VIIIB	IB	IIB	IIIA	IVA	VA	VIA	VIIA	Noble Gases
								2 He 4.003
			5 B 10.81	6 C 12.011	7 N 14.007	8 O 15.999	9 F 18.998	10 Ne 20.179
			13 Al 26.982	14 Si 28.086	15 P 30.974	16 S 32.06	17 Cl 35.453	18 Ar 39.948
28 Ni 58.70	29 Cu 63.546	30 Zn 65.38	31 Ga 69.72	32 Ge 72.59	33 As 74.922	34 Se 78.96	35 Br 79.904	36 Kr 83.80
46 Pd 106.4	47 Ag 107.868	48 Cd 112.41	49 In 114.82	50 Sn 118.69	51 Sb 121.75	52 Te 127.60	53 I 126.905	54 Xe 131.30
78 Pt 195.09	79 Au 196.966	80 Hg 200.59	81 Tl 204.37	82 Pb 207.19	83 Bi 208.2	84 Po (209)	85 At (210)	86 Rn (~222)

67 Ho 164.930	68 Er 167.26	69 Tm 168.934	70 Yb 173.04	71 Lu 174.96
99 Es (254)	100 Fm (257)	101 Md (258)	102 No (259)	103 Lr (260)

Chemical Principles

Third Edition

Richard E. Dickerson
California Institute of Technology

Harry B. Gray
California Institute of Technology

Gilbert P. Haight, Jr.
University of Illinois

The Benjamin/Cummings Publishing Company, Inc.
Menlo Park, California · Reading, Massachusetts
London · Amsterdam · Don Mills, Ontario · Sydney

Sponsoring editor: Mary Forkner
Production editor: Betsey Rhame
Cover designer: Stephen Osborn Associates
Book designer: Marjorie Spiegelman

Figures 1-3, 1-5, 1-7, 1-8, 2-1, 4-3, 4-4, 5-2, 5-3, 16-1, 16-2, 16-3, 16-4, 21-9, 21-19, 23-4, and 23-5 are adapted or reprinted with permission from *Chemistry, Matter, and the Universe* by Richard E. Dickerson and Irving Geis. Benjamin/Cummings Publishing Co., Inc., Menlo Park, Calif., 1976.

Figures 13-8, 13-9, 13-11, 13-15, 13-16, 13-17, 13-28, 13-29, 13-30, 13-31, 13-32, 13-33, 13-34, 13-35, 13-36, 13-37, 13-38, 13-39, 14-3, 14-4, and 14-7 are reprinted with permission from *Chemical Bonds* by Harry B. Gray. Benjamin/Cummings Publishing Co., Inc., Menlo Park, Calif., 1973.

Library of Congress Cataloging in Publication Data
Dickerson, Richard Earl.
  Chemical principles.

  Includes bibliographies and index,
  1. Chemistry. I. Gray, Harry B., joint author.
II. Haight, Gilbert Pierce, joint author. III. Title.
QD31.2.D52 1979   540   77-87336
ISBN 0-8053-2398-8

ISBN 0-8053-2398-8
  EFGHIJKL-DO-8987654321

The Benjamin/Cummings Publishing Company, Inc.
2727 Sand Hill Road
Menlo Park, California 94025

Preface

This edition of *Chemical Principles,* like its predecessors, is designed to be used in a general university chemistry course which must provide both an overview of chemistry for nonspecialists and a sound foundation for later study for science or chemistry majors. Hence there are several survey chapters introducing different areas of chemistry, including inorganic, nuclear, organic, and biochemistry, and an attempt is made throughout the book to place chemistry in its historical and cultural setting. At the same time, the quantitative aspects of chemistry are presented in a manner consistent with their importance, in a way that will make it easy to build upon them in later courses.

This is the first complete revision of *Chemical Principles* since the first edition was published in 1969. The authors have rethought and replanned the entire book, especially the first thirteen chapters, trying to make it a better pedagogical tool without losing the special viewpoints and flavor that made the earlier editions so successful. The history and the anecdotal asides that help to make the subject palatable have been retained, but they have been better segregated from the factual material for which a student will be held responsible.

THIRD EDITION REORGANIZATION AND LEVEL

The basic material of the first six chapters has been rearranged into a more logical and more easily absorbed order. These chapters, although not formally set off from the rest of the book, actually make up one study unit on quantitative chemistry: atoms and moles, stoichiometry, heats of reaction, gas laws and the kinetic theory, chemical equilibrium and acid-base equilibrium. They have been rethought and rewritten as a block by one of the authors, with more in-text examples and new end-of-chapter problems. The mole concept, balancing of equations, and stoichiometry in general now appear in the first two chapters where they will be most useful as preparation for the laboratory. At the same time, stoichiometry, which can be one of the dullest topics in chemistry, is presented along with heats of reaction as two illustrations of a fundamental physical principle: the conservation of mass and energy. A long but essential chapter on equilibrium has been broken down into two more accessible chapters: one on principles of equilibrium, and the second on acid-base equilibria in aqueous solutions. It is hoped that these five chapters will give the student a solid foundation in vocabulary and problem-solving skills without which further progress in chemistry is impossible. After a solid dose of "basic training," Chapter 6 provides a little historical relief, with the story of how we arrived at the knowledge contained in the first five chapters.

Chapters 7 through 14 make up a second study unit on atomic structure, chemical periodicity, and chemical structure and bonding. This too has been carefully reconsidered and revised as a whole by a single author, thus the Third Edition revision helps unify the text. In response to user requests, the material on chemical periodicity and inorganic oxidation-reduction chemistry has been unified in Chapters 9 and 10.

The treatment of molecular orbitals and chemical bonding in the Second Edition had been liked by most users, but had been considered a little too high-level and difficult to get into. Now we have divided this material into two chapters, Chapter 12 on principles of the molecular orbital theory and applications to simple diatomic molecules, and Chapter 13 on polyatomic molecules and molecular spectroscopy. We have also provided a new Chapter 11 as an introduction to bonding theories, as far as one can go with electron pairing and electron pair repulsion short of quantum mechanics. The Valence Shell Electron Pair Repulsion (VSEPR) theory, which has been surprisingly neglected in this country, provides an intuitively simple and nonmathematical way of explaining the shapes of molecules. These three chapters plus the subsequent one on bonding in solids and liquids will give the student a secure grounding in the principles of bonding, molecular structure, and spectroscopy.

Chapters 15–19 make up a third study unit on thermodynamics and equilibrium. The material on the first and second laws is essentially the same as in previous editions, but has been divided into two more digestible chapters. The statistical description of entropy has been simplified. A new chapter has been added on phase equilibria, Chapter 18. Since this is quantitative material and frequently is difficult for the beginning student, we have significantly increased the number of worked examples in the text, revised the chapter-end problems, and added new ones.

The last four chapters cover special topics that may not be included in all introductory courses: coordination chemistry, organic and biochemistry, chemical kinetics, and nuclear chemistry. After much agonized debate about principles and pedagogy, we finally decided to place these chapters at the end, where they can be used or not as the individual instructor chooses. (We hope that they will all be used.) They have all been revised and rewritten where necessary, especially the chapter on organic and biochemistry.

PEDAGOGY

Each chapter begins with a list of key concepts. This provides students with a brief overview of the chapter material, both before they start the chapter and after they finish, as a quick check on their retention of key ideas. Throughout the text of each chapter, we have concentrated on expanding the solutions to problems worked in the chapter. Problem examples relevant to each concept are presented, and solutions proceed step by step. Chapters conclude with a summary in which key terms, introduced in the chapter, are called out in boldface type. Each summary is followed by 20 to 40 self-study questions and a series of problems arranged by subject.

The Third Edition contains over 100 more end-of-chapter problems than its predecessor. Moreover, new problems have been written to parallel the development of each chapter, and all problems have been titled and grouped by subject matter. Following the more quantitative chapters, the problems have been paired, with first an odd-numbered problem and then an even-numbered problem testing the same skills. Answers to the odd-numbered problems are given in Appendix 6. Hence the even-numbered problems can be assigned as homework, and if the students cannot work a problem, they can try the preceding odd-numbered problem first as practice, checking their solution against the Appendix.

SI UNITS

After considerable debate, the authors have decided to "bite the bullet" with regard to SI units. There is a traditional attachment to the calorie as the unit of heat, and it will be a long time before the calorie is eliminated from the scientific literature. Nevertheless, the sheer logic of SI units, their ease of use, and the way that they make obvious the connection between heat, work, and energy, all argue for a changeover now to what will be the standard units of the next generation of scientists. SI units and the logic behind them are explained in Appendix 1. The calorie is mentioned in this book because every scientist will still have to know what a calorie is, but all calculations are carried out in joules. Thermodynamic tables in Appendix 3 and elsewhere in the book have all been converted to joules. At the same time, we have refused to become overly doctrinaire and throw the baby out with the bathwater. The standard atmosphere (101,325 pascals) has been considered to be as reasonable a derived unit in gas law calculations as is the electronic charge (0.16022 attocoulomb) for expressing the charge on an ion. The careful reader will even discover angstrom units lurking here and there, and we offer no apologies. Our goal has been to train intelligent scientists and laypeople who can read, understand, and use the literature.

SUPPLEMENTS

All of the supplemental aides to the Third Edition have been revised by their authors on the basis of the new manuscript: *Programmed Reviews* by Lassila, Barrow, Kenney, Litle, and Thompson; *Relevant Problems* by Butler and Grosser; a new *Study Guide* covering the entire text by Tom Taylor; and an *Instructors' Manual* by Ben Chastain. Some or all of these may be useful adjuncts to the main textbook in your course.

ACKNOWLEDGMENTS

We are grateful to the many reviewers who read the Third Edition revision with care: Marcetta Darensbourg, Leo E. Kallan, Curtis B. Anderson, Paul M. Treichel, Jean Lassila, George Miller, Caroline Eastman, Lawrence E. Wilkins, Paul Hunter, and Peter Linde. We would like to thank Ben Chastain and Mildred Johnson for yeoman service in reading every line of the new edition, and offering detailed suggestions based on their experience with the Second Edition. Gloria Joyce deserves our thanks for reducing some of our more convoluted prose to comprehensibility. And we thank Laura Dagen, Sue Brittenham, and Dee Barr, who typed every page of the revision and managed to remain good-natured in the process. Mary Forkner, as sponsoring editor, provided us with feedback from reviewers and users that led to the present book. Betsey Rhame carried out the remarkable task of producing a book that is not only complete and attractive, but right on schedule. Lastly, we offer belated thanks and recognition

Chapter 6   **Are Atoms Real? From Democritus to Dulong and Petit**   **225**

Chapter 7   **The Periodic Table**   **257**

1

Atoms, Molecules, and Ions

According to convention there is a sweet and a bitter, a hot and a cold, and according to convention there is order. In truth there are atoms and a void.

Democritus (400 B.C.)

In the trial scene in *Alice in Wonderland,* the White Rabbit, called to the witness stand, asks, "Where shall I begin, please?" The answer is straightforward: "Begin at the beginning, and go on till you come to the end, then stop." But we shall begin in the middle, with a description of what atoms and molecules are like, *before* saying anything about how we know that atoms exist. When we examine the evidence for atomic and molecular structure in later chapters, you will have at least an idea of the goal of the effort. The result, we hope, will be to make this textbook more comprehensible than most of Lewis Carroll's books. (The White Rabbit's evidence did not fare very well: "If any one of them can explain it," said Alice, "I'll give him sixpence. *I* don't believe there's an atom of meaning in it.")

1-1 THE STRUCTURE OF ATOMS

An **atom** (ours, not Carroll's) consists of a positively charged **nucleus**, surrounded by one or more negatively charged particles called **electrons**. The positive charges equal the negative charges, so the atom has no overall charge; it is electrically neutral. Most of an atom's mass is in its nucleus; the mass of an electron is only 1/1836 the mass of the lightest nucleus, that of hydrogen. Although the nucleus is heavy, it is quite small compared with

Table 1-1

Fundamental Particles of Matter

Particle	Charge	Mass (amu)
Proton	+1	1.00728
Neutron	0	1.00867
Electron	−1	0.000549

the overall size of an atom. The radius of a typical atom is around 1 to 2.5 angstroms (Å), whereas the radius of a nucleus is about 10^{-5} Å.* If an atom were enlarged to the size of the earth, its nucleus would be only 200 feet in diameter and could easily rest inside a small football stadium.

The nucleus of an atom contains **protons** and **neutrons.** Protons and neutrons have nearly equal masses, but they differ in charge. A neutron has no charge, whereas a proton has a positive charge that exactly balances the negative charge on an electron. Table 1-1 lists the charges of these three fundamental particles, and gives their masses expressed in **atomic mass units.** The atomic mass unit (amu) is defined as exactly one-twelfth the mass of a carbon atom that has six protons and six neutrons in its nucleus. With this scale, protons and neutrons have masses that are close to, but not precisely, 1 amu each. [As a matter of information at this point, there are 6.022×10^{23} amu in 1 gram (g). This number is known as **Avogadro's number,** N, and later in the chapter we will show one of the ways this number can be calculated.]

The number of protons in the nucleus of an atom is known as the **atomic number,** Z. It is the same as the number of electrons around the nucleus, because an atom is electrically neutral. The **mass number** of an atom is equal to the total number of heavy particles: protons and neutrons. When two atoms are close enough to combine chemically—to form *chemical bonds with one another*—each atom "sees" mainly the outermost electrons of the other atom. Hence these outer electrons are the most important factors in the chemical behavior of atoms. Neutrons in the nucleus have little effect on chemical behavior, and the protons are significant only because they determine how many electrons surround the nucleus in a neutral atom. All atoms with the same atomic number behave in much the same way chemically and are classified as the same chemical **element.** Each element

*One angstrom unit (Å) equals 10^{-10} meters (m), or 10^{-8} centimeters (cm). If you are not familiar with metric units, see Appendix 1 for more information on the *Système International* (SI), a simplified version of the metric system adopted by scientists throughout the world. We shall generally use SI units in this book. If you are not familiar with the use of exponential numbers (scientific notation), read Appendix 4.

has its own name and a one- or two-letter symbol (usually derived from the element's English or Latin name). For example, the symbol for carbon is C, and the symbol for calcium is Ca. The symbol for sodium is Na—the first two letters of its Latin (and German) name, *natrium*—to distinguish it from nitrogen, N, and sulfur, S. On the inside back cover of this book is an alphabetical list of the elements and their symbols.

Example 1

What is the atomic symbol for bromine, and what is its atomic number? Why isn't the symbol for bromine just the first letter of its name? What other element preempts the symbol B? (Use table inside back cover.)

Solution

Bromine's atomic number is 35, and its symbol is Br; B is the symbol for boron.

1-2 ISOTOPES

Although all atoms of an element have the same number of protons, the atoms may differ in the number of neutrons they have (Table 1-2). These differing atoms of the same element are called **isotopes**. Four isotopes of helium (He) are shown in Figure 1-1. All atoms of chlorine (Cl) have 17 protons, but there are chlorine isotopes having 15 to 23 neutrons. Only two chlorine isotopes exist in significant amounts in nature, those with 18 neutrons (75.53% of all chlorine atoms found in nature), and those with 20 neutrons (24.47%). To write the symbol for an isotope, place the atomic number as a subscript and the mass number (protons plus neutrons) as a superscript to the left of the atomic symbol. The symbols for the two naturally occurring isotopes of chlorine then would be $^{35}_{17}Cl$ and $^{37}_{17}Cl$. Strictly speaking, the subscript is unnecessary, since all atoms of chlorine have 17 protons. Hence the isotope symbols are usually written without the subscript: ^{35}Cl and ^{37}Cl. In discussing these isotopes, we use the terms *chlorine-35* and *chlorine-37*. For an atom to be stable, it generally must have a few more neutrons than protons. Nuclei that have too many of either kind of fundamental particle are unstable, and break down radioactively in ways that are discussed in Chapter 23.

Example 2

How many protons, neutrons, and electrons are there in an atom of uranium-238? Write the symbol for this isotope.

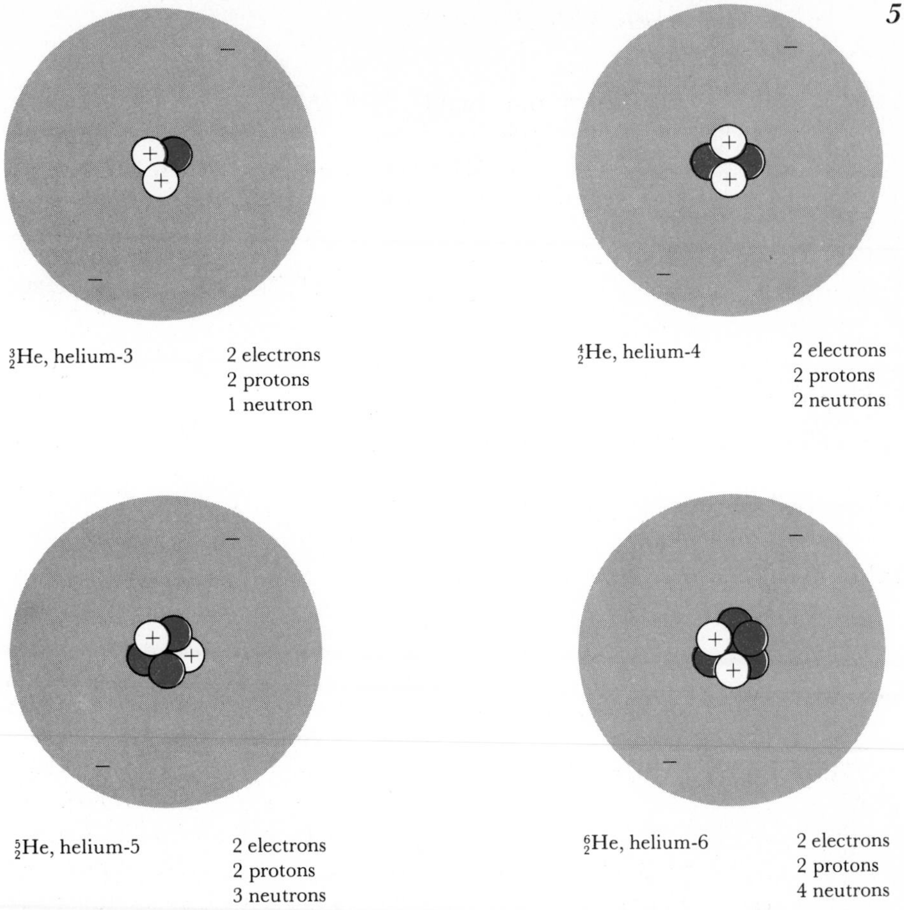

$^3_2$He, helium-3   2 electrons
2 protons
1 neutron

$^4_2$He, helium-4   2 electrons
2 protons
2 neutrons

$^5_2$He, helium-5   2 electrons
2 protons
3 neutrons

$^6_2$He, helium-6   2 electrons
2 protons
4 neutrons

Figure 1-1   Four isotopes of helium (He). All atoms of helium have two protons (hence two electrons), but the number of neutrons can vary. Most helium atoms in nature have two neutrons (helium-4), and fewer than one helium atom per million in nature have one neutron (helium-3). The other helium isotopes, helium-5, helium-6, and helium-8 (not shown) are unstable and are seen only briefly in nuclear reactions (see Chapter 23). The size of the nucleus is grossly exaggerated here. If the nucleus were of the size shown, the atom would be half a kilometer across.

Solution   The atomic number of uranium (see the inside back cover) is 92, and the mass number of the isotope is given as 238. Hence it has 92 protons, 92 electrons, and $238 - 92 = 146$ neutrons. Its symbol is $^{238}_{92}$U (or ^{238}U).

The total mass of an atom is called its **atomic weight**, and this is almost but not exactly the sum of the masses of its constituent protons, neutrons, and electrons.* When protons, neutrons, and electrons combine to form an

*The terms *atomic weight* and *molecular weight* are universally used by working scientists, and will be used in this book, even though these technically are masses rather than weights.

Table 1-2

Composition of Typical Atoms and Ions

	Electrons	Protons	Neutrons	Atomic number	Atomic weight (amu)	Total charge (electron units)
Hydrogen atom, ^1_1H or H	1	1	0	1	1.008	0
Deuterium atom, ^2_1H or D	1	1	1	1	2.014	0
Tritium atom, ^3_1H or T	1	1	2	1	3.016	0
Hydrogen ion, H^+	0	1	0	1	1.007	+1
Helium atom, ^4_2He	2	2	2	2	4.003	0
Helium nucleus or alpha particle, He^{2+} or α	0	2	2	2	4.002	+2
Lithium atom, ^7_3Li	3	3	4	3	7.016	0
Carbon atom, $^{12}_6C$	6	6	6	6	12.000	0
Oxygen atom, $^{16}_8O$	8	8	8	8	15.995	0
Chlorine atom, $^{35}_{17}Cl$	17	17	18	17	34.969	0
Chlorine atom, $^{37}_{17}Cl$	17	17	20	17	36.966	0
Naturally occurring mixture of chlorine	17	17	18 or 20	17	35.453	0
Uranium atom, $^{234}_{92}U$	92	92	142	92	234.04	0
Uranium atom, $^{235}_{92}U$	92	92	143	92	235.04	0
Uranium atom, $^{238}_{92}U$	92	92	146	92	238.05	0
Naturally occurring mixture of uranium	92	92	varied	92	238.03	0

atom, some of their mass is converted to energy and is given off. (This is the source of energy in nuclear fusion reactions.) Because the atom cannot be broken down into its fundamental particles unless the energy for the missing mass is supplied from outside it, this energy is called the **binding energy** of the nucleus.

Example 3

Calculate the mass that is lost when an atom of carbon-12 is formed from protons, electrons, and neutrons.

Solution

Since the atomic number of every carbon atom is 6, carbon-12 has 6 protons and therefore 6 electrons. To find the number of neutrons, we subtract the number of protons from the mass number: $12 - 6 = 6$ neutrons. We can use the data in Table 1-1 to calculate the total mass of these particles:

Protons: 6×1.00728 amu $=$ 6.04368 amu
Neutrons: 6×1.00867 amu $=$ 6.05202 amu
Electrons: 6×0.00055 amu $=$ 0.00330 amu

Total particle mass: 12.09900 amu

But by the definition of the scale of atomic mass units, the mass of one carbon-12 atom is exactly 12 amu. Hence 0.0990 amu of mass has disappeared in the process of building the atom from its particles.

Example 4

Calculate the expected atomic weight of the isotope of chlorine that has 20 neutrons. Compare this with the actual atomic weight of this isotope as given in Table 1-2.

Solution

The chlorine isotope has 17 protons and 20 neutrons:

Protons: 17×1.00728 amu $= 17.1238$ amu
Neutrons: 20×1.00867 amu $= 20.1734$ amu
Electrons: 17×0.00055 amu $=$ 0.0094 amu

Total particle mass: 37.3066 amu
Actual observed atomic weight: 36.966 amu

Mass loss: 0.341 amu

Each isotope of an element is characterized by an atomic number (total number of protons), a mass number (total number of protons and neutrons), and an atomic weight (mass of atom in atomic mass units). Since mass losses upon formation of an atom are small, the mass number is usually the same as the atomic weight rounded to the nearest integer. (For example, the atomic weight of chlorine-37 is 36.966, which is rounded to 37.) If there are several isotopes of an element in nature, then of course the experimentally observed atomic weight (**the natural atomic weight**) will be the *weighted average* of the isotope weights. The average is weighted according to the percent abundance of the isotopes. Chlorine occurs in nature as 75.53% chlorine-35 (34.97 amu) and 24.47% chlorine-37 (36.97 amu), so the weighted average of the isotope weights is

$$(0.7553 \times 34.97 \text{ amu}) + (0.2447 \times 36.97 \text{ amu}) = 35.46 \text{ amu}$$

The atomic weights given inside the back cover of this book are all weighted averages of the isotopes occurring in nature, and these are the figures we shall use henceforth—unless we are specifically discussing one isotope. All isotopes of an element behave the same way chemically for the most part. Their behavior will differ in regard to mass-sensitive properties such as diffusion rates, which we'll look at later in this book.

Example 5

Magnesium (Mg) has three significant natural isotopes: 78.70% of all magnesium atoms have an atomic weight of 23.985 amu, 10.13% have an atomic weight of 24.986 amu, and 11.17% have an atomic weight of 25.983 amu. How many protons and neutrons are present in each of these three isotopes? How do we write the symbols for each isotope? Finally, what is the weighted average of the atomic weights?

Solution

There are 12 protons in all magnesium isotopes. The isotope whose atomic weight is 23.985 amu has a mass number of 24 (protons and neutrons), so 24 − 12 protons gives 12 neutrons. The symbol for this isotope is ^{24}Mg. Similarly, the isotope whose atomic weight is 24.986 amu has a mass number of 25, 13 neutrons, and ^{25}Mg as a symbol. The third isotope (25.983 amu) has a mass number of 26, 14 neutrons, and ^{26}Mg as a symbol. We calculate the average atomic weight as follows:

$$(0.7870 \times 23.985) + (0.1013 \times 24.986) + (0.1117 \times 25.983) = 24.31 \text{ amu}$$

Example 6

Boron has two naturally occurring isotopes, ^{10}B and ^{11}B. We know that 80.22% of its atoms are ^{11}B, atomic weight 11.009 amu. From the natural atomic weight given on the inside back cover, calculate the atomic weight of the ^{10}B isotope.

Solution

If 80.22% of all boron atoms are ^{11}B, then 100.00 − 80.22, or 19.78%, are the unknown isotope. We can use W to represent the unknown atomic weight in our calculation:

$$(0.8022 \times 11.009) + (0.1978 \times W) = 10.81 \text{ amu (natural atomic weight)}$$

$$W = \frac{10.81 - 8.831}{0.1978} = 10.01 \text{ amu}$$

1-3 MOLECULES

The formation of atoms from fundamental particles, interesting as this might be to the physicist, is far from being the ultimate stage in the organization of matter. As we mentioned earlier, when atoms are close enough to one another that the outer electrons of one atom can interact with the other atoms, then attractions can be set up between atoms, strong enough to hold them together in what is termed a **chemical bond**. In the simplest cases the bond arises from the sharing of two electrons between a pair of atoms, with one electron provided by each of the bonded atoms. Bonds based on electron sharing are known as **covalent bonds**, and two or more atoms held together as a unit by covalent bonds are known as a **molecule**. One of the principal triumphs of the theory of quantum mechanics in chemistry (see Chapter 8) has been its ability to predict the kinds of atoms that will bond together, and the three-dimensional structures and reactivities of the molecules that result. (A major section of this book, Chapters 8–14, is devoted to chemical bonding theories.)

In **molecular diagrams**, a covalent, electron-sharing bond is represented by a straight line connecting the bonded atoms. In the water molecule, one atom of oxygen (O) is bonded to two hydrogen (H) atoms. The diagram for the molecule can be drawn two ways:

$$H-O-H \qquad \text{or} \qquad H \overset{\displaystyle O}{\diagup \diagdown} H$$

The second version acknowledges the fact that a water molecule is not linear; the two H—O bonds make an angle of 105° with one another.

Molecules of hydrogen gas, hydrogen sulfide, ammonia, methane, and methyl alcohol (methanol) have the following bond structures:

$$H-H \qquad H \overset{\displaystyle S}{\diagup \diagdown} H \qquad H \overset{\displaystyle \overset{H}{|}}{\underset{\diagup \diagdown}{N}} H$$

hydrogen   hydrogen sulfide   ammonia

$$H-\overset{\displaystyle \overset{H}{|}}{\underset{\underset{H}{|}}{C}}-H \qquad H-\overset{\displaystyle \overset{H}{|}}{\underset{\underset{H}{|}}{C}}-O-H$$

methane   methyl alcohol

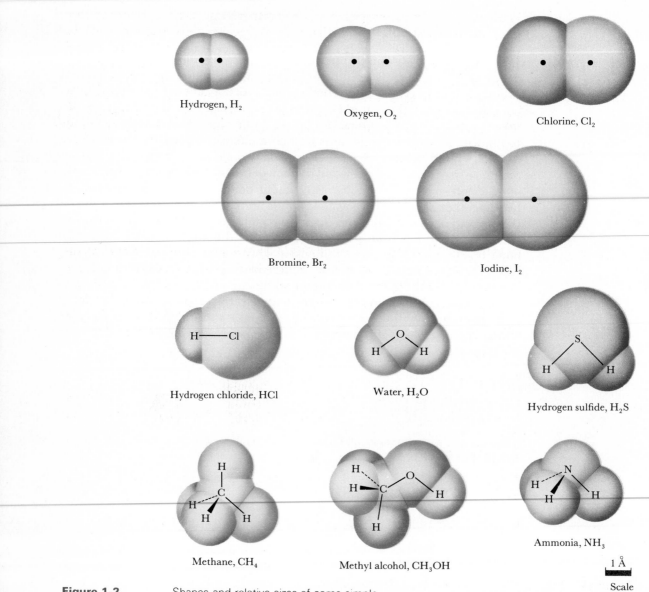

Hydrogen, H_2

Oxygen, O_2

Chlorine, Cl_2

Bromine, Br_2

Iodine, I_2

Hydrogen chloride, HCl

Water, H_2O

Hydrogen sulfide, H_2S

Methane, CH_4

Methyl alcohol, CH_3OH

Ammonia, NH_3

1 Å
Scale

Figure 1-2 Shapes and relative sizes of some simple molecules. Two bonded atoms appear to interpenetrate because their electron clouds overlap. By convention, a tapered bond in a drawing represents a bond pointing out toward the observer, with the wide end of the taper closest, and a dashed line is used for a bond that points back behind the plane of the page.

These diagrams show only the connections between atoms in the molecules. They do not show the three-dimensional geometries (or shapes) of the molecules. Figure 1-2 shows the shapes and the relative bulk of several molecules. Note that the bond angle in molecules having more than two atoms can vary. The angle in the water molecule is 105°, and the angle in hydrogen sulfide is 92°; the four atoms connected to the central carbon in methane and methyl alcohol are directed to the four corners of a tetrahedron. The bond structure in straight-chain octane, one of the components of gasoline, is

$$\begin{matrix} & H & H & H & H & H & H & H & H & \\ & | & | & | & | & | & | & | & | & \\ H- & C- & C- & C- & C- & C- & C- & C- & C & -H \\ & | & | & | & | & | & | & | & | & \\ & H & H & H & H & H & H & H & H & \end{matrix}$$

Each of the molecular diagrams shown can be condensed to a **molecular formula**, which tells how many atoms of each element are in the molecule, but provides little or no information as to how the atoms are connected. The molecular formula for hydrogen is H_2; water, H_2O; hydrogen sulfide, H_2S; ammonia, NH_3; methane, CH_4; methyl alcohol, CH_3OH or CH_4O; and octane, C_8H_{18}. The formula for octane can also be written

$$CH_3-CH_2-CH_2-CH_2-CH_2-CH_2-CH_2-CH_3$$

The sum of the atomic weights of all the atoms in a molecule is its **molecular weight**. Using the atomic weights on the inside back cover, we can calculate molecular weights. The molecular weight of hydrogen, H_2, is

$$2 \times 1.0080 \text{ amu} = 2.0160 \text{ amu}$$

A water molecule, H_2O, has two atoms of hydrogen and one atom of oxygen, so:

$$(2 \times 1.0080 \text{ amu}) + (15.9994 \text{ amu}) = 18.0154 \text{ amu}$$

Example 7

Calculate the molecular weight of methyl alcohol.

Solution

The molecular formula is CH_3OH or CH_4O. Then:

1 carbon:	1×12.011 amu = 12.011 amu
4 hydrogens:	4×1.008 amu = 4.032 amu
1 oxygen:	1×15.999 amu = 15.999 amu
Total molecular weight:	32.04 amu

(If you wonder why the last figure has been dropped, see the discussion of significant figures in Appendix 4.)

In Example 7 notice that the natural atomic weight of carbon is not 12.000 amu but 12.011 amu, since carbon occurs as a mixture of 98.89% carbon-12 and 1.11% carbon-13, with trace amounts of carbon-14.

Example 8

What is the molecular weight of pure octane?

Solution

Since the molecular formula is C_8H_{18}, the molecular weight is

$$(8 \times 12.011) + (18 \times 1.008) = 114.23 \text{ amu}$$

1-4 FORCES BETWEEN MOLECULES

Although the strongest attractions of an atom are for other atoms to which it is bonded in a molecule, two molecules themselves exert small but appreciable attractions on one another. Molecules are slightly "sticky." These forces, caused by momentary fluctuations in electron distributions around the atoms, are known as **van der Waals attractions** (after Dutch physicist Johannes van der Waals). They are responsible for the existence of three states (or phases) of matter at different temperatures: solids, liquids, and gases. **Temperature** is just a measure of the heat energy or energy of motion that a collection of molecules possesses. At low temperatures, the molecules have little energy of motion. The van der Waals attractions hold them together in an orderly, close-packed crystalline array or lattice (Figure 1-3c). This is the **solid** state. If more energy is fed into the crystal so the temperature rises, the molecules will vibrate about their average or equilibrium positions in the crystal. Enough energy will cause the ordered structure of the molecular crystal to break up, and the molecules will be free to slide past one another, although they are still touching (Figure 1-3b). This is the **liquid** state, and the transition temperature between solid and liquid is called the **melting point,** T_m. The liquid is still held together by van der Waals attractions, although the molecules have too much energy of motion to be locked into a rigid array. If still more energy is given the liquid, the molecules will begin to move fast enough to overcome the van der Waals attractions, separate entirely from one another, and travel in independent molecular trajectories through space (Figure 1-3a). This is the **gas** phase, and the transition temperature between liquid and gas is called the **boiling point,** T_b. Changes in phase are treated in more detail in Chapter 18.

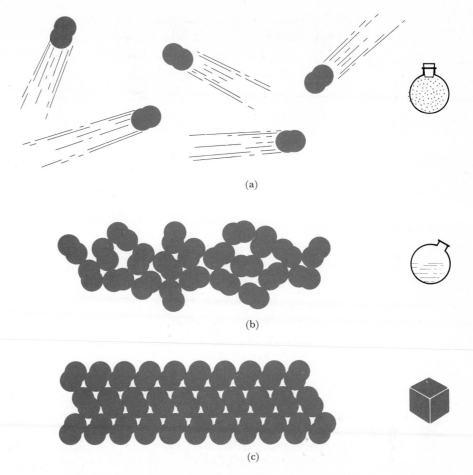

(a)

(b)

(c)

Figure 1-3   The three states of matter: (a) In a gas the individual molecules move freely through space, colliding and rebounding. A gas adapts to the shape of its container and can easily be expanded or compressed. (b) Molecules in a liquid are in contact, but free to slide past one another. A liquid also adapts to the shape of its container, but it has a relatively fixed volume. (c) In a crystalline solid, molecules are packed into a regular array, giving the solid both a fixed volume and a definite shape. Work must be done to break or deform a crystal. Adapted from R. E. Dickerson and I. Geis, *Chemistry, Matter, and the Universe*, W. A. Benjamin, Menlo Park, Calif., 1976.

The melting and boiling points of some simple molecules are compared in Table 1-3. In general, larger molecules have higher melting and boiling points, since they have larger surface areas for van der Waals attractions. Thus at 1 atm. pressure H_2 boils at $-252.5°C$, CH_4 boils at $-164.0°C$, but C_8H_{18} must be heated to $+125.7°C$ before the molecules will separate from one another and go into the gas phase.

A second kind of force between molecules also influences melting and boiling points: the **polarity** of the molecules. If two atoms that are connected by an electron-pair covalent bond do not have the same attraction for

Table 1-3
Melting and Boiling Points of Some Simple Molecular Substances

Substance	Molecular formula	T_m (°C)	T_b (°C)
Gases			
Hydrogen	H_2	−259.1	−252.5
Oxygen	O_2	−218.4	−183.0
Methane	CH_4	−182.5	−164.0
Hydrogen sulfide	H_2S	−85.5	−60.7
Chlorine	Cl_2	−101.0	−34.6
Ammonia	NH_3	−77.7	−33.4
Liquids			
Bromine	Br_2	−7.2	+58.8
Methanol	CH_3OH	−93.9	+65.0
Water	H_2O	0	+100
n-Octane	C_8H_{18}	−56.8	+125.7
Solids			
Iodine	I_2	+113.5	+184.4
Sucrose (cane sugar)	$C_{12}H_{22}O_{11}$	+185	decomposes

electrons, then the electron pair will shift toward the atom with the greater electron pulling power. This will give that atom a slight excess of negative charge (represented by δ^- rather than by just a minus sign, which would imply a full electron charge), and will confer a slight positive charge (δ^+) on the atom that lost out in the tug-of-war for the electron pair. Because the electron-attracting power (**electronegativity**) of oxygen is greater than that of hydrogen, the oxygen atom in a molecule of water or methyl alcohol is slightly negative, and the hydrogen atoms are slightly positive (Figure 1-4). Such a molecule is termed *polar* because it behaves like a tiny electric dipole; that is, the negative charge on the oxygen attracts other nearby positive charges, and the positive charge on each hydrogen attracts other negative charges. This is another attractive force between molecules, in addition to van der Waals attractions. Because of the forces binding its molecules, methanol melts and boils at much higher temperatures than methane, which is similar to it in molecular size. Methanol is a liquid at room temperature, whereas methane is a gas. In water, the attractions between hydrogen and oxygen from different molecules are so strong that they are given the name of **hydrogen bonds.** Hydrogen bonds are especially important in proteins and other giant molecules in living organisms. If it were not for polarity and hydrogen bonding, water would melt and boil at lower temperatures even than H_2S (Table 1-3). It would be a gas at room temperature, rather than the earth's most common liquid.

Example 11

How many molecules of H_2 and Cl_2 would be present in the experiment of Example 10?

Solution   In 496.0 moles of any substance, there will be 496.0 moles $\times$ 6.022 $\times$ 10^{23} molecules mole^{-1}, which equals 2.99 $\times$ 10^{26} molecules.

As a sobering example of just how large Avogadro's number is, 1 mole of coconuts, each 14 centimeters (cm) in diameter, would fill a volume as large as the entire planet earth. The use of moles in chemical calculations is the subject of the next chapter, but the idea has been introduced here because we need to know how to scale up from the molecular to the laboratory level.

1-6   IONS

The idea of a covalent bond suggests equal sharing of the electron pair by the bonded atoms, but the brief discussion of polarity in Section 1-4 indicated that the sharing is not always equal. The relative electronegativity or electron-attracting power of atoms is of great importance in explaining chemical behavior, and is treated in detail in Chapters 9 and 10. Sodium atoms (and all metals in general) have a weak hold on electrons, whereas chlorine atoms are very electronegative. Hence in common table salt (sodium chloride, NaCl), each sodium atom, Na, loses one electron (e^-) to form a sodium ion, Na^+. Each chlorine atom picks up one electron to become a chloride ion, Cl^-:

$$Na \rightarrow Na^+ + e^- \quad \text{and} \quad \tfrac{1}{2}Cl_2 + e^- \rightarrow Cl^-$$

We write $\tfrac{1}{2}Cl_2$ because free chlorine gas exists as **diatomic** (two-atom) molecules, not as free chlorine atoms. Solid sodium chloride (Figure 1-5) has sodium and chloride ions packed into a three-dimensional lattice in such a way that each positive Na^+ ion is surrounded on four sides and top and bottom by negative Cl^- ions, and each Cl^- is similarly surrounded by six nearest neighbor Na^+ ions. This is a particularly stable arrangement of positive and negative charges.

Metals in general lose one to three electrons easily to become positively charged ions, or **cations**:

$$
\begin{array}{lll}
Li & \rightarrow Li^+ \ + \ e^- & \text{lithium ion} \\
Na & \rightarrow Na^+ \ + \ e^- & \text{sodium ion} \\
K & \rightarrow K^+ \ + \ e^- & \text{potassium ion} \\
Mg & \rightarrow Mg^{2+} + 2e^- & \text{magnesium ion} \\
Ca & \rightarrow Ca^{2+} + 2e^- & \text{calcium ion} \\
Al & \rightarrow Al^{3+} + 3e^- & \text{aluminum ion}
\end{array}
$$

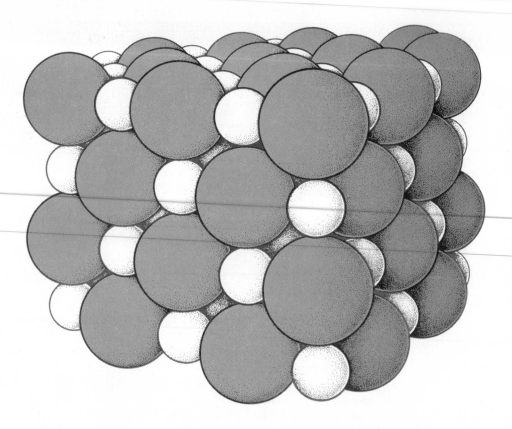

Figure 1-5   Common table salt (sodium chloride, NaCl) is built from closely packed sodium ions, Na$^+$ (small spheres), and chloride ions, Cl$^-$ (large, colored spheres). Each ion of one charge is surrounded by six ions of the opposite charge at the four compass points and above and below. This is a particularly stable arrangement of charges, and it occurs in many salts. From Dickerson and Geis, *Chemistry, Matter, and the Universe*. The Benjamin/ Cummings Publishing Co., Menlo Park, Ca., © 1976.

Some nonmetals, in contrast, pick up electrons to become negatively charged ions, or **anions**:

$$\tfrac{1}{2}F_2 + e^- \rightarrow F^- \qquad \text{fluoride ion}$$
$$\tfrac{1}{2}Cl_2 + e^- \rightarrow Cl^- \qquad \text{chloride ion}$$
$$\tfrac{1}{2}O_2 + 2e^- \rightarrow O^{2-} \qquad \text{oxide ion}$$
$$S + 2e^- \rightarrow S^{2-} \qquad \text{sulfide ion}$$

Table 1-4

Some Simple Ions of Elements

Cations				Anions		
+1	**+2**	**+3**	**+4**	**−3**	**−2**	**−1**
Li^+	Be^{2+}	Al^{3+}	Sn^{4+}	N^{3-}	O^{2-}	F^-
Na^+	Mg^{2+}	Sc^{3+}	Mn^{4+}	P^{3-}	S^{2-}	Cl^-
K^+	Ca^{2+}	Y^{3+}	U^{4+}		Se^{2-}	Br^-
Rb^+	Sr^{2+}	Ga^{3+}	Th^{4+}			I^-
Cs^+	Ba^{2+}	In^{3+}	Ce^{4+}			
Cu^+	Mn^{2+}	Tl^{3+}				
Ag^+	Fe^{2+}	Sb^{3+}				
Tl^+	Co^{2+}	Bi^{3+}				
	Ni^{2+}	V^{3+}				
	Cu^{2+}	Cr^{3+}				
	Zn^{2+}	Fe^{3+}				
	Cd^{2+}	Co^{3+}				
	Hg^{2+}					
	Sn^{2+}					
	Pb^{2+}					

Other simple ions made from single atoms are shown in Table 1-4. The charge on a simple, single-atom ion such as Al^{3+} or S^{2-} is its **oxidation state** or **oxidation number**. It is the number of electrons that must be added to reduce (or removed to oxidize) the ion to the neutral species:

Reduction: $Al^{3+} + 3e^- \rightarrow Al$
Oxidation: $S^{2-} \rightarrow S + 2e^-$

Pulling electrons away from an atom or removing them altogether is **oxidation**. Adding electrons to an atom or merely shifting them toward it is **reduction**.

Example 12

Is chlorine oxidized or reduced in forming the chloride ion? What is the oxidation state of the ion?

Solution   Chlorine is reduced, since one electron per chlorine atom is added to form the ion. The chloride ion, Cl^-, is in the -1 oxidation state.

Example 13

When metals are converted into their ions, are they oxidized or reduced? What is the oxidation state of the aluminum ion?

Solution

Metals are oxidized to their ions, since electrons are removed. The aluminum ion, Al^{3+}, is in the $+3$ oxidation state.

If two or more oxidation states for a metal ion are possible, they are differentiated by writing the oxidation state in Roman numerals after the name of the atom. An older nomenclature, still in use, identifies the higher oxidation state by the ending -*ic* and the lower by -*ous*. Hence,

Fe^{2+}	iron(II) or ferrous	Fe^{3+}	iron(III) or ferric
Cu^{+}	copper(I) or cuprous	Cu^{2+}	copper(II) or cupric
Sn^{2+}	tin(II) or stannous	Sn^{4+}	tin(IV) or stannic

Example 14

When the ferric ion is converted to the ferrous ion, is this an oxidation or reduction? Write the equation for the process.

Solution

The equation is $Fe^{3+} + e^{-} \rightarrow Fe^{2+}$. The process is a reduction since an electron is added.

The modern nomenclature with Roman numerals is easier to use because it does not require you to remember what the two oxidation states of a metal are, in order to know what a compound is from its name.

A **salt** is a compound made up of positive and negative ions. Because a salt must be electrically neutral, the total charge on its positive and negative ions must be zero. Since each ion of Sn^{2+} has a charge of $+2$, twice as many chloride ions with -1 charge each are required to produce a zero net charge. Hence the salt of Sn^{2+} and Cl^{-} ions has the overall composition $SnCl_2$, rather than SnCl or $SnCl_3$. It is called stannous chloride or tin(II) chloride. The formula for stannic chloride or tin(IV) chloride is $SnCl_4$.

In addition to these simple ions, compound or complex ions can be formed between a metal or nonmetal and oxygen, chlorine, ammonia (NH_3), the hydroxide ion (OH^{-}), or other chemical groups. The sulfate ion, SO_4^{2-}, has four oxygens at the corners of a tetrahedron around the central sulfur atom, and an overall charge of -2. The nitrate ion, NO_3^{-}, has three oxygen atoms in an equilateral triangle around the nitrogen, and a -1 charge. The ammonium ion, NH_4^{+}, has four hydrogens at the corners of a tetrahedron, and a $+1$ charge. These ions are thought of as units because they form salts

Table 1-5

Some Common Complex Ions

Cations			Anions			
+1	+2	+3	−4	−3	−2	−1
NH_4^+ Ammonium	$Cu(NH_3)_4^{2+}$ Tetraammine-copper(II)	$Co(NH_3)_6^{3+}$ Hexaammine-cobalt(III)	$Fe(CN)_6^{4-}$ Hexacyano-ferrate(II) (ferrocyanide)	PO_4^{3-} Phosphate	CO_3^{2-} Carbonate	OH^- Hydroxide
$Ag(NH_3)_2^+$ Diammine-silver(I)	VO^{2+} Vanadyl			AsO_4^{3-} Arsenate	SO_4^{2-} Sulfate	NO_3^- Nitrate
$(CH_3)_4N^+$ Tetramethyl-ammonium	UO_2^{2+} Uranyl			AsO_3^{3-} Arsenite	SO_3^{2-} Sulfite	NO_2^- Nitrite
NO^+ Nitrosyl	$Ni(NH_3)_6^{2+}$ Hexaammine-nickel(II)			BO_3^{3-} Borate	CrO_4^{2-} Chromate	BF_4^- Fluoroborate
NO_2^+ Nitryl				$Fe(CN)_6^{3-}$ Hexacyano-ferrate(III) (ferricyanide)	$Cr_2O_7^{2-}$ Dichromate	CN^- Cyanide
					$C_2O_4^{2-}$ Oxalate	ClO^- Hypochlorite
					$S_2O_3^{2-}$ Thiosulfate	ClO_2^- Chlorite
					$PtCl_4^{2-}$ Tetrachloro-platinate(II)	ClO_3^- Chlorate
					$PtCl_6^{2-}$ Chloroplatinate or hexachloro-platinate(IV)	ClO_4^- Perchlorate
						BrO_3^- Bromate
						IO_3^- Iodate
						MnO_4^- Perman-ganate
						SCN^- Thiocyanate
						$C_2H_3O_2^-$ Acetate
						I_3^- Triiodide

the way single-atom ions do, and go through many chemical reactions unchanged. Silver nitrate, $AgNO_3$, is a salt containing equal numbers of Ag^+ and NO_3^- ions. Ammonium sulfate is a salt with twice as many ammonium ions, NH_4^+, as sulfate ions, SO_4^{2-}, and the chemical formula $(NH_4)_2SO_4$. Other typical complex ions are shown in Table 1-5.

Table 1-6

Common Coordination Numbers

Element	Coordination number	Examples
Fe	6	$Fe(CN)_6^{4-}$, $Fe(CN)_6^{3-}$
Co	4, 6	$CoCl_4^{2-}$, $Co(NH_3)_6^{3+}$, $Co(H_2O)_6^{2+}$
Ni	4, 6	$Ni(CN)_4^{2-}$, $Ni(NH_3)_6^{2+}$
Cu	4, 6	$CuCl_4^{2-}$, $Cu(H_2O)_6^{2+}$
Zn	4	$Zn(CN)_4^{2-}$
Pt	4, 6	$PtCl_4^{2-}$, $PtCl_6^{2-}$
B	3, 4	BO_3^{3-}, BF_4^-
C	3, 4	CO_3^{2-}, CH_4, CF_4
N	3, 4	NO_3^-, NH_4^+
Si	4, 6	SiO_4^{2-}, SiF_6^{2-}
S	4, 6	SO_4^{2-}, SF_6
Cl	1, 2, 3, 4	ClO^-, ClO_2^-, ClO_3^-, ClO_4^-
As	3, 4	AsO_3^{3-}, AsO_4^{3-}
Sb	6	$Sb(OH)_6^-$, $SbCl_6^-$
I	3, 4, 6	IO_3^-, IO_4^-, IO_6^{5-}

When a central atom is surrounded by several equally spaced atoms, the number of surrounding atoms is called the **coordination number**. The most important factor is size. Nitrogen in the nitrate ion, NO_3^-, has room for three oxygen atoms around it, and hence a coordination number of 3 for oxygen. The sulfur atom is larger than a nitrogen atom, and can accommodate one more oxygen atom in the sulfate ion, SO_4^{2-}. Hence the coordination number of sulfur for oxygen is 4.

The most common coordination numbers are 2, 3, 4, and 6. (See Table 1-6.) An ion or molecule with a central atom having a coordination number of 2 can be either linear, as carbon dioxide with $O-C-O$ in a straight line, or bent, as in water, H_2O. Possible structures for ions or molecules with coordination numbers of 3, 4, and 6 are shown in Figure 1-6.

It is not strictly correct to talk about molecular formulas and molecular weights of salts, since there are no molecules in salts—only ordered lattices of ions. No one sodium ion in the sodium chloride structure shown in Figure 1-5 "belongs" to a particular chloride ion. It is correct, however, to speak of the **chemical formula** of a salt, and the **formula weight** that corresponds to it. Since the chemical formula for sodium chloride is NaCl, the formula weight of sodium chloride is the sum of the atomic weights of one atom of sodium and one atom of chlorine:

1 sodium:	22.990 amu
1 chlorine:	35.453 amu
Total:	58.443 amu

Coordination number 3

Trigonal planar

BO_3^{3-}

Trigonal pyramidal

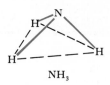

NH_3

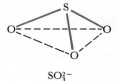

SO_3^{2-}

Coordination number 4

Square planar

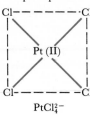

$PtCl_4^{2-}$

Tetrahedral

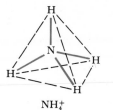

NH_4^+

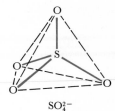

SO_4^{2-}

Coordination number 6

Octahedral

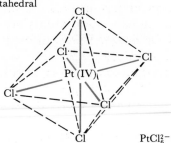

$PtCl_6^{2-}$

Figure 1-6   Geometry of atoms around central atoms with coordination numbers 3, 4, and 6. If L is any peripheral atom and M is the central atom, then the bond angle L—M—L is 120° for trigonal planar, 109.5° for tetrahedral, and typically around 109.5° for trigonal pyramidal geometries. Square planar and octahedral geometries have two L—M—L angles, 90° and 180°.

It is conventional to call this the "molecular weight" of sodium chloride, and no confusion results as long as you realize what a salt structure is like. A mole of sodium chloride is 58.443 g. It will contain 6.022×10^{23} sodium ions and 6.022×10^{23} chloride ions. Even though they are not paired off into molecules, the ratio is strictly one to one.

Example 15

What is the molecular weight of ammonium sulfate?

Solution The chemical formula of ammonium sulfate is $(NH_4)_2SO_4$, so the molecular weight (actually the formula weight) is

2 nitrogens:	2×14.007 amu =	28.014 amu
8 hydrogens:	8×1.008 amu =	8.064 amu
1 sulfur:	1×32.06 amu =	32.06 amu
4 oxygens:	4×15.999 amu =	63.996 amu
Total:		132.13 amu

The simple anions are named by adding *-ide* to the name of the element, as in the fluoride (F^-), chloride (Cl^-), oxide (O^{2-}), and sulfide (S^{2-}) ions. Where more than one complex anion of an element with oxygen can be formed, the suffixes *-ate* and *-ite* are used for the higher and lower oxidation states, respectively. Thus,

Sulfate ion	SO_4^{2-}	Sulfite ion	SO_3^{2-}
Nitrate ion	NO_3^-	Nitrite ion	NO_2^-
Arsenate ion	AsO_4^{3-}	Arsenite ion	AsO_3^{3-}

If more than two such anions exist, then the prefixes *hypo-* ("under") and *per-* ("beyond") are used:

Perchlorate ion	ClO_4^-
Chlorate ion	ClO_3^-
Chlorite ion	ClO_2^-
Hypochlorite ion	ClO^-

Melting Points and Boiling Points of Salts

A salt crystal represents a particularly stable balance of positive and negative charges, with each type of ion being kept out of the way of others of like charge. Melting a salt crystal means upsetting this delicate balance of charges, and allowing mutually repelling ions to come closer together from time to time as the ions flow past one another. This disruption of structure requires large amounts of energy to accomplish, so the melting points of salts are higher than those of molecular solids. The melting points of two salts, sodium chloride (NaCl) and potassium sulfate (K_2SO_4), are compared in Table 1-7 with those of the elements from which the salts are made.

Table 1-7

Melting and Boiling Points of Two Salts and Their Component Elements

Substance	Chemical formula	T_m (°C)	T_b (°C)
Sodium metal	Na	97.8	882.9
Chlorine gas	Cl_2	−101.0	−34.6
Sodium chloride (salt)	NaCl	801	1413
Potassium metal	K	64	774
Sulfur	S	119	445
Oxygen gas	O_2	−218	−183
Potassium sulfate (salt)	K_2SO_4	1069	1689

Metallic sodium melts at 97.8°C, and solid chlorine melts at −101°C, but their combination, sodium chloride (common table salt), requires a temperature of 801°C before it will melt. Boiling or vaporizing a salt is even more difficult. The ions remain ions in the liquid state, tumbling past one another as in any other liquid; but before the gas phase can be attained, Na^+ and Cl^- ions must pair off into neutral NaCl molecules. To accomplish this pairing, electrons have to be pulled away from Cl^- ions, which have a strong attraction for them, and pushed toward Na^+ ions, which do not want them. The NaCl bond in sodium chloride vapor is extremely polar, with the electron pair skewed strongly toward the chlorine atom, but the separation still is not as complete as in Na^+ and Cl^- ions. Much energy is required to push electrons where they are not wanted and to make NaCl molecules from Na^+ and Cl^- ions, so high temperatures are required before this can happen. Hence the very high boiling points of salts in comparison with molecular compounds, as illustrated in Table 1-7.

1-7 IONS IN SOLUTION

Although salts are hard to melt and even harder to vaporize, many can be dissolved easily in a polar liquid such as water. The reason for this is simple. The water molecules help to dismantle the salt crystal, since the partial positive and negative charges on the polar water molecules (Figure 1-4) provide a substitute for the positive and negative charges that were present in the crystal lattice. Figure 1-7 illustrates what happens when a crystal such as sodium chloride is dissolved in water. Each positively charged Na^+ is surrounded by water molecules with their negatively charged oxygens turned toward it, and each negatively charged Cl^- ion is surrounded by water molecules with their positively charged hydrogens closest. The ions

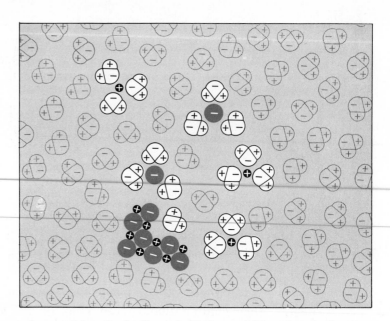

Figure 1-7

Breakup of a salt crystal by water molecules, with hydration of ions. Each salt ion in solution is surrounded by polar water molecules with the opposite charge to that of the ion turned toward it. This electrostatic hydration energy compensates for the loss of attractions between ions in the salt crystal. From Dickerson and Geis, *Chemistry, Matter, and the Universe.*

from the salt crystal are said to be **hydrated**. If the stability that hydration gives the ions in solution is greater than the stability of the crystal lattice, then the salt will dissolve. Sodium chloride is a familiar example of a soluble salt. In contrast, if the hydration energy is too small, then the crystal will be the more stable form, and it will not dissolve in water. Silver chloride (AgCl) and barium sulfate ($BaSO_4$) are examples of insoluble salts. When a salt crystal dissolves, it does not simply come apart into ions; it is taken apart by the molecules of the liquid in which it is dissolved (the solvent). This is why salts will not dissolve in nonpolar liquids such as gasoline (octane, C_8H_{18}); there are no charges on the solvent molecules to make up for the loss of charge attractions within the crystalline salt.

Salt solutions conduct electricity, and this property was extremely important early in the development of theories of chemical bonding. Electrical conduction in metals takes place by means of moving electrons; the metal ions remain in place. Crystalline salts do not conduct electricity at all, but if the salt is melted, then positive ions can migrate one way through the liquid and negative ions can move the other way in the presence of an electric field. This mobility of ions is even greater if the salt is dissolved in water and the ions consequently are hydrated.

Some of the first concrete ideas about the nature of chemical bonding came from the electrolysis experiments of the English scientist Michael Faraday (1791–1867). (**Electrolysis** means "breaking apart with electricity.") If sodium chloride is melted (above 801°C) and if two electrodes (the **cathode** and the **anode**) are inserted into the melt as shown in Figure 1-8 and an electric current is passed through the molten salt, then **chemical reactions** take place at the electrodes: Sodium ions migrate to the cathode, where electrons enter the melt, and are reduced to sodium metal:

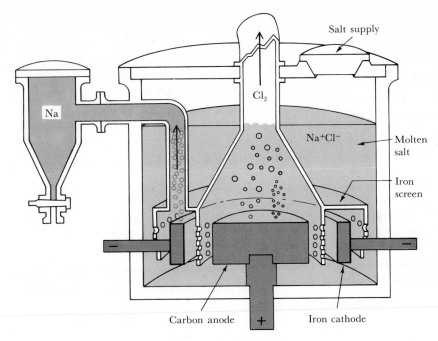

Figure 1-8 A commercial electrolysis cell for the production of metallic sodium and chlorine gas from molten NaCl. Liquid sodium floats to the top of the melt above the cathode and is drained off into a storage tank. Chlorine gas bubbles out of the melt above the anode. From Dickerson and Geis, *Chemistry, Matter, and the Universe.*

$$Na^+ + e^-(\text{from cathode}) \rightarrow Na$$

Chloride ions migrate the other way, toward the anode, give up their electrons to the anode, and are oxidized to chlorine gas:

$$Cl^- \rightarrow \tfrac{1}{2}Cl_2 + e^-(\text{to anode})$$

The overall reaction is the breakdown of sodium chloride into its elements:

$$Na^+ + Cl^- \rightarrow Na + \tfrac{1}{2}Cl_2$$

Sodium ions are reduced and chloride ions are oxidized. Electrolysis can also be carried out by passing electric current through solutions of salts (Figure 1-9). If a solution of sodium chloride in water is electrolyzed, chlorine gas is given off at the anode as in the case of molten sodium chloride, but the cathode product is hydrogen gas rather than metallic sodium:

$$Na^+ + Cl^- + H_2O \rightarrow Na^+ + \tfrac{1}{2}Cl_2 + \tfrac{1}{2}H_2 + OH^- \qquad (1\text{-}1)$$

This is the same result that would be obtained if liquid sodium chloride was first electrolyzed to give metallic sodium:

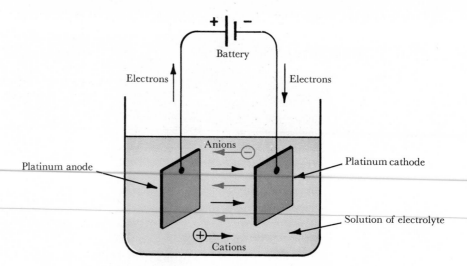

Figure 1-9   Schematic diagram of an electrolysis cell. For current to be carried, the fluid must contain mobile ions, either as a molten salt or as hydrated ions in solution. A substance capable of carrying current by migration of ions is called an *electrolyte*. If the electrolyte is a solution of $CuCl_2$, which dissociates (breaks apart) to give Cu^{2+} and Cl^- ions, then as current is passed through the cell, Cu^{2+} ions migrate to the cathode and are reduced to metallic copper, and Cl^- ions migrate to the anode, where they are oxidized to Cl_2 gas. Platinum electrodes are used because they are chemically inert and will not react.

$$Na^+ + Cl^- \rightarrow Na + \tfrac{1}{2}Cl_2 \tag{1-2}$$

and the sodium was then dumped into water:

$$Na + H_2O \rightarrow Na^+ + \tfrac{1}{2}H_2 + OH^- \tag{1-3}$$

Equation 1-1 is just the sum of equations 1-2 and 1-3, since the sodium metal that is produced in equation 1-2 is used up in equation 1-3. There is nothing mysterious about the different cathode products during electrolysis of sodium chloride in a melt or in solution. If water is present, some of the H_2O molecules will be dissociated into H^+ and OH^- ions. Because H^+ has a greater affinity for electrons than Na^+ does, the H^+ ions will take electrons away from metallic sodium, making the anode product H_2 rather than Na, and leaving Na^+ ions in solution. In contrast, Cu^{2+} ions have a greater affinity for electrons than H^+ ions do, so the anode product of electrolysis of $CuCl_2$ is metallic copper, whether the process is carried out in the melt or in solution (Figure 1-9). Typical products of electrolysis of solutions and melts are given in Table 1-8. Electrochemical reactions and cells are discussed in detail in Chapter 19. At the moment, we are focusing on what electrochemical reactions tell us about chemical bonding.

Table 1-8

Products of Electrolysis

Electrolyte	Cathode product	Anode product
Sulfuric acid (H_2SO_4) in H_2O	H_2	O_2
Sodium sulfate (Na_2SO_4) in H_2O	H_2	O_2
Sodium chloride (NaCl) in H_2O	H_2	Cl_2
Potassium iodide (KI) in H_2O	H_2	I_2
Copper sulfate ($CuSO_4$) in H_2O	Cu	O_2
Silver nitrate ($AgNO_3$) in H_2O	Ag	O_2
Mercuric nitrate [$Hg(NO_3)_2$] in H_2O	Hg	O_2
Lead nitrate [$Pb(NO_3)_2$] in H_2O	Pb	O_2 and some PbO_2
Molten lye (NaOH); not in H_2O	Na	O_2

Faraday found that there was a quantitative relationship between the amount of electricity passed through an electrolytic cell and the amount of chemical change produced. He formulated **Faraday's laws of electrolysis**, which in terms of the modern theory of atoms and ions can be expressed as follows:

1. Passing the same quantity of electricity through a cell always leads to the same amount of chemical change for a given reaction. The weight of an element deposited or liberated at an electrode is proportional to the amount of electricity that is passed through.

2. It takes 96,485 coulombs of electricity to deposit or liberate 1 mole of a substance that gains or loses one electron during the cell reaction. If n electrons are involved in the reaction, then $96,485n$ coulombs of electricity are required to liberate a mole of product.

The quantity 96,485 coulombs of electricity has become known as 1 **faraday** in his honor, and has been given the symbol $\mathscr{F}$. Faraday's laws become self-evident when you realize that $1\mathscr{F}$ is simply the charge on 1 mole of electrons, or 6.022×10^{23} electrons. The scale-up factor of 6.022×10^{23} from molecules to moles is paralleled by the same scale-up factor from 1 electron charge to $1\mathscr{F}$ of charge. At the time, of course, Faraday knew neither the value of Avogadro's number nor the charge on an electron. His experiments did tell him, however, that charges on ions came in multiples of a fundamental unit, such that 96,485 coulombs corresponded to a mole of these units. The word *electron* first appeared in 1881, when the British physicist G. J. Stoney coined it to denote this fundamental unit of ionic charge. Its application to a real negatively charged particle came a decade later.

Example 16

Write equations for the reactions that occur when current is passed through molten NaCl. How many grams of sodium and chlorine are released when $1\mathscr{F}$ of charge is passed through the cell?

Solution

The cathode reaction is $Na^+ + e^- \rightarrow Na$, and the anode reaction is $Cl^- \rightarrow \frac{1}{2}Cl_2 + e^-$. When 1 mole of electrons $(1\mathscr{F})$ passes through molten NaCl, each electron reduces one sodium ion, so 1 mole of sodium atoms is produced. Hence 22.990 g of Na are deposited at the cathode. At the anode, 1 mole of electrons is removed from 1 mole of chloride ions, leaving 1 mole of chlorine atoms, which combine pairwise to make $\frac{1}{2}$ mole of Cl_2 molecules. Hence the weight of chlorine gas released is 35.453 g (the atomic weight of Cl, half the molecular weight of Cl_2).

Example 17

How many grams of magnesium metal and chlorine gas are released when $1\mathscr{F}$ of electricity is passed through an electrolytic cell containing molten magnesium chloride, $MgCl_2$?

Solution

The cathode reaction is this: $Mg^{2+} + 2e^- \rightarrow Mg$, and the anode reaction is this: $2Cl^- \rightarrow Cl_2 + 2e^-$. Since two electrons are required to reduce each ion of Mg^{2+}, 1 mole of electrons will be sufficient to reduce half a mole of magnesium ions, depositing 12.153 g of magnesium. (The atomic weight of magnesium is 24.305 g $mole^{-1}$.) As in Example 16, 1 mole of Cl^- ions is oxidized, liberating half a mole or 35.453 g of Cl_2 gas.

Example 18

The main commercial source of aluminum metal is the electrolysis of molten salts of Al^{3+}. How many faradays of charge, and how many coulombs, must be passed through the melt to deposit 1 kg of metal?

Solution

One kilogram of aluminum is 1000 g/26.98 g $mole^{-1}$, or 37.06 moles. Since each atom of aluminum deposited requires three electrons, 37.06 moles will require 3×37.06, or 111.2, moles of electrons. Hence $111.2\mathscr{F}$ or 10,730,000 coulombs will be needed.

Example 19

Electron flow at the rate of 1 coulomb per second (coulomb sec^{-1}) is a current of 1 ampere (A). Currents in industrial electrolytic production of aluminum are ordinarily in the range of 20,000 to 50,000 A. If a cell is operated at 40,000 A (40,000 coulombs sec^{-1}), how long will it take to produce the kilogram of aluminum metal mentioned in Example 18?

Solution

The time required will be

$$\frac{10{,}730{,}000 \text{ coulombs}}{40{,}000 \text{ coulombs sec}^{-1}} = 268 \text{ sec or 4.5 min}$$

Faraday's laws are represented diagrammatically in Figure 1-10. We have been using these laws with a prior knowledge of the charges on different ions, and the knowledge that 96,485 coulombs is the total charge on 6.022×10^{23} electrons. History actually operated in reverse: Faraday and others used electrolysis experiments to find out what the charges on ions were. The reasoning used is illustrated in Table 1-9. If twice as much electricity is required to liberate a mole of copper as a mole of silver (assuming

Table 1-9

Deduction of Ionic Charge by Electrolysis

Product of electrolysis	Electrode	Faradays per mole of atoms deposited	Ion in solution
Silver (Ag)	Cathode	1[a]	Ag$^+$
Chlorine (Cl$_2$)	Anode	1	Cl$^-$
Copper (Cu)	Cathode	2	Cu^{2+}
Hydrogen (H$_2$)	Cathode	1	H$^+$
Iodine (I$_2$)	Anode	1	I$^-$
Oxygen (O$_2$)[b]	Anode	2	O^{2-}
Zinc (Zn)	Cathode	2	Zn^{2+}

[a]For example, electrolysis of silver nitrate solution for 1 hour by using a current of 0.5 A deposits 2.015 g of silver; $2.015/107.9 = 0.0187$ mole of silver.

$$\frac{(0.5 \text{ coulomb sec}^{-1}) \times 3600 \text{ sec}}{96.485 \text{ coulombs } \mathscr{F}^{-1}} = 0.0187 \, \mathscr{F}$$

[b]Actually, oxygen (O$_2$) is produced by a complicated electrode reaction. The species O^{2-} can exist in molten oxides, but in water O^{2-} becomes 2OH$^-$ by reaction with a water molecule.

(a) $Cu^{2+} + 2Cl_2 \rightarrow Cu + Cl_2$

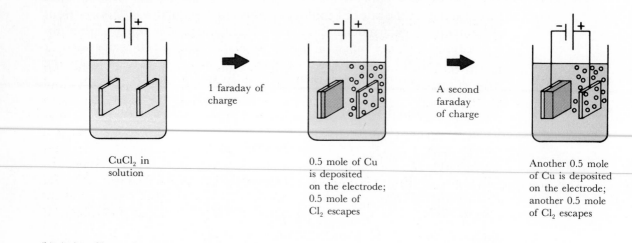

| CuCl₂ in
solution | 1 faraday of
charge | 0.5 mole of Cu
is deposited
on the electrode;
0.5 mole of
Cl₂ escapes | A second
faraday
of charge | Another 0.5 mole
of Cu is deposited
on the electrode;
another 0.5 mole
of Cl₂ escapes |

(b) $Ag^+ + Cl^- \rightarrow Ag + \frac{1}{2}Cl_2$

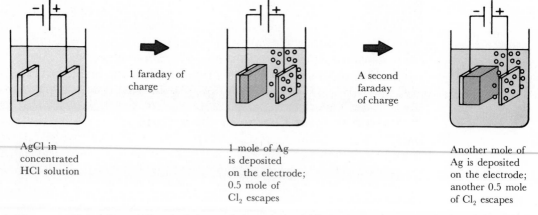

| AgCl in
concentrated
HCl solution | 1 faraday of
charge | 1 mole of Ag
is deposited
on the electrode;
0.5 mole of
Cl₂ escapes | A second
faraday
of charge | Another mole of
Ag is deposited
on the electrode;
another 0.5 mole
of Cl₂ escapes |

Figure 1-10   Illustrations of Faraday's laws of electrolysis. (a) Two electrons are required to reduce each ion of Cu^{2+}, or 2 moles of electrons ($2\mathscr{F}$) for each mole of copper. Each faraday is enough to oxidize 1 mole of Cl^- ions to $\frac{1}{2}$ mole of Cl_2 gas. (b) Only $1\mathscr{F}$ of charge is required to reduce 1 mole of Ag^+ ions to metallic silver, since the ionic charge on Ag^+ is only $+1$. Chlorine gas is liberated at the same rate per faraday as before.

that you know the atomic weights of the two metals and can calculate the weights of a mole of each), then the copper ion must have twice the charge of the silver ion. In Table 1-9, the number of faradays of charge required to liberate 1 mole of an element is the same as the number of charges, positive or negative, on the ion.

1-8 GASEOUS IONS

It was proposed even as far back as the time of Ben Franklin and John Dalton that the forces between particles of matter must be electrical in some way. But because like charges repel one another, it was believed, wrongly, that bonds could not exist between identical atoms, whereas we now know that most common gases occur as diatomic molecules: H_2, N_2, O_2, F_2, Cl_2, and so on. This one blunder led to nearly a half-century of confusion about molecular structure and atomic weights, during which it was thought that hydrogen gas was H instead of H_2, water was HO instead of H_2O, and oxygen had an atomic weight of 8 rather than 16. Electron pairs as the "glue" that holds atoms together in covalent bonds were not even proposed systematically until 1913, by G. N. Lewis; they were not explained theoretically for still another 20 years. Faraday's experiments showed that the charges on ions did occur naturally in fundamental pieces or units such that a mole of these charges equaled $1\,\mathscr{F}$, and Stoney named this elemental unit the *electron*. But Stoney's electron was not necessarily a particle that could be isolated and studied.

The people who first showed that electrons were real particles that could be added to or removed from atoms were physicists studying the effects of electricity on gases. They found that if they set up an electrical potential of 10,000 volts between two electrodes in a sealed tube (a *Crookes* tube) containing gas at low pressure, they observed a glow discharge (Figure 1-11). This discharge is what makes neon signs glow. The electrical potential strips electrons from atoms of the gas, sending the electrons streaming toward the anode and positive ions toward the cathode. These moving electrons (the cathode rays) can be detected by watching the flashes of light on a zinc sulfide screen placed in the tube. If a lightweight pinwheel is set up in the path of the electrons inside the tube, the electrons even make the pinwheel rotate. On their way to the anode, the cathode rays strike other atoms of the gas, causing the emission of light in a glow discharge. The color of the glow discharge will vary, depending on the gas used inside the tube.

If a metal plate with a slit is placed in front of the cathode, then the electrons in the cathode ray will be confined to a thin beam. This beam is deflected by electric and magnetic fields in a way that indicates that the particles in the beam carry a negative charge. The relative amount of bending of the **canal rays** (positive ions) and **cathode rays** (negative electrons) shows that the cathode ray particles are extremely light, whereas the positive ions are roughly as heavy as the original atoms from which they came. The exact nature of the canal rays depends on what gas is used in the tube, but the cathode rays are the same for all gases. J. J. Thomson (1856–1940) suggested that the particles in the cathode rays might in fact be Stoney's "electrons," and in 1897 he found a way to use the deflection of the

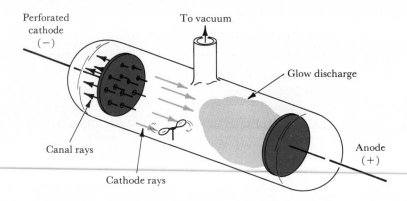

Figure 1-11   A Crookes tube. When a high voltage (about 10,000 volts) is applied across two electrodes in a sealed glass tube containing gas at low pressure, the voltage induces the
breakdown of gas molecules into electrons and positive ions. The electrons stream toward
the anode and are known as *cathode rays*, and the positive ions stream toward the
cathode and are termed *canal rays*. If the cathode is perforated, the positive ions will
pass through it and cause a glow where they strike the glass walls. If a lightweight
pinwheel is placed in the path of the cathode rays, they can cause it to rotate. As the
electrons of the cathode rays move toward the anode, they strike other gas molecules
and set up a glow discharge that is familiar in neon signs.

beam by electric and magnetic fields to calculate the charge-to-mass ratio
(e/m) of the particles. He found that

$$\frac{e}{m} = 1.76 \times 10^8 \text{ coulombs g}^{-1}$$

Example 20

Assume that Thomson's cathode ray particles are in fact the same as Stoney's
and Faraday's electrons, and that $1\mathscr{F}$ is a mole of electrons. Calculate the
mass of one electron.

Solution   The charge on one electron is

$$e = \frac{1\mathscr{F}}{N} = \frac{96,485 \text{ coulombs mole}^{-1}}{6.022 \times 10^{23} \text{ electrons mole}^{-1}}$$

$$= 1.602 \times 10^{-19} \text{ coulomb}$$

$$m = \frac{1.602 \times 10^{-19} \text{ coulomb}}{1.76 \times 10^8 \text{ coulomb g}^{-1}} = 0.910 \times 10^{-27} \text{ g}$$

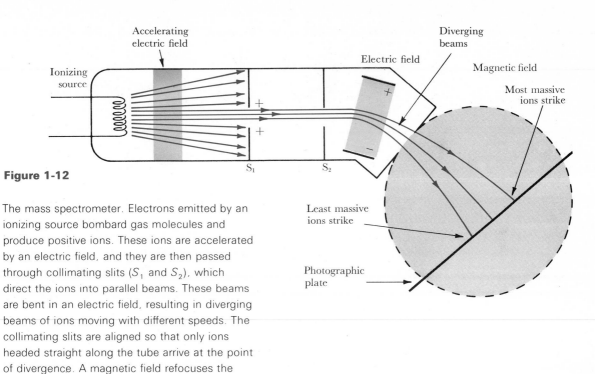

Figure 1-12

The mass spectrometer. Electrons emitted by an ionizing source bombard gas molecules and produce positive ions. These ions are accelerated by an electric field, and they are then passed through collimating slits (S_1 and S_2), which direct the ions into parallel beams. These beams are bent in an electric field, resulting in diverging beams of ions moving with different speeds. The collimating slits are aligned so that only ions headed straight along the tube arrive at the point of divergence. A magnetic field refocuses the beams in such a way that all ions of the same charge-to-mass ratio strike the same spot on the photographic plate.

A modern descendant of the Crookes tube and Thomson's apparatus is the mass spectrometer (Figure 1-12). It is a valuable research tool for measuring the mass per unit charge of any substance that can be given a positive charge. Mass spectrometry offers the most direct measurement of atomic weights of elements, and it is the method by which isotopes can be both identified and separated. By looking at the masses of the fragments into which molecules are broken down during electron bombardment in the spectrometer, organic chemists can obtain useful information about the molecular structure of a substance. (During the development of the atomic bomb in World War II, mass spectrometry was used to separate fissionable ^{235}U from ^{238}U, although the extremely low pressures that mass spectrometry requires were not practical for large-scale production.)

Although the ratio of electron charge to electron mass was measured in 1897 by Thomson, the charge itself was not measured until 1911, when Robert A. Millikan (1868–1953) obtained the charge by the ingenious experiment illustrated in Figure 1-13. He used x rays to irradiate a spray of tiny oil droplets between two chargeable plates. Electrons from ionization of the air around the drops adhered to the drops, giving them one, two, or more electron charges. Millikan measured first the rate of free fall of the charged

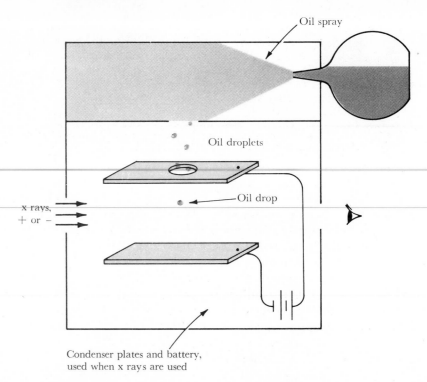

Oil spray

Oil droplets

x rays,
+ or −

Oil drop

Condenser plates and battery,
used when x rays are used

Figure 1-13   Millikan's oil-drop experiment. Tiny droplets of oil are introduced between two plates that
can be given an electrostatic charge. A drop of oil is allowed to fall freely through the
air, and its path is monitored. The radius of the drop is calculated from the terminal
velocity of its fall and the viscosity of air. The air is ionized by x rays, and negatively
charged particles (electrons) stick to the oil drops. The charge on a drop can be
determined from the voltage that must be applied across the condenser plates to make
the drop hang motionless, with electrostatic and gravitational forces in balance.

drops through air of known viscosity. Then he measured the voltage across
the plates that was sufficient to suspend the drops motionless between the
plates. He calculated that the charge on any one drop was always an integral
multiple of 1.6022×10^{-19} coulomb, and he concluded correctly that this
was the charge on a single electron.

Example 20 actually was presented the wrong way around. We can
solve it only because we know the value of Avogadro's number, whereas
in fact Millikan's results furnished one way of *calculating* Avogadro's number.

Example 21

Assume that you do not know the value of Avogadro's number, but that
you recognize that the faraday is the charge necessary to reduce 1 mole of

Na$^+$ ions, with one of Millikan's electrons combining with each ion. Calculate the number of ions in a mole, or Avogadro's number.

Solution

$$N = \frac{96,485 \text{ coulombs mole}^{-1}}{1.6022 \times 10^{-19} \text{ coulomb ion}^{-1}} = 6.022 \times 10^{23} \text{ ions mole}^{-1}$$

Summary

An **atom** consists of a positively charged **nucleus** surrounded by enough negatively charged **electrons** to yield zero net charge. The nucleus is constructed from positively charged **protons** and neutral **neutrons**, each of mass approximately 1 amu. The mass of an electron is approximately 1/1836 the mass of a proton; the charge on an electron is equal but opposite in sign to the charge on a proton. The total number of protons in the nucleus (and electrons in a neutral atom) is the **atomic number**, Z. The total number of both protons and neutrons is the **mass number**, and the mass of the atom, in atomic mass units, is its **atomic weight**. The atomic weight is always slightly less than the sum of masses of the particles that go into making an atom, because mass is converted to energy and lost when the atom is formed.

All atoms with the same number of protons, and therefore the same atomic number, are classified as the same **element** and represented by a one- or two-letter symbol. Atoms of the same element with varying numbers of neutrons are called **isotopes** of the element. Isotopes are identified by placing the mass number as a superscript to the left of the symbol of the element (e.g., ^{37}Cl). The atomic number is sometimes added as a subscript (e.g., $^{37}_{17}$Cl), although it is actually not necessary since the element's name and atomic number are known from the symbol. Each isotope of an element has its own atomic weight, and the **natural atomic weight** is the weighted average of these isotopic values, the weighting being according to the natural abundance of each isotope.

A collection of atoms held together by chemical bonds is a **molecule** Usually, but not always, the bonding in a molecule can be explained in terms of electron pairs, each holding two atoms together. Such an electron-pair bond is a **covalent bond**. The sum of the atomic weights of all the atoms in a molecule is its **molecular weight**. Although atoms in different molecules are not directly bonded to one another, all molecules are slightly "sticky," and are attracted to other molecules. These **van der Waals attractions** will make the molecules of a gas adhere to one another to form a liquid if the temperature falls low enough, and make the molecules of a liquid fit together in a regular crystalline array in a solid if the temperature falls lower still. The temperatures at which these two transitions occur are the **boiling point**, T_b, and the **melting point**, T_m, respectively.

If two atoms differ in their intrinsic electron-pulling power or **electronegativity**, then the electron pair of the bond between them will be shifted toward the atom with the greater attraction, giving it a negative charge and the other atom a positive charge. The bond, and molecules that contain such bonds, are said to be **polar**. Polar molecules can attract one another, and they can also attract positively and negatively charged ions. Melting and boiling points of polar molecules are higher than would be expected from van der Waals attractions alone, because their polarity provides a second type of intermolecular attraction.

Atomic and molecular weights are measured on a scale of **atomic mass units (amu)**, where 1 amu is defined as exactly one-twelfth of the mass of a ^{12}C atom. A quantity of a chemical substance (atoms, molecules, or ions) equal to the atomic weight expressed in grams is defined as 1 **mole** of that substance. One mole of any substance—atoms, molecules or ions—contains the same number of particles of that substance. This property makes the mole a useful means of counting out particles merely by weighing them. The units of atomic and molecular weights are either grams per mole or amu per molecule (or atom).

Some atoms, those of metals in particular, have a weak hold on their electrons and can lose one, two, or more electrons to become positively charged **ions**, or **cations**. Many nonmetals or groups of atoms can acquire one or more negative charges to become negatively charged ions, or **anions**. A **salt** is a compound of the relative number of cations and anions that will produce zero overall charge. Common table salt, NaCl, contains equal numbers of Na^+ and Cl^- ions. The pulling away or outright removal of electrons is termed **oxidation**, and the addition to or shifting of electrons toward an atom is **reduction**. Since electrons are never created or destroyed in chemical reactions, whenever one substance is oxidized, some other substance must be reduced.

Simple anions made by adding electrons to single atoms have names ending in *-ide*, as *chloride*, Cl^-, and *sulfide*, S^{2-}, ions. For complex ions of a nonmetal atom with oxygen, the higher and lower oxidation state ions are differentiated by the suffixes *-ate* and *-ite*. The oxidation state of a metal cation (see Chapter 10) is indicated by a Roman numeral after the name of the metal, as in Fe^{3+}, iron(III), or by the suffixes *-ic* and *-ous*.

Although salts do not have separate molecules and, strictly speaking, cannot have molecular weights, they do have chemical formulas that express their overall composition in the simplest possible way. The weight of 1 mole of these atoms is the **formula weight** of the salt, but it is customary to refer to this as the salt's "molecular weight." Thus magnesium chloride has one Mg^{2+} ion for every two Cl^- ions, a net charge of zero, a chemical formula of $MgCl_2$, and a molecular weight of 95.211 g mole^{-1}.

The **coordination number** in a complex ion or molecule is the number of atoms or chemical groups bonded directly to the central atom. These bonding groups can be simple ions such as O^{2-} and Cl^- or molecules such as ammonia (NH_3) and water (H_2O). The maximum coordination number for a given central atom depends on the size of the atom and the size of its surrounding groups. The most common coordination numbers are 2, 3, 4, and 6.

Salts have higher melting and boiling points than molecular substances, because heat energy must be supplied to break apart the stable crystal lattice, and even more heat energy is required to force positive and negative ions to pair off and share electrons in neutral molecules that can go into a gas phase. Salts often dissolve readily in water, however, because polar attractions by the water molecules can compensate for the attractions of other ions in the crystal. Ions surrounded by polar water molecules in solution are said to be **hydrated**. Gasoline and similar nonpolar liquids cannot dissolve salts because they cannot hydrate (or **solvate**, if the solvent is other than water) the ions.

If a current of electricity is passed through molten salt or a salt solution, the current is carried by ions migrating in opposite directions. At the **cathode**, where electrons enter the salt medium, metal cations can be reduced to pure metal. At the **anode**, where electrons flow out of the salt and back into the external circuit, anions can be oxidized to liberate pure nonmetallic elements. This is the process of **electrolysis**. Faraday found a quantitative relationship between the amount of charge passed through a cell and the amount of chemical change produced: 96,485 coulombs of charge will bring about 1 mole of a change that involves one electron per ion. The quantity, 96,485 coulombs, is simply the charge on 1 mole of electrons and is called 1 faraday ($\mathscr{F}$) of charge.

Electrons as separate particles were studied by physicists interested in low-pressure gas discharges under high voltages. *Cathode rays* consist of a beam of electrons stripped away from the gas atoms. J. J. Thomson showed, by means of deflecting magnetic and electrostatic fields, that the cathode rays were made of negatively charged particles, and he measured the charge-to-mass ratio of the particles. R. A. Millikan completed the process, in his oil-drop experiment, by successfully measuring the charge on the electron. This, combined with Faraday's results, led to the calculation of **Avogadro's number**, the number of electrons in a faraday of charge, or the number of particles in a mole of any substance. The mass spectrometer, a descendant of Thomson's gas-discharge tubes, is a modern analytical tool and a means of finding the charge-to-mass ratio for any atomic or molecular species that can be given a charge.

Self-Study Questions

1. What particles are found in the nucleus of an atom? What charge does each kind of particle have? Does the nucleus have an overall charge, and is it positive or negative?
2. In view of your answer to Question 1, how is it that an atom is electrically neutral? What makes it so?
3. Which kind of particle is most responsible for the observed chemical behavior of an atom? Why is this so?

4. What are the differences between atomic weight, atomic number, and mass number? Which of these is most directly related to the chemical behavior of an atom, and why? What are all atoms that have the same value of this number called?

5. What are isotopes? Do all isotopes of an element have the same (a) atomic weight? (b) atomic number? (c) mass number?

6. How is the symbol for an element chosen? How can it be related to the name of the element?

7. In the isotope barium-138, is 138 the atomic weight, the atomic number, or the mass number? How many protons, electrons, and neutrons are found in an atom of barium-138? Which of the following ways of writing this isotope is correct: (a) $_{138}Ba$; (b) ^{138}Ba; (c) Ba_{138}?

8. The observed atomic weight of barium in nature is 137.34 g mole^{-1}. Give *three* reasons why this weight is not identical with the 138 of Question 7.

9. What are *mass loss* and *binding energy,* in terms of atomic weights?

10. How is the atomic weight of an element as listed on the inside back cover of this book related to the atomic weights of the isotopes of the element?

11. What is a molecule? What is an electron-pair bond between two atoms in a molecule called?

12. What are the molecular formulas for hydrogen, chlorine, oxygen, water, methane, methanol, hydrogen sulfide, and ammonia?

13. What are the shapes of the molecules mentioned in Question 12? Sketch the geometry of the bonds around the central atom in water, hydrogen sulfide, ammonia, and methane.

14. How is the molecular weight of a molecule related to the atomic weights of the atoms contained in it?

15. How many molecules of chlorine are there in 1 mole of chlorine gas? How many atoms of chlorine are there in 1 mole of chlorine gas?

16. How many moles of hydrogen gas molecules can be made from 1 mole of hydrogen atoms?

17. Why do gases condense into liquids when the temperature falls low enough? Why do liquids freeze to solids at even lower temperatures?

18. Why are the melting and boiling points of polar molecules higher than those of nonpolar molecules of similar size?

19. Why are the melting and boiling points of salts much higher than those even of polar molecules?

20. What is electronegativity? How does a difference in electronegativity between two bonded atoms affect the bond?

21. What is a mole of a substance? Which has more particles: a mole of hydrogen or a mole of methane?

22. If the hydrogen and methane molecules of Question 21 were torn apart to atoms, which would produce the greater number of atoms? How many atoms would result from a mole of methane?

23. How many moles of oxygen *atoms* are needed to make a mole of water molecules? How many moles of oxygen gas *molecules* does this correspond to?

24. Which atom is more electronegative, Na or Cl? How does this determine the charge on the ions in NaCl, sodium chloride?

25. What is a salt? Sodium chloride is a salt, and it contains an equal number of anions and cations. Why do some salts have unequal numbers of anions and cations?

26. Are there molecules in a salt? What meaning does *molecular weight* have for a salt? How many sodium ions are there in a mole of sodium chloride?

27. What should the charge be on a *hydride ion,* judging from the suffix *-ide?* An example is lithium hydride, LiH. Does this fit your answer just given?

28. The formula CuCl stands for copper(I) chloride. Why doesn't BeO stand for beryllium(I) oxide? What does BeO stand for?

29. Why is Cu^{2+} the cupr*ic* ion, whereas Fe^{2+} (which has the same charge) is the ferr*ous* ion? What are the modern names for these ions?

30. What are oxidation and reduction?

31. Oxygen is more electronegative than any element except fluorine. Why, then, does the combination of another element with oxygen amount to an oxidation of that element, by the way oxidation has been defined in this chapter? When copper is oxidized by oxygen to Cu^{2+} in CuO, what substance is reduced?

32. What do the prefixes *per-* and *hypo-* signify in the oxyanions (oxygen-containing anions) of chlorine?

33. How do molten salts conduct electricity? How do salt solutions conduct electricity?

34. What is electrolysis? At which electrode do electrons flow into the material being electrolyzed? At which electrode does reduction take place?

35. Toward which electrode do anions migrate? Can you see why ions of that charge should have been named *an*ions?

36. The Greek prefix *cata-* means "away," as in *catapult,* and *ana-* means "back." In terms of electron flow, why are the cathode and anode in an electrolysis cell appropriately named?

37. What is the relationship between quantity of charge passed through an electrolysis cell and amount of chemical reaction carried out? What is $1\mathscr{F}$ of charge, and what does it mean at the atomic level?

38. How many faradays of electricity are required to reduce 10 moles of ferric ions to metallic iron? How many coulombs is this? How long would a current of 10 A have to flow?

39. How did J. J. Thomson show that cathode rays were made up of discrete particles? How did he show that these particles were negatively charged? Could he measure the charge per particle?

40. How did Millikan measure the charge on the electron?

41. How can Avogadro's number be calculated from the results of Faraday's and Millikan's experiments?

Problems

Atomic symbols

1. Six elements have names that begin with the letter *R*. How are their atomic symbols derived from their names in a way that avoids confusion among them?

2. Four elements have names beginning with the letter *G*. How are these distinguished by atomic symbols? One symbol does not itself begin with *G*. Which one is it, and where do you think it might have come from?

Expected atomic weights

3. How many protons and neutrons are present in an atom of aluminum-27? How many electrons are there? What is the expected atomic weight? Compare this with the observed value on the inside back cover of this book, bearing in mind that aluminum has only one naturally occurring isotope. How do you explain the discrepancy in atomic weights?

4. What is the expected atomic weight of niobium-93? How many protons, neutrons, and electrons are there per atom? The only natural isotope is ^{93}Nb. How does your value for atomic weight compare with the observed value given on the inside back cover? Why the discrepancy?

Atomic structure

5. How many electrons, protons, and neutrons are present in an atom of nitrogen-14? How is its symbol written with both subscripts and superscripts? What would an atom of nitrogen-14 become if one more neutron were added to its nucleus? How would this affect the number of electrons around the nucleus? What would nitrogen-14 become if one more proton were added to the nucleus? How would this affect the total number of electrons?

6. How many protons, neutrons, and electrons are found in an atom of sulfur-32? How is the atomic symbol written (subscripts and superscripts)? What would this atom become if one neutron were removed from the nucleus, and how would the removal affect the number of electrons present? What would it become if one proton were removed instead, and how would this removal affect the number of electrons?

Natural atomic weights

7. Zinc, Zn, has five stable isotopes, with natural abundance and individual atomic weights as follows:

a) 48.89%   63.9291 amu
b) 27.81%   65.9260 amu
c) 4.11%   66.9271 amu
d) 18.54%   67.9249 amu
e) 0.62%   69.9253 amu

How many protons and neutrons are present in each isotope? What is the symbol for each isotope? (Use superscripts only.) What is the average atomic weight of naturally occurring zinc?

8. From the data on the inside back cover of this book, would you expect naturally occurring strontium to consist of one isotope or more than one? Use the Table of Isotopes in the *CRC Handbook of Chemistry and Physics* (or similar reference source) to find out how many isotopes strontium has, and their percent natural abundance. From these data, calculate the average molecular weight of strontium and compare your answer with the weight given on the inside back cover of this book.

9. Naturally occurring copper is a mixture of two isotopes, one in 69.09% abundance and of atomic weight 62.9298 amu atom^{-1}. What is the atomic weight of the other isotope, given the natural atomic weight listed on the inside back cover?

10. Gallium has two isotopes, of atomic weights 68.9257 and 70.9249. How many protons and neutrons are present in each type of atom? What is the percent natural abundance of each isotope, given the average atomic weight listed on the inside back cover?

Avogadro's number

11. Which has more atoms, 1 g of boron or 1 g of aluminum? How many atoms are present in each? Which has more atoms, 1 mole of boron or 1 mole of aluminum?

12. Which has more atoms, 1 mole of neon or 1 mole of sodium? How many atoms are present in each? How many grams of sodium will have as many atoms as 1 g of neon?

13. How many atoms are there in 0.00745 g of tungsten wire?

14. What is the weight in grams of 7.63×10^{20} atoms of arsenic?

Molecular weight

15. What is the molecular weight of carbon tetrabromide, CBr_4? How many molecules are there in 8.50 g of CBr_4 vapor?

16. How many nitrogen atoms are present in 25.00 g of N_2O_4? How many moles of N_2 gas would this quantity yield if the compound were broken down into N_2 and O_2?

17. What is the molecular weight of oxalic acid, $(COOH)_2$ or $HOOC—COOH$? How many grams of carbon would be present in 100 g of oxalic acid?

18. What is the molecular weight of the amino acid alanine, $C_3H_7O_2N$?

Conservation of mass

19. The common ore of copper is Cu_2O; it is converted to metallic copper through heating with carbon from coke or charcoal:

$$2Cu_2O + C \rightarrow 4Cu + CO_2$$

What is the percent copper by weight in Cu_2O? How many grams of copper are present in a kilogram of ore? How many grams of carbon will be required to react completely with the kilogram of ore? How many grams, and how many moles, of carbon dioxide will be given off? Show from your calculations that the total weight of ore and carbon is identical to that of copper and CO_2 after the reaction.

20. Iron ore is smelted by heating it with charcoal in the reaction

$$2Fe_2O_3 + 3C \rightarrow 4Fe + 3CO_2$$

What is the percent iron by weight in this iron ore? How much carbon will be required to smelt a metric ton (1000 kg) of iron ore? How many kilograms of iron will result, and how much carbon dioxide will be given off? Show that there is no change in the total mass of reactants and products during the reaction. (*Q:* How did the ancient Anatolians discover iron? *A:* They smelt it.)

Ionic compounds

21. Write chemical formulas for the following compounds:
    a) lithium sulfide
    b) zinc phosphate
    c) calcium sulfate
    d) uranyl oxalate
    e) zinc hexacyanoferrate(III)
    f) ferric sulfate
    g) chromic fluoride

22. Write chemical formulas for the following compounds:
    a) calcium phosphate
    b) mercuric acetate
    c) ammonium hexacyanoferrate(II)
    d) magnesium nitrite
    e) stannous chloride
    f) potassium permanganate
    g) barium carbonate
    h) silver arsenite

Faraday's laws

23. How many faradays and coulombs of electricity are required to reduce 0.782 g of Cu^{2+} to metallic copper?

24. How many coulombs of electricity are required to reduce 0.300 mole of iron(III) to iron(II)?

Electrolysis

25. How many moles of metallic aluminum can be obtained by passing a current of 1.00 A through molten $AlCl_3$ for 7 hours?

26. The quantity of charge passed through a circuit is sometimes measured by determining the mass of solid silver deposited by the electrolysis of an Ag^+ solution. If a cathode increases in mass by 0.197 g, how many coulombs have passed through the electrolysis cell?

27. When aqueous HCl solution is electrolyzed, for how much time must a current of 0.020 A flow to liberate 0.015 mole of hydrogen gas at the cathode?

28. When molten NaCl is electrolyzed, the products are Na and Cl_2, but when aqueous NaCl is electrolyzed, the products are H_2 and Cl_2. Why the difference? Write equations for the electrode reactions with aqueous NaCl. If a current of 4.0 A flows through the salt solution, how long will it take to produce 1.0 mole of H_2 gas? How many moles of Cl_2 gas will be produced in the same time?

29. A certain quantity of electricity is passed through a solution of silver nitrate in water, causing 2.00 g of silver to be deposited at the cathode. How many grams of lead will be deposited if the same quantity of electricity is passed through a solution of $PbCl_2$?

30. Molten $ZnCl_2$ is electrolyzed by passing a current of 3.0 A through an electrolysis cell for a certain length of time. In this process, 24.5 g of Zn are deposited on the cathode. What is the chemical equation for the reaction at the cathode? At the anode? How long does the process take? What weight of chlorine gas is liberated at the anode?

31. One of the major purposes for building the large dams on the Columbia River was to provide cheap hydroelectric power for the electrolytic production of aluminum. The power plant at each dam produces approximately 2×10^8 A of electricity at a voltage high enough to decompose molten salts of aluminum. What is the daily production of metallic aluminum in kilograms if all of the electricity from one dam is used? How many dams would be needed for a daily production of 3000 metric tons (1 metric ton = 1000 kg) of aluminum?

Suggested Reading

R. E. Dickerson and I. Geis, *Chemistry, Matter, and the Universe,* W. A. Benjamin, Menlo Park, Calif., 1976. A somewhat more elementary introduction to chemistry than this book, with more emphasis on the descriptive and pictorial, and less mathematics. Extensively illustrated by Irving Geis. Strong emphasis on chemistry of life.

J. P. Hunt, *Metal Ions in Aqueous Solution,* W. A. Benjamin, Menlo Park, Calif., 1963. Continuation of the discussion of ionic solutions.

W. F. Kieffer, *The Mole Concept in Chemistry,* Reinhold, New York, 1962. The concept of the mole, and its application to chemical calculations.

H. M. Leicester, *The Historical Background of Chemistry,* Dover, New York, 1971. A very readable introduction to the subject, including the Greeks, the Arabic and medieval alchemists, the rise of the new chemistry after Lavoisier and Dalton, and the progression of new chemical ideas down to the era of quantum chemistry.

Lucretius, *The Nature of the Universe* (De Rerum Natura), Penguin, London, 1967. A good prose translation of the best record that we have of the atomic theory of Democritus and the Epicureans.

2

Conservation of Mass and Energy

When you can measure what you are
speaking about, and express it in numbers,
you know something about it; but when
you cannot measure it, when you cannot
express it in numbers your knowledge is of
a meagre and unsatisfactory kind; it may
be the beginning of knowledge, but you
have scarcely, in your thoughts, advanced
to the stage of science.

William Thomson, Lord Kelvin (1891)

The concept of atoms goes back to the Greek philosophers. Democritus (470–360? B.C.) proposed that all matter is made up of separate, indestructible atoms, that different kinds of atoms have different structures and behavior, and that the observed properties of substances arise because of the way their individual atoms arrange themselves and combine with one another. His theories are essentially a primitive version of the material in Chapter 1. Why, then, did the ancient Greeks not use the theories of Democritus and go on to develop atomic energy? Why did 2000 years pass before modern science began to develop?

The answer, in large part, is that the Greeks did not think quantitatively about atoms, and they were not experimentalists. Their science was a philosophical explanation of the universe, rather than a pragmatic tool with which to manipulate the world around them. Cheap human labor kept them from having to worry about developing a scientific technology. The Greek scientist Heron of Alexandria invented several steam-driven mechanisms that could have led directly to the steam engine, but he used them only as toys and novelties.

The atomic theory of Democritus was sterile because it did not lead to quantitative predictions that could be tested. It failed to develop beyond abstract concepts because it did not have the feedback from successful and unsuccessful experiments in the real world to challenge and improve it.

A scientific theory, to be useful, must be quantitative. It must predict: "If I do *this,* then *that* will happen, and to an extent that I can calculate in advance." Such a prediction is testable. It can be seen to be correct, increasing our confidence in the theory behind it; or what frequently is more important, it can be seen to be incorrect, causing us to revise and improve the theory. Scientific theories grow by continual destruction and rebuilding. A theory that predicts nothing that could possibly be tested conveys no information, and is worthless.

The importance of precise measurement of mass in chemical reactions escaped the Greek philosophers. It also escaped the medieval European alchemists, metallurgists, and iatrochemists (medicinal chemists). The great French chemist Antoine Lavoisier (1743–1794) was the first to realize that mass was the fundamental quantity conserved during chemical reactions. The total mass of all products formed must be precisely the same as the total mass of the starting materials. With this principle Lavoisier demolished the long-accepted phlogiston theory of heat (see Chapter 6) by showing that when a substance burns, it combines with another element, oxygen, rather than decomposing and giving off a mysterious universal substance called phlogiston. The principle of the conservation of mass is the cornerstone of all chemistry. More than just total mass is conserved; the same number of each type of atom must be present both before and after a chemical reaction, no matter how intricately these atoms may combine and rearrange into molecules.

Energy also must be conserved in chemical reactions. To the chemist, this means that the heat that is absorbed or given off in a particular chemical reaction (the **heat of reaction**) must be the same no matter how that reaction is carried out—in one step or in several. For example, the heat given off when hydrogen gas and graphite (a form of carbon) are burned must be the same as that given off when hydrogen and carbon are used to make synthetic gasoline, and this gasoline is then used as fuel for an automobile engine. If the heat given off in the two variations of the reaction were not the same, then the more efficient reaction could be run in the forward direction and the less efficient one could be run in reverse. The result would be a cyclical no-fuel furnace that would pour out endless quantities of heat at no cost to the operator. Perpetual-motion schemes of all types vanish as soon as one becomes quantitative about heat, energy, and work. This is the basis of thermodynamics, which is covered in detail in Chapters 15–17.

In this chapter we shall look at the consequences, for chemistry, of two principles:

1. Atoms are neither created nor destroyed during chemical reactions (**conservation of mass**).

2. Heats of reaction are additive. If two reactions can be added to give a third, then the heat of the third reaction is equal to the sum of the heats of the first two reactions (**conservation of energy**).

Both these principles may seem obvious at first, but they are also quite powerful tools in explaining chemical behavior.

2-1 ATOMIC WEIGHTS, MOLECULAR WEIGHTS, AND MOLES

As soon as chemists realized that **mass**—not volume, density or some other measurable property—was the fundamental property that was conserved during chemical reactions, they began to try to establish a correct scale of atomic masses (atomic weights) for all the elements. How they did this is described in Chapter 6; the result of their years of work is the table of natural atomic weights on the inside back cover of this book. As we saw in Chapter 1, the molecular weights of molecular compounds and the formula weights of nonmolecular compounds (such as salts) are found by adding the atomic weights of all the constituent atoms.

Central to all chemical calculations is the concept of the mole. As defined in Chapter 1, a mole of any substance is the quantity that contains as many particles of the substance as there are atoms in exactly 12 g of carbon-12. Thus a mole of a substance is a quantity in grams that is numerically equal to its molecular weight expressed in atomic mass units. The number of particles in a mole is called Avogadro's number, and the experiments of Millikan and Faraday described at the end of Chapter 1 are one means of establishing its value:

$$N = 6.022 \times 10^{23} \text{ particles mole}^{-1}$$

Moles are a way of manipulating atoms or molecules in bundles of 6.022×10^{23}. The molecular weights of H_2, O_2, and H_2O were worked out in Chapter 1. If we know that two molecules of hydrogen gas, H_2, react with one molecule of oxygen gas, O_2, to produce two molecules of water, H_2O, then we can predict that 2 moles of H_2, or 4.032 g, will react with 1 mole of O_2, or 31.999 g, to yield 2 moles of water, or 36.031 g (Figure 2-1). The check addition, $4.032 + 31.999 = 36.031$, verifies the conservation of mass during the reaction. The chemist measures substances in grams, by weighing them. Yet it is more meaningful to convert these quantities from grams to moles, because then one is working with relative molecular proportions, scaled up by a uniform factor of N.

2-2 CHEMICAL ANALYSES: PERCENT COMPOSITION AND EMPIRICAL FORMULAS

Chemical analysis involves breaking a substance down into its elements and then measuring the relative amount of each element present, either in grams per 100 g of original compound, or as a percent by weight. One way of

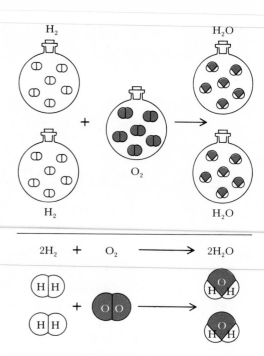

Figure 2-1

Two molecules of hydrogen gas combine with one molecule of oxygen to yield two molecules of water. Avogadro's principle tells us that equal volumes of different gases contain equal numbers of molecules, at a specified temperature and pressure. Hence two volumes of H_2 gas will combine with one volume of O_2, to produce two volumes of water vapor, and two moles of H_2 will combine with one mole of O_2 to produce two moles of water vapor. From Dickerson and Geis, *Chemistry, Matter, and the Universe.*

doing this, if the compound is a **hydrocarbon** (made up only of carbon and hydrogen), is to burn a known amount of the substance in oxygen, and measure the quantities of CO_2 (carbon dioxide) and H_2O that result.

Example 1

When 25.00 g of an unknown hydrocarbon is burned, 68.58 g of CO_2 and 56.15 g of H_2O are produced. How many grams of carbon and hydrogen did the original sample contain?

Solution

The atomic weight of carbon is 12.011 g mole^{-1}, and the molecular weight of CO_2 is 44.010 g mole^{-1}. First we find the percentage of carbon in carbon dioxide:

$$\frac{12.011}{44.010} \times 100 = 27.29\% \text{ carbon}$$

If 27.29% of CO_2 is carbon, then the quantity of carbon in 68.58 g of CO_2 will be

$$27.29\% \times 68.58 \text{ g} = 18.72 \text{ g carbon}$$

A similar calculation for hydrogen in water yields

$$\frac{2 \times 1.008}{18.015} \times 100 = 11.19\% \text{ hydrogen}$$

$$11.19\% \times 56.15 \text{ g} = 6.283 \text{ g hydrogen}$$

As a check: 18.72 g + 6.283 g = 25.00 g.

Example 2

How many grams of carbon and hydrogen per 100.0 g of sample are there in the hydrocarbon of Example 1?

Solution

$$\frac{100.0 \text{ g}}{25.00 \text{ g}} \times 18.72 \text{ g carbon} = 74.88 \text{ g carbon per 100 g sample}$$

$$\frac{100.0 \text{ g}}{25.00 \text{ g}} \times 6.28 \text{ g hydrogen} = 25.12 \text{ g hydrogen per 100 g sample}$$

Example 3

What is the percent composition by weight of the hydrocarbon of Example 1?

Solution

$$\frac{18.72 \text{ g carbon}}{25.00 \text{ g total}} \times 100\% = 74.88\% \text{ carbon}$$

$$\frac{6.28 \text{ g hydrogen}}{25.00 \text{ g total}} \times 100\% = 25.12\% \text{ hydrogen}$$

Once we know the percent composition by weight, we can use atomic weights to obtain the relative number of atoms of each type in a compound.

Example 4

Calculate the relative number of carbon and hydrogen atoms in the compound of Example 3.

Solution

It is easiest to work with 100.0 g of the substance, so the percent elemental composition figures become grams of the respective elements. First we divide each amount of carbon and hydrogen by their atomic weights:

$$\frac{74.88 \text{ g carbon}}{12.011 \text{ g mole}^{-1}} = 6.234 \text{ moles carbon}$$

$$\frac{25.12 \text{ g hydrogen}}{1.008 \text{ g mole}^{-1}} = 24.92 \text{ moles hydrogen}$$

These are the relative numbers of moles of carbon and hydrogen, and it is here that the concept of moles becomes useful. These numbers must also be the relative numbers of *atoms* of carbon and hydrogen. For every 6.234 atoms of carbon in the unknown hydrocarbon, there are 24.92 atoms of hydrogen. If we look for a common factor for these two numbers, we see that they are in a 1:4 ratio. Dividing both numbers by the lower number, 6.234, we find that for every one atom of carbon there are 24.92/6.234 = 3.997 or four atoms of hydrogen.

Example 5

A common liquid is 11.19% hydrogen and 88.81% oxygen by weight. What are the relative numbers of hydrogen and oxygen atoms?

Solution

Again working with 100.0 g of the substance, we calculate the number of moles of each element:

$$\frac{11.19 \text{ g hydrogen}}{1.008 \text{ g mole}^{-1}} = 11.10 \text{ moles hydrogen}$$

$$\frac{88.81 \text{ g oxygen}}{15.999 \text{ g mole}^{-1}} = 5.551 \text{ moles oxygen}$$

Dividing both numbers by the smaller, to search for a common factor, we find that there are two atoms of hydrogen for every atom of oxygen.

Example 6

A common laboratory solvent, a hydrocarbon, is made up of 92.26% carbon and 7.74% hydrogen. What are the relative numbers of carbon and hydrogen atoms in the substance?

Solution

The answer is that one carbon is found for each hydrogen.

An elemental analysis, by itself, is not enough to decide the correct molecular formula of a compound. The formula for methane is CH_4, which would fit the results of the calculation in Example 4. But the analytical results would also be compatible with the molecules C_2H_8, C_3H_{12}, or

C_4H_{16}, if they could exist. The substance in Example 5 might be water, H_2O, but it could also be H_4O_2 or some higher multiple. If you recognize, correctly, that only CH_4 and H_2O are chemically sensible, then you are bringing to bear new chemical information that is not present in the analytical data alone. Most chemists would assume that the molecule in Example 6 was benzene, C_6H_6. But it could also be acetylene, C_2H_2 (except for the fact that acetylene is a gas at room temperature, and the unknown hydrocarbon was said to be a common laboratory solvent, which would exclude acetylene) or any of the five other less common hydrocarbon molecules shown in Figure 2-2.

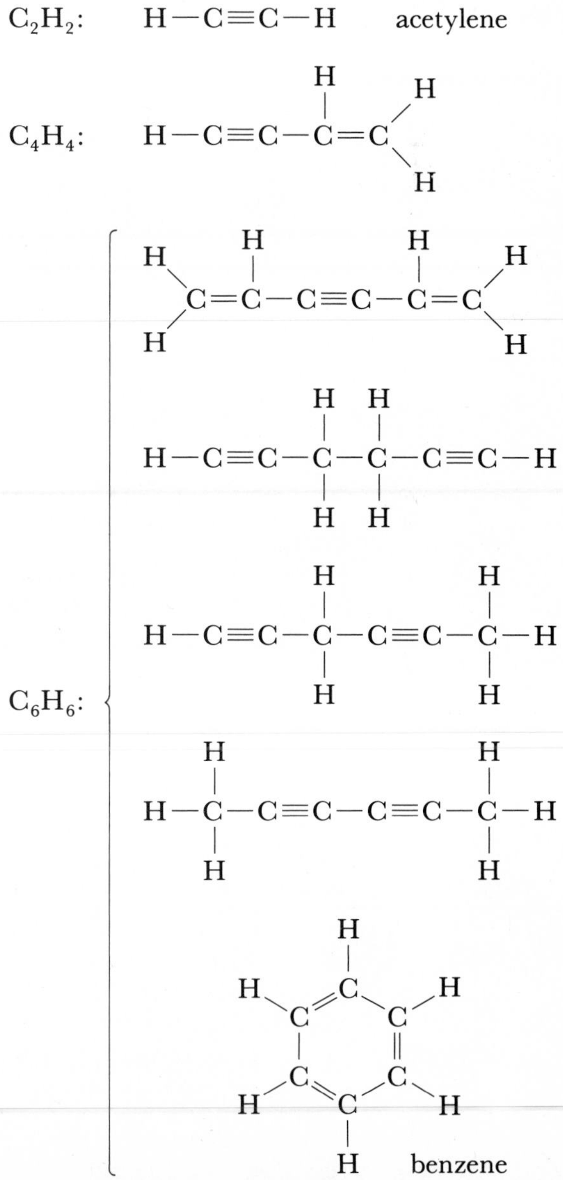

Figure 2-2

Seven different molecules with the empirical formula CH. A simple elemental analysis could not distinguish between them. An approximate molecular weight could distinguish between C_2H_2, C_4H_4, and C_6H_6, but even more information would be required to identify the particular C_6H_6 molecule present.

A chemical formula that gives the relative number of each type of atom, as integers with no common factor, is called the **empirical formula** of the substance. It is the empirical formula that results from an elemental analysis of a substance, not the molecular formula, which could be the same as the empirical formula or could be some integral multiple of it. The empirical formula is the same as the molecular formula for methane, CH_4, and for water, H_2O; the empirical formulas of acetylene and benzene are both CH, but the molecular formulas are C_2H_2 and C_6H_6, respectively.

It frequently happens that some simple physical measurement can give a rough approximation of the molecular weight of a substance. Gas densities (Chapter 3), freezing-point depression, and osmotic pressure measurements (Chapter 18) are useful in this regard. If such an approximate molecular weight is available, then it can be used along with the empirical formula to decide the true molecular formula.

Example 7

Glucose is 40.00% carbon by weight, 6.71% hydrogen, and 53.29% oxygen. What is its empirical formula, and what is its molecular formula?

Solution

Working with 100.0 g of glucose, we first find the number of moles of each element:

$$\frac{40.00 \text{ g carbon}}{12.011 \text{ g mole}^{-1}} = 3.330 \text{ moles carbon}$$

$$\frac{6.71 \text{ g hydrogen}}{1.008 \text{ g mole}^{-1}} = 6.66 \text{ moles hydrogen}$$

$$\frac{53.29 \text{ g oxygen}}{15.999 \text{ g mole}^{-1}} = 3.331 \text{ moles oxygen}$$

This is obviously a molar ratio of one carbon to two hydrogens to one oxygen, so the empirical formula is CH_2O. With the information provided we have no way of knowing whether this, or some multiple of this, is the true molecular formula.

Example 8

From other experiments we know that glucose has a molecular weight of approximately 175 g mole^{-1}. Use this information and the results of Example 7 to find the molecular formula and the exact molecular weight of glucose.

Solution   The weight corresponding to the empirical formula is

$$12.011 + (2 \times 1.008) + 15.999 = 30.026 \text{ g mole}^{-1}$$

The approximate molecular weight is roughly six times this value, so the precise molecular weight is 6×30.026 g mole^{-1} = 180.16 g mole^{-1}, and the molecular formula is $C_6H_{12}O_6$.

2-3   CHEMICAL EQUATIONS

When propane gas, C_3H_8, is burned in oxygen, the products are carbon dioxide and water. This can be written as a chemical equation:

$$C_3H_8 + O_2 \rightarrow CO_2 + H_2O \tag{2-1}$$

If chemistry were not a quantitative science, then this description of the reaction, identifying both the reactants and the products, would be adequate. But we expect more from a chemical equation. *How many* molecules of oxygen are required per molecule of propane, and how many molecules of carbon dioxide and water result? Equation 2-1 is an unbalanced equation. When we add numerical coefficients (placed to the left of the formula) that tell how many of each kind of molecule are involved, then there will be the same number of each kind of atom on the left and right sides of the equation, since atoms are neither created nor destroyed in a chemical reaction. The result will be a **balanced equation.**

To balance equation 2-1, we note first that the 3 carbon atoms on the left will lead to 3 molecules of CO_2 as products, each requiring 2 oxygen atoms, or 6 oxygens in all. Similarly, the 8 hydrogen atoms in propane will produce 4 molecules of water, requiring 4 more oxygen atoms. This total of 10 oxygens on the right must come from 5 molecules of O_2. The correct coefficients for the four substances in equation 2-1 are therefore 1, 5, 3, and 4:

$$C_3H_8 + 5O_2 \rightarrow 3CO_2 + 4H_2O \tag{2-2}$$

Each side of this balanced equation contains 3 carbon atoms, 8 hydrogen atoms, and 10 oxygen atoms.

Example 9

Trinitrotoluene (TNT), $C_7H_5N_3O_6$, combines violently with oxygen to produce CO_2, water, and N_2. Write a balanced chemical equation for the explosion.

Solution   The unbalanced equation is

$$C_7H_5N_3O_6 + O_2 \rightarrow CO_2 + H_2O + N_2$$

Since there are odd numbers of hydrogen and nitrogen atoms on the left, and even numbers on the right, it will be easier to balance the equation on the basis of two molecules of TNT:

$$2C_7H_5N_3O_6 + O_2 \rightarrow CO_2 + H_2O + N_2$$

The 14 carbons, 10 hydrogens, and 6 nitrogens then will result in 14 carbon dioxide, 5 water, and 3 nitrogen molecules:

$$2C_7H_5N_3O_6 + O_2 \rightarrow 14CO_2 + 5H_2O + 3N_2$$

Now all atoms are balanced on left and right sides of the equation except for oxygen. Of the 33 oxygens on the right, 12 are provided on the left by the 2 starting molecules of TNT, and 21 must be supplied by $10\frac{1}{2} O_2$. The final, balanced equation is

$$2C_7H_5N_3O_6 + 10\tfrac{1}{2}O_2 \rightarrow 14CO_2 + 5H_2O + 3N_2$$

Example 9 led to an equation with a fractional coefficient for oxygen. This can be removed by multiplying all coefficients on both sides by 2:

$$4C_7H_5N_3O_6 + 21O_2 \rightarrow 28CO_2 + 10H_2O + 6N_2 \tag{2-3}$$

but this is not necessary, since there is no reason why all coefficients must be integers. It would even be correct to base the equation on a single molecule of TNT:

$$C_7H_5N_3O_6 + 5\tfrac{1}{4}O_2 \rightarrow 7CO_2 + 2\tfrac{1}{2}H_2O + 1\tfrac{1}{2}N_2 \tag{2-4}$$

A balanced chemical equation such as equation 2-3 has several levels of meaning. Most simply, it describes the starting materials and the products. It also tells us that the number of each kind of atom entering the reaction is the same as the number leaving. Each type of atom individually is conserved during the reaction. Equation 2-3 is also a statement that for every 4 molecules of TNT, 21 molecules of oxygen are required, and the products are 28 molecules of CO_2, 10 molecules of water, and 6 molecules of N_2. Scaling the reaction up by a factor of 6.022×10^{23} to go from molecules to moles, 4 moles of TNT react with 21 moles of O_2 to produce 28 moles of CO_2, 10 moles of H_2O, and 6 moles of N_2. The individual molecular weights are

$C_7H_5N_3O_6$	227.13 g mole^{-1}
O_2	31.999 g mole^{-1}
CO_2	44.010 g mole^{-1}
H_2O	18.015 g mole^{-1}
N_2	28.013 g mole^{-1}

Hence equation 2-3 also tells us that 4×227.13 g $= 908.52$ g of TNT requires 21×31.999 g $= 671.98$ g of oxygen for complete reaction. It also tells us that the products will be 28×44.010 g $= 1232.3$ g of CO_2, 10 $\times$

18.015 g = 180.15 g of H_2O, and 6 × 28.013 g = 168.08 g of N_2. We can verify that mass is indeed conserved:

Reactants		Products	
908.52 g	TNT	1232.3 g	CO_2
671.98 g	O_2	180.15 g	H_2O
		168.08 g	N_2
1580.5 g	Total	1580.5 g	Total

What a balanced chemical equation does *not* tell us is the molecular mechanism or course of events by which the reaction takes place. Equation 2-3 should not be construed as suggesting that 4 TNT molecules must collide simultaneously with 21 oxygen molecules. Even three-body collisions are so much rarer than two-body collisions that they can be dismissed from consideration in most chemical reactions. An elaborate series of individual steps could take place, as long as the overall net reaction was described correctly by equation 2-3 or equation 2-4.

The reactants and products need not be molecules:

$$CaCO_3 + 2HCl \rightarrow CaCl_2 + CO_2 + H_2O \qquad (2\text{-}5)$$

Equation 2-5 describes the reaction of $CaCO_3$, calcium carbonate (limestone), and HCl, hydrochloric acid, to produce an aqueous solution of calcium chloride, $CaCl_2$, and carbon dioxide. The equation is balanced, because the number of each type of atom is the same on both sides. The molar meaning is clear: 1 mole or 100.09 g of calcium carbonate requires 2 moles or 72.92 g of hydrochloric acid for complete reaction, and the products will be 1 mole each of calcium chloride (110.99 g mole^{-1}), carbon dioxide (44.01 g mole^{-1}), and water (18.02 g mole^{-1}). You can verify from these figures that mass is conserved during the reaction. The molecular interpretation is less straightforward, since calcium carbonate is a salt and not a molecular compound. Equation 2-5 should not be taken as meaning that one *molecule* of calcium carbonate reacts with two *molecules* of HCl. Although HCl exists as discrete molecules in the gas phase, in solution the molecules dissociate into H^+ and Cl^- ions. A better approximation of what actually happens at the molecular level is

$$CaCO_3(s) + 2H^+(aq) \rightarrow Ca^{2+}(aq) + CO_2(g) + H_2O(l) \qquad (2\text{-}6)$$

The letters in parentheses describe the physical state of each species (*s*, solid; *aq*, a hydrated ion in aqueous solution; *g*, gas; *l*, liquid). This equation says that solid calcium carbonate reacts with two hydrated protons (hydrogen ions) in aqueous solution to produce hydrated calcium ions, gaseous carbon dioxide, and liquid water. The chloride ions remain as hydrated chloride ions in solution before and after the reaction; hence they are omitted from the equation. Equation 2-5, like other balanced chemical equations, is most useful in describing the amounts of materials involved, rather than the molecular mechanism of reaction. Equation 2-6, although a better description of what is going on at the level of atoms and ions, is less useful in keeping track of the quantities of matter involved.

Example 10

Metallic sodium reacts with water to produce hydrogen gas and sodium hydroxide solution (a mixture of Na^+ and OH^- ions). Write (a) a balanced equation for the overall reaction and (b) an equation that more accurately describes the actual atomic or ionic species present.

Solution   The balanced equation is

$$Na + H_2O \rightarrow \tfrac{1}{2}H_2 + NaOH$$

or

$$2Na + 2H_2O \rightarrow H_2 + 2NaOH$$

A better description of what is actually present would be

$$2Na(s) + 2H_2O(l) \rightarrow H_2(g) + 2Na^+(aq) + 2OH^-(aq)$$

2-4   CALCULATIONS OF REACTION YIELDS

Balanced chemical equations are primarily used to calculate the expected yield (quantity of product) from a reaction, and to determine whether any of the reactants will remain unused when the other reactants are depleted.

Example 11

How many grams of hydrogen are needed to combine with 100.0 g of carbon to make benzene, C_6H_6? How many moles and how many grams of benzene will be produced?

Solution   The balanced equation for the reaction is

$$6C + 3H_2 \rightarrow C_6H_6$$

The number of moles of carbon present is

$$\frac{100.0 \text{ g carbon}}{12.011 \text{ g mole}^{-1}} = 8.326 \text{ moles carbon}$$

The balanced equation tells us that half as many moles of H_2 are needed as moles of C, so we need 4.163 moles of H_2.

$$4.163 \text{ moles} \times 2.016 \text{ g mole}^{-1} = 8.393 \text{ g hydrogen}$$

The molecular weight of benzene is

$$(6 \times 12.011 \text{ g}) + (6 \times 1.008 \text{ g}) = 78.11 \text{ g mole}^{-1}$$

One-sixth as many moles of benzene are produced as moles of carbon used up, or $8.326/6 = 1.388$ moles of benzene. Hence the amount of benzene produced is

$$1.388 \text{ moles} \times 78.11 \text{ g mole}^{-1} = 108.4 \text{ g benzene}$$

As a check on arithmetic, note that 100.0 g of carbon and 8.4 g of hydrogen combine to produce 108.4 g of benzene. Again, mass is conserved during a chemical reaction.

Example 12

How many grams of silver sulfide (Ag_2S) can be formed by the reaction $2Ag + S \rightarrow Ag_2S$, if we start with 10.00 g of silver (Ag) and 1.00 g of sulfur (S)? Which starting material, if any, will be left over, and how much?

Solution

$$\begin{aligned} \text{Reaction:} \quad & 2Ag \;+\; S \;\rightarrow\; Ag_2S \\ \text{Masses:} \quad & 215.7 \text{ g} + 32.06 \text{ g} = 247.8 \text{ g} \end{aligned}$$

The quantity of sulfur required to react with 10.00 g of silver is

$$\frac{32.06 \text{ g S}}{215.7 \text{ g Ag}} \times 10.00 \text{ g Ag} = 1.486 \text{ g S}$$

But we only have 1.00 g sulfur, so not all the silver can react. Turning the problem around, the amount of silver needed to react with the entire 1.00 g of sulfur is

$$\frac{215.7 \text{ g Ag}}{32.06 \text{ g S}} \times 1.00 \text{ g S} = 6.73 \text{ g Ag}$$

Hence there will be $10.00 - 6.73 = 3.27$ g of silver left over. The quantity of silver sulfide produced will be

$$\frac{247.8 \text{ g Ag}_2S}{32.06 \text{ g S}} \times 1.00 \text{ g S} = 7.73 \text{ g Ag}_2S$$

Notice that Example 12 was worked a different way: First the total masses of all reactants and products were written under the balanced chemical equation, and then the ratios of these masses were worked out to find the desired answers. The problem can also be solved using moles, and the choice is one of convenience.

Alternative Solution

First we find the number of moles of silver and of sulfur:

$$\frac{10.00 \text{ g Ag}}{107.9 \text{ g mole}^{-1}} = 0.0927 \text{ mole Ag}$$

$$\frac{1.00 \text{ g S}}{32.06 \text{ g mole}^{-1}} = 0.0312 \text{ mole S}$$

Since 2 moles of silver are required for every mole of sulfur, and there are more than twice as many moles of silver as moles of sulfur, some of the silver must be left behind when all the sulfur has been used. The 0.0312 mole of sulfur will combine with 0.0624 mole of silver, and form 0.0312 mole of Ag_2S. Left behind are $0.0927 - 0.0624 = 0.0303$ mole of silver. Translating these quantities back to grams, we have

$$0.0303 \text{ mole Ag} \times 107.9 \text{ g mole}^{-1} = 3.27 \text{ g Ag left over}$$

$$0.0312 \text{ mole Ag}_2S \times 247.8 \text{ g mole}^{-1} = 7.73 \text{ g Ag}_2S \text{ produced}$$

We arrived at the same answers in the first solution to the example. The mole method is surer, but slower. The ratio method is faster, but you can go astray more easily if you are not absolutely sure of what you are doing. Use the mole method until you are proficient in chemical calculations.

2-5 SOLUTIONS AS CHEMICAL REAGENTS

Liquid solutions are convenient media for chemical reactions. Rapid mixing of the liquid means that potential reactants are brought close to one another frequently, so collisions and chemical reactions can take place much faster than they would in a crystalline solid. Moreover, a given number of molecules in a liquid is confined to a smaller space than the same number of molecules in a gas, so reactant molecules in a liquid have more of a chance to come in contact. Water is an especially good solvent for chemical reactions because its molecules are polar. The H_2O molecules, and the H^+ and OH^- ions into which water dissociates to a small extent, can help to polarize bonds in other molecules, weaken bonds, and encourage chemical attack. It is no accident that life evolved in the oceans rather than in the upper atmosphere or on dry land. If life had been forced to evolve using solid-state crystal reactions, the 4.5 billion years of earth's history to date might barely have been time enough for the process to begin.

Concentration Units: Molarity and Molality

In solutions involving a liquid and a gas or solid, the liquid component is called the **solvent,** and the other component is called the **solute.** If the solution is made up of two liquids, the distinction is less clear, but the substance present in the greater amount is usually considered as the solvent. The most common way of expressing concentration in solution is **molarity,** or the number of moles of solute *per liter of solution.** The symbol M is read as "moles per liter of solution," as in $1.5M$ NaCl. The symbol c is used to denote concentration in moles per liter, as is the chemical symbol in brackets [H], although such brackets are sometimes used to represent concentration in any units. Hence the expression c_{NaCl} would be read as "the concentration of sodium chloride in moles per liter of solution." This is *not* the solution that would result from adding 1 mole of NaCl to a liter of water, since the total volume after mixing would be a little more than 1 liter. Sodium and chloride ions take up room, even when dissolved in water. The proper procedure in making a $1.0M$ solution would be to dissolve the salt in less than a liter of water, and then slowly add more water, with mixing, until the total volume reached 1 liter.

For many salts, we can use the approximation that *volumes are additive,* or that the volume of a solution will be equal to the original volume of the solvent plus that of the crystals that were dissolved.

Example 13

If 264 g of ammonium sulfate, $(NH_4)_2SO_4$, are dissolved in 1.000 liter of water, what will the approximate final volume and the approximate molarity of the solution be, assuming additivity of volumes? The density of crystalline $(NH_4)_2SO_4$ is 1.76 g ml^{-1}.

---

*The SI unit of length is the meter (m), divided into 10 decimeters (dm) or 100 centimeters (cm). The unit of volume is the cubic meter (m^3). For laboratory work the cubic meter is too large to be convenient, so it is customary to use the *liter,* which in the SI is *defined* as 1 dm^3, and the milliliter (ml), which equals 1 cm^3 (or sometimes cc). By strict logic the liter is an extraneous unit in SI, but it is too convenient, and its use too deeply ingrained, to be eliminated. Scientists in the past tended to use milliliters for liquid volumes, and cubic centimeters for volumes of solids. Hence the volume of a sodium chloride *solution* would be measured in milliliters, but the density of rock salt (sodium chloride crystals) would be reported in grams per cubic centimeter, or g cm^{-3}. We shall use only milliliters in this chapter, but thereafter shall feel free to use cubic centimeters wherever that unit seems more natural. Remember that 1 m^3 = 1000 liters, 1 liter = 1000 ml, and 1 ml = 1 cm^3. For more information on SI, see Appendix 1.

Solution   The volume of solid ammonium sulfate added is

$$\frac{264 \text{ g}}{1.76 \text{ g ml}^{-1}} = 150 \text{ ml or } 0.150 \text{ liter}$$

The final solution volume then will be $1.000 + 0.150 = 1.150$ liters. The number of moles of solute is

$$\frac{264 \text{ g}}{132 \text{ g mole}^{-1}} = 2.00 \text{ moles ammonium sulfate}$$

The molarity then is

$$\frac{2.00 \text{ moles}}{1.150 \text{ liters}} = 1.74 \text{ moles liter}^{-1}, \text{ or } 1.74M \text{ (NH}_4)_2\text{SO}_4$$

The approximation of additivity of volumes must be used with care. In this example, the true molarity of such a solution is $1.80M$, so the approximation is only 3.3% in error. But for liquids whose molecules interact strongly, such as ethyl alcohol and water, the total volume may shrink after mixing because of molecular attractions. Additivity of volumes should be used only as a rough guide to molarity.

An alternative expression of concentration, **molality,** is based on the amount of solvent used rather than the solution that results. The molality of a solute is the number of moles of solute *in 1 kg of solvent* (not of solution). The density of water is 1.00 g ml^{-1}, so 1 kg of water occupies a volume of 1 liter. Hence the ammonium sulfate solution of Example 13 is a 2.00 molal solution, since it was made up from 2.00 moles of solute in a kilogram (1 liter) of water. For solvents other than water, we must use the density of the liquid to convert from kilograms to liters.

Example 14

Suppose 5.00 g of acetic acid, $C_2H_4O_2$, are dissolved in 1 liter of ethanol. Calculate the molality of the resulting solution. The density of ethanol is 0.789 g ml^{-1}. Can you calculate the molarity from the information given?

Solution   The molecular weight of acetic acid is $60.05 \text{ g mole}^{-1}$, so the number of moles is

$$\frac{5.00 \text{ g}}{60.05 \text{ g mole}^{-1}} = 0.0833 \text{ mole acetic acid}$$

The number of kilograms of solvent used is

$$1.000 \text{ liter} \times 0.789 \text{ kg liter}^{-1} = 0.789 \text{ kg ethanol}$$

Notice that 1 g ml^{-1} is the same as 1 kg liter^{-1}, since there are 1000 g in a kilogram and 1000 ml in a liter. The molality then is

$$\frac{0.0833 \text{ mole solute}}{0.789 \text{ kg solvent}} = 0.106 \text{ mole kg}^{-1}$$

The solution is therefore 0.106 molal. The molarity of the solution cannot be calculated because we know neither the volume of the acetic acid nor whether volumes are additive when acetic acid is dissolved in ethanol.

The symbol m is used for concentration expressed as molality. We would write the results of Example 14 as

$$m_{\text{acetic acid}} = 0.106 \text{ mole kg}^{-1}$$

Dilution Problems

If we **dilute** a solution (add more solvent), the number of moles of solute does not change. If c is the molarity (not molality) of the solution and V is the volume in liters, then the number of moles of solute is

$$c \,(\text{moles liter}^{-1}) \times V(\text{liters}) = cV(\text{moles})$$

If we use the subscript 1 to represent a solution before it is diluted with more solvent, and the subscript 2 for the diluted solution, then

$$\text{Moles of solute} = c_1 V_1 = c_2 V_2$$

Example 15

To what volume must 5.00 ml of 6.00M HCl be diluted to make the concentration 0.100M?

$$V_2 = \frac{c_1}{c_2} \times V_1 = \frac{6.00M}{0.100M} \times 5.00 \text{ ml} = 300 \text{ ml}$$

Solution

This does not mean that 300 ml of water must be added, but that the total volume of solution must be brought to 300 ml.

Example 16

If 175 ml of a 2.00M solution are diluted to 1.00 liter, what will the molarity be?

Solution

$$c_2 = \frac{V_1}{V_2} \times c_1 = \frac{175 \text{ ml}}{1000 \text{ ml}} \times 2.00M = 0.350M$$

Acid–Base Neutralization

Probably the most familiar definition of acids and bases is that by the Swedish physicist and chemist Svante Arrhenius (1859–1927): An **acid** is a substance that increases the hydrogen ion concentration, $[H^+]$, when added to water, and a **base** is a substance that increases the hydroxide ion concentration, $[OH^-]$, when added to water. Some of the more common acids and bases are listed in Tables 2-1 and 2-2. The first 11 acids in Table 2-1, from HF to HNO_3, dissociate in aqueous solution to release one proton or hydrogen ion:

$$HNO_3 \rightarrow H^+(aq) + NO_3^-(aq)$$

nitric acid   nitrate ion

Table 2-1

Common Acids

HF	Hydrofluoric
HCl	Hydrochloric
HClO	Hypochlorous
$HClO_2$	Chlorous
$HClO_3$	Chloric
$HClO_4$	Perchloric
HBr	Hydrobromic
$HBrO_3$	Bromic
HI	Hydriodic
HNO_2	Nitrous
HNO_3	Nitric
H_2CO_3	Carbonic
H_2SO_3	Sulfurous
H_2SO_4	Sulfuric
H_3PO_2	Hypophosphorous
H_3PO_3	Phosphorous
H_3PO_4	Phosphoric
H_3BO_3	Boric
HCOOH	Formic,
CH_3COOH	Acetic,

Table 2-2

Common Bases

LiOH	Lithium hydroxide
NaOH	Sodium hydroxide
KOH	Potassium hydroxide
$Mg(OH)_2$	Magnesium hydroxide
$Ca(OH)_2$	Calcium hydroxide
$Ba(OH)_2$	Barium hydroxide
NH_3	Ammonia

The abbreviation (aq) is a reminder that the ions are hydrated, but it is really not necessary since *every* ion in aqueous solution is hydrated, and we shall omit it in the rest of this discussion. Remember that the water molecules are always present, surrounding each ion and helping to stabilize it in solution.

Carbonic, sulfurous, and sulfuric acids release two protons in two stages, and the three phosphorus-containing acids produce three protons:

$$H_2CO_3 \rightarrow H^+ + HCO_3^- \rightarrow 2H^+ + CO_3^-$$

carbonic · bicarbonate · carbonate
acid · ion · ion

$$H_2SO_4 \rightarrow H^+ + HSO_4^- \rightarrow 2H^+ + SO_4^{2-}$$

sulfuric · bisulfate · sulfate
acid · ion · ion

$$H_3PO_4 \rightarrow H^+ + H_2PO_4^- \rightarrow 2H^+ + HPO_4^{2-} \rightarrow 3H^+ + PO_4^{3-}$$

phosphoric · dihydrogen · monohydrogen · phos-
acid · phosphate · phosphate · phate
· ion · ion · ion

Carbonic acid is classed as a **weak acid** because its loss of protons is only partial; the species present in aqueous solution are a mixture of carbonate and bicarbonate ions and a small amount of undissociated carbonic acid. In contrast, sulfuric acid is a **strong acid** because the loss of the first of the two H^+ is complete in aqueous solution. (Acid-dissociation equilibria are considered in detail in Chapter 5.) Nitric and hydrochloric acids are common strong acids, and phosphoric acid is weak. Organic acids such as formic and acetic release a proton from their —COOH carboxyl groups:

$$CH_3-COOH \rightarrow CH_3-COO^- + H^+$$

acetic acid · acetate ion

It is common to use the abbreviations HAc for acetic acid and Ac^- for the acetate ion.

Hydroxide bases such as sodium hydroxide and magnesium hydroxide dissolve in water to release hydroxide ions:

$$NaOH \rightarrow Na^+ + OH^-$$
$$Mg(OH)_2 \rightarrow Mg^{2+} + 2OH^-$$

The hydroxide ions are already present in solid $NaOH$, just as chloride ions are present in $NaCl$. Ammonia, NH_3, is also a base, but it has no hydroxide ions of its own. Instead, it produces them by reacting with water molecules:

$$NH_3 + H_2O \rightarrow NH_4^+ + OH^-$$

ammonia   ammonium   hydroxide
                              ion   ion

Ammonia is sometimes written as ammonium hydroxide, NH_4OH, to make it resemble the metal hydroxide bases (such as sodium hydroxide, $NaOH$). But this is incorrect; there is no such substance as ammonium hydroxide; there is only ammonia.

Acids and bases are useful because the H^+ and OH^- ions that they produce can attack molecules in solution and bring about chemical changes that would be difficult or slow in their absence. When acids and bases react with one another, the H^+ and OH^- ions combine to form water molecules. This is called **neutralization:**

$$H^+ + OH^- \rightarrow H_2O$$

The easiest way to determine how much of an acid or base is present is to find out how much of a base or acid of known concentration is required to neutralize it completely. This is the process of acid–base **titration.** One **equivalent** (equiv) of an acid is the quantity of acid that will release 1 mole of protons or H^+ in neutralizing a base, and 1 equiv of a base is the quantity that will produce 1 mole of OH^- ions. Complete neutralization occurs when the same number of equivalents of acid and base react with one another. For acids that release one proton per molecule, such as HCl and HNO_3, the equivalent is the same as the mole, and 1 equivalent weight is the same as the molecular weight. But since H_2SO_4 is capable of releasing two H^+ ions, 1 mole of H_2SO_4 corresponds to 2 equiv, and the equivalent weight of sulfuric acid in acid–base neutralizations is half the molecular weight. The equivalent weight of phosphoric acid, H_3PO_4, or the quantity that will produce 1 mole of H^+ ions, is one-third the molecular weight. Similarly, the mole and the equivalent are identical for $NaOH$, KOH, and NH_3, but the equivalent weight of $Ca(OH)_2$ is half its molecular weight.

We can appreciate the usefulness of the concept of equivalents by looking at the neutralization of phosphoric acid by magnesium hydroxide:

$$2H_3PO_4 + 3Mg(OH)_2 \rightarrow Mg_3(PO_4)_2 + 6H_2O$$

Molecular weight:	98.0 g	58.3 g	262.9 g	18.0 g
Equivalent weight:	32.7 g	29.2 g		

One mole or 98.0 g of phosphoric acid will not neutralize 1 mole or 58.3 g of magnesium hydroxide, but 1 equiv or 32.7 g of phosphoric acid will neutralize 1 equiv or 29.2 g of magnesium hydroxide. This is the same answer that would be obtained by using the balanced equation shown. Since 2 moles of acid react with 3 moles of base as shown, $2 \times 98.0 = 196$ g of phosphoric acid will neutralize $3 \times 58.3 = 175$ g of magnesium hydroxide. These numbers are just the numbers obtained by using equivalents, but scaled up by a factor of 6.

Example 17

Use equivalents to find the number of grams of nitric acid, HNO_3, needed to neutralize 100.0 g of barium hydroxide, $Ba(OH)_2$.

Solution

The molecular weight of HNO_3 is 63.01 g mole^{-1}; of $Ba(OH)_2$, 171.34 g mole^{-1}. The corresponding equivalent weights are $63.01/1 = 63.01$ g equiv^{-1} for HNO_3, and $171.34/2 = 85.67$ g equiv^{-1} for $Ba(OH)_2$. The number of equivalents of barium hydroxide is

$$\frac{100.0 \text{ g}}{85.67 \text{ g equiv}^{-1}} = 1.167 \text{ equiv of } Ba(OH)_2$$

The same number of equivalents of nitric acid is needed:

$$1.167 \text{ equiv} \times 63.01 \text{ g equiv}^{-1} = 73.53 \text{ g nitric acid}$$

Alternative Solution

This example could also be solved by using the balanced chemical equation.

The balanced equation is

$$2HNO_3 + Ba(OH)_2 \rightarrow Ba(NO_3)_2 + 2H_2O$$

The number of moles of barium hydroxide at the start is

$$\frac{100.0 \text{ g}}{171.3 \text{ g mole}^{-1}} = 0.5838 \text{ mole barium hydroxide}$$

The balanced equation tells us that twice as many moles of nitric acid are required: 1.167 moles HNO_3. In grams, this is

$$1.167 \text{ moles} \times 63.01 \text{ g mole}^{-1} = 73.53 \text{ g nitric acid}$$

The use of equivalents eliminates the need to work out a balanced equation for the reaction.

The **normality** of a solution, represented by N, is the number of equivalents of solute per liter of solution. A $1.00M$ solution of phosphoric acid is $3.00N$, and a $0.010M$ solution of $Ca(OH)_2$ is $0.020N$.

Example 18

If 4.00 g of sodium hydroxide are dissolved in water and the volume is brought up to 500 ml, find the molarity and the normality of the solution.

Solution   Since the molecular weight of NaOH is 40.0 g mole^{-1},

$$\frac{4.00 \text{ g}}{40.0 \text{ g mole}^{-1}} = 0.100 \text{ mole NaOH}$$

$$\frac{0.100 \text{ mole NaOH}}{0.500 \text{ liter solution}} = 0.200 \text{ mole liter}^{-1}, \text{ or } 0.200M \text{ NaOH}$$

Because 1 mole of NaOH releases 1 mole of OH^- ions, the molarity and normality are the same. The solution is $0.200N$.

Example 19

If 10.0 g of sulfuric acid (H_2SO_4) are mixed slowly with enough water to make a final volume of 750 ml, what are the molarity and normality of the resulting solution?

Solution $$\frac{10.0 \text{ g}}{98.1 \text{ g mole}^{-1}} = 0.102 \text{ mole sulfuric acid}$$

$$\frac{0.102 \text{ mole}}{0.750 \text{ liter}} = 0.136M \text{ sulfuric acid}$$

Since each mole of sulfuric acid contributes 2 equiv, the solution is $2 \times 0.136 = 0.272$ normal in sulfuric acid, or $0.272N$ H_2SO_4.

Acid–Base Titration

Chemists frequently use **titrations** to compare relative concentrations of chemical equivalents in acid–base solutions (Figure 2-3). When enough acid solution from a burette (shown in the figure) has been added to neutralize the base in the sample being analyzed, the number of equivalents of acid and base involved must be the same. The point of neutralization is called the **equivalence point**. An acid–base **indicator** such as litmus or phenolphthalein can be used to determine the equivalence point. From the volume

Figure 2-3

An acid—base titration. The solution in the flask contains an unknown number of equivalents of base (or acid). The burette is calibrated to show volume to the nearest 0.001 cm³. It is filled with a solution of strong acid (or base) of known concentration. Small increments are added from the burette until, at the end point, one drop or less changes the indicator color permanently. (An indication that the equivalence point is being approached is the appearance—and disappearance on stirring—of the color that the indicator assumes beyond neutralization.) At the equivalence point, the total amount of acid (or base) is recorded from the burette readings. The number of equivalents of acid and base must be equal at the equivalence point.

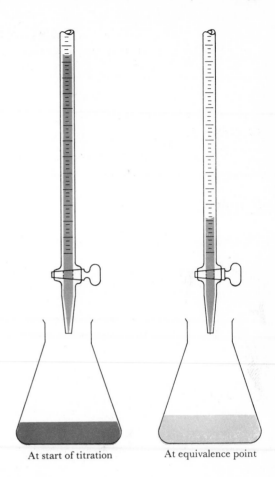

At start of titration   At equivalence point

of acid solution used and its normality, we can calculate the number of equivalents of base in the unknown sample. If N_A and N_B are the normalities of acid and base solutions, and V_A and V_B are the volumes of each at neutrality, then

$$\text{Number of equivalents} = N_A V_A = N_B V_B \tag{2-7}$$

Example 20

If 25.00 ml of phosphoric acid (H_3PO_4) are just enough to neutralize 30.25 ml of a sodium hydroxide solution, what is the ratio of the normalities of the two solutions? What is the ratio of molarities?

Solution

$$\frac{N_A}{N_B} = \frac{V_B}{V_A} = \frac{30.25 \text{ ml}}{25.00 \text{ ml}} = 1.210$$

Since the normality of the acid is three times its molarity, and the normality of the base is the same as its molarity, the molarity ratio is

$$\frac{c_A}{c_B} = \frac{N_A/3}{N_B} = 0.403$$

Example 21

In a titration, 25.00 ml of a solution of calcium hydroxide, $Ca(OH)_2$, require 10.81 ml of 0.100N HCl for neutralization. Calculate (a) the normality of the $Ca(OH)_2$ solution, (b) the molarity, and (c) the number of grams of $Ca(OH)_2$ present in the sample.

Solution

The normality of the solution of $Ca(OH)_2$ is

$$N_B = \frac{V_A}{V_B} \times N_A = \frac{10.81 \text{ ml}}{25.00 \text{ ml}} \times 0.100N = 0.0432N$$

Since 1 mole of calcium hydroxide yields 2 equiv OH^-, the molarity is half the normality, or 0.0216M $Ca(OH)_2$. The number of moles of $Ca(OH)_2$ is

0.0216 mole liter^{-1} × 0.02500 liter = 0.000541 mole

Since the molecular weight of $Ca(OH)_2$ is 74.1 g mole^{-1}, the weight present is

0.000541 mole × 74.1 g mole^{-1} = 0.0401 g $Ca(OH)_2$

Example 22

An organic chemist synthesizes a new acid. He dissolves 0.500 g in a convenient volume of water and finds that it requires 15.73 ml of 0.437N NaOH for neutralization. What is the equivalent weight of the new compound as an acid? If it is known that the acid contains three ionizing —COOH groups, what is the molecular weight?

Solution

The number of equivalents of base is

0.01573 liter × 0.437 equiv liter^{-1} = 0.00687 equiv

The equivalent weight is found from

$$\frac{0.500 \text{ g}}{0.00687 \text{ equiv}} = 72.8 \text{ g equiv}^{-1}$$

If the equivalent weight of the acid is 72.8 g, and each mole yields 3 equiv, then the molecular weight is 3 × 72.8 g = 218 g.

2-6 HEATS OF REACTION: CONSERVATION OF ENERGY

So far this chapter has been devoted to the consequences of the conservation of mass, and little has been said about energy. But the principle that heats of reaction are additive, that energy is conserved in a process whether the process is carried out in one step or in several, is an important one. Heat and work are both forms of energy, and are measured in the same units. If you do work on an object or collection of objects, you can increase the energy or make the system heat up, depending on how the work is done. Lifting a heavy object is a conversion of work to potential energy, and friction is a conversion of work to heat. Conversely, energy can be reconverted to work when a heavy object falls, and heat is converted to work in an automobile engine. Of these three—heat, work, and energy—the chemist usually is more concerned with heat: the heat that may be absorbed or given off when a chemical reaction takes place.

By Newton's laws of motion, the force on an object is the product of its mass and acceleration:

$$\text{Force} = \text{mass} \times \text{acceleration}$$
$$F = m \times a$$

The force that must be applied to a 1.00 kg mass to give it an acceleration of 1 meter per second per second (1 m sec^{-2}) is defined as a force of 1 **newton** (N). Hence 1 N = 1 kg m sec^{-2}. (SI units are based on length in meters and mass in kilograms.)

Example 23

When a pitcher whips a 5.00 ounce (oz) baseball around an arc 5.00 m in circumference in order to accelerate it from zero to 90 miles hr^{-1}, what average acceleration does he give to the ball during the pitch, and what average force does he exert on it during his windup?

Solution

Assume uniform acceleration on the ball from the time the windup begins until the ball leaves the pitcher's hand. For uniform acceleration from rest, $v = at$ and $s = \frac{1}{2}at^2$, where v is velocity, a is acceleration, t is time, and s is distance. Eliminating time from the two expressions yields $a = v^2/2s$. If $v = 40.2$ m sec^{-1} (90 miles hr^{-1}) and $s = 5.00$ m, then

$$a = \frac{(40.2)^2}{2 \times 5.00} = 162 \text{ m sec}^{-2}$$

Since the mass (m) is 0.142 kg (5.00 oz), the average force applied to the ball during the swing is

$$F = m \times a = 0.142 \times 162 \text{ kg m sec}^{-2} = 23.0 \text{ N}$$

The work that is done on an object is the product of the force exerted on it along the direction of motion, and the distance through which the force is applied:

$$\text{Work} = \text{force} \times \text{distance}$$
$$W = F \times s$$

The work done when 1 N of force is exerted on an object for a distance of 1 m is defined as 1 **joule** (J). Hence $1 \text{ J} = 1 \text{ N m} = 1 \text{ kg m}^2 \text{ sec}^{-2}$. For a thrown object, all the work is converted into kinetic energy (energy of motion); in other circumstances, part or all the work can end as heat.

Example 24

How much work is done on the baseball in the pitch described in Example 23? How much kinetic energy does the ball have as it leaves the pitcher's hand?

Solution

The work done on the baseball is

$$W = F \times s = 23.0 \text{ N} \times 5.00 \text{ m} = 115 \text{ N m or } 115 \text{ J}$$

The ball ends with a kinetic energy of 115 J.

As a check on these results, we can calculate the kinetic energy directly:

$$E = \tfrac{1}{2} mv^2 = \tfrac{1}{2} \times 0.142 \times (40.2)^2 \text{ kg m}^2 \text{ sec}^{-2} = 115 \text{ J}$$

where E is energy.

The advantage of the joule as a unit of heat is that it makes immediately apparent the connection between heat, work, and energy. An older unit of energy that arose from heat measurements is the **calorie**. One calorie (cal) is defined as the quantity of heat required to raise the temperature of 1 g of pure water by $1\,°C$ (from $14.5\,°C$ to $15.5\,°C$, to be exact). This definition had no obvious connection with work, and in fact the calorie was defined in the nineteenth century, before anyone realized that heat and work were alternative forms of the same thing: energy. We will use only joules in this book, but you should be aware of calories since most of the preexisting literature uses that unit. The calorie is approximately four times as big as a joule: **1 cal = 4.184 J**. Heats of reaction of mole quantities of substances are typically in the range of kilojoules (kJ) or kilocalories (kcal), where 1 kJ = 1000 J and 1 kcal = 1000 cal.

As an illustration of heats of reaction and the principle of additivity of heat, let us look at the decomposition of hydrogen peroxide, H_2O_2. When an aqueous solution of hydrogen peroxide reacts to form oxygen gas and liquid water, heat is given off. The amount of heat will vary somewhat with the temperature at which the reaction occurs, but at $25\,°C$, the commonly accepted standard "room temperature" for measuring and tabulating heats of reaction, each mole of H_2O_2 that decomposes produces 94.7 kJ of heat. (If this energy could be used with perfect efficiency, it would be enough to accelerate 823 baseballs as described in Example 24.)

The heat involved in a chemical reaction carried out at constant pressure (or at least with the final pressure brought back to the starting value) is known as the change in the **enthalpy** of the reacting system, ΔH (read as "delta *H*"). As we shall see in Chapter 15, the **energy** change, ΔE, corresponds to the heat of the reaction if the reaction is carried out at constant volume, as in the bomb calorimeter shown in Figure 2-4. Enthalpy can be

Figure 2-4

A schematic representation of a bomb calorimeter used for the measurement of heats of combustion. The weighed sample is placed in a crucible, which in turn is placed in the bomb. The sample is burned completely in oxygen under pressure. The sample is ignited by an iron wire ignition coil that glows when heated. The calorimeter is insulated by means of the jacket, which can be equipped with a special thermal regulator (not shown). The temperature of the calorimeter fluid is measured with the thermometer. From the change in temperature, the heat of reaction can be calculated.

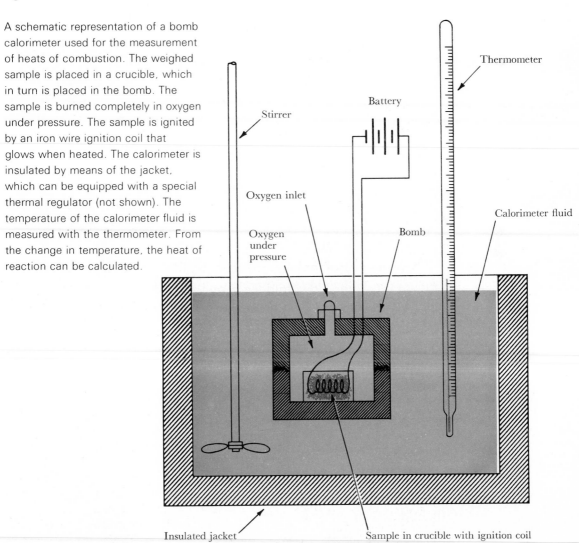

regarded as a "corrected" energy, the correction being for any work that the chemicals might do in pushing against the atmosphere if they expand. The difference between ΔE and ΔH is small but significant, but it is not important to us now. If heat is given off during the reaction, then the enthalpy of the reacting system of chemical falls; ΔH, the change in enthalpy, is negative. Such a reaction is called **exothermic**. In an **endothermic** reaction heat is absorbed and the enthalpy of the reaction mixture rises. For the hydrogen peroxide reaction, we can write

$$H_2O_2(aq) \rightarrow H_2O(l) + \tfrac{1}{2}O_2(g) \qquad \Delta H = -94.7 \text{ kJ} \qquad (2\text{-}8)$$

This heat is released when 1 mole of hydrogen peroxide decomposes to 1 mole of water and $\tfrac{1}{2}$ mole of oxygen gas, or for 1 mole of the reaction as just written. If all the coefficients of the reaction are doubled, then the heat of reaction must be doubled also, since it then refers to twice as much reaction:

$$2H_2O_2(aq) \rightarrow 2H_2O(l) + O_2(g) \qquad \Delta H = -189.4 \text{ kJ} \qquad (2\text{-}9)$$

"One mole of the reaction as written" now means 2 moles of hydrogen peroxide decomposing to 2 moles of water and 1 mole of oxygen, because the coefficients in the equation have all been doubled.

The heat of a reaction also depends on the physical state of the reactants and products. If hydrogen peroxide were to decompose to give oxygen gas and water vapor instead of liquid, part of the 94.7 kJ would be diverted to evaporating H_2O:

$$H_2O(l) \rightarrow H_2O(g) \qquad \Delta H = +44.0 \text{ kJ} \qquad (2\text{-}10)$$

and less heat would be given off by decomposition of peroxide:

$$H_2O_2(aq) \rightarrow H_2O(g) + \tfrac{1}{2}O_2(g) \qquad \Delta H = -50.7 \text{ kJ} \qquad (2\text{-}11)$$

An important assumption is hidden here: that heats of reaction are additive (Figure 2-5). Equation 2-9 plus equation 2-10 gives equation 2-11, and so we have assumed that the heat of the third reaction will be the sum of the first two:

$$\Delta H = -94.7 + 44.0 \text{ kJ} = -50.7 \text{ kJ}$$

The additivity of reaction heats follows directly from the first law of thermodynamics (Chapter 15): The change in energy or enthalpy between two states depends only on the nature of those states, and not on how the change is carried out. A collection of chemicals in a given state has a certain energy and a certain enthalpy, neither of which depends in any way on how the chemicals were brought to that state. (That is, the past history of the chemicals can affect their present energy and enthalpy, but we do not need to know that history to measure the values of E and H.) Hence the difference between enthalpies of reactants and products, or the heat of reaction, can depend only on the nature of the starting and ending states, and not on the particular way that the reaction is carried out. This is sometimes called **Hess'**

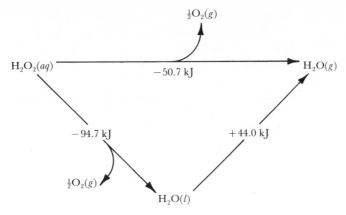

Figure 2-5

The change in enthalpy for the reaction $H_2O_2(aq) \rightarrow H_2O + \frac{1}{2}O_2(g)$ can be obtained without actually measuring the enthalpy change for the reaction by adding ΔH for the two reactions

$$H_2O(aq) \rightarrow H_2O(l) + \frac{1}{2}O_2(g) \qquad \Delta H = -94.7 \text{ kJ}$$
$$H_2O(l) \rightarrow H_2O(g) \qquad \Delta H = 44.0 \text{ kJ}$$

law of heat summation, which is a rather dignified name for a natural consequence of the first law of thermodynamics.

The additivity of heats of reaction makes a great amount of experimentation in thermochemistry (the chemistry of heat and energy) unnecessary. We need not measure and tabulate the enthalpy change of every conceivable chemical reaction. For example, if we know the heat of vaporization of liquid water (equation 2-10) and the heat of decomposition of hydrogen peroxide to liquid water (equation 2-9), then we never need to measure the heat of decomposition of hydrogen peroxide to water vapor; the answer can be calculated ahead of time. If a certain reaction is inconvenient to carry out, there may be a set of easier reactions whose sum is the reaction in question. After we have carried out the individual experiments, we can add the enthalpy changes in the same way as the chemical equations, to find the heat of the difficult-to-measure reaction.

Suppose that someone proposed a scheme for making diamonds by oxidizing methane:

$$CH_4(g) + O_2(g) \rightarrow C(di) + 2H_2O(l)$$

[The notation (s) is not sufficient for carbon, since diamond (di) must be differentiated from graphite (gr).] You would like to find whether the reaction will liberate heat that must be allowed for in the design of the reaction vessel. This particular synthesis has never been carried out (and probably never will be), yet you can give your misguided friend his answer from a knowledge of the heats of easier reactions. The **heat of combustion** of a substance containing C, N, O, and H is the heat, per mole of substance, of the reaction with enough oxygen to produce CO_2, N_2, and liquid H_2O. Heats of combustion are easy to measure, and were among the first reaction heats to be measured and tabulated systematically. Extensive tables of heats of combustion can be found in books such as the *CRC Handbook of Chemistry and Physics* or *Lange's Handbook of Chemistry*. The heats of combustion of methane and diamond are

$$CH_4(g) + 2O_2(g) \rightarrow CO_2(g) + 2H_2O(l) \qquad \Delta H = -890 \text{ kJ} \qquad (2\text{-}12)$$
$$C(di) + O_2(g) \rightarrow CO_2(g) \qquad \Delta H = -395 \text{ kJ} \qquad (2\text{-}13)$$

The desired diamond-synthesizing reaction is produced by subtracting the second reaction from the first, or by adding the first reaction to the *reverse* of the second, and the heat of reaction is found in the same way:

$$CH_4(g) + 2O_2(g) \rightarrow CO_2(g) + 2H_2O(l) \qquad \Delta H = -890 \text{ kJ} \qquad (2\text{-}12)$$
$$CO_2(g) \rightarrow C(di) + O_2(g) \qquad \Delta H = +395 \text{ kJ} \qquad (2\text{-}14)$$
$$\overline{CH_4(g) + O_2(g) \rightarrow C(di) + 2H_2O(l) \qquad \Delta H = -495 \text{ kJ} \qquad (2\text{-}15)}$$

Notice that when a reaction is turned around and run in reverse, the heat of reaction changes sign, since a process that gave off 395 kJ in one direction must absorb 395 kJ in reverse.

Heats of Formation

Because of the additivity of heats of reaction, not all heats have to be tabulated—only those for the minimum set of reactions from which all others can be obtained. The set that has been agreed upon by scientists and engineers is made up of the **heats of formation** of compounds from their pure elements in **standard states**. For solids and liquids this standard state is the most common form of the element at 25°C or 298 K and 1 atmosphere (atm)* external pressure; gases are defined similarly but at 1 atm partial pressure.† The standard state for thermodynamic measurements involving carbon is graphite (*gr*), not diamond (*di*). The heats of formation for all the compounds involved in the diamond synthesis are

$$C(gr) + 2H_2(g) \rightarrow CH_4(g) \qquad \Delta H = -74.8 \text{ kJ} \qquad (2\text{-}16)$$
$$C(gr) \rightarrow C(di) \qquad \Delta H = +1.9 \text{ kJ} \qquad (2\text{-}17)$$
$$H_2(g) + \tfrac{1}{2}O_2(g) \rightarrow H_2O(l) \qquad \Delta H = -285.8 \text{ kJ} \qquad (2\text{-}18)$$

A table of standard heats of formation of compounds from their pure elements is given in Appendix 3. In that table, the subscript 298 refers to the temperature (298 K), and the zero superscript signifies that reactants and products are all in their standard states.

To illustrate how heats of general reactions are found from heats of formation, let us look again at the diamond synthesis, equation 2-15. That

---

*The conversion to the absolute or Kelvin temperature scale is considered in Chapter 3, as is the atmosphere as a unit of pressure. (By SI convention, no degree sign is used for the Kelvin scale.)

†The partial pressure of a gas in a mixture is the pressure that the gas would show if all the other gases were removed and it were the only gas present.

reaction can be obtained by adding equation 2-17 to twice equation 2-18 and the reverse of equation 2-16:

$$
\begin{array}{llll}
\text{C}(gr) \rightarrow \text{C}(di) & \Delta H = & +1.9 \text{ kJ} & (2\text{-}17) \\
2\text{H}_2(g) + \text{O}_2(g) \rightarrow 2\text{H}_2\text{O}(l) & \Delta H = & -571.6 \text{ kJ} & 2 \times (2\text{-}18) \\
\text{CH}_4(g) \rightarrow \text{C}(gr) + 2\text{H}_2(g) & \Delta H = & +74.8 \text{ kJ} & -(2\text{-}16) \\
\hline
\text{CH}_4(g) + \text{O}_2(g) \rightarrow \text{C}(di) + 2\text{H}_2\text{O}(l) & \Delta H = & -494.9 \text{ kJ} & (2\text{-}15)
\end{array}
$$

The heat of reaction is found in exactly the same manner:

$$\Delta H = (+1.9) + 2(-285.8) - (-74.8) = -494.9 \text{ kJ}$$

As you can see, the consequences of not keeping track of signs and coefficients can be disastrous. The surest method is to write out each equation, with its heat of reaction, in such a way that the sum of the individual equations will be the desired reaction. If all the coefficients of an equation are multiplied by an arbitrary number n, then the heat of formation must also be multiplied by n, and if a formation equation is turned around and run in reverse, then the sign ΔH must be changed. If the individual equations add to give the desired reaction, then the individual heats add to give the corresponding overall heat of reaction.

A convenient shortcut is to think of the heat of formation of a compound as if it were, in a sense, the enthalpy of the compound itself. (Warning: This is possible only because the heats of formation of the elements are zero by definition.) Then the heat of a reaction becomes the sum of heats of formation of all the products, minus the heats of formation of all the reactants, each being multiplied by the coefficient of that substance in the balanced equation.

Example 25

What is the standard heat of the reaction by which ferric oxide is reduced by carbon to iron and carbon monoxide in a blast furnace?

Solution

The reaction is as follows, with the standard heat of formation per mole written below each compound:

$$
\begin{array}{lcccc}
& \text{Fe}_2\text{O}_3(s) + & 3\text{C}(gr) \rightarrow & 2\text{Fe}(s) + & 3\text{CO}(g) \\
\Delta H \text{ (kJ mole}^{-1}) & -822.1 & 0.0 & 0.0 & -110.5
\end{array}
$$

The standard heat of formation of the elements from themselves, of course, is zero by definition. For the reaction as written,

$$\Delta H = 2(0.0) + 3(-110.5) - (-822.1) - 3(0.0) = +490.6 \text{ kJ}$$

These results are consistent with the fact that much heat must be supplied to reduce iron ore to iron. Note, however, that 490.6 kJ is the net heat that would be absorbed if the reaction were run at 298 K, not the 1800 K of a blast furnace. Yet this calculated figure is also the heat absorbed if ferric oxide and carbon are heated from 298 K to 1800 K, allowed to react, and the products are cooled again to room temperature. The enthalpy change or heat of a reaction depends only on the initial and final states of the participants, and not on whether the temperature remained constant or went to blast-furnace levels in between. All that matters is that the temperature is brought back down to 298 K at the end.

As another example of the principle, the net heat evolved when water is made from hydrogen and oxygen will be the same whether a mixture of H_2 and O_2 at 298 K explodes violently and the resulting water is cooled back to 298 K, or whether the same mixture reacts slowly in the presence of finely divided platinum as a catalyst, never increasing its temperature. So, in referring to heats of reaction, when we say that the values are correct for the process carried out "at 1 atm pressure and 298 K," we require only that the reactants begin at these conditions, and that the products end there. This is why tables of heats of formation under standard conditions (Appendix 3) are useful.

Example 26

What is the heat of combustion of liquid benzene?

Solution

The balanced equation, and the individual heats of formation, are

$$2C_6H_6(l) + 15O_2(g) \rightarrow 12CO_2(g) + 6H_2O(l)$$

$$\Delta H \text{ (kJ mole}^{-1}) \quad +49.0 \qquad 0.0 \qquad -393.5 \qquad -285.8$$

For the reaction as written, with 2 moles of benzene, the heat is

$$\Delta H = 12(-393.5) + 6(-285.8) - 2(+49.0) - 15(0.0) = -6540 \text{ kJ}$$

Hence the heat of combustion of liquid benzene, per mole, is half this amount, or $-3270 \text{ kJ mole}^{-1}$.

2-7   CONSERVATION PRINCIPLES

It may seem odd that scientists, who are presumably interested in the changes that matter can undergo, should spend so much time searching for conservation principles, which tell us what does *not* change. Everyone knows that all things change; seasons come and go, living things grow and die, mountains erode away to topsoil. These observations are not science. The first high priests in primitive society were not the philosophers who observed that spring turned into summer and ultimately into winter, but the clever observers who noticed that spring always came back again, and who learned how to compute when it would happen. This kind of knowledge is useful to the tribe.

True science begins when we understand the limitations on random change. A rock thrown into the air does not rise forever; it eventually begins to fall because of limitations on its kinetic energy. A tank of gasoline cannot propel an automobile forever because of limitations on the chemical energy of the fuel. A chemical plant cannot turn out more products than raw materials received, because of limitations on mass. The search for an understanding of change in the universe is a search for the limits between which we can predict that change will be confined. Then we can anticipate what will happen, and make this knowledge work to our advantage. It is safer to be chased by a crazed locomotive engineer than by an irate motorist, because you understand the limitations on motion of the railway engine and can use that knowledge to get out of its way.

The first two conservation principles to be recognized in science were the principles of conservation of mass and of energy. In the physics of motion, momentum is often conserved. In nuclear reactions, mass and energy can be interconverted, but their sum must be constant. Nuclear energy is produced only at the cost of a disappearance of mass, by Einstein's relationship, $E = mc^2$, where E equals energy, m equals mass, and c^2 equals the speed of light squared. Charge also is conserved during nuclear reactions. When a carbon-14 nucleus decays to nitrogen-14, an electron is emitted (this is called *beta* radiation):

$$^{14}_{6}\text{C} \rightarrow {}^{14}_{7}\text{N} + {}^{0}_{-1}e$$

The superscripts tell the mass number or total number of heavy particles—protons and neutrons—and this number is the same in carbon-14 and in nitrogen-14. This is an example of conservation of heavy particles, related to the conservation of mass. The subscripts give the charge on the particles and this charge can increase from $+6$ to $+7$ only if a negative charge escapes into the surroundings. This -1 charge is carried by the electrons of the beta radiation. Total charge is conserved.

Even more mysterious quantities are conserved or nearly conserved in nuclear and fundamental-particle reactions: **parity**, **strangeness**, and **charm**. We find these properties mysterious because we never heard of them before we learned of the necessity that they be conserved. Mass and energy were familiar before it was realized that they had to be conserved. But who ever heard of the charm of a fundamental particle before its conservation principle was enunciated? In what seems like an inversion of logical order, it was noticed that particles seem to fall into classes, with rigid rules about how the members of these classes interact with one another; then a new property was invented such that, if this property is conserved during particle reactions, the observed behavior is accounted for.

Still, this is not too different from the way that electrostatic charge was discovered (Figure 2-6). Experimenters in the eighteenth century observed that if they rubbed a glass rod vigorously with silk, it accumulated a charge that could be transferred in part to a lightweight pith ball hanging from a thread. After the ball acquired the charge from the glass rod, it was thereafter repelled by the rod. In contrast, they found that if they rubbed a hard rubber rod with wool, the rod *attracted* the pith ball that had been electrified by touching it with the glass rod. To explain this behavior, scientists hypothesized that there were two different kinds of charge, and that like charges repelled, whereas unlike charges attracted. Benjamin Franklin coined the terms *positive* and *negative* for these two charges. Charge, in Franklin's day, was no less mysterious than *charm* and *strangeness* are today.

Figure 2-6

Two types of electrical charged were detected as early as the eighteenth century. A glass rod rubbed with silk acquires one type of charge (a), a hard rubber rod rubbed with wool acquires the other type (b). When the charge is transferred to a pith ball (c), the repulsion of like charges (d) and attraction of unlike ones (e) is shown by the deflection of the ball.

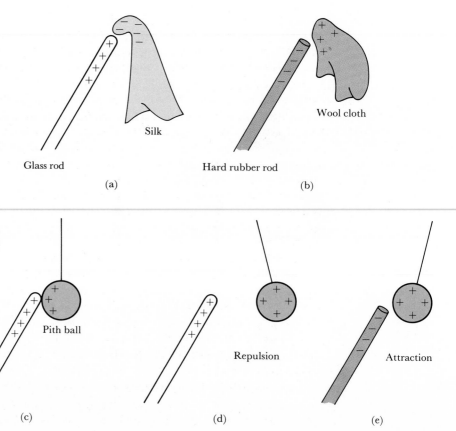

The idea behind all these conservation principles is the same: When we know what cannot happen, then we are better able to predict what will happen. Conservation of mass and conservation of energy are the two most important principles to the chemist, and all the calculations in this chapter have been consequences of them.

Summary

Conservation of mass, to a chemist, means the conservation of the total number of each type of atom during a chemical reaction. Exactly as many of each type of atom must be present in the products of the reaction as were provided by the reactants. A chemist counts out molecules by weighing a substance, and then converting weight in grams into the number of **moles**. A mole of any substance contains the same number of particles, $N = 6.022 \times 10^{23}$; it is the quantity in grams that is numerically the same as the molecular weight of the substance in atomic mass units. To obtain the number of moles in a sample, divide the weight in grams by the molecular weight of the material in grams per mole.

An **elemental analysis** of a substance will give the percentage of each element by weight, or the number of grams of each element in 100 g of the material under analysis. Dividing each of these gram quantities by the atomic weight of the element produces a set of numbers giving the relative amounts of each kind of atom. The simplest formula that specifies the relative number of each kind of atom is called the **empirical formula**. If the compound is a salt, then it has no discrete molecules, although the empirical formula is treated as if it were a molecular formula when we write chemical equations. If the compound is made up of molecules, then the true molecular formula and the molecular weight may be either the same as or integral multiples of the empirical formula and the formula weight. The information from an elemental analysis is insufficient to determine the true molecular formula, but if a rough molecular weight is known from some other physical measurement, then it can be used to choose the proper multiple of the empirical formula as the true molecular formula.

A chemical equation is **balanced** when the correct molecular or empirical formula is given for each reactant and product, and when the number of each kind of atom on the left side of the equation equals the number of that kind of atom on the right side. The coefficients of the reactants and products need not be integers, but it is sometimes more convenient if they are. If all the coefficients on both sides of a balanced equation are multiplied or divided by the same number, the equation remains balanced.

A balanced chemical equation expresses the relative amounts of reactants and products that are involved in the reaction. At the atomic level, it describes the relative numbers of atoms and molecules participating.

It provides no information, however, about the mechanism by which the reaction takes place. A process could occur by one step or by a series of steps, and it could still have the same overall chemical equation. From calculations based on the balanced chemical equation, we can choose the proper relative quantities of reactants, predict the quantities of products to be formed if reaction is complete, and decide whether one or another of the reactants is present in excess in a given starting situation.

Chemical reactions are especially convenient in solution since the liquids can easily be measured, handled, and mixed. The **molarity** of a reactant or product, c, is the number of moles per liter of solution. The abbreviation M is used for the units of molarity or moles per liter. In contrast, the **molality**, m, is the number of moles per kilogram of solvent. If water is the solvent, with a density of 1 g ml^{-1}, then molality corresponds to the number of grams per liter of solvent, but this is still not the same as molarity, since the total volume in general will change slightly when solute and solvent are mixed. For dilute solutions the volume increase when solute is added becomes negligible, and molarity and molality, in aqueous solutions, become effectively equivalent. When calculating concentrations during dilutions, it is helpful to remember that addition of solvent leaves the number of moles of solute unchanged:

$$\text{Moles of solute} = cV = \text{constant}$$

In Arrhenius' classical definition, an acid is a substance that increases the hydrogen ion concentration, $[H^+]$, when added to water, and a base is a substance that increases the hydroxide ion concentration, $[OH^-]$, when added to water. One mole of acid may be capable of releasing $1, 2,$ or 3 moles of H^+ upon complete dissociation. The **equivalent weight** of an acid is that quantity that is capable of liberating 1 mole of H^+ if it dissociates completely, so the equivalent weight of an acid such as H_3PO_4 is one-third its molecular weight. Similarly, if 1 mole of a base is capable of producing 2 moles of OH^- in solution, as in $Ca(OH)_2$, then its equivalent weight is half its molecular weight.

Even though an acid is capable in principle of producing 1 or more moles of H^+ per mole of acid, its actual degree of dissociation may be smaller because of its strong hold on the hydrogen ions. Acids that fail to dissociate completely in water are called **weak acids**, and **strong acids** are those that dissociate completely. Partial dissociation does not affect the definition of equivalent weights. Carbonic acid, H_2CO_3, still has an equivalent weight half its molecular weight, even though in pure water it is only partially dissociated. If some of the H^+ from carbonic acid is used up by combining with OH^- from a base, more H^+ then will dissociate until nothing but CO_3^{2-} ions are left.

The reaction of H^+ from an acid with OH^- from a base is known as **neutralization**:

$$H^+ + OH^- \rightarrow H_2O$$

One **equivalent** (equiv), or 1 equivalent weight, of an acid will combine with, and neutralize, 1 equiv of base. The amount of an acid in an unknown sample can be measured by finding out how much base solution of a known concentration is required to neutralize it. This is called **titration**, and it is a common analytical procedure. The **normality**, N, of an acid or base solution is the number of equivalents per liter of solution. For an acid that would release three protons per molecule at complete dissociation, the normality is three times the molarity, since the equivalent weight is only one-third the molecular weight. The product of normality (N) times volume (V) is the number of acid–base equivalents, and at neutrality they must be the same for acid and base:

$$N_A V_A = N_B V_B$$

The heat that is given off during a chemical reaction carried out at constant volume is a measure of the drop in **energy**, E, of the reacting system of chemicals. If the reaction is carried out at constant pressure rather than constant volume, then the heat given off corresponds to the fall in a slightly different quantity, the **enthalpy**, H. If heat is emitted, then the change in enthalpy, ΔH, is negative, and the reaction is said to be **exothermic**. If heat is absorbed, then ΔH is positive and the reaction is **endothermic**.

Heat and work are alternate forms of energy. If a mass (m) of 1 kg is given an acceleration (a) of 1 m sec^{-2}, the force (F) required is 1 **newton** (N):

$$F = m \times a = 1\,\text{N} = 1\,\text{kg m sec}^{-2}$$

If such a force is exerted on an object for a distance (s) of 1 m, then the work (W) done on the object is 1 **joule** (J):

$$W = F \times s = 1\,\text{J} = 1\,\text{N m} = 1\,\text{kg m}^2\,\text{sec}^{-2}$$

To put the joule in perspective, a pitched baseball traveling 90 miles hr^{-1} has an energy of 115 J. The older unit of heat, the **calorie** (cal), is approximately four times as large: 1 cal = 4.184 J.

The first law of thermodynamics states that the change in energy or enthalpy between two states depends only on the nature of those states, and not on how the change was carried out. Hence the heat of a chemical reaction depends only on the overall reaction, and not on whether it is carried out in one step or several. This means that heats of reaction are additive: If reaction A plus reaction B gives reaction C, then the heat of reaction C is found by adding the heats of reactions A and B. This leads to a great economy in tabulating heats of reaction: We need only measure the heat of reaction for the set of reactions from which all other reactions can be obtained. The set chosen is made up of the reactions of formation of all compounds from their elements in standard states. The **standard state** of a liquid or solid is the most common form at 298 K, under an external pressure of 1 atm. A similar definition applies to gases, but with 1 atm partial pressure. Standard **heats of formation** of compounds from their elements are tabulated as in Appendix 3 for a wide range of substances.

We can use these tabulated heats of formation as if they were the absolute enthalpies of the substances involved. The heat of reaction for a balanced chemical equation is the sum of heats of formation of all the product molecules, less the heats of formation of the reactants. Care must be taken, of course, to see that each heat of formation is multiplied by the number of times the substance appears in the balanced equation.

Self-Study Questions

1. Why is the expression of the amounts of reactants and products in moles more closely related to the balanced chemical equation than is the expression of quantities in grams?
2. In what way is the principle of conservation of mass in chemical reactions more than just a conservation of total mass before and after reaction?
3. What is the empirical formula of a substance? How is it obtained from an elemental analysis expressed in percent by weight?
4. How is the empirical formula related to the molecular formula? Can the molecular formula be decided from an elemental analysis?
5. How can the empirical formula turn an imprecise molecular weight into an exact molecular weight?
6. What is "balanced" in a balanced chemical equation? How does such an equation illustrate the conservation of mass?
7. Can a balanced equation have fractional coefficients, such as in $3\frac{1}{2} H_2O$? How can an equation that refers to half-molecules be meaningful?
8. How is a balanced equation useful in predicting amounts of products? How can it be used to calculate the relative amounts of reactants that are needed?
9. What is the difference between molarity and molality as units of solution concentration?
10. In dilute aqueous solution, molarity and molality approach one another. Why is this so? Why is it correct only for aqueous solutions?
11. What is normality? Is it more closely related to molality or molarity? For what type of chemical reactions is normality useful as a concentration unit?
12. What are Arrhenius' definitions of an acid and a base?
13. How do strong and weak acids differ?
14. For what kinds of acid are the molecular weight and equivalent weight the same? Why is the molecular weight always an *integral* multiple of the equivalent weight?
15. What is neutralization? What is the product of the neutralization reaction?
16. What is titration? How is it used in chemical analysis? What information can be obtained from an acid–base titration?

17. How are acid–base equivalents useful in titration calculations? What is the relationship between equivalents of acid and of base at the equivalence point of a titration, or at neutrality?

18. What is the SI unit of heat, work, or energy? How is it related to the force exerted on an object?

19. If acceleration is given by the equation $a = v^2/2s$, as stated in Example 23, then prove that the kinetic energy of motion is $E = \frac{1}{2}mv^2$, if this energy is just the work that was exerted on the moving object to set it in motion from rest.

20. What is the difference between energy, E, and enthalpy, H? Under what conditions is the heat of reaction the same as the change in energy? When is it the same as the change in enthalpy?

21. Why must we specify the state of reactants and products when calculating the heat of a reaction?

22. What does the first law of thermodynamics say about E and H that is relevant to heats of reaction? How does one deduce from it that heats of reaction are additive? What does this additivity mean?

23. In what sense is the additivity of heats of reaction a great labor-saving principle?

24. Why are heats of formation particularly appropriate for tabulation? In what way can all other heats of reaction be obtained from heats of formation?

25. Why is it legitimate to regard heats of formation from elements as if they were absolute enthalpy values of the compounds, when calculating heats of reaction? What happens to the elements on the left side of all of the heat-of-formation equations? In what way does this illustrate the principle of conservation of mass?

26. What does the requirement "at constant pressure and 298 K" mean in terms of the applicability of standard heat of formation tables such as in Appendix 3? Why are such tables useful even for reactions that do not remain uniformly at 298 K thoughout?

Problems

Percent composition

1. What is the weight percent of each element in $K_2Cr_2O_7$? What is its molecular weight?

2. The formula for nitroglycerine is $C_3H_5O_9N_3$. Calculate the molecular weight and the weight percent of each element.

Empirical formula

3. One of the components of Portland cement contains 52.7% calcium (Ca), 12.3% silicon (Si), and 35.0% oxygen (O). What is its empirical formula?

4. A compound is 22.8% sodium (Na), 21.5% boron (B), and 55.7% oxygen (O). What is its empirical formula, and what is its formula weight?

Molecular formula

5. The molecular weight of an oxide of phosphorus (P) is approximately 280 amu. Elemental analysis shows that the compound contains 43.6% phosphorus by weight. Determine the molecular formula and exact molecular weight of the oxide.

6. When 5.00 g of an unknown hydrocarbon, containing only carbon and hydrogen atoms, are burned as fuel, 14.65 g of carbon dioxide, CO_2, are produced. What is the empirical formula of the hydrocarbon? If other experiments establish an approximate molecular weight for the hydrocarbon of 25 to 35 g mole^{-1}, what are the true molecular weight and the molecular formula? Write a balanced chemical equation for the burning of the hydrocarbon.

Atomic weight calculation

7. A metal, M, forms an oxide having the formula M_2O_3, containing 68.4% metal by weight. Calculate the atomic weight of M.

8. When 5.00 g of element A react completely with 15.0 g of element B, compound AB is formed. When 3.00 g of element A react with 18.0 g of element C, compound AC_2 is formed. The atomic weight of element B is 60.0. Calculate the atomic weights of elements A and C.

Balanced equations

9. The following is a balanced chemical equation:

$$Zn + H_2SO_4 \rightarrow H_2 + ZnSO_4$$

State the qualitative meaning of the equation. What is the meaning in terms of the relative weights of compounds involved? What do we mean by saying that the equation is balanced?

10. Balance the following equations:
   a) $Fe_2O_3 + Al \rightarrow Fe + Al_2O_3$
   b) $Na_2SO_3 + HCl \rightarrow$
   $$NaCl + SO_2 + H_2O$$
   c) $Mg_3N_2 + H_2O \rightarrow$
   $$Mg(OH)_2 + NH_3$$
   d) $Pb + PbO_2 + H_2SO_4 \rightarrow$
   $$PbSO_4 + H_2O$$
   State what each equation means in terms of the appearance and disappearance of moles of reactants and products.

11. Write a balanced equation for each of the following: (a) the reaction of sodium with water to produce hydrogen and sodium hydroxide; (b) the reaction of calcium hydroxide and carbon dioxide to form calcium carbonate and water; (c) the reaction of carbon monoxide with hydrogen to form methane and water; (d) the reaction of aluminum nitrate with ammonium hydroxide to produce aluminum hydroxide and ammonium nitrate.

12. Write balanced equations for each of the following: (a) the reaction of aluminum with hydrogen chloride to produce aluminum chloride and hydrogen; (b) the reaction of ammonia with oxygen to produce nitric oxide (NO) and water; (c) the reaction of zinc with phosphorus to produce zinc phosphide, Zn_3P_2; (d) the reaction of nitric acid with zinc hydroxide to produce zinc nitrate and water.

Product yield

13. Carbon burns in air to produce carbon dioxide. What weight of carbon dioxide results from burning 100 g of carbon?

14. Laughing gas, N_2O, causes hysteria and unconsciousness when inhaled. It is

made from ammonium nitrate by the reaction

$$NH_4NO_3 \rightarrow H_2O + N_2O$$

Balance the equation, and calculate the weight of N_2O produced from 7.50 g of ammonium nitrate.

15. Potassium dichromate, $K_2Cr_2O_7$, reacts with oxalic acid, $H_2C_2O_4$, and sulfuric acid, H_2SO_4, according to the following equation:

$$3H_2C_2O_4 + K_2Cr_2O_7 + 5H_2SO_4 \rightarrow \\ 2KHSO_4 + Cr_2(SO_4)_3 \\ + 6CO_2 + 7H_2O$$

(a) Is this equation balanced? If not, balance it. (b) If 450 ml of 0.200M potassium dichromate solution react with excess oxalic and sulfuric acids, how many moles of CO_2 will form? How many grams?

16. Sodium sulfide reacts with sulfuric acid to produce sodium sulfate and hydrogen sulfide. Assume that excess sulfuric acid is allowed to react with 10.0 g of sodium sulfide and calculate the following:
a) The moles of sodium sulfide used
b) The moles of hydrogen sulfide liberated
c) The number of grams of hydrogen sulfide liberated

17. Calcium phosphide, Ca_3P_2, reacts with water to produce phosphine gas (PH_3) and calcium hydroxide. Write a balanced equation for this process. How many grams of PH_3 gas will 1.75 g of calcium phosphide produce?

18. When vanadium oxide, VO, reacts with iron oxide, Fe_2O_3, the products are V_2O_5 and FeO. If no other reactants or products are involved, write a balanced equation for the reaction. If 6.50 g of vanadium oxide react with an excess of iron oxide, how many grams of V_2O_5 are produced? How many grams of V_2O_5 can be formed from 2.00 g of VO and 5.75 g of Fe_2O_3?

19. How many moles of XeF_6 can be made from 0.0320 g of xenon and 0.0304 g of fluorine? How many grams of XeF_6 result? Is an excess of either reactant left behind, and if so, how much?

20. How many moles of ethanol, C_2H_5OH, can be made from 25 g of carbon? If a 5-carat diamond could be converted completely to ethanol, how many liters of gin would result? (One carat equals 0.2053 g, 80-proof gin is 40% ethanol by volume, and the density of ethanol is 0.790 g ml^{-1}.)

21. (a) One mole of propane gas, C_3H_8, is burned with 5 moles of O_2 according to the reaction

$$C_3H_8 + 5O_2 \rightarrow 3CO_2 + 4H_2O$$

Is some of either reactant left unused when the other reactant is gone? If so, which one is left, and how many grams remain behind? (b) One gram of propane gas is now burned with 5 g of O_2 according to the same reaction. Is some of either reactant left unused when the other is gone? If so, which one, and how much is left behind?

22. When metallic silver and powdered sulfur are heated together, black, solid Ag_2S forms. A mixture containing 1.73 g of Ag and 0.540 g of S is heated to bring about reaction. Is any of one starting material left behind after the other is completely converted to Ag_2S? How many grams of which element remain behind?

Molarity

23. Potassium perchlorate, $KClO_4$, has a solubility of about 7.5 g liter^{-1} of solution in water at $0°C$. What is the molarity of a saturated solution at $0°C$?

24. An amount of 50.0 ml of ether, $C_4H_{10}O$, which has a density of 0.714 g ml^{-1}, is dissolved in enough ethyl alcohol to make 100 ml of solution. Calculate the molarity of ether in the solution.

Molarity, molality, normality

25. A solution of sulfuric acid was prepared from 95.94 g of H_2O and 10.66 g of H_2SO_4. The volume of the solution that resulted was 100.00 ml. Calculate the molality, molarity, normality, and density of the solution.

Molarity and normality

26. How many grams of solute are required to prepare the following solutions: (a) 2 liters of $2.5M$ sulfuric acid; (b) 0.5 liter of $1.0N$ H_3PO_4 (assuming that the dissociation produces HPO_4^{2-} ions, and that the solution is to be used in acid–base neutralization reactions; and (c) 1.0 liter of sodium hydroxide solution that will titrate, milliliter for milliliter, with $0.5N$ HCl?

Dilution

27. How many milliliters of water must be *added* to 200 ml of $5.00M$ HNO_3 to make a solution that is $2M$ HNO_3?

28. What volume of a solution of $1.53M$ sulfuric acid must be diluted with water to obtain 25 ml of $0.0500M$ sulfuric acid?

29. When 25.0 ml of $0.400M$ H_2SO_4 and 50.0 ml of $0.850M$ H_2SO_4 are mixed,

what is the molarity of sulfuric acid in the final solution?

30. Calculate the volume of (a) $2.10M$ KOH needed to make 500 ml of $0.0100M$ KOH; (b) $18M$ sulfuric acid needed to make 2.0 liters of $0.100M$ H_2SO_4; (c) the solution made by diluting 2.0 ml of $6N$ HNO_3 to give $0.01N$ HNO_3.

Concentration

31. Calculate the molarity and normality of a solution that contains 0.0156 g of $Ba(OH)_2$ in 245 ml of solution.

32. Sodium carbonate, Na_2CO_3, dissolves in water and forms carbonate ions, CO_3^{2-}, each of which can accept two protons to form carbonic acid, H_2CO_3. What is the number of equivalents of base per mole of sodium carbonate? What is the normality of a solution made by adding 1.35 g of sodium carbonate to enough water to make 50.0 ml of solution?

33. The density of 65% nitric acid is 1.40 g ml^{-1}. How many milliliters of nitric acid would be required to prepare 500 ml of a $0.50N$ solution?

34. How much water must be added to 100.0 ml of concentrated hydrochloric acid to make a $0.1000N$ solution? Concentrated hydrochloric acid contains 37.00% by weight of HCl and has a density of 1.190 g ml^{-1}.

Neutralization reaction

35. Write a balanced equation for the reaction of magnesium hydroxide with H_3PO_4 that produces $Mg_3(PO_4)_2$ and water.

Neutralization

36. Equal volumes of $0.050M$ $Ba(OH)_2$ and $0.040M$ HCl are allowed to react. Calculate the molarity of each of the ions present after the reaction.

37. What volume of $0.200M$ H_2SO_4 will neutralize 20.0 ml of $0.120N$ NaOH?

38. Calculate the weight in grams of $Mg(OH)_2$ required to react with 20.0 ml of $0.103N$ H_3PO_4 and convert it completely to PO_4^{3-}.

Titration

39. What volume of $0.2N$ HCl is needed to titrate 20 ml of $0.35N$ NaOH?

40. If 0.350 g of calcium hydroxide is dissolved in water, how many milliliters of $0.100M$ HCl are required to titrate the solution to neutrality?

41. A volume of 35.8 ml of $0.100M$ sodium hydroxide is needed to titrate 20.0 ml of sulfuric acid solution. What are the normality and molarity of the sulfuric acid?

42. Find the number of grams of lysergic acid, $C_{15}H_{15}N_2COOH$, in a solution that requires 8.6 ml of $0.10M$ NaOH solution to titrate to an equivalence point. (Only one H per lysergic acid molecule reacts with NaOH.)

Acid identification

43. A student was given an unknown acid, which was either acetic (CH_3COOH), pyruvic ($CH_3COCOOH$), or propionic (CH_3CH_2COOH). He prepared a solution of the unknown acid by dissolving 0.100 g of the acid in 50.0 ml of water. He then titrated the solution to an end point with 11.3 ml of $0.100M$ NaOH. Identify the unknown acid. (Only one H per acid molecule reacts with NaOH.)

Molarity of NH₄Cl solution

44. A strong base will react with an ammonium salt to liberate ammonia. In addition to providing a qualitative test for NH_4^+, this reaction may be used to determine the molarity of an NH_4Cl solution. To 20.0 ml of an NH_4Cl solution, 50.0 ml of $0.500M$ NaOH were added. The ammonia was expelled by heating the solution. The remaining solution was titrated to an end point with 15.0 ml of $0.500M$ HCl. Write balanced equations for the reactions that occurred, and calculate the molarity of the NH_4Cl solution. What weight of NH_3 was liberated?

Formic acid titration

45. Formic acid, HCOOH, can be produced by distilling ants. What weight of formic acid dissolved in 4.32 ml of aqueous solution will require 3.72 ml of $0.0173N$ NaOH for complete neutralization? Assume that only one of the hydrogen atoms in formic acid dissociates.

Oxalic acid titration

46. Oxalic acid, $(COOH)_2$, is a moderately poisonous constituent of rhubarb leaves. Calculate the volume of $0.114N$ NaOH required to react completely with 0.273 g of the acid if the equivalent weight of oxalic acid is 45.0 g. If this equivalent weight is correct, what does this indicate about the number of hydrogen atoms that dissociate per molecule of acid?

Calibration of base concentration

47. The molecular weight of potassium acid phthalate ($HOOC-C_6H_4-COOK$, abbreviated KHP) is 204.2. If 1.673 g of KHP are dissolved in 80 ml of water, and this solution requires 34.50 ml of a NaOH solution to give phenolphthalein indicator a slight pink color, what is the molarity of the NaOH?

48. When titrating $0.15M$ HCl with a magnesium hydroxide solution of unknown concentration, 35.0 ml of the acid are required to neutralize 25.0 ml of the base. Calculate the molarity of the base.

Equivalent weight of acid

49. When dissolved in water, a 0.375-g sample of a weak acid requires 28.8 ml of $0.1250M$ sodium hydroxide to cause phenolphthalein to turn a pale pink. What is the equivalent weight of the acid?

50. A chemist dissolves 0.300 g of an unknown acid in a convenient volume of water. He finds that 14.60 ml of $0.426N$ NaOH are required to neutralize the acid. What is the equivalent weight of the acid?

Identification by titration

51. A 0.162-g sample of an unknown acid requires 12.7 ml of $0.0943M$ solution of NaOH for neutralization. What is the equivalent weight of the acid? The acid is either $CH_3-C_6H_4-COOH$ or $CH_3-CH_2-C_6H_4-COOH$. Which possibility is more likely?

52. In a laboratory exercise, the ammonia in 0.250 g of a compound $Cu(NH_3)_xSO_4$ is neutralized by 25.37 ml of $0.201N$ HCl when titrated to an end point. (a) What is the percentage of NH_3 in the compound by weight? (b) What is the value of the subscript, x, in the formula?

Work and gravitational energy

53. At the surface of the earth, around $40°$ north latitude, the downward acceleration of gravity (g) is 9.806 m sec^{-2}. What force does an 80-kg (176-lb) man exert on the floor by standing on it? How much work is required to raise the man from the first floor to the second, if the distance between floors is 3.00 m? *If* all the heat from burning methane could be converted into work to raise the man (this is not thermodynamically possible), how many grams of methane would have to burn to provide enough energy to raise the man to the second floor? (See equation 2-12.)

Work and energy

54. When the brakes are applied in an automobile, energy of motion is converted into heat at the brake shoes and drum. How much energy is required to bring a 1200-kg automobile to a stop from an initial speed of 30 m sec^{-1} (67 miles hr^{-1})? If the heat of combustion of heptane, C_7H_{16}, one of the principal components of gasoline, is 4800 kJ mole^{-1}, and if the conversion of heat to work were perfect (which it is not), how many moles and how many grams of heptane would have been required to accelerate the car from rest to 30 m sec^{-1}?

Enthalpy change and the form of the reaction equation

55. Calculate the enthalpy change or the heat of the following reactions at 298 K by using data in Appendix 3:

a) $2HI(g) \rightarrow H_2(g) + I_2(s)$
b) $HI(g) \rightarrow \frac{1}{2}H_2(g) + \frac{1}{2}I_2(s)$
c) $2HI(g) \rightarrow H_2(g) + I_2(g)$
d) $H_2(g) + I_2(s) \rightarrow 2HI(g)$
e) $2HI(g) \rightarrow H_2(g) + 2I(g)$
f) $2HI(g) \rightarrow 2H(g) + 2I(g)$
g) $3HI(g) \rightarrow H_2(g) + I_2(s) + HI(g)$
What is the heat of sublimation of I_2 or the conversion from solid to gas?

Endothermic and exothermic reactions

56. Is the conversion of the mineral rutile, TiO_2, to Ti_3O_5 in the open atmosphere exothermic or endothermic? What is the heat of the reaction at 298 K? Write a balanced equation for it.

57. Sulfur has two crystalline forms, rhombic and monoclinic. With the data in Appendix 3, calculate the heat of conversion from rhombic to monoclinic sulfur. What is the heat of formation of $SO_2(g)$ from $O_2(g)$ and monoclinic S at 298 K?

58. Given the following reactions,

$$2P(s) + 3Cl_2(g) \rightarrow 2PCl_3(l)$$
$$\Delta H^0 = -635.1 \text{ kJ}$$

$$PCl_3(l) + Cl_2(g) \rightarrow PCl_5(s)$$
$$\Delta H^0 = -137.3 \text{ kJ}$$

calculate the heat of formation of solid PCl_5 at 25°C. What is the heat of vaporization of $PCl_3(l)$ at this temperature?

Heat of reaction

59. What is the heat of the reaction that produces sodium carbonate from Na_2O and carbon dioxide gas at 298 K?

60. Nitrogen trichloride, NCl_3, is an unstable yellow oil that explodes at 95°C with the release of N_2, Cl_2, and 230 kJ mole^{-1} of heat. Write a balanced equation for the reaction, including ΔH. Calculate the amount of heat released by the decomposition of 10.0 g of NCl_3 at 95°C.

61. Calculate the heat of the following reaction at 298 K; ΔH^0 for $PbO_2(s)$ is -276.6 kJ.

$$4Al(s) + 3PbO_2(s) \rightarrow$$
$$3Pb(s) + 2Al_2O_3(s)$$

What does the algebraic sign in your answer indicate?

62. Nitrogen dioxide, NO_2, is one of the atmospheric pollutants produced by automobiles. It is formed when the high temperature in the internal combustion engine causes atmospheric N_2 and O_2 to combine to produce NO, which then reacts with more O_2 to give NO_2. This pollutant ultimately is converted to HNO_3. A proposed equation for this conversion at 298 K is

$$O_3(g) + 2NO_2(g) + H_2O(g) \rightarrow$$
$$2HNO_3(aq) + O_2(g)$$

Calculate the change in enthalpy for the conversion reaction. Necessary data can be obtained from Appendix 3.

Heat of combustion

63. The heat of combustion of $CH_3OH(l)$ to carbon dioxide gas and liquid water at 298 K is 715 kJ mole^{-1}, whereas that of formic acid, $HCOOH(l)$, is 261 kJ mole^{-1}. Calculate the heat of the reaction at 298 K

$$CH_3OH(l) + O_2(g) \rightarrow$$
$$HCOOH(l) + H_2O(l)$$

64. A combustion bomb containing 5.40 g aluminum metal and 15.97 g Fe_2O_3 is placed in an ice calorimeter that initially contained 8.000 kg of ice and 8.000 kg of liquid water. The reaction

$$2Al(s) + Fe_2O_3(s) \rightarrow Al_2O_3(s) + 2Fe(s)$$

is begun by remote control. It is observed that the calorimeter contains 7.746 kg of ice and 8.254 kg of water. The heat of fusion of ice is 335 J g^{-1}. What is the enthalpy change for the reaction at 0°C?

Suggested Reading

S. W. Benson, *Chemical Calculations,* Wiley, New York, 1971, 3rd ed.

W. F. Kieffer, *The Mole Concept in Chemistry,* Reinhold, New York, 1962.

L. K. Nash, *Stoichiometry,* Addison-Wesley, Reading, Mass., 1966.

M. J. Sienko, *Chemistry Problems,* W. A. Benjamin, Menlo Park, Calif., 1972, 2nd ed. Chapters 2 through 5 are especially relevant here.

C. A. Van der Werf, *Acids, Bases, and the Chemistry of the Covalent Bond,* Reinhold, New York, 1961.

R. C. Weast, Ed., *CRC Handbook of Chemistry and Physics,* Chemical Rubber Co., Cleveland. New editions annually. Probably the most-used handbook of chemical data.

3

Gas Laws and the Kinetic Theory

Scientific research consists in seeing what everyone else has seen, but thinking what no one else has thought.

A. Szent-Gyorgyi (b. 1893)

The word *gas* comes from *gaos,* a Dutch form of the word *chaos.* Gases were the last substances to be understood chemically. Solids and liquids were easy to identify and differentiate, but the idea of different kinds of "ayres" came only slowly. Carbon dioxide was not prepared from limestone until 1756. Hydrogen was discovered in 1766; nitrogen, in 1772; and oxygen, in 1781. Although gases were late in being identified, they were the first substances whose physical properties could be explained in terms of simple laws. It is fortunate that when matter in this most elusive state is subjected to changes in temperature and pressure, it behaves according to rules much simpler than those that solids and liquids follow. Moreover, one of the best tests of the atomic theory is its ability to account for the behavior of gases. This is the story of the present chapter.

Given any trapped sample of gas, we can measure its mass, its volume, its pressure against the walls of a container, its viscosity, its temperature, and its rate of conducting heat and sound. We can also measure the rate at which it effuses through an orifice into another container, and the rate at which it diffuses through another gas. In this chapter we shall show that these properties are not independent of one another, that they can all be related by a simple theory which assumes that gases consist of moving and colliding particles.

3-1 AVOGADRO'S LAW

One of the most important hypotheses in the development of atomic theory was made by Amedeo Avogadro (1776–1856) in 1811. He proposed that equal volumes of all gases, at a specified temperature and pressure, contain equal numbers of molecules. This means that the density of a gas—its weight per unit volume—in grams per milliliter must be proportional to the molecular weight of the gas. If Avogadro's ideas had not been ignored for another 50 years, the process of arriving at a dependable set of atomic weights for the elements would have taken much less time. (The entire story is told in Chapter 6.) It was a belated tribute to an unjustly ignored scientist to call the number of molecules per mole **Avogadro's number**.

   If we accept Avogadro's principle, then the number of molecules, n, and also the number of moles, is proportional to V, the volume of the gas:

$$n = \text{number of moles} = kV \qquad (\text{at constant } P \text{ and } T) \qquad (3\text{-}1)$$

In this equation, k is a proportionality constant; it changes with temperature and pressure. We shall be looking for other such relationships for gases—relationships connecting the pressure, P, the volume, V, the temperature, T, and the number of moles in a sample, n.

3-2 THE PRESSURE OF A GAS

If a glass tube, closed at one end, is filled with mercury (Hg), and the open end is inverted in a pool of mercury as in Figure 3-1a, the level of mercury in the tube will fall until the mercury column stands about 760 millimeters (mm) above the surface of the pool. The pressure produced at the pool surface by the weight of the mercury column is balanced exactly by the pressure of the surrounding atmosphere. Because there is a balance of opposing pressures, more mercury will not flow into or out of the tube. A device such as this (called a *barometer*) can measure atmospheric pressure, as Italian mathematician and physicist Evangelista Torricelli (1608–1647) first realized. He showed that it was the *pressure* at the bottom of the mercury column that mattered, and not the total weight of mercury; thus, the height of mercury in a barometer tube is independent of the size or shape of the tube. Atmospheric pressure at sea level supports a column of mercury 760 mm high. Because mercury columns were used so often in early pressure experiments, the "millimeter of mercury" became a common unit of pressure. Pressure is force per unit area ($P = F/A$), and the SI unit of pressure is the pascal (Pa), defined as 1 newton per square meter (N m^{-2}). (The newton, as you may recall from Chapter 2, is the force that will impart an acceleration of 1 m sec^{-2} to a 1-kg mass; $1 \text{ N} = 1 \text{ kg m sec}^{-2}$.)

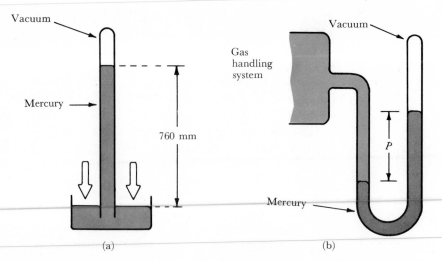

Figure 3-1

Measuring gas pressure. (a) Torricellian barometer. When a mercury-filled tube is inverted in a dish of mercury, the level in the tube falls, thereby leaving a vacuum at the top of the tube. Only a trace of mercury vapor is present in the space at the top. The height of the column is determined by the pressure of the atmosphere on the mercury in the reservoir. (b) In a gas-handling system, pressure P (in millimeters of Hg) is determined by measuring the difference in heights of the two mercury columns of a manometer. If the system is evacuated completely, the levels are equal.

Example 1

The density of liquid mercury is 13.596 g cm^{-3}. How would you express a pressure of 1 mm Hg in pascals?

Solution

Imagine a sheet of mercury 1 m square and 1 mm thick. Converting first to centimeters for convenience, we would find its volume to be

$$0.100 \text{ cm} \times 100 \text{ cm} \times 100 \text{ cm} = 1000 \text{ cm}^3$$

and its mass to be

$$1000 \text{ cm}^3 \times 13.596 \text{ g cm}^{-3} = 13{,}596 \text{ g or } 13.596 \text{ kg}$$

Since force is equal to mass times acceleration, and the acceleration of gravity at sea level, g, is 9.806 m sec^{-2}, the force exerted by the mercury on its supporting table would be

$$F = mg$$
$$F = 13.596 \text{ kg} \times 9.806 \text{ m sec}^{-2} = 133.32 \text{ kg m sec}^{-2}, \text{ or } 133.32 \text{ N}$$

Since the area of the sheet is 1 m², the pressure on the table would be

$$P = \frac{F}{A}$$

$$P = \frac{133.32 \text{ N}}{1 \text{ m}^2} = 133.32 \text{ N m}^{-2} = 133.32 \text{ pascals (Pa)}$$

Example 2

Standard sea level pressure is exactly 760 mm Hg. Express this in pascals.

Solution

From Example 1, we know that a pressure of 1 mm Hg is equal to 133.3 Pa. Therefore,

$$760 \text{ mm Hg} \times 133.32 \text{ Pa mm}^{-1} = 101{,}323 \text{ Pa}$$

The pascal is too small a unit to be convenient for measuring gas pressures, just as the cubic meter is too large to be convenient for measuring liquid volumes. Hence we shall follow a long-established tradition of measuring gas pressures in standard atmospheres, where

$$1 \text{ atmosphere (atm)} = 101{,}325 \text{ Pa} = 760 \text{ mm Hg}$$

The atmosphere then becomes an auxiliary or secondary unit to the strict SI, like the liter for volume, and the charge on the electron for ionic charges.

Example 3

At 8000 ft in the Colorado Rockies, the pressure of the atmosphere is approximately three-quarters what it is at sea level. Express this pressure in standard atmospheres, in pascals, and in millimeters of mercury.

Solution

The pressure is 0.750 atm, 76,000 Pa, or 570 mm Hg.

3-3 BOYLE'S LAW RELATING PRESSURE AND VOLUME

Robert Boyle (1627–1691), who gave us the first operational definition of an element (see Chapter 6), was also interested in phenomena occurring in evacuated spaces. When devising vacuum pumps for removing air from vessels, he noticed a property familiar to anyone who has used a hand pump for inflating a tire or football, or who has squeezed a balloon without breaking it: As air is compressed, it pushes back with increased vigor. Boyle called this the "spring of the air," and measured it with the simple device shown in Figure 3-2a and b.

Boyle trapped a quantity of air in the closed end of the J-tube as in Figure 3-2a, and then compressed it by pouring increasing amounts of mercury into the open end (b). At any point in the experiment, the total pressure on the enclosed gas is the atmospheric pressure *plus* that produced

Figure 3-2   Dependence of volume of gas sample on pressure. (a) The simple J-tube apparatus used by Boyle to measure pressure and volume. When the height of the column is equal in the open and closed parts of the tube, the pressure exerted on the gas sample is equal to atmospheric pressure. (b) The pressure on the gas is increased by adding mercury to the tube. (c) The gas burette, a device employing the same principle as the J-tube apparatus. The gas is at atmospheric pressure. (d) The pressure on the gas is increased by raising the mercury reservoir. In (a) and (b) the cross section of the J-tube is assumed constant, so the height of the gas sample is a measure of volume. In (c) and (d) the volume of the gas is measured by the calibrated burette.

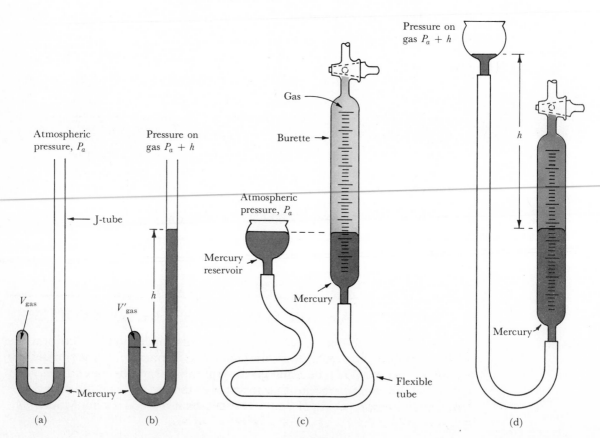

Table 3-1

Boyle's Original Data Relating Pressure and Volume for Atmospheric Air[a]

	Volume (index marks along uniform bore tubing)[b]	Pressure (inches of mercury)[c]	$P \times V$
A	48	$29\frac{2}{16}$	1400
	46	$30\frac{9}{16}$	1406
	44	$31\frac{15}{16}$	1408
	42	$33\frac{8}{16}$	1410
	40	$35\frac{5}{16}$	1412
	38	37	1408
	36	$39\frac{5}{16}$	1416
	34	$41\frac{10}{16}$	1420
	32	$44\frac{3}{16}$	1416
	30	$47\frac{1}{16}$	1414
	28	$50\frac{5}{16}$	1410
	26	$54\frac{5}{16}$	1412
	24	$58\frac{13}{16}$	1414
	23	$61\frac{5}{16}$	1411
	22	$64\frac{1}{16}$	1411
	21	$67\frac{1}{16}$	1410
	20	$70\frac{11}{16}$	1415
	19	$74\frac{2}{16}$	1410
	18	$77\frac{14}{16}$	1403
	17	$82\frac{12}{16}$	1410
	16	$87\frac{14}{16}$	1407
	15	$93\frac{1}{16}$	1398
	14	$100\frac{7}{16}$	1408
	13	$107\frac{13}{16}$	1395
B	12	$111\frac{9}{16}$	1342

[a]Reprinted by permission from J. B. Conant, *Harvard Case Histories in Experimental Science*, Harvard University Press, Cambridge, 1957, Vol. 1, p. 53.

[b]End data points A and B correspond to those labels on Figure 3-3.

[c]The height, h, in Figure 3-2b, plus $29\frac{1}{8}$ inches for atmospheric pressure.

by the excess mercury, which has height h in the open tube. Boyle's original pressure–volume data on air are given in Table 3-1. Although he did not take special pains to keep the temperature of the gas constant, it probably varied only slightly. Boyle did note that the heat from a candle flame produced a drastic alteration in the behavior of air.

Analysis of Data

After a scientist obtains data such as those in Table 3-1, he then attempts to infer a mathematical equation relating the two mutually dependent quantities that he has measured. One technique is to plot various powers of each quantity against one another until a straight line is obtained. The general equation for a straight line is

$$y = ax + b \qquad (3-2)$$

in which x and y are variables and a and b are constants. If b is zero, the line passes through the origin.

Figure 3-3 shows several possible plots of the data for pressure, P, and volume, V, given in Table 3-1. The plots of P versus $1/V$ and V versus $1/P$ are straight lines through the origin. A plot of the logarithm of P versus the logarithm of V is also a straight line with negative slope of -1. From these plots the equivalent equations are deduced:

$$P = \frac{a}{V} \qquad (3-3a)$$

$$V = \frac{a}{P} \qquad (3-3b)$$

and

$$\log V = \log a - \log P \qquad (3-3c)$$

These equations represent variants of the usual formulation of Boyle's law: *For a given number of moles of gas molecules, the pressure is inversely proportional to the volume if the temperature is held constant.*

When the relationship between two measured quantities is as simple as this one, it can be deduced numerically as well. If each value of P is multiplied by the corresponding value of V, the products all are nearly the same for a single sample of gas at constant temperature (Table 3-1). Thus,

$$PV = a \simeq 1410 \qquad (3-3d)$$

Equation 3-3d represents the hyperbola obtained by plotting P versus V (Figure 3-3a). This experimental function relating P and V now can be checked by plotting PV against P to see if a horizontal straight line is obtained (Figure 3-3e).

Boyle found that for a given quantity of any gas at constant temperature, the relationship between P and V is given reasonably precisely by

$$PV = \text{constant} \qquad \text{(at constant } T \text{ and } n) \qquad (3-4)$$

For comparing the same gas sample at constant temperature under different pressure and volume conditions, Boyle's law can be written conveniently as

$$P_1 V_1 = P_2 V_2 \qquad (3-5)$$

with the subscripts 1 and 2 representing the different conditions.

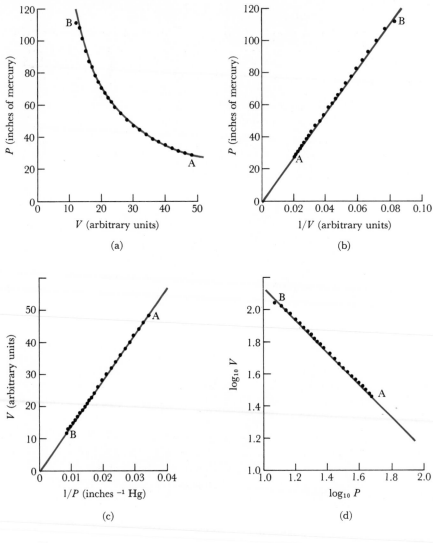

Figure 3-3

Plots of Boyle's data of Table 3-1 on various scales. (a) *P* versus *V*, which gives a hyperbola. (b) *P* versus $1/V$. (c) *V* versus $1/P$. (d) log *V* versus log *P*. (e) *PV* versus *P*. A and B mark the same end data points in each plot. If the data are plotted as *P* versus $1/V$ (or *V* versus $1/P$), the curve obeys the linear equation $y = ax + b$, in which *P* is the *y*-coordinate and $1/V$ is the *x* coordinate. The proportionality constant *a* can be determined from the slope of the straight line in (b), or from the height of the horizontal line in (e). Note how sensitive plot (e), with its expanded vertical scale, is to errors in the data (and possibly unsuspected trends).

Example 4

Plastic bags of peanuts or potato chips purchased in Aspen, Colorado, are frequently puffed up because the air sealed inside at sea level has expanded under the lower surrounding pressure at the 8000-ft elevation. If 100 cm³ of air are sealed inside a bag at sea level, what volume will the air occupy at the same temperature in Aspen? (Assume that the bag is so wrinkled that it does not limit gas expansion, and see Example 3 for missing data.)

Solution   Use Boyle's law in the form of equation 3-5, with subscript 1 representing sea level and subscript 2 representing 8000 ft. Then $P_1 = 1.000$ atm, $V_1 = 100$ cm³, $P_2 = 0.750$ atm, and V_2 is to be calculated:

$$P_1 V_1 = P_2 V_2$$
$$1.000 \text{ atm} \times 100 \text{ cm}^3 = 0.750 \text{ atm} \times V_2$$
$$V_2 = 133 \text{ cm}^3$$

3-4   CHARLES' LAW RELATING VOLUME AND TEMPERATURE

We know that air expands on heating, thereby decreasing its density. For this reason, balloons rise when inflated with warm air. About 100 years after Boyle derived his law, Jacques Charles (1746–1823), in France, measured the effect of changing temperature on the volume of an air sample. This measurement can be made quite easily with the device shown in Figure 3-4.

Figure 3-4

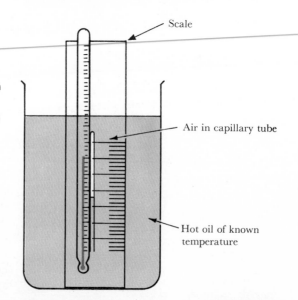

Experimental determination of the relationship between volume and temperature of a gas. The apparatus consists of a small capillary tube and a thermometer mounted on a ruled scale and immersed in a hot oil bath. As the system cools, the oil rises in the tube, and the length of air space and the temperature are measured at intervals. For a tube of constant bore, the length of the air space is a measure of the gas volume. So long as the bottom of the air space in the capillary is maintained at the same depth below the surface of the oil bath, the pressure in the capillary will be constant.

Scale

Air in capillary tube

Hot oil of known temperature

Figure 3-5

A plot of data obtained with the apparatus in Figure 3-4, showing that volume is proportional to the absolute temperature. Just such a plot employing the Celsius scale of temperature was originally used to locate the absolute zero of temperature. Notice how easily a small error in the slope of the line through the data points could produce a large error in the value of absolute zero. It should be clear that, if at all possible, such long extrapolations should be avoided.

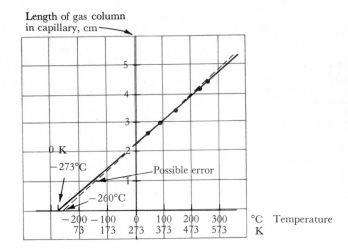

Some sample data are plotted in Figure 3-5; these show that a graph of V versus T is a straight line with an extrapolated intercept of $-273°$ on the Celsius scale of temperature, or $-460°$ on the Fahrenheit scale. Charles expressed his law as

$$V = c(t + 273)$$

where V is the volume of gas sample, t is the temperature on the Celsius scale, and c is a proportionality constant.

Later, Lord Kelvin (1824–1907) suggested that the intercept of $-273°C$ represented an absolute minimum of temperature below which it is not possible to go. Scientists now use Kelvin's absolute scale of temperature with $0 \text{ K} = -273.15°C$ and $0°C = 273.15 \text{ K}$. Charles' law is expressed as

$$V = cT \quad \text{(at constant } P \text{ and } n) \tag{3-6}$$

where T is the absolute Kelvin temperature (i.e., $T = t + 273.15$). Equation 3-6 indicates that *at constant pressure the volume of a given number of moles of gas is directly proportional to the absolute temperature.* For light gases such as hydrogen and helium, Charles' law is so accurate that gas thermometers often replace mercury thermometers for precise temperature measurement (Figure 3-6). A mercury thermometer calibrated to read $0°C$ in a water–ice mixture and $100°C$ in boiling water is inaccurate by as much as 0.1 degree (deg) at intermediate points, whereas a hydrogen thermometer is much more accurate throughout this region.

If the same gas sample is being compared at constant pressure but different temperatures and volumes, then Charles' law can be written as

$$\frac{V_1}{T_1} = \frac{V_2}{T_2} \quad \text{or} \quad \frac{V_1}{V_2} = \frac{T_1}{T_2} \tag{3-7}$$

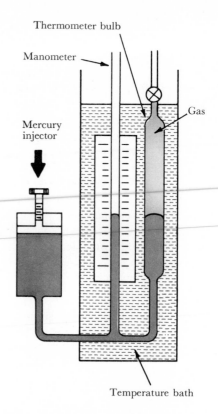

Figure 3-6

A simple gas thermometer. The gas volume is a measure of the absolute temperature. The scale can be calibrated with the freezing point (0°C) and boiling point (100°C) of water. The mercury is injected into or removed from the apparatus to maintain constant atmospheric pressure.

This way of writing the expression emphasizes the fact that the ratio of volumes matches the ratio of absolute temperatures, if pressure and the number of moles are constant.

Example 5

The same plastic bag of peanuts mentioned in Example 4 is laid on a windowsill in the sun, where its temperature increases from 20°C to 30°C. If the original volume is 100.0 cm³, what is the final volume after warming?

Solution

$V_1 = 100.0$ cm³; $T_1 = 20$°C, or 293.15 K; and $T_2 = 30$°C, or 303.15 K. To calculate V_2, we use equation 3-7:

$$\frac{V_1}{T_1} = \frac{V_2}{T_2}$$

Substituting our data and solving for V_2, we get

$$\frac{100.0 \text{ cm}^3}{293.15} = \frac{V_2}{303.15}$$

$$V_2 = \frac{303.15}{293.15} \times 100.0 \text{ cm}^3 = 103.4 \text{ cm}^3$$

Notice that the absolute Kelvin temperature (K) must be used, *not* the Celsius temperature.

3-5 THE COMBINED GAS LAW

The three gas equations that we have encountered so far all may be written in terms of the proportionality of volume to another quantity:

Avogadro's law: $V \propto n$ (at constant P and T)

Boyle's law: $V \propto \dfrac{1}{P}$ (at constant T and n)

Charles' law: $V \propto T$ (at constant P and n)

(The symbol $\propto$ means "is proportional to.") Therefore, the volume must be proportional to the product of these three terms, or

$$V \propto \frac{nT}{P} = R\frac{nT}{P}$$

or

$$PV = nRT \tag{3-8}$$

where R is the proportionality constant. Equation 3-8 is known as the **ideal gas law.** It contains all of our earlier laws as special cases and, in addition, predicts more relationships that can be tested. For example, the French chemist and physicist Joseph Gay-Lussac verified the prediction that, at constant volume, the pressure of a fixed amount of gas is proportional to its absolute temperature. (In effect, equation 3-8 is a definition of the ideal gas; the differences between real gases and the hypothetical ideal gas are discussed in Section 3-8.)

The gas law is often useful when expressed in the form of ratios of starting and final variables. Suppose a fixed amount of gas at constant temperature is compressed from P_1 to P_2, with volumes V_1 and V_2. Then $P_1 = nRT/V_1$, $P_2 = nRT/V_2$, and the pressure ratio and volume ratio are related by

$$\frac{P_2}{P_1} = \frac{V_1}{V_2} \quad (T, n \text{ constant}) \tag{3-9}$$

This is Boyle's law in ratio form. In a similar way, as we have seen, Charles' law states that the ratio of starting volume to final volume matches the temperature ratio at constant pressure:

$$\frac{V_2}{V_1} = \frac{T_2}{T_1} \qquad (P, n \text{ constant}) \qquad (3\text{-}7)$$

Increasing the number of moles of gas at constant temperature and pressure by a certain factor increases the volume by the same factor:

$$\frac{V_2}{V_1} = \frac{n_2}{n_1} \qquad (P, T \text{ constant}) \qquad (3\text{-}10)$$

And increasing the number of moles of gas at constant temperature in a tank of fixed volume by a certain factor increases the pressure inside the tank by the same factor:

$$\frac{P_2}{P_1} = \frac{n_2}{n_1} \qquad (T, V \text{ constant}) \qquad (3\text{-}11)$$

You should be able to derive these equations and also the analogous equation that expresses Gay-Lussac's observations about pressure and temperature at constant volume easily from the ideal gas law.

Example 6

An old-fashioned diving bell is simply a cylinder closed at the top and open at the bottom, like an inverted drinking glass, with benches around the inside for the divers to sit on. Air pressure alone keeps the water out. A diver sitting in a bell that has an air volume of 8000 liters wants to drive the water level down by increasing the volume to 10,000 liters, because his feet are getting wet. If he filled the bell with 650 moles of an O_2–N_2 mixture to begin with, how many more moles of gas will he have to release into the bell to obtain his desired volume increase?

Solution

This is a problem in changes of volume with number of moles; pressure and temperature remain constant. Therefore we use equation 3-10, with $V_1 = 8000$ liters, $V_2 = 10,000$ liters, and $n_1 = 650$ moles. We need to calculate n_2. Substituting our data in equation 3-10 and solving for n_2, we get

$$\frac{10,000 \text{ liters}}{8000 \text{ liters}} = \frac{n_2}{650 \text{ moles}}$$

$$n_2 = \frac{10,000 \text{ liters}}{8000 \text{ liters}} \times 650 \text{ moles} = 813 \text{ moles}$$

Hence $813 - 650 = 163$ moles of gas must be added.

Notice that it makes no difference to the problem whether the gas is pure or a mixture of N_2 and O_2. Within the limits of validity of the ideal gas expression, all gases behave the same way with respect to pressure, volume, and temperature, if measured in moles rather than in grams. The ideal gas expression itself can be written as a ratio, in a form useful for considering simultaneous pressure, temperature, and volume changes in a fixed quantity of gas:

$$\frac{P_1 V_1}{T_1} = \frac{P_2 V_2}{T_2} \qquad (3\text{-}12)$$

Example 7

When a weather balloon is filled with hydrogen gas at 1.000 atm pressure and 25°C, it has a diameter of 3.00 m and a volume of 14,100 liters. At high altitude the atmospheric pressure drops to half its sea-level value; the temperature is −40°C. What then is the volume of the balloon? What is its diameter?

Solution

We have $P_1 = 1.000$ atm; $V_1 = 14,100$ liters; $T_1 = 25$°C, or 298.15 K; $P_2 = 0.500$ atm; $T_2 = -40$°C, or 233.15 K. Rearranging equation 3-12 to solve for V_2, we get

$$V_2 = \frac{T_2}{T_1} \times \frac{P_1}{P_2} \times V_1 = \frac{233.15 \text{ K}}{298.15 \text{ K}} \times \frac{1.000 \text{ atm}}{0.500 \text{ atm}} \times 14,100 \text{ liters}$$

$$= 22,100 \text{ liters}$$

Assuming that the balloon is spherical, we can find its diameter, d, by using the formula $V = \frac{4}{3}\pi r^3$ or $V = 4\pi r^3/3$. The diameter then is 3.48 m.

Notice that in Example 7 the temperature drop alone would have brought about a volume decrease to $233.15/298.15 = 0.782$ of the initial volume, and the pressure drop alone would have brought about a twofold volume increase. The actual increase, by a factor of 1.56, is the product of these two effects.

The numerical value of the gas constant, R, in the ideal gas law (equation 3-8) depends on the units in which pressure and volume are measured—assuming that only the absolute, or Kelvin, temperature scale is used. If pressure is in atmospheres and volume is in liters, then $R = 0.082054$ liter atm K^{-1} mole^{-1}. But, as you can see from Appendix 1, R also can be expressed as 8.3143 J K^{-1} mole^{-1}. We shall show in Chapter 15 that the product PV has the units of work or energy.

Example 8

How much volume will 75.0 g of hydrogen gas occupy at 1.000 atm pressure and 298 K?

Solution   To answer this question we must use the full ideal gas equation, equation 3-8, with $P = 1.000$ atm, $R = 0.08205$ liter atm K^{-1} mole^{-1}, $T = 298$ K, and $n = 75.0$ g/2.016 g mole^{-1} = 37.2 moles. Thus

$$V = \frac{nRT}{P} = \frac{37.2 \text{ moles} \times 0.08205 \text{ liter atm K}^{-1} \text{ mole}^{-1} \times 298 \text{ K}}{1.000 \text{ atm}}$$

$$= 910 \text{ liters}$$

Example 9

A 1000-liter tank is filled to a pressure of 10.00 atm at 298 K, requiring 11.5 kg of gas. How many moles of gas are present? What is the molecular weight of the gas? Assuming the gas to be a pure element, can you identify it?

Solution $V = 1000$ liters, $P = 10.00$ atm, and $T = 298$ K. Hence,

$$n = \frac{PV}{RT} = \frac{10.00 \text{ atm} \times 1000 \text{ liters}}{0.08205 \text{ liter atm K}^{-1} \text{ mole}^{-1} \times 298 \text{ K}}$$

$$= 409 \text{ moles}$$

We convert the weight of the gas in kilograms to grams (11.5 kg = 11,500 g) and then calculate the molecular weight (mol wt):

$$\text{Mol wt} = \frac{11,500 \text{ g}}{409 \text{ moles}} = 28.1 \text{ g mole}^{-1}$$

Since we are assuming that the gas is an element, it must be N_2.

Example 10

An 8-liter boiler is designed to withstand pressures up to 1000 atm. If 1.50 kg of water vapor is in the boiler, to what temperature can it be heated before the boiler explodes?

Solution $P = 1000$ atm, $V = 8.00$ liters. To find the number of moles, we divide the number of grams of water (1.50 kg = 1500 g) by the molecular weight of water. Hence, $n = 1500$ g/18.0 g mole^{-1} = 83.3 moles. Therefore,

$$T = \frac{PV}{nR} = \frac{1000 \times 8.00}{83.3 \times 0.08205} \text{ K} = 1170 \text{ K or } 897°C$$

Standard Temperature and Pressure

It is frequently useful to compare volumes of gases involved in physical and chemical processes. Such comparisons are interpreted most easily if the gases are at the same temperature and pressure, although generally it is inconvenient to make all measurements under such carefully controlled conditions; 0°C (273 K) and 1.000 atm have been designated arbitrarily as **standard temperature and pressure (STP)**. If we know the volume of a sample of gas at any condition, we can easily calculate the volume it would have as an ideal gas at STP by employing the combined gas law. This calculated volume is useful even if the substance itself becomes a liquid or solid at STP.

Example 11

In an experiment, 300 cm³ of steam are at 1.000 atm pressure and 150°C. What is the ideal volume at STP?

Solution

We can use equation 3-7, substituting the subscript STP for the subscript 2. Rearranging the equation to solve for V_{STP}, we get

$$V_{STP} = V_1 \frac{T_{STP}}{T_1}$$

$$= 300 \text{ cm}^3 \times \frac{273 \text{ K}}{423 \text{ K}} = 194 \text{ cm}^3$$

This is the volume that the steam would occupy at STP *if* it behaved like an ideal gas instead of condensing.

At STP, 1 mole of an ideal gas occupies 22.414 liters, as can be seen from the ideal gas law:

$$\text{Volume per mole} = \frac{V}{n} = \frac{RT}{P}$$

$$= \frac{0.082054 \text{ liter atm K}^{-1} \text{ mole}^{-1} \times 273.15 \text{ K}}{1.0000 \text{ atm}}$$

$$= 22.414 \text{ liters mole}^{-1}$$

This volume is often called a **standard molar volume**.

Ideality and Nonideality

The equations describing the various gas laws are exact mathematical expressions. Measurements of volume, pressure, and temperature more accurate than those of Boyle and Charles show that gases only *approach* the behavior that the equations express. Gases depart radically from so-called ideal behavior when under high pressure or at temperatures near the boiling point of the corresponding liquids. Thus, the gas laws, or more precisely the ideal gas laws, accurately describe the actual behavior of a real gas only at low pressures and at temperatures far above the boiling point of the substance in question. In Section 3-8 we shall return to the problem of how to correct the simple ideal gas law for the behavior of real gases.

3-6 THE KINETIC MOLECULAR THEORY OF GASES

At STP, 1 mole of carbon dioxide gas occupies 22.2 liters (ideally 22.4 liters), whereas the same amount of dry ice (solid CO_2) has a volume of only about 28 cm^3 (assuming a density of dry ice of 1.56 g cm^{-3}). This greater volume of the gas, plus the fact that a gas is compressed or expanded so easily, suggests strongly that much of a gas is empty space. But how does a system that is mostly empty space exert pressure on its surroundings? Experiments such as the one in Figure 3-7 indicate that gas molecules move, and move in straight lines. They also collide with the walls of the container, with one another, and with any other objects that may be in the container with the gas (Figure 3-8). As we shall see, the collision with the container walls produces pressure. It is unnecessary to assume any forces between molecules and container to account for pressure.

We can explain many observed properties of gases by a simple theory of molecular behavior that was developed in the latter half of the nineteenth century by Ludwig Boltzmann (1844–1906), James Clerk Maxwell (1831–1879), and others. This **kinetic molecular theory** has three assumptions:

1. A gas is composed of molecules that are extremely far from one another in comparison with their own dimensions. They can be considered as essentially shapeless, volumeless points, or small, hard spheres.

2. These gas molecules are in a state of constant random motion, which is interrupted only by collisions of the molecules with each other and with the walls of the container.

3. The molecules exert no forces on one another or on the container other than through the impact of collision. Furthermore, these collisions are **elastic;** that is, no energy is lost as friction during collision.

Our experience with colliding bodies such as a tennis ball bouncing on pavement is that some kinetic energy is lost on collision: The energy is transformed into heat as a result of what we call **friction**. A bouncing tennis ball gradually "dies down" and comes to rest because its collisions with the pavement are subject to friction and are therefore **inelastic**. If molecular collisions involved friction, the molecules gradually would slow down and lose kinetic energy, thereby hitting the walls with decreasing change of momentum, so the pressure would drop slowly to zero. This process does not

Figure 3-7

An experiment to test whether gas molecules move in straight lines. Two flasks are joined by a straight tube with a side arm and stopcock. The bottom flask contains material, such as iodine, that can be vaporized by heating. The flasks are evacuated and the material heated, thereby producing vapor. Molecules leave the solid in random directions, and condensation occurs uniformly over the entire surface of the bottom flask. However, only molecules moving vertically can pass through the collimating holes in the connecting tube and into the top flask. These molecules pass straight through and form a single spot directly opposite the source material. A high vacuum (low pressure) is required to prevent molecular collisions from randomizing molecular motion in the connection tube and upper flask.

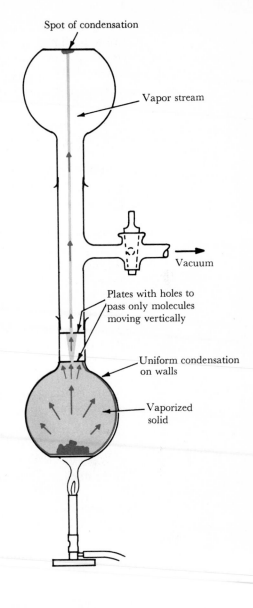

Spot of condensation

Vapor stream

Vacuum

Plates with holes to pass only molecules moving vertically

Uniform condensation on walls

Vaporized solid

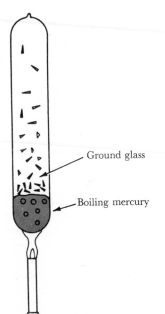

Ground glass

Boiling mercury

Figure 3-8

An experiment demonstrating collisions of gas molecules with large solid particles. Particles of ground glass are kept suspended like dust particles in air by bombardment with moving mercury molecules. The heavy molecules (mainly monatomic Hg) leaving the surface of the boiling mercury have high kinetic energy, some of which is transferred to the glass particles on collision.

take place; therefore, we must postulate that **molecular collisions are frictionless, that is, perfectly elastic**. In other words, the total kinetic energy of colliding molecules remains constant.

The Phenomenon of Pressure and Boyle's Law

This simple model is adequate to explain pressure and to provide a molecular explanation of Boyle's law. Consider a container, which we will make cubical for simplicity, with a side of length l (Figure 3-9b). Suppose that the container is evacuated completely except for one molecule of mass m that moves with a velocity v having components v_x, v_y, and v_z parallel to the x, y, and z edges of the box (see Figure 3-9a).*

Let us look first at what happens when the molecule rebounds from a collision with one of the YZ walls, which are perpendicular to the x axis.

Pressure is force per unit area, and force is the rate of change of momentum (mass times velocity) with time. When a molecule bounces off the shaded wall in Figure 3-9b, it exchanges momentum of $2mv_x$ with the wall; for the particle begins with momentum in the x direction of $-mv_x$ and ends with momentum $+mv_x$. The velocity components in the y and z directions are not changed during a collision with the YZ wall and do not enter into the calculation. No matter how many collisions the molecule has with an XY or an XZ wall along the way, if the x component of velocity is v_x, the molecule will return to collide with the original YZ wall in a time $2l/v_x$. If the molecule transfers momentum of $2mv_x$ every $2l/v_x$ sec, then the rate of change of momentum with time, or the force, F_x, is

$$F_x = \frac{2mv_x}{2l/v_x} = \frac{mv_x^2}{l}$$

The force per unit area, or the pressure, is

$$P_x = \frac{mv_x^2}{l \cdot l^2} = \frac{mv_x^2}{l^3} = \frac{mv_x^2}{V} \tag{3-13}$$

---

*If the idea of the breakdown of a vector such as velocity into its three components, v_x, v_y, and v_z, is unfamiliar, there is another explanation that, although less exact, leads to the same answer. This is to assume that since the motions of a molecule in the x, y, and z directions are unrelated, we can think of the molecules as being divided into three groups: one third moving in the x direction, one third in the y direction, and one third in the z direction. The pressure from one molecule on the YZ wall is then $P_x = mv^2/V$ (analogous to equation 3-13). The pressure from all the molecules moving in a direction perpendicular to that wall is $N/3$ times this value (where N is the total number of molecules), or

$$P_x = \frac{N}{3} \frac{\overline{mv^2}}{V}$$

as in equation 3-21. The rest of the proof is the same. (The bar over v^2 indicates an average over all molecules.)

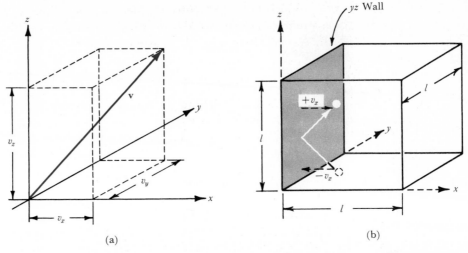

Figure 3-9

(a) The velocity of a molecule of gas resolved into components. We determine the components of the velocity vector **v** by dropping perpendiculars from the head and tail of the vector to the coordinate axes. (b) Collision of molecule with wall showing change in direction of x component of velocity.

since the area of the wall is l^2, and the total volume of the box is $V = l^3$. Similarly, for the other walls,

$$P_y = \frac{mv_y^2}{V} \tag{3-14}$$

$$P_z = \frac{mv_z^2}{V} \tag{3-15}$$

If the box now contains N molecules rather than just one,

$$P_x = N\frac{m\overline{v_x^2}}{V} \tag{3-16}$$

$$P_y = N\frac{m\overline{v_y^2}}{V} \tag{3-17}$$

$$P_z = N\frac{m\overline{v_z^2}}{V} \tag{3-18}$$

in which the quantities $\overline{v^2}$ are the *averages* over all molecules of the squares of the velocity components, since we cannot assume that all molecules have the same velocity.

The total velocity of a molecule is related to its velocity components by

$$v^2 = v_x^2 + v_y^2 + v_z^2 \tag{3-19}$$

If the motions of the individual molecules are truly random and unrelated, the average of the square of the velocity component in each direction will be the same. There will be no preferred direction of motion in the gas:

$$\overline{v_x^2} = \overline{v_y^2} = \overline{v_z^2} = \tfrac{1}{3}\overline{v^2} \tag{3-20}$$

(As before, the bars over v^2 indicate averages over all molecules.) An immediate consequence of this randomness of motion is that the pressure will be the same on all walls, a fact that certainly agrees with our observations of real gases. Rewriting equations 3-16, 3-17, and 3-18 in terms of $\overline{v^2}$ gives

$$P_x = \frac{N}{3}\frac{m\overline{v^2}}{V} \qquad P_y = \frac{N}{3}\frac{m\overline{v^2}}{V} \qquad P_z = \frac{N}{3}\frac{m\overline{v^2}}{V}$$

and

$$P_x = P_y = P_z = P = \frac{N}{3}\frac{m\overline{v^2}}{V} \tag{3-21}$$

or

$$PV = \frac{N}{3}m\overline{v^2} \tag{3-22}$$

This last expression looks very much like Boyle's law. Boyle's law maintains that the product of pressure and volume for a gas is constant *at constant temperature;* our derivation from the simple kinetic molecular theory states that the *PV* product is constant for a given mean velocity of gas molecules. If the theory is correct, the mean velocity of the molecules of a gas cannot depend on either pressure or volume, but only on temperature. The mean molecular kinetic energy is represented by the symbol $\overline{\epsilon}$ (where ϵ is the Greek letter epsilon) and is expressed as $\tfrac{1}{2}mv^2$; furthermore, if N is Avogadro's number, the kinetic energy, E_k, of 1 mole of molecules is $N\overline{\epsilon}$. For a mole of gas, the *PV* product of Boyle's law is proportional to the kinetic energy per mole:

$$E_k = N\overline{\epsilon} = \frac{1}{2}N m\overline{v^2} \tag{3-23}$$

Multiplying and dividing the right-hand term by 3 and rearranging gives

$$E_k = \left(\frac{3}{2}\right)\left(\frac{1}{3}\right) N m\overline{v^2} = \left(\frac{3}{2}\right)\left(\frac{N}{3}\right) m\overline{v^2} \tag{3-24}$$

Comparison with equation 3-22 shows that

$$PV = \frac{2}{3}E_k \tag{3-25}$$

The combination of this derivation from the kinetic theory and the observed ideal gas law (equation 3-8) tells us that the kinetic energy per mole is directly proportional to the temperature. Or, reversing the statement:

Absolute temperature, T, is an indication of the kinetic energy of gas molecules and ultimately of their mean square velocity. For 1 mole of an ideal gas, $PV = RT$. Substituting the value for PV given in equation 3-25 gives

$$E_k = \frac{3}{2}RT \tag{3-26}$$

But $E_k = N\bar{\epsilon}$, in which $\bar{\epsilon} = \frac{1}{2}m\overline{v^2}$; therefore,

$$T = \frac{2}{3}\frac{N}{R}\frac{1}{2}m\overline{v^2} = \frac{M\overline{v^2}}{3R} \tag{3-27}$$

in which the molecular weight is $M = Nm$. In short, *temperature is a measure of the motion of molecules.* If we heat a gas and raise its temperature, we do so by increasing the velocity of its molecules. When a gas (or any other substance) cools, its molecular motion diminishes. This molecular motion need not be confined to movement of whole molecules from one place to another, which is the picture that we have drawn for an ideal gas. It also can include *rotations* of entire molecules or of groups on a molecule, and *vibrations* of molecules.

We now can see more clearly what happens when the kinetic energy of macroscopic objects is dissipated as heat. When a speeding car skids to a halt, its braking is achieved by converting its energy of motion into frictional heat. But this conversion means changing the motion of the large object—the automobile—into increased relative motion of the molecules of the brake shoes and drum, the tires, and the pavement. Instead of having rubber molecules in the tires vibrating relatively slowly but moving rapidly as a unit, we have a heated tire with molecules moving more rapidly relative to one another but without a net direction of motion. The motions of the molecules have become less directional and more randomized.

This behavior is typical of all real processes. It is easy to go from coherent motion (the rolling tire) to incoherent motion (the hot but stationary tire); it is not possible to go the other way without paying a price. As we shall see in Chapter 16, in any real process the disorder of the object under examination, plus all of the surroundings with which it interacts, always will increase. In other words, in this world things always get messier. This notion is simply the second law of thermodynamics. The quantity that measures this disorder, and which we shall learn to use later in chemical situations, is called **entropy**, S.

3-7 PREDICTIONS OF THE KINETIC MOLECULAR THEORY

The test of any theory is not its beauty or its internal consistency, but its usefulness in predicting the behavior of real systems correctly. By this criterion, the kinetic molecular theory is a good one, as we shall see.

Molecular Size

Some simple calculations with solid and gaseous carbon dioxide illustrate an important difference between molecular environments in gases and in condensed phases of matter.

Example 12

The density of dry ice, or solid carbon dioxide, is 1.56 g cm^{-3}. Find the volume per mole and the volume per molecule.

Solution

The molecular weight of carbon dioxide is 44.01 amu, so 1 mole of dry ice weighs 44.01 g. To find the volume of 1 mole, divide weight by density:

$$V = \frac{44.01 \text{ g}}{1.56 \text{ g cm}^{-3}} = 28.2 \text{ cm}^3$$

Dividing the volume per mole by Avogadro's number yields the volume per molecule:

$$V = \frac{28.2 \text{ cm}^3}{6.022 \times 10^{23} \text{ molecules}} = 4.68 \times 10^{-23} \text{ cm}^3 \text{ molecule}^{-1}$$

Since 1 cm = 10^8 Å, the volume per molecule is 46.8 Å^3.

The cube root of 46.8 is 3.6, so we should expect the carbon dioxide molecule to fit inside a cube approximately 3.6 Å on edge; therefore the carbon dioxide molecule is about 3.6 Å in diameter. This is roughly what would be expected from other experiments for the size of the carbon dioxide molecule, giving us confidence in the correctness of the picture of dry ice as built up from closely packed molecules of carbon dioxide. Now let's try a comparable calculation for gaseous carbon dioxide.

Example 13

The density of carbon dioxide gas at STP is 1.977 g liter^{-1}. Find the volume per mole and the volume per molecule.

Solution

One mole or 44.01 g of carbon dioxide gas has a volume of

$$V = \frac{44.01 \text{ g}}{1.977 \text{ g liter}^{-1}} = 22.26 \text{ liters or } 22,260 \text{ cm}^3$$

(Notice the small deviation from ideal gas behavior.) The volume per molecule is

$$V = \frac{22,260 \text{ cm}^3}{6.022 \times 10^{23} \text{ molecules}} \times \frac{10^{24} \text{ Å}^3}{1 \text{ cm}^3} = 36,900 \text{ Å}^3 \text{ molecule}^{-1}$$

Carbon dioxide gas has a molar volume that is 790 times the volume of the solid. The *volume per molecule* in the gas phase corresponds to a cube that is 33.4 Å on a side (Figure 3-10). Only one part in 790 of the gas volume is actually filled by molecules.

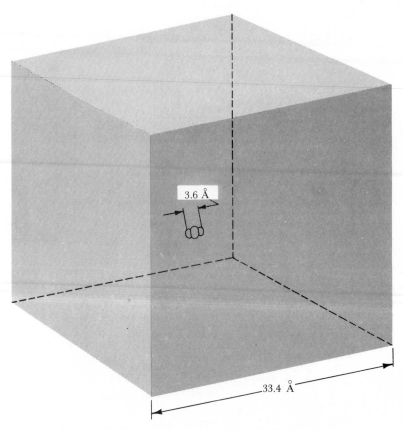

Figure 3-10   The relative size of a CO_2 molecule and the volume per molecule available to it in carbon dioxide gas at STP. Of course, one molecule is not confined to this volume, nor are other molecules excluded from it.

Molecular Speeds

With nothing more than the elementary kinetic theory presented here, we can calculate the **root-mean-square (rms) speed**, v_{rms}, which is the square root of the average of the squares of the speeds of individual molecules. From equation 3-27, v_{rms} is

$$\sqrt{\overline{v^2}} = v_{rms} = \sqrt{\frac{3RT}{M}} \tag{3-28}$$

where R is the gas constant, T is the absolute temperature, and M is the molecular weight. This equation is a good example of the absolute necessity of keeping track of units. The gas constant, R, must be expressed as 8.314 J K^{-1} $mole^{-1}$, *not* as 0.08205 liter atm K^{-1} $mole^{-1}$, if the speed is to be expressed in m sec^{-1}. Since 1 J = 1 kg m^2 sec^{-2}, the units of $3RT/M$ are

$$\frac{(kg\ m^2\ sec^{-2}\ K^{-1}\ mole^{-1})(K)}{(kg\ mole^{-1})} = m^2\ sec^{-2}$$

and v_{rms} is in the desired units. At STP the expression is

$$v_{rms} = \frac{2610}{M^{1/2}}\ m\ sec^{-1} \tag{3-29}$$

Example 14

Use equation 3-29 to calculate the root-mean-square speeds at STP of molecules of H_2, N_2, O_2, and HBr.

Solution

The molecular weights of these four gases are 2.02, 28.01, 32.00, and 80.91 g $mole^{-1}$, respectively. Hence the rms speeds are 1840 m sec^{-1} for H_2, 493 m sec^{-1} for N_2, 461 m sec^{-1} for O_2, and 290 m sec^{-1} for HBr. In more familiar units, these are speeds of 4140, 1109, 1037, and 653 miles hr^{-1}, respectively.

Notice that heavier molecules are slower moving at a given temperature. Molecules that have greater mass do not have to move as rapidly as lighter molecules do to have the same kinetic energy, and it is kinetic energy that is directly related to temperature.

Although the root-mean-square speed of nitrogen molecules at STP may be 493 m sec^{-1}, this does not mean that all nitrogen molecules travel at this speed. There is a *distribution* of speeds, from zero to values considerably above 493 m sec^{-1}. As individual gas molecules collide and exchange energy, their speeds will vary. The actual distribution of speeds in nitrogen gas at 1 atm pressure and three different temperatures is shown in Figure 3-11. These curves portray a *Maxwell–Boltzmann distribution* of speeds. The equations for these curves can be derived from the kinetic theory of gases by

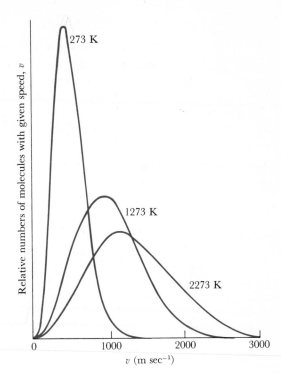

Figure 3-11   The distribution of speeds among molecules in nitrogen gas at three different temperatures. At higher temperatures the average speed is greater, there are fewer molecules that have precisely this average speed, and there is a broader distribution of speed among molecules.

using statistical or probability arguments. At higher temperatures, the root-mean-square speed increases, as expected from equation 3-28. But Figure 3-11 shows that the distribution of speeds also becomes more diffuse. There is a greater range of speeds, and fewer molecules have a speed close to the average value.

From the size of a molecule, the speed with which it travels, and the number of other molecules per unit of volume around it, we can calculate the **mean free path** (the distance a molecule travels between two successive collisions) and the **collision frequency**. Molecules such as O_2 or N_2 travel an average of 1000 Å between collisions, and they experience approximately 5 billion collisions per second at STP (Figure 3-12).

Dalton's Law of Partial Pressures

If each molecule in a gas travels independently of every other except at moments of collisions, and if collisions are elastic, then in a mixture of different gases the total kinetic energy of all the different gases will be the sum of the kinetic energies of the individual gases:

$$E = E_1 + E_2 + E_3 + E_4 + \cdots$$

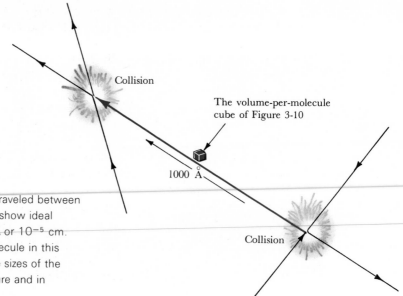

Figure 3-12

The mean free path, or distance traveled between collisions, for gas molecules that show ideal behavior at STP, is about 1000 Å or 10^{-5} cm. We can imagine the size of a molecule in this drawing by comparing the relative sizes of the cube 33.4 Å on a side in this figure and in Figure 3-10.

Since each gas molecule moves independently, the pressure that each gas exerts on the walls of the container can be derived separately (equation 3-25). For example,

$$p_1 = \frac{2E_1}{3V} \qquad p_2 = \frac{2E_2}{3V} \qquad p_3 = \frac{2E_3}{3V} \qquad (3\text{-}30)$$

This pressure exerted by one component of a gas mixture is called its **partial pressure**, p. Each of these equations can be rewritten to give kinetic energy in terms of pressure:

$$E_1 = \tfrac{3}{2}p_1 V \qquad E_2 = \tfrac{3}{2}p_2 V \qquad E_3 = \tfrac{3}{2}p_3 V$$

Substituting in the energy expression and canceling the $\tfrac{3}{2}V$ terms from both sides of the equation produces

$$P = p_1 + p_2 + p_3 + p_4 + \cdots = \sum_j p_j \qquad (3\text{-}31)$$

The special sign at the right is a **summation sign**, which is a shorthand way of writing the instructions: Sum all the terms of the type p_j for all the different values of j. It will be used frequently.

The *total pressure, P,* then, is the *sum of the partial pressures* of the individual components of the gas mixture, each considered as if it were the only gas present in the given volume. John Dalton (1766–1844) proposed his *law of partial pressures* during the gas investigations that eventually led him to the theory of atoms.

An important measure of concentrations in a mixture of gases (and in solutions and solids as well) is the **mole fraction**, X. The mole fraction of the jth component in a mixture of substances is defined as the number of moles (n) of the given substance divided by the total number of moles of all substances:

$$X_j = \frac{n_j}{n_1 + n_2 + n_3 + n_4 + \cdots} = \frac{n_j}{\sum_i n_i} \qquad (3\text{-}32)$$

Another version of Dalton's law is the statement that the partial pressure of one component in a mixture of gases is its concentration in mole fraction times the total pressure. If there are n_j moles of gas j present in a mixture, the partial pressure of that gas is calculable from the ideal gas law:

$$p_j = n_j \frac{RT}{V} = \frac{n_j}{n} \times n \times \frac{RT}{V} \qquad \left(n = n_1 + n_2 + n_3 + \cdots = \sum_i n_i \right)$$

Since $n_j/n = X_j$ is the mole fraction, and $nRT/V = P$ is the *total* pressure, Dalton's law becomes

$$p_j = X_j P \qquad (3\text{-}33)$$

Example 15

A gas mixture at $100°C$ and 0.800 atm pressure contains 50% helium, He, and 50% xenon, Xe, by weight. What are the partial pressures of the individual gases?

Solution

First find the number of moles of helium and xenon in any given sample. A convenient sample choice is 100 g. Then the number of moles of each gas is

$$n_{He} = \frac{50.0 \text{ g}}{4.00 \text{ g mole}^{-1}} = 12.5 \text{ moles He}$$

$$n_{Xe} = \frac{50.0 \text{ g}}{131.3 \text{ g mole}^{-1}} = 0.381 \text{ mole Xe}$$

The next step is to calculate the mole fraction, X_j, of each component:

$$X_{He} = \frac{12.5}{12.5 + 0.381} = 0.970$$

$$X_{Xe} = \frac{0.381}{12.5 + 0.381} = 0.030$$

According to Dalton's law, the partial pressure of each component is expressed as $p_j = X_j P$. Thus, we have

$$p_{He} = 0.970\,P = 0.970(0.800) = 0.776 \text{ atm}$$
$$p_{Xe} = 0.030\,P = 0.030(0.800) = 0.024 \text{ atm}$$

Notice that no total volume was specified, and a convenient but arbitrary sample size was used for calculation purposes. Why is the answer independent of volume? Will the answer change if the temperature is changed?

Often gases are collected over liquids such as water or mercury, as in Figure 3-13. Dalton's law must be applied in such cases to account for partial evaporation of the liquid into the space occupied by the gas.

Example 16

Oxygen gas generated in an experiment is collected at 25°C in a bottle inverted in a trough of water (Figure 3-13). The external laboratory pressure is 1.000 atm. When the water level in the originally full bottle has fallen to the level in the trough, the volume of collected gas is 1750 ml. How many moles of oxygen gas have been collected?

Solution

If the water levels inside and outside the bottle are the same, then the total pressure inside the bottle equals 1.000 atm. But at 25°C the vapor pressure of water (or the pressure of water vapor in equilibrium with the liquid) is 23.8 mm Hg or 0.0313 atm, so the partial pressure of oxygen gas is only 1.000 − 0.031, or 0.969 atm. The mole fraction of oxygen gas in the bottle is 0.969 and not 1.000, and the partial pressure of oxygen also is 0.969 atm. The number of moles is

$$n = \frac{PV}{RT} = \frac{0.969 \text{ atm} \times 1750 \text{ cm}^3}{82.054 \text{ cm}^3 \text{ atm K}^{-1} \text{ mole}^{-1} \times 298 \text{ K}}$$

$$= 0.0694 \text{ mole}$$

What would the answer have been had the pressure of water vapor been neglected?

Example 17

On a humid day at 43.3°C in Galveston, Texas, the vapor pressure of water is 0.087 atm. What is the water content of the atmosphere, expressed as a mole fraction? Assuming that dry air is 20 mole percent O_2 and 80 mole percent N_2, what is the water content in percent by weight?

Solution

The answers are 0.087 and 5.62%.

Figure 3-13

When oxygen gas is collected by displacing water from an inverted bottle, the presence of water vapor in the collecting bottle must be recognized when calculating the amount of oxygen collected. The correction is made easily by using Dalton's law of partial pressures.

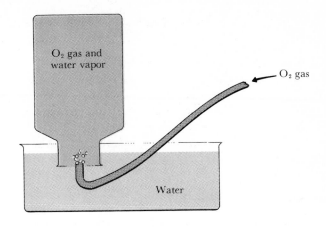

O₂ gas and water vapor

O₂ gas

Water

Figure 3-14

Effusion is the flow of gas from a small hole in a container into an outside region of equal pressure. According to Graham's law, the rates of effusion of two gases at equal temperature are inversely proportional to the square roots of their molecular weights or, by kinetic molecular theory, proportional to their velocities.

Other Predictions of the Kinetic Molecular Theory

Derivations from the kinetic molecular theory that are not much more complicated in principle than the ones we have seen for the gas pressure furnish us with a host of other predictions about the behavior of gases. These predictions have been tested by many scientists and have encouraged confidence in the theory. A derivation of the probability of a molecule hitting a hole in the wall of a container leads to Graham's law of effusion, which predicts that the rate of leakage of a gas from a small hole in a tank will be inversely proportional to the square root of the molecular weight (Figure 3-14).

Thomas Graham (1805–1869) observed, in 1846, that the rates of effusion of gases are inversely proportional to the square roots of their densities. Since, by Avogadro's hypothesis, the density of a gas is proportional to its molecular weight, Graham's observation agrees with the kinetic theory, which predicts that the rate of escape is proportional to molecular velocity or inversely proportional to the square root of the molecular weight (equation 3-29):

$$\frac{\text{Rate}_2}{\text{Rate}_1} = \frac{v_2}{v_1} = \left(\frac{M_1}{M_2}\right)^{1/2} \tag{3-34}$$

However, the law begins to fail at high densities, in which molecules collide several times with one another as they escape through the hole. The law also fails when there are holes large enough so the gas has a hydrodynamic flow toward the hole, thereby leading to the formation of a jet of escaping gas. But so long as isolated molecules escape by going through the hole during their random motions through a stationary gas, the kinetic molecular theory prediction is exact.

Example 18

A given volume of oxygen gas effuses through a small orifice into a vacuum in 1 min. The same volume of an unknown gas takes 1 min and 34 sec to effuse at the same temperature. What is the approximate molecular weight of the unknown gas? If its empirical formula is CH, what is its molecular formula, and what is its molecular weight?

Solution

From equation 3-34 we can derive

$$\frac{t_2}{t_1} = \frac{\text{Rate}_1}{\text{Rate}_2} = \left(\frac{M_2}{M_1}\right)^{1/2}$$

the subscript 1 representing oxygen and the subscript 2 representing the unknown gas. Substituting the observed data, and rearranging the equation to solve for M_2, we get

$$M_2 = 32.00 \text{ g mole}^{-1} \times \left(\frac{1.57}{1.00}\right)^2 = 78.9 \text{ g mole}^{-1}$$

This is an approximate value, subject to errors in measuring gas flow. Since the formula weight of the unknown gas is 13.02 g (we know this from the empirical formula, CH), there must be six formula weights in the true molecular weight. Since $6 \times 13.02 = 78.12$, the gas is C_6H_6.

Example 19

Calculate the relative effusion rates of the two isotopic forms of uranium hexafluoride, $^{238}UF_6$ and $^{235}UF_6$. All fluorine is ^{19}F.

Solution

The molecular weight of $^{238}UF_6$ is 352.0 amu, and that of $^{235}UF_6$ is 349.0 amu. The ratio of effusion rates then is

$$\frac{\text{Rate}(235)}{\text{Rate}(238)} = \left(\frac{352.0}{349.0}\right)^{1/2} = 1.0043$$

Although there is only 0.43% difference in effusion rates of the hexafluorides of the two isotopes of uranium, scientists used this difference to separate fissionable ^{235}U from ^{238}U during the construction of the first atomic bombs at the end of World War II. No other separation method proved workable at the time. The scientists used UF_6 because it is a gaseous compound of uranium, but the small separation ratio meant that the gas had to be passed through many thousands of porous barriers in the special gas diffusion plant at Oak Ridge, Tennessee, to achieve a useful degree of separation.

The kinetic molecular theory also allows us to predict gaseous diffusion, viscosity, and thermal conductivity, the three so-called transport properties. Each phenomenon can be treated mathematically as the diffusion of some molecular property down a gradient. In gaseous diffusion, mass diffuses from regions of high to low concentration, or down a concentration gradient. Viscosity of a fluid arises because slowly moving molecules diffuse into (and retard) rapidly moving fluid layers, and faster molecules diffuse into (and accelerate) the slow regions. This is a transport of momentum down a velocity gradient. Thermal conductivity is the scattering of rapidly moving molecules into regions of slower ones. It can be described as a transport of kinetic energy down a temperature gradient. In all three cases, the kinetic molecular theory predicts the diffusion coefficient, with best accuracy at low gas pressures and high temperatures. These are just the conditions for which the simple ideal gas law is most applicable.

In summary, the elementary kinetic molecular theory, as outlined here, provides a correct explanation for the behavior of ideal gases. It gives us confidence in the reality of molecules, and encourages us to look for molecular modifications of the simple theory that will account for deviations from ideal gas behavior.

3-8 REAL GASES AND DEVIATIONS FROM THE IDEAL GAS LAW

If gases were ideal, the quotient PV/RT would always equal 1 for 1 mole of gas. Actually all real gases deviate, to some extent, from ideal behavior; the quantity $Z = PV/RT$, called the *compressibility coefficient,* is one measure of this deviation. Z is plotted against pressure for several gases at 273 K in Figure 3-15, and for one gas at several temperatures in Figure 3-16. We can interpret the behavior of real gases as a combination of intermolecular attractions (which are effective over comparatively long distances) and repulsions caused by the finite sizes of molecules (which become significant only when molecules are crowded together at high pressures). At low pressures—but still too high for ideal behavior—intermolecular attractions make the molar volume unexpectedly low, and the compressibility coefficient is less than 1. However, at sufficiently high pressures the crowding of molecules begins to predominate, and the molar volume of the gas is greater than it would be if the molecules were point masses. The higher the temperature (Figure 3-16), the less significant the intermolecular attraction will be in comparison with the kinetic energy of the moving molecules, and the lower will be the pressure at which the bulk factor dominates and Z rises above 1.

An equation such as the ideal gas law, $PV = nRT$, is known as an **equation of state** because it describes the state of a system in terms of the measurable variables P, V, T, and n (Figure 3-17). Other equations of state that have been proposed describe the behavior of real gases better than the ideal gas

Figure 3-15

Deviations from the ideal gas law for several gases at 273 K, in terms of the compressibility factor $Z = PV/RT$. The dip of Z below 1.0 at low pressures is caused by intermolecular attractions; the rise above 1.0 at high pressures is produced by the shorter range intermolecular repulsions as the molecules, of finite bulk, are crowded closely together.

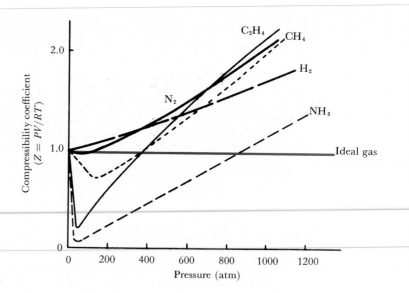

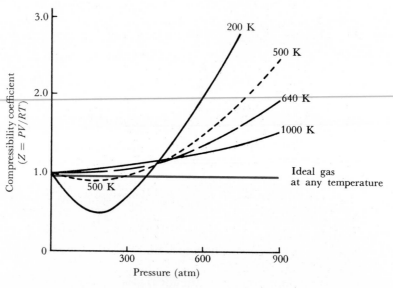

Figure 3-16 PV/RT for 1 mole of methane gas at several temperatures. Note that PV/RT is less than 1.0 at low pressures and greater than 1.0 at high pressures. Ideal gas behavior is approached at high temperatures.

Figure 3-17

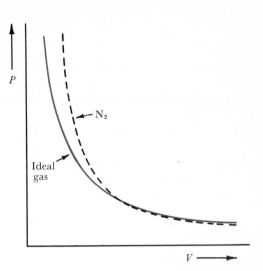

Pressure—volume curves for nitrogen and an ideal gas at constant temperature. At low pressures, the molar volume of N_2 gas is less than that of an ideal gas because of intermolecular attraction. At high pressures, the nonzero volume of individual N_2 molecules makes the gas volume greater than ideal.

law. The best known of these equations is the one introduced, in 1873, by Johannes van der Waals. Van der Waals assumed that, even for a real gas, there is an ideal pressure, P^*, and an ideal volume, V^*, that would apply to the ideal expression $P^* V^* = nRT$; but because of the imperfections of the gas, these were not the same as the measured pressure, P, and measured volume, V. The ideal volume, he reasoned, should be less than the measured volume because the individual molecules have a finite volume instead of being point masses, and the portion of the container's volume that is occupied by other molecules is unavailable to any given molecule. Therefore, the "ideal" volume should be less than the measured volume by a constant, b, that is related to molecular size by $V^* = V - b$.

Moreover, a gas molecule subject to attractions from other gas molecules strikes the walls with less force than it would if these attractions were absent. For as the molecule approaches the wall, there are more gas molecules behind it in the bulk of the gas than there are between it and the wall (Figure 3-18). The number of collisions with the wall in a given time is proportional to the density of the gas, and each collision is softened by a back-attraction factor, which itself is proportional to the density of molecules doing the attracting. Therefore, the correction factor to P is proportional to the square of the gas density, or inversely proportional to the square of the volume: $P^* = P + a/V^2$, where a is related to the attractions between molecules. The complete van der Waals equation is

$$\left(P + \frac{a}{\overline{V}^2}\right)(\overline{V} - b) = RT$$

Here $\overline{V}$ is the *volume per mole*, or: $\overline{V} \equiv V/n$. The equation is simpler when written this way. Similarly, the ideal gas law can be written $P\overline{V} = RT$ as easily as $PV = nRT$. The constants a and b are chosen empirically to provide

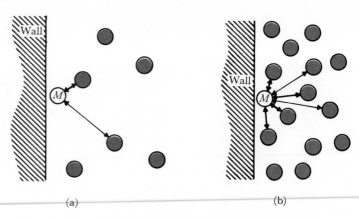

(a)   (b)

Figure 3-18   Reduction of pressure of a real gas as a result of inter-molecular attractions. (a) Gas at low density. (b) Gas at high density. A molecule M in a high-density gas hits the wall with a smaller impact than in a low-density gas because the attractions of its nearest neighbors reduce the force of its impact.

Table 3-2

Measures of Molecular Size Obtained from the Kinetic Theory

	Van der Waals constants		Spherical molecular diameters, d (Å)		
Gas	a (liter^2 atm mole^{-2})	b (cm^3 mole^{-1})	From van der Waals[a]	From gas viscosity	From density of liquid or solid[b]
Hg	8.09	17.0	2.38	3.60	3.26
He	0.0341	23.70	2.48	2.00	—
H$_2$	0.2444	26.61	2.76	2.18	—
H$_2$O	5.464	30.49	2.88	2.72	3.48
O$_2$	1.360	31.83	2.90	2.96	3.75
N$_2$	1.390	39.12	3.14	3.16	4.00
CO$_2$	3.592	42.67	3.24	4.60	4.54

[a]This is a bad approximation of d for all gases except Hg and He.

[b]We are assuming that the molecules are spheres, which when most closely packed fill 74% of the space available. If M is the molecular weight, N is Avogadro's number, and D is the density, the molecular volume is

$$V_m = \frac{\pi}{6} d^3 = 0.74 \frac{M}{ND}$$

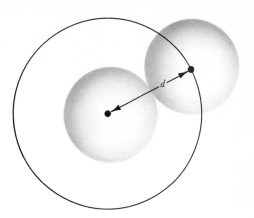

Figure 3-19   This center of no other molecule can come closer than a molecular diameter to the center of a given molecule. The volume around each molecule from which other molecules are excluded then is $\frac{4}{3}\pi d^3$, or eight times the molecular volume of $\frac{\pi}{6}d^3$.

the best relationship of the equation to the actual PVT behavior of a gas. Even so, the molecular size calculated from this purely experimental b agrees well with the ones obtained by other means (Table 3-2), and gives us confidence that we have the right explanation for deviations from ideality.

Experimentally obtained values of a and b for several gases are given in Table 3-2, along with several calculations of molecular diameters. We might suppose that the constant b is simply the excluded volume per mole, V_m (as in Figure 3-19): $b = 8NV_m = \frac{4}{3}\pi Nd^3$. However, collision is a two-molecule process and this calculation overcounts the excluded volume by a factor of 2. The molecular diameters in Table 3-2 were obtained from the b values by having b equal to $4NV_m$, in which $V_m = \pi d^3/6$ is the volume of one molecule.

The van der Waals equation is applicable over a much wider range of temperatures and pressures than is the ideal gas law; it is even compatible with the condensation of a gas to a liquid.

Summary

We have seen four experimentally derived principles or laws of gas behavior, which all gases obey approximately, especially under conditions of low pressure and high temperature.

   1. Avogadro's law: At fixed pressure and temperature, the volume of any gas is proportional to the number of moles present.
   2. Boyle's law: At constant temperature, the volume of a sample of gas is inversely proportional to the pressure on the gas.
   3. Charles' law: At constant pressure, the volume of a sample of gas is proportional to the temperature of the gas on the absolute, or Kelvin, scale.

4. Gay-Lussac's law: At constant volume, the pressure exhibited by a sample of gas is proportional to the temperature of the gas on the absolute scale.

Although real gases only approximate this behavior, we can define an **ideal gas** as one that follows the preceding laws exactly under all conditions. All the foregoing observations can be combined in one expression, the **ideal gas law**:

$$PV = nRT$$

If P is pressure in atmospheres, V is volume in liters, T is the absolute Kelvin temperature, and n is the number of moles, then the proportionality constant R, known as the gas constant, has the value

$$R = 0.08205 \text{ liter atm } K^{-1} \text{ mole}^{-1}$$

In practical calculations, the ideal gas law is often most useful in one of the various ratio forms given in equations 3-9 through 3-12.

Standard temperature and pressure, or **STP**, is defined as 273.15 K (0°C) and exactly 1 atm. Gas properties are frequently converted to STP conditions for comparison purposes, even for gases such as H_2O that liquefy at STP. One mole of any ideal gas at STP has a volume of 22.414 liters, and this is called a **standard molar volume**.

The **kinetic molecular theory of gases** successfully explains the behavior of ideal gases with a minimum of starting assumptions, and also provides a framework for understanding the deviations of real gases from ideal gas behavior. In its simplest form, the kinetic theory assumes that a gas is made up of noninteracting point molecules in a state of constant motion, colliding elastically with one another and with the walls of a container. In extending the theory to cover real gases, we recognize that molecules have finite volume and exert attractive forces on one another.

In the kinetic theory, pressure is simply the result of collision of molecules with the container walls, and transfer of momentum. The product of pressure and volume is equal to two-thirds the kinetic energy of motion of the molecules (equation 3-25). Combining this fact with the observed ideal gas law, we come to the important conclusion that the kinetic energy of motion of the molecules is directly proportional to absolute temperature (equation 3-26), or that *temperature is simply a consequence of molecular motion.*

Comparing the densities of substances in the gaseous and condensed phases, we find that the average space available to a gas molecule at STP is roughly three orders of magnitude (or 10^3) greater than the volume of the molecule itself. The **root-mean-square (rms) speed** of a molecule is inversely proportional to the square root of its molecular weight (equation 3-28), and this speed is on the order of several thousand miles per hour at STP. The heavier the molecule, the more slowly it moves.

The actual molecular velocities in a gas vary in a distribution around this rms value, with some velocities nearly zero and others very much faster than average. Velocities of individual molecules vary as the molecules collide with one another and rebound. Nevertheless, the *distribution* of molecular speeds remains constant at constant temperature. An ideal gas at STP has a **mean free path**, or average distance between collisions, on the order of 1000 Å, and a collision rate of around 5×10^9 collisions sec^{-1}.

Dalton's law of partial pressures says that each component of a gas mixture behaves as if it were the only gas present. The **mole fraction** X_j, is the number of moles of gas j present divided by the total number of moles of all gases. The partial pressure of gas j is the mole fraction of j times the total pressure: $p_j = X_j P$. The sum of partial pressures of all components is the total pressure.

The kinetic theory of gases predicts that the rate of effusion of a gas through a small orifice will be inversely proportional to the square root of molecular velocity (equation 3-34), and this prediction is borne out by experiment. The theory is also successful in accounting quantitatively for diffusion of gases, viscosity, and conductivity of heat.

Real gases deviate from ideal behavior because molecules are not volumeless, shapeless points, and because real molecules attract one another. Molecular attractions become impossible to neglect when the molecules are moving more slowly, at low temperatures; molecular volumes become significant when the gas is compressed. Hence the gases approximate ideal behavior most closely at high temperature and low pressure.

Van der Waals modified the ideal gas law to take both of the preceding factors into account. The van der Waals equation, $(P + a/\bar{V}^2)(\bar{V} - b) = RT$ for 1 mole of gas, has an experimental constant, b, that is related to molecular volume, and another constant, a, that is related to molecular attractions or "stickiness." From the van der Waals constant b we can obtain approximate molecular diameters, and these values agree roughly with estimates of diameters obtained from densities of solids or from gas viscosities.

Self-Study Questions

1. Why should gases obey simpler laws than those governing liquids or solids?
2. Early hydraulic engineers found that no suction pump could lift water more than approximately 34 ft. Can you explain this phenomenon from the information in this chapter?
3. How did Boyle design his experiment to test the "spring of the air" theory?
4. Why is a plot of experimental data that produces a straight line useful or desirable?
5. How is an absolute scale of temperature defined in terms of gas behavior?

6. Under what conditions does Boyle's law apply? When is Charles' law applicable? How are these laws derived from the complete ideal gas law?

7. What does STP signify, and why is it useful?

8. What molecular explanation can you give for the deviation of real gases from ideal gas behavior? Under what conditions will real gases most resemble ideal behavior?

9. What experimental evidence is there that each of the three assumptions of the kinetic molecular theory of gases is valid?

10. What can we say that the product, PV, for an ideal gas is proportional to the kinetic energy, E_k?

11. Why can we say that the temperature is proportional to the square of the speed of the molecules (actually to the mean square speed)?

12. If the molecules in a liter of hydrogen gas and those in a liter of oxygen gas are moving with the same mean square speed, which gas is hotter?

13. What fraction of a typical gas is occupied by the volume of the molecules of which it is composed? What direct physical measurements can tell you this?

14. Why is the gas volume of 22.414 liters significant?

15. How does the speed of sound in air at sea level compare with the root-mean-square speed of the molecules in the air?

16. Which would you expect to be greater, the average speed or the root-mean-square speed? Can you explain your answers by using the definitions of the two speeds?

17. What does Dalton's law of partial pressure indicate about the behavior of gases in a mixture?

18. Why does the compressibility coefficient of a real gas deviate above and below 1.00 as it does?

19. How is a measure of molecular size obtained from the van der Waals equation?

Problems

Atmospheric pressure

1. One atmosphere of pressure will push a column of mercury to a height of 760 mm when the cross-sectional area of the column is 1.00 cm². What would be the height of the column of mercury supported by the atmosphere if its cross-sectional area were 0.500 cm²? How high would the mercury column be if the area were 2 cm²?

2. Standard atmospheric pressure is quoted by the U.S. Weather Bureau as 29.92 in. How can pressure have units of inches? Show how this measure is related to other units of pressure discussed in this chapter.

Boyle's law

3. A gas at an initial pressure of 0.921 atm is allowed to expand at constant temperature until the pressure falls to 0.197

atm. What is the ratio of final volume to initial volume?

4. An ideal gas occupies 76.0 liters at 1.00 atm pressure. What pressure will reduce the volume to 10.0 liters, if T = constant?

5. A sample of neon gas occupies 75.0 ml at 1.00 atm pressure. If the temperature is unchanged, what volume will it occupy at (a) 5.00 atm, (b) 0.100 atm, and (c) 1000 Pa?

6. An experiment is being carried out at 75°C in a 5.00-liter flask that contains an evacuated glass bulb of 400 cm³ volume. If the bulb breaks, what will be the new pressure in the flask, assuming no temperature change?

Charles' law

7. If a sample of gas at 25°C occupies 2.34 liters, what will be its volume at 400°C if the pressure is unchanged?

8. A sample of gas is heated from 25°C to 50°C at constant pressure. Will the gas volume double? Why, or why not? What will be the ratio of final volume to initial volume?

Ideal gas law

9. At STP, 10.3 g of a gas occupies 453 in.³ What is the volume of this sample at 1.25 atm and 100°C?

10. The temperature of a 0.0100-g sample of chlorine gas, Cl_2, in a 10-ml sealed glass container is increased, in an oven, from 20°C to 250°C. What is the initial pressure at 20°C? What is the pressure at 250°C?

Molecules

11. How many molecules of an ideal gas are there in 1.000 ml of the gas if the temperature is −80°C and the pressure 1.000 Pa?

12. What pressure will be exerted by 5.0 × 10^{13} molecules of an ideal gas in 1.000 ml at 0°C? Express your answer in atmospheres and in pascals.

Partial pressures

13. A 2.00-liter flask at 27°C contains 4.40 g of carbon dioxide and 1.00 g of nitrogen gas. What is the pressure inside the flask? What are the partial pressures of each of the two components? (Express pressures in atmospheres.)

14. One liter each of oxygen, nitrogen, and hydrogen gas, all originally at 1.00 atm pressure, are forced into a single 2.00-liter container. What is the resulting pressure if temperature is unchanged? What are the partial pressures of the components?

15. A gas mixture contains half argon and half helium by weight, with a total pressure of 1.11 atm. What is the partial pressure of each gas in the mixture?

16. A mixture of gases contains 0.5 mole of oxygen, 0.1 mole of hydrogen, and 0.8 mole of nitrogen. The total pressure is 0.80 atm. What is the partial pressure of each gas?

17. The concentration of carbon monoxide, CO, in cigarette smoke is 20,000 parts per million (ppm) by volume. Calculate the partial pressure of carbon monoxide in 1 liter of cigarette smoke which exerts a total pressure of 1.00 atm.

18. A mixture of 3.86 g of CCl_4 (carbon tetrachloride) and 1.92 g of C_2H_4 (ethylene) at 450°C exerts how many atmospheres of pressure inside a 30-ml metal bomb? How much pressure is contributed by ethylene?

Gas density

19. What is the density of XeF_6 gas at STP in grams per liter? What will be its density at 25°C and 1.30 atm?

20. The density of a gas at STP is 1.62 g liter^{-1}. What will be its density at 302 K and 0.950 atm?

Molecular weight

21. If 0.750 g of a gas occupies 4.62 liters at 0.976 atm and 20°C, what is the molecular weight of the gas? What might the gas be?

22. A 1.12-liter sample of a gas weighs 0.400 g when measured at 0°C and 0.500 atm. The gas is 25.0% hydrogen by weight and 75.0% carbon. What is the molecular weight of the gas? What are its empirical formula and molecular formula?

Molecular formula

23. A 250-ml sample of a compound with the empirical formula CH_2 weighs 0.395 g at 0.921 atm and 27°C. What are the molecular weight and formula of the compound?

24. A sample of 0.524 g of a compound fills a volume of 129 ml at 25°C and 0.991 atm. Chemical analysis shows that it is 23.5% carbon, 2.0% hydrogen, and 74.5% fluorine by weight. What are its molecular weight and molecular formula?

Empirical formula

25. A 0.490-g sample of a compound is heated through the successive evolution of the following gases, all at 1.00 atm pressure: 280 ml of water vapor at 182°C, 112 ml of ammonia vapor at 273°C, 0.0225 g of water at 400°C, and 0.200 g of SO_3 at 700°C. At the end of the heating, 0.090 g of FeO remains. Deduce the empirical formula for the compound.

Molecular proportions

26. Both solid LiH and CaH_2 react with water to produce hydrogen gas and the corresponding hydroxide, LiOH or $Ca(OH)_2$. A 0.850 g sample of a mixture of LiH and CaH_2 produces 1.200 liters of H_2 at STP. What percentage of the starting mixture was LiH? (Give both a mole percent and a percent by weight.)

Vapor pressure

27. One liter of dry air at 1.00 atm and 86°C is placed in contact with 1.00 ml of liquid water at the same temperature. The volume of the gas phase remains constant throughout the experiment. The vapor pressure of water at this temperature is 0.593 atm and its density is 0.970 g ml^{-1}. When equilibrium has been established,
a) What is the partial pressure of air in the vessel?
b) What is the partial pressure of water vapor in the vessel?
c) What is the total pressure in the vessel?
d) How many moles of water will have evaporated?
e) What volume of liquid water, if any, will remain?

28. One gram of methane, CH_4, is burned to produce CO_2 gas and liquid H_2O. At 25°C, the pressure exerted by the products is 0.987 atm. The vapor pressure of water at 25°C is 0.0313 atm. Calculate the volume of dry CO_2 produced in the reaction.

Energy and temperature

29. Calculate the kinetic energy per mole of ideal gas molecules at a temperature of 25°C. If you use the value of the gas constant $R = 0.08205$ liter atm K^{-1} mole^{-1}, the kinetic energy will be in units of liter atm, which is unorthodox but perfectly acceptable. The calculation is easier with $R = 8.3144$ J K^{-1} mole^{-1}. How does this kinetic energy per mole compare in magnitude with the energy of a chemical bond, which typically is on the order of 350 kJ mole^{-1}? What would happen if kinetic energy of motion and bond energy were more similar in magnitude?

30. Calculate the kinetic energy per mole at 300°C for the following gas molecules, assuming ideal behavior: (a) H_2, (b) CH_4, (c) HBr. Why is this problem simpler than it looks? Calculate the rms velocities of the three molecules at 300°C, and compare their relative magnitudes. What general principle can you draw from them? Why is the second half of this problem more complex than the first?

Molecular volumes

31. Liquid benzene, C_6H_6, has a density of 0.879 g ml^{-1}. If benzene vapor behaves as an ideal gas, what will the vapor density be at STP? Calculate the volume per molecule, in cubic angstroms, for the liquid and gaseous states. By what factor does the volume per molecule increase when the liquid evaporates?

32. At its normal boiling point of -164°C, liquid methane, CH_4, has a density of 0.466 g ml^{-1}. If methane vapor behaved as an ideal gas at that temperature, what would be its density at 1 atm pressure? What are the volumes per molecule for liquid and for gas?

Molecular speeds

33. What is the rms speed of oxygen molecules at 25°C? To what temperature must the gas be raised to increase the speed by a factor of 10 while maintaining a constant volume? By what factor would the pressure increase during this temperature increase at constant volume?

34. The speed of sound waves in an ideal gas is given by the formula

$$\text{Speed of sound} = c = \sqrt{\frac{\gamma RT}{M}}$$

This is remarkably like the equation for the rms speed of the molecules themselves, except that γ is a constant that has the theoretical value of 5/3 for monatomic gases such as He and Ne, and a value near 1.41 for diatomic gases such as N_2 and O_2. Calculate the speed of sound in pure nitrogen gas at 1.00 atm and 25°C, and compare this speed with the rms speed of the nitrogen molecules themselves.

35. Earth's atmosphere is approximately 80% nitrogen gas and 20% oxygen. The speed of sound in air can be calculated by using an average molecular weight in the expression given in the preceding problem. Calculate the speed of sound in air at 1.00 atm pressure and 25°C. Will sound travel faster, or slower, than this in helium gas? What will the speed of sound be in air at 35,000 ft, where the temperature is -40°C?

36. How does the speed of sound in a gas

at 25°C and 1.00 atm pressure compare with that in the same gas at 25°C and 50.0 atm pressure? How do the rms molecular speeds compare? Does your answer seem sensible?

Dalton's law

37. One liter of hydrogen gas is collected over water at 10°C and 1.053 atm. The vapor pressure of water at this temperature is 0.0121 atm. If the hydrogen then is separated from the water and dried at constant temperature, what will be the new volume of the dry hydrogen gas? If the water vapor that was removed from the hydrogen is stored at 100°C and 0.0159 atm, what will be its volume?

38. Ultraviolet light from the sun converts some of the oxygen, O_2, in the upper atmosphere to ozone, O_3. If a sample at constant temperature and volume is irradiated until 5% of the O_2 is converted to O_3, what will the final pressure be, assuming an initial pressure of 0.526 atm?

Graham's law

39. A sample of an unknown gas is shown by analysis to contain only sulfur and oxygen. The gas requires 28.3 sec to effuse through an orifice into a vacuum, whereas an identical number of O_2 molecules passes through the same orifice in 20.0 sec. Determine the molecular weight and formula of the gas.

40. In the same time required for 6 liters of carbon dioxide to effuse through a porous barrier, only 5 liters of an unknown gas will pass through. Estimate the molecular weight of the unknown gas.

Van der Waals gas

41. What is the volume per mole of an ideal gas at STP, or 1 atm pressure and 273.15 K? The van der Waals equation is a better description of the behavior of real gases. What pressure does this equation predict for a mole of O_2 kept at the volume that you just calculated for an ideal gas, and at 273.15 K? What is the percent difference between the ideal and van der Waals pressure predictions?

42. Use the van der Waals equation to calculate the molar volume of carbon dioxide at STP. [*Note:* To avoid solving a cubic equation, you can use the method of successive approximations. Solve for $\overline{V}$ in the term $(\overline{V} - b)$, using the ideal gas value of the molar volume in the denominator of the $a/\overline{V}^2$ term. Then, if you are not happy with your approximate answer, you can repeat the process, using the approximate answer in the denominator of $a/\overline{V}^2$, and continue until the answers cease changing from one cycle to the next.]

Postscript to Gas Laws and Atomic Theory

When the dust has settled after a new discovery, it is all too easy to forget how much controversy and effort went into its development. Thomas Thomson (1773–1852) was Regius Professor of Chemistry at the University of Glasgow, and was the man to whom John Dalton turned for help in publicizing his new theory of atoms. In 1830, Thomson published his *History of Chemistry,* which is particularly interesting because many of the participants in the atomic revolution in chemistry were alive, active, and friends of Thomson. In the last chapter of his *History,* Thomson describes the circumstances of the birth of the atomic theory:

66In the year 1804, on the 26th of August, I spent a day or two at Manchester, and was much with Mr. Dalton. At that time he explained to me his notions respecting the composition of bodies. I wrote down at the time the opinions which he offered . . . [A brief account of the atomic theory followed.]

Mr. Dalton informed me that the atomic theory first occurred to him during his investigation of olefiant gas [acetylene, C_2H_2] and carburetted hydrogen gas [ethylene, C_2H_4], at that time imperfectly understood, and the constitution of which was first fully developed by Mr. Dalton himself. It was obvious from the experiments which he made upon them, that the constituents of both were carbon and hydrogen, and nothing else. He found further, that if we reckon the carbon in each the same, then carburetted hydrogen gas contains exactly twice as much hydrogen as olefiant gas does. This determined him to state the ratios of these constituents in numbers, and to consider the olefiant gas as a compound of one atom of carbon and one atom of hydrogen; and carburetted hydrogen of one atom of carbon and two atoms of hydrogen. The idea thus conceived was applied to carbonic oxide, water, ammonia, etc.; and numbers representing the atomic weights of oxygen, azote, etc., deduced from the best analytical experiments which chemistry then possessed. Let not the reader suppose that this was an easy task. Chemistry at that time did not possess a single analysis which could be considered as even approaching to accuracy. . . .

In the third edition of my *System of Chemistry,* published in 1807, I introduced a short sketch of Mr. Dalton's theory, and thus made it known to the chemical world. . . . These facts gradually drew the attention of chemists to Mr. Dalton's views. There were, however, some of our most eminent chemists who were very hostile to the atomic theory. The most conspicuous of these was Sir Humphry Davy. In the autumn of 1807 I had a long conversation with him at the Royal Institution, but could not convince him that there was any truth in the hypothesis. A few days after, I dined with him at the Royal Society Club, at the Crown and Anchor, in the Strand. Dr. Wollaston was present at the dinner. After dinner every member of the club left the tavern, except Dr. Wollaston, Mr. Davy, and myself, who stayed behind and had tea. We sat about an hour and a half together, and our whole conversation was about the atomic theory. Dr. Wollaston was a convert as well as myself; and we tried to convince Davy of the inaccuracy of his opinions; but, so far from being convinced, he went away, if possible, more prejudiced against it than ever. Soon after, Davy met Mr. David Gilbert, the late distinguished president of the Royal Society; and he amused him with a caricature description of the atomic theory, which he exhibited in so ridiculous a light, that Mr. Gilbert was astonished how a man of sense or science could be taken in with such a tissue of absurdities. . . . [Wollaston finally convinced Gilbert after a long recital of the chemical evidence.]

Mr. Gilbert went away a convert to the truth of the atomic theory; and he had the merit of convincing Davy that his former opinions on the subject were wrong. What arguments he employed I do not know; but they must have been convincing ones, for Davy ever after became a strenuous supporter of the atomic theory. The only alteration which he made was to substitute *proportion* for Dalton's word, *atom.* Dr. Wollaston substituted for it the term *equivalent.* The object of these substitutions was to avoid all theoretical annunciations. But, in fact, these terms, *proportion, equivalent,* are neither of them so convenient as the term *atom;* and unless we adopt the hypothesis with which Dalton set out, namely, that the ultimate particles of bodies are *atoms* incapable of further division, and that chemical combination consists in the union of these atoms with each other, we lose all the new light which the atomic theory throws upon chemistry, and bring our notions back to the obscurity of the days of Bergman and of Berthollet. **"**

Suggested Reading

J. Hildebrand, *An Introduction to Molecular Kinetic Theory,* Van Nostrand Reinhold, New York, 1963.

T. L. Hill, *Lectures on Matter and Equilibrium,* W. A. Benjamin, Menlo Park, Calif., 1966. Written at the honors freshman level. The first four chapters, on states of matter, gases, and intermolecular forces, are particularly useful.

W. Kauzmann, *Kinetic Theory of Gases,* W. A. Benjamin, Menlo Park, Calif., 1966. Thorough, clear. Chapters 1 and 2, on equations of state of gases, are especially relevant. Chapter 4 continues the discussion of the distribution of molecular velocities, but requires calculus.

4

Will It React?
An Introduction
to Chemical
Equilibrium

And so, nothing that to our world appears,
Perishes completely, for nature ever
Upbuilds one thing from another's ruin;
Suffering nothing yet to come to birth
But by another's death.

Lucretius (95–55 B.C.)

The main question asked in Chapter 2 was "If a given set of substances will react to give a desired product, how much of each substance is needed?" Our basic assumptions were that matter cannot be arbitrarily created or destroyed, and that atoms going into a reaction must come out again as products.

In this chapter we ask a second question: "Will a reaction occur, eventually?" Is there a tendency or a drive for a given reaction to take place, and if we wait long enough will we find that reactants have been converted spontaneously into products? This question leads to the ideas of **spontaneity** and of **chemical equilibrium.** A third question, "Will a reaction occur in a reasonably short time?" involves chemical kinetics, which will be discussed in Chapter 22. For the moment, we will be satisfied if we can predict which way a chemical reaction will go by itself, ignoring the time factor.

4-1 SPONTANEOUS REACTIONS

A chemical reaction that will occur on its own, given enough time, is said to be **spontaneous.** In the open air, and under the conditions inside an automobile engine, the combustion of gasoline is spontaneous:

$$C_7H_{16} + 11O_2 \rightarrow 7CO_2 + 8H_2O$$

(The reaction is exothermic, or heat emitting. The enthalpy change, which was defined in Chapter 2, is large and negative: $\Delta H = -4812$ kJ mole^{-1} of heptane at 298 K. The heat emitted causes the product gases to expand, and it is the pressure from these expanding gases that drives the car.) In contrast, the reverse reaction under the same conditions is not spontaneous:

$$7CO_2 + 8H_2O \not\rightarrow C_7H_{16} + 11O_2$$

No one seriously proposes that gasoline can be obtained spontaneously from a mixture of water vapor and carbon dioxide.

Explosions are examples of rapid, spontaneous reactions, but a reaction need not be as rapid as an explosion to be spontaneous. It is important to understand clearly the difference between rapidity and spontaneity. If you mix oxygen and hydrogen gases at room temperature, they will remain together without appreciable reaction for years. Yet the reaction to produce water is genuinely spontaneous:

$$2H_2 + O_2 \rightarrow 2H_2O$$

We know that this is true because we can trigger the reaction with a match, or with a catalyst of finely divided platinum metal.

The preceding sentence suggests why a chemist is interested in whether a reaction is spontaneous, that is, whether it has a natural tendency to occur. If a desirable chemical reaction is spontaneous but slow, it may be possible to speed up the process. Increasing the temperature will often do the trick, or a catalyst may work. We will discuss the function of a catalyst in detail in Chapter 22. But in brief, we can say now that a catalyst is a substance that helps a naturally spontaneous reaction to go faster by providing an easier pathway for it. Gasoline will burn rapidly in air at a high enough temperature. The role of a spark plug in an automobile engine is to provide this initial temperature. The heat produced by the reaction maintains the high temperature needed to keep it going thereafter. Gasoline will combine with oxygen at room temperature if the proper catalyst is used, because the reaction is naturally spontaneous but slow. But no catalyst will ever make carbon dioxide and water recombine to produce gasoline and oxygen at room temperature and moderate pressures, and it would be a foolish chemist who spent time trying to find such a catalyst. In short, an understanding of spontaneous and nonspontaneous reactions helps a chemist to see the limits of what is *possible*. If a reaction is possible but not currently realizable, it may be worthwhile to look for ways to carry it out. If the process is inherently impossible, then it is time to study something else.

4-2 EQUILIBRIUM AND THE EQUILIBRIUM CONSTANT

The speed with which a reaction takes place ordinarily depends on the concentrations of the reacting substances. This is common sense, since most reactions take place when molecules collide, and the more molecules there are per unit of volume, the more often collisions will occur.

The industrial fixation of atmospheric nitrogen is very important in the manufacture of agricultural fertilizers (and explosives). One of the steps in nitrogen fixation, in the presence of a catalyst, is

$$N_2 + O_2 \rightarrow 2NO \tag{4-1}$$

If this reaction took place by simple collision of one molecule of N_2 and one molecule of O_2, then we would expect the rate of collision (and hence the rate of reaction) to be proportional to the concentrations of N_2 and O_2:

Rate of NO production $\propto [N_2][O_2]$

or

$$R_1 = k_1[N_2][O_2] \tag{4-2}$$

The proportionality constant k_1 is called the **forward-reaction rate constant,** and the bracketed terms $[N_2]$ and $[O_2]$ represent concentrations in moles per liter. This rate constant, which we will discuss in more detail in Chapter 22, usually varies with temperature. Most reactions go faster at higher temperatures, so k_1 is larger at higher temperatures. But k_1 does *not* depend on the concentrations of nitrogen and oxygen gases present. All of the concentration dependence of the overall forward reaction rate, R_1, is contained in the terms $[N_2]$ and $[O_2]$. If this reaction began rapidly in a sealed tank with high starting concentrations of both gases, then as more N_2 and O_2 were consumed, the forward reaction would become progressively slower. The rate of reaction would decrease because the frequency of collision of molecules would diminish as fewer N_2 and O_2 molecules were left in the tank.

The reverse reaction can also occur. If this reaction took place by the collision of two molecules of NO to make one molecule of each starting gas,

$$2NO \rightarrow N_2 + O_2 \tag{4-3}$$

then the rate of reaction again would be proportional to the concentration of each of the colliding molecules. Since these molecules are of the same compound, NO, the rate would be proportional to the *square* of the NO concentration:

Rate of NO removal $\propto [NO][NO]$

or

$$R_2 = k_2[NO]^2 \tag{4-4}$$

where R_2 is the overall reverse reaction rate and k_2 is the reverse-reaction rate constant. If little NO is present when the experiment begins, this reaction will occur at a negligible rate. But as more NO accumulates by the forward reaction, the faster it will be broken down by the reverse reaction.

Thus as the forward rate, R_1, decreases, the reverse rate, R_2, increases. Eventually the point will be reached at which the forward and reverse reactions exactly balance:

$$R_1 = R_2$$
$$[N_2][O_2]k_1 = k_2[NO]^2 \tag{4-5}$$

This is the condition of **equilibrium**. Had you been monitoring the concentrations of the three gases, N_2, O_2, and NO, you would have found that the composition of the reacting mixture had reached an equilibrium state and thereafter ceased to change with time. This does not mean that the individual reactions had stopped, only that they were proceeding at equal rates; that is, they had arrived at, and thereafter maintained, a condition of balance or equilibrium.

The condition of equilibrium can be illustrated by imagining two large fish tanks, connected by a channel (Figure 4-1). One tank initially contains 10 goldfish, and the other contains 10 guppies. If you watch the fish swimming aimlessly long enough, you will eventually find that approximately 5 of each type of fish are present in each tank. Each fish has the same chance of blundering through the channel into the other tank. But as long as there are more goldfish in the left tank (Figure 4-1a), there is a greater probability that a goldfish will swim from left to right than the reverse. Similarly, as long as the number of guppies in the right tank exceeds that in the left, there will be a net flow of guppies to the left, even though there is nothing in the left tank to make the guppies prefer it. Thus the rate

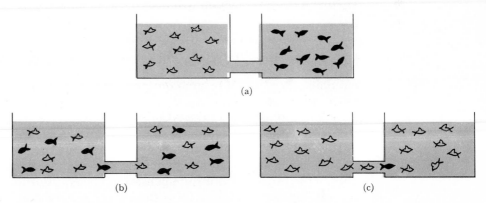

(a)

(b) (c)

Figure 4-1   Illustration of dynamic equilibrium: two fish tanks connected by a channel. (a) Start of experiment, with 10 goldfish in the left tank and 10 guppies in the right. (b) Equilibrium state, with 5 of each kind of fish in each tank. (c) If we were to observe one single fish (here a guppy among goldfish), we would find that it spends half its time in each tank. The equilibrium of the tank in (b) is a dynamic, averaged state and not a static condition. The fish do not stop swimming when they have become evenly mixed.

of flow of guppies is proportional to the concentration of guppies present. A similar statement can be made for the goldfish.

At equilibrium (Figure 4-1b), on an average there will be 5 guppies and 5 goldfish in each tank. But they will not always be the same 5 of each fish. If 1 guppy wanders from the left tank into the right, then it or a different guppy may wander back a little later. Thus at equilibrium we find that the fish have not stopped swimming, only that over a period of time the total number of guppies and goldfish in each tank remains constant. If we were to fill each tank with 9 goldfish and then throw in 1 guppy, we would see that, in its aimless swimming, it would spend half its time in one tank and half in the other (Figure 4-1c).

In the NO reaction we considered, there will be a constant concentration of NO molecules at equilibrium, but they will not always be the same NO molecules. Individual NO molecules will react to re-form N_2 and O_2, and other reactant molecules will make more NO. As with the goldfish, only on a head-count or concentration basis have changes ceased at equilibrium.

The equilibrium condition for the NO-producing reaction, equation 4-1, can be rewritten in a more useful form:

$$\frac{[NO]^2}{[N_2][O_2]} = \frac{k_1}{k_2} = K_{eq} \qquad (4\text{-}6)$$

in which the ratio of forward and reverse rate constants is expressed as a simple constant, the **equilibrium constant, K_{eq}**. This equilibrium constant will vary as the temperature varies, but it is independent of the concentrations of the reactants and products. It tells us the ratio of products to reactants at equilibrium, and is an extremely useful quantity for determining whether a desired reaction will take place spontaneously.

4-3   GENERAL FORM OF THE EQUILIBRIUM CONSTANT

We derived the equilibrium-constant expression for the NO reaction by assuming that we knew the way that the forward and reverse steps occurred at the molecular level. If the NO reaction proceeded by simple collision of two molecules, the derivation would be perfectly correct. The actual mechanism of this reaction is more complicated. But it is important, and fortunate for chemists, that we do not have to know the reaction mechanism to write the proper equilibrium constant. The equilibrium-constant expression can always be written from the balanced chemical equation, with no other information, even when the forward and reverse rate expressions are more complicated than the balanced equation would suggest. (We shall prove this in Chapter 16.) In our NO example, the forward reaction actually takes place by a series of complicated chain steps. The reverse reaction takes place by a complementary set of reactions, so that these complications cancel one another in the final ratio of concentrations that gives us the equilibrium constant. The details of the mechanism are "invisible" to the equilibrium-constant expression, and irrelevant to equilibrium calculations.

A general chemical reaction can be written as

$$aA + bB \rightleftarrows cC + dD \tag{4-7}$$

In this expression, A and B represent the reactants; C and D, the products. The letters a, b, c, and d represent the *number* of moles of each substance involved in the balanced reaction, and the double arrows indicate a state of equilibrium. Although only two reactants and two products are shown in the general reaction, the principle is extendable to any number. The correct equilibrium-constant expression for this reaction is

$$K_{eq} = \frac{[C]^c[D]^d}{[A]^a[B]^b} \tag{4-8}$$

It is the ratio of product concentrations to reactant concentrations, with each concentration term raised to a power given by the number of moles of that substance appearing in the balanced chemical equation. Because it is based on the quantities of reactants and products present at equilibrium, equation 4-8 is called the **law of mass action**.

Example 1

Give the equilibrium-constant expression for the reaction

$$CO + H_2O \rightleftarrows CO_2 + H_2$$

Solution

The equilibrium constant is given by

$$K_{eq} = \frac{[CO_2][H_2]}{[CO][H_2O]}$$

Since all four substances have a coefficient of 1 in the balanced equation, their concentrations are all raised to the first power in the equilibrium-constant expression.

Example 2

What is the equilibrium-constant expression for the formation of water from hydrogen and oxygen gases? The reaction is

$$2H_2 + O_2 \rightleftarrows 2H_2O$$

Solution

$$K_{eq} = \frac{[H_2O]^2}{[H_2]^2[O_2]}$$

Since two moles of hydrogen and water are involved in the chemical equation, their concentrations are squared in the K_{eq} expression.

Example 3

Give the equilibrium-constant expression for the dissociation (breaking up) of water into hydrogen and oxygen. The reaction is

$$2H_2O \rightleftarrows 2H_2 + O_2$$

Solution

$$K_{eq} = \frac{[H_2]^2[O_2]}{[H_2O]^2}$$

An important general point emerges here. This reaction is the reverse of that of Example 2, and the equilibrium-constant expression is the inverse, or reciprocal, of the earlier one. *If a balanced chemical reaction is reversed, then the equilibrium-constant expression must be inverted,* since what once were reactants now are products, and vice versa.

Example 4

The dissociation of water can just as properly be written as

$$H_2O \rightleftarrows H_2 + \tfrac{1}{2}O_2$$

What then is the equilibrium-constant expression?

Solution

$$K_{eq} = \frac{[H_2][O_2]^{1/2}}{[H_2O]}$$

Notice that when the reaction from Example 3 is divided by 2, resulting in the Example 4 reaction, the equilibrium constant is the square root of the old value, or the old K_{eq} to the one-half power. Similarly, if the reaction is doubled, the K_{eq} must be squared. In general, it is perfectly proper to multiply all the coefficients of a balanced chemical reaction by any positive or negative number, n, and the equation will remain balanced. (Multiplying all the coefficients of an equation by -1 is formally the same as writing the equation in reverse. Write out a simple equation and prove to yourself that this is so.) *But if all the coefficients of an equation are multiplied by n, then the new equilibrium-constant expression is the old one raised to the nth power.* Hence, when working with equilibrium constants, one must keep the corresponding chemical reactions clearly in mind.

Example 5

The reaction for the formation or the breakdown of ammonia can be written in a number of ways:

a) $N_2 + 3H_2 \rightleftarrows 2NH_3$
b) $\frac{1}{2}N_2 + \frac{3}{2}H_2 \rightleftarrows NH_3$
c) $\frac{1}{3}N_2 + H_2 \rightleftarrows \frac{2}{3}NH_3$
d) $NH_3 \rightleftarrows \frac{1}{2}N_2 + \frac{3}{2}H_2$

(Each of these expressions might be appropriate, depending on whether you were focusing on nitrogen, ammonia, hydrogen, or the dissociation of ammonia.) What are the equilibrium-constant expressions for each formulation, and how are the equilibrium constants related?

Solution

a) $K_a = \dfrac{[NH_3]^2}{[N_2][H_2]^3}$ c) $K_c = \dfrac{[NH_3]^{2/3}}{[N_2]^{1/3}[H_2]}$

b) $K_b = \dfrac{[NH_3]}{[N_2]^{1/2}[H_2]^{3/2}}$ d) $K_d = \dfrac{[N_2]^{1/2}[H_2]^{3/2}}{[NH_3]}$

$$K_a = K_b^2 = K_c^3 = K_d^{-2} = \frac{1}{K_d^2}$$

Notice that there is nothing wrong with fractional powers in the equilibrium-constant expression.

4-4 USING EQUILIBRIUM CONSTANTS

Equilibrium constants have two main purposes:

1. To help us tell whether a reaction will be spontaneous under specified conditions.
2. To enable us to calculate the concentration of reactants and products that will be present once equilibrium has been reached.

We can illustrate how equilibrium constants can be used to achieve these ends, and also the fact that an equilibrium constant is indeed *constant,* with real data from one of the most intensively studied of all reactions, that between hydrogen and iodine to yield hydrogen iodide:

$$H_2(g) + I_2(g) \rightleftarrows 2HI(g) \tag{4-9}$$

If we mix hydrogen and iodine in a sealed flask and observe the reaction, the gradual fading of the purple color of the iodine vapor tells us that iodine is being consumed. This reaction was studied first by the German chemist Max Bodenstein in 1893. Table 4-1 contains the data from Bodenstein's experiments. The experimental data are in the first three columns. In the fourth column, we have calculated the simple ratio of product and reactant concentrations, $[HI]/[H_2][I_2]$, to see if it is constant. It clearly is not, for as the hydrogen concentration is decreased and the iodine concentration is increased, this ratio varies from 2.60 to less than 1. The law of mass action (Section 4-3) dictates that the equilibrium-constant expression should contain the *square* of the HI concentration, since the reaction involves 2 moles of HI for every mole of H_2 and I_2. The fifth column shows that the ratio $[HI]^2/[H_2][I_2]$ is constant within a mean deviation of approximately 3%.* Therefore, this ratio is the proper equilibrium-constant expression, and the average value of K_{eq} for these six runs is 50.53.

The equilibrium constant can be used to determine whether a reaction under specified conditions will go spontaneously in the forward or in the reverse direction. The ratio of product concentration to reactant concentration, identical to the equilibrium constant in form but not necessarily at equilibrium conditions, is called the **reaction quotient, Q**:

$$Q = \frac{[HI]^2}{[H_2][I_2]} \qquad \text{(not necessarily at equilibrium)} \tag{4-10}$$

If there are too many reactant molecules present for equilibrium to exist, then the concentration terms in the denominator will make the reaction quotient, Q, smaller than K_{eq}. The reaction will go forward spontaneously to make more product. However, if an experiment is set up so that the reaction quotient is greater than K_{eq}, then too many product molecules are present for equilibrium and the reverse reaction will proceed spontaneously. Therefore, a comparison of the actual concentration ratio or reaction quotient with the equilibrium constant allows us to predict in which direction a reaction will go spontaneously under the given set of circumstances:

$$\begin{array}{ll} Q < K_{eq} & \text{(forward reaction spontaneous)} \\ Q > K_{eq} & \text{(reverse reaction spontaneous)} \\ Q = K_{eq} & \text{(reactants and products at equilibrium)} \end{array} \tag{4-11}$$

*These are Bodenstein's original numbers. Modern data can be much more accurate, with less deviation in K_{eq}. The mean deviation is the average of the deviations of individual calculated K_{eq} from the average K_{eq}.

Table 4-1

Experimental Measurements of Equilibrium Concentrations[a]

Experimental data			Calculations from data		
$[H_2]$	$[I_2]$	$[HI]$	$\dfrac{[HI]}{[H_2][I_2]}$	$\dfrac{[HI]^2}{[H_2][I_2]} = K_{eq}$	Deviation from average K_{eq}
18.14	0.41	19.38	2.60	50.50	−0.03
10.96	1.89	32.61	1.57	51.34	+0.81
4.57	8.69	46.28	1.16	53.93	+3.40
2.23	23.95	51.30	0.96	49.27	−1.26
0.86	67.90	53.40	0.91	48.83	−1.70
0.65	87.29	52.92	0.93	49.35	−1.18
				6)303.22	6)8.28
				50.53	1.38

Average $K_{eq} = 50.53$ $\dfrac{1.38}{50.53} \times 100 = 2.7\%$ mean deviation

[a]For the reaction $H_2(g) + I_2(g) \rightleftarrows 2HI(g)$, at 448°C in a sulfur vapor constant-temperature bath. Concentrations are in mole per liter $\times 10^{+3}$ (e.g., the first hydrogen concentration is 18.14×10^{-3} mole liter^{-1}).

Example 6

If 1.0×10^{-2} mole each of hydrogen and iodine gases are placed in a 1-liter flask at 448°C with 2.0×10^{-3} mole of HI, will more HI be produced?

Solution

The reaction quotient under these conditions is

$$Q = \frac{(2.0 \times 10^{-3})^2}{(1.0 \times 10^{-2})^2} = 0.040$$

This is smaller than the equilibrium value of 50.53 in Table 4-1, which tells us that excess reactants are present. Hence, equilibrium will not be reached until more HI has been formed.

Example 7

If only 1.0×10^{-3} mole each of H_2 and I_2 had been used, together with 2.0×10^{-3} mole of HI, would more HI have been produced spontaneously?

Solution   You can verify that the reaction quotient is $Q = 4.0$. Because this is less than K_{eq}, the forward reaction is still spontaneous.

Example 8

If the conditions of Example 7 are changed so that the HI concentration is increased to 2.0×10^{-2} mole liter^{-1}, what happens to the reaction?

Solution   The reaction quotient now is $Q = 400$. This is greater than K_{eq}. There are now too many product molecules and too few reactant molecules for equilibrium to exist. Thus the reverse reaction occurs more rapidly than the forward reaction. Equilibrium is reached only by converting some of the HI to H_2 and I_2, so the reverse reaction is spontaneous.

Example 9

If the conditions of Example 7 are changed so that the HI concentration is 7.1×10^{-3} mole liter^{-1}, in which direction is the reaction spontaneous?

Solution   Under these conditions,

$$Q = \frac{(7.1 \times 10^{-3})^2}{(1.0 \times 10^{-3})^2} = 50.4 = K_{eq}$$

Since Q equals K_{eq} within the limits of accuracy of the data, the system as described is at equilibrium, and neither the forward nor the backward reaction is spontaneous. (Both reactions are still taking place at the molecular level, of course, but they are balanced so their net effects cancel.)

The second use for equilibrium constants is to calculate the concentrations of reactants and products that will be present at equilibrium.

Example 10

If a 1-liter flask contains 1.0×10^{-3} mole each of H_2 and I_2 at 448°C, what amount of HI is present when the gas mixture is at equilibrium?

Solution   The K_{eq} expression is treated as an ordinary algebraic equation, and solved for the HI concentration:

$$\frac{[HI]^2}{(1.0 \times 10^{-3})^2} = K_{eq} = 50.53$$

$$[HI]^2 = 50.53 \times 1.0 \times 10^{-6}$$

$$[HI] = 7.1 \times 10^{-3} \text{ mole liter}^{-1}$$

You can verify that in Example 7 the HI concentration was less than this equilibrium value; in Example 8 it was more; and in Example 9 it was just this value.

Example 11

One-tenth of a mole, 0.10 mole, of hydrogen iodide is placed in an otherwise empty 5.0 liter flask at 448°C. When the contents have come to equilibrium, how much hydrogen and iodine will be in the flask?

Solution

From the stoichiometry of the reaction, the concentrations of H_2 and I_2 must be the same. For every mole of H_2 and I_2 formed, 2 moles of HI must decompose. Let y equal the number of moles of H_2 or I_2 *per liter* present at equilibrium. The initial concentration of HI before any dissociation has occurred is

$$[HI]_0 = \frac{0.10 \text{ mole}}{5.0 \text{ liters}} = 0.020 \text{ mole liter}^{-1}$$

Begin by writing a balanced equation for the reaction, then make a table of concentrations at the start and at equilibrium:

	H_2	$+ \, I_2$	$\rightleftarrows$ 2HI
Start (moles liter^{-1}):	0	0	0.020
Equilibrium:	y	y	$0.020 - 2y$

The HI concentration of 0.020 mole liter^{-1} has been decreased by $2y$ for every y moles of H_2 and I_2 that are formed. The equilibrium-constant expression is

$$50.53 = \frac{(0.020 - 2y)^2}{y^2}$$

We immediately see that we can take a shortcut by taking the square root of both sides:

$$7.11 = \frac{0.020 - 2y}{y}$$

$$9.11y = 0.020$$

$$y = 0.0022 \text{ mole liter}^{-1}$$

For 5 liters, $5 \times 0.0022 = 0.011$ mole of H_2 and of I_2 will be present at equilibrium. Only $(0.020 - 0.0044) \times 5 = 0.080$ mole of HI will be left in the 5-liter tank, and the fraction of HI dissociated at equilibrium is

$$\frac{2y}{[HI]_0} = \frac{0.0044}{0.020} = 0.22, \text{ or } 22\% \text{ dissociation}$$

Shortcuts such as taking the square root in the preceding example are not always possible, yet part of the skill of solving equilibrium problems lies in recognizing shortcuts when they occur and using them. The key is often a good intuition about what quantities are large and small relative to one another, and this intuition comes from thoughtful practice and understanding of the chemistry involved. You should remember that these are chemical problems, not mathematical ones.

In many cases a quadratic equation must be solved.

Example 12

If 0.00500 mole of hydrogen gas and 0.0100 mole of iodine gas are placed in a 5.00 liter tank at 448°C, how much HI will be present at equilibrium?

Solution

The initial concentrations of H_2 and I_2 are

$$[H_2]_0 = \frac{0.00500 \text{ mole}}{5.00 \text{ liters}} = 0.00100 \text{ mole liter}^{-1}$$

$$[I_2]_0 = \frac{0.0100 \text{ mole}}{5.00 \text{ liters}} = 0.00200 \text{ mole liter}^{-1}$$

This time, let the unknown variable y be the moles per liter of H_2 or I_2 that have reacted at equilibrium:

	H_2	$+$	I_2	$\rightleftarrows$	2HI
Start (moles liter^{-1}):	0.00100		0.00200		0.0
Equilibrium:	$0.00100 - y$		$0.00200 - y$		$2y$

The equilibrium expression is

$$50.53 = \frac{(2y)^2}{(0.00100 - y)(0.00200 - y)}$$

The square-root shortcut is now impossible because the starting concentrations of H_2 and I_2 are unequal. Instead we must reduce the equation to a quadratic expression:

$$46.53y^2 - 0.1516y + 1.011 \times 10^{-4} = 0$$

A general quadratic equation of the form $ay^2 + by + c = 0$ can be solved by the quadratic formula,

$$y = \frac{-b \pm \sqrt{b^2 - 4ac}}{2a}$$

Thus for this problem

$$y = \frac{0.1516 \pm \sqrt{0.02298 - 0.01881}}{93.06}$$

$y = 2.32 \times 10^{-3}$ and 0.935×10^{-3} mole liter^{-1}

The first solution is physically impossible since it shows more H_2 reacting than was originally present. The second solution is the correct answer: $y = 0.935 \times 10^{-3}$ mole liter^{-1}. Therefore, the equilibrium concentrations are

$$[H_2] = 0.00100 - 0.000935 = 0.065 \times 10^{-3} \text{ mole liter}^{-1}$$
$$[I_2] = 0.00200 - 0.000935 = 1.065 \times 10^{-3} \text{ mole liter}^{-1}$$
$$[HI] = 2(0.935 \times 10^{-3}) = 1.87 \times 10^{-3} \text{ mole liter}^{-1}$$

4-5 UNITS AND EQUILIBRIUM CONSTANTS

As we have seen, the square brackets around a chemical symbol, as in $[N_2]$, represent concentrations, usually but not exclusively in units of moles liter^{-1}. Concentrations expressed as moles liter^{-1} are often given the special symbol c, as in c_{N_2}, the concentration of N_2 in moles liter^{-1}. The equilibrium constant with concentrations measured in these units is denoted by K_c.

An equilibrium constant as we have defined it thus far may itself have units. In Example 1, K_{eq} is unitless since the moles^2 liter^{-2} of the numerator and denominator cancel. In Example 2, the units of K_{eq} are moles^{-1} liter since concentration occurs to the second power in the numerator and to the third power in the denominator. In Example 3 the units of K_{eq} are the inverse: moles liter^{-1}. The units demanded by Example 4, moles$^{1/2}$ liter$^{-1/2}$, may seem strange but they are perfectly respectable.

Example 13

What are the units for the equilibrium constants in the four reactions of Example 5?

Solution

Constant	Units
K_a	moles^{-2} liter^2
K_b	moles^{-1} liter
K_c	moles$^{-2/3}$ liter$^{2/3}$
K_d	moles liter^{-1}

The question of units for K_{eq} becomes important as soon as we realize that we can measure concentration in units other than moles liter^{-1}. The partial pressure in atmospheres is a convenient unit when dealing with gas mixtures, and the equilibrium constant then is identified by K_p. Since the numerical values of K_p and K_c in general will be different, one must be sure what the units are when using a numerical constant.

Example 14

One step in the commercial synthesis of sulfuric acid is the reaction of sulfur dioxide and oxygen to make sulfur trioxide:

$$2SO_2 + O_2 \rightleftarrows 2SO_3$$

At 1000 K, the equilibrium constant for this reaction is $K_p = 3.50 \text{ atm}^{-1}$. If the total pressure in the reaction chamber is 1.00 atm and the partial pressure of unused O_2 at equilibrium is 0.10 atm, what is the ratio of concentrations of product (SO_3) to reactant (SO_2)?

Solution

$$K_p = \frac{p_{SO_3}^2}{p_{SO_2}^2 p_{O_2}} \qquad (p_j = \text{partial pressure of } j)$$

$$\text{Ratio} = \frac{p_{SO_3}}{p_{SO_2}} = \sqrt{K_p \times p_{O_2}} = \sqrt{3.50 \times 0.10} = 0.59$$

The equilibrium mixture has 0.59 mole of SO_3 for every 1 mole of SO_2.

The ideal gas law permits us to convert between atmospheres and moles liter^{-1}, and between K_p and K_c:

$$PV = nRT \tag{3-8}$$

$$P = \frac{n}{V}RT = cRT \tag{4-12}$$

In the general chemical reaction written earlier,

$$aA + bB \rightleftarrows cC + dD \tag{4-7}$$

Δn (read "delta n"), the *increase* in number of moles of gas during the reaction, is

$$\Delta n = c + d - a - b \tag{4-13}$$

The equilibrium-constant expression in terms of partial pressures is

$$K_p = \frac{p_C^c p_D^d}{p_A^a p_B^b} \tag{4-14}$$

With the ideal gas law applied to each gas component, we can convert this expression to K_c:

$$K_p = \frac{(cRT)^c (cRT)^d}{(cRT)^a (cRT)^b} = \frac{c^c c^d}{c^a c^b} (RT)^{\Delta n} = K_c (RT)^{\Delta n} \qquad (4\text{-}15)$$

(Do not confuse the two uses of the symbol c in equation 4-15: one is for concentration in moles liter^{-1} and the other for the number of moles of substance C.)

Example 15

What is the numerical value of K_c for the reaction of Example 14?

Solution

Three moles of reactant gases are converted into only 2 moles of product, so $\Delta n = -1$. Hence at 1000 K,

$$K_p = 3.50 \text{ atm}^{-1} = K_c (RT)^{-1}$$
$$K_c = K_p \times RT$$
$$= 3.50 \text{ atm}^{-1} \times 0.08205 \text{ liter atm K}^{-1} \text{ mole}^{-1} \times 1000 \text{ K}$$
$$= 287 \text{ moles}^{-1} \text{ liter}$$

Although the numerical answers that result when different units are used may differ, the physical reality must be the same.

Example 16

What is the concentration of oxygen in Example 14, in moles liter^{-1}? Solve Example 14 again using K_c from Example 15.

Solution

$$K_c = \frac{c_{SO_3}^2}{c_{SO_2}^2 c_{O_2}} = 287 \text{ moles}^{-1} \text{ liter}$$

$$c_{O_2} = \frac{p_{O_2}}{RT} = \frac{0.10 \text{ atm}}{82.05 \text{ liter atm mole}^{-1}} = 0.00122 \text{ mole liter}^{-1}$$

$$\text{Ratio} = \frac{c_{SO_3}}{c_{SO_2}} = \sqrt{K_c \times c_{O_2}} = \sqrt{287 \times 0.00122} = 0.59$$

This is the same ratio of SO_3 to SO_2 as was obtained when atmospheres were used. The choice is one of convenience.

4-6 EQUILIBRIA INVOLVING GASES WITH LIQUIDS OR SOLIDS

All the examples considered so far have involved only one physical state, a gas, and are examples of **homogeneous equilibria**. Equilibria that involve two or more physical states (such as a gas with a liquid or a solid) are called **heterogeneous equilibria**. If one or more of the reactants or products are solids or liquids, how does this affect the form of the equilibrium constant?

The answer, in short, is that any pure solids or liquids that may be present at equilibrium have the same effect on the equilibrium no matter how much solid or liquid is present. The concentration of a pure solid or liquid can be considered constant, and for convenience all such constant terms are brought to the left side of the equation and incorporated into the equilibrium constant itself. As an example, limestone (calcium carbonate, $CaCO_3$), breaks down into quicklime (calcium oxide, CaO) and carbon dioxide, CO_2:

$$CaCO_3(s) \rightleftharpoons CaO(s) + CO_2(g)$$

The simple equilibrium-constant expression is

$$K'_{eq} = \frac{[CaO(s)][CO_2(g)]}{[CaCO_3(s)]}$$

As long as any solid limestone and quicklime are in contact with the gas, their effect on the equilibrium is unchanging. Hence the terms $[CaCO_3]$ and $[CaO]$ remain constant and can be merged with K'_{eq}:

$$K_{eq} = K'_{eq} \frac{[CaCO_3(s)]}{[CaO(s)]} = [CO_2(g)]$$

This form of the equilibrium-constant expression tells us that, *at a given temperature,* the concentration of carbon dioxide gas above limestone and calcium oxide is a fixed quantity. (This is true only as long as *both* solid forms are present.) Measuring concentration in units of atmospheres, we get

$$K_p = p_{CO_2}$$

with the experimental value 0.236 atm at 800°C.

We can see what this means experimentally by considering a cylinder to which $CaCO_3$ and CaO have been added. The cylinder has a movable piston, as shown in Figure 4-2. If the piston is fixed at one position, then $CaCO_3$ will decompose until the pressure of CO_2 above the solids is 0.236 atm (if the temperature is 800°C). If you try to decrease the pressure by raising the piston, then more $CaCO_3$ will decompose until the pressure again rises to 0.236 atm. Conversely, if you try to increase the pressure by lowering the piston, some of the CO_2 gas will react with CaO and become $CaCO_3$, decreasing the amount of CO_2 gas present until the pressure once more is 0.236 atm. The only way to increase p_{CO_2} is to raise the temperature, which increases the value of K_p itself to 1 atm at 894°C and to 1.04 atm at 900°C.

An even simpler example is the vaporization of a liquid such as water:

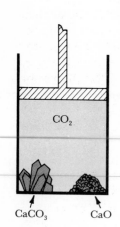

Figure 4-2

Solid $CaCO_3$ and CaO are placed in a cylinder with a movable piston, which is initially pressed against the solids to ensure that no extraneous gases are present. At equilibrium, enough $CaCO_3$ will have decomposed so that the pressure of CO_2 gas above the solid phases is a fixed value that varies with temperature but is independent of how much of each of the solids is present.

$$H_2O(l) \rightleftarrows H_2O(g)$$

This process can be treated as a chemical reaction in a formal sense even though bonds within molecules are not made or broken. Imagine that the cylinder shown in Figure 4-2 is half-filled with water rather than with $CaCO_3$ and CaO, and that the piston is initially brought down to the surface of the water. As the piston is raised, liquid will evaporate until the pressure of water vapor is a constant value that depends only on temperature. This is the **equilibrium vapor pressure** of water at that temperature. At 25°C, the vapor pressure of water is 0.0313 atm. At 100°C, the vapor pressure reaches 1 atm and, as we shall see in Chapter 18, this is just the definition of the normal boiling point of water. The pressure of water vapor above the liquid in the cylinder does not depend on whether the water in the cylinder is 1 cm or 10 cm deep; the only requirement is that some water be present and capable of evaporating to make up any decrease in vapor pressure. Only when the piston is raised to the point where no more liquid exists can the pressure of water vapor fall below 0.0313 atm, if the cylinder is at 25°C. Similarly, if the piston is lowered, some of the vapor condenses, keeping the pressure at 0.0313 atm. Only when all vapor has condensed and the piston is resting on the surface of the liquid can the pressure inside the cylinder be raised above 0.0313 atm.

The formal equilibrium treatment of the evaporation of water would be

$$K'_{eq} = \frac{[H_2O(g)]}{[H_2O(l)]}$$

$[H_2O(l)]$ = constant, as long as liquid is present

$$K_{eq} = K'_{eq}[H_2O(l)] = [H_2O(g)]$$

In pressure units, the expression would be

$$K_p = p_{H_2O(g)}$$

From a practical standpoint, what the preceding discussion means is that the concentration terms for pure solids and liquids are simply eliminated from the equilibrium-constant expression. (They are present, implicitly, in the K_{eq}.)

Example 17

If the hydrogen iodide reaction previously discussed in this chapter is carried out at room temperature, then iodine is present as deep purple crystals rather than as vapor. What then is the form of the equilibrium-constant expression, and does the equilibrium depend on the amount of iodine crystals present?

Solution   The reaction is

$$H_2(g) + I_2(s) \rightleftarrows 2HI(g)$$

and the equilibrium-constant expression is:

$$K_{eq} = \frac{[HI]^2}{[H_2]}$$

As long as some $I_2(s)$ crystals are present, the quantity is immaterial as far as equilibrium is concerned.

Example 18

Tin(IV) oxide reacts with carbon monoxide to form metallic tin and CO_2, by the reaction

$$SnO_2(s) + 2CO(g) \rightleftarrows Sn(s) + 2CO_2(g)$$

What is the equilibrium-constant expression?

Solution $$K_{eq} = \frac{[CO_2]^2}{[CO]^2}$$

Example 19

What is the equilibrium-constant expression for the following reaction leading to liquid water?

$$CO_2(g) + H_2(g) \rightleftarrows CO(g) + H_2O(l)$$

What would the expression be if the product were water vapor?

Solution   If the product is $H_2O(l)$, the equilibrium-constant expression is

$$K_{eq} = \frac{[CO]}{[CO_2][H_2]}$$

If the product is $H_2O(g)$, the equilibrium-constant expression is

$$K_{eq} = \frac{[CO][H_2O]}{[CO_2][H_2]}$$

The preceding example shows that as long as liquid water is present the gas-phase concentration is fixed at the vapor pressure of water at that temperature. Hence the water contribution, being constant, can be lumped into K_{eq}.

4-7 FACTORS AFFECTING EQUILIBRIUM: LE CHATELIER'S PRINCIPLE

Equilibrium represents a balance between two opposing reactions. How sensitive is this balance to changes in the conditions of a reaction? What can be done to change the equilibrium state? These are very practical questions if, for example, one is trying to increase the yield of a useful product in a reaction.

Under specified conditions, the equilibrium-constant expression tells us the ratio of product to reactants when the forward and backward reactions are in balance. This equilibrium constant is not affected by changes in concentration of reactants or products. However, if products can be withdrawn continuously, then the reacting system can be kept constantly off-balance, or short of equilibrium. More reactants will be used and a continuous stream of new products will be formed. This method is useful when one product of the reaction can escape as a gas, be condensed or frozen out of a gas phase as a liquid or solid, be washed out of the gas mixture by a spray of a liquid in which it is especially soluble, or be precipitated from a gas or solution.

For example, when solid lime (CaO) and coke (C) are heated in an electric furnace to make calcium carbide (CaC_2),

$$CaO(s) + 3C(s) \rightleftharpoons CaC_2(s) + CO(g)\uparrow$$

the reaction, which at 2000–3000°C has an equilibrium constant of close to 1.00, is tipped toward calcium carbide formation by the continuous removal of carbon monoxide gas. In the industrial manufacture of titanium dioxide for pigments, $TiCl_4$ and O_2 react as gases:

$$TiCl_4(g) + O_2(g) \rightleftharpoons TiO_2(s)\downarrow + 2Cl_2(g)$$

The product separates from the reacting gases as a fine powder of solid TiO_2, and the reaction is thus kept moving in the forward direction. When ethyl acetate or other esters used as solvents and flavorings are synthesized from carboxylic acids and alcohols,

$$CH_3COOH + HOCH_2CH_3 \rightleftharpoons CH_3COOCH_2CH_3 + H_2O$$

$$\text{acetic acid} \qquad \text{ethanol} \qquad\qquad \text{ethyl acetate}$$

the reaction is kept constantly off-balance by removing the water as fast as it is formed. This can be done by using a drying agent such as Drierite ($CaSO_4$), by running the reaction in benzene and boiling off a constant-boiling benzene–water mixture, or by running the reaction in a solvent in which the water is completely immiscible and separates as droplets in a second phase. A final example: Since ammonia is far more soluble in water than either hydrogen or nitrogen is, the yield of ammonia in the reaction

$$N_2(g) + 3H_2(g) \rightleftharpoons 2NH_3(g)$$

can be raised to well over 90% by washing the ammonia out of the equilibrium mixture of gases with a stream of water, and recycling the nitrogen and hydrogen.

Temperature

All the preceding methods will upset an equilibrium (in our examples, in favor of desired products) without altering the equilibrium constant. A chemist can often enhance yields of desired products by increasing the equilibrium constant so that the ratio of products to reactants at equilibrium is larger. The equilibrium constant is usually temperature dependent. In general, both forward and reverse reactions are speeded up by increasing the temperature, because the molecules move faster and collide more often. If the increase in the rate of the forward reaction is greater than that of the reverse, then K_{eq} increases with temperature and more products are formed at equilibrium. If the reverse reaction is favored, then K_{eq} decreases. Thus K_{eq} for the hydrogen–iodine reaction at 448°C is 50.53, but at 425°C it is 54.4, and at 357°C it increases to 66.9. Production of HI is favored to some extent by an increase in temperature, but its dissociation to hydrogen and iodine is favored much more.

The hydrogen iodide–producing reaction is exothermic or heat emitting:

$$H_2(g) + I_2(g) \rightleftarrows 2HI(g)$$
$$\Delta H_{298} = -10.2 \text{ kJ per 2 moles of HI}$$

(If you check this figure against Appendix 3, remember that this reaction involves gaseous iodine, not solid.) If the external temperature of this reaction is lowered, the equilibrium is shifted in favor of the heat-emitting or forward reaction; conversely, if the temperature is increased, the reverse reaction, producing H_2 and I_2, is favored. The equilibrium shifts so as to counteract to some extent the effect of adding heat externally (raising the temperature) or removing it (lowering the temperature).

The temperature dependence of the equilibrium point is one example of a more general principle, known as *Le Chatelier's principle: If an external stress is applied to a system at chemical equilibrium, then the equilibrium point will change in such a way as to counteract the effects of that stress.* If the forward half of an equilibrium reaction is exothermic, then K_{eq} will decrease as the temperature increases; if it is endothermic, K_{eq} will increase. Only for a heat-absorbing reaction can the equilibrium yield of products be improved by increasing the temperature. A good way to remember this is to write the reaction explicitly with a heat term:

$$H_2(g) + I_2(g) \rightleftarrows 2HI(g) + \text{heat (given off)}$$

Then it is clear that adding heat, just like adding HI, shifts the reaction to the left. (See Figure 4-3.)

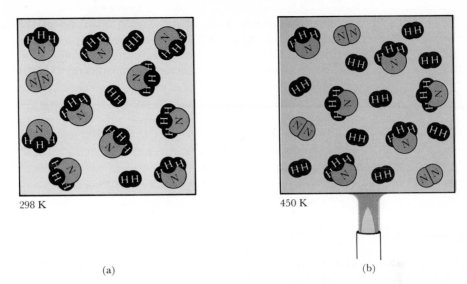

(a)

(b)

Figure 4-3   Le Chatelier's principle and temperature. The dissociation of ammonia,

$$2NH_3(g) \rightarrow 3H_2(g) + N_2(g)$$

is endothermic or heat-absorbing. (a) Ammonia equilibrium at room temperature. (b) The temperature increase produced by adding heat is partially counteracted by using some of the heat to dissociate NH_3 molecules and form N_2 and H_2. From Dickerson and Geis, *Chemistry, Matter, and the Universe.*

Pressure

Le Chatelier's principle is true for other kinds of stress, such as pressure changes. The equilibrium constant, K_{eq}, is not altered by a pressure change at constant temperature. However, the relative amounts of reactants and products will change in a way that can be predicted from Le Chatelier's principle.

The hydrogen–iodine reaction involves an equal number (2) of moles of reactants and product. Therefore, if we double the pressure at constant temperature, the volume of the mixture of gases will be halved. All concentrations in moles liter^{-1} will be doubled, but their *ratio* will be the same. In Example 12, doubling the concentrations of the reactants and product does not change the equilibrium constant:

$$K_{eq} = \frac{(1.87 \times 10^{-3} \text{ mole liter}^{-1})^2}{(0.065 \times 10^{-3} \text{ mole liter}^{-1})(1.065 \times 10^{-3} \text{ mole liter}^{-1})}$$

$$= \frac{(3.74 \times 10^{-3})^2}{(0.13 \times 10^{-3})(2.13 \times 10^{-3})} = 50.51$$

Thus the hydrogen–iodine equilibrium is not sensitive to pressure changes. Notice that in this case K_{eq} does not have units, since the concentration units in the numerator and denominator cancel.

In contrast, the dissociation of ammonia is affected by changes in pressure because the number of moles (2) of reactant does not equal the total number of moles (4) of products:

$$2NH_3(g) \rightleftharpoons N_2(g) + 3H_2(g)$$

The equilibrium constant for this reaction at 25°C is

$$K_{eq} = \frac{[N_2][H_2]^3}{[NH_3]^2} = 2.5 \times 10^{-9} \text{ mole}^2 \text{ liter}^{-2}$$

One set of equilibrium conditions is

$$N_2 = 3.28 \times 10^{-3} \text{ mole liter}^{-1}$$
$$H_2 = 2.05 \times 10^{-3} \text{ mole liter}^{-1}$$
$$NH_3 = 0.106 \text{ mole liter}^{-1}$$

(Can you verify that these concentrations satisfy the equilibrium condition?) If we now double the pressure at constant temperature, thereby halving the volume and doubling each concentration,

$$N_2 = 6.56 \times 10^{-3} \text{ mole liter}^{-1}$$
$$H_2 = 4.10 \times 10^{-3} \text{ mole liter}^{-1}$$
$$NH_3 = 0.212 \text{ mole liter}^{-1}$$

the ratio of products to reactants, the reaction quotient, is no longer equal to K_{eq}:

$$Q = \frac{(6.56 \times 10^{-3})(4.10 \times 10^{-3})^3}{(0.212)^2} = 1.0 \times 10^{-8} \text{ mole}^2 \text{ liter}^{-2}$$

Since Q is greater than K_{eq}, too many product molecules are present for equilibrium. The reverse reaction will run spontaneously, thereby forming more NH_3 and decreasing the amounts of H_2 and N_2. Consequently, part of the increased pressure is offset when the reaction shifts in the direction that lowers the total number of moles of gas present. In general, a reaction that reduces the number of moles of gas will be favored by an increase in pressure, and one that produces more gas will be disfavored. (See Figure 4-4.)

Example 20

If the hydrogen–iodine reaction were run at a temperature at which the iodine was a solid, would an increase in pressure shift the equilibrium reaction toward more HI, or less? What would be the effect of pressure on K_{eq}?

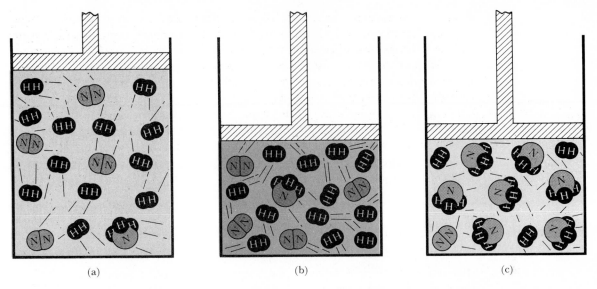

(a)   (b)   (c)

Figure 4-4   Le Chatelier's principle and pressure. (a) At initial equilibrium 17 molecules (moles) of gas are present: 12 of H_2, 4 of N_2, and 1 of NH_3. (b) When the gas is compressed into a smaller volume a stress is created, which is evidenced by a higher pressure. (c) This stress can be relieved and the pressure reduced if some of the molecules of H_2 and N_2 combine to form more NH_3, since the total number of gas molecules is thereby reduced. From Dickerson and Geis, *Chemistry, Matter, and the Universe.*

Solution   Since the reaction of 2 moles of gaseous HI now yields 1 mole of gaseous H_2 and 1 mole of solid I_2, the stress of increased pressure is relieved by dissociating HI to H_2 and I_2. However, K_{eq} will be unchanged by the pressure increase.

Catalysis

What effect does a catalyst have on a reaction *at equilibrium?* None. A catalyst cannot change the value of K_{eq}, but it can increase the speed with which equilibrium is reached. This is the main function of a catalyst. It can take the reaction *only* to the same equilibrium state that would be reached eventually without the catalyst.

Catalysts are useful, nevertheless. Many desirable reactions, although spontaneous, occur at extremely slow rates under ordinary conditions. In automobile engines, the main smog-producing reaction involving oxides of nitrogen is

$$N_2 + O_2 \rightleftarrows 2NO$$

(Once NO is present, it reacts readily with more oxygen to make brown NO_2.) At the high temperature of an automobile engine, K_{eq} for this reaction is so large that appreciable amounts of NO are formed. However, at 25°C, $K_{eq} = 10^{-30}$. (Using only the previous two bits of information and Le Chatelier's principle, predict whether the reaction as written is endothermic or exothermic. Check your answer using data from Appendix 3.) The amount of NO present in the atmosphere at equilibrium at 25°C should be negligible. NO should decompose spontaneously to N_2 and O_2 as the exhaust gases cool. But any Southern Californian can verify that this is not what happens. Both NO and NO_2 are indeed present, because the gases of the atmosphere are not at equilibrium.

The rate of decomposition of NO is extremely slow, although the reaction is spontaneous. One approach to the smog problem has been to search for a catalyst for the reaction

$$2NO \rightleftarrows N_2 + O_2$$

that could be housed in an exhaust system and could break down NO in the exhaust gases as they cool. Finding a catalyst is possible; a practical problem arises from the gradual poisoning of the catalyst by gasoline additives, such as lead compounds. This is the reason why new cars with catalytic converters only use lead-free gasoline.

A proof of the assertion that a catalyst cannot change the equilibrium constant is illustrated in Figure 4-5. If a catalyst *could* shift the equilibrium point of a reacting gas mixture and produce a volume change, then this expansion and contraction could be harnessed by mechanical means and made to do work. We would have a true perpetual-motion machine that would deliver power without an energy source. From common sense and experience we know this to be impossible. This "common sense" is stated scientifically as the first law of thermodynamics, which will be discussed in Chapter 15. A mathematician would call this a *proof by contradiction:* If we assume that a catalyst can alter K_{eq}, then we must assume the existence of a perpetual-motion machine. However, a perpetual-motion machine cannot exist; therefore our initial assumption was wrong, and we must conclude that a catalyst cannot alter K_{eq}.

In summary, K_{eq} is a function of temperature, but it is not a function of reactant or product concentrations, total pressure, or the presence or absence of catalysts. The relative amounts of substances at equilibrium can be changed by applying an external stress to the equilibrium mixture of reactants and products, and the change is one that will relieve this stress. This last statement, Le Chatelier's principle, enables us to predict what will happen to a reaction when external factors are changed, without having to make exact calculations.

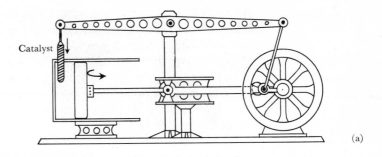

(a)

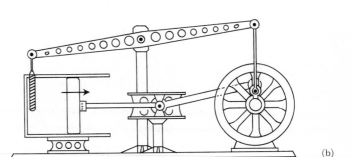

(b)

Figure 4-5

The ammonia perpetual-motion engine. A mixture of NH_3, H_2, and N_2 is contained in a chamber by the piston at the left, and the hatched cylinder suspended from the left end of the rocker arm contains a mythical catalyst that would shift the equilibrium point of the reaction

$$2NH_3(g) \rightleftarrows N_2(g) + 3H_2(g)$$

to the right. In (a) and (b), as the catalyst is introduced, ammonia dissociates to nitrogen and hydrogen. The total volume of gas increases and the piston is pushed to the right. In (c) and (d), as the catalyst is withdrawn, N_2 and H_2 reassociate to form ammonia; hence the volume shrinks and the piston is driven to the left. This self-contained, two-stage process provides an unlimited supply of power at the flywheel on the right, without an external input of energy. For practical difficulties, see the text.

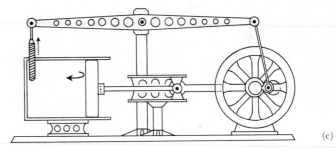

(c)

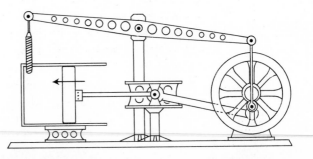

(d)

Summary

A **spontaneous reaction** is one that will take place, given enough time, without outside assistance. Some spontaneous reactions are rapid, but time is not an element in the definition of spontaneity. A reaction can be almost infinitely slow and still be spontaneous.

The net reaction that we observe is the result of competition between forward and reverse steps. If the forward process is faster, then products accumulate, and we say that the reaction is spontaneous in the forward direction. If the reverse process is faster, then reactants accumulate, and we say that the reverse reaction is the spontaneous one. If both forward and reverse processes take place at the same rate, then no net change is observed in any of the reaction components. This is the condition of **chemical equilibrium**.

The ratio of products to reactants, each concentration term being raised to a power corresponding to the coefficient of that substance in the balanced chemical equation, is called the **equilibrium constant, K_{eq}.** (See equation 4-8.) It can be used to predict whether a given reaction under specified conditions will be spontaneous, and to calculate the concentrations of reactants and products at equilibrium. The **reaction quotient, Q,** has a form that is identical with that of the equilibrium constant, K_{eq}, but Q applies under nonequilibrium conditions as well. For a given set of conditions, if Q is smaller than K_{eq}, the forward reaction is spontaneous; if Q is greater than K_{eq}, the reverse reaction is spontaneous; and if $Q = K_{eq}$, the system is at equilibrium.

The equilibrium constant can be used with any convenient set of concentration units: moles liter^{-1}, pressure in atmospheres, or others. Its numerical value will depend on the units of concentration, so one must be careful to match the proper values of K_{eq} and units when solving problems. If gas concentrations are expressed in moles liter^{-1}, the equilibrium constant is designated by K_c; if in atmospheres, by K_p. Just as partial pressure of the jth component of a gas mixture is related to moles per liter by $p_j = c_j RT$, so K_p and K_c are related by $K_p = K_c(RT)^{\Delta n}$, in which Δn is the net change in number of moles of gas during the reaction.

When some of the reactants or products are pure solids or liquids, they act as infinite reservoirs of material as long as some solid or liquid is left. Their effect on equilibrium depends only on their presence, not on how much of the solid or liquid is present. Their effective concentrations are constant, and can be incorporated into K_{eq}. In practice, this simply means omitting concentration terms for pure solids and liquids from the equilibrium-constant expression. Evaporation of a liquid can be treated formally as a chemical reaction with the liquid as reactant and vapor as product. These conventions for writing concentration terms for a liquid permit us to write the equilibrium constant for evaporation as $K_p = p_j$, where p_j is the equilibrium vapor pressure of substance j.

Le Chatelier's principle states that if stress is applied to a system at equilibrium the amounts of reactants and products will shift in such a manner as to minimize the stress. This means that for a heat-absorbing, or endothermic, reaction, K_{eq} increases as the temperature is increased, since carrying out more of the reaction is a way of absorbing some of the added heat. Similarly, cooling increases K_{eq} for a heat-emitting or exothermic reaction. Although the equilibrium constant K_{eq} is independent of pressure, and changing the total pressure on a reacting system does not alter K_{eq} directly, an increase in pressure does cause the reaction to shift in the direction that decreases the total number of moles of gas present.

A catalyst has no effect at all on K_{eq} or the conditions of equilibrium. All that a catalyst can do is to make the system reach equilibrium faster than it would have done otherwise. Catalysts can make inherently spontaneous but slow reactions into rapid reactions, but they cannot make non-spontaneous reactions take place of their own accord.

Self-Study Questions

1. What is a spontaneous reaction? Must a spontaneous reaction be rapid? Illustrate with an example other than those given in this chapter.
2. How does a catalyst affect a spontaneous reaction? What does it do to the equilibrium point in a reaction?
3. What is meant by the rate constant for a chemical reaction? What is an equilibrium constant? How does the equilibrium constant depend on the concentrations of reactants and products? How does it depend on the relative numbers of molecules that are involved in a chemical reaction?
4. How, in principle, is the equilibrium constant for a reaction related to the forward and reverse rate constants? Can the rate-constant expressions for forward and reverse reactions be written from a knowledge of only the balanced total chemical reaction? Why, or why not? Can the equilibrium-constant expression be written from a knowledge of only the balanced total chemical reaction? Why, or why not?
5. How is the equilibrium-constant expression changed if all quantities in a balanced chemical reaction are doubled? How is it changed if they are halved? What is the effect on the equilibrium constant of writing the reaction in reverse?
6. What is the reaction quotient, and how is it related to the equilibrium constant? How can comparisons of the two quantities help us to decide whether a reaction under given conditions is spontaneous, spontaneous in reverse, or at equilibrium?
7. How does the numerical value of an equilibrium constant depend on the choice of units for expressing concentrations? For gases, how does one convert from partial pressures to moles liter^{-1}? How are K_p and K_c related?
8. For what kinds of reactions is the equilibrium constant a unitless number? How then are K_p and K_c related?

9. Why is it legitimate to omit concentration terms from the equilibrium-constant expression for those components that are pure solids or liquids?
10. How can the equilibrium-constant concept be used to explain vapor pressure?
11. What is Le Chatelier's principle? How does it help to predict the outcome of changes of temperature on a chemical reaction?
12. How can Le Chatelier's principle explain the effects of pressure changes on chemical equilibrium?
13. How does a catalyst change K_{eq}? Justify your answer by reference to the existence or nonexistence of perpetual-motion machines. What makes a catalyst useful?

Problems

Equilibrium expressions

1. Write the correct equilibrium-constant expressions for the following reactions:
   a) $2H_2S(g) \rightleftarrows 2H_2(g) + S_2(g)$
   b) $2H_2S(g) \rightleftarrows 2H_2(g) + 2S(s)$
   c) $PCl_3(g) + Cl_2(g) \rightleftarrows PCl_5(g)$
   d) $Na_2CO_3(s) \rightleftarrows Na_2O(s) + CO_2(g)$
   e) $2NO_2(g) \rightleftarrows 2NO(g) + O_2(g)$

2. Write the correct equilibrium-constant expressions for the following reactions:
   a) $4NH_3(g) + 7O_2(g) \rightleftarrows 4NO_2(g) + 6H_2O(g)$
   b) $2NO_2(g) + 7H_2(g) \rightleftarrows 2NH_3(g) + 4H_2O(l)$
   c) $NH_4Cl(s) \rightleftarrows NH_3(g) + HCl(g)$
   d) $H_2(g) + (CN)_2(g) \rightleftarrows 2HCN(g)$
   e) $2ZnS(s) + 3O_2(g) \rightleftarrows 2ZnO(s) + 2SO_2(g)$

3. Write the correct equilibrium-constant expressions for the following reactions:
   a) $N_2O_4(g) \rightleftarrows 2NO_2(g)$
   b) $N_2O_4(g) \rightleftarrows N_2(g) + 2O_2(g)$
   c) $2NO_2(g) \rightleftarrows N_2O_4(g)$
   d) $NO_2(g) \rightleftarrows \frac{1}{2}N_2O_4(g)$
   e) $N_2O_5(s) \rightleftarrows 2NO_2(g) + \frac{1}{2}O_2(g)$

4. Write the correct equilibrium-constant expressions for the following reactions:
   a) $Cl_2(g) + H_2O(g) \rightleftarrows 2HCl(g) + \frac{1}{2}O_2(g)$

   b) $HCl(g) + \frac{1}{4}O_2(g) \rightleftarrows \frac{1}{2}Cl_2(g) + \frac{1}{2}H_2O(g)$
   c) $4HCl(g) + O_2(g) \rightleftarrows 2Cl_2(g) + 2H_2O(l)$
   d) $C(s) + CO_2(g) \rightleftarrows 2CO(g)$
   e) $NH_4SH(s) \rightleftarrows NH_3(g) + S(s) + H_2(g)$

K_p and K_c

5. At 1476 K, the equilibrium constant for the reaction

$$CO(g) + \frac{1}{2}O_2(g) \rightleftarrows CO_2(g)$$

is given by

$$K_p = 2.5 \times 10^5$$

a) What are the proper units (in atmospheres) for K_p as just given?
b) What is the numerical value of K_p for the reaction

$$2CO_2(g) \rightleftarrows 2CO(g) + O_2(g)$$

What are the units of this K_p?
c) What is the numerical value of K_c for the foregoing reaction, and what are its units?
d) In an equilibrium mixture of the three gases, CO_2 and CO are present in equimolar quantities. What is the concentration of O_2, in both atmospheres and moles liter^{-1}?

6. At 1476 K, the equilibrium constant for the reaction

$$C(s) + CO_2(g) \rightleftarrows 2CO(g)$$

is given by

$$K_p = 1.67 \times 10^3$$

a) What are the proper units (in atmospheres) for K_p as just given?
b) What is the numerical value of K_c for this reaction, and what are its units?
c) What is the form of the equilibrium-constant expression for the reaction

$$C(s) + O_2(g) \rightleftarrows CO_2(g)$$

d) What is the numerical value of K_p for this last reaction? (Missing but essential information is given in Problem 5.)
e) What general principle does part d suggest, regarding the equilibrium-constant expression for a sum of two reactions?

Reaction quotient

7. The equilibrium constant for the reaction

$$N_2(g) + O_2(g) \rightleftarrows 2NO(g)$$

at 2130°C is 2.5×10^{-3}. Why is it unnecessary to be told whether the value is for K_p or K_c? The reaction absorbs heat. For the following conditions, determine whether the reaction is spontaneous forward or in reverse, or whether it is at equilibrium:
a) A 1-liter box contains 0.02 mole of NO, 0.01 mole of O_2, and 0.02 mole of N_2 at 2130°C.
b) A 20-liter box contains 0.01 mole of N_2, 0.001 mole of O_2, and 0.02 mole of NO at 2130°C.
c) A 1-liter box contains 1.00 mole of N_2, 16 moles of O_2, and 0.2 mole of NO at 2500°C.

8. Gaseous nitrosyl bromide, NOBr, is formed by the reaction

$$2NO(g) + Br_2(g) \rightleftarrows 2NOBr(g)$$

At 25°C, the equilibrium constant for this reaction is $K_p = 116$ atm^{-1}. The reaction as written is exothermic, or heat emitting. When the following amounts of the three gases are introduced into a 1-liter flask, is the reaction spontaneous in the forward or reverse direction, or is it at equilibrium?
a) 0.045 atm NOBr, 0.01 atm NO, and 0.10 atm Br_2 at 25°C
b) 0.045 atm NOBr, 0.10 atm NO, and 0.01 atm Br_2 at 25°C
c) 0.108 atm NOBr, 0.10 atm NO, and 0.01 atm Br_2 at 0°C

Calculation of K_{eq}

9. When the sulfur trioxide reaction was run at 1000 K, a sample of the equilibrium gas mixture showed 0.562 atm of SO_2, 0.101 atm of O_2, and 0.332 atm of SO_3. Calculate the equilibrium constant, K_p, for the formation of 1 mole of SO_3 from SO_2 and O_2. Calculate K_p for the dissociation of 2 moles of SO_3 into 1 mole of O_2 and 2 moles of SO_2. How are these two equilibrium constants related?

10. If pure CO_2 is placed in a sealed tank and the temperature is raised to 2000 K, the CO_2 will be 1.6% dissociated into CO and O_2; that is, of every 1000 molecules of CO_2 at the beginning, 16 will decompose to CO and 984 will remain as CO_2. Assuming that all O_2 present comes from such decomposition, and that the total pressure is 1 atm, calculate K_p and K_c for the reaction

$$CO_2(g) \rightleftarrows CO(g) + \tfrac{1}{2}O_2(g)$$

11. 3.00 moles of NO, 5.00 moles of ClNO,

and 2.00 moles of Cl_2 are placed in a 25-liter tank at a temperature of 503 K. After the reaction has come to equilibrium, there are 6.12 moles of ClNO in the tank.

a) Write a balanced equation for the reaction producing 1 mole of ClNO from the other two components.

b) Calculate K_p and K_c for this reaction at 503 K, with proper units for each.

12. If carbon dioxide is passed over a bed of graphite at 1050°C, the product-gas stream (assumed to be at equilibrium) is 0.74 mole percent CO_2 and 99.26 mole percent CO.

a) Write a balanced equation for the production of 2 moles of CO from carbon and CO_2.

b) If the total pressure is 2.00 atm, calculate K_p and K_c for the reaction.

Equilibrium concentrations

13. Experiments have shown that at 60°C and 1 atm total pressure, the equilibrium ratio of NO_2 to N_2O_4 in moles in a closed vessel is exactly 2:1.

a) Calculate the equilibrium constant, K_c, for the dissociation of 1 mole of N_2O_4 into 2 moles of NO_2.

b) By measuring the intensity of the brown color of NO_2, an experimenter determines that 0.30 mole of NO_2 is present in a 1-liter flask in which equilibrium has been established with N_2O_4 at 60°C. Calculate the number of moles of N_2O_4 present.

14. The equilibrium constant for the reaction

$$2HCl(g) \rightleftarrows H_2(g) + Cl_2(g)$$

at 25°C is 6.2×10^{-54}.

a) Without any calculations, what does the numerical value of this constant tell you about the dissociation of HCl?

b) Calculate the concentration in moles liter^{-1} of hydrogen gas in equilibrium with 0.010 mole of HCl in a 1-liter container, assuming that the container had originally been filled with pure HCl.

c) If 0.0050 mole of Cl_2 and 0.0050 mole of H_2 are placed together in a 1-liter tank and allowed to come to equilibrium at 25°C, what will be the final concentration of Cl_2 gas, in moles liter^{-1}? (If you have trouble, compare this part with part b.)

15. The equilibrium constant for the reaction

$$CO(g) + H_2O(g) \rightleftarrows H_2(g) + CO_2(g)$$

is $K_{eq} = 5.10$ at 800 K. What are the units of K_{eq}, and how does its numerical value compare with K_p and K_c? Explain the results of the comparison. If a tank is charged with 1 atm of CO and 10.00 atm of H_2O, and these gases are allowed to come to equilibrium, what will be the partial pressures of H_2 and H_2O?

16. The reaction

$$SO_2(g) + \tfrac{1}{2}O_2(g) \rightleftarrows SO_3(g)$$

has an equilibrium constant of 33.4 atm$^{-1/2}$ at 530°C. Calculate the total pressure in a reaction tank that has been charged with 10.00 atm each of SO_2 and O_2, which are allowed to come to equilibrium. (This problem is simple in principle, but requires skillful use of the method of successive approximations to avoid having to solve a cubic equation.)

Heterogeneous equilibria

17. The decomposition of calcium carbonate,

$$CaCO_3(s) \rightleftarrows CaO(s) + CO_2(g)$$

has at 1013 K an equilibrium constant of

$K_c = 0.0060$. If concentration is measured in moles liter^{-1}, what are the units of this K_c? Suppose that 0.100 mole of calcium carbonate is placed in a 10-liter tank at 1013 K; calculate the equilibrium values of

a) The concentration of CO_2 in moles liter^{-1}

b) The number of moles of CO_2 present in the tank

c) The fraction of the initial calcium carbonate that has been converted to calcium oxide

d) The pressure of carbon dioxide in the tank

18. At 613 K, $K_c = 0.064$ for the reaction

$$Fe_2O_3(s) + 3H_2(g) \rightleftharpoons 2Fe(s) + 3H_2O(g)$$

a) What are the units of K_c?

b) If the reaction is carried out in such a way that the partial pressure of hydrogen at equilibrium is 1.00 atm, what is the concentration of hydrogen in moles liter^{-1}?

c) What is the concentration of water vapor at equilibrium, in moles liter^{-1} and in atmospheres?

Le Chatelier's principle

19. At room temperature the equilibrium constant, K_p, for dissociation of N_2O_4 into NO_2 is 5.83×10^{-3}, and at 10°C above room temperature it is 1.26×10^{-2}. Is the dissociation of N_2O_4 exothermic, or endothermic? How did you decide?

20. The heat of the reaction

$$2NO(g) \rightleftharpoons N_2(g) + O_2(g)$$

is $\Delta H_{298} = 180$ kJ.

a) Is nitric oxide, NO, more stable at high or low temperature?

b) In fact, nitric oxide decomposes more rapidly at high temperatures. Why is this not inconsistent with your answer to part a?

21. The following reaction is endothermic, or heat absorbing:

$$PCl_3(g) + Cl_2(g) \rightleftharpoons PCl_5(g)$$

What will be the effect on the equilibrium of each of the following disturbances?

a) Increase in total pressure

b) Addition of equal molar quantities of PCl_3 and PCl_5 at constant volume

c) Addition of chlorine gas at constant volume

d) Doubling the volume

e) An increase in temperature

22. At 25°C and 20 atm, the reaction

$$N_2(g) + 3H_2(g) \rightleftharpoons 2NH_3(g)$$

has a ΔH of -92.5 kJ. What will be the effect on equilibrium of each of the following changes?

a) Increasing the temperature to 300°C while holding the pressure at 20 atm

b) Increasing pressure to 30 atm while keeping the temperature at 25°C

c) Removing half the ammonia and allowing the system to come to equilibrium again

d) Adding a Cr_2O_3 catalyst for ammonia synthesis

Suggested Reading

A. J. Bard, *Chemical Equilibrium*, Harper & Row, New York, 1966.

M. J. Sienko, *Chemistry Problems*, W. A. Benjamin, Menlo Park, Calif., 1972, 2nd ed.

5

Solution Equilibria: Acids and Bases

*The words "acid" and "base" are
functional terms, and not labels.
They describe what a substance does,
rather than what it* is.

R. von Handler (b. 1931)

Almost all the reactions that a chemist is concerned with take place in solution rather than in gaseous or solid phases. Most of these reactions occur in *aqueous* solution, where water is the solvent. There are good reasons for this preference for liquid media. Molecules must come into contact to react, and the rates of migration of atoms or molecules within crystals usually are too slow to be useful. In contrast, molecules of gases are mobile, but gas volumes are inconveniently large, and many substances cannot be brought into the gas phase without decomposing. Solutions of reacting molecules in liquids offer an optimum combination of compactness, ease of handling, and rapidity of mixing of different substances.

As we saw in Chapter 1, water has special virtues as a solvent. It is polar, in the sense illustrated in Figure 5-1. The oxygen atom draws the electrons of the O—H bonds toward itself, acquiring a slight negative charge and leaving small positive charges on the two hydrogen atoms. Water therefore can interact with other polar molecules. Moreover, water molecules dissociate to a small extent into H^+ and OH^- ions, a property that is important in acid–base reactions. This chapter is concerned with reactions and equilibria in aqueous solution, especially those involving acids and bases.

5-1 EQUILIBRIA IN AQUEOUS SOLUTIONS

If reactants and products in a chemical reaction are in solution, the form of the equilibrium-constant expression is the same as for gas reactions, but

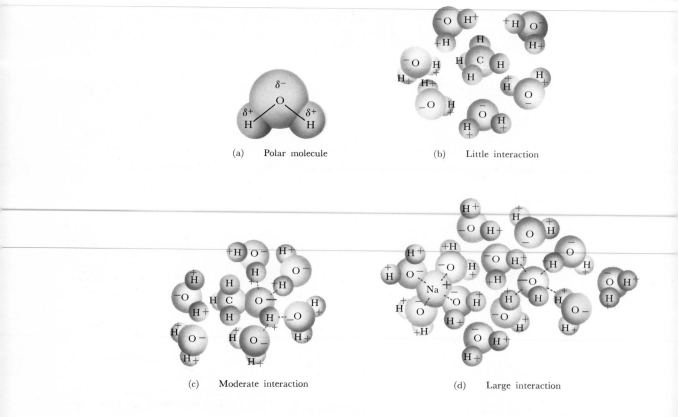

(a)   Polar molecule

(b)   Little interaction

(c)   Moderate interaction

(d)   Large interaction

Figure 5-1   Water is a polar molecule, with excess electrons and a partial negative charge on the oxygen atom, and an electron deficiency and a partial positive charge on each hydrogen atom. (b) The methane molecule, CH_4, is nonpolar: Its electrons are distributed evenly over the molecule. It has no local regions of positive and negative charge to attract water molecules, so water is a poor solvent for methane. (c) Methanol, CH_3OH, is polar, although less so than water. It has an excess of electrons and a small negative charge on the oxygen atom, and a small positive charge on the attached hydrogen atom. Methanol interacts well with water molecules by electrostatic forces, making it soluble in water. (d) Sodium hydroxide, NaOH, dissociates into positive and negative ions. These ions interact strongly with the polar water molecules, so NaOH is extremely soluble in water. Each Na^+ and OH^- ion has a cluster of water molecules surrounding it, with their negative charges closest to the sodium ions and their positive charges closest to the hydroxide ions. The ions are said to be hydrated.

the logical units of concentration are moles per liter of solution (units of molarity).

$$aA + bB \rightleftarrows cC + dD \tag{4-7}$$

$$K_{eq} = \frac{[C]^c[D]^d}{[A]^a[B]^b} \tag{4-8}$$

Some reactions in aqueous solution involve water as a participant. A well-studied example is the **hydrolysis** ("splitting by water") of the ethyl acetate molecule to yield acetic acid and ethyl alcohol (ethanol):

$$\underset{\text{ethyl acetate}}{CH_3-\overset{\overset{\displaystyle O}{\|}}{C}-O-C_2H_5} + \underset{\text{water}}{H_2O} \rightleftarrows \underset{\text{acetic acid}}{CH_3-\overset{\overset{\displaystyle O}{\|}}{C}-OH} + \underset{\text{ethyl alcohol}}{HO-C_2H_5} \tag{5-1}$$

Because all the other participant molecules themselves are polar, they dissolve well in water, which is therefore a good dispersing agent. In addition, water plays a direct role as a reactant molecule.

The equilibrium-constant expression for this reaction, in principle, is

$$K'_{eq} = \frac{[CH_3COOH][C_2H_5OH]}{[CH_3COOC_2H_5][H_2O]} \tag{5-2}$$

However, since water is present in such excess in its role as solvent, the water concentration is virtually unchanged during the reaction. In dilute solutions this is approximately the concentration of water in its pure state.

$$[H_2O] = \frac{1000 \text{ g liter}^{-1}}{18.0 \text{ g mole}^{-1}} = 55.6 \text{ moles liter}^{-1} \tag{5-3}$$

This constant water concentration can be brought over to the left side of equation 5-2 and incorporated into the equilibrium constant, as we saw for condensed phases in Chapter 4, so the equilibrium-constant expression becomes

$$K_{eq} = K'_{eq}[H_2O] = \frac{[CH_3COOH][C_2H_5OH]}{[CH_3COOC_2H_5]} \tag{5-4}$$

Other reactions in aqueous solution involve ions; an example is the precipitation of silver ions with chloride ions, in the form of insoluble silver chloride:

$$Ag^+ + Cl^- \rightleftarrows AgCl(s)$$

In this process water is not a direct reactant or product, but it does interact with the ions to keep them in solution. Any ion in aqueous solution is **hydrated**, or surrounded by polar water molecules as in Figure 5-1d. If the central ion is positive (a cation), then the negatively charged oxygen atoms of the water molecules are pointed toward it; if the central ion is negative (an anion), the positively charged hydrogen atoms of the water molecules are closest.

Each hydrated ion thus is stabilized by an immediate environment of charges opposite in sign to its own charge. When a salt crystal dissolves in water, the attractions between ions of opposite charge in the crystal are broken. In compensation, similar attractions are set up between ions and the hydrating water molecules. Solubility of salt crystals is the result of a balance or competition between crystal forces and hydration forces. This is why salts do not dissolve in nonpolar solvents such as benzene, which cannot offer hydrating attractions.

5-2   IONIZATION OF WATER AND THE pH SCALE

Water itself ionizes to a small extent:

$$H_2O(l) \rightleftarrows H^+ + OH^- \tag{5-5}$$

Each ion is surrounded with polar water molecules (as Na^+ and OH^- are in Figure 5-1d). The hydrated state of the proton, H^+, is sometimes represented as H_3O^+, meaning $H^+ \cdot H_2O$. But this is an unnecessary and even misleading notation. A more accurate representation of a hydrated proton would be $H_9O_4^+$, or $H^+ \cdot (H_2O)_4$, to represent the cluster:

$$
\left[
\begin{array}{c}
H \quad\quad\quad\quad H \\
| \quad\quad\quad\quad\quad | \\
O \cdots\quad\quad\cdots O \\
\diagup \quad \diagdown \quad \diagup \quad \diagdown \\
H \quad H \quad H \quad H \\
\diagdown \quad\quad \diagup \\
O \\
| \\
H \\
\vdots \\
O \\
\diagup \quad \diagdown \\
H \quad\quad H
\end{array}
\right]^+
$$

We will assume that H^+ and OH^-, like all other ions, are hydrated in aqueous solution, and we will therefore represent them simply as H^+ and OH^-.

Table 5-1

Measured Temperature Dependence of the Ion-Product Constant for Water, K_w

$$K_w = [H^+][OH^-]$$

$T\,(°C):$	0	25	40	60
$K_w:$	0.115×10^{-14}	1.008×10^{-14}	2.95×10^{-14}	9.5×10^{-14}

The equilibrium-constant expression for the dissociation of water is

$$K'_{eq} = \frac{[H^+][OH^-]}{[H_2O]} \tag{5-6}$$

The constant $[H_2O]$ form can be combined with K'_{eq} as before, producing

$$K_w = 55.6 K'_{eq} = [H^+][OH^-] \tag{5-7}$$

This new equilibrium constant, K_w, is called the **ion-product constant** for water. Like most equilibrium constants, K_w varies with temperature. Some experimental values of the ion-product constant are given in Table 5-1.

Example 1

From the data in Table 5-1 and Le Chatelier's principle, predict whether the dissociation of water liberates or absorbs heat.

Solution Since a higher temperature favors dissociation, dissociation is an endothermic or heat-absorbing process. From Appendix 3, $\Delta H_{(diss\ of\ H_2O)} = +55.90$ kJ mole^{-1}. This is the energy required to break one O—H bond, thereby leaving both electrons with the oxygen atom.

It is customary to take $K_w = 1.00 \times 10^{-14}$ as being accurate enough for room-temperature equilibrium calculations. (It is also customary in acid base equilibrium calculations to write K_w as if it were an exact number, 10^{-14}, rather than 1.00×10^{-14}.) This means that in pure water, where the concentrations of hydrogen and hydroxide ions are equal,

$$[H^+] = [OH^-] = 10^{-7} \text{ mole liter}^{-1} \tag{5-8}$$

Since large powers of 10 are clumsy to deal with, a logarithmic notation has been devised, called the **pH scale** (Figure 5-2). (The symbol pH stands for "negative power of hydrogen ion concentration.") The pH is the negative logarithm of $[H^+]$:

$$pH = -\log_{10}[H^+] \tag{5-9}$$

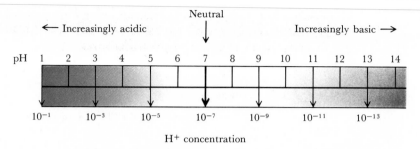

Figure 5-2   The pH scale. From Dickerson and Geis, *Chemistry, Matter, and the Universe.*

If the hydrogen ion concentration is 10^{-7} mole liter^{-1}, then

$$pH = -\log_{10}(10^{-7}) = +7$$

By an analogous definition,

$$pOH = -\log_{10}[OH^-] \tag{5-10}$$

and the pOH of pure water is also $+7$. The equilibrium constant K_w also can be expressed in logarithmic terms:

$$pK_w = -\log_{10}K_w = +14 \tag{5-11}$$

Finally, the equilibrium expression for dissociation of water,

$$[H^+][OH^-] = K_w = 10^{-14} \tag{5-12}$$

can be written

$$pH + pOH = 14 \tag{5-13}$$

In an acid solution, $[H^+]$ is greater than 10^{-7}, so the pH is less than 7. The ion-product equilibrium still holds, and $[OH^-]$ can be found from the expression

$$[OH^-] = \frac{K_w}{[H^+]} = \frac{10^{-14}}{[H^+]} \tag{5-14}$$

or

$$pOH = pK_w - pH = 14 - pH \tag{5-15}$$

The approximate pH values of some common solutions are given in Table 5-2.

Example 2

From Table 5-2, what is the hydrogen ion concentration of orange juice? What is the hydroxide ion concentration?

Table 5-2

Acidity of Some Common Solutions

Substance	pH
Commercial concentrated HCl (37% by weight)	~ −1.1
1*M* HCl solution	0.0
Gastric juice	1.4
Lemon juice	2.1
Orange juice	2.8
Wine	3.5
Tomato juice	4.1
Black coffee	5.0
Urine	6.0
Rainwater	6.5
Milk	6.9
Pure water at 24°C	7.0
Blood	7.4
Baking soda solution	8.5
Borax solution	9.2
Limewater	10.5
Household ammonia	11.9
1*M* NaOH solution	14.0
Saturated NaOH solution	~15.0

Solution

Since the pH of orange juice is 2.8, the hydrogen ion concentration is

$$[H^+] = 10^{-2.8} = 10^{+0.2} \times 10^{-3} = 1.6 \times 10^{-3}$$
$$= 0.0016 \text{ mole liter}^{-1}$$

The hydroxide ion concentration can be obtained by either of two equivalent methods:

$$[OH^-] = \frac{10^{-14}}{1.6 \times 10^{-3}} = 6.3 \times 10^{-12} \text{ mole liter}^{-1}$$

or

$$pOH = 14 - pH = 11.2$$
$$[OH^-] = 10^{-11.2} = 10^{+0.8} \times 10^{-12} = 6.3 \times 10^{-12} \text{ mole liter}^{-1}$$

Example 3

What is the ratio of hydrogen ions to hydroxide ions in pure water? In orange juice?

Solution   In pure water the ratio is 10^{-7} to 10^{-7} or 1 to 1. In orange juice, from Table 5-2, the ratio is 1.6×10^{-3} to 6.3×10^{-12} or 250,000,000 to 1.

To maintain equilibrium, the added H^+ ions from the juice have pushed the water dissociation reaction in the direction of undissociated H_2O, thereby removing OH^- ions from the solution. Orange juice is not a particularly strong acid, and the enormous fluctuation of ionic ratios even in this example illustrates the usefulness of power-of-ten and logarithmic (pH, pOH, pK) notation.

5-3   STRONG AND WEAK ACIDS

Arrhenius defined an acid (Chapter 2) as a substance that increases the hydrogen ion concentration of an aqueous solution, and a base as a substance that increases the hydroxide ion concentration. A more general definition was proposed in 1923 by Johannes Brønsted and T. M. Lowry. The Brønsted–Lowry definition can be applied to nonaqueous solutions as well: An acid is any substance that is capable of giving up a hydrogen ion, or proton, and a base is any substance that can combine with and therefore remove a hydrogen ion. Now that we understand that water molecules exist in equilibrium with their dissociated H^+ and OH^- ions, we can see that the two definitions are equivalent when water is the solvent. Arrhenius and Brønsted acids are both hydrogen-ion-releasing substances. If a Brønsted base combines with hydrogen ions, it shifts the equilibrium of equation 5-5 in favor of dissociation until balance is restored. More hydroxide ions are formed in the process, so in water a Brønsted base is an Arrhenius base as well.

In aqueous solution, acids are classified as either strong or weak. Strong acids are completely dissociated or ionized, and they include hydrogen acids such as hydrochloric acid (HCl) and hydroiodic acid (HI), and oxyacids (oxygen-containing acids) such as nitric acid (HNO_3), sulfuric acid (H_2SO_4), and perchloric acid ($HClO_4$). Each of these acids loses one proton in solution, and the acid-dissociation constant, K_a, is so large ($> 10^3$) that too little undissociated acid remains to be measured. (HSO_4^- loses a second proton and is a weak acid.)

Weak acids have measurable ionization constants in aqueous solution, because they do not dissociate completely. Examples (at $25\,°C$) are

Sulfuric: $HSO_4^- \rightleftharpoons H^+ + SO_4^{2-}$ $K_a = \dfrac{[H^+][SO_4^{2-}]}{[HSO_4^-]}$
(2nd ionization)

$$= 1.2 \times 10^{-2} \quad (5\text{-}16)$$

Hydrofluoric: $\quad HF \rightleftarrows H^+ + F^- \qquad\qquad K_a = \dfrac{[H^+][F^-]}{[HF]}$

$$= 3.5 \times 10^{-4} \quad (5\text{-}17)$$

Acetic: $\quad CH_3COOH \rightleftarrows CH_3COO^- + H^+ \quad K_a = \dfrac{[H^+][CH_3COO^-]}{[CH_3COOH]}$

$$= 1.76 \times 10^{-5} \quad (5\text{-}18)$$

Hydrocyanic: $\quad HCN \rightleftarrows H^+ + CN^- \qquad\qquad K_a = \dfrac{[H^+][CN^-]}{[HCN]}$

$$= 4.9 \times 10^{-10} \quad (5\text{-}19)$$

The distinction between strong and weak acids is somewhat artificial. The ionization of HCl is not simply a dissociation; it is, rather, the result of successful competition of H_2O molecules with Cl^- ions for the proton, H^+:

$$HCl + xH_2O \rightleftarrows H^+ \cdot (H_2O)_x + Cl^- \qquad\qquad (5\text{-}20)$$

In the Brønsted–Lowry theory, *any proton donor is an acid,* and *any proton acceptor is a base* (Figure 5-3). Therefore, HCl is an acid, and Cl^- is its **conjugate base**. Since HCl loses a proton readily it is a strong acid, and since Cl^- has so little affinity for the proton it is a weak base. In contrast, HCN is a very weak acid, because relatively few HCN molecules release their proton. Its conjugate base, CN^-, is a strong base by virtue of its high affinity for a proton.

Water is a somewhat stronger base than Cl^-, and when it is present in excess, as in an aqueous solution of HCl, it takes virtually all the protons from HCl, leaving it completely ionized. CN^- is a much stronger base than H_2O, so only a small fraction of the protons from HCN become bound to the water molecules. In other words, HCN is only slightly ionized in aqueous solution, as its K_a of 4.9×10^{-10} indicates.

Because water is present in great excess, any acid whose conjugate base is weaker than H_2O (i.e., has a lesser affinity for protons than has H_2O) will be ionized almost completely in aqueous solution. We cannot distinguish between the behavior of HCl and of $HClO_4$ (perchloric acid) in water solution. Both are completely dissociated and are therefore strong acids. However, for a solvent with a lesser attraction for protons than water, we do find differences between HCl and $HClO_4$. With diethyl ether as a solvent, perchloric acid is still a strong acid, but HCl is only partially ionized and hence is a weak acid. Diethyl ether does not solvate a proton as strongly as water does (Figure 5-4). (**Solvation** is a generalization of the concept of hydration, which applies to solvents other than water.) The equilibrium point in the reaction

$$HCl + xC_2H_5OC_2H_5 \rightleftarrows H^+ \cdot (C_2H_5OC_2H_5)_x + Cl^- \qquad\qquad (5\text{-}21)$$

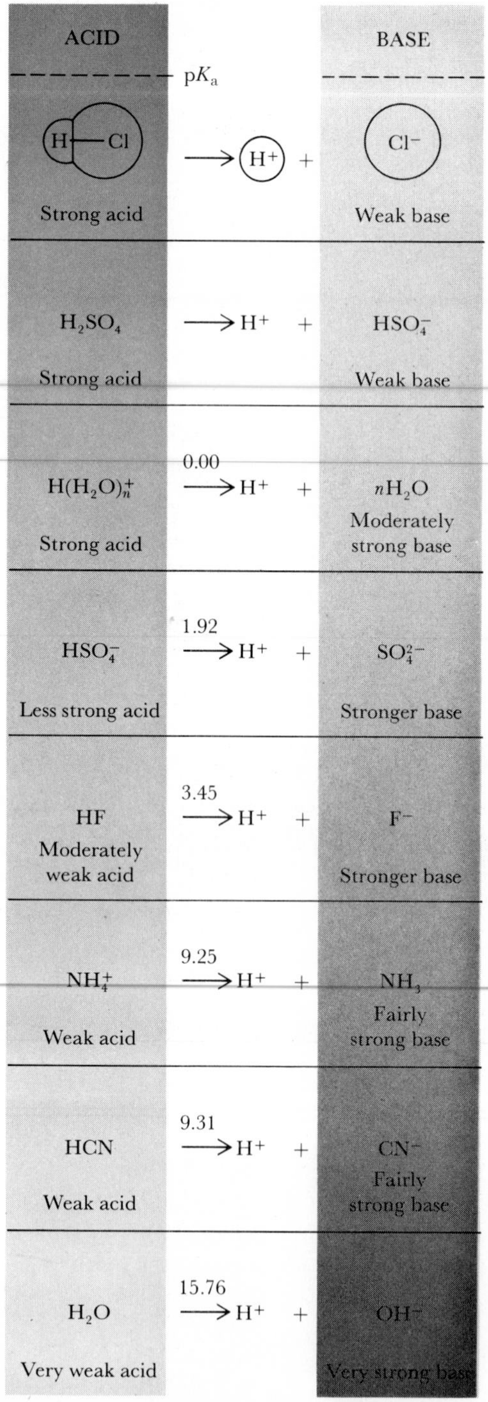

Figure 5-3   Brønsted-Lowry acids and bases. In the theory of Brønsted and Lowry, an acid is any substance that releases protons in solution, and a base is any substance that removes protons by combining with them. HCl is a strong acid because it readily releases H^+ ions. Cl^- is a weak base because it has a small tendency to combine with H^+. HCl and Cl^- are termed a conjugate pair of acid and base. From Dickerson and Geis, *Chemistry, Matter, and the Universe.*

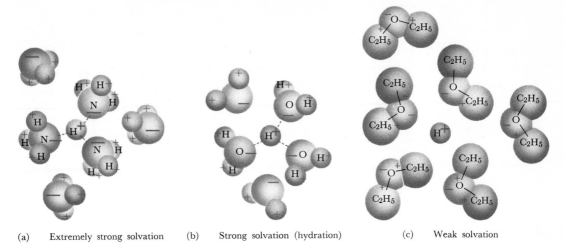

(a)   Extremely strong solvation   (b)   Strong solvation (hydration)   (c)   Weak solvation

Figure 5-4   Comparison of relative strengths of solvation of a hydrogen ion in (a) liquid ammonia,
(b) water, and (c) diethyl ether. The binding between proton and solvent ammonia
molecules is extremely strong, and liquid ammonia will take protons from and make
strong acids of substances that in aqueous solution are only weak acids. In contrast,
diethyl ether is such an ineffectual proton-solvating molecule that many substances that
are strong acids in water can retain their proton and be only partially dissociated weak
acids in ethyl ether. The + and − represent partial charges arising from local defi-
ciencies and excesses of electrons.

lies far to the left, so HCl is only partially dissociated in ether. Only in an
extremely strong acid, such as perchloric acid, does the anion have so little
attraction for the proton that it will release it to ether as an acceptor solvent.
Clearly, by using solvents other than water, we can see differences in
acidity (or proton affinity) that are masked in aqueous solution. This
masking of relative acid strengths by solvents such as water is known as the
leveling effect.

The dissociation constants for a number of acids in aqueous solution are
listed in Table 5-3, with estimates of the K_a for strong acids that are "leveled" by the solvent in aqueous solution. The dissociation of protonated
solvent, H_3O^+, into hydrated protons and H_2O, represents merely a shuf-
fling of protons from one set of water molecules to another, and must have a
K_{eq} of 1.00. In liquid ammonia as a solvent, all acids whose conjugate bases
are weaker than NH_3 would be leveled by the solvent and would be totally
ionized strong acids. Thus hydrofluoric acid and acetic acid are both strong
acids in liquid ammonia.

The leveling effect of solvent and the origin of strong and weak acids
are summarized in Figure 5-4. The distinction between strong and weak
acids depends on the solvent as much as it does on the inherent properties of
the acids themselves. Nevertheless, in aqueous solution the distinction is

Table 5-3

Dissociation Constants of Some Acids at 25°C[a]

Acid[b]	HA	A⁻	K_a	pK_a
Perchloric	$HClO_4$	ClO_4^-	$\sim 10^{+8}$	~ -8
Permanganic	$HMnO_4$	MnO_4^-	$\sim 10^{+8}$	~ -8
Chloric	$HClO_3$	ClO_3^-	$\sim 10^{+3}$	~ -3
Nitric	HNO_3	NO_3^-		
Hydrobromic	HBr	Br^-		
Hydrochloric	HCl	Cl^-		
Sulfuric (1)	H_2SO_4	HSO_4^-		
Hydrated proton or protonated solvent	H^+	H_2O(solvent)	1.00	0.00
Trichloroacetic	CCl_3COOH	CCl_3COO^-	2×10^{-1}	0.70
Oxalic (1)	$HOOC-COOH$	$HOOC-COO^-$	5.9×10^{-2}	1.23
Dichloroacetic	$CHCl_2COOH$	$CHCl_2COO^-$	3.32×10^{-2}	1.48
Sulfurous (1)	H_2SO_3	HSO_3^-	1.54×10^{-2}	1.81
Sulfuric (2)	HSO_4^-	SO_4^{2-}	1.20×10^{-2}	1.92
Phosphoric (1)	H_3PO_4	$H_2PO_4^-$	7.52×10^{-3}	2.12
Bromoacetic	$CH_2BrCOOH$	CH_2BrCOO^-	2.05×10^{-3}	2.69
Malonic (1)	$HOOC-CH_2-COOH$	$HOOC-CH_2-COO^-$	1.49×10^{-3}	2.83
Chloroacetic	$CH_2ClCOOH$	CH_2ClCOO^-	1.40×10^{-3}	2.85
Nitrous	HNO_2	NO_2^-	4.6×10^{-4}	3.34
Hydrofluoric	HF	F^-	3.53×10^{-4}	3.45
Formic	$HCOOH$	$HCOO^-$	1.77×10^{-4}	3.75
Benzoic	C_6H_5COOH	$C_6H_5COO^-$	6.46×10^{-5}	4.19
Oxalic (2)	$HOOC-COO^-$	$^-OOC-COO^-$	6.4×10^{-5}	4.19
Acetic	CH_3COOH	CH_3COO^-	1.76×10^{-5}	4.75
Propionic	CH_3CH_2COOH	$CH_3CH_2COO^-$	1.34×10^{-5}	4.87
Malonic (2)	$HOOC-CH_2-COO^-$	$^-OCC-CH_2-COO^-$	2.03×10^{-6}	5.69
Carbonic (1)	$CO_2 + H_2O$	HCO_3^-	4.3×10^{-7}	6.37
Sulfurous (2)	HSO_3^-	SO_3^{2-}	1.02×10^{-7}	6.91
Hydrogen sulfide (1)	H_2S	HS^-	9.1×10^{-8}	7.04
Phosphoric (2)	$H_2PO_4^-$	HPO_4^{2-}	6.23×10^{-8}	7.21
Ammonium ion	NH_4^+	NH_3	5.6×10^{-10}	9.25
Hydrocyanic	HCN	CN^-	4.93×10^{-10}	9.31
Silver ion	$Ag^+ + H_2O$	$AgOH$	9.1×10^{-11}	10.04
Carbonic (2)	HCO_3^-	CO_3^{2-}	5.61×10^{-11}	10.25
Hydrogen peroxide	H_2O_2	HO_2^-	2.4×10^{-12}	11.62
Hydrogen sulfide (2)	HS^-	S^{2-}	1.1×10^{-12}	11.96
Phosphoric (3)	HPO_4^{2-}	PO_4^{3-}	2.2×10^{-13}	12.67
Water[c]	H_2O	OH^-	1.8×10^{-16}	15.76

[a]HA is the acid form, with acid strength decreasing down the table. A⁻ is the conjugate base, with base strength increasing down the table. The equilibrium is HA $\rightleftarrows$ H⁺ + A⁻ and the equilibrium-constant expression is

$$K_a = \frac{[H^+][A^-]}{[HA]} \qquad pK_a = -\log_{10} K_a$$

[b]The notation (1) indicates a first dissociation or proton-transfer reaction; (2) indicates a second dissociation; and (3) indicates a third dissociation.

[c]This K_a value for water explicitly uses $[H_2O] = 55.6$ moles liter⁻¹ in the denominator, for the sake of consistency with the other entries in the table. The standard K_w is obtained by noting that $55.6 \times 1.8 \times 10^{-16} = 1.0 \times 10^{-14} = K_w$.

real. As long as the discussion is confined to aqueous solutions (as ours will be from now on), we shall find it useful to think about and to treat the two classes of acids separately.

5-4 STRONG AND WEAK BASES

In Arrhenius' terminology a base is a substance that decreases the hydrogen ion concentration of a solution. Sodium hydroxide, potassium hydroxide, and similar compounds are bases because they dissolve and dissociate completely in aqueous solution to yield hydroxide ions:

$$NaOH \rightleftarrows Na^+ + OH^-$$
$$KOH \rightleftarrows K^+ + OH^- \tag{5-22}$$

These excess hydroxide ions then disturb the water dissociation equilibrium, and combine with some of the protons normally found in pure water:

$$H^+ + OH^- \rightleftarrows H_2O \qquad [H^+] = \frac{K_w}{[OH^-]} < 10^{-7} \tag{5-23}$$

In the more generalized Brønsted–Lowry definition, the hydroxide ion itself is the base, because it is the substance that combines with the proton. The Na^+ and K^+ ions merely provide the positive ions that are necessary for overall electrical neutrality for the chemical compound.

The commonly encountered hydroxides of alkali metals (Li, Na, K) all dissolve and dissociate completely to produce the same Brønsted–Lowry base, OH^-. These hydroxides all are *strong bases,* analogous to strong acids such as HCl and HNO_3. Other substances such as ammonia and many organic nitrogen compounds also can combine with protons in solution and act as Brønsted–Lowry bases. These compounds are generally *weaker* bases than the hydroxide ion, because they have a smaller attraction for protons. For example, when ammonia competes with OH^- for protons in an aqueous solution, it is only partially successful. It can combine with only a portion of the H^+ ions, thus

$$NH_3 + H^+ \rightleftarrows NH_4^+ \tag{5-24}$$

will have a measurable equilibrium constant.

There is no logical reason why this reaction cannot be described by an acid-dissociation constant, as in Table 5-3. The ammonium ion, NH_4^+, is the Brønsted–Lowry **conjugate acid** of the base NH_3. There is no reason why, in an acid–base pair, it is the acid that must be neutral and the base charged, as in HCl/Cl^- and HCN/CN^-. The NH_4^+ ion is just as respectable an acid as HCl or HCN, and although weaker than HCl, it is actually stronger than HCN. Thus, we can describe the ammonia reaction as an acid dissociation:

$$NH_4^+ \rightleftarrows NH_3 + H^+ \qquad K_a = 5.6 \times 10^{-10} \quad \text{(from Table 5-3)} \tag{5-25}$$

or, if we want to focus on the basic behavior of NH_3,

$$NH_3 + H^+ \rightleftarrows NH_4^+ \qquad K_{eq} = \frac{1}{K_a} = 1.8 \times 10^{+9} \tag{5-26}$$

However, chemical language has become trapped by the older acid–base terminology introduced by Arrhenius, and you should be aware of this. Arrhenius thought of a base as a substance that releases OH^- ions into aqueous solution. For alkali metal hydroxides such as NaOH the process was straightforward:

$$NaOH \rightleftarrows Na^+ + OH^- \tag{5-27}$$

But what about NH_3? Where do the hydroxide ions come from? Arrhenius assumed that when ammonia dissolved in water the reaction was

$$NH_3 + H_2O \rightleftarrows NH_4OH \rightleftarrows NH_4^+ + OH^- \tag{5-28}$$

This brought NH_3 into line by postulating an intermediate—ammonium hydroxide—that dissociated like any other hydroxide. Sodium hydroxide is a strong base that dissociates completely; ammonium hydroxide would be a weak base that dissociates only partially. Arrhenius defined a base-dissociation constant, K_b, as

$$BOH \rightleftarrows B^+ + OH^- \qquad K_b = \frac{[B^+][OH^-]}{[BOH]} \tag{5-29}$$

where B usually represents a metal. For ammonia, K_a and K_b would be related by

$$K_b = \frac{[NH_4^+][OH^-]}{[NH_3]} = \frac{[NH_4^+][OH^-][H^+]}{[NH_3][H^+]} = \frac{K_w}{K_a} \tag{5-30}$$

$$K_b = \frac{10^{-14}}{5.6 \times 10^{-10}} = 1.8 \times 10^{-5} \tag{5-31}$$

Unfortunately for Arrhenius' theory, there is no evidence that ammonium hydroxide, NH_4OH, exists as a real compound. It is more accurate to say that the polar ammonia molecule is hydrated like any other polar molecule: $NH_3 \cdot (H_2O)_x$. Ammonia, NH_3, combines directly with a proton and with water molecules:

$$\begin{array}{ll} NH_3 + H^+ + xH_2O \rightleftarrows NH_4^+ & \text{(in acid solutions)} \\ NH_3 + xH_2O \rightleftarrows NH_4^+ + OH^- & \text{(in basic solutions)} \end{array} \tag{5-32}$$

Nevertheless, Arrhenius' notation is too deeply embedded in the fabric of chemistry to dislodge, and we often will use K_b for weak bases rather than K_a for their conjugate acids. In general, the completely dissociated strong bases that we shall encounter will be hydroxide compounds, and the weak bases will be ammonia and organic nitrogen compounds such as those listed in Table 5-4. K_b always can be found from K_a and K_w and the expression

$$K_a \times K_b = K_w \tag{5-33}$$

Table 5-4

Dissociation Constants of Some Weak Bases[a] at 25°C

Base	B	BH+	K_b	pK_b
Aniline	$-NH_2$	$-NH_3^+$	4.3×10^{-10}	9.37
Pyridine	N	N^+-H	1.8×10^{-9}	8.75
Imidazol			9.1×10^{-8}	7.05
Hydrazine	N_2H_4	$N_2H_5^+$	9.8×10^{-7}	6.01
Ammonia	NH_3	NH_4^+	1.79×10^{-5}	4.75
Trimethylamine	$(CH_3)_3N$	$(CH_3)_3NH^+$	6.4×10^{-5}	4.19
Methylamine	CH_3-NH_2	$CH_3-NH_3^+$	3.7×10^{-4}	3.34
Dimethylamine	$(CH_3)_2NH$	$(CH_3)_2NH_2^+$	5.4×10^{-4}	3.27

[a]If B represents the base, the equilibrium equation is $B + H_2O \rightleftarrows BH^+ + OH^-$, in which BH^+ is the conjugate acid. Base strengths increase down the table, and conjugate acid strengths decrease. The equilibrium-constant expression is

$$K_b = \frac{[BH^+][OH^-]}{[B]} \qquad pK_b = -\log_{10} K_b$$

5-5 SOLUTIONS OF STRONG ACIDS AND BASES: NEUTRALIZATION AND TITRATION

When an amount of strong acid is added to water, the effect is that of adding the same amount of hydrogen ions, since the acid is totally dissociated.

Example 4

What is the hydrogen ion concentration of a 0.0100M nitric acid solution? What is the pH?

Solution $[H^+] = 0.010$ mole liter^{-1}

$pH = -\log_{10}(10^{-2}) = 2.00$

The solution is quite acidic.

Example 5

What are the hydrogen ion concentration and the pH of a 0.0050M sodium hydroxide solution?

Solution   The hydroxide ion contribution from completely dissociated NaOH is

$[OH^-] = 0.0050$ mole liter^{-1}

This large quantity of hydroxide ions will repress the normal dissociation of water and enhance the reaction to the left:

$$H_2O \longleftrightarrow H^+ + OH^-$$

The hydrogen ion concentration is found from the water equilibrium expression:

$$[H^+] = \frac{K_w}{[OH^-]} = \frac{10^{-14}}{0.0050} = 2.0 \times 10^{-12} \text{ mole liter}^{-1}$$

$$pH = -\log_{10}(2.0) - \log_{10}(10^{-12}) = -0.30 + 12.0 = 11.7$$

The solution is quite basic.

Example 6

If we mix equal volumes of the solutions of the previous two examples, what will be the pH of the resulting solution?

Solution   If equal volumes are mixed, then the concentration of each solute will be halved, since the final volume is twice the volume of each starting solution. The final solution would be 0.0050M in nitric acid and 0.0025M in sodium hydroxide. But acid and base will react and neutralize one another until one or the other is used up:

$$H^+ + NO_3^- + Na^+ + OH^- \rightleftarrows H_2O + NO_3^- + Na^+$$

or simply

$$H^+ + OH^- \rightleftarrows H_2O$$

since sodium and nitrate ions take no part in the neutralization reaction. In this case, sodium hydroxide is in shorter supply. When all the base has been neutralized, we still have

$$0.0050 - 0.0025 = 0.0025 \text{ mole liter}^{-1} \text{ excess nitric acid}$$
$$[H^+] = 0.0025 = 2.5 \times 10^{-3} \text{ mole liter}^{-1}$$
$$pH = -\log_{10}(2.5) + 3.0 = 2.6$$

Example 7

How many milliliters of $0.10M$ HCl must we add to 200 ml of $0.0050M$ KOH to bring the pH down to 10.0?

Solution

Without HCl, the pH of the potassium hydroxide solution would be 11.7, as in Example 5. Let y equal the number of milliliters of HCl solution needed to yield a pH of 10.0. Since 0.0050 mole liter^{-1} is the same as 0.0050 millimoles ml^{-1}, the total number of millimoles (mmoles) of KOH is

$$n_{KOH} = 0.0050 \text{ mmole ml}^{-1} \times 200 \text{ ml} = 1.0 \text{ mmole}$$

The total number of millimoles of HCl that must be added is

$$n_{HCl} = 0.10 \text{ mmole ml}^{-1} \times y \text{ ml} = 0.10y \text{ mmole}$$

Since the final solution is basic, $n_{KOH} > n_{HCl}$. The net amount of hydroxide ions left over after partial neutralization by HCl is

$$n_{base} = n_{KOH} - n_{HCl} = 1.0 - 0.10y$$

The final volume is

$$V = 200 + y \text{ ml}$$

and therefore the final hydroxide in concentration is

$$[OH^-] = \frac{n_{base}}{V} = \frac{1.0 - 0.10y}{200 + y}$$

A pH of 10.0 means a pOH of 4.0 and $[OH^-] = 10^{-4}$ mole liter^{-1}, thus

$$\frac{1.0 - 0.10y}{200 + y} = 10^{-4}$$

and

$$y = 9.8 \text{ ml of } 0.10 \text{ } M \text{ HCl to be added.}$$

Titration and Titration Curves

If we add equal numbers of equivalents of a strong acid and a strong base, they will neutralize one another completely, and the pH will be 7.0. As we saw in Chapter 2, this makes possible the titration method of measuring quantities of acid or base.

Example 8

One hundred fifty milliliters of HCl solution of unknown concentration are titrated with $0.10M$ NaOH. Eighty milliliters of base solution are required to neutralize the acid. How many moles of HCl were present originally, and what was the acid-solution concentration?

Solution

The number of millimoles of base used is

$$n_{\text{NaOH}} = 0.10 \text{ mmole ml}^{-1} \times 80 \text{ ml} = 8.0 \text{ mmoles}$$

This must be the same as the number of millimoles of acid originally present, if neutralization was complete.

Thus the original concentration of HCl was

$$[\text{HCl}]_0 = \frac{8.0 \text{ mmoles}}{150 \text{ ml}} = 0.053 \text{ mmole ml}^{-1} \text{ or mole liter}^{-1}$$

A common way of determining the equivalence point of titration (the point at which neutralization occurs) is with an acid–base **indicator**. Indicators are weak organic acids or bases that have different colors in their ionized and neutral states (or in two ionized states). If their color change occurs in the neighborhood of pH 7, and if we add a few drops of indicator solution to the solution being titrated, we see this color change at the end point of the titration. We will discuss some common indicators in the section on weak acids. The matching of indicator color-change point and the end point of a titration does not have to be very exact, because the pH swings drastically through several units as neutralization becomes complete. This can make life easy for the analytical chemist, and it is worth looking more closely at the behavior of pH during titration. To illustrate what we have just said, let us calculate the titration curve for a typical strong acid and strong base.

Example 9

Fifty milliliters of $0.10M$ HNO_3 are titrated with $0.10M$ KOH, in an experimental arrangement such as that shown in Figure 2-3. Calculate the pH of the solution as a function of the volume of KOH solution added (v, in milliliters).

Solution

It is easiest to treat this calculation in three parts: before neutralization, at neutralization (equivalence point), and after neutralization. Before the equivalence point, calculate how much base has been added, assume that all of this base was used to neutralize some of the acid, and calculate how much acid would remain unneutralized, as a function of the volume of base solution added.

Table 5-5

Titration of 50 ml of 0.10*M* Nitric Acid by 0.10*M* Potassium Hydroxide

Before equivalence point

Base solution added, v (ml)	$\dfrac{50 - v}{50 + v}$	[H$^+$]	pH
0	1.00	0.100	1.00
10	$\frac{40}{60}$	0.067	1.18
20	$\frac{30}{70}$	0.043	1.37
30	$\frac{20}{80}$	0.025	1.60
40	$\frac{10}{90}$	0.011	1.95
45	$\frac{5}{95}$	0.0053	2.28
48	$\frac{2}{98}$	0.0020	2.69
49	$\frac{1}{99}$	0.0010	3.00
49.9	$\frac{0.1}{99.9}$	0.0001	4.00
49.99	$\frac{0.01}{99.99}$	0.00001	5.00

After equivalence point

Base solution added, v (ml)	$\dfrac{v - 50}{v + 50}$	[OH$^-$]	pOH	pH
50.01	$\frac{0.01}{100.01}$	0.00001	5.00	9.00
50.1	$\frac{0.1}{100.1}$	0.0001	4.00	10.00
51	$\frac{1}{101}$	0.0010	3.00	11.00
52	$\frac{2}{102}$	0.0020	2.71	11.29
55	$\frac{5}{105}$	0.0048	2.32	11.68
60	$\frac{10}{110}$	0.0091	2.04	11.96
70	$\frac{20}{120}$	0.0167	1.78	12.22
80	$\frac{30}{130}$	0.023	1.64	12.36
90	$\frac{40}{140}$	0.029	1.54	12.46
100	$\frac{50}{150}$	0.033	1.48	12.52

Original: $n_{HNO_3} = 50$ ml $\times$ 0.10 mmole ml^{-1} = 5.0 mmoles

Added: $n_{KOH} = v$ ml $\times$ 0.10 mmole ml^{-1}

Net acid: $n_{acid} = 5.0$ mmoles $-$ 0.10v mmole

Total volume: $V = 50 + v$ ml

Hydrogen ion concentration: $[H^+]_{net} = \dfrac{5.0 - 0.10v}{50 + v} = \dfrac{50 - v}{50 + v}(0.10)$ mmole ml^{-1}

The calculation of [H$^+$] for various values of v is shown in Table 5-5, and these calculations are plotted with open circles at the left of Figure 5-5. At the equivalence point, the amounts of acid and base are equal and the

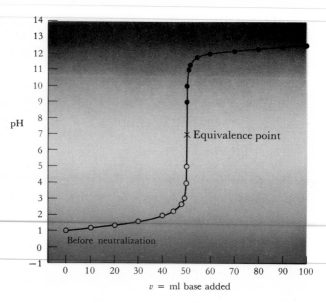

Figure 5-5   Titration curve for typical strong acid and base. Fifty milliliters of $0.10M$ HNO_3 are titrated with increasing amounts of $0.10M$ KOH. Data are given in Table 5-5. Notice how rapidly the pH changes in the region of the end point, or of exact neutralization of acid by base. Any acid—base indicator that changes color between pH 4 and pH 10 could be used to detect the equivalence point in this titration.

pH is 7.0. After the equivalence point, we only need to calculate how much base was added in excess of that required to neutralize the acid, and use this to find $[OH^-]$, pOH, and pH:

Original:	$n_{HNO_3} = 5.0$ mmoles (as before)
Added:	$n_{KOH} = v$ ml $\times$ 0.10 mmole ml^{-1}
Net base:	$n_{base} = 0.10v - 5.0$ mmoles
Final volume:	$V = 50 + v$ ml

Hydroxide ion concentration:

$$[OH^-] = \frac{0.10v - 5.0}{50 + v} = \frac{v - 50}{v + 50}(0.10) \text{ mmole ml}^{-1}$$

This calculation for several values of v and the corresponding pH values are listed in Table 5-5 and are plotted with solid circles on the right of Figure 5-5. It now is obvious why the choice of an indicator is not too critical in such a titration. Any indicator that changes color between pH 4 and pH 10 will do.

Titrating a weak acid with a strong base, or a weak base with a strong acid, is more complicated because the weak component is only partially dissociated. Dissociation equilibria of the type discussed in the next section must be used. We will not be concerned in this chapter with such titrations, but they are treated in Appendix 5, with an example of a titration curve corresponding to Figure 5-5.

5-6 EQUILIBRIA WITH WEAK ACIDS AND BASES

Because weak acids are only partially dissociated in water, the contribution of a weak acid such as acetic acid to the hydrogen ion concentration is less than the total concentration of added acid. The equilibrium-constant expression for dissociation of the acid must be used explicitly. These general principles can be illustrated with a concrete example, that of calculating the pH of a solution of $0.0100M$ acetic acid. As we saw in Example 5-4 for nitric acid, a strong acid, a $0.0100M$ solution has a pH of 2.00. Because acetic acid is a weak acid and only partially dissociated, a $0.0100M$ solution will have a hydrogen ion concentration of less than $0.0100M$, and a pH greater than 2.0.

It is common to represent the acetate ion, CH_3COO^-, simply by Ac^-, and the undissociated acetic acid molecule, CH_3COOH, by HAc as if it were a simple inorganic acid. (The forms OAc^- and $HOAc$ also are used, to indicate that acetic acid is an oxyacid with the dissociating proton attached to an oxygen atom.) The dissociation of HAc is incomplete:

$$HAc \rightleftarrows H^+ + Ac^-$$

and the equilibrium expression describing dissociation is

$$K_a = \frac{[H^+][Ac^-]}{[HAc]} = 1.76 \times 10^{-5} \quad \text{(from Table 5-3)}$$

We know the initial overall concentration, c_0, of acetic acid:

$$c_0 = 0.0100 \text{ mole liter}^{-1}$$

and we know that at equilibrium some of this acetic acid remains undissociated and some of it has ionized to acetate ions, Ac^-:

$$c_0 = [HAc] + [Ac^-] \quad \text{(mass-balance equation)}$$

This is called a **mass-balance equation**, because it states that *total* acetate is neither created nor destroyed during dissociation. We also know that the concentrations of hydrogen ions and acetate ions are equal, since dissociation of HAc is the only source of H^+. (It is legitimate to neglect H^+ from the dissociation of water, since acetic acid represses water dissociation even below its normal small extent.) Thus

$$[H^+] = [Ac^-] \quad \text{(charge-balance equation)}$$

This is known as a **charge-balance equation**, because it states that the total positive charge in the solution must equal the total negative charge. We now can use these data about conservation of acetate and neutrality of the solution to simplify the equilibrium-constant expression. Let the hydrogen ion concentration that we are seeking be $[H^+] = y$, and eliminate $[Ac^-]$ at once using the charge-balance equation:

$$K_a = \frac{y^2}{[HAc]} \qquad \text{(equilibrium equation)}$$

$$c_0 = [HAc] + y \qquad \text{(mass-balance equation)}$$

The second equation tells us that the concentration of undissociated HAc equals the original overall concentration, c_0, minus the amount that has dissociated, y:

$$[HAc] = c_0 - y$$

The equilibrium expression then is

$$K_a = \frac{y^2}{c_0 - y} \qquad (5\text{-}34)$$

Substituting the value of K_a from Table 5-3, we get

$$1.76 \times 10^{-5} = \frac{y^2}{0.0100 - y}$$

or

$$y^2 + 1.76 \times 10^{-5}y - 1.76 \times 10^{-7} = 0$$

This is a quadratic equation, which can be solved with the quadratic formula. If $ay^2 + by + c = 0$, then

$$y = \frac{-b \pm \sqrt{b^2 - 4ac}}{2a} \qquad (5\text{-}35)$$

For this problem, $a = 1$, $b = 1.76 \times 10^{-5}$, and $c = -1.76 \times 10^{-7}$.

$$y = \frac{-1.76 \times 10^{-5} \pm \sqrt{3.10 \times 10^{-10} + 7.04 \times 10^{-7}}}{2}$$

or

$$y = \frac{-1.76 \times 10^{-5} \pm 8.39 \times 10^{-4}}{2}$$

Only the positive answer is reasonable, because one cannot have a negative concentration. Thus the answer is

$$y = 4.11 \times 10^{-4} \text{ mole liter}^{-1}$$

Under certain physical conditions you can take a shortcut to avoid the quadratic formula. In this example, since you know that the acid is only slightly dissociated, you can try neglecting y in the denominator of the equilibrium expression for K_a, thereby assuming that it is small in comparison with 0.0100 mole liter^{-1}, and that the concentration of undissociated acetic acid is virtually the same as the total acetic acid present. This assumption gives

$$1.76 \times 10^{-5} = \frac{y^2}{0.0100}$$

and an approximate answer of

$$y = 4.2 \times 10^{-4} = 0.00042 \text{ mole liter}^{-1}$$

This is close to the correct answer of 0.000411 mole liter^{-1}. You can make a quick improvement by using this approximate value in the undissociated acetate concentration in the denominator:

$$1.76 \times 10^{-5} = \frac{y^2}{0.0100 - 0.00042}$$

$$y = 4.11 \times 10^{-4} \text{ mole liter}^{-1}$$

Repetition of the foregoing process until the answer remains constant from one cycle to the next is called the **method of successive approximation**. If your intuition for how much dissociation the acid undergoes is good enough, you can often solve an equilibrium problem by an approximate solution and a quick correction in less time than it takes to solve the quadratic formula. If your original guess is not so good, two or three cycles of approximation may be required before you arrive at an unchanging value for y.

   As our results show, acetic acid is indeed only slightly dissociated at 0.0100M concentration. Of the initial 0.0100 mole liter^{-1}, 0.000411 mole has dissociated, and 0.0096 mole remains as dissolved but undissociated HAc molecules. The percent dissociation is

$$\frac{4.11 \times 10^{-4} \text{ mole}}{0.0100 \text{ mole}} \times 100 = 4.11\%$$

Since the hydrogen ion concentration is $[H^+] = 4.11 \times 10^{-4} \, M$, the pH of this solution is 3.39.

   What happens if we dilute the acetic acid solution? Does a greater or lesser percent of the acetic acid then dissociate? Does the pH increase or decrease?

Example 10

What are the pH and percent dissociation in a solution of 0.00100M acetic acid?

Solution   The equilibrium expression is as before:

$$K_a = \frac{y^2}{c_0 - y}$$

$$1.76 \times 10^{-5} = \frac{y^2}{0.00100 - y}$$

Neglecting y in comparison with c_0, the approximate solution is

$$1.76 \times 10^{-5} = \frac{y^2}{0.00100}$$

$$y = 1.33 \times 10^{-4} \text{ mole liter}^{-1}$$

and the solution obtained by using this value to correct the undissociated HAc concentration is

$$y = 1.24 \times 10^{-4} \text{ mole liter}^{-1}$$

Using this second value to correct c_0 in another cycle of approximation makes no change in y, so the process can be halted. Now the pH is 3.91 instead of 3.39, and the percent dissociation is

$$\frac{1.24 \times 10^{-4} \text{ mole}}{0.00100 \text{ mole}} \times 100 = 12.4\%$$

Although the actual hydrogen ion concentration is lower (witness the larger pH), a greater fraction of the HAc present is dissociated into ions. This is Le Chatelier's principle again. If a solution containing HAc, H^+, and Ac^- is diluted, thereby lowering its total concentration of all ions and molecules, the equilibrium will attempt to reestablish itself, as reactions change, in the direction that will increase the total concentration of solute particles of one kind or another. Compare this behavior with the effect of increasing the pressure on the ammonia gas equilibrium in Chapter 4.

Indicators

An indicator is a weak acid (or a weak base) that has sharply different colors in its dissociated and undissociated states. Methyl orange (Figure 5-6) is a complex organic compound that is red in its neutral, un-ionized form and yellow when ionized. It can be represented as the weak acid HIn:

$$HIn \rightleftarrows H^+ + In^-$$

    red   yellow

Adding acid shifts the indicator equilibrium to the left, and adding base shifts it to the right. Hence methyl orange is red in acids and yellow in bases.

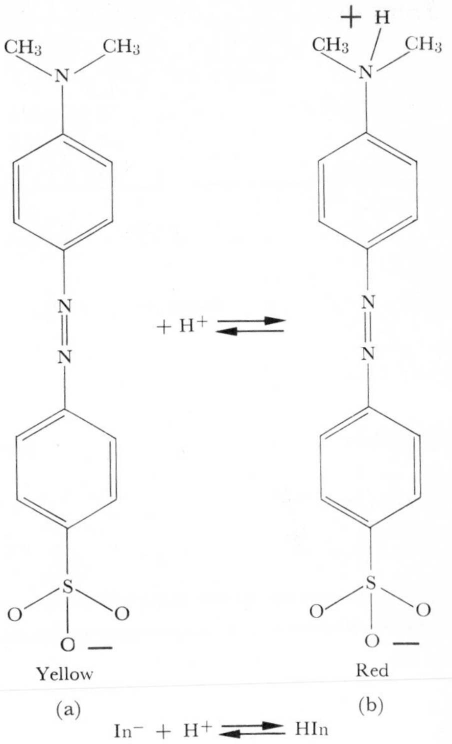

Figure 5-6

The basic form (a) and acid form (b) of the indicator methyl orange. The different colors of the two structures, yellow and red, give methyl orange its usefulness in displaying the pH of a solution into which it has been introduced. The complex structure can be symbolized by an ion, In^-, which can combine with a proton as shown at the bottom of the figure.

$$In^- + H^+ \rightleftharpoons HIn$$

The intensity of color from indicators such as methyl orange is so great that the colors can be seen easily even when the amount added to a solution is too small to have an appreciable influence on the pH of the solution. Nevertheless, the ratio of dissociated to undissociated indicator depends on the hydrogen ion concentration

$$K_a = \frac{[H^+][In^-]}{[HIn]} \tag{5-36}$$

and

$$\frac{[In^-]}{[HIn]} = \frac{K_a}{[H^+]} \tag{5-37}$$

$$\log_{10}\left(\frac{[In^-]}{[HIn]}\right) = pH - pK_a \tag{5-38}$$

For methyl orange, $K_a = 1.6 \times 10^{-4}$ and $pK_a = 3.8$. The neutral (red) and dissociated (yellow) forms of the indicator are present at equal concentrations when the pH $= 3.8$. The eye is sensitive to color changes over a range of concentration ratios of approximately 100, or over two pH units. Below pH 2.8, a solution containing methyl orange is red, and above approximately 4.8 it is clearly yellow. As you can see from Figure 5-5, an indicator change over two pH units is quite satisfactory for strong acid–base titrations.

Methyl orange could be used for the titration in Figure 5-5, even though its pK_a is far from the titration equivalence point of 7.0, only because the change in pH at the equivalence point is so large. For titrations of weak acids, this would not be true, and it would be better to pick an indicator with a pK_a closer to the expected equivalence point. Other indicators are shown in Figure 5-7, along with the pH range in which their color changes occur. Phenolphthalein is a particularly convenient and common indicator, which changes from colorless to pink in the range of pH 8 to 10.

Contribution to [H⁺] from Dissociation of Water

Nothing has been said in the discussions of either strong acids or weak acids about a contribution to the hydrogen ion concentration from the dissociation of water itself. It has been tacitly assumed that all H^+ comes from the acid. This is a valid assumption for all but the most dilute solutions of very weak acids such as HCN. The correction for water dissociation seldom is necessary, so it will not be covered in this chapter. A complete treatment is given in Appendix 5.

Figure 5-7   Some common acid–base indicators, with the pH ranges in which their color changes occur. The choice of an indicator for an acid–base titration depends on the expected pH at the equivalence point of the titration and the width of swing of pH values as the equivalence point is passed.

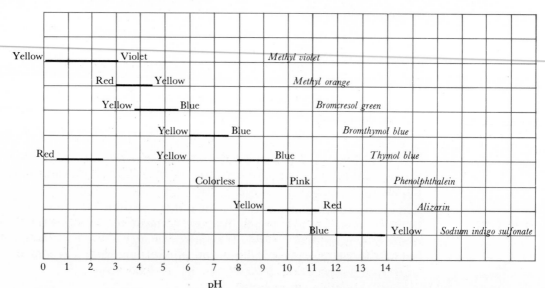

5-7 WEAK ACIDS AND THEIR SALTS

What will happen to a weak acid such as acetic acid if we add some sodium acetate (NaAc), which is the salt of a strong base (NaOH) and acetic acid? The salt will dissolve and dissociate completely into sodium and acetate ions. From Le Chatelier's principle, we would expect these added acetate ions to force the weak acetic acid equilibrium system in the direction of less dissociation. This is exactly what happens. The acid-equilibrium expression is the same:

$$K_a = \frac{[H^+][Ac^-]}{[HAc]} \tag{5-39}$$

However, two sources of acetate ions now exist: NaAc and HAc. The acetate ion supplied by sodium acetate is measured by c_s, the total molarity of the salt, since dissociation is complete. Acetate concentration from acetic acid is measured by the hydrogen ion concentration, since every dissociation of HAc to produce Ac^- also produces a proton. Therefore, the total acetate ion concentration is

$$[Ac^-]_{total} = [Ac^-]_{NaAc} + [Ac^-]_{HAc} = c_s + [H^+] \tag{5-40}$$

(Again, we have neglected any protons contributed by the dissociation of water.) The concentration of un-ionized acetic acid is the overall acid concentration, c_a, less the acetate from dissociation:

$$[HAc] = c_a - [Ac^-]_{HAc} = c_a - [H^+] \tag{5-41}$$

If we represent the hydrogen ion concentration by y, we have

$$K_a = \frac{y(c_s + y)}{(c_a - y)} \tag{5-42}$$

When the added salt concentration, c_s, is zero, this is the simple weak acid–dissociation equilibrium expression that we have seen previously in equation 5-34.

Example 11

What are the pH and percent dissociation of a solution of $0.010M$ acetic acid in the presence of (a) no NaAc, (b) $0.0050M$ NaAc, and (c) $0.010M$ NaAc?

Solution

From Le Chatelier's principle, we would expect the dissociation of HAc to be repressed as more NaAc is added. The pH should increase and the percent dissociation should decrease. (a) This problem was already solved in Section 5-6, yielding pH 3.39 and 4.11% dissociation.
(b) For $c_s = 0.0050$ mole liter^{-1},

$$1.76 \times 10^{-5} = \frac{y(0.0050 + y)}{0.010 - y} \quad \text{(from equation 5-42)}$$

This is most easily solved by successive approximations. As a first approximation we can assume that y will be smaller than 0.0050 or 0.010, and we can therefore neglect it when it is added to or subtracted from these quantities:

$$y_1 = 1.76 \times 10^{-5} \times \frac{0.010}{0.0050} = 3.52 \times 10^{-5} = 0.000035 \text{ mole liter}^{-1}$$

As a second approximation, we can use this trial value of y to "correct" 0.0050 to 0.005035, and 0.010 to 0.009965, and solve the equation again:

$$y_2 = 1.76 \times 10^{-5} \times \frac{0.009965}{0.005035} = 3.48 \times 10^{-5} \text{ mole liter}^{-1}$$

A third approximation is unnecessary, and the answer should be rounded to 3.5×10^{-5} mole liter^{-1}:

$$\text{pH} = 5 - \log_{10} 3.5 = 5 - 0.54 = 4.46$$

$$\text{Percent dissociation} = \frac{3.5 \times 10^{-5}}{0.010} \times 100 = 0.35\%$$

(c) For $c_s = 0.010$ mole liter^{-1},

$$y = [\text{H}^+] = 1.76 \times 10^{-5} \text{ mole liter}^{-1}$$
$$\text{pH} = 4.75$$
$$\text{Percent dissociation} = 0.18\%$$

Notice that the acetic acid now dissociates so little that even the first approximation is adequate.

Results for these and a few other sodium acetate concentrations are listed in Table 5-6 and are plotted in Figure 5-8. The first salt added has a

Table 5-6

Effect of Adding Sodium Acetate to 0.010*M* Acetic Acid Solution

c_s = concentration of sodium acetate in moles liter^{-1}

c_s:	0.0	0.001	0.002	0.005	0.010	0.020
pH:	3.4	3.8	4.1	4.5	4.8	5.1
Percent dissociation of acetic acid:	4.1	1.5	0.84	0.35	0.18	0.09

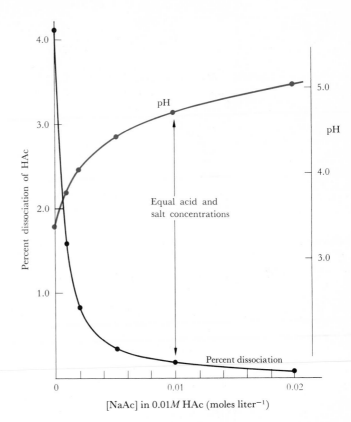

Figure 5-8

The effect of added sodium acetate on the dissociation of acetic acid. Data plotted here are listed in Table 5-6, and were calculated as explained in the text. The first salt added represses acetic acid dissociation to a great extent and causes a rapid increase in pH. Later additions are not as effective.

large effect on the degree of dissociation and pH; later additions of salt cause less change. When acid and salt are present in equal concentrations, the pH is equal to the pK_a of the acid.

Buffers

If the concentrations of a solution of a weak acid and a salt of the acid anion are reasonably high, then the solution is resistant to changes in hydrogen ion concentration.

Example 12

A solution is $0.050M$ in HAc and $0.050M$ in NaAc. Calculate the change in pH when 0.0010 mole of hydrochloric acid (HCl) is added to a liter of solution, assuming that the volume increase upon adding the HCl is negligible. Compare this to the pH if the same amount of HCl is added to a liter of pure water.

Solution   Before adding HCl the acetic acid equilibrium is

$$K_a = \frac{[H^+][Ac^-]}{[HAc]} = \frac{y(0.050)}{(0.050)}$$

Thus

$$y = K_a = 1.76 \times 10^{-5} \text{ mole liter}^{-1}$$
$$pH = pK_a = 4.75$$

(Again, we were justified in ignoring y in the $[Ac^-]$ and $[HAc]$ terms because the value is small compared to 0.050.)

The added protons from HCl combine with acetate ions to form more acetic acid:

$$Ac^- + H^+ \text{ (from HCl)} \rightarrow HAc$$

Thus to a good approximation, all the added protons are used up, and the new acetic acid and acetate concentrations are

$$[HAc] = 0.050 + [H^+]_{HCl} = 0.051 \text{ mole liter}^{-1}$$
$$[Ac^-] = 0.050 - [H^+]_{HCl} = 0.049 \text{ mole liter}^{-1}$$

$$K_a = \frac{y(0.049)}{(0.051)}$$

$$y = 1.76 \times 10^{-5} \times \frac{0.051}{0.049} = 1.83 \times 10^{-5} \text{ mole liter}^{-1}$$

$$pH = 5 - 0.26 = 4.74$$

The pH changes from 4.75 to 4.74, a difference of only 0.01 unit. In the absence of HAc and NaAc, the same concentration of HCl would produce a pH of 3.0.

This resistance to pH change is called **buffering action**, and the solution of HAc and NaAc is an acetate buffer. Buffers are used widely for pH control in laboratory chemistry, in the chemical industry, and in living organisms. A carbonate buffer system in your bloodstream, involving the reaction

$$H^+ + HCO_3^- \rightleftarrows H_2CO_3 \rightleftarrows CO_2 + H_2O \qquad (5\text{-}43)$$

maintains the blood pH around 7.4. When a biochemist studies enzyme activity in the laboratory, he must use a buffer system to maintain a constant pH during the experiments, otherwise his results may have little meaning. One of the sillier disputes in commercial advertising is that between two pharmaceutical companies as to whether buffers added to aspirin to combat an acid reaction in the stomach are a benefit or an adulterant.

In general, if the concentration of strong acid added to a buffer solution is y moles liter^{-1}, the equilibrium equation becomes

$$K_a = \frac{[H^+][A^-]}{[HA]} = \frac{[H^+](c_s - y)}{c_a + y} \tag{5-44}$$

in which c_s and c_a are the salt and buffering acid concentrations, respectively. After addition of the foreign acid, the hydrogen ion concentration is

$$[H^+] = K_a \frac{(c_a + y)}{(c_s - y)} \tag{5-45}$$

and the pH is

$$pH = pK_a + \log_{10} \frac{(c_s - y)}{(c_a + y)} \tag{5-46}$$

If base is added, hydrogen ions are removed, and the same expressions can be used with a negative value of y.

Example 13

A formic acid buffer is prepared with 0.010 mole liter^{-1} each of formic acid (HCOOH) and sodium formate (HCOONa). What is the pH of the solution? What is the pH if 0.0020 mole liter^{-1} of solid sodium hydroxide (NaOH) is added to a liter of buffer? What would be the pH of the sodium hydroxide solution without buffer? What would the pH have been after adding sodium hydroxide if the buffer concentrations had been 0.10 mole liter^{-1} instead of 0.010?

Solution

The answers are

Buffer:	pH = 3.75
After adding NaOH:	pH = 3.92
Without buffer:	pH = 11.30
With stronger buffer:	pH = 3.77

In the preceding example, you can see the dramatic effect of the formate buffer in keeping the solution acidic in spite of the added base, and the importance of reasonably high buffer concentrations if the buffering capacity of the solution is not to be exceeded.

5-8 SALTS OF WEAK ACIDS AND STRONG BASES: HYDROLYSIS

A sodium chloride solution is neutral, with a pH of 7.0. This is reasonable, because sodium hydroxide is a strong base and hydrochloric acid is a strong acid, and if equal amounts of each were added, neutralization would be complete. In contrast, sodium acetate is the salt of a strong base and a weak acid. Intuitively we would expect a sodium acetate solution to be somewhat basic, and it is. Some of the acetate ions from the salt combine with water to form undissociated acetic acid and hydroxide ions:

$$Ac^- + H_2O \rightleftharpoons HAc + OH^- \tag{5-47}$$

This sometimes is called a **hydrolysis reaction**, the implication being that H_2O breaks up crystals of sodium acetate. It does, when the salt crystal dissolves in water, but this is not the point. In solution the acetate ion acts as a base. It is as good a Brønsted base as ammonia, and the ammonium ion is a perfectly good acid, like HAc.

$$NH_3 + H_2O \rightleftharpoons NH_4^+ + OH^-$$

We should not let the different charges on the acetate ion (-1) and ammonia (0) obscure the similarity of their acid–base behavior.

The equilibrium constant for acetate hydrolysis has the same form as any other base dissociation:

$$K_b = \frac{[HAc][OH^-]}{[Ac^-]} \tag{5-48}$$

$$K_b = \frac{[NH_4^+][OH^-]}{[NH_3]} \tag{5-30}$$

where, as usual, the virtually unchanging water concentration is incorporated into the equilibrium constant. This constant sometimes is written K_h for "hydrolysis constant," but the added nomenclature is unnecessary. It is a simple base-equilibrium constant of the kind we have seen before, except that acetate ion is the base.

As always, K_b is related to the corresponding acid-dissociation constant, K_a, by

$$K_b = \frac{[HAc][OH^-]}{[Ac^-]} = \frac{[HAc][OH^-][H^+]}{[Ac^-][H^+]} = \frac{K_w}{K_a} \tag{5-49}$$

Thus

$$K_b = \frac{10^{-14}}{1.76 \times 10^{-5}} = 5.68 \times 10^{-10}$$

(Recall the ammonia–water equilibrium expressions at the end of Section 5-4.) This value is all we need to calculate the pH of a sodium acetate solution.

Example 14

What is the pH of a solution of $0.010M$ NaAc?

Solution   Acetate ions from NaAc combine with H_2O to produce undissociated HAc molecules and OH^- ions (equation 5-47). The equilibrium expression is

$$5.68 \times 10^{-10} = \frac{[HAc][OH^-]}{[Ac^-]}$$

Let the hydroxide ion concentration be y. Since every reaction of an acetate ion with water produces one hydroxide ion and one undissociated HAc molecule, the concentration of each of the latter two species must be y moles liter^{-1}. The remaining acetate ions are those originally present from NaAc minus those that have combined with water:

$$[Ac^-] = 0.010 - y$$

and we arrive at the familiar expression

$$K_b = \frac{y^2}{c_s - y} = \frac{y^2}{0.010 - y} = 5.68 \times 10^{-10} \tag{5-50}$$

This is even easier to solve than the weak-acid problems. Since the equilibrium constant is so small, y will be correspondingly small and can be neglected in the denominator in comparison to 0.010. The result is

$$y^2 = 0.010 \times 5.68 \times 10^{-10} = 5.7 \times 10^{-12}$$

$$y = 2.4 \times 10^{-6} \text{ mole liter}^{-1} = [OH^-]$$

$$[H^+] = \frac{K_w}{[OH^-]} = \frac{10^{-14}}{2.4 \times 10^{-6}} = 4.2 \times 10^{-9}$$

$$pH = 9 - 0.6 = 8.4$$

(As before, we have neglected any contribution to the hydrogen ion concentration from water molecules. Our procedure is accurate enough for most situations, including the purposes of this chapter. The full derivation is found in Appendix 5.)

5-9 POLYPROTIC ACIDS: ACIDS THAT LIBERATE MORE THAN ONE HYDROGEN ION

If water is the solvent, sulfuric acid, H_2SO_4, loses one proton as a strong acid with an immeasurably large dissociation constant.

$$H_2SO_4 \rightarrow H^+ + HSO_4^-$$

It also can lose a second proton as a weak acid with a measurable dissociation constant. Acids that can liberate more than one proton are called **polyprotic acids**.

$$HSO_4^- \rightleftarrows H^+ + SO_4^{2-} \qquad K_{a_2} = 1.20 \times 10^{-2} \qquad pK_{a_2} = 1.92$$

For carbonic acid, H_2CO_3, both dissociations are weak:

$$H_2CO_3 \rightleftarrows H^+ + HCO_3^- \qquad K_{a_1} = 4.3 \times 10^{-7} \qquad pK_{a_1} = 6.37$$
$$HCO_3^- \rightleftarrows H^+ + CO_3^{2-} \qquad K_{a_2} = 5.61 \times 10^{-11} \qquad pK_{a_2} = 10.25$$

The relative values of K_{a_1} and K_{a_2} for a given acid are intuitively reasonable. One would expect HCO_3^-, which already has a negative charge, to be less ready than neutral H_2CO_3 to lose another proton.

Phosphoric acid, H_3PO_4, has three dissociations:

$$H_3PO_4 \rightleftarrows H^+ + H_2PO_4^- \qquad pK_{a_1} = 2.12$$
$$H_2PO_4^- \rightleftarrows H^+ + HPO_4^{2-} \qquad pK_{a_2} = 7.21$$
$$HPO_4^{2-} \rightleftarrows H^+ + PO_4^{3-} \qquad pK_{a_3} = 12.67$$

Thus, in an aqueous solution of phosphoric acid there will be seven ionic and molecular species present: H_3PO_4, $H_2PO_4^-$, HPO_4^{2-}, PO_4^{3-}, H_2O, H^+, and OH^-. Life might appear impossibly complicated, were we not able to make some approximations.

At a pH equal to the pK_a for a particular dissociation, the two forms of the dissociating species are present in equal concentrations. For the second dissociation of phosphoric acid, for which $pK_{a_2} = 7.21$,

$$K_{a_2} = \frac{[H^+][HPO_4^{2-}]}{[H_2PO_4^-]}$$

$$\log \frac{[HPO_4^{2-}]}{[H_2PO_4^-]} = pH - pK_{a_2}$$

When $pH = pK_{a_2}$, we have the ratio

$$\frac{[HPO_4^{2-}]}{[H_2PO_4^-]} = 1.00$$

Hence, in a neutral solution, $H_2PO_4^{2-}$ and HPO_4^- are present in about the same concentrations. Very little undissociated H_3PO_4 will be found, since from the first dissociation constant,

$$K_{a_1} = \frac{[H^+][H_2PO_4^-]}{[H_3PO_4]}$$

$$\log \frac{[H_2PO_4^-]}{[H_3PO_4]} = pH - pK_{a_1} = 7.00 - 2.12 = 4.88$$

$$\frac{[H_2PO_4^-]}{[H_3PO_4]} = 10^{4.88} = 7.6 \times 10^4 = 76,000$$

Similarly, little PO_4^{3-} will exist:

$$\log \frac{[PO_4^{3-}]}{[HPO_4^{2-}]} = pH - pK_{a_3} = 7.00 - 12.67 = -5.67$$

$$\frac{[PO_4^{3-}]}{[HPO_4^{2-}]} = 10^{-5.67} = 2.1 \times 10^{-6} = \frac{1}{480,000}$$

The only phosphate species that we have to consider near pH = 7 are $H_2PO_4^-$ and HPO_4^{2-}. Similarly, in strong acid solutions near pH = 3, only H_3PO_4 and $H_2PO_4^-$ are important. As long as the pK_a's of successive dissociations are separated by three or four units (as they almost always are), matters are simplified.

There is still another simplification. When a polyprotic acid such as carbonic acid, H_2CO_3, dissociates, most of the protons present come from the first dissociation:

$$H_2CO_3 \rightleftarrows H^+ + HCO_3^- \qquad pK_{a_1} = 6.37$$

Since the second dissociation constant is smaller by four orders of magnitude (and the pK_{a_2} larger by four units), the contribution of hydrogen ions from the second dissociation will be only one ten-thousandth as large. Correspondingly, the second dissociation has a negligible effect on the concentration of the product of the first dissociation, HCO_3^-.

Example 15

At room temperature and 1 atm CO_2 pressure, water saturated in CO_2 has a carbonic acid concentration of approximately 0.040 mole liter^{-1}. Calculate the pH and the concentrations of all carbonate species for a 0.040 M H_2CO_3 solution.

Solution

Considering initially only the first dissociation:

$$K_{a_1} = 4.3 \times 10^{-7} = \frac{y^2}{0.040 - y} \qquad \text{in which } y = [H^+]$$

From our experience with acetic acid, which has an even larger K_a, we should expect to be able to neglect y in the denominator. The extent of dissociation of an acid with such a small K_a will be very small:

$$y^2 = 4.3 \times 10^{-7} \times 0.040 = 1.7 \times 10^{-8}$$
$$y = 1.3 \times 10^{-4} \text{ mole liter}^{-1}$$

This is the concentration of both hydrogen ion and bicarbonate ion, HCO_3^-:

$$[H^+] = 1.3 \times 10^{-4} \text{ mole liter}^{-1}$$
$$[HCO_3^-] = 1.3 \times 10^{-4} \text{ mole liter}^{-1}$$
$$[H_2CO_3] = 0.040 - 0.00013 = 0.040 \text{ mole liter}^{-1}$$
$$pH = 4 - 0.12 = 3.88$$

Consequently, carbonated beverages have an acidity somewhere between those of wine and tomato juice (see Table 5-2). For the second dissociation:

$$HCO_3^- \rightleftarrows H^+ + CO_3^{2-}$$

$$K_{a_2} = 5.6 \times 10^{-11} = \frac{[H^+][CO_3^{2-}]}{[HCO_3^-]}$$

Since this second dissociation has only a minor effect on the first one, we can assume that the hydrogen ion and bicarbonate ion concentrations are effectively the same:

$$[CO_3^{2-}] = \frac{[HCO_3^-]}{[H^+]} \times K_{a_2} = K_{a_2} = 5.6 \times 10^{-11} \text{ mole liter}^{-1}$$

Note the rather surprising result that the concentration of the second dissociation product is equal to the second dissociation constant!

Example 16

Calculate the sulfide ion concentration in a solution saturated in H_2S (0.10 mole liter^{-1}) (a) if the solution is made from distilled water and (b) if the solution is made pH = 3.0 with HCl. Use K_a values in Table 5-3.

Solution

In distilled water, the first dissociation is

$$K_{a_1} = 9.1 \times 10^{-8} = \frac{y^2}{0.10}$$

The dissociation constant is so small that y in the denominator can be neglected immediately. Dissociation will be extremely slight:

$$y = [H^+] = [HS^-] = 9.5 \times 10^{-5} \text{ mole liter}^{-1}$$
$$pH = 5 - 0.98 = 4.02$$

From the second dissociation:

$$[S^{2-}] = \frac{[HS^-]}{[H^+]} \times K_{a_2} = K_{a_2} = 1.1 \times 10^{-12} \text{ mole liter}^{-1}$$

As with the H_2CO_3 example, the anion produced by the second dissociation has a concentration equal to the second dissociation constant.

In contrast, in HCl solution at pH = 3.0:

$$K_{a_1} = \frac{[H^+][HS^-]}{[H_2S]} = \frac{1.0 \times 10^{-3}[HS^-]}{0.10} = 9.1 \times 10^{-8}$$

$$[HS^-] = 9.1 \times 10^{-6} \text{ mole liter}^{-1}$$

$$K_{a_2} = \frac{[H^+][S^{2-}]}{[HS^-]} = \frac{1.0 \times 10^{-3}[S^{2-}]}{9.1 \times 10^{-6}} = 1.1 \times 10^{-12}$$

$$[S^{2-}] = \frac{9.1 \times 10^{-6} \times 1.1 \times 10^{-12}}{1.0 \times 10^{-3}} = 1.0 \times 10^{-14}$$

The acid has repressed the dissociation of H_2S, making the sulfide ion concentration only one-hundredth of what it is in pure water. As we shall see in the next section, we can use acids to exert a fine control on sulfide concentration in analytical methods by controlling the pH.

5-10 EQUILIBRIA WITH SLIGHTLY SOLUBLE SALTS

When most solid salts dissolve in water, they dissociate almost completely into hydrated positive and negative ions. The solubility of a salt in water represents a balance between the attraction of the ions in the crystal lattice and the attraction between these ions and the polar water molecules. This balance may be a delicate one, easily changed in going from one compound to an apparently similar one, or from one temperature to another. It is not possible to give hard-and-fast rules as to whether a compound is soluble, or even to account for all observed behavior.

One important factor certainly is the electrostatic attraction between ions. Crystals made up of small ions that can be packed closely together are generally harder to pull apart than crystals made up of large ions. Therefore, for a given cation, fluorides (F^-) and hydroxides (OH^-) are less soluble than nitrates (NO_3^-) and perchlorates (ClO_4^-). Chlorides are intermediate in size, and their behavior is difficult to predict from general principles.

The charge on the ions also is important. More highly charged ions such as phosphates (PO_4^{3-}) and carbonates (CO_3^{2-}) interact strongly with cations and are less soluble than the singly charged nitrates and perchlorates.

The terms *soluble* and *insoluble* are relative, and the degree of solubility can be related to an equilibrium constant. For a "slightly soluble" salt such as silver chloride, an equilibrium exists between the dissociated ions and the solid compound:

$$AgCl(s) \rightleftarrows Ag^+ + Cl^- \tag{5-51}$$

The equilibrium expression for this reaction is

$$K_{eq} = \frac{[Ag^+][Cl^-]}{[AgCl(s)]} \tag{5-52}$$

As long as solid AgCl remains, its effect on the equilibrium does not change. As with the H_2O concentration in the water dissociation equilibrium, the concentration of the solid salt can be incorporated into the equilibrium constant:

$$K_{sp} = K_{eq}[AgCl(s)] = [Ag^+][Cl^-] \tag{5-53}$$

This new equilibrium constant, K_{sp}, is called the *solubility-product constant*. For substances in which the ions are not in a 1:1 ratio, the form of the solubility-product expression is analogous to our previous equilibrium expressions:

$$PbCl_2 \rightleftarrows Pb^{2+} + 2Cl^- \qquad K_{sp} = [Pb^{2+}][Cl^-]^2$$
$$Al(OH)_3 \rightleftarrows Al^{3+} + 3OH^- \qquad K_{sp} = [Al^{3+}][OH^-]^3$$
$$Ag_2CrO_4 \rightleftarrows 2Ag^+ + CrO_4^{2-} \qquad K_{sp} = [Ag^+]^2[CrO_4^{2-}]$$
$$Ba_3(PO_4)_2 \rightleftarrows 3Ba^{2+} + 2PO_4^{3-} \qquad K_{sp} = [Ba^{2+}]^3[PO_4^{3-}]^2$$

Solubility equilibria are useful in predicting whether a precipitate will form under specified conditions, and in choosing conditions under which two chemical substances in solution can be separated by selective precipitation.

The solubility-product constant of a slightly soluble compound can be calculated from its solubility in moles liter^{-1}.

Example 17

The solubility of AgCl in water is 0.000013 mole liter^{-1} at 25°C. What is its solubility-product constant, K_{sp}?

Solution   The equilibrium expression is

$$AgCl \rightleftarrows Ag^+ + Cl^-$$

The concentrations of Ag^+ and Cl^- are equal because for each mole of solid AgCl that dissolves, 1 mole each of Ag^+ and Cl^- ions is produced. Hence the concentration of each ion is equal to the overall solubility, s, of the solid in moles liter^{-1}:

$$[Ag^+] = [Cl^-] = s = 1.3 \times 10^{-5} \text{ mole liter}^{-1}$$
$$K_{sp} = [Ag^+][Cl^-] = s^2 = 1.7 \times 10^{-10}$$

Example 18

At a certain temperature the solubility of $Fe(OH)_2$ in water is 7.7×10^{-6} mole liter^{-1}. Calculate its K_{sp} at that temperature.

Solution   The equilibrium equation is

$$Fe(OH)_2 \rightleftarrows Fe^{2+} + 2OH^-$$

and the solubility-product expression is

$$K_{sp} = [Fe^{2+}][OH^-]^2$$

Since one mole of dissolved $Fe(OH)_2$ produces one mole of Fe^{2+} and *two* moles of OH^-,

$$[Fe^{2+}] = s = 7.7 \times 10^{-6} \text{ mole liter}^{-1}$$
$$[OH^-] = 2s = 1.54 \times 10^{-5} \text{ mole liter}^{-1}$$
$$K_{sp} = 7.7 \times 10^{-6} \times (1.54 \times 10^{-5})^2 = 1.8 \times 10^{-15}$$

The solubility-product constants of a number of substances are listed in Table 5-7. Substances are grouped by anion and listed in the order of decreasing K_{sp}; anions are listed roughly in the order of decreasing solubility. Once the solubility-product constant is known, it can be used to calculate the solubility of a compound at a specified temperature.

Table 5-7

Solubility-Product Constants, K_{sp}, at 25 °C

Fluorides		Chromates (cont.)		Hydroxides (cont.)	
BaF_2	2.4×10^{-5}	Ag_2CrO_4	1.9×10^{-12}	$Ni(OH)_2$	1.6×10^{-16}
MgF_2	8×10^{-8}	$PbCrO_4$	2×10^{-16}	$Zn(OH)_2$	4.5×10^{-17}
PbF_2	4×10^{-8}			$Cu(OH)_2$	1.6×10^{-19}
SrF_2	7.9×10^{-10}	Carbonates		$Hg(OH)_2$	3×10^{-26}
CaF_2	3.9×10^{-11}	$NiCO_3$	1.4×10^{-7}	$Sn(OH)_2$	3×10^{-27}
		$CaCO_3$	4.7×10^{-9}	$Cr(OH)_3$	6.7×10^{-31}
Chlorides		$BaCO_3$	1.6×10^{-9}	$Al(OH)_3$	5×10^{-33}
$PbCl_2$	1.6×10^{-5}	$SrCO_3$	7×10^{-10}	$Fe(OH)_3$	6×10^{-38}
$AgCl$	1.7×10^{-10}	$CuCO_3$	2.5×10^{-10}	$Co(OH)_3$	2.5×10^{-43}
$Hg_2Cl_2{}^a$	1.1×10^{-18}	$ZnCO_3$	2×10^{-10}	Sulfides	
Bromides		$MnCO_3$	8.8×10^{-11}	MnS	7×10^{-16}
$PbBr_2$	4.6×10^{-6}	$FeCO_3$	2.1×10^{-11}	FeS	4×10^{-19}
$AgBr$	5.0×10^{-13}	Ag_2CO_3	8.2×10^{-12}	NiS	3×10^{-21}
$Hg_2Br_2{}^a$	1.3×10^{-22}	$CdCO_3$	5.2×10^{-12}	CoS	5×10^{-22}
Iodides		$PbCO_3$	1.5×10^{-15}	ZnS	2.5×10^{-22}
PbI_2	8.3×10^{-9}	$MgCO_3$	1×10^{-15}	SnS	1×10^{-26}
AgI	8.5×10^{-17}	$Hg_2CO_3{}^a$	9.0×10^{-15}	CdS	1.0×10^{-28}
$Hg_2I_2{}^a$	4.5×10^{-29}			PbS	7×10^{-29}
Sulfates		Hydroxides		CuS	8×10^{-37}
$CaSO_4$	2.4×10^{-5}	$Ba(OH)_2$	5.0×10^{-3}	Ag_2S	5.5×10^{-51}
Ag_2SO_4	1.2×10^{-5}	$Sr(OH)_2$	3.2×10^{-4}	HgS	1.6×10^{-54}
$SrSO_4$	7.6×10^{-7}	$Ca(OH)_2$	1.3×10^{-6}	Bi_2S_3	1.6×10^{-72}
$PbSO_4$	1.3×10^{-8}	$AgOH$	2.0×10^{-8}	Phosphates	
$BaSO_4$	1.5×10^{-9}	$Mg(OH)_2$	8.9×10^{-12}	Ag_3PO_4	1.8×10^{-18}
		$Mn(OH)_2$	2×10^{-13}	$Sr_3(PO_4)_2$	1×10^{-31}
Chromates		$Cd(OH)_2$	2.0×10^{-14}	$Ca_3(PO_4)_2$	1.3×10^{-32}
$SrCrO_4$	3.6×10^{-5}	$Pb(OH)_2$	4.2×10^{-15}	$Ba_3(PO_4)_2$	6×10^{-39}
$Hg_2CrO_4{}^a$	2×10^{-9}	$Fe(OH)_2$	1.8×10^{-15}	$Pb_3(PO_4)_2$	1×10^{-54}
$BaCrO_4$	8.5×10^{-11}	$Co(OH)_2$	2.5×10^{-16}		

^{a}As Hg_2^{2+} ion. $K_{sp} = [Hg_2^{2+}][X^-]^2$

Example 19

What is the solubility of lead sulfate, $PbSO_4$, in water at 25 °C?

Solution   The dissociation reaction is

$$PbSO_4 \rightleftarrows Pb^{2+} + SO_4^{2-}$$

Let the unknown solubility be s moles liter^{-1}. Then since each mole of dissolved $PbSO_4$ produces 1 mole of each ion,

$$[Pb^{2+}] = [SO_4^{2-}] = s$$

The solubility-product equation is

$$K_{sp} = [Pb^{2+}][SO_4^{2-}] = s^2 = 1.3 \times 10^{-8} \quad \text{(from Table 5-7)}$$
$$s = 1.1 \times 10^{-4} \text{ mole liter}^{-1}$$

Example 20

In Table 5-7 we see that cadmium carbonate, $CdCO_3$, and silver carbonate, Ag_2CO_3, have approximately the same solubility-product constants. Compare their molar solubilities in water (at 25°C).

Solution   For cadmium carbonate,

$$K_{sp} = [Cd^{2+}][CO_3^{2-}] = s^2 = 5.2 \times 10^{-12}$$
$$s = 2.3 \times 10^{-6} \text{ mole liter}^{-1}$$

For Ag_2CO_3 the expression is slightly different. If the solubility again is s moles liter^{-1}, since each mole of salt produces 2 moles of Ag^+ ions,

$$[Ag^+] = 2s$$
$$[CO_3^{2-}] = s$$
$$K_{sp} = [Ag^+]^2[CO_3^{2-}] = (2s)^2 \times s = 4s^3 = 8.2 \times 10^{-12}$$
$$s = 1.3 \times 10^{-4} \text{ mole liter}^{-1}$$

Although cadmium carbonate and silver carbonate have nearly the same solubility-product constants, their solubilities in moles liter^{-1} differ by a factor of 100 because the form of the solubility-product expression is different. The solubility of Ag_2CO_3 is sensitive to the square of the metal-ion concentration, because two silver ions per carbonate ion are necessary to build the solid crystal.

Common-Ion Effect

In the preceding example, the solubility of silver carbonate in pure water was calculated to be 1.3×10^{-4} mole liter^{-1}. Will silver carbonate be more soluble or less soluble in silver nitrate solution? Le Chatelier's principle leads us to predict that a new, outside source of silver ions would shift the silver carbonate equilibrium reaction in the direction of less dissociation:

$$Ag_2CO_3 \rightleftharpoons 2Ag^+ + CO_3^{2-} \tag{5-54}$$

or that silver carbonate would be less soluble in a silver nitrate solution than in pure water. This decrease in the solubility of one salt in a solution of another salt that has a common cation or anion is called the **common-ion effect**.

Example 21

What is the solubility at $25°C$ of calcium fluoride, CaF_2, (a) in pure water, (b) in $0.10M$ calcium chloride, $CaCl_2$, and (c) in $0.10M$ sodium fluoride, NaF?

Solution

(a) If the solubility in pure water is s, then

$$[Ca^{2+}] = s$$
$$[F^-] = 2s$$
$$K_{sp} = s \times 4s^2 = 4s^3 = 3.9 \times 10^{-11}$$
$$s = 2.1 \times 10^{-4} \text{ mole liter}^{-1}$$

(b) In $0.10M$ $CaCl_2$, the calcium ion concentration is the sum of the concentration of calcium ions from calcium chloride and from calcium fluoride, whose solubility we are seeking:

$$[Ca^{2+}] = 0.10 + s$$
$$[F^-] = 2s$$
$$K_{sp} = (0.10 + s)(2s)^2 = 3.9 \times 10^{-11}$$

This is a cubic equation, but a moment's thought about the chemistry involved will eliminate the need to solve it as such. With such a small solubility-product constant, you can predict that the solubility of calcium fluoride will be very small in comparison with 0.10 mole liter^{-1}. (You already should realize from (a) and Le Chatelier's principle that in this problem s will be less than 2.1×10^{-4} mole liter^{-1}.) If our prediction is valid, we can simplify the solubility-product equation and calculate the approximate solubility:

$$0.10 \times (2s)^2 = 3.9 \times 10^{-11}$$

$$s^2 = \frac{3.9 \times 10^{-11}}{4 \times 0.10} = 9.75 \times 10^{-11}$$

$$s = 0.99 \times 10^{-5} = 9.9 \times 10^{-6} \text{ mole liter}^{-1}$$

Therefore the approximation is justified. Only 4.7% as much CaF_2 will dissolve in $0.10M$ $CaCl_2$ as in pure water:

$$\frac{9.9 \times 10^{-6}}{2.1 \times 10^{-4}} \times 100 = 4.7\%$$

(c) In $0.10M$ NaF,

$$[Ca^{2+}] = s \quad \text{and} \quad [F^-] = 0.10 + 2s$$

since fluoride ions come from NaF as well as from CaF_2. The solubility-product equation is

$$K_{sp} = s(2s + 0.10)^2 = 3.9 \times 10^{-11}$$

Again, thinking about the chemical meaning will avoid the necessity of solving a cubic equation. The $2s$ term will be very small compared to 0.10 mole liter^{-1}, therefore,

$$s(0.10)^2 = 3.9 \times 10^{-11}$$
$$s = 3.9 \times 10^{-9} \text{ mole liter}^{-1}$$

This approximation is even more valid than the previous one, since from the calculation

$$\frac{3.9 \times 10^{-9}}{2.1 \times 10^{-4}} \times 100 = 0.0019\%$$

only 0.0019 % as much CaF$_2$ will dissolve in 0.10M NaF as in pure water. Fluoride is more effective than calcium as a common ion because it has a second-power effect on the solubility equilibrium.

The common-ion method of controlling solubility often is used with solutions of sulfide ion, S^{2-}, because many metals form insoluble sulfides, and the sulfide ion concentration can be controlled by adjusting the pH.

Example 22

What is the maximum possible concentration of Ni^{2+} ion in water at 25°C that is saturated with H$_2$S and maintained at pH 3.0 with HCl?

Solution

From the solubility-product equilibrium equation we predict that too much nickel ion will cause the precipitation of nickel sulfide, NiS:

$$K_{sp} = [\text{Ni}^{2+}][\text{S}^{2-}] = 3 \times 10^{-21}$$

The only new twist to this problem is finding the sulfide ion concentration from the H$_2$S equilibrium. Hydrogen sulfide dissociates in two steps, each with an equilibrium constant:

$$\begin{array}{ll}
\text{H}_2\text{S} \rightleftarrows \text{H}^+ + \text{HS}^- & K_{a_1} = 9.1 \times 10^{-8} \\
\text{HS}^- \rightleftarrows \text{H}^+ + \text{S}^{2-} & K_{a_1} = 1.1 \times 10^{-12} \\
\hline
\text{H}_2\text{S} \rightleftarrows 2\text{H}^+ + \text{S}^{2-} & K_{a_{1.2}} = K_{a_1} \times K_{a_2}
\end{array}$$

Because the overall dissociation is the sum of two dissociation steps, the overall equilibrium constant, $K_{a_{1.2}}$, is the product of K_{a_1} and K_{a_2}:

$$K_{a_{1.2}} = \frac{[\text{H}^+][\text{HS}^-]}{[\text{H}_2\text{S}]} \times \frac{[\text{H}^+][\text{S}^{2-}]}{[\text{HS}^-]} = \frac{[\text{H}^+]^2[\text{S}^{2-}]}{[\text{H}_2\text{S}]}$$

$$K_{a_{1.2}} = 9.1 \times 10^{-8} \times 1.1 \times 10^{-12} = 1.0 \times 10^{-19}$$

Saturated H_2S is approximately $0.10M$ at $25°C$, and the very small value of $K_{a_{1,2}}$ means that dissociation of H_2S is very slight. Hence we can write

$$[H_2S] = 0.10 \text{ mole liter}^{-1} \quad \text{and} \quad [H^+]^2[S^{2-}] = 1.0 \times 10^{-20}$$

in a saturated H_2S solution. This "ion product" for saturated H_2S is a useful relationship to remember.

In this problem, the pH has been adjusted to 3.0 with hydrochloric acid, so

$$[H^+] = 1.0 \times 10^{-3} \text{ mole liter}^{-1}$$

Therefore, the sulfide ion concentration can be calculated from

$$[S^{2-}] = K_{a_{1,2}} \times \frac{[H_2S]}{[H^+]^2} = 1.0 \times 10^{-19} \times \frac{0.10}{(1.0 \times 10^{-3})^2}$$

which gives

$$[S^{2-}] = 1.0 \times 10^{-14} \text{ mole liter}^{-1}$$

Since NiS will precipitate if the solubility product is exceeded, the highest value that the nickel ion concentration can have is

$$[Ni^{2+}] = \frac{K_{sp}}{[S^{2-}]} = \frac{3 \times 10^{-21}}{1 \times 10^{-14}} = 3 \times 10^{-7} \text{ mole liter}^{-1}$$

Separation of Compounds by Precipitation

Solubility-product constants can be used to devise methods for separating ions in solution by selective precipitation. The entire traditional qualitative-analysis scheme is based on the use of these equilibrium constants to determine the correct precipitating ions and the correct strategy.

Example 23

A solution is $0.010M$ in barium chloride, $BaCl_2$, and $0.020M$ in strontium chloride, $SrCl_2$. Can either Ba^{2+} or Sr^{2+} be precipitated selectively with concentrated sodium sulfate, Na_2SO_4, solution? Which ion will precipitate first? When the second ion just begins to precipitate, what is the residual concentration of the first ion, and what fraction of the original amount of the first ion is left in solution? (For simplicity, assume that the Na_2SO_4 solution is so concentrated that the volume change in the Ba–Sr solution can be neglected.)

Solution

The upper limit on barium sulfate solubility is given by

$$K_{sp} = [Ba^{2+}][SO_4^{2-}] = 1.5 \times 10^{-9}$$

With 0.010 mole liter^{-1} of Ba^{2+}, precipitation of barium sulfate will not occur until the sulfate ion concentration increases to

$$[SO_4^{2-}] = \frac{1.5 \times 10^{-9}}{0.010} = 1.5 \times 10^{-7} \text{ mole liter}^{-1}$$

Strontium sulfate will precipitate when the sulfate concentration is

$$[SO_4^{2-}] = \frac{K_{sp(SrSO_4)}}{[Sr^{2+}]} = \frac{7.6 \times 10^{-7}}{0.020} = 3.8 \times 10^{-5} \text{ mole liter}^{-1}$$

Therefore, barium will precipitate first. When the sulfate concentration has risen to 3.8×10^{-5} mole liter^{-1} and strontium sulfate just begins to precipitate, the residual barium concentration left in solution will be

$$[Ba^{2+}] = \frac{1.5 \times 10^{-9}}{3.8 \times 10^{-5}} = 3.9 \times 10^{-5} \text{ mole liter}^{-1}$$

This quantity is

$$\frac{3.9 \times 10^{-5}}{0.010} \times 100 = 0.39\%$$

or 0.39% of the original Ba^{2+} present. Thus 99.6% of the barium has been precipitated before any strontium begins to precipitate.

Summary

In this chapter we have applied the concepts of chemical equilibrium to ions in aqueous solution, especially to acid–base and precipitation reactions. We have used the equilibrium-constant expression from Chapter 4, with concentrations expressed in units of moles per liter (moles liter^{-1}). Since the concentration of water is effectively constant, especially in dilute solutions, we have incorporated all water concentration terms, $[H_2O]$, into the equilibrium constants.

Water itself ionizes with an equilibrium or **ion-product constant** at room temperature of $K_w = [H^+][OH^-] = 10^{-14}$. To avoid the inconvenience of large exponential numbers, a negative exponent notation is used, whereby $pH = -\log_{10}[H^+]$, $pOH = -\log_{10}[OH^-]$, and $pK_{eq} = -\log_{10} K_{eq}$. In this notation, the dissociation of water can be represented by $pH + pOH = pK_w = 14$. For pure water, $[H^+]$ and $[OH^-]$ must be the same, and equal to 10^{-7} mole liter^{-1}, so the pH and pOH each are equal to 7. The pH is a convenient measure of acidity, since in acid solutions the pH is less than 7, and in basic solutions it is greater than 7.

According to the Brønsted–Lowry theory of acids and bases, any substance that gives up a proton is an *acid,* and any substance that can combine with a proton and remove it from solution is a *base.* When an acid loses its proton, it becomes the **conjugate base**. A strong acid such as HCl has a weak conjugate base, Cl⁻, and a weak acid such as HAc or NH_4^+ has a relatively strong conjugate base, Ac⁻ or NH_3. Any acid whose conjugate base is sufficiently weaker than H_2O (has a lesser affinity for H^+) will be dissociated completely in aqueous solution, hence it is classified as a **strong acid**. Acids that dissociate only partially in aqueous solution are **weak acids**

Strong acids and bases are simple to deal with, since their dissociation is complete in aqueous solution. When a strong acid is added to water, the increase in hydrogen ion concentration equals the concentration of added acid. **Neutralization** occurs when H^+ from an acid combines with OH^- from a base to form water molecules. The amount of acid present in a sample can be determined by finding out how much base of known strength is required to make the solution neutral as measured by an acid–base **indicator**. This is called **titration**, and it is a useful analytical procedure.

The equilibrium expression for a weak acid, equation 5-34, is obtained with the help of two conservation expressions: a mass-balance equation that says the total amount of acid anion is constant, and a charge-balance equation that says the solution must remain neutral as a whole. This simple expression can be solved as a quadratic equation or by the method of successive approximations, and it is valid as long as the solution is so acid that the contribution to $[H^+]$ from the dissociation of water can be neglected. If this is not the case, then a more complete expression (Appendix 5) must be used. Acid–base indicators themselves are weak acids or weak bases whose dissociated and undissociated forms have different colors.

A **buffer** is a mixture of a weak acid and its salt with a strong base, or alternatively, of a weak base and its salt with a strong acid. The equilibrium between the acid and salt form of the substances shifts to counteract the effect of added acid or base, making the buffered solution resistant to pH change. The pH in such solutions can be calculated from equations 5-42 and 5-46.

Hydrolysis is the interaction of the salt of a weak acid (or weak base) with water to form undissociated acid (or base) and OH^- (or H^+) ions. What is sometimes described as a **hydrolysis constant** is actually nothing more than the dissociation constant for the conjugate of the weak acid or base. The base constant, K_b, and the acid constant, K_a, are related by $K_a K_b = K_w$.

Some acids can release more than one proton in successive dissociations. These are called **polyprotic acids**. As long as the successive dissociation constants, K_1, K_2, and so on, differ by factors of 10^{-4} or 10^{-5}, the successive dissociations can be treated as separate events.

Most of the general comments just made about solving acid–base equilibrium problems are applicable to solubility equilibria, for situations in which ions combine to form an insoluble salt. Solubility-product calculations are more useful to indicate whether precipitation will occur under certain conditions, what the upper limit on concentration of an ion in solution may be, and whether two ions can be separated in solution by selective precipitation.

Self-Study Questions

1. In what sense is water a polar molecule? How does its polarity help it to be a better solvent? For what kinds of substances is water an especially good solvent?

2. Why is water a better solvent for methanol than for methane? Why is table salt more soluble in water than in benzene?

3. What is hydration of an ion? What kinds of ions are hydrated? What is solvation, and how is it related to hydration? How does hydration affect the solubility of salts?

4. What is the ion-product constant for the dissociation of water? Why does the concentration of water not appear explicitly in the expression?

5. What is the pH scale, and why is it useful? Does a strong acid have a high pH or a low pH? Does it have a high pOH or a low pOH? If the pH of a solution is 3, what are the pOH and the hydrogen and hydroxide ion concentrations?

6. When the pH changes by two units, by what factor does the hydrogen ion concentration change? When the pH changes by one unit, by what factor does the hydroxide ion concentration change?

7. What is an acid, and what is a base, in the theories of Arrhenius and of Brønsted and Lowry? What are conjugate acids and bases? Give two examples of conjugate acid–base pairs, in one of which the acid is charged and the base is neutral, and in the other of which the acid is neutral and the base is charged.

8. What is the difference between strong and weak acids? Why is it correct to say that a strong acid is one whose acid anion is a weaker base than a water molecule? What is the *leveling effect*? In methanol as a solvent, will some acids that are strong in aqueous solution become weak acids, or will some of the acids that are weak in aqueous solution become strong acids? Explain in terms of competition for the proton.

9. How are K_a and K_b, the acid- and base-dissociation constants, defined? Why does their product always equal K_w? Illustrate with an example other than the ones used in this chapter.

10. What is meant by an *equivalent* of an acid or a base in titration? (See Chapter 2 if you have forgotten.) How many equivalent weights of acid are there per mole of hydrochloric acid? Phosphoric acid? Sulfuric acid? How many moles of sodium hydroxide would be required to neutralize 1 mole of sulfuric acid?

11. Why is the choice of an indicator relatively uncritical in the titration of a strong acid by a strong base? Why is it more critical when a weak acid is titrated?

12. What makes some weak acids or bases suitable as indicators? What do they "indicate," and how?

13. Is the pH of a solution of a weak base greater or smaller than the pH of a

solution of the same concentration of a strong base, assuming the same number of equivalents of base per mole? Why?

14. What are mass-balance and charge-balance equations, and how are they involved in deriving equilibrium-constant expressions? What is the physical meaning of these balance equations in terms of atoms, ions, and charge?

15. What is the method of successive approximations, and how can it be applied to solving problems involving weak acids or weak bases?

16. Will a solution of ammonium chloride be acidic, neutral, or basic? As the concentration of the solution is increased, what will happen to the pH?

17. When is the contribution from dissociation of water important in weak-acid equilibria? When may it be neglected?

18. What will be the effect on the degree of dissociation of an aqueous ammonia solution if we add some ammonium chloride? What do you think would happen to the degree of dissociation if instead we added a substance that formed a complex ion such as $Cu(NH_3)_4^{2+}$ with ammonia molecules? How is Le Chatelier's principle involved here?

19. How does a buffer counteract attempts to change the pH of a solution? What are the two components of a typical buffer solution?

20. Under what conditions will the pH of a buffer mixture be equal to the pK_a of the buffer acid? How could you set up the buffer mixture so the pH was one unit less than the pK_a?

21. What mass and charge balances are involved in deriving the general expression for equilibrium in a solution of a weak acid and its salt with a strong base? How does the general expression reduce to the simpler expressions that we have used for weak acids, buffers, and hydrolysis?

22. What is hydrolysis? What is the hydrolysis constant, and how is it related to K_a, K_b, and K_w?

23. What are polyprotic acids? How do their successive equilibrium dissociation constants usually compare?

24. If a dissociation has a pK_a of 4, what will be the ratio of undissociated to dissociated forms in a neutral solution? What would this same ratio be if the pK_a were 9 instead?

25. What is the relationship between solubility (s) and solubility-product constants (K_{sp}) for $CaCO_3$? For CaF_2? For $Ca_3(PO_4)_2$?

26. What is the common-ion effect, and how does it influence solubility equilibria?

27. How can a knowledge of solubility-product constants be used to make analytical separations of ions in solution?

28. How can pH be used to control the concentration of sulfide ion, S^{2-}, in solution? As the pH is increased, does the sulfide ion concentration increase, or does it decrease? Give a physical explanation for your answer.

Problems

Calculation of pH

1. What is the pH of a 0.01M NaOH solution?

2. What is the pH of a $1.0 \times 10^{-10}M$ HCl solution? Why is the answer not pH 10?

3. What is the pH of each of the following solutions: (a) 0.001M HCl; (b) 0.02M HCl; (c) 0.001M NaOH; (d) 0.001M Ba(OH)$_2$?

4. What is the pH of each of the following solutions: (a) 0.10M NaOH; (b) 0.020M HCl; (c) 0.050M NaCl; (d) pure water?

5. What are pH and pOH in a 0.120M solution of acetic acid?

6. What is the pH of each of the following solutions: (a) 0.10M acetic acid; (b) 0.10M chloroacetic acid; (c) 0.01M chloroacetic acid; (d) 0.10M hydrogen cyanide?

Changes in pH

7. A solution contains H$^+$, NO$_2^-$, and HNO$_2$ in equilibrium. In which direction will the pH change if each of the following is added to the original solution: (a) HCl; (b) HCN; (c) NaCN; (d) NaCl; (e) H$_2$O?

8. In which direction will the pH of pure water change if each one of the following is added to it: (a) HCN; (b) HCl; (c) NaCN; (d) NaCl; (e) NH$_4$Cl; (e) NH$_3$; (f) NH$_4$CN?

pH and K_a

9. If a 0.10M acetic acid solution is 1.3% ionized, what is the pH of the solution? What is K_a for the acid? Compare your answer with the value in Table 5-3.

10. If a 0.10M HF solution is 5.75% ionized, what is the pH of the solution? What is K_a for HF? Compare your value with that in Table 5-3.

pH

11. From the data in Table 5-3, calculate the dissociation constant for ammonium hydroxide. Is undissociated NH$_4$OH really present in the solution? If not, what is the reaction for the production of ammonium ion and OH$^-$? What is the pH of a 0.0100M solution of ammonia?

12. A detergent box must bear a warning label if its contents will form a solution of pH greater than 11, because a strong base degrades protein structure. Should a box bear such a label if the H$^+$ concentration of a solution of its contents is 2.5×10^{-12} mole liter^{-1}?

Weak acids

13. The ionization constant for arsenious acid, HAsO$_2$, is 6.0×10^{-10}. What is the pH of a 0.10M solution of arsenious acid? What is the pOH of a 0.10M solution?

14. What are the CN$^-$ ion concentration and the pOH in a 1.00M aqueous solution of HCN?

15. What is the equilibrium concentration of NO$_2^-$ ion in a 0.25M aqueous solution of nitrous acid? What is the pH? What is the percent ionization of HNO$_2$?

Weak bases

16. A solution of ammonia has a hydrogen

ion concentration of 8.0×10^{-9} mole liter^{-1}. What is the pOH of this solution?

17. Pyridine is an organic base that reacts with water as follows:

$$C_5H_5N + H_2O \rightleftarrows C_5H_5NH^+ + OH^-$$

The base-dissociation constant for this reaction, K_b, is 1.58×10^{-8}. What is the concentration of $C_5H_5NH^+$ ion in a solution that was initially $0.10M$ in pyridine? What is the pH of the solution?

18. Hydrazine is a weak base that dissociates in water as follows:

$$N_2H_4 + H_2O \rightleftarrows N_2H_5^+ + OH^-$$

The equilibrium constant for this dissociation at $25°C$ is 2.0×10^{-6}. Write the equilibrium-constant expression for the reaction. If the initial hydrazine concentration is $0.010M$, what is the concentration of hydrazinium ion, $N_2H_5^+$? What is the pH?

Hydrolysis

19. What is the pH of a $0.18M$ solution of ammonium chloride?

20. What is the pH of a $0.025M$ solution of sodium acetate?

21. The hypobromite ion, OBr^-, is the conjugate base of the weak hypobromous acid, $HOBr$. When $0.100M$ sodium hypobromite, $NaOBr$, is dissolved in water, the pH of the solution is 10.85. Write the equation for the hydrolysis of $NaOBr$, and the equilibrium-constant expression for the reaction. Calculate the value of the hydrolysis constant and of the acid-dissociation constant for $HOBr$.

22. The phenolate ion, $C_6H_5O^-$, is the anion of the weak acid phenol, C_6H_5OH.

The anion undergoes hydrolysis according to the equation

$$C_6H_5O^- + H_2O \rightleftarrows C_6H_5OH + OH^-$$

A $0.0100M$ solution of sodium phenolate has a pH of 11.0. Write the expression for the hydrolysis constant. Calculate the numerical values of the hydrolysis constant and the acid-dissociation constant for phenol.

23. The pH of a $0.100M$ sodium nitrite solution is 8.15. Calculate the hydrolysis constant, K_b, for NO_2^-. Calculate the dissociation constant for nitrous acid.

24. What is the pH of a $1.0M$ solution of sodium cyanide?

Buffers

25. A buffer solution is made with $0.30M$ sodium cyanide and $0.30M$ HCN. What is the pH of the buffer solution?

26. What is the pH of a buffer that is $0.20M$ in NH_3 and $0.40M$ in NH_4Cl?

27. A buffer solution is made from equal volumes of $0.10M$ acetic acid and $0.10M$ sodium acetate. What is the pH of the buffer?

28. What is the pH of a solution made from equal volumes of $0.20M$ propionic acid and $0.20M$ sodium propionate?

pH control

29. If 0.010 mole of HCl gas is dissolved in 1 liter of pure water, what is the final pH? If the same amount of HCl is dissolved instead in 1 liter of the buffer solution of the preceding problem, what is the final pH?

30. A solution is $0.10M$ in formic acid and $0.010M$ in sodium formate. What is the

pH of the solution? If 0.10 g of sodium hydroxide is added to a liter of the buffer, what is the final pH? What will the final pH be if 1.0 g of NaOH is added instead?

31. If 20 ml of a solution of $0.6M$ ammonia are mixed with 10 ml of a $1.8M$ ammonium chloride solution, what is the final pH? If 1 ml of a $1.0M$ HCl solution is added, what will the pH become?

32. If the buffer solution of the preceding problem had been prepared from $0.06M$ ammonia and $0.18M$ ammonium chloride, would the same HCl solution change the pH more or less than in the first situation? Why? What would the pH be before and after addition of HCl?

Titration

33. Novocain (Nvc) is a weak organic base that reacts with water as follows:

$$Nvc + H_2O \rightleftarrows NvcH^+ + OH^-$$

The base-dissociation constant for this reaction is $K_b = 9.0 \times 10^{-6}$. Suppose that a $0.010M$ solution of Novocain is titrated with nitric acid. (a) What is the pH of the Novocain solution at the beginning of titration, before any acid has been added? (b) At the equivalence point of the titration, the solution behaves just like a solution of $0.010M$ $NvcH^+NO_3^-$. What is the pH of this solution? (c) The indicator bromcresol green has a pK_a of 5.0. Is this indicator suitable for the titration?

34. If $0.10M$ pyridine solution ($K_b = 1.58 \times 10^{-8}$) is titrated with HCl, what is the pH of the solution when the ratio of equivalents of H^+ added to initial equivalents of pyridine is 0.50? What is the pH when this ratio is 1.00? From the information in Figure 5-5, which indi-

cator would be most suitable for this titration—methyl violet, methyl orange, bromthymol blue, or alizarin?

Dissociation constant

35. A student titrated a spoonful of an unknown monoprotic acid with an NaOH solution of unknown concentration. After adding 5.00 ml of base, he found the pH of the solution to be 6.0. The equivalence point came when he added 7.00 ml of additional base. Calculate the dissociation constant of the acid.

Equilibrium constants

36. Determine equilibrium constants for the following reactions, using data in this chapter:
   a) $NO_2^- + HF \rightleftarrows HNO_2 + F^-$
   b) $CH_3COOH + F^- \rightleftarrows$
       $HF + CH_3COO^-$
   c) $CH_3COOH + SO_3^{2-} \rightleftarrows$
       $HSO_3^- + CH_3COO^-$
   d) $NH_3 + HSO_3^- \rightleftarrows SO_3^{2-} + NH_4^+$
   Arrange the Brønsted acids in these equations in order of increasing acid strength.

Polyprotic acids

37. What are the concentrations of H^+, HCO_3^-, CO_3^{2-}, and $CO_2(aq)$ in a saturated solution of carbonic acid ($0.034M$ in CO_2)? What is the pH of this solution?

38. What are the concentrations of H^+, HSO_4^-, SO_4^{2-}, and H_2SO_4 in a $0.30M$ solution of sulfuric acid? What is the pH?

39. What are the concentrations of H^+, HSO_3^-, SO_3^{2-}, and H_2SO_3 in a $0.050M$ solution of SO_2?

40. What are the concentrations of H^+, $H_2AsO_4^-$, $HAsO_4^{2-}$, AsO_4^{3-}, and H_3AsO_4 in a $0.30M$ solution of arsenic acid?

K_{sp} calculation

41. The solubility of silver phosphate, Ag_3PO_4, in water is 0.0067 g liter^{-1} at 20°C. What is the solubility product, K_{sp}, for this salt? What is the solubility of silver phosphate, in moles liter^{-1}, in a solution containing a total of 0.10 mole liter^{-1} of Ag^+?

42. If a solution containing 0.16 mole liter^{-1} of Pb^{2+} is made 0.10 M in chloride ion, 99.0% of the Pb^{2+} is removed as $PbCl_2$. What is K_{sp} for $PbCl_2$?

Solubilities

43. With data in Table 5-7, calculate the solubility in moles liter^{-1} of MgF_2 in pure water. What is the solubility in $0.050M$ NaF?

44. What is the solubility of CoS in pure water, in moles liter^{-1}? What is the solubility of CoS in $0.10M$ sodium sulfide solution?

Common-ion effect

45. What is the silver ion concentration in a solution of silver chromate in pure water? In $0.10M$ chromate solution?

46. Calculate the calcium ion concentration in a saturated solution of calcium fluoride. What will the calcium ion concentration be if the solution is also $0.05M$ in NaF?

Precipitation

47. A solution is made $0.10M$ in Mg^{2+}, $0.10M$ in NH_3, and $1.0M$ in NH_4Cl. Will $Mg(OH)_2$ precipitate?

48. How many grams of ammonium chloride must be added to 100 ml of $0.050M$ ammonium hydroxide to prevent the precipitation of ferrous hydroxide, $Fe(OH)_2$, when the NH_4Cl–NH_4OH mixture is added to 100 ml of $0.020M$ $FeCl_2$? Assume that the addition of solid NH_4Cl produces no volume change.

pH and precipitation control

49. The K_{sp} of calcium phosphate, $Ca_3(PO_4)_2$, is 1.3×10^{-32}, and the third ionization constant of phosphoric acid is 2.2×10^{-13}. Suppose that 0.31 g of calcium phosphate is added to 100 ml of water, and the pH of the solution is adjusted until all of the calcium phosphate dissolves. What is this pH? (Assume that HPO_4^{2-} is the only other species formed in the solution, and that $CaHPO_4$ is soluble.)

50. In the precipitation of metal sulfides, selective precipitation can be achieved by adjusting the hydrogen ion concentration. At what pH does ZnS begin to precipitate from a $0.077M$ solution of H_2S containing $0.08M$ Zn^{2+}? (Necessary data are in the tables in this chapter.)

Solubility and pH

51. What is the solubility of AgOH in a buffer at pH 13?

52. In a water solution saturated with H_2S, the "ion product" for H_2S is as given in Example 22. Calculate the solubility of FeS at pH 9 and at pH 2. Can you see how its behavior might be useful in analytical separations?

53. Calculate the solubility of $Mg(OH)_2$ in aqueous solution at pH 2 and pH 12.

How is this behavior useful in chemical separations?

54. Frequently, communities partially soften their water by adding slaked lime, $Ca(OH)_2$, to the water supply. The slaked lime reacts with HCO_3^-,

$$Ca(OH)_2(s) + 2HCO_3^- \rightarrow$$
$$CaCO_3(s) + CO_3^{2-} + 2H_2O(l)$$

to produce a mole of CO_3^{2-}, which further reacts with Ca^{2+} originally in the water to precipitate $CaCO_3$. Thus Ca^{2+} ion is added to remove Ca^{2+} ion. A malfunctioning water-softening plant once delivered saturated $Ca(OH)_2$ solution to the home owners' taps in Charleston, Illinois. Calculate the pH of a saturated solution of $Ca(OH)_2$ at 0°C. Is it unsafe for human consumption? (See Problem 12.) The solubility of $Ca(OH)_2$ can be found in a handbook.

Precipitation

55. Three suggestions are made for ways to remove silver ions from solution: (a) Make the solution $0.010M$ in NaI. (b) Buffer the solution at pH 13. (c) Make the solution $0.0010M$ in Na_2S. What will be the equilibrium silver ion concentration in each case? Which course of action is most effective in removing Ag^+ ions?

Suggested Reading

A. J. Bard, *Chemical Equilibrium*, Harper & Row, New York, 1966.

J. N. Butler, *Ionic Equilibrium*, Addison-Wesley, Reading, Mass., 1964.

J. N. Butler, *Solubility and pH Calculations*, Addison-Wesley, Reading, Mass., 1964.

A. F. Clifford, *Inorganic Chemistry of Qualitative Analysis*, Prentice-Hall, Englewood Cliffs, N.J., 1961.

E. S. Gould, *Inorganic Reactions and Structure*, Holt, Rinehart and Winston, New York, 1962, 2nd ed.

K. B. Harvey and G. B. Porter, *Introduction to Physical Inorganic Chemistry*, Addison-Wesley, Reading, Mass., 1963.

E. J. King, *Acid–Base Equilbria*, Macmillan, New York, 1965.

E. J. King, *Qualitative Analysis and Electrolyte Solutions*, Harcourt Brace Jovanovich, New York, 1959.

T. Moeller and R. O'Conner, *Ions in Aqueous Systems*, McGraw-Hill, New York, 1972.

L. Pauling, *The Nature of the Chemical Bond*, Cornell University Press, Ithaca, N.Y., 1960, 3rd ed.

M. J. Sienko, *Chemistry Problems*, W. A. Benjamin, Menlo Park, Calif., 1972, 2nd ed.

M. J. Sienko and R. A. Plane, *Physical Inorganic Chemistry*, W. A. Benjamin, Menlo Park, Calif., 1963.

6

Are Atoms Real?
From Democritus
to Dulong and Petit

The atomic hypothesis provides a convenient form of speech, which successfully describes many of the facts in a metaphorical manner. But the handy way in which the atomic hypothesis lends itself to the representation of the characteristic features of a chemical change falls short of constituting a proof that atoms have any real existence.

Alexander Smith, Professor of Chemistry, University of Chicago (1910)

In the first five chapters you encountered some of the most fundamental ideas in chemistry: atoms, molecules, moles, conservation of mass and energy, behavior of gases, kinetic theory, equilibrium, and acid–base chemistry in solution. All these ideas have been presented in a very dogmatic way, without proof of any kind. It is time now to stop being a believer and to become a skeptic. How do we know that the material in the first five chapters is true? How do we know, for example, that the molecular formula for water is H_2O? After all, the best chemists in the world thought it was HO for a full 58 years after the atomic theory was proposed in 1802 by John Dalton. Why did they change their minds? What gives us the right to assert that one atom of carbon is approximately 12 times as heavy as one atom of hydrogen? It is not easy to think of ways to weigh out equal numbers of atoms without the mole concept, and this concept depends on the existence of a reliable set of atomic weights, which brings us back in a circle to the relative weights of carbon and hydrogen. How can the circle be broken?

How are the atomic numbers for elements obtained? Why should atoms with the same atomic number but different atomic weights (isotopes) have so nearly identical chemical properties that they are given the same symbol and classed as one element? What evidence is there that the negative charges in an atom are on the outside, and the positive charges are grouped in a tiny central nucleus that contains virtually all the mass of the atom? And what do we mean by the radius of an atom? Is not the size of an atom as difficult to measure as its weight? What laboratory measurements can be related to such microscopic dimensions, and how can we be sure that the relationship is correct?

How, in fact, do we know that atoms exist at all? How do we really know that everything said so far is not the product of the chemist's hyperactive imagination? Perhaps Professor Smith, the author of our chapter-opening quote, was right. Alchemists explained chemical reactions in terms of mythological figures or planets (the distinction was not clear in their own minds) that they associated with the reagents: gold with the sun, copper with Venus, iron with Mars, tin with Jupiter, and lead with Saturn. In what way are atoms more successful models than Greek gods? And how are hydrogen, helium, lithium, beryllium, and so on really more satisfactory as "fundamental materials" than the earth, air, fire and water of Empedocles in ancient Greece?

We have already mentioned Faraday's experiments with ions and electrolysis, and Thomson's and Millikan's measurements of electron charge and mass, in Chapter 1. The tremendous achievement of Mendeleev and Meyer in building the periodic table of the elements is the subject of Chapter 7. The work of Rutherford, Bohr, Schrödinger, and others in developing the modern theory of atomic structure and bonding is described in Chapter 8. In this chapter we shall go back even farther, and focus on two men who revolutionized chemistry: Antoine Lavoisier (1743–1794), who demonstrated that the fundamental quantity in any chemical reaction is *mass,* and John Dalton (1766–1844), who proposed that the fundamental units in chemical reactions are *atoms.* Dalton was not the first to propose the idea of atoms in principle, but he was the first to show in a convincing way that atoms do exist, and that they are a useful basis for understanding chemical reactions.

This chapter is an exercise in both chemical history and chemical understanding—the two frequently go hand in hand. One of the guiding principles of this book is that knowing how chemical concepts evolved helps to make them more comprehensible and more interesting. Such historical material is usually presented in postscripts at the ends of chapters. In effect, this chapter is one long postscript to the first five chapters. As you travel through the rest of this book, study and learn the material in the chapters, and relax and enjoy the postscripts.

6-1 THE CONCEPT OF AN ELEMENT

One of the oldest ideas in science is that of fundamental materials out of which everything else is made. Empedocles (500 B.C.), in Greece, performed what may be the first recorded chemical analysis. He noted that when wood burns, smoke or air rises first and is followed by flame or fire. Water vapor

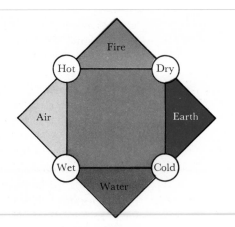

Figure 6-1

The Greeks of the fifth century B.C. pictured all material substances as composed of different proportions of the four basic elements: earth, air, fire, and water. These elements shared, in pairs, the properties of heat or cold and wetness or dryness: Earth was cold and dry, water was cold and wet, air was hot and wet, and fire was hot and dry.

will condense on a cool surface held near the flame. After combustion, the remains are ash or earth. Empedocles interpreted combustion as a breaking down of the burning substance into its four elements: earth, air, fire, and water. He and later writers generalized these into the four elements of which all substances were composed in varying proportions (Figure 6-1). Originally, at least, these ideas were not meant to be flights of metaphysical invention, but were attempts to explain observations. Later, among the Greek, Arabic, and medieval alchemists, the ideas become imbued with mysticism. Then earth, air, fire, and water were abandoned as fundamental elements, but varying sets of what we now would call elements or simple compounds were chosen by different alchemists as the fundamental materials of nature.

Aristotle (384–322 B.C.) gave a definition of an element that, even now, can hardly be improved:

"Everything is either an element or composed of elements. . . . An element is that into which other bodies can be resolved, and which exists in them either potentially or actually, but which cannot itself be resolved into anything simpler, or different in kind."

However, this definition doesn't answer the question of how to recognize an element when you encounter one. Robert Boyle (1627–1691) gave a more practical definition: *An element is a substance that will always gain weight when undergoing chemical change.* This statement must be understood in the sense in which it was intended. For example, when iron rusts, the iron oxide produced weighs more than the original iron. Yet the weight of the iron *and* the oxygen that combines with it is exactly the same as the weight of the iron oxide. Conversely, when the red powder of mercuric oxide is heated, oxygen gas is emitted, and the silvery liquid mercury that remains weighs less than the original red powder. But if this decomposition takes place in an enclosed flask, one sees that there is no *overall* loss of weight during the reaction. (It was a century after Boyle that Lavosier made careful weighing experiments demonstrating the conservation of mass in such reactions.)

By Boyle's definition, mercuric oxide could not be an element, because it can be decomposed into parts, each of which is lighter than the original substance. Mercury could provisionally be called an element, at least until the day when someone else succeeded in separating it into components. Until the present century of spectroscopy and other laboratory techniques, it was easy to prove that a substance was *not* an element, but impossible to prove that one was. As the famous German chemist Justus von Liebig wrote, in 1857, "The elements count as simple substances not because we know that they are so, but because we do not know that they are not."

The elements called the *rare earths* provide an example of the difficulties of proving by purely chemical means that a substance is an element. In 1839, the Swedish chemist Carl Mosander extracted a new element from cerium nitrate and named it *lanthanum* (from the Greek *lanthanein*, "to lie hidden"). Two years later he showed that his lanthanum-containing preparation contained a second element which he christened *didymium* (from the Greek *didymos*, or "twin"). In 1879, François Lecoq de Boisbaudran isolated another substance, *samarium*, from the didymium preparation, and all these were accepted as chemical elements. But didymium vanished from the rolls of chemistry in 1885, when the Austrian Carl von Welsbach separated it into two new elements, *neodymium* ("new twin") and *praseodymium* ("green twin"). It is only because we now have the periodic table, and understand the principles behind its construction, that we can say that there can be *no* other new elements between hydrogen, $_1$H, and element 105.

What kinds of substances are elements? The first to be recognized correctly as such were the metals. Gold, silver, copper, tin, iron, platinum, lead, zinc, mercury, nickel, tungsten, and cobalt all are metals. In fact, all but 22 of the 105 known elements have metallic properties. Five of the nonmetals (helium, neon, argon, krypton, and xenon) were discovered in the mixture of gases that remained when all the nitrogen and oxygen in air were removed. Chemists thought that these "noble" gases were inert until 1962, when it was shown that xenon combines with fluorine, the most chemically active nonmetal. The other chemically active nonmetals are either gases (such as hydrogen, nitrogen, oxygen, and chlorine) or brittle, crystalline solids (such as carbon, sulfur, phosphorus, arsenic, and iodine). Only one nonmetallic element, bromine, is liquid under ordinary conditions.

6-2 COMPOUNDS, COMBUSTION, AND THE CONSERVATION OF MASS

Most eighteenth-century chemists were devoted to preparing and describing pure compounds, and to decomposing them to the elements from which they are formed. The great advances of the time were in the chemistry of gases. In 1756, Joseph Black completely changed chemists' ideas about gases when he showed, in his M.D. thesis at Edinburgh, that marble (which we know to be primarily calcium carbonate, $CaCO_3$) could be decomposed to quicklime (calcium oxide, CaO) and a gas (carbon dioxide, CO_2), and that *the process could be reversed.* This demonstration proved that there were different kinds of gases, and that they could take part in chemical reactions just as well as liquids and solids could. One of Black's contemporaries, John Robinson, wrote the following:

"He had discovered that a cubic inch of marble consisted of about half its weight of pure lime and as much air as would fill a vessel holding six wine gallons. . . . . What could be more singular than to find so subtle a substance as air existing in the form of a hard stone, and its presence accompanied by such a change in the properties of the stone?"

In the following years, Henry Cavendish discovered hydrogen (1766), Daniel Rutherford found nitrogen (1772), and Joseph Priestley invented carbonated water and identified nitrous oxide ("laughing gas"), nitric oxide, carbon monoxide, sulfur dioxide, hydrogen chloride, ammonia, and oxygen. In 1781, Cavendish proved that water is a combination of only hydrogen and oxygen, after he had witnessed Priestley explode the two gases in what Priestley later recalled as "a random experiment to entertain a few philosophical friends." The discovery of oxygen (Figure 6-2) led Antoine

Figure 6-2

Priestley's apparatus for preparing oxygen gas. Mercuric oxide in a small pan floating on the surface of the mercury bath is decomposed to liquid mercury and oxygen by solar heat. The arrangement of the mercury bath and bell jar prevents loss of the gas evolved.

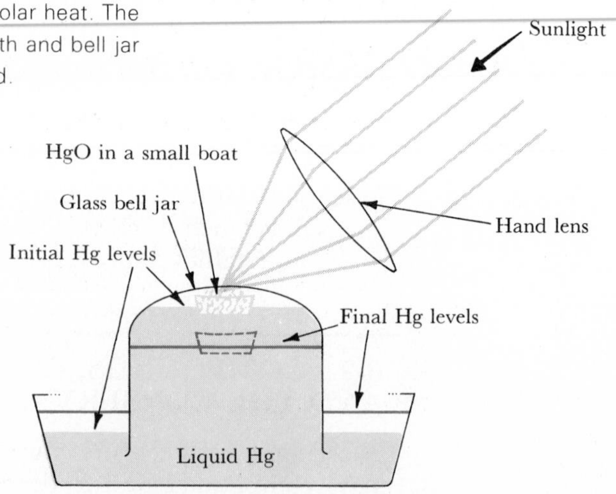

Lavoisier to overthrow the predominant idea of eighteenth-century chemistry, the phlogiston theory. The process by which this theory was shattered illustrates the great importance of quantitative measurements in chemistry.

Phlogiston

When Empedocles watched wood burn, he was impressed with the idea that something *left* the wood; only a light fluffy ash remained. It became generally accepted that combustion was the decomposition of a substance accompanied by a loss of weight. Metal oxides are usually less dense and less compact than the metals from which they come. Even when it became known that the oxide was heavier than the original metal, a confusion between density (weight per unit volume) and weight itself compounded the error. The Germans Johann Becher and Georg Stahl proposed, in 1702, that all combustible material contains an element called **phlogiston**, which escapes when the material burns. According to their theory,

1. Metals, when heated, lose phlogiston and become **calces**. (A calx is a crumbly residue.)

2. Calces, when heated with charcoal, reabsorb phlogiston and become metals again. The charcoal is necessary because the original phlogiston has become scattered through the surrounding atmosphere and lost.

3. Charcoal must therefore be very rich in phlogiston.

By this theory, a lit match goes out when it is placed in a closed bottle because the air in the bottle becomes saturated with phlogiston; respiration in living organisms is a purification process in which phlogiston is removed; a mouse under a bell jar eventually dies when the air around him has absorbed all the phlogiston it can.

Think about these ideas for a while. So long as you make no weighing experiments, this theory explains combustion as well as our present ideas do, and seems to agree with common-sense observations about the appearance of metals and calces. Jean Rey, in France, had demonstrated that tin gains weight when it burns, but chemists, unaccustomed to attaching much importance to weight, overlooked the significance of Rey's work. In 1723, Stahl gave a clever explanation for Rey's finding:

"The fact that metals when transformed into their calces increase in weight, does not disprove the phlogiston theory, but, on the contrary, confirms it, because phlogiston is lighter than air, and, in combining with substances, strives to lift them, and so decrease their weight; consequently, a substance which has lost phlogiston must be heavier than before."

It is no wonder that hydrogen, when it was discovered, was hailed as the first preparation of pure phlogiston! Again there was a confusion between the two ideas of weight and of density (in terms of buoyancy).

Conservation of Mass

Lavoisier discovered that mercuric calx lost weight when it was heated and free mercury and a gas were produced. He measured the volume of gas released. Then he showed that when mercury was reconverted to calx, the same volume of this gas was reabsorbed and there was a weight increase equal to the earlier loss. On the basis of careful weighing experiments such as these, Lavoisier proposed that combustible materials burn by *adding* oxygen, thus increasing in weight. (*Oxygen* was his name for the gas. Priestley called it *dephlogisticated air* since it could apparently absorb even more phlogiston than atmospheric air could.) Lavoisier demonstrated that the products obtained from burning wood, sulfur, phosphorus, charcoal, and other substances were gases whose weight always exceeded that of the solids that burned. His rebuttal to the metallurgical explanations of Becher and Stahl was as follows:

1. Metals combine with oxygen from the air to form calces, which are oxides.

2. Hot charcoal removes oxygen from calces to form a metal and a gas, O_2 (at that time called *fixed air*).

3. Charcoal, therefore, does not combine with the *metal;* rather, it removes the oxygen that had previously been combined with the metal in the calx.

The key to this theory was the chemical balance. Lavoisier was the first chemist to realize the importance of the principle of the conservation of mass. In his *Traité Elémentaire de Chimie,* he wrote:

"We must lay it down as an incontestable axiom, that in all the operations of art and nature, nothing is created; an equal quantity of matter exists both before and after the experiment. . . . Upon this principle, the whole art of performing chemical experiments depends."

Lavoisier was a businessman first and a chemist second. His full-time occupation was as a member of the *Ferme Générale,* an agency that collected taxes on a commission basis for the French government before the revolution. One of his biographers has called his conservation of mass dictum the "principle of the balance sheet," and has claimed to see its origin in his role as tax collector. Be that as it may, in 1794, his connection with the *Ferme Générale* cost him his life.*

---

*Hailed before a revolutionary tribunal because of his past aristocratic associations, Lavoisier heard Coffinhal, president of the tribunal, reject a plea for clemency: "The Republic has no need of chemists and savants. The course of justice shall not be interrupted." This was surely one of the most serious governmental cutbacks in the history of science.

Lavoisier published his textbook, *Traité Elémentaire de Chimie,* in 1789, and it would be difficult to overemphasize the impact that it had on chemistry. In addition to setting forth the principle of conservation of mass in chemical reactions and overthrowing the phlogiston theory, the book contained in an appendix what is essentially our present system of nomenclature. For a generation, therefore, chemistry became "the French science" (the phrase lingered longer in France than elsewhere).

6-3 DOES A COMPOUND HAVE A FIXED COMPOSITION?

After Lavoisier, chemists began an intensive study of quantities in chemical reaction, that is, masses. The distinction between compounds and mixtures or solutions gradually became clear. A feud developed between those who claimed that the ratios of elements in compounds were fixed and those who believed that a continuous range of proportions was possible. The French chemist Berthollet cited alloys of metals in support of the idea of variable composition. But J. L. Proust, in Madrid, insisted that compounds had fixed composition, and correctly recognized alloys as solid solutions, not compounds:

"The properties of true compounds are invariable as is the ratio of their constituents. Between pole and pole, they are found identical in these two respects; their appearance may vary owing to the manner of aggregation, but their [chemical] properties never. No differences have yet been observed between the oxides of iron from the South and those from the North. The cinnabar of Japan is constituted according to the same ratio as that of Spain. Silver is not differently oxidized or muriated in the muriate of Peru than in that of Siberia."

This principle has been called the **law of constant composition.** The dispute between Berthollet and Proust had the good effect of sending chemists to the laboratory to prove the ideas of one or the other camps, and incidentally to compile rapidly a body of knowledge about chemical composition.* Of course, Proust was right; yet there are solid crystalline materials in which, because of defects in the crystal structure, the same ratio of atoms is not quite that predicted by the ideal chemical formula. For example, iron sulfide can vary from $Fe_{1.1}S$ to $FeS_{1.1}$, depending on how the sample is prepared. Such substances are called *nonstoichiometric solids,* although it has been suggested that they be called "berthollides" after the loser in the debate just discussed.

---

*The orthodox viewpoint is that they went to the laboratory to decide between two conflicting theories. Let us be honest: Scientists are people, and science is seldom conducted in such a nonpartisan vacuum.

Equivalent Proportions

Between 1792 and 1802, an obscure German chemist named Jeremias Richter made an important discovery that was almost completely ignored by his contemporaries. His idea was the one of **equivalent proportions**: The same relative amounts of two elements that combine with one another will also combine with a third element (assuming that the reactions are possible at all). This concept is easy to understand from a few examples:

1 g of hydrogen combines with 8 g of oxygen to form water.
1 g of hydrogen combines with 3 g of carbon to form methane.
1 g of hydrogen combines with 35.5 g of chlorine to form hydrogen chloride.
1 g of hydrogen combines with 25 g of arsenic to form arsine.

The chemical reactions and formulas (which were not known at the time) are, in fact,

$$2H_2 + O_2 \rightarrow 2H_2O$$
$$2H_2 + C \rightarrow CH_4$$
$$H_2 + Cl_2 \rightarrow 2HCl$$
$$3H_2 + 2As \rightarrow 2AsH_3$$

Using modern atomic weights, verify that the preceding statements about weights involved in the reactions are true.

Richter's law of equivalent proportion states that if carbon and oxygen combine they should do so in the ratio of 3 to 8 by weight. This is true for what we now know to be CO_2. If they react, carbon and chlorine should do so in the ratio of 3 to 35.5, and this is true for the liquid that we now know as carbon tetrachloride, CCl_4. In a similar way, arsenic forms $AsCl_3$ and As_2O_3, and chlorine and oxygen form Cl_2O.

Combining Weights

A **combining weight** can be defined for each element as the weight of the element that combines with 1 g of hydrogen. If no hydrogen compound exists, it is the weight that combines with 8 g of oxygen or with one combining weight of some other element that does form a hydrogen compound. In this way a branching network of reactions can lead to a table of combining weights for all the elements. Richter's principle, if true, assures us that there will be no contradictions within the table. Such a set of combining weights is shown in Figure 6-3 and Table 6-1.

There is one serious flaw to this scheme, which is why no one took Richter very seriously. The flaw is that many elements appear to have *more than one* combining weight. Carbon forms a second oxide (we know it now as carbon monoxide, CO), in which the ratio of carbon to oxygen is only 3 to 4. Either the combining weight of carbon has risen to 6, or that of oxygen

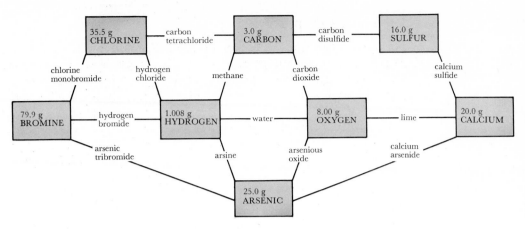

Figure 6-3   Weights of elements that combine with one another in forming the compounds indicated. We can predict from the diagram that 25.0 g of arsenic, for example, will combine with 35.5 g of chlorine or 16.0 g of sulfur. This, in fact, does occur. Arsenic and chlorine react in the predicted weight ratio to give arsenic trichloride ($AsCl_3$), whereas arsenic and sulfur react in the predicted ratio to give the compound arsenous sulfide (As_2S_3).

has fallen to 4. In ethane (C_2H_6) the combining weight of carbon is 4, in ethylene (C_2H_4) it is 6, and in acetylene (C_2H_2) it is 12. The expected oxide of sulfur, SO, does not appear, and in the two most common oxides (SO_2 and SO_3) sulfur has combining weights of 8 and $5\frac{1}{3}$ (Figure 6-4).

Table 6-1

Combining Weights Derived from Simple Compounds

Element	Combining weights (and sources)[a]			
H	1 (by definition)			
O	8 (H_2O)			
C	3 (CH_4, CO_2)	4 (C_2H_6)	6 (CO, C_2H_4)	12 (C_2H_2)
Cl	$35\frac{1}{2}$ (HCl, CCl_4)			
Br	80 (HBr, CBr_4, ClBr)			
As	25 (AsH_3, As_2O_3, $AsBr_3$)			
S	16 (H_2S, CS_2)	8 (SO_2)	$5\frac{1}{3}$ (SO_3)	
Ca	20 (CaO, CaS, Ca_3As_2)			
N	$4\frac{2}{3}$ (NH_3)	$3\frac{1}{2}$ (NO_2)	7 (NO)	14 (N_2O)

[a]Although the compounds were known in Dalton's time, the chemical formulas were not. Apparent anomalies in combining weights are in color.

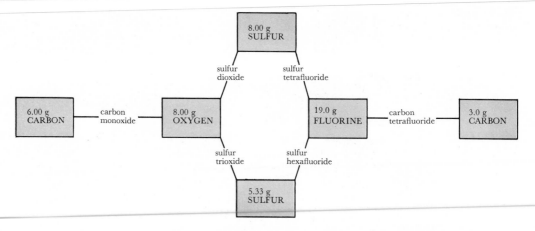

Figure 6-4

Variable combining weights for sulfur and carbon. Note that in this figure and in Figure 6-3 there are three combining weights for sulfur: 5.33 g, 8.00 g, and 16.0 g. These weights are in the ratios of 2:3:6. The two combining weights for carbon are in the ratio of 1:2.

Nitrogen is particularly troublesome. In ammonia it has a combining weight of $4\frac{2}{3}$, and in the three oxides known since Priestley's time its combining weights are $3\frac{1}{2}$, 7, and 14. If you know chemical formulas, the combining weights are easy to calculate, and you should be able to check them. But if you know *only* the combining weights, could you deduce the formulas? The significance of the ratios of elements in compounds was obscured even more by the habit of reporting composition in percent by weight; it was John Dalton who developed the trick of writing them as ratios to one common element and setting up combining weight tables, which we still do. When Humphry Davy reported that the three oxides of nitrogen contained 29.50%, 44.05%, and 63.30% nitrogen by weight, no one noticed that the nitrogen was combining in the relative ratios of 1 to 2 to 4. (These percentages are Davy's experimental values. What are the correct percentages?) By 1802, it was established that compounds had fixed compositions, and that there could be several such definite compositions between the same two elements. Yet no one knew why, or where to go from there.

6-4   JOHN DALTON AND THE THEORY OF ATOMS

John Dalton, a science (or "natural philosophy") teacher in the Manchester, England, schools, was compelled by such data as those in Section 6-3 to propose a theory of atoms, which he presented to the Literary and Philosophical Society of Manchester in 1802 and published three years later. His theory was as follows:

1. All matter is made up of atoms. These are the ultimate particles, and are indivisible and indestructible.

2. All atoms of a given element are identical, both in weight and in chemical properties.

3. Atoms of different elements have different weights and different chemical properties.

4. Atoms of different elements can combine in simple whole numbers to form compounds.

5. When a compound is decomposed, the recovered atoms are unchanged and can form the same or new compounds.

Dalton also emphasized weights, as had Lavoisier; furthermore, Dalton invented a convenient symbolism for atoms, shown in Figure 6-5. Dalton's symbol for hydrogen represents more than merely an unspecified amount of

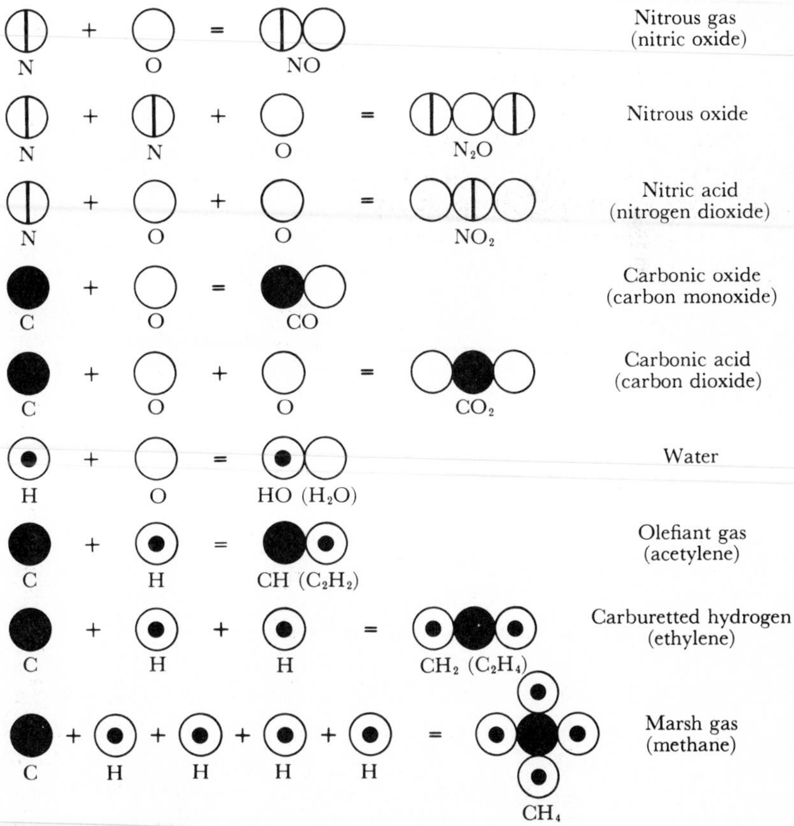

Figure 6-5

Dalton's original symbolism for reactions that form simple compounds. Modern symbols appear beneath them. The names to the right are Dalton's. When either the formula or the name of the product is different today, the modern formula or name is given in parentheses.

hydrogen. It represents either one *atom* of hydrogen or some standard weight of hydrogen containing a standard number of atoms (such as the atomic weight containing Avogadro's number of atoms). Chemical formulas and equations are therefore not merely symbolic, but quantitative.

The Greek Atomic Theory

The idea of atoms was far from new. Democritus and the Epicureans in Greece had proposed an atomic theory, about 400 B.C., that contained virtually all of Dalton's ideas on the subject. The original writings are lost, but we know of this theory from attacks by its opponents and from a long poem written, in 55 B.C., by a Roman Epicurean, Lucretius. (The poem is entitled *De Rerum Natura,* "On the Nature of Things.") After Lucretius, the ideas of atomism drifted in and out of alchemy for nearly 1900 years without making a significant impact on it. Isaac Newton and Lavoisier both believed in atoms, but more as philosophical concepts or figures of speech that helped in thinking about reactions than as a theory requiring experiment.

There is an important point here that cannot be overstressed. A theory in science is important *if, and only if,* it makes the understanding of the behavior of the real world clearer. Describing bronze as a substitutional alloy of tin and copper is superior to describing it as the confluence of Jupiter and Venus, in alchemical terminology, because the tin–copper theory suggests experiments by which the properties of bronze might be explained, predicted, and even improved, whereas the "celestial marriage" theory leads nowhere. But perhaps it is less apparent that Democritus' atomic theory, and even Newton's, was not much of an improvement on this celestial marriage idea; it was Dalton's measurements, explanations, and predictions that made atomic theory valuable.

Fixed Ratios

Dalton took the table of combining weights as his point of departure and asked why the ratios of elements in compounds should be fixed. His answer was that *a compound consists of a large number of identical molecules, each of which is built up from the same small number of atoms, arranged in the same way.* Yet Dalton still needed to know how many atoms of carbon and oxygen combined in each molecule of an oxide of carbon, and how many hydrogen and oxygen atoms combined in a water molecule. Lacking any other guide, he proposed a "rule of simplicity" that started him off well but eventually landed him in serious trouble. The most stable two-element molecule, he reasoned, would be the simple diatomic one, AB. If only one compound of two elements were known, it would be an AB compound. Next most stable would be the triatomic molecules, AB_2 and A_2B. If only two or three compounds of two elements were known, they would be of these three types. This rule was one of those principles of economy, like the minimization of energy in mechanics or the principle of least action in physics, which are sometimes right, and sometimes wrong. Dalton was wrong.

Dalton began by mistakenly assuming from his rule of simplicity that water had a diatomic formula, HO. This made the atomic weight of oxygen equal to its combining weight of 8 (all relative to 1 for hydrogen). He then turned to the oxides of carbon and nitrogen; the possible choices are shown in Table 6-2. (All atomic weights in this discussion are based on the true numerical values, not on Dalton's values. He was a notoriously poor experimentalist. The atomic weight of oxygen, even on his own terms, began at 6.5 and slowly worked up to 8.) One oxide of carbon had a carbon-to-oxygen ratio of 0.75, and the other had a ratio of 0.375. If the first oxide were CO—he assumed that one of them had to be—then, as Table 6-2 shows, the other would be CO_2. Thus, the atomic weight of carbon would be 6. If the second oxide were CO, the first would have to be C_2O. (Can you prove this?) Then carbon would have an atomic weight of 3. Since oxide A was more stable to decomposition, he argued that this one must be CO, and correctly chose possibility 1. For the oxides of nitrogen, he similarly ruled out possibilities 1 and 3 because the five-atom molecules clashed with his rule of simplicity; and he again made the correct assignment of an atomic weight of 7 for nitrogen. (Correct, that is, relative to 8 for oxygen.)

Table 6-2

The Possible Choices of Chemical Formulas Open to Dalton for the Oxides of Carbon and Nitrogen

	C/O Mass ratio	Possibility 1	Possibility 2	
Oxide A	3/4	CO	C_2O	
Oxide B	3/8	CO_2	CO	
Atomic weight of C (assuming O = 8)		6	3	

	N/O Mass ratio	Possibility 1	Possibility 2	Possibility 3
Oxide A	$3\frac{1}{2}/8$	NO	NO_2	NO_4
Oxide B	7/8	N_2O	NO	NO_2
Oxide C	14/8	N_4O	N_2O	NO
Atomic weight of N (assuming O = 8)		$3\frac{1}{2}$	7	14

Dalton should have sensed trouble as soon as he came to ammonia. He assumed by the rule of simplicity that the molecular formula for ammonia was NH. However, since $4\frac{2}{3}$ g of nitrogen combines with 1 g of hydrogen, this assumption would have meant an atomic weight of $4\frac{2}{3}$ for nitrogen, a value in conflict with the number 7 calculated from the oxides. As an alternative, he could have kept the atomic weight of 7 and worked out the formula for ammonia:

Hydrogen: $\dfrac{1 \text{ g of hydrogen}}{1 \text{ g mole}^{-1}} = 1$ mole of hydrogen atoms

Nitrogen: $\dfrac{4\frac{2}{3} \text{ g of nitrogen}}{7 \text{ g mole}^{-1}} = 0.667$ mole of nitrogen atoms

With the molar ratio of hydrogen to nitrogen (and therefore the ratio of atoms as well) being 1:0.667, or 3:2, the chemical formula would have to be N_2H_3, N_4H_6, or some higher multiple. Such a result would have shaken Dalton's faith in the rule of simplicity, and might have forced him to go back and find the right track. Yet he was undone by the poor quality of his experimental data. His initial value for the combining weight of oxygen was 6.5, which he raised to 7 in 1808. Davy increased it to 7.5, and Proust finally arrived at the correct figure (given Dalton's assumptions) of 8. Dalton refused to believe their values (a stubborn attitude for such a poor experimentalist), and all the nitrogen calculations described here were carried out by Dalton with a nitrogen atomic weight of 5 rather than 7.

Law of Multiple Proportions

It is easy to be critical of a man who has gone astray because of bad data. But the real achievement of the atomic theory, which made people accept it almost at once, was not the calculation of atomic weights. It was that the atomic theory explained perfectly a principle that had lain unnoticed in the published literature for over 15 years, relating elements that combine to form more than one compound. This was Dalton's **law of multiple proportions**.

The law of multiple proportions states that if two elements combine to form more than one compound, then the amounts of one element that combine with a fixed amount of the other will differ by factors that are the ratios of small whole numbers. (Or, that you can multiply the amounts by a suitable constant and produce a set of integers.) Since we have been using combining weights, perhaps a more meaningful statement is that if an element shows several combining weights, these weights will differ among themselves by ratios of small whole numbers. For example, the combining weights of carbon in Table 6-1 differ in the ratios of 3 to 4 to 6 to 12, or, more revealingly, in the ratios of 1/4 to 1/3 to 1/2 to 1. The combining weights for sulfur are in the ratios of 1 to 1/2 to 1/3, and the ones for nitrogen are

1/3 to 1/4 to 1/2 to 1 in NH_3, NO_2, NO, and N_2O. Dalton's explanation of these simple fractions was that one, or two, or any small number of atoms could combine with one of another kind, but that a molecule with 1.369 . . . atoms combined with 1 atom of another was physically impossible according to atomic theory. The combining weights differ by small whole number fractions because the atoms combine in small whole numbers.

A search through the chemical literature showed that this law was the universal rule. It is one thing to prove your theory with new data that you have collected, but it is much more impressive to prove it with everyone else's; this is what Dalton did. The acceptance of the atomic theory was rapid and almost unanimous.

6-5   EQUAL NUMBERS IN EQUAL VOLUMES: GAY-LUSSAC AND AVOGADRO

As chemists tried to deduce formulas for more and more compounds, the flaws in Dalton's atomic weights and in his rule of simplicity became more and more obvious. No one could come up with a dependable method of deciding on chemical formulas. Of the three pieces of molecular information —combining weights of the elements, atomic weights of the elements, and molecular formulas—any one could be calculated if the other two were known. Yet only one, the combining weight, was directly measurable. Dalton's wrong assumptions about formulas led to wrong atomic weights, which led back again to wrong formulas for new compounds. Between 1850 and 1860 more than 13 different formulas were assigned to acetic acid, the common acid of vinegar. The confusion was so great that some chemists despaired for the atomic theory. Jean Dumas wrote:

"If I were in charge, I would efface the word *atom* from science, for I am persuaded that it goes far beyond experience, and chemistry must never go beyond experience."

The great German chemist Friedrich Wöhler complained, even as early as 1835, that

". . . organic chemistry just now is enough to drive one mad. It gives me the impression of a primeval tropical forest, full of the most remarkable things, a monstrous and boundless thicket, with no way to escape, into which one may well dread to enter."

However, the key to the dilemma was already in the chemical literature, and had been since 1811. The first step was provided by Gay-Lussac, and the second by Avogadro.

Gay-Lussac

In 1808, Joseph Gay-Lussac (1778–1850) began a series of experiments with the *volumes* of reacting gases. He found that equal volumes of HCl gas and ammonia form neutral, solid ammonium chloride. An initial excess of either gas is left over at the end of the reaction. Two volumes of hydrogen react with one of oxygen to form two volumes of steam; three volumes of hydrogen react with one of nitrogen to yield two volumes of ammonia; and one volume of hydrogen reacts with one of chlorine to produce two volumes of HCl gas. In these and other experiments, in which the gas reactions were usually explosions triggered by a spark in an enclosed container, Gay-Lussac always found that gases react [in simple whole number units of volume units] provided that after the explosion the products are brought back to the temperature and pressure of the initial gases.

Gay-Lussac was a cautious man and a protégé of Berthollet, who as we have seen did not believe in compounds with fixed compositions. Gay-Lussac drew no conclusions in his *Memoire,* but the possibility of a connection with Dalton's atomic theory was apparent.

Avogadro

Dalton used Gay-Lussac's data to "prove" that equal volumes of gas do not have equal numbers of molecules, another wrong turn, like his rule of simplicity. Dalton's argument is illustrated in Figure 6-6a. The Italian physicist Amedeo Avogadro (1776–1856) saw another path. He began by assuming that equal volumes of gas (at the same temperature and pressure) contain *equal numbers of molecules.* As Figure 6-6b shows, this assumption requires that gases of the reactive elements such as hydrogen, oxygen, chlorine, and nitrogen be composed of two-atom molecules instead of single, isolated atoms. If Avogadro had been believed when he published his ideas in 1811, a half-century of confusion in chemistry would have been avoided. To many people, though, his ideas seemed like one shaky assumption (equal numbers in equal volumes) buttressed by an even shakier one (diatomic molecules). At that time, ideas of chemical bonding were based almost entirely on electrical attraction and repulsion, and it was difficult for scientists to understand how two *identical* atoms could do anything but repel each other. And if they did attract, why did they not form larger molecules, such as H_3 and H_4? Jöns Jakob Berzelius (1779–1848) used data on the vapors of sulfur and phosphorus to undercut Avogadro. Yet Berzelius did not realize that these were examples of just such higher aggregates (S_8 and P_4). Avogadro himself did not help matters; he mixed terminology so much that it sometimes appeared as if he were splitting hydrogen atoms ("elementary molecules") rather than separating atoms in a diatomic molecule ("integral molecules").

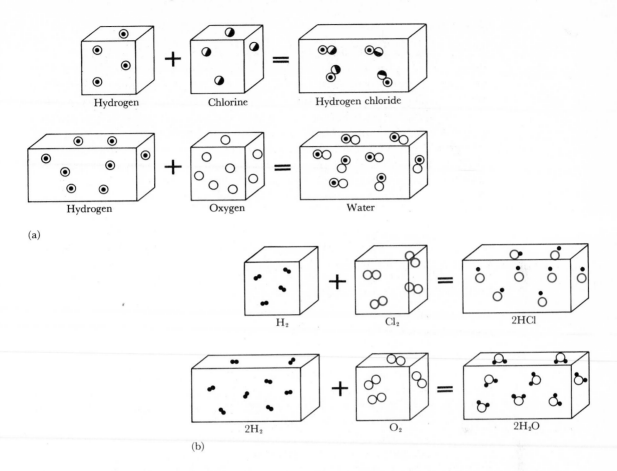

(a)

(b)

Figure 6-6   Gay-Lussac's results on the combining volumes of gases and the explanations by (a) Dalton and (b) Avogadro. Gay-Lussac found that one volume of hydrogen and one of chlorine produce two volumes of HCl gas, and that two volumes of hydrogen react with one of oxygen to produce two volumes of steam. (a) Dalton agreed that if the volume of HCl is twice the volume of either hydrogen or chlorine there must be half as many molecules per volume unit in HCl. Similarly, if there are n molecules of hydrogen per unit of volume, and if each of these molecules produces a water molecule in the same total volume, there will also be n molecules per volume unit in water. But only half the volume of oxygen is required, so the density of oxygen must be $2n$ molecules per volume unit. Thus, in hydrogen chloride, hydrogen, chlorine, water, and oxygen, the numbers of molecules per volume unit are $n/2$, n, n, n, and $2n$. (b) Avogadro proposed that each molecule of hydrogen, chlorine, or oxygen contains two atoms. All the participants in the HCl reaction would therefore have the same number of molecules per volume unit of gas. Applying this same assumption to the water reaction led to a new formula for water, H_2O, and ultimately to a complete revision of Dalton's atomic weight scale.

6-6   CANNIZZARO AND A RATIONAL METHOD OF CALCULATING ATOMIC WEIGHTS

By 1860, the confusion about atomic weights was so widespread that nearly every chemist of any repute had his own private method of writing chemical formulas. August Kekulé (the inventor of the Kekulé structure for benzene) called a conference in Karlsruhe, Germany, to try to reach some kind of an agreement. The man who settled the entire issue was the Italian Stanislao Cannizzaro (1826–1910), who based a rigorous method for finding atomic weights on the long-ignored work of his countryman Avogadro.

Cannizzaro's reasoning, based on Avogadro's principle that equal volumes of gas contain equal numbers of molecules, was as follows:

1. Assume that the atomic weight of hydrogen is 1.0, and that hydrogen gas is made up of diatomic molecules, as Gay-Lussac's gas volume experiments suggest.

2. Assume that Avogadro was correct in deducing that oxygen gas is also diatomic, O_2, and hence that the molecular formula for water is H_2O, not HO. (See Figure 6-6b.) Since the combining weight of oxygen in water is 8.0, the atomic weight of oxygen must be 16.0, and the molecular weight of O_2 must be 32.0.

3. If equal volumes of all gases contain equal numbers of molecules, then the molecular weight (M) of a gas is proportional to the density (D) of the gas: $M = k\,D$. Use H_2 and O_2 to evaluate the proportionality constant, k:

Gas	Density, D (g liter^{-1})	Molecular weight, M (g mole^{-1})	Constant, k (liters mole^{-1})
H_2	0.0894	2.0	22.37
O_2	1.427	32.0	22.42
		Average value:	22.4

(The fact that k is the same for both H_2 and O_2 indicated that Cannizzaro was on the right track.)

4. Evaluate the molecular weights of a series of compounds containing the elements whose atomic weights are to be determined. Starting with percent composition by weight from chemical analyses, and molecular weights calculated from gas densities, calculate the weight of each element *per molecular unit*. Look over these weights for a given element to see if the numbers can be interpreted as integral multiples of some common factor, which may then be the atomic weight.

In the data given in Table 6-3, carbon occurs only in multiples of 12, hydrogen in multiples of 1, and chlorine in multiples of 35.3. Hence the atomic

Table 6-3

Experimental Calculation of Molecular Formulas

Compound	Density, D (g liter^{-1})	Molecular weight, $M = kD$	Elemental composition percent by weight			Weight per molecular unit			Probable formula
			C	H	Cl	C	H	Cl	
Methane	0.714	16.0	74.8	25.0	—	12.0	4.03	—	CH_4
Ethane	1.335	29.9	79.8	20.2	—	23.9	6.04	—	C_2H_6
Benzene	3.47	77.8	92.3	7.7	—	71.8	6.00	—	C_6H_6
Chloroform	5.32	119.1	10.05	0.844	89.10	12.0	1.01	106.2	$CHCl_3$
Ethyl chloride	2.87	64.3	37.2	7.8	55.0	23.9	5.02	35.4	C_2H_5Cl
Carbon tetra-chloride	6.81	152.6	7.8	—	92.2	11.9	—	141.0	CCl_4
			Greatest common factor =			12.0	1.0	35.3	

weight of carbon cannot be greater than 12, although it could be an integral submultiple of 12, such as 6, 4, or 3.

The atomic weights obtained by Cannizzaro's method are either the true atomic weights or integral multiples of them. If just ethane, benzene, and ethyl chloride had been included in the table, then the conclusion might have been drawn that the atomic weight of carbon was 24. If information from other carbon compounds had been added to the table, and just *one* analysis gave a weight of 6 for carbon, then the lower value would have to have been accepted, making the probable formulas C_2H_4, C_4H_6, $C_{12}H_6$, C_2HCl_3, and so on. However, no matter how many carbon compounds were analyzed by Cannizzaro's method, the weight per molecular unit always came out to be an integral multiple of 12. Hence this value was accepted as the atomic weight of carbon.

Cannizzaro's achievement was the last link in the chain of logic that began with Proust and the law of constant composition. The battle was over, save for the computing. Scientists could find accurate atomic weights for any element that appeared in compounds having measurable vapor densities. With these atomic weights, the percent composition of a new compound would lead unambiguously to the chemical formula. The *mole* was defined as we have stated in Chapter 1, that is, as the number of grams of a compound equal to its molecular weight on Cannizzaro's scale (which is the one we use today, with improvements in accuracy). It was realized that a mole of any compound would have the same number of molecules. Although the value of that number was not then known, it was named *Avogadro's number, N,* in belated recognition of his contribution.

With the hindsight that comes from knowledge of the ideal gas law, we can see that Cannizzaro's constant k is simply RT/P:

$$PV = nRT = \frac{wRT}{M}$$

$$PM = DRT$$

$$k = \frac{M}{D} = \frac{RT}{P}$$

where P = pressure w = weight in grams
        V = volume M = molecular weight
        n = number of molecules D = density = w/V
        R = constant k = constant = M/D
        T = temperature

The gas density values that were used in the previous demonstration of
Cannizzaro's argument are those at standard temperature and pressure or
STP: 1.00 atm and 273 K. Thus,

$$k = \frac{0.08205 \times 273}{1.00} \text{ liters mole}^{-1} = 22.4 \text{ liters mole}^{-1}$$

Example 1

At STP, the following vapor densities and percent compositions are observed for three compounds of C, H, and S:

Compound	Density (g liter^{-1})	Percent by weight		
		C	H	S
x	3.40	16.0	0.0	84.0
y	2.14	25.0	8.4	66.6
z	2.77	38.7	9.7	51.6

Assuming atomic weights of 12 for C and 1 for H, as just found, calculate
the probable atomic weight of sulfur, S, and the probable formulas for
molecules x, y, and z.

Solution

For each compound, first calculate the molecular weight (mol wt) from the
gas density, and the weight of each element per molecule:

Compound	Mol wt	Weight per molecule			Probable formula
		C	H	S	
x	76.1	12.2	0.0	63.9	CS_2
y	48.0	12.0	4.0	32.0	CH_4S
z	62.1	24.0	6.0	32.0	C_2H_6S

Greatest common factor for sulfur = 32.0

The probable atomic weight of sulfur is therefore 32.0 g mole^{-1}.

Example 2

Suppose that another compound, w, has a vapor density of 1.38 g liter^{-1}, and an analysis of 38.7% C, 9.4% H, and 51.6% S by weight. How would this force you to revise your conclusions for Example 1?

Solution

The molecular weight of compound w would be 31.0 g mole^{-1}, and the percent composition would indicate 12.0 g or 1 mole of C per mole, 2.9 g or 3 moles of H, and only 16.0 g of sulfur. Hence the revised atomic weight of S would be 16.0 g mole^{-1} and the formula for molecule w would be CH_3S. The formula for x then would be CS_4; for y, CH_4S_2; and for z, $C_2H_6S_2$. Needless to say, a compound such as w has never been observed, and the atomic weight of 32.0 for sulfur is valid.

By calculations such as these, a consistent set of atomic weights was obtained for the lighter elements that can be found in gaseous molecules.

6-7 ATOMIC WEIGHTS FOR THE HEAVY ELEMENTS: DULONG AND PETIT

One problem remained: What does one do about the heavy elements, especially metals, that cannot be prepared readily in gaseous compounds? The problem can be illustrated by considering lead and silver.

Example 3

The combining weight of lead (amount per 8.00 g of oxygen) in lead oxide is 51.8 g. What is the atomic weight of lead?

Solution

We know that 103.6 of lead combine with 16 g, or 1 mole, of oxygen atoms, but we can go no further without knowing the chemical formula for lead oxide. Hence, we are caught in the same vicious circle from which Cannizzaro escaped for the lighter atoms. *If* the formula is PbO, then the atomic weight of lead is 103.6. But if the formula is Pb_2O, the atomic weight is 51.8, and if PbO_2, 207.2. Can you show that, in general, if the formula for lead oxide is Pb_xO_y, the atomic weight of lead will be 103.6 (y/x)? The problem has several solutions.

Example 4

Silver oxide is 93.05% silver by weight. What is the atomic weight of silver?

Solution

If we take, for simplicity, a specimen sample of 100 g, there will be 93.05 g of silver for every 6.95 g of oxygen. The combining weight of silver (amount per 8.00 g of oxygen) is then 93.05 g $\times$ (8.00 g/6.95 g) = 107.1 g. One mole of oxygen atoms combines with twice this amount, or 214.2 g, of silver. The choice of atomic weight now is limited to a set of multiples or fractions of 107.1 g, depending on what we assume the formula to be.

Formula:	Ag_2O	Ag_3O_2	AgO	Ag_2O_3	AgO_2
Atomic weight:	107.1	142.8	214.2	321.3	428.4

We need some means of deciding among these values. Without further information a choice cannot be made.

Pierre Dulong (1785–1838) and Alexis Petit (1791–1820) had discovered a method of estimating atomic weights of the heavier elements in 1819, but it had been overlooked in the general confusion that attended chemistry at that time. They made a systematic study of all physical properties that could possibly have a correlation with atomic weight, and found a good one in the heat capacities of solids. The **gram heat capacity** of a substance is the number of joules of heat needed to raise the temperature of 1 g of the substance by 1°C. It is an easily measured property. The product of the gram heat capacity and the atomic weight of an element is the heat required to raise the temperature of 1 mole by 1°C, or the **molar heat capacity**. Dulong and Petit noticed that for many solid elements whose atomic weight was known the molar heat capacity was very close to 25 J deg^{-1} mole^{-1} (Table 6-4). This indicates that the process of heat absorption must be related more strongly to the number of atoms of matter present than to the mass of matter. Later work on the theory of heat capacities of solids has shown that there should be such a constant molar heat capacity for simple solids. Dulong and Petit gave no explanation, however.

Since they advanced no reason for this phenomenon, at the time it was regarded by most chemists as being as dubious as Dalton's rule of simplicity (which was wrong) or Avogadro's principle of equal volumes/equal number of molecules (which was right). It was not until Cannizzaro prepared the way with light atoms that the method of Dulong and Petit was appreciated for heavy atoms.

We now can choose among the possible precise values of atomic weight derived from analytical data by using an approximate value obtained by assuming, with Dulong and Petit, that the molar heat capacity is approximately constant for all solids, 25 J deg^{-1} mole^{-1}.

Table 6-4

Dulong and Petit's Data on the Molar Heat Capacities of Solid Elements[a,b]

Element	Gram heat capacity (J deg^{-1} g^{-1})	Atomic weight	Molar heat capacity (J deg^{-1} mole^{-1})
Bi	0.120	212.8	25.64
Au	0.125	198.9	24.79
Pt	0.133	188.6	25.04
Sn	0.215	117.6	25.30
Zn	0.388	64.5	25.01
Ga	0.382	64.5	24.60
Cu	0.397	63.31	25.14
Ni	0.433	59.0	25.56
Fe	0.460	54.27	24.98
Ca	0.627	39.36	24.67
S	0.787	32.19	25.30

[a]Reproduced by permission from J. B. Conant, *Harvard Case Histories in Experimental Science,* Harvard University Press, Cambridge, 1957, Vol. 1, p. 305.

[b]All data are the original values (with atomic weights changed to a scale in which O = 16.0, and calories converted to joules). If you use modern atomic weights rather than Dulong and Petit's as given in this table, does this make the molar heat capacities more nearly constant?

Example 5

The heat capacities of lead and silver, as tabulated by Dulong and Petit, are 0.123 and 0.233 J deg^{-1} g^{-1}, respectively. Use this information to choose the proper atomic weights for Examples 3 and 4.

Solution

$$\text{Approximate atomic weight of lead} = \frac{25}{0.123} = 203$$

$$\text{Approximate atomic weight of silver} = \frac{25}{0.233} = 107$$

The correct choices from the previous examples must be 207.2 for lead and 107.1 for silver; the chemical formulas then are PbO_2 and Ag_2O.

Example 6

A common cobalt-containing mineral, linnaeite, contains 58.0% cobalt and 42.0% sulfur by weight. The heat capacity of cobalt metal is 0.434 J deg^{-1}

g^{-1}. Assuming that you know the atomic weight of sulfur to be 32.06, compute the atomic weight of cobalt and write the correct empirical formula for linnaeite.

Solution

The correct answers are 59 and Co_3S_4.

6-8   COMBINING CAPACITIES, "VALENCE," AND OXIDATION NUMBER

With Dalton's atomic theory, and with the contributions of Avogadro, Dulong and Petit, and Cannizzaro, it became possible to deduce atomic weights for elements from chemical analyses and physical data such as vapor densities and heat capacities. These deductions have given us the table of atomic weights shown on the inside back cover. The next great task of chemistry was to explain the formulas that could be derived.

The most primitive concept in chemical bonding is probably the idea of **combining capacity**, sometimes called "valence." The combining capacity of an element in a given compound is defined as the ratio of its true atomic weight to its combining weight in that compound:

$$\text{Combining capacity} = \frac{\text{atomic weight}}{\text{combining weight}}$$

Hydrogen has a combining capacity of 1, by definition. Oxygen has a combining capacity of 2 in H_2O and most other compounds, but a combining capacity of 1 in hydrogen peroxide, H_2O_2. Using the data in Table 6-1, we can see that Cl and Br have combining capacities of 1; Ca, 2; and As, 3; carbon shows several combining capacities: 4, 3, 2, and 1. Sulfur has a combining capacity of 2 in H_2S, 4 in SO_2, and 6 in SO_3. Nitrogen has a combining capacity of 3 in ammonia, 4 in NO_2, 2 in NO, and 1 in N_2O. Notice that in these two-element compounds the total combining capacity of one element exactly balances the total combining capacity of the other. In SO_3, one sulfur atom with a combining capacity of 6 balances three oxygen atoms having a capacity of 2 each. The formulation of the concept of combining capacity or "valence" was the first step toward a theory of chemical bonding. The second step was to assign plus and minus signs to these combining capacities so that the *sum* of the signed capacities for a molecule is zero. Hydrogen was assigned the value $+1$; therefore the value for oxygen had to be -2 so the sum for water, H_2O, would be zero. The formula for sulfuric acid, H_2SO_4, requires that sulfur be associated with the value $+6$:

$$
\begin{array}{lrl}
\text{H:} & 2 \times +1 = & +2 \\
\text{O:} & 4 \times -2 = & -8 \\
\text{S:} & 1 \times +6 = & \underline{+6} \\
\text{Sum:} & & 0
\end{array}
$$

These signed combining capacities are just the oxidation numbers that we encountered in Chapter 1. They are important in theories of chemical bonding because they describe how electrons are shifted toward or away from atoms in a molecule.

Summary

This chapter has chronicled the process by which scientists deduced that chemical compounds are made up of specific numbers of atoms of different kinds having specific atomic weights, and slowly worked out a set of reliable atomic weights. The theory of atoms originated as a philosophical concept rather than as a means of manipulating substances and reactions. Antoine Lavoisier laid the foundation by establishing that **mass** was the fundamental quantity in chemical reactions. John Dalton turned the philosophy into reality by showing that the atomic theory would account for the experimental observations that were summarized in the **laws of equivalent proportions and multiple proportions**

The task of deciding upon a set of consistent atomic weights was not an easy one, and Dalton himself went astray. The circular argument involving assumed atomic weights and assumed molecular formulas was not broken until 1860, when Cannizzaro applied a principle that had been discovered in 1811 by Avogadro but had been ignored: Under the same conditions of temperature and pressure, equal volumes of all gases contain equal numbers of molecules. Since this meant that gas density was proportional to molecular weight, the way was open for establishing the standard atomic weight scale that we still use today. The quantitative foundations of modern chemistry had been laid.

Self-Study Questions

1. A. D. Risteen is quoted by J. W. Mellor as saying, as late as 1895, "I cannot see what warrant there is for assuming that when a weight A of one substance combines with another whose weight is B, the weight of the resulting compound is universally and necessarily A + B." Ignoring the subtlety of mass–energy conversion that is important in nuclear buildup, can you prove that Risteen was mistaken? Why should mass be conserved, rather than some other physical properties, such as volume, density, or temperature?

2. How was the idea of the conservation of mass fatal to the phlogiston theory?

3. Use the simple atomic theory to explain the following observed behavior to a skeptic:
   a) The law of constant composition

b) The law of equivalent proportions

c) The law of multiple proportions

4. What is Dalton's rule of simplicity? How did it lead to trouble?

5. Suppose that in 1812 you had been offered two conflicting hypotheses:

a) The gaseous elements exist as single atoms. Not all gases necessarily have the same number of molecules per liter at a given temperature and pressure.

b) All gases at the same pressure and temperature have the same number of molecules per liter. The common gaseous elements have atoms that associate in pairs.

Which alternative would you have chosen? Can you really justify your choice, other than by twentieth-century hindsight?

6. What did Cannizzaro do with Avogadro's hypothesis that made it acceptable when Cannizzaro presented it 50 years after Avogadro had done so?

7. Why is the atomic weight of oxygen 16 on Cannizzaro's scale, but 8 on Dalton's?

8. How can the combining weight of an element be calculated from atomic weights and chemical formulas?

9. How can a chemical formula be deduced if one knows only the atomic weights and the combining weights of the elements in the compound?

10. How can the atomic weight be deduced from the combining weight of an element and the formula of the compound in which it appears?

11. Cannizzaro's method can establish an upper limit for an atomic weight but never a lower limit. Why?

12. Show that if the combining weight of silver in silver oxide is 107.1 g and the formula is Ag_xO_y then the atomic weight of silver is given by 214.2 (y/x).

13. Why does the constancy of the product of gram heat capacity and atomic weight for metals suggest that heat absorption is a property of the number of particles of matter rather than a property of the number of grams?

14. What is the combining capacity of the sulfur atom in the following compounds: H_2S, SO, SO_2, SO_3, S_8?

Problems

Mass and volume

1. The following reaction is carried out at 110°C, where all reactants and products are gases.

$$CO_2 + H_2 \rightarrow CO + H_2O$$

Calculate the molecular weights of reactants and products, and show that mass is conserved during the reaction—that the total mass after reaction is identical with the mass present at the start of the reaction. Is volume also conserved in this reaction? Which of these conservation principles is more fundamental in chemistry?

2. The following reaction leads to synthesis of ammonia gas:

$$3H_2 + N_2 \rightarrow 2NH_3$$

Is mass conserved in this reaction? Show with molecular weights that it is, or that it is not. Is volume conserved in this reaction? (What principle did you use in answering the last question?)

Combining weights

3. If 1.00 g of a metal reacts with 0.348 g of oxygen, what is the combining weight of the metal? How much metal would combine with 1 mole of oxygen atoms? If the empirical formula for the metal oxide is M_2O, what is the atomic weight of the metal? Can you identify it from the periodic table on the inside front cover?

4. Iron reacts with oxygen to produce oxides of different compositions, depending on experimental conditions. The percentages of iron in three of these oxides are 77.73%, 72.36%, and 69.94%. What is the combining weight of iron in each of these three compounds? How do these weights illustrate the law of multiple proportions? Using atomic weights from the periodic table, calculate the empirical formula of each oxide.

Avogadro's principle

5. If 10 liters of hydrogen gas, H_2, and 10 liters of oxygen gas, O_2, are allowed to react, what will be the volume of water vapor produced if the reaction is carried out at constant temperature and pressure? What volume of which reactant will remain unreacted? (Assume a temperature high enough to make water a gas.)

6. If 6 liters of propane gas, C_3H_8, are burned in oxygen, how many liters of oxygen are required? How many liters of CO_2 and water vapor are produced? What is the ratio of the density in grams per liter of propane and oxygen gases at the same temperature and pressure?

Atomic weights

7. Four compounds of carbon, hydrogen, and an unknown element X have the vapor densities under conditions of Table 6-3 and percent composition by weight as shown in the table below. Assuming the known atomic weights of carbon and hydrogen, what is the probable atomic weight of element X and what are the molecular formulas of compounds A through D? From the atomic weights in the periodic table, can you identify element X?

Table for Problem 7

Compound	Density (g liter^{-1})	Weight percent composition		
		C	H	X
A	4.3	12.7	3.2	84.1
B	7.8	6.9	1.2	91.9
C	11.3	4.8	0.4	95.8
D	14.8	3.6	–	96.4

8. There are three compounds of carbon and another element, Y. Compound A is 86.4% Y by weight and has a vapor density of 3.92 g liter^{-1} under conditions of Table 6-3. Compound B is 82.6% Y and has a density of 6.16 g liter^{-1}, and compound C is 61.4% Y with a density of 2.77 g liter^{-1}. What is the largest possible atomic weight for Y? If this value is correct, what are the molecular formulas for A, B, and C? What other values for the molecular weight of Y are possible? From the periodic table inside the front cover, can you identify element Y? What are the most likely values for the molecular weights of the compounds?

Dulong and Petit

9. When a 1.00-g sample of a copper oxide reacts with hydrogen gas, the products are water and 0.799 g of copper metal. The gram heat capacity of copper is 0.385 J deg^{-1} g^{-1}. What is the atomic weight of copper?

10. Suppose 3.70 g of metal combines with 1.94 g of oxygen gas. The gram heat capacity of the metal is 0.586 J deg^{-1} g^{-1}. What is the combining weight of the metal, and what is its atomic weight? Identify the metal from the periodic table.

11. A 1.00-g sample of uranium reacts with 0.0126 g of hydrogen gas. If the gram heat capacity of uranium metal is 0.113 J deg^{-1} g^{-1}, what is the atomic weight of uranium? What is the combining capacity of uranium?

12. The gram heat capacity of an unknown metal, M, is 0.123 J deg^{-1} g^{-1}. The metal forms an oxide in which 8.50 g of metal combines with 1 g of O_2. Calculate the combining weight, atomic weight, and combining capacity of the metal. What is the formula of the metal oxide? Can you identify the metal from the periodic table?

Molecular weights and formulas

13. A compound contains 40.0% carbon, 53.5% oxygen, and 6.67% hydrogen. At STP, the vapor density of the compound is 2.67 g liter^{-1}. Calculate the empirical formula, molecular weight, and molecular formula.

14. A compound contains 65.45% carbon, 29.06% oxygen, and 5.49% hydrogen by weight. A 3.30-g sample of the compound yields 672 ml of vapor, measured at STP. Determine the empirical formula, molecular weight, molecular formula, and number of molecules in 3.30 g.

Gas volumes

15. Toluene, a compound of hydrogen and carbon only, can be burned to produce carbon dioxide and water vapor. When this occurs, one volume of toluene vapor reacts with nine volumes of O_2 to give seven volumes of CO_2 and four volumes of water vapor. What is the simplest or empirical formula for toluene?

16. Two volumes of an unknown gas react with three volumes of O_2 at 110°C to give two volumes of CO_2 and four volumes of water vapor, and no other products. What is the empirical formula of the unknown gas, and what is its molecular formula?

Postscript: Joseph Priestley and Benjamin Franklin

Joseph Priestley (1733–1804) is, after Lavoisier, one of the most interesting personalities of this period in chemistry. A Unitarian minister in Leeds and Birmingham in the north of England, and later a suspected sympathizer with the French Revolution, he was constantly under attack for his heretical religious and political views. He published widely, both in chemistry (which brought him fame) and in politics and religion (which brought him notoriety). Although an innovator and a careful experimentalist, he did not have the theoretical grasp of chemical principles that his younger French colleague had. Priestley's oxygen experiments inspired Lavoisier to initiate the research that demolished the phlogiston theory, but Priestley himself continued to hold stubbornly to the phlogiston theory until his death.

In 1791, on the second anniversary of the storming of the Bastille, rioting against suspected "republicans" broke out in Birmingham. The city was under mob rule for three days. Priestley's church, home, laboratory, apparatus, and manuscripts were burned, and he had to flee in disguise to Worcester. After that, he spent three unhappy years in London, and finally emigrated to the United States. He was offered both a professorship and a church, but declined both, choosing to live in relative seclusion for his last decade.

Priestley and Benjamin Franklin were in continual correspondence, the former about his gas chemistry, the latter about his voltaic piles and Leyden jars. In a letter to Benjamin Vaughan, in 1788, Franklin wrote:

❝ Remember me affectionately to good Dr. Price and to the honest heretic Dr. Priestley. I do not call him *honest* by way of distinction, for I think all the heretics I have known have been virtuous men. They have the virtue of fortitude, or they would not venture to their own heresy; and they cannot afford to be deficient in any of the other virtues, as that would give advantage to their enemies; and they have not, like orthodox sinners, such a number of friends to excuse or justify them. Do not, however, mistake me. It is not to my good friend's heresy that I impute his honesty. On the contrary, it is his honesty that has brought him the character of heretic. ❞

Eleven years before the tragedy at Birmingham, Franklin wrote a letter to Priestley that is both perceptive in its forecasts and pessimistic in its time scale, and fully as relevant now as it was in 1780.

❝ I always rejoice to hear of your still being employed in experimental researches into nature, and of the success you meet with. The rapid progress *true* science now makes, occasions my regretting sometimes that I was born too soon. It is impossible to imagine the height to which may be carried, in a thousand years, the power of man over matter. We may perhaps learn to deprive large masses of their gravity, and give them absolute levity, for the sake of easy transport. Agriculture may diminish its labor and double its produce; all diseases may by sure means be prevented or cured, not excepting that of old age, and our lives lengthened at pleasure even beyond the antediluvian standard. O that moral science were in as fair a way of improvement, that men would cease to be wolves to one another, and that human beings would at length learn what they now improperly call humanity! **99**

Suggested Reading

O. T. Benfey, Ed., *Classics in the Theory of Chemical Combination,* Dover, New York, 1963.

J. B. Conant and L. K. Nash, *Harvard Case Histories in Experimental Science,* Harvard University Press, Cambridge, 1957. Eight critical developments in the experimental sciences, presented as case studies for non-scientists. Extensive reproductions of the original papers. Excellent for understanding the human side of scientific progress. Includes Boyle, the phlogiston theory, the nature of heat, the atomic theory, Pasteur and fermentation, and the nature of electricity.

H. M. Leicester, *The Historical Background of Chemistry,* Dover, New York, 1971. A very readable introduction to the subject, including the Greeks, the Arabic and medieval alchemists, the rise of the new chemistry after Lavoisier and Dalton, and the progression of new chemical ideas to the era that we shall call the "quantum revolution."

Lucretius, *The Nature of the Universe* (De Rerum Natura), Penguin, London, 1967. A good prose translation by R. E. Latham. The best record that we have of the atomic theory of Democritus and the Epicureans.

D. McKie, *Antoine Lavoisier: Scientist, Economist, Social Reformer,* Macmillan, New York, 1962. A good account of a scientist and man of public affairs in the midst of a revolution.

J. W. Mellor, *A Comprehensive Treatise on Inorganic and Theoretical Chemistry,* Wiley, New York, 1922. A multivolume treatise, of which the first seven chapters of Volume 1 form a good history of chemistry. More details, more quotations from original sources than in the book by Leicester.

T. Thomson, *The History of Chemistry,* Colburn and Bentley, London, 1830. Particularly interesting for its account of alchemy and its immediate successors, and because an account of Dalton and the atomic theory constitute the *last* chapter of the book. Written by a friend and scientific mentor of Dalton.

7

The Periodic Table

It often matters vastly with what others,
In what arrangements the primordial germs
Are bound together, and what motions, too,
They exchange among themselves,
*   for these same atoms*
Do put together sky, and sea, and lands,
Rivers and suns, grains, trees and
*   breathing things.*
But yet they are commixed in different ways
With different things, with motions each its own.

Lucretius (55 B.C.)

I n this chapter we shall examine the correlations that exist between the physical and chemical properties of the elements and their compounds. These correlations lead directly to a fundamental classification scheme for matter, the periodic table. To Ernest Rutherford, who once remarked that there are two kinds of science—physics and stamp collecting—the periodic table would be the ultimate stamp album. If this were the final chapter of our book, his impression would be confirmed. But we organize the elements of the universe into the periodic table so chemistry can begin, not end. Once we have established the classification scheme, we must set out to explain it, in terms of electrons and the other subatomic particles from which atoms are constructed. This explanation is the task of later chapters. But before we begin to theorize about the world, let's see what it's really like.

7-1   EARLY CLASSIFICATION SCHEMES

Very early in the development of chemistry, chemists recognized that certain elements have similar properties. In the earliest classification scheme, there were only two divisions, metals and nonmetals. Metallic elements have a certain lustrous appearance, they are malleable (can be hammered into thin sheets) and ductile (can be drawn into wires), they conduct heat and electricity, and they form compounds with oxygen that are basic. Nonmetallic elements have no one characteristic appearance, they generally do not conduct heat and electricity, and they form acidic oxides.

Döbereiner's Triads

In 1829, the German chemist Johann Döbereiner observed several groups of three elements (*triads*) with similar chemical properties. In every case the atomic weight of one element in the triad was nearly the average of the other two. For example, each element in the triad *chlorine, bromine,* and *iodine* forms colored vapors containing diatomic molecules. Each element combines with metals and has a combining weight equal to its atomic weight. Each element forms ions with oxygen that have a single negative charge: ClO^-, ClO_3^-, BrO_3^-, and IO_3^-. The atomic weight of bromine (80) is approximately the average of those of chlorine (35.5) and iodine (127). Table 7-1 lists the similarities of elements in this and other triads.

In addition to recognizing the triads given in Table 7-1, Döbereiner observed a peculiar triad of the metals *iron, cobalt,* and *nickel,* all of which have similar properties and almost the same atomic weights. The metals are used in structural materials (steel) and may be ferromagnetic like iron; in their $+2$ and $+3$ states they form complex ions that are colored.

This discovery of families of elements (the number 3 per family proved to be insignificant) provided an incentive to those who were attempting to find a rational means of classifying the elements.

Newlands' Law of Octaves

Between 1850 and 1865 many new elements were discovered, and chemists made considerable progress in the determination of atomic weights. Thus, more accurate atomic weights were made available for old elements, and reasonably accurate values were presented for new elements. In 1865, the English chemist John Newlands (1839–1898) explored the problem of the periodic recurrence of similar behavior of elements. He arranged the lightest of the known elements in order of increasing atomic weight as follows:

H	Li	Be	B	C	N	O
F	Na	Mg	Al	Si	P	S
Cl	K	Ca	Cr	Ti	Mn	Fe

Newlands noticed that the eighth element (fluorine, F) resembled the first (hydrogen, H), the ninth resembled the second, and so forth. His observation that every eighth element had similar properties led him to compare his chemical octaves with musical octaves, and he himself called it his *law of octaves*. Periodicity by octaves in chemistry suggested to him a fundamental harmony like the one in music. The comparison, although appealing, is invalid. Had Newlands known of the noble gases, his periodicity of properties would have been by nines rather than by eights. He never would have used his musical analogy, and he might have been spared some of the ridicule and indifference that he suffered. (See the Postscript for more on Newlands.)

Table 7-1

Döbereiner's Triads

Triad elements and atomic weights	Elemental form	Principal compounds	Special properties
(I) Cl, Br, I; 35.5, 80, 127	Colored diatomic molecules: Cl_2 (yellow), Br_2 (brown), I_2 (violet)	Form simple salts containing -1 ions Cl^-, Br^-, I^-. Form oxyanions containing one to four oxygen atoms: ClO_4^-, ClO_3^-, BrO_3^-, IO_3^-, ClO^-, IO_4^-. Hydrogen compounds are molecular: HCl, HBr, HI.	Free elements react vigorously with electron donors to form negative ions Cl^-, Br^-, I^-: $2Na + Cl_2 \rightarrow 2Na^+ + 2Cl^-$ $I_2 + S^{2-} \rightarrow 2I^- + S$ Salts (like NaCl) are very soluble in water. Halide salts of Li, Na, and K give neutral solutions. Hydrogen compounds are strong acids and ionize completely in water: $HBr + H_2O \rightarrow H_3O^+ + Br^-$
(II) S, Se, Te; 32, 79, 127.6	Colored crystalline nonmetals (Te somewhat metallic): S_8 (yellow), Se_8 (red)	Form simple salts with -2 ions: S^{2-}, Se^{2-}, Te^{2-}, and very smelly compounds with hydrogen: H_2S, H_2Se, H_2Te. Form oxyanions with up to four oxygen atoms: SO_3^{2-},[a] SO_4^{2-}, SeO_4^{2-}. Form dioxides and trioxides: SO_2, SO_3,[a] SeO_2, TeO_2, TeO_3.	Salts, except those with triads III and IV, are slightly soluble in water: CuS, ZnS, HgS. Soluble salts (Na_2S) give basic solutions: $S^{2-} + H_2O \rightarrow HS^- + OH^-$ Hydrogen compounds are weak acids.
(III) Ca, Sr, Ba; 40, 88, 137	Reactive metals	Form salts containing $+2$ ions: Ca^{2+}, Sr^{2+}, Ba^{2+} in $BaSO_4$, $CaCO_3$, $SrCl_2$, and so on.	Salts give bright colors in flame: Ca (orange), Sr (red), Ba (green). Sulfates and carbonates are insoluble. Metals replace hydrogen slowly from water.
(IV) Li, Na, K; 7, 23, 39	Very reactive metals	Form salts containing $+1$ ions: Li^+, Na^+, K^+ in Li_2CO_3, NaCl, K_3PO_4, and so on.	Almost all salts are soluble; metals and salts give brightly colored flames: Li (red), Na (yellow), K (purple). Metals react violently with water to produce hydrogen and soluble ionic hydroxides: $2Na + 2H_2O \rightarrow H_2 + 2Na^+ + 2OH^-$

[a]Note the importance of charge: SO_3^{2-} is very different from SO_3 (no charge).

Newlands' effort was admittedly a step in the right direction. However, three serious criticisms can be directed at his classification scheme:

1. There were no places in his table for new elements, which were being discovered rapidly. Moreover, in the later parts of the table, there were several places where two elements were forced into the same position. (See the Postscript.)

2. There was no scholarly evaluation of the work on atomic weights and no selection of probable best values.

3. Certain elements did not seem to belong where they were placed in the scheme. For example, chromium (Cr) is not sufficiently similar to aluminum (Al), nor is manganese (Mn), a metal, to phosphorus (P), a nonmetal. Iron (Fe), a metal, and sulfur (S), a nonmetal, do not resemble each other either.

7-2 THE BASIS FOR PERIODIC CLASSIFICATION

The development of the periodic table as we now know it is credited mainly to the Russian chemist Dmitri Mendeleev (1834–1907), although the German chemist Lothar Meyer worked out essentially the same system independently and almost simultaneously. So far as we know, neither man was aware of Newlands' work. Mendeleev's periodic table (Figure 7-1), presented in 1869, followed Newlands' plan of arranging the elements in order of increasing atomic weight, but with the following substantial improvements:

1. Long periods were instituted for the elements now known as transition metals. These long periods are shown folded in half in his original table, with each full period taking two lines. This innovation removed the necessity of placing metals such as vanadium (V), chromium, and manganese under nonmetals such as phosphorus, sulfur, and chlorine.

2. If the properties of an element suggested that it did not fit in the arrangement according to atomic weight, a space was left in the table. For example, no element existed that would fit in the space below silicon (Si). Thus, a space was left for a new element, which was named *ekasilicon*.

3. A scholarly evaluation of atomic weight data was made. For example, as a result of this work the combining capacity of chromium in its highest oxide was changed from 5 to the correct value of 6. The combining weight of chromium was known to be 8.66 g. Hence, instead of 43.3 (5 × 8.66), the revised atomic weight of chromium became 52.0 (6 × 8.66).

Indium (In), with a combining weight of 38.5, had been assigned a combining capacity of 2 and therefore an atomic weight of 77, and had been placed between arsenic (As) and selenium (Se). Since their properties were consistent with placement below phosphorus and sulfur, which were next to one another, arsenic and selenium also had to be side by side in

Row	Group I $\overline{R_2O}$	Group II $\overline{RO}$	Group III $\overline{R_2O_3}$	Group IV RH_4 RO_2	Group V RH_3 R_2O_5	Group VI RH_2 RO_3	Group VII RH R_2O_7	Group VIII $\overline{}$ RO_4
1	H = 1							
2	Li = 7	Be = 9.4	B = 11	C = 12	N = 14	O = 16	F = 19	
3	Na = 23	Mg = 24	Al = 27.3	Si = 28	P = 31	S = 32	Cl = 35.5	
4	K = 39	Ca = 40	— = 44	Ti = 48	V = 51	Cr = 52	Mn = 55	Fe = 56, Co = 59, Ni = 59, Cu = 63
5	(Cu = 63)	Zn = 65	— = 68	— = 72	As = 75	Se = 78	Br = 80	
6	Rb = 85	Sr = 87	?Yt = 88	Zr = 90	Nb = 94	Mo = 96	— = 100	Ru = 104, Rh = 104, Pd = 106, Ag = 108
7	(Ag = 108)	Cd = 112	In = 113	Sn = 118	Sb = 122	Te = 125	I = 127	
8	Cs = 133	Ba = 137	?Di = 138	?Ce = 140				
9								
10			?Er = 178	?La = 180	Ta = 182	W = 184		Os = 195, Ir = 197, Pt = 198, Au = 199
11	(Au = 199)	Hg = 200	Tl = 204	Pb = 207	Bi = 208			
12				Th = 231		U = 240		

Figure 7-1   The periodic table of Mendeleev as it appeared when published in English, in 1871. The elements appear in order of increasing atomic weight. Note the space left under Si for an unknown (at that time) element of atomic weight 72, and the incorrect atomic weights (for example, In). The letter R in the column headings is the general symbol for an element in the table. Elements in parentheses indicate the continuation of a period from the preceding row.

Mendeleev's scheme. A reevaluation showed that indium had an atomic weight of 114.8 and a combining capacity of 3, which is consistent with its position below aluminum and gallium (Ga) in the present table.

The atomic weight of platinum (Pt) had been thought to be greater than that of gold (Au). Mendeleev thought otherwise, because of the chemistry of the two metals and the places that they should occupy in his table. New determinations inspired by Mendeleev showed 198 for platinum and 199 for gold, thereby placing platinum ahead of gold and under palladium (Pd), which of all the other elements most resembles platinum.

4. On the basis of the known periodic behavior summarized in the table, predictions of the properties of the undiscovered elements were made. These predictions later proved to be amazingly accurate, as one can see by comparing the predicted properties of ekasilicon and the properties reported for the element called germanium (Ge), which now occupies the ekasilicon space. This comparison is given in Table 7-2.

From the table it is evident how Mendeleev was able to predict accu-

Table 7-2

Mendeleev's Predictions for the Element Ekasilicon (Germanium)

Properties	Silicon (Si) and its compounds	Mendeleev's predictions for ekasilicon (Es)	Germanium (Ge) and its compounds	Tin (Sn) and its compounds
Atomic weight	28	72	72.6	119
Appearance	Gray, diamondlike	Gray metal	Gray metal	White metal or gray nonmetal
Melting point (°C)	1410	High	958	232
Density (g cm^{-3})	2.32	5.5	5.36	7.28 or 5.75
Action of acid and alkali	Acid resistant; slow attack by alkali	Acid and alkali resistant	Not attacked by HCl or lye (NaOH); attacked by HNO_3	Slow attack by conc. HCl; attacked by HNO_3; inert to lye (NaOH)
Oxide formula and density (g cm^{-3})	SiO_2, 2.65	EsO_2, 4.7	GeO_2, 4.70	SnO_2, 7.0
Sulfide formula and properties	SiS_2, decomposes in water	EsS_2, insoluble in water, soluble in ammonium sulfide solution	GeS_2, insoluble in water, soluble in ammonium sulfide solution	SnS_2, insoluble in water, soluble in ammonium sulfide solution
Chloride formula	$SiCl_4$	$EsCl_4$	$GeCl_4$	$SnCl_4$
Boiling point of chloride (°C)	57.6	100	83	114
Density of chloride (g cm^{-3})	1.50	1.9	1.88	2.23
Preparation of element	Reduction of K_2SiF_6 with sodium	Reduction of EsO_2 or K_2EsF_6 with sodium	Reduction of K_2GeF_6 with sodium	Reduction of SnO_2 with carbon

rately the physical and chemical properties of the missing element. Its position in the periodic table was below silicon and above tin (Sn). The physical properties of germanium are just about the average between those observed for silicon and for tin. To predict the chemical properties for ekasilicon Mendeleev also used information from the known relative properties of phosphorus, arsenic, and antimony (Sb) in the column to the right in the periodic table.

Correlations such as this guided the search for new elements and compounds and stimulated investigation when known data did not conform with other correlations. One consequence of this research was that we gained improved values for atomic weights and densities.

The Periodic Law

Mendeleev summarized his discoveries in the **periodic law**: The properties of chemical elements are not arbitrary, but vary with the atomic weight in a systematic way.

After most of the elements had been discovered and their atomic weights carefully determined, several discrepancies persisted. For example, the order of increasing atomic weight within Mendeleev's Group VIII (Figure 7-1) was found to be Fe, Co, Ni, Cu in the fourth period (row 4), Ru, Rh, Pd, Ag in the fifth (row 6), and Os, Ir, Pt, Au in the sixth (row 10). Yet Ni resembles Pd and Pt more than Co does. Again, Te has a higher atomic weight than I, but I clearly belongs with Br and Cl, and Te resembles Se and S in chemical properties. When the noble gases were discovered, it was revealed that Ar had a higher atomic weight than K, whereas all the other noble gases had lower atomic weights than the adjacent alkali metals. In these three instances, increasing atomic weight clearly is *not* acceptable as a means of placing elements in the periodic table. Therefore, the elements were assigned atomic numbers from 1 to 92 (now 105). (The atomic numbers of the elements *approximately* increase with their atomic weights.) When the elements are arranged according to increasing atomic number, chemically similar elements lie in vertical columns (families or groups) of the periodic table.

In 1912, Henry G. J. Moseley (1887–1915) observed that the frequencies of x rays emitted from elements could be correlated better with atomic numbers than with atomic weights. The relationship between an element's atomic number and the frequency (or energy) of x rays emitted from the element is a consequence of atomic structure. As we shall see in Chapter 8, the electrons in an atom are arranged in *energy levels*. When an element is bombarded by a powerful beam of electrons, electrons from the innermost levels or shells (closest to the nucleus) can be ejected from the atoms. When outer electrons drop into these shells to fill the vacancies, energy is emitted as x radiation. The x-ray spectrum of an element (the collection of frequencies of x rays emitted) contains information about the electronic energy levels of the atom. The important point for our present purpose is that the energy of a level varies with the charge on the nucleus of the atom. The greater the nuclear charge, the more tightly the innermost electrons are bound. More energy is required to knock off one of these electrons; consequently, there is more energy emitted when an electron falls back into a vacancy in the shells. Moseley discovered that the frequency of x rays emitted (designated by the Greek letter ν, nu) varies with atomic number, Z, according to

$$\nu = c(Z - b)^2$$

in which c and b are characteristic of a given x-ray line and are the same for all elements.

In April 1914, Moseley published the results of his work on 39 elements from $_{13}$Al to $_{79}$Au. (Recall that the atomic number is indicated by a sub-

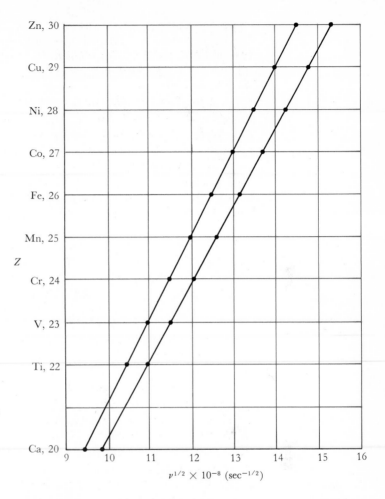

Figure 7-2

Moseley's plot of the square root of x ray frequency against atomic numbers for the elements calcium through zinc. The two lines come from two different, identifiable frequencies in each atom's spectrum.

script to the left of the symbol of an element.) A portion of his data is plotted in Figure 7-2. Moseley wrote the following:

"The spectra of the elements are arranged on horizontal lines spaced at equal distances. The order chosen for the elements is the order of the atomic weights, except in the cases of Ar, Co, and Te, where this clashes with the order of the chemical properties. Vacant lines have been left for an element between Mo and Ru, an element between Nd and Sm, and an element between W and Os, none of which are yet known. . . . This is equivalent to assigning to successive elements a series of successive characteristic integers. . . . Now if either the elements were not characterized by these integers, or any mistake had been made in the order chosen or in the number of places left for unknown elements, these regularities (the straight lines) would at once disappear. We can therefore conclude from the evidence of the x-ray spectra alone, without using any theory of atomic structure, that these integers are really characteristic of the elements. . . . Now Rutherford has proved that the most important constituent of an atom is its central positively charged nucleus, and van den Broek has put forward the view that the charge carried by this nucleus is in all cases an integral multiple of the charge on the hydrogen nucleus. There is every reason to suppose that the integer which controls the x-ray spectrum is the same as the number of electrical units in the nucleus, and these experiments therefore give the strongest possible support to the hypothesis of van den Broek."*

The three undiscovered elements mentioned by Moseley were later found to be elements 43 (technetium, Tc), 61 (promethium, Pm), and 75 (rhenium, Re). A confusing "double element" was cleared up in 1923, when D. Coster and G. Hevesy showed that one of the unoccupied horizontal lines of Moseley's chart belonged to the new element hafnium (Hf, 72). Moseley's work was perhaps the most fundamental single step in the development of the periodic table. It proved that atomic number (or the charge on the nucleus), and not atomic weight, was the essential property in explaining chemical behavior.

7-3   THE MODERN PERIODIC TABLE

The easiest way to understand the periodic table is to build it. Although this may seem to be a difficult task, surprisingly little knowledge of chemistry is required to understand that the form on the inside front cover of this book is inevitable. If we arrange elements by atomic number, as Moseley did, then certain chemical properties repeat at definite intervals (Figure 7-3, top). The chemically inert gases (at least thought to be inert until 1962, when chemists produced compounds containing xenon bound to fluorine and oxygen), He, Ne, Ar, Kr, Xe, and Rn, have atomic numbers 2, 10, 18, 36, 54, and 86, or numerical intervals of 2, 8, 8, 18, 18, and 32. Each of these gases precedes an extremely reactive, soft metal that tends to form a +1 ion: the **alkali metals** Li, Na, K, Rb, Cs, and Fr. And each gas is preceded by a reactive element that can gain an electron to form a −1 ion: hydrogen and the **halogens** F, Cl, Br, I, and At. These key elements are shown in color in the row at the top of Figure 7-3.

These chemical similarities are best represented by dividing the list of 105 elements into seven rows or **periods** (Figure 7-3). However, the first period has only 2 elements, the next two have 8, the next two, 18, and the sixth and probably the seventh periods have 32. How can we align 8 entries over 18, and 18 over 32?

The **alkaline earth metals**, Be, Mg, Ca, Sr, and Ba, are so similar in

*By the time these lines were published, Moseley was in the British army, and less than a year later he was dead, at the age of 27, on a hillside on Gallipoli.

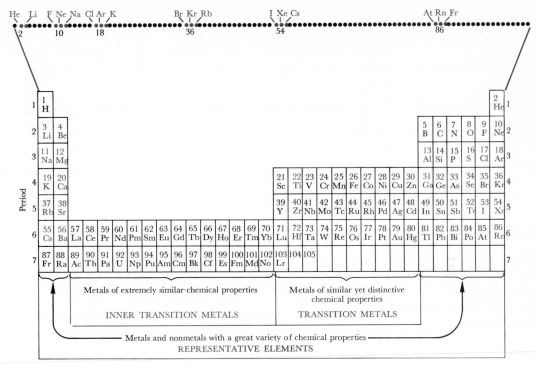

Figure 7-3   When elements are listed in order of increasing atomic number, as in the strip at the top, the recurrence of similar chemical properties suggests the folding into the "super-long" form of the periodic table shown below the strip. Elements can be classified into three categories, based on the extent to which chemical and physical properties change from one position in the table to the next.

chemical properties that we need little imagination to place them as shown. Nonmetals are at the right end of each period, and O, S, Se, and Te constitute a series of elements with a combining capacity of 2 and an increase in metallic behavior from O to Te: O is a nonmetal, and Te exists in the unspecific intermediate zone known as the **semimetals** or **metalloids**. Elements N, P, As, Sb, and Bi comprise a group whose characteristics are the ability to gain three electrons in certain compounds and a gradation from the nonmetallic N and P, to the semimetallic As, to the metallic Sb and Bi. The elements C, Si, Ge, Sn, and Pb all have a combining capacity of 4. For these elements, the border zone between metals and nonmetals is located at an earlier period; C is a nonmetal, Si and Ge are semimetals, and Sn and Pb are metals. Finally, the series B, Al, Ga, In, and Tl form +3 ions; B is semimetallic and the others are metallic. The properties of Al and Ga are more similar than those of Al and Sc. To bring Al above Ga, it is necessary to shift the 8-element periods to the extreme right above the 18-element period below.

The "superfluous" elements in Periods 4 and 5 ($_{21}$Sc to $_{30}$Zn, and $_{39}$Y to $_{48}$Cd) constitute a series of metals, all of which exhibit a great variety of ionic states, the $+2$ and $+3$ states seeming to be the most common. Their properties do not change from one element to another nearly as much as the properties in the series B, C, N, O, and F change. We call these "super-fluous" elements **transition metals**. (We defer the question of *what* is in transition to Chapters 9 and 10.) When we look for chemical parallels between the fifth and sixth periods, we find that $_{40}$Zr and $_{72}$Hf are virtually identical in behavior. Again, our preferred arrangement is to place the elements in Period 5 beyond $_{38}$Sr as far as possible to the right atop Periods 6 and 7. The extra elements in Period 6, $_{57}$La to $_{70}$Yb, are practically identical in chemical behavior. These elements are called the **rare earths**, or **lanthanides**. Their partners in the seventh period ($_{89}$Ac to $_{102}$No) are known as the **actinides**. Because the lanthanides are so similar in chemical properties, they are found together in nature and are extremely difficult to separate.

In summary, the elements can be classified into three groups (Figure 7-3): the **representative elements**, with diverse properties; the **transition metals**, more similar but yet clearly distinguishable; and the **inner transition metals** (lanthanides and actinides), with very similar properties. The representative elements are called representative because they show a broader range of properties than are found in the other elements, and because they are the elements with which we are most familiar.

(The radioactivity and nuclear instability of the actinides, especially uranium, have given them a historical significance that their chemical properties perhaps would not have justified. An old-time chemist is a person who still thinks of uranium as an obscure heavy element used in yellow pottery glazes and stained glass. It is ironic that a nuclear war would be fought with the raw material of stained-glass windows.)

There is a more compact form of the periodic table that indicates more clearly the relative variability of properties of neighboring elements (Figure 7-4). Trends in chemical behavior are often easier to understand if only the representative elements are examined, with the transition metals set to one side as a special case and the inner transition metals virtually ignored. In this table, the vertical columns are called **groups**, and those of the representative elements are numbered IA through VIIA and 0. The groups of the transition elements are numbered in a way to remind you that they should be inserted in the representative element table. The numbering includes Groups IIIB to VIIB, then three columns all labeled collectively Group VIIIB, then Groups IB and IIB. Group IIIB follows Group IIA in the representative elements, and Group IIB precedes Group IIIA. This kind of numbering is clearer in the standard, "long form" of the periodic table on the inside front cover. We can see that the standard form is a compromise between the compactness of Figure 7-4 and the completeness of Figure 7-3. The lanthanides and actinides have been of so little relative importance that they have not been given group numbers.

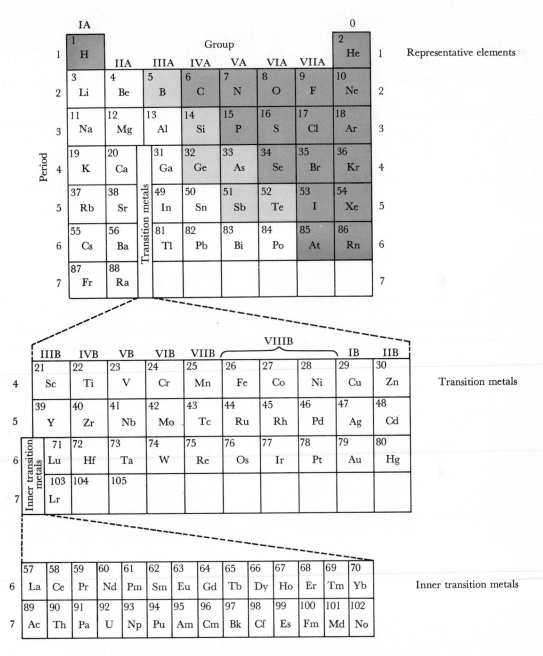

Figure 7-4   This compact, folded form of the periodic table emphasizes the natural division of elements into three categories: the extremely variable representative elements, the more similar transition metals, and the quite similar inner transition metals. Nonmetals are in color, and semimetals in a lighter color. The standard long version of the periodic table on the inside front cover is a compromise between this table and the one in Figure 7-3.

7-4   PERIODICITY OF CHEMICAL PROPERTIES AS ILLUSTRATED BY BINARY HYDRIDES AND OXIDES

In this section we shall see how the periodic table enables us to predict the molecular formulas and the chemical properties of compounds of metals and nonmetals with hydrogen and oxygen.

Binary Hydrides

The number of hydrogen atoms that combine with one atom of a given representative element in the first three periods of the periodic table varies (as shown in Figure 7-5) from 1 to 4 and back to 1 again across each period. This number is equal to either the group number or 8 minus the group number, whichever is smaller. This fact alone offers a clue to the way in which H is bonded in each hydrogen compound.

Compounds of metals with hydrogen—called *hydrides*—are mostly ionic. In alkali hydrides such as KH or NaH, there is a transfer of negative charge to each hydrogen atom. Alkali hydrides have the NaCl crystal structure (Chapter 1), but BeH_2, MgH_2, and AlH_3 manifest a new phenomenon, "bridging" hydrogen. In this arrangement each H atom in the crystal is equidistant between two metal atoms and appears to form a hydrogen bridge between them. Whenever H has a net negative charge, this extra charge can apparently be used to make a second bond to another atom, if there is enough potential bonding capacity in the other atom. The negatively charged H is present in NaH, but not the capacity for multiple bonds. However, Be, Mg, and Al satisfy both demands, and bridge structures are formed. The boron–hydrogen compound B_2H_6 (Figure 7-6) is an example of hydrogen bridging *within* a molecule, and the other known boron

Figure 7-5

Periodicity of combining ratios of the lightest elements in compounds with hydrogen. The number of hydrogen atoms combined with one atom of these elements is usually the group number or 8 minus the group number, whichever is smaller.

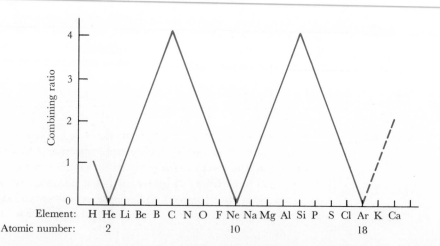

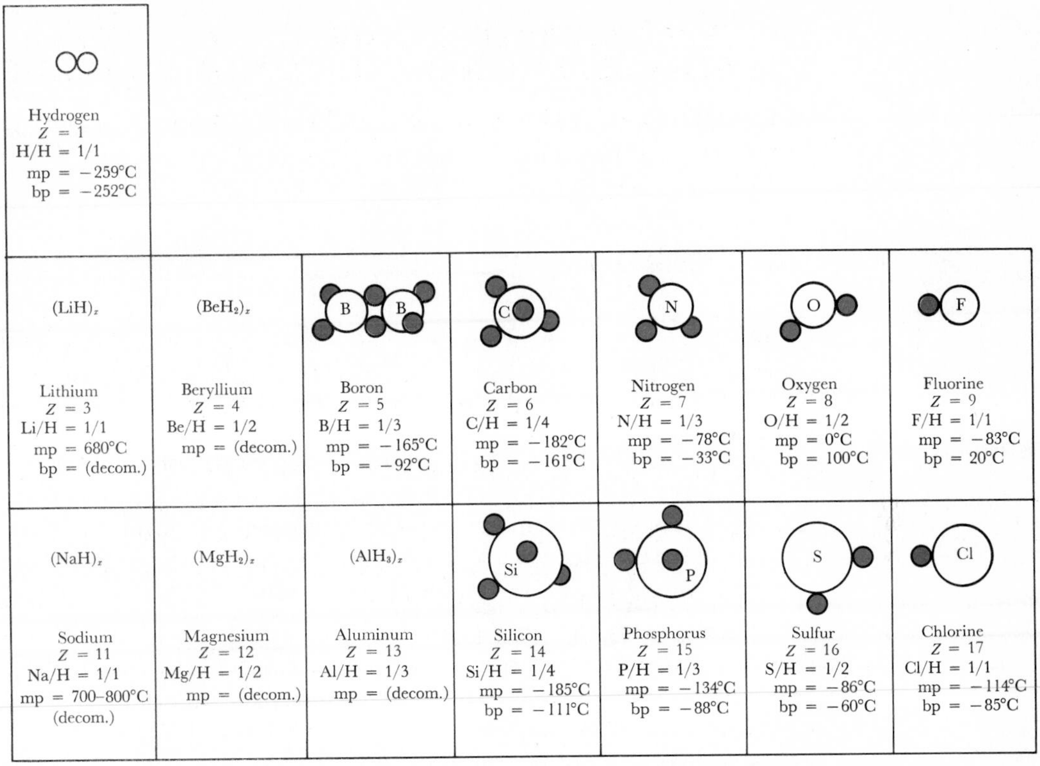

Figure 7-6 Hydrogen compounds of the elements in the first three rows of the periodic table. Combining ratios of the elements (with hydrogen) increase to four (in CH_4 and SiH_4) and then decrease. The hydrides of Li, Be, Na, Mg, and Al are solids at room temperature ($25\,°C$) and have infinitely extended network structures. The simple molecules LiH, BeH_2, NaH, and MgH_2 exist only at low pressures and high temperatures in the gas phase. AlH_3 is not an isolated molecule; it can exist only in the polymeric form $(AlH_3)_x$. The remaining hydrogen compounds are liquids or gases at room temperature. These compounds consist of discrete molecules having the composition and structure depicted schematically in the figure. The structure of the interesting B_2H_6 molecule is discussed in detail in Chapter 21. (Note: Z stands for atomic number; mp, melting point; bp, boiling point.)

hydrides (such as B_4H_{10}, B_5H_9, B_5H_{11}, B_6H_{10}, and $B_{10}H_{14}$) all make extensive use of such hydrogen bridges.

Compounds of hydrogen with elements in the right half of the periods are small molecular compounds in which the number of hydrogen atoms in a molecule is dictated by the number of covalent bonds that the other atom can form. Molecules of such compounds are held together in crystals only by weak forces between molecules; thus their melting and boiling points are very low (Figure 7-6).

Ionic hydrides react with water to produce basic solutions:

$$Na^+H^- + H_2O \rightarrow H_2 + OH^- + Na^+$$

Conversely, the halogen compounds are acidic:

$$HCl + H_2O \rightarrow H_3O^+ + Cl^-$$

Example 1

What are the formulas of the hydrides of cesium (Cs) and selenium (Se)? Which hydride has the higher melting point? Write a balanced equation for the reaction of cesium hydride with water.

Solution

Cesium is in Group IA. Therefore the formula of its hydride is CsH. Selenium is in Group VIA, and its hydride is H_2Se. Cesium hydride is an ionic hydride with a much higher melting point than that of H_2Se, which is a molecular substance with a low melting point (and boiling point). The reaction of CsH with H_2O yields H_2, Cs^+, and OH^-:

$$CsH(s) + H_2O(l) \rightarrow H_2(g) + Cs^+ + OH^-$$

Binary Oxides

The representative elements form oxides with the formulas expected from the elements' positions in the periodic table; in the third period these oxides are Na_2O, MgO, Al_2O_3, SiO_2, P_2O_5, SO_3, and Cl_2O_7. Oxides of elements at the lower left of the table are strong bases. They have a large negative charge on the O atom, and are ionic. The melting points of these ionic oxides are typically around 2000°C, and many decompose before melting. They react with water to make basic solutions:

$$Na_2O + H_2O \rightarrow 2Na^+ + 2OH^-$$

At the other extreme, oxides of elements at the upper right of the table are strong acids:

$$Cl_2O_7 + 3H_2O \rightarrow 2H_3O^+ + 2ClO_4^- \quad \text{(perchloric acid)}$$
$$SO_3\ \ \ + 3H_2O \rightarrow 2H_3O^+ + SO_4^{2-} \quad \text{(sulfuric acid)}$$

Cl_2O_7 is explosively unstable, and SO_3 reacts vigorously with water to produce acidic solutions. The acids have been represented here as completely ionized or dissociated, but this can be as misleading as writing them in their undissociated forms: $HClO_4$ and H_2SO_4. As we saw in Chapter 5, H_2SO_4 is only partially dissociated in water.

Between acidic and basic oxides lies a diagonal band of oxides that are amphoteric: BeO, Al_2O_3, and Ga_2O_3; GeO_2 through PbO_2; and Sb_2O_5 and

Bi_2O_5. (Amphoteric oxides show both acidic and basic behavior.) They are virtually insoluble in water but can be dissolved by either acids or bases:

$$BeO + 2H_3O^+ \rightarrow Be^{2+} + 3H_2O$$
$$BeO + 2OH^- + H_2O \rightarrow Be(OH)_4^{2-}$$

The notation in the first equation is conventional but inconsistent. The hydration of the proton is denoted by the symbol H_3O^+. However, the Be^{2+} cation is also very strongly hydrated, especially so because of its small size. It should be written as $Be(H_2O)_n^{2+}$, or at least $Be^{2+}(aq)$. But so long as the hydration of cations is understood, it need not be spelled out every time.

The amphoteric and basic oxides are solids with high melting points. For instance, Al_2O_3 is the abrasive known as corundum, or emery; SiO_2 is quartz. Only the oxides of C, N, S, and the halogens are normally liquids or gases. The contrast between C and Si in carbon dioxide and quartz is analogous to the contrast between C and N in diamond and nitrogen gas. The difference between C and Si arises because C can make double bonds to O and therefore form a molecular compound of limited size. However, Si must make single bonds with four different O atoms; hence, it must assume a three-dimensional network structure in which tetrahedrally arranged Si atoms are connected by bridging O atoms.

In all the oxides discussed so far, the chemical formula can be predicted from the group number. But there are other oxides whose formulas cannot be predicted from group numbers. For example, C can form CO as well as CO_2. The compound N_2O_5 is not the only nitrogen oxide: NO_2, N_2O_3, and NO are others. Sulfur can form SO_2, S_2O_3, S_2O, and S_2O_7, as well as SO_3. But in these compounds the element does not make full use of its potential combining capacity. Thus, the general trends in properties are best illustrated by the oxides that we have been examining.

Example 2

Write a balanced equation for the reaction of calcium oxide with water.

Solution

Calcium (Ca) is in Group IIA. Its oxide is therefore CaO. The reaction is

$$CaO(s) + H_2O(l) \rightarrow Ca^{2+} + 2OH^-$$

The equation shows that CaO is a strong base, as its reaction with water yields OH^- ions.

Summary

Physical and chemical properties of elements are periodic functions, not of atomic weight, but of atomic number. Moseley suggested, and it later was verified, that the atomic number is the total positive charge on the nucleus, equal to the total number of electrons around the nucleus in a neutral atom.

Particularly stable, inert elements occur at intervals of 2, 8, 8, 18, 18, and 32 in atomic number. These intervals, and only the most basic knowledge of similarities among elements, led to the formulation of a **periodic table**, in which similar elements are in vertical columns or **groups**, and in which chemical properties change in an orderly manner along horizontal rows or **periods**. The full, extended periodic table can be folded into a compact form that illustrates the division of the elements into three categories: the diversified **representative elements**, the more similar **transition metals**, and the virtually identical **inner transition metals**

The correlation between combining capacity and group number in the periodic table may be illustrated by the hydrides and oxides of the representative elements. Elements at the lower left of the periodic table are metals. Their hydrides and oxides are ionic, and aqueous solutions of these compounds are basic. Elements at the upper right corner of the table are nonmetals. Their hydrogen compounds and oxides are small molecules with covalent bonds; they are gases or liquids, and are acidic. Between the two extremes at upper right and lower left in the table there is a gradation of properties. As the elements pass from nonmetals through semimetals to metals, their hydrogen compounds go from acidic, to neutral or inert, to basic (although there are many complications to this overall trend), and the oxides proceed in a more regular manner from acidic, to amphoteric, to basic.

Self-Study Questions

1. In what ways was Mendeleev's classification of the elements superior to Newlands'?
2. Why did Mendeleev's periodic classification lead to a reexamination of combining capacities?
3. How did Mendeleev predict the properties of ekasilicon?
4. What is incorrect about Mendeleev's periodic law?
5. How did Moseley deduce the existence of undiscovered elements?
6. What are the identifying characteristics of the following groups of elements: halogens, alkali metals, noble gases, alkaline earths?
7. What is a *group* in the periodic table? What is a *period?* How many elements are in each of the first six periods?
8. What are the differences among the elements in the three categories of representative elements, transition metals, and inner transition metals?
9. What does the letter *A* or *B* after a group number tell you about the category of the elements in that group?
10. How does metallic character vary within Groups IIIA, IVA, or VA? How does metallic character vary across a period?

11. How does the combining capacity of the representative elements vary with group number in the hydrogen compounds and the oxides?

12. What is the difference in bonding in the hydrogen compounds NaH, MgH_2, and NH_3?

13. Why is there such a difference between the melting points or boiling points of CO_2 and SiO_2? Is there any similarity between this phenomenon and the difference in melting points of carbon and nitrogen?

14. How do the chemical properties of the oxides change from left to right across a period of the table?

15. Which elements are out of their proper sequence in Newlands' table in the Postscript? Why do you think that they are misplaced as they are? (Glucinium, G, was an early name for beryllium, Be.)

Problems

Elements

1. Efforts are presently being made to synthesize or discover new elements of very high atomic number [G. T. Seaborg, "From Mendeleev to Mendelevium and Beyond," *Chemistry*, **43**, 6 (1970)]. Which existing element would be most like element 111? Like 112? Like 118?

2. Predict the empirical formulas of chlorides of elements 111, 112, and 118.

3. What are the group numbers of (a) N; (b) Al; (c) Cl; (d) Rb?

4. What are the group numbers of (a) Ga; (b) Sb; (c) Ba; (d) P?

Periodic properties

5. In the elements Si, Ge, Sn, and Pb, do the nonmetallic properties increase or decrease in the series?

6. Imagine that you are taking chemistry prior to the discovery of strontium ($Z = 38$). Considering strontium's position in the periodic table, predict the following properties: (a) the chemical formula of its most common oxide; (b) the chemical formula of its most common chloride; (c) the chemical formula of its most common hydride; (d) the solubility of its hydride in water, and the acidity or basicity of the resulting solution; (e) the principal ion formed in aqueous solution.

Hydrogen compounds

7. Write a balanced equation for the reaction of hydrogen iodide with water.

8. Write a balanced equation for the reaction of calcium hydride, CaH_2, with water.

9. What would you predict as the formulas of the hydrogen compounds of the following elements: Ca, Te, Ge, W? Which compounds will be ionic? In which will the hydrogens behave as cations? In which will they be anions? Which aqueous solution will be most basic?

Postscript to the Classification of the Elements

The story of John A. R. Newlands is a melancholy illustration of the fact that in science a good idea alone is not enough. The idea must be substantiated with enough evidence to gain acceptance. Newlands' story is also an example of the dangers of poor nomenclature.

Newlands, the son of a Scottish minister, was a graduate of Glasgow University. From his mother, who was of Italian descent, he inherited a love of music and the fervor that led him to join Garibaldi in the struggle for Italian independence in 1860. On his return to England he completed his chemical studies and established himself as a private analytical chemist for industry. His reputation in chemistry was based on his expertise in sugar chemistry, but his lifelong hobby was chemical periodicity.

The high point of his work on periodicity was to occur on March 1, 1866, when he presented his "law of octaves" before the Chemical Society in London. He expected acclaim, but received only indifference and heavy-handed humor. The paper on which his talk was based was rejected by the *Journal of the Chemical Society*. The account of the meeting was reported in *Chemical News* [**13**, 113 (1866)] as follows:

66 Mr. John A. R. Newlands read a paper entitled 'The Law of Octaves and the Causes of Numerical Relations among the Atomic Weights.' The author claims the discovery of a law according to which the elements analogous in their properties exhibit peculiar relationships, similiar to those subsisting in music between a note and its octave. Starting from the atomic weights on Cannizzaro's system, the author arranges the known elements in order of succession, beginning with the lowest atomic weight (hydrogen) and ending with thorium (=231.5); placing, however, nickel and cobalt, platinum and iridium, cerium and lanthanum, etc., in positions of absolute equality or in the same line. The fifty-six elements so arranged are said to form the compass of eight octaves, and the author finds that chlorine, bromine, iodine, and fluorine are thus brought into the same line, or occupy corresponding places in his scale. Nitrogen and phosphorus, oxygen and sulfur, etc., are also considered as forming true octaves. The author's supposition will be exemplified in Table II, shown to the meeting, and here subjoined:

Table II—Elements Arranged in Octaves

H 1	F 8	Cl 15	Co & Ni 22	Br 29	Pd 36	I 43	Pt & Ir 50
Li 2	Na 9	K 16	Cu 23	Rb 30	Ag 37	Cs 44	Os 51
G 3	Mg 10	Ca 17	Zn 24	Sr 31	Cd 38	Ba & V 45	Hg 52
B 4	Al 11	Cr 18	Y 25	Ce & La 32	U 39	Ta 46	Tl 53
C 5	Si 12	Ti 19	In 26	Zr 33	Sn 40	W 47	Pb 54
N 6	P 13	Mn 20	As 27	Di & Mo 34	Sb 41	Nb 48	Bi 55
O 7	S 14	Fe 21	Se 28	Ro & Ru 35	Te 42	Au 49	Th 56

Dr. Gladstone made objection on the score of its having been assumed that no elements remain to be discovered. The last few years had brought forth thallium, indium, cesium, and rubidium, and now the finding of one more would throw out the whole system. The speaker believed there was as close an analogy subsisting between the metals named in the last vertical column as in any of the elements standing on the same horizontal line.

Professor G. F. Foster humorously inquired of Mr. Newlands whether he had ever examined the elements according to the order of their initial letters? For he believed that any arrangement would present occasional coincidences, but he condemned one which placed so far apart manganese and chromium, or iron from nickel and cobalt.

Mr. Newlands said that he had tried several other schemes before arriving at that now proposed. One founded upon the specific gravity of the elements had altogether failed, and no relation could be worked out of the atomic weights under any other system than that of Cannizzaro.❞

And so dies a good story. The questioner did not ask about "chords and arpeggios," as is sometimes said, but only about alphabetical order. The disbelief was apparent, however, and the unfortunate musical analogy made Newlands' ideas look more like numerology than science. The lack of space for new elements and the crowding of two elements into one space were serious flaws. Perhaps the main feature that made Mendeleev's scheme superior was the introduction of the long periods after the first two eight-element ones. Mendeleev buttressed his table with a host of chemical evidence, including his famous predictions for new elements and their chemistry. He clearly deserves his reputation as the creator of the periodic table.

Yet we should not forget Newlands, struggling to have his contribution recognized. He published note after note in *Chemical News,* first elaborating on his table, and then welcoming Mendeleev's table in 1869 as the vindication of his own. Seven years after the *Journal of the Chemical Society* rejected his 1866 paper, he was given a reason, of sorts, by the Society president, Dr. Odling. The paper had not been published, he said, because they "had made it a rule not to publish papers of a purely theoretical nature, since it was likely to lead to correspondence of a controversial character."

Newlands collected all his papers and published them as a book in 1884, and documented his claims to priority in the pages of *Chemical News* and in an account to the German Chemical Society. Perhaps in an outburst of conscience, the Royal Society of Great Britain awarded him the Davy medal in 1887, five years after it had presented the same award to Mendeleev.

Suggested Reading

J. L. Hall and D. A. Keyworth, *Brief Chemistry of the Elements,* W. A. Benjamin, Menlo Park, Calif., 1971.

J. W. Mellor, *Comprehensive Treatise on Inorganic and Theoretical Chemistry,* Macmillan, New York, 1922. See especially Chapter VI.

R. L. Rich, *Periodic Correlations,* W. A. Benjamin, Menlo Park, Calif., 1965.

R. T. Sanderson, *Chemical Periodicity,* Van Nostrand Reinhold, New York, 1960.

M. E. Weeks and H. M. Leicester, *Discovery of the Elements,* Chemical Education Publishing Co., Easton, Pa., 1968, 7th ed.

8

Quantum Theory and Atomic Structure

*The continuity of all dynamical effects was
formerly taken for granted as the basis of all
physical theories and, in close correspondence
with Aristotle, was condensed in the well-
known dogma—Natura non facit saltus—
nature makes no leaps. However, present-day
investigation has made a considerable breach
even in this venerable stronghold of physical
science. This time it is the principle of thermodynamics with which that theorem has been
brought into collision by new facts, and unless
all signs are misleading, the days of its
validity are numbered. Nature does indeed seem
to make jumps—and very extraordinary ones.*

Max Planck (1914)

Physics seemed to be settling down quite satisfactorily in the late nineteenth century. A clerk in the U.S. Patent Office wrote a now-famous
letter of resignation in which he expressed a desire to leave a dying
agency, an agency that would have less and less to do in the future since
most inventions had already been made. In 1894, at the dedication of a
physics laboratory in Chicago, the famous physicist A. A. Michelson suggested that the more important physical laws all had been discovered, and
"Our future discoveries must be looked for in the sixth decimal place."
Thermodynamics, statistical mechanics, and electromagnetic theory had
been brilliantly successful in explaining the behavior of matter. Atoms
themselves had been found to be electrical, and undoubtedly would follow
Maxwell's electromagnetic laws.

Then came x rays and radioactivity. In 1895, Wilhelm Röntgen (1845–
1923) evacuated a Crookes tube (Figure 1-11) so the cathode rays struck the
anode without being blocked by gas molecules. Röntgen discovered that a
new and penetrating form of radiation was emitted by the anode. This
radiation, which he called *x rays,* traveled with ease through paper, wood,
and flesh but was absorbed by heavier substances such as bone and metal.
Röntgen demonstrated that x rays were not deflected by electric or magnetic fields and therefore were not beams of charged particles. Other
scientists suggested that the rays might be electromagnetic radiation like

light, but of a shorter wavelength. The German physicist Max von Laue proved this hypothesis 18 years later when he diffracted x rays with crystals.

In 1896, Henri Becquerel (1852–1908) observed that uranium salts emitted radiation that penetrated the black paper coverings of photographic plates and exposed the photographic emulsion. He named this behavior **radioactivity**. In the next few years, Pierre and Marie Curie isolated two entirely new, and radioactive, elements from uranium ore and named them *polonium* and *radium*. Radioactivity, even more than x rays, was a shock to physicists of the time. They gradually realized that radiation occurred during the breakdown of atoms, and that atoms were not indestructible but could decompose and decay into other kinds of atoms. The old certainties, and the hopes for impending certainties, began to fall away.

The radiation most commonly observed was of three kinds, designated alpha (α), beta (β), and gamma (γ). Gamma radiation proved to be electromagnetic radiation of even higher frequency (and shorter wavelength) than x rays. Beta rays, like cathode rays, were found to be beams of electrons. Electric and magnetic deflection experiments showed the mass of α radiation to be 4 amu and its charge to be $+2$; α particles were simply nuclei of helium, ^4_2He.

The next certainty to slip away was the quite satisfying model of the atom that had been proposed by J. J. Thomson.

8-1 RUTHERFORD AND THE NUCLEAR ATOM

In Thomson's model of the atom all the mass and all the positive charge were distributed uniformly throughout the atom, with electrons embedded in the atom like raisins in a pudding. Mutual repulsion of electrons separated them uniformly. The resulting close association of positive and negative charges was reasonable. Ionization could be explained as a stripping away of some of the electrons from the pudding, thereby leaving a massive, solid atom with a positive charge.

In 1910, Ernest Rutherford (1871–1937) disproved the Thomson model, more or less by accident, while measuring the scattering of a beam of α particles by extremely thin sheets of gold and other heavy metals. (His experimental arrangement is shown in Figure 8-1.) He expected to find a relatively small deflection of particles, as would occur if the positive charge and mass of the atoms were distributed throughout a large volume in a uniform way (Figure 8-2a). What he observed was quite different, and wholly unexpected. In his own words:

"In the early days I had observed the scattering of α particles, and Dr. Geiger in my laboratory had examined it in detail. He found in thin pieces of heavy metal that the scattering was usually small, of the order of one degree. One day Geiger came to me and said, 'Don't you think that

Figure 8-1

The experimental arrangement for Rutherford's measurement of the scattering of α particles by very thin metal foil. The source of the α particles was radioactive polonium, encased in a lead block that protected the surroundings from radiation and confined the α particles to a beam. The gold foil used was about 6×10^{-5} cm thick. Most of the α particles passed through the gold leaf with little or no deflection, *a*. A few were deflected at wide angles, *b*, and occasionally a particle rebounded from the foil, *c*, and was detected by a screen or counter placed on the same side of the foil as the source.

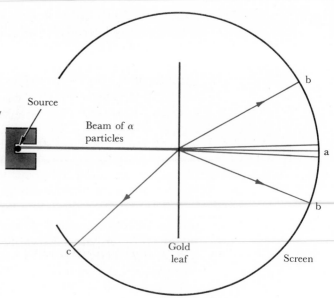

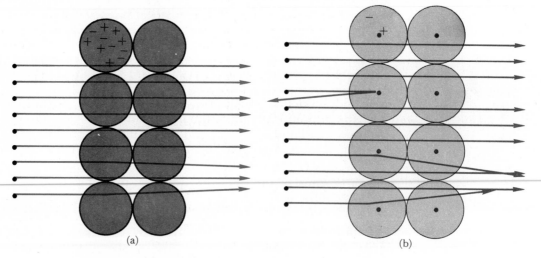

(a)   (b)

Figure 8-2   The expected outcome of the Rutherford scattering experiment, if one assumes (a) the Thomson model of the atom, and (b) the model deduced by Rutherford. In the Thomson model, mass is spread throughout the atom, and the negative electrons are embedded in the positive mass in a uniform manner. There would be little deflection of the beam of positively charged α particles. In the Rutherford model, all the positive charge and virtually all the mass are concentrated in a very small nucleus. Most α particles pass through undeflected. But close approach to a nucleus will make the α particle swerve sharply in its path, and head-on collision with the nucleus will make the α particle rebound in the direction from which it came.

young Marsden, whom I am training in radioactive methods, ought to begin a small research?' Now I had thought that too, so I said, 'Why not let him see if any α particles can be scattered through a large angle?' I may tell you in confidence that I did not believe they would be, since we knew that the α particle was a very fast massive particle, with a great deal of energy, and you could show that if the scattering was due to the accumulated effect of a number of small scatterings, the chance of an α particle's being scattered backwards was very small. Then I remember two or three days later Geiger coming to me in great excitement and saying, 'We have been able to get some of the α particles coming backwards.' . . . It was quite the most incredible event that has ever happened to me in my life. It was almost as incredible as if you fired a 15-inch shell at a piece of tissue paper and it came back and hit you."

Rutherford, Geiger, and Marsden calculated that this observed back-scattering was precisely what would be expected if virtually all the mass and positive charge of the atom were concentrated in a dense nucleus at the center of the atom (Figure 8-2b). They also calculated the charge on the gold nucleus as 100 ± 20 (actually 79), and the radius of the gold nucleus as something less than 10^{-12} cm (actually nearer to 10^{-13} cm).

The picture of the atom that emerged from these scattering experiments was of an extremely dense, positively charged nucleus surrounded by negative charges—electrons. These electrons inhabited a region with a radius 100,000 times that of the nucleus. The majority of the α particles passing through the metal foil were not deflected because they never encountered the nucleus. However, particles passing close to such a great concentration of charge would be deflected; and those few particles that happened to collide with the small target would be bounced back in the direction from which they had come.

The validity of Rutherford's model has been borne out by later investigations. An atom's nucleus is composed of protons and neutrons (Figure 8-3). Just enough electrons are around this nucleus to balance the nuclear charge. But this model of an atom cannot be explained by classical physics. What keeps the positive and negative charges apart? If the electrons were stationary, electrostatic attraction would pull them toward the nucleus to form a miniature version of Thomson's atom. Conversely, if the electrons were moving in orbits around the nucleus, things would be no better. An electron moving in a circle around a positive nucleus is an oscillating dipole when the atom is viewed in the plane of the orbit; the negative charge appears to oscillate up and down relative to the positive charge. By all the laws of classical electromagnetic theory, such an oscillator should broadcast energy as electromagnetic waves. But if this happened, the atom would lose energy, and the electron would spiral into the nucleus. By the laws of classical physics, the Rutherford model of the atom could not be valid. Where was the flaw?

Particle	Neutron	Proton	Electron
Charge	No charge	One positive charge or 1.6021×10^{-19} coulomb	One negative charge or 1.6021×10^{-19} coulomb
Mass	1.67×10^{-27} kg	1.67×10^{-27} kg	9.11×10^{-31} kg

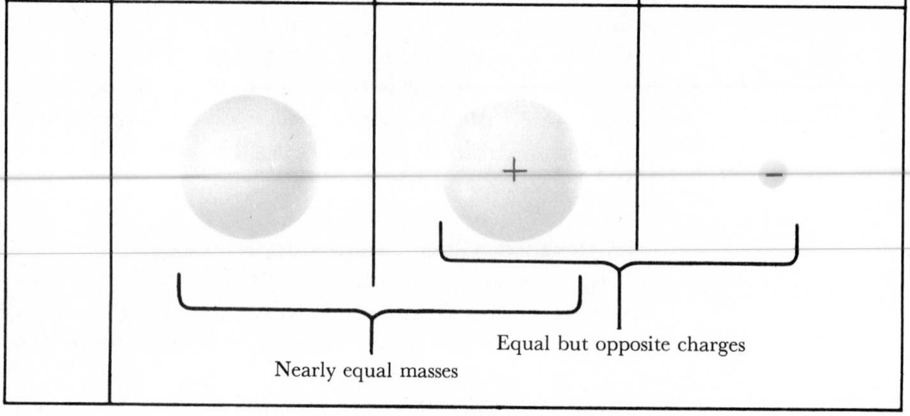

Equal but opposite charges

Nearly equal masses

Figure 8-3   A comparison of properties of a neutron, a proton, and an electron. The mass of a proton is 1836 times as great as that of an electron. However, the electrostatic force of attraction between two particles is independent of their masses and dependent only on their charges and how far they are apart.

8-2   THE QUANTIZATION OF ENERGY

Other flaws that were just as disturbing as Rutherford's impossibly stable atoms were appearing in physics at this time. By the turn of the century scientists realized that radio waves, infrared, visible light, and ultraviolet radiation (and x rays and γ rays a few years later) were **electromagnetic waves** with different wavelengths. These waves all travel at the same speed, c, which is 2.9979×10^{10} cm sec^{-1} or 186,000 miles sec^{-1}. (This speed seems almost instantaneous until you recall that the slowness of light is responsible for the 1.3-sec delay each way in radio messages between the earth and the moon.) Waves such as these are described by their wavelength (designated by the Greek letter lambda, λ), amplitude, and frequency (ν, the Greek letter nu), which is the number of cycles of a moving wave that passes a given point per unit of time (Figure 8-4). The speed of the wave, c, which is constant for all electromagnetic radiation, is the product of the frequency (the number of cycles per second or hertz, Hz) and the length of each cycle (the wavelength):

$$c = \nu\lambda \tag{8-1}$$

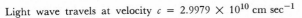

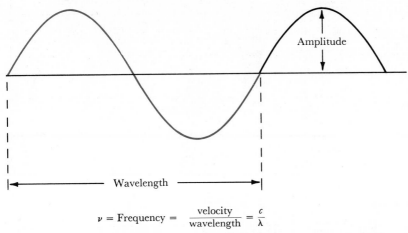

(a)

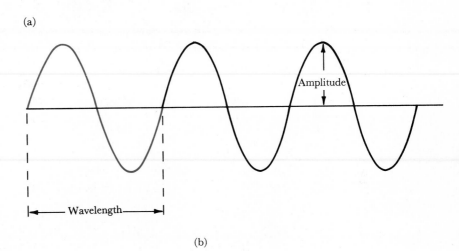

(b)

Figure 8-4   Electromagnetic waves. (a) The profile of a traveling wave at an instant of time. Amplitude, wavelength (λ), velocity (c), and frequency (ν) are shown. The wave number, $\bar{\nu}$, measured in waves per centimeter, or cm^{-1}, is the reciprocal of the wavelength. (b) A wave of shorter wavelength and hence higher frequency. (The product, $\lambda\nu$, is constant.)

The reciprocal of the wavelength is called the wave number, $\bar{\nu}$:

$$\bar{\nu} = 1/\lambda$$

Its units are commonly waves per centimeter, or cm^{-1}.

The electromagnetic spectrum as we know it is shown in Figure 8-5a. The scale is logarithmic rather than linear in wavelength; that is, it is in

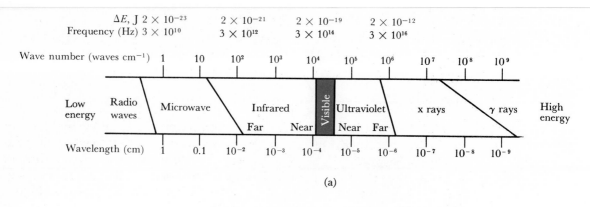

(a)

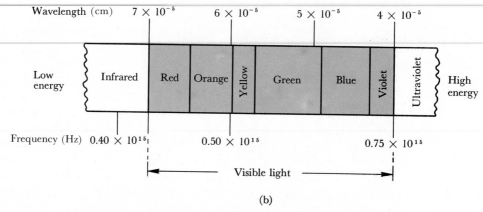

(b)

Figure 8-5   The spectrum of electromagnetic radiation. The visible region is only a small part of the entire spectrum. (a) Overall spectrum. (b) Visible region.

increasing powers of 10. On this logarithmic scale, the portion of the electromagnetic radiation that our eyes can see is only a small sector halfway between radio waves and gamma rays. The visible part of the spectrum is shown in Figure 8-5b.

Example 1

Light of wavelength 5000 Å (or 5×10^{-5} cm) falls in the green region of the visible spectrum. Calculate the wave number, $\bar{\nu}$, corresponding to this wavelength.

Solution   The wave number is equal to the reciprocal of the wavelength, so

$$\bar{\nu} = 1/\lambda$$

$$= 1/5 \times 10^{-5} \text{ cm} = 0.2 \times 10^5 \text{ cm}^{-1}$$
$$= 2 \times 10^4 \text{ cm}^{-1}$$

The Ultraviolet Catastrophe

Classical physics gave physicists serious trouble even when they used it to try to explain why a red-hot iron bar is red. Solids emit radiation when they are heated. The ideal radiation from a perfect absorber and emitter of radiation is called **blackbody radiation**. The spectrum, or plot of relative intensity against frequency, of radiation from a red-hot solid is shown in Figure 8-6a. Since most of the radiation is in the red and infrared frequency regions, we see the color of the object as red. As the temperature is increased, the peak of the spectrum moves to higher frequencies, and we see the hot object as orange, then yellow, and finally white when enough energy is radiated through the entire visible spectrum.

The difficulty in this observation is that classical physics predicts that the curve will keep rising to the right rather than falling after a maximum. Thus there should be much more blue and ultraviolet radiation emitted than is actually observed, and all heated objects should appear blue to our eyes. This complete contradiction of theory by facts was called the **ultraviolet catastrophe** by physicists of the time.

In 1900, Max Planck provided an explanation for this paradox. To do this he had to discard a hallowed tenet of science—that variables in nature change in a continuous way (nature does not make jumps). According to classical physics, light of a certain frequency is emitted because charged objects—atoms or groups of atoms—in a solid vibrate or oscillate with that frequency. We could thus theoretically calculate the intensity curve of the spectrum if we knew the relative number of oscillators that vibrate with each frequency. All frequencies are thought to be possible, and the energy associated with a particular frequency depends only on how many oscillators are vibrating with that frequency. There should be no lack of high-frequency oscillators in the blue and ultraviolet regions.

Planck made the revolutionary suggestion that the energy of electromagnetic radiation comes in packages, or **quanta**. The energy of one package of radiation is proportional to the frequency of the radiation:

$$E = h\nu \tag{8-2}$$

The proportionality constant, h, is known as **Planck's constant** and has the value 6.6262×10^{-34} J sec. By Planck's theory, a group of atoms cannot emit a *small* amount of energy at a *high* frequency; high frequencies can be emitted only by oscillators with a *large* amount of energy, as given by $E = h\nu$. The probability of finding oscillators with high frequencies is therefore slight because the probability of finding groups of atoms with such unusually large vibrational energies is low. Instead of rising, the spectral curve falls at high frequencies, as in Figure 8-6.

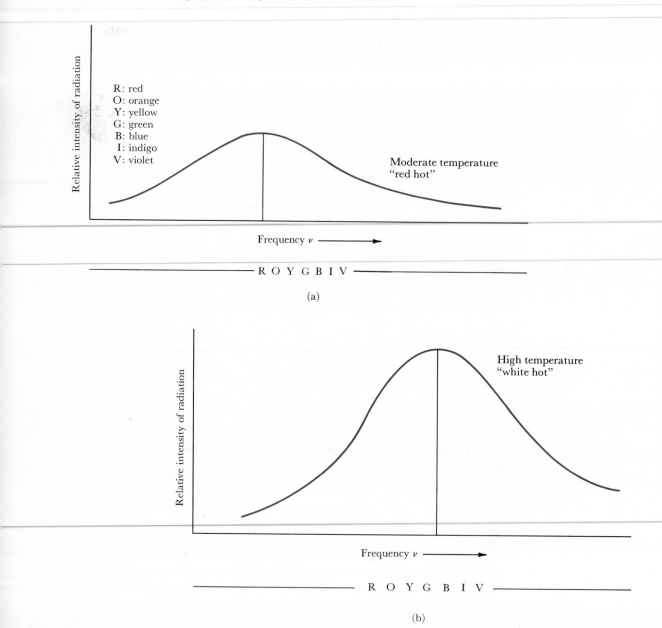

Figure 8-6   The radiation from a hot object is referred to as *blackbody radiation*. (a) For a moderately hot object, most of the radiation is in the red region of the spectrum, and the object is said to be "red hot." (b) As the temperature is increased, the object glows orange, then yellow as the maximum of the radiation curve moves to higher frequencies, and finally white when the radiation occurs at all visible wavelengths in significant amounts. At even higher temperatures, there is less radiation in the red region, and the glow assumes a bluish tinge.

Was Planck's theory correct, or was it only an *ad hoc* explanation to account for one isolated phenomenon? Science is plagued with theories that explain the phenomenon for which they were invented, and thereafter never explain another phenomenon correctly. Was the idea that electromagnetic energy comes in bundles of fixed energy that is proportional to frequency only another one-shot explanation?

The Photoelectric Effect

Albert Einstein (1879–1955) provided another example of the quantization of energy, in 1905, when he successfully explained the **photoelectric effect**, in which light striking a metal surface can cause electrons to be given off. (Photocells in automatic doors use the photoelectric effect to generate the electrons that operate the door-opening circuits.) For a given metal there is a minimum frequency of light below which no electrons are emitted, no matter how intense the beam of light. To classical physicists it seemed non-sensical that for some metals the most intense beam of red light could not drive off electrons that could be ejected by a faint beam of blue light.

Einstein showed that Planck's hypothesis explained such phenomena beautifully. The energy of the quanta of light striking the metal, he said, is greater for blue light than for red. As an analogy, imagine that the low-frequency red light is a beam of Ping-Pong balls and that the high-frequency blue light is a beam of steel balls with the same velocity. Each impact of a quantum of energy of red light is too small to dislodge an electron; in our analogy, a steady stream of Ping-Pong balls cannot do what one rapidly moving steel ball can. These quanta of light were named **photons**. Because of the successful explanation of both the blackbody and photoelectric effects, physicists began recognizing that light behaves like particles as well as like waves.

Example 2

Consider once again the green light in Example 1. The relationship $E = h\nu$ allows us to calculate the energy of one green photon. What is this energy in kilojoules? In kilojoules per mole of green photons?

Solution

Let us assume that we know the wavelength to two significant digits, 5.0×10^{-5} cm. The frequency, ν, of this green light is

$$c = \lambda\nu = 3.0 \times 10^{10} \text{ cm sec}^{-1} = (5.0 \times 10^{-5} \text{ cm})\, \nu$$

$$\nu = \frac{3.0 \times 10^{10} \text{ cm sec}^{-1}}{5.0 \times 10^{-5} \text{ cm}}$$

$$= 0.60 \times 10^{15} \text{ sec}^{-1} \quad (\text{or } 0.60 \times 10^{15} \text{ Hz})$$

The energy of one green photon, then, is

$$E = h\nu$$

$$= (6.63 \times 10^{-34} \text{ J sec})(0.60 \times 10^{15} \text{ sec}^{-1})$$
$$= 4.0 \times 10^{-19} \text{ J, or } 4.0 \times 10^{-22} \text{ kJ}$$

This is the energy of *one* green photon. To obtain the energy of a mole of green photons, we must multiply by Avogadro's number:

$$E = (4.0 \times 10^{-22} \text{ kJ photon}^{-1})(6.02 \times 10^{23} \text{ photons mole}^{-1})$$
$$= 2.4 \times 10^2 \text{ kJ mole}^{-1}$$

The Spectrum of the Hydrogen Atom

The most striking example of the quantization of light, to a chemist, appears in the search for an explanation of atomic spectra. Isaac Newton (1642–1727) was one of the first scientists to demonstrate with a prism that white light is a spectrum of many colors, from red at one end to violet at the other. We know now that the electromagnetic spectrum continues on both sides of the small region to which our eyes are sensitive; it includes the infrared at low frequencies and the ultraviolet at high frequencies.

All atoms and molecules absorb light of certain characteristic frequencies. The pattern of absorption frequencies is called an **absorption spectrum** and is an identifying property of any particular atom or molecule. The absorption spectrum of hydrogen atoms is shown in Figure 8-7. The lowest-energy absorption corresponds to the line at 82,259 cm^{-1}. Notice that the absorption lines are crowded closer together as the limit of 109,678 cm^{-1} is approached. Above this limit absorption is continuous.

If atoms and molecules are heated to high temperatures, they emit light of certain frequencies. For example, hydrogen atoms emit red light when

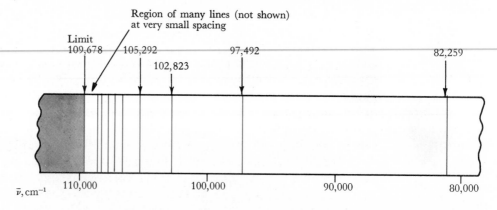

Figure 8-7   The electromagnetic absorption spectrum of hydrogen atoms in the ultraviolet region. The scale is in wave numbers (cm^{-1}). The lines in this spectrum represent ultraviolet radiation that is absorbed by hydrogen atoms when a mixture of all wavelengths is passed through a gas sample.

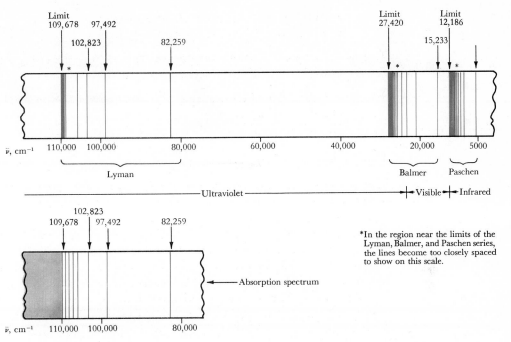

Figure 8-8   The emission spectrum from heated hydrogen atoms. The emission lines occur in series, named for their discoverers: Lyman, Balmer, Paschen; the Brackett and Pfund series are farther to the right in the infrared region. The lines become more closely spaced to the left in each series, until they finally merge at the series limit.

they are heated. An atom that possesses excess energy (e.g., an atom that has been heated) emits light in a pattern known as its **emission spectrum**. A portion of the emission spectrum of atomic hydrogen is shown in Figure 8-8. Note that the lines occur at the same wave numbers in the two types of spectra.

If we look more closely at the emission spectrum in Figure 8-8, we see that there are three distinct groups of lines. These three groups or series are named after the scientists who discovered them. The series that starts at 82,259 cm^{-1} and continues to 109,678 cm^{-1} is called the **Lyman series** and is in the ultraviolet portion of the spectrum. The series that starts at 15,233 cm^{-1} and continues to 27,420 cm^{-1} is called the **Balmer series** and covers a large portion of the visible and a small part of the ultraviolet spectrum. The lines between 5332 cm^{-1} and 12,186 cm^{-1} are called the **Paschen series** and fall in the near-infrared region. The Balmer spectra of hydrogen from several stars are shown in Figure 8-9.

J. J. Balmer proved, in 1885, that the wave numbers of the lines in the

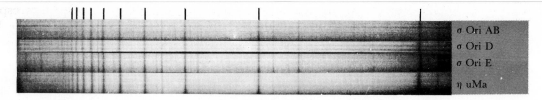

Figure 8-9 Balmer hydrogen-atom spectra from several stars. The three σ Ori spectra are from a group of stars, σ Orionis, located just below the belt in the constellation Orion. η UMa is η Ursa Majoris, the end of the handle in the Big Dipper. Note the universality of the atomic hydrogen spectrum. Hydrogen is hydrogen, no matter where you find it. Balmer lines are marked above the spectra. The other lines are primarily from helium. (Courtesy John Oke, California Institute of Technology.)

Balmer spectrum of the hydrogen atom are given by the empirical relationship

$$\bar{\nu} = R_H \times \left(\frac{1}{4} - \frac{1}{n^2} \right) \qquad n = 3, 4, 5, \ldots \tag{8-3}$$

Later, Johannes Rydberg formulated a general expression that gives all of the line positions. This expression, called the **Rydberg equation**, is

$$\bar{\nu} = R_H \times \left(\frac{1}{n_1^2} - \frac{1}{n_2^2} \right) \tag{8-4}$$

In the Rydberg equation n_1 and n_2 are integers, with n_2 greater than n_1; R_H is called the **Rydberg constant** and is known accurately from experiment to be 109,677.581 cm^{-1}.

Example 3

Calculate $\bar{\nu}$ for the lines with $n_1 = 1$ and $n_2 = 2, 3,$ and 4.

Solution $n_1 = 1, n_2 = 2$ line:

$$\bar{\nu} = 109{,}678\left(\frac{1}{1^2} - \frac{1}{2^2} \right) = 109{,}678\left(1 - \frac{1}{4} \right) = 82{,}259 \text{ cm}^{-1}$$

$n_1 = 1, n_2 = 3$ line:

$$\bar{\nu} = 109{,}678\left(\frac{1}{1^2} - \frac{1}{3^2} \right) = 109{,}678\left(1 - \frac{1}{9} \right) = 97{,}492 \text{ cm}^{-1}$$

$n_1 = 1, n_2 = 4$ line:

$$\bar{\nu} = 109{,}678\left(\frac{1}{1^2} - \frac{1}{4^2} \right) = 109{,}678\left(1 - \frac{1}{16} \right) = 102{,}823 \text{ cm}^{-1}$$

We see that the wave numbers obtained in Example 3 correspond to the first three lines in the Lyman series. Thus we expect that the Lyman series corresponds to lines calculated with $n_1 = 1$ and $n_2 = 2, 3, 4, 5, \ldots$. We can check this by calculating the wave number for the line with $n_1 = 1$ and $n_2 = \infty$.

$n_1 = 1, n_2 = \infty$ line:

$$\bar{\nu} = 109{,}678(1 - 0) = 109{,}678 \text{ cm}^{-1}$$

The wave number 109,678 cm^{-1} corresponds to the highest emission line in the Lyman series.

The wave number for $n_1 = 2$ and $n_2 = 3$ is

$$\bar{\nu} = 109{,}678 \left(\frac{1}{4} - \frac{1}{9} \right) = 15{,}233 \text{ cm}^{-1}$$

This corresponds to the first line in the Balmer series. Thus, the Balmer series corresponds to the $n_1 = 2, n_2 = 3, 4, 5, 6, \ldots$ lines. You probably would expect the lines in the Paschen series to correspond to $n_1 = 3, n_2 = 4, 5, 6, 7, \ldots$. They do. Now you should wonder where the lines are with $n_1 = 4, n_2 = 5, 6, 7, 8, \ldots$, and $n_1 = 5, n_2 = 6, 7, 8, 9, \ldots$. They are exactly where the Rydberg equation predicts they should be. The $n = 4$ series was discovered by Brackett and the $n = 5$ series was discovered by Pfund. The series with $n = 6$ and higher are located at very low frequencies and are not given special names.

The Rydberg formula, equation 8-4, is a summary of observed facts about hydrogen atomic spectra. It states that the wave number of a spectral line is the difference between two numbers, each inversely proportional to the square of an integer. If we draw a set of horizontal lines at a distance R_H/n^2 down from a baseline, with $n = 1, 2, 3, 4, \ldots$, then each spectral line in any of the hydrogen series is observed to correspond to the distance between two such horizontal lines in the diagram (Figure 8-10). The Lyman series occurs between line $n = 1$ and those above it; the Balmer series occurs between line $n = 2$ and those above it; the Paschen series occurs between line $n = 3$ and those above it; and the higher series are based on lines $n = 4$, 5, and so on. Is the agreement between this simple diagram and the observed wave numbers of spectral lines only a coincidence? Does the idea of a wave number of an emitted line being the difference between two "wave-number levels" have any physical significance, or is this just a convenient graphical representation of the Rydberg equation?

8-3 BOHR'S THEORY OF THE HYDROGEN ATOM

In 1913, Niels Bohr (1885–1962) proposed a theory of the hydrogen atom that, in one blow, did away with the problem of Rutherford's unstable atom and gave a perfect explanation of the spectra we have just discussed.

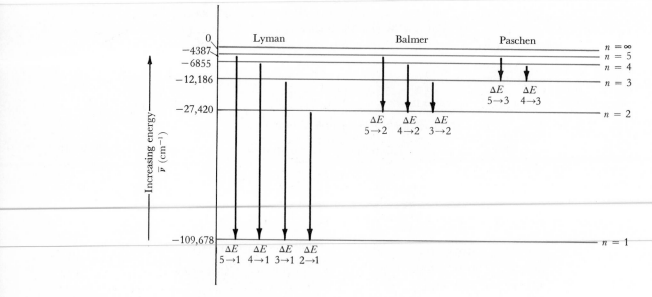

Figure 8-10

An energy-level diagram that accounts for the observed hydrogen spectrum. This diagram can be considered as a graphic representation of the Rydberg equation: $\bar{\nu} = R_H[(1/n_1^2) - (1/n_2^2)]$. However, Bohr atttributed more meaning to it. He proposed that these levels represent the only possible energy states of a hydrogen atom: $E \propto 1/n^2$ He also postulated that an atom generates a spectral line when it goes from one energy state to another of lower energy, and that the wave number of the emitted line is determined by the change in energy: $\Delta E = hc\bar{\nu}$. Only lines with $n = 1, 2, 3, 4$, and 5 and the limit at $n = \infty$ are shown.

There are two ways of proposing a new theory in science, and Bohr's work illustrates the less obvious one. One way is to amass such an amount of data that the new theory becomes obvious and self-evident to any observer. The theory then is almost a summary of the data. This is essentially the way Dalton reasoned from combining weights to atoms. The other way is to make a bold new assertion that initially does not seem to follow from the data, and then to demonstrate that the consequences of this assertion, when worked out, explain many observations. With this method, a theorist says, "You may not see why, yet, but please suspend judgment on my hypothesis until I show you what I can do with it." Bohr's theory is of this type.

Bohr answered the question of why the electron does not spiral into the nucleus by simply postulating that *it does not.* In effect, he said to classical physicists: "You have been misled by your physics to expect that the electron would radiate energy and spiral into the nucleus. Let us assume that it

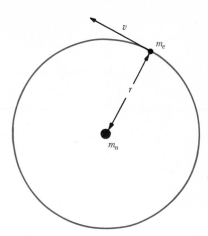

Figure 8-11

Bohr's picture of the hydrogen atom. A single electron of mass m_e moves in a circular orbit with velocity v at a distance r from a nucleus of mass m_n. To explain the spectrum of Figure 8-8, or the Rydberg equation diagram of Figure 8-10, Bohr had to postulate that the angular momentum of the electron, $m_e vr$, was restricted to integral multiples of the quantity $h/2\pi$. The integers are the numbers n of Figure 8-10.

does not, and see if we can account for more observations than by assuming that it does." The observations that he explained so well are the wavelengths of lines in the atomic spectrum of hydrogen.

Bohr's model of the hydrogen atom is illustrated in Figure 8-11: an electron of mass m_e moving in a circular orbit at a distance r from a nucleus. If the electron has a velocity of v, it will have an **angular momentum** of $m_e vr$. (To appreciate what angular momentum is, think of an ice skater spinning on one blade like a top. The skater begins spinning with his arms extended. As he brings his arms to his sides, he spins faster and faster. This is because, in the absence of any external forces, angular momentum is conserved. As the mass of the skater's arms comes closer to the axis of rotation, or as r decreases, the velocity of his arms must increase in order that the product mvr remain constant.) Bohr postulated, as the first basic assumption of his theory, that in a hydrogen atom there could only be orbits for which *the angular momentum is an integral multiple of Planck's constant divided by 2π:*

$$m_e vr = n\left(\frac{h}{2\pi}\right)$$

There is no obvious justification for such an assumption; it will be accepted only if it leads to the successful explanation of other phenomena. Bohr then showed that, with no more new assumptions, and with the laws of classical mechanics and electrostatics, his principle leads to the restriction of the energy of an electron in a hydrogen atom to the values

$$E = -\frac{k}{n^2} \qquad n = 1, 2, 3, 4, \ldots \tag{8-5}$$

The integer n is the same integer as in the angular momentum assumption, $m_e vr = n(h/2\pi)$; k is a constant that depends only on Planck's constant, h, the mass of an electron, m_e, and the charge on an electron, e:

$$k = \frac{2\pi^2 m_e e^4}{h^2} = 13.595 \text{ electron volts (eV)* atom}^{-1}$$

$$= 1312 \text{ kJ mole}^{-1}$$

The radius of the electron's orbit also is determined by the integer n:

$$r = n^2 a_0 \qquad\qquad (8\text{-}6)$$

The constant, a_0, is called the **first Bohr radius** and is given in Bohr's theory by

$$a_0 = \frac{h^2}{4\pi^2 m_e e^2} = 0.529 \text{ Å}$$

The first Bohr radius is often used as a measure of length called the **atomic unit**, a.u.

The energy that an electron in a hydrogen atom can have is **quantized**, or limited to certain values, by equation 8-5. The integer, n, that determines these energy values is called the **quantum number**. When an electron is removed (ionized) from an atom, that electron is described as excited to the quantum state $n = \infty$. From equation 8-5, we see that as n approaches ∞, E approaches zero. Thus, the energy of a completely ionized electron has been chosen as the zero energy level. Because energy is required to remove an electron from an atom, an electron that is bound to an atom must have less energy than this, and hence a negative energy. The relative sizes of the first five hydrogen-atom orbits are compared in Figure 8-12.

Example 4

For a hydrogen atom, what is the energy, relative to the ionized atom, of the **ground state**, for which $n = 1$? How far is the electron from the nucleus in this state? What are the energy and radius of orbit of an electron in the **first excited state**, for which $n = 2$?

Solution   The answers are

$$E_1 = -\frac{k}{1^2} = -1312 \text{ kJ mole}^{-1}$$

$$E_2 = -\frac{k}{2^2} = -328.0 \text{ kJ mole}^{-1}$$

$$r_1 = 1^2 \times 0.529 \text{ Å} = 0.529 \text{ Å}$$

---

*An electron volt is equal to the amount of energy an electron gains as it passes from a point of low potential to a point one volt higher in potential ($1 \text{ eV} = 1.6022 \times 10^{-19} \text{ J}$).

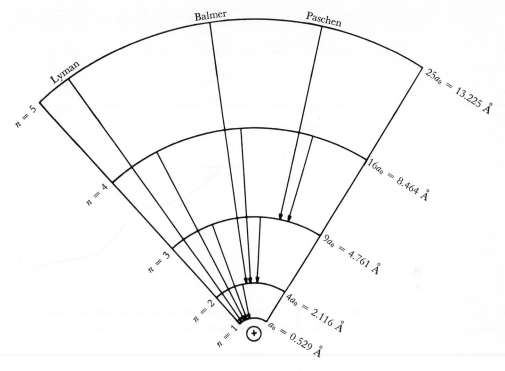

Figure 8-12 Relative sizes of the first five Bohr orbits for atomic hydrogen. The transitions from one orbit to a lower one are indicated as in Figure 8-10. Each arc of a circle represents part of a circular orbit for an electron around the positive nucleus at the bottom of the diagram. The radius of the nth orbit is calculated from $r = n^2 a_0$, in which a_0 is the first Bohr radius; $a_0 = (h^2 / 4\pi^2 m_e e^2) = 0.529$ Å.

$$r_2 = 2^2 \times 0.529 \text{ Å} = 2.12 \text{ Å}$$

Example 5

Using the Bohr theory, calculate the ionization energy of the hydrogen atom.

Solution

The **ionization energy**, **IE**, is that energy required to remove the electron, or to go from quantum state $n = 1$ to $n = \infty$. This energy is

$$IE = E_\infty - E_1 = 0.00 - (-1312 \text{ kJ mole}^{-1})$$
$$= +1312 \text{ kJ}$$

Example 6

Diagram the energies available to the hydrogen atom as a series of horizontal lines. Plot the energies in units of k for simplicity. Include at least the first eight quantum levels and the ionization limit. Compare your result with Figures 8-10 and 8-13.

Try this one yourself.

In the second part of his theory, Bohr postulated that absorption and emission of energy occur when an electron moves from one quantum state to another. The energy emitted when an electron drops from state n_2 to a lower quantum state n_1 is the difference between the energies of the two states:

$$\Delta E = E_1 - E_2 = -k\left(\frac{1}{n_1^2} - \frac{1}{n_2^2}\right) \tag{8-7}$$

The light emitted is assumed to be quantized in exactly the way predicted from the blackbody and photoelectric experiments of Planck and Einstein:

$$|\Delta E| = h\nu = hc\bar{\nu} \tag{8-8}$$

If we divide equation 8-7 by hc to convert from energy to wave number units, we obtain the Rydberg equation,

$$\bar{\nu} = \frac{k}{hc}\left(\frac{1}{n_1^2} - \frac{1}{n_2^2}\right) \tag{8-9}$$

With the Bohr theory, we now can calculate the Rydberg constant from first principles:

$$R_{\mathrm{H}} = \frac{k}{hc} = \frac{2\pi^2 m_e e^4}{h^3 c} = 109{,}737.3 \text{ cm}^{-1}$$

Recall that the experimental value of R_{H} is 109,677.581 cm^{-1}.

The graphic representation of the Rydberg equation, Figure 8-10, now is seen to be an energy-level diagram of the possible quantum states of the hydrogen atom. We can see why light is absorbed or emitted only at specific wave numbers. The absorption of light, or the heating of a gas, provides the energy for an electron to move to a higher orbit. Then the *excited* hydrogen atom can emit energy in the form of light quanta when the electron falls back to a lower-energy orbit. From this emission come the different series of spectral lines:

1. The Lyman series of lines arises from transitions from the $n = 2, 3,$ 4, . . . levels to the ground state ($n = 1$).
2. The Balmer series arises from transitions from the $n = 3, 4, 5, . . .$ levels to the $n = 2$ level.

3. The Paschen series arises from transitions from the $n = 4, 5, 6, \ldots$ levels to the $n = 3$ level.

An excited hydrogen atom in quantum state $n = 8$ may drop directly to the ground state and emit a photon in the Lyman series, or it may drop first to $n = 3$, emit a photon in the Paschen series, and then drop to $n = 1$ and emit a photon in the Lyman series. The frequency of each photon depends on the energy difference between levels:

$$\Delta E = E_a - E_b = h\nu$$

By cascading down the energy levels, the electron in one excited hydrogen atom can successively emit photons in several series. Therefore, all series are present in the emission spectrum from hot hydrogen. However, when measuring the absorption spectrum of hydrogen gas at lower temperatures we find virtually all the hydrogen atoms in the ground state. Therefore, almost all the absorption will involve transitions from $n = 1$ to higher states, and only the Lyman series will be observed.

Energy Levels of a General One-Electron Atom

Bohr's theory can also be used to calculate the ionization energy and spectral lines of any atomic species possessing only one electron (e.g., He^+, Li^{2+}, Be^{3+}). The energy of a Bohr orbit depends on the square of the charge on the atomic nucleus (Z is the atomic number):

$$E = -\frac{Z^2 k}{n^2}$$

where

$$k = 13.595 \text{ eV or } 1312 \text{ kJ mole}^{-1}$$
$$n = 1, 2, 3, 4, \ldots \infty$$

The equation reduces to equation 8-5 in the case of atomic hydrogen ($Z = 1$).

Example 7

Calculate the third ionization energy of a lithium atom.

Solution

A lithium atom is composed of a nucleus of charge $+3$ ($Z = 3$) and three electrons. The first ionization energy, IE_1, of an atom with more than one electron is the energy required to remove one electron. For lithium,

$$Li(g) \rightarrow Li^+(g) + e^- \qquad \Delta E = IE_1$$

The energy needed to remove an electron from the unipositive ion, Li^+, is defined as the second ionization energy, IE_2, of lithium,

$$Li^+(g) \rightarrow Li^{2+}(g) + e^- \qquad \Delta E = IE_2$$

and the third ionization energy, IE_3, of lithium is the energy required to remove the one remaining electron from Li^{2+}. For lithium, $Z = 3$ and $IE_3 = (3)^2(13.595 \text{ eV}) = 122.36 \text{ eV}$. (The experimental value is 122.45 eV.)

The Need for a Better Theory

The Bohr theory of the hydrogen atom suffered from a fatal weakness: It explained nothing *except* the hydrogen atom and any other combination of a nucleus and one electron. For example, it could account for the spectra of He^+ and Li^{2+}, but it did not provide a general explanation for atomic spectra. Even the alkali metals (Li, Na, K, Rb, Cs), which have a single valence electron outside a closed shell of inner electrons, produce spectra that are at variance with the Bohr theory. The lines observed in the spectrum of Li could be accounted for only by assuming that each of the Bohr levels beyond the first was really a collection of levels of different energies, as in Figure 8-13: two levels for $n = 2$, three levels for $n = 3$, four for $n = 4$, and so on. The levels for a specific n were given letter symbols based on the appearance of the spectra involving these levels: s for "sharp," p for "principal," d for "diffuse," and f for "fundamental."

Arnold Sommerfeld (1868–1951) proposed an ingenious way of saving the Bohr theory. He suggested that orbits might be elliptical as well as circular. Furthermore, he explained the differences in stability of levels with the same principal quantum number, n, in terms of the ability of the highly elliptical orbits to bring the electron closer to the nucleus (Figure 8-14). For a point nucleus of charge $+1$ in hydrogen, the energies of all levels with the same n would be identical. But for a nucleus of $+3$ screened by an inner shell of two electrons in Li, an electron in an outer circular orbit would experience a net attraction of $+1$, whereas one in a highly elliptical orbit would penetrate the screening shell and feel a charge approaching $+3$ for part of its traverse. Thus, the highly elliptical orbits would have the additional stability illustrated in Figure 8-13. The s orbits, being the most elliptical of all in Sommerfeld's model, would be much more stable than the others in the set of common n.

The Sommerfeld scheme led no further than the alkali metals. Again an impasse was reached, and an entirely fresh approach was needed.

8-4   PARTICLES OF LIGHT AND WAVES OF MATTER

At the beginning of the twentieth century, scientists generally believed that all physical phenomena could be divided into two distinct and exclusive classes. The first class included all phenomena that could be described by laws of classical, or Newtonian, mechanics of motion of discrete particles.

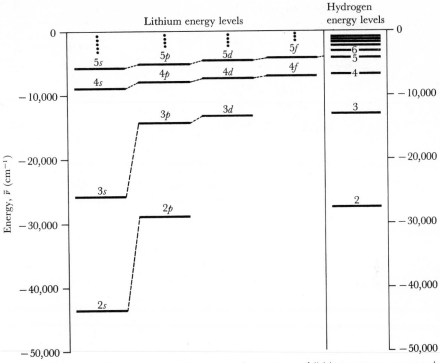

Figure 8-13

The energy levels required to explain the observed spectrum of lithium atoms compared with the hydrogen levels at the right. Levels *n* = 1 are off scale at the bottom. For quantum number *n*, there are *n* levels, traditionally identified by the letters *s, p, d, f, g*, etc. The levels farthest to the right for each quantum number (2*p*, 3*d*, 4*f*, . . .) approach the corresponding hydrogen levels, whereas all other levels of the same quantum number are more stable. Sommerfeld accounted for this stability by postulating elliptical orbits, in which the *s* orbits were most elliptical and the 2*p*, 3*d*, 4*f*, . . . orbits were nearly circular (Figure 8-14).

Figure 8-14

The Sommerfeld orbits for hydrogen. For a point nucleus, all orbits of the same principal quantum number, *n*, will have the same energy. For a nucleus surrounded by a shielding cloud of electrons, the more elliptical orbits that penetrate this cloud will experience (for part of their path) a stronger attraction from the nucleus and will be more stable. Hence, the 4*p* level, for example, is lower in energy than the 4*f* (Figure 8-13).

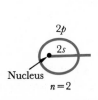

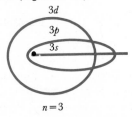

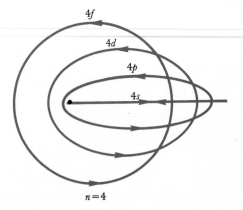

The second class included all phenomena showing the continuous properties of waves.

One outstanding property of matter, apparent since the time of Dalton, is that it is built of discrete particles. Most material substances appear to be continuous: water, mercury, salt crystals, gases. But if our eyes could see the nuclei and electrons that constitute atoms, and the fundamental particles that make up nuclei, we would discover quickly that every material substance in the universe is composed of a certain number of these basic units and therefore is quantized. *Objects appear continuous only because of the minuteness of the individual units.*

In contrast, light was considered to be a collection of waves traveling through space at a constant speed; any combination of energies and frequencies was possible. However, Planck, Einstein, and Bohr showed that light, when observed under the right conditions, also behaves as though it occurs in particles, or quanta.

In 1924, the French physicist Louis de Broglie (b. 1892) advanced the complementary hypothesis that all matter possesses wave properties. De Broglie pondered the Bohr atom, and asked himself where, in nature, quantization of energy occurs most naturally. An obvious answer is in the vibration of a string with fixed ends. A violin string can vibrate with only a selected set of frequencies: a fundamental tone with the entire string vibrating as a unit, and overtones of shorter wavelengths. A wavelength in which the vibration fails to come to a **node** (a place of zero amplitude) at both ends of the string would be an impossible mode of vibration (Figure 8-15). The vibration of a string with fixed ends is quantized by the **boundary conditions** that the ends cannot move.

Can the idea of standing waves be carried over to the theory of the Bohr atom? Standing waves in a circular orbit can exist only if the circumference of the orbit is an integral number of wavelengths (Figure 8-15c, d). If it is not, waves from successive turns around the orbit will be out of phase and will cancel. The value of the wave amplitude at 10° around the orbit from a chosen point will not be the same as at 370° or 730°, yet all these represent the same point in the orbit. Such ill-behaved waves are not **single-valued** at any point on the orbit: Single-valuedness is a boundary condition on acceptable waves.

For single-valued standing waves around the orbit, the circumference is an integer, n, times the wavelength:

$$2\pi r = n\lambda$$

But from Bohr's original assumption about angular momentum,

$$2\pi r = n\left(\frac{h}{m_e v}\right)$$

Therefore, the idea of standing waves leads to the following relationship between the mass of the electron, m_e, its velocity, v, and its wavelength, λ:

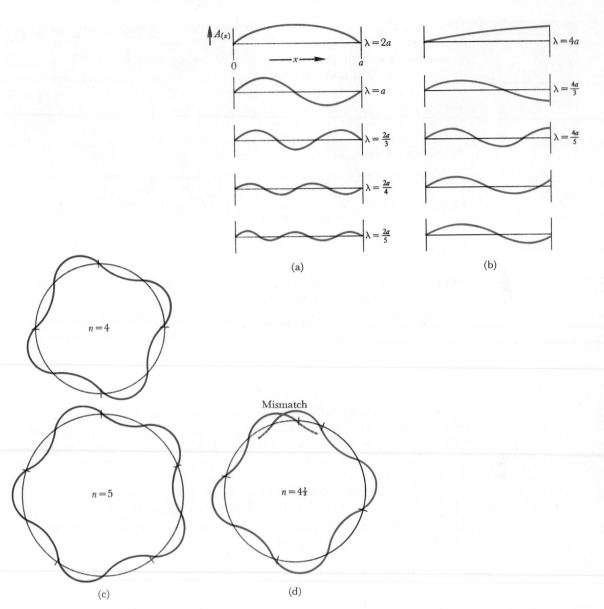

(a)

(b)

(c)

(d)

Figure 8-15 (a) Acceptable and (b) unacceptable standing waves or modes of vibration in a violin string, as determined by the boundary conditions that the ends of the string be motionless. The boundary conditions limit the possible wavelengths of vibration to $\lambda = 2a/n$, in which a is the length of the string and n is an integer $n = 1, 2, 3, \ldots$. (c) Acceptable and (d) unacceptable electron waves in a Bohr orbit. The boundary conditions for a standing wave in the circular orbit are that the circumference be an integral number of wavelengths: $2\pi r = n\lambda$. This requirement and Bohr's angular momentum postulate, $m_e v r = n(h/2\pi)$, lead directly to de Broglie's relationship between the mass of a particle, its velocity, and its wavelength: $\lambda = h/m_e v$.

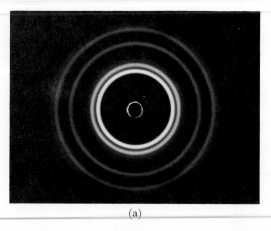
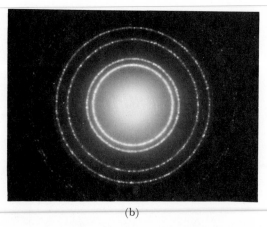

(a)   (b)

Figure 8-16   Difffraction of waves by aluminum foil. (a) X rays of wavelength 0.71 Å. (b) Electrons of energy 600 eV, or wavelength 0.50 Å. The similarity of these two patterns is strong evidence for the wave properties of particles. (Courtesy Film Studio, Education Development Center.)

$$\lambda = \frac{h}{m_e v} \qquad\qquad (8\text{-}10)$$

De Broglie proposed this relationship as a general one. With every particle, he said, there is associated a wave. The wavelength depends on the mass of the particle and how fast it is moving. If this is so, the same sort of diffraction from crystals that von Laue observed with x rays should be produced with electrons.

In 1927, C. Davisson and L. H. Germer demonstrated that metal foils diffract a beam of electrons exactly as they diffract an x-ray beam, and that the wavelength of a beam of electrons is given correctly by de Broglie's relationship (Figure 8-16). Electron diffraction is now a standard technique for determining molecular structure.

Example 8

A typical electron diffraction experiment is conducted with electrons accelerated through a potential drop of 40,000 volts, or with 40,000 eV of energy. What is the wavelength of the electrons?

Solution

First convert the energy, E, from electron volts to joules:

$$E = 40,000 \text{ eV} \times \frac{1.6022 \times 10^{-19} \text{ J}}{1 \text{ eV}} = 6.409 \times 10^{-15} \text{ J}$$

(This and several other useful conversion factors, plus a table of the values of frequently used physical constants, are in Appendix 2.) Since the energy is $E = \frac{1}{2}m_e v^2$, the velocity of the electrons is

$$v = \left(\frac{2E}{m_e}\right)^{1/2} = \left(\frac{2 \times 6.409 \times 10^{-15} \text{ kg m}^2 \text{ sec}^{-2}}{9.110 \times 10^{-31} \text{ kg}}\right)^{1/2}$$

$$= (1.407 \times 10^{16} \text{ m}^2 \text{ sec}^{-2})^{1/2} = 1.186 \times 10^8 \text{ m sec}^{-1}$$

(In the expression $E = \frac{1}{2}m_e v^2$, if the mass is in kilograms and the velocity is in m sec^{-1}, then the energy is in joules: 1 J equals 1 kg m^2 sec^{-2} of energy. We used this conversion of units in the preceding step. The mass of the electron, $m_e = 9.110 \times 10^{-31}$ kg, is found in Appendix 2.) The momentum of the electron, $m_e v$, is

$$m_e v = 9.110 \times 10^{-31} \text{ kg} \times 1.186 \times 10^8 \text{ m sec}^{-1}$$
$$= 10.80 \times 10^{-23} \text{ kg m sec}^{-1}$$

Finally, the wavelength of the electron is obtained from the de Broglie relationship:

$$\lambda = \frac{h}{m_e v} = \frac{6.6262 \times 10^{-34} \text{ J sec}}{10.80 \times 10^{-23} \text{ kg m sec}^{-1}}$$

$$= 0.06135 \times 10^{-10} \frac{\text{kg m}^2 \text{ sec}^{-2} \text{ sec}}{\text{kg m sec}^{-1}} = 0.06135 \times 10^{-10} \text{ m}$$

$$= 0.06135 \text{ Å}$$

So 40-kilovolt (kV) electrons produce the diffraction effects expected from waves with a wavelength of six-hundredths of an angstrom.

Such calculations are all very well, but the question remains: Are electrons waves or are they particles? Are light rays waves or particles? Scientists worried about these questions for years, until they gradually realized that they were arguing about language and not about science. Most things in our everyday experience behave either as what we would call waves or as what we would call particles, and we have created idealized categories and used the words *wave* and *particle* to identify them. The behavior of matter as small as electrons cannot be described accurately by these large-scale categories. Electrons, protons, neutrons, and photons are not waves, and they are not particles. Sometimes they act as if they were what we commonly call waves, and in other circumstances they act as if they were what we call particles. But to demand, "Is an electron a wave *or* a particle?" is pointless.

This wave–particle duality is present in all objects; it is only because of the scale of certain objects that one behavior predominates and the other is negligible. For example, a thrown baseball has wave properties, but a wavelength so short that we cannot detect it.

Example 9

A 200-g baseball is thrown with a speed of 30 m sec^{-1}. Calculate its de Broglie wavelength.

Solution

The answer is $\lambda = 1.1 \times 10^{-34}$ m $= 1.1 \times 10^{-24}$ Å.

Example 10

How fast (or rather, how slowly) must a 200-g baseball travel to have the same de Broglie wavelength as a 40-kV electron?

Solution

The wavelength of a 40-kV electron is 0.0613 Å.

$$v = \frac{h}{\lambda m} = \frac{6.6262 \times 10^{-34} \text{ kg m}^2 \text{ sec}^{-2} \text{ sec}}{0.06130 \times 10^{-10} \text{ m} \times 0.200 \text{ kg}}$$

$$= 0.540 \times 10^{-21} \text{ m sec}^{-1} = 1.70 \times 10^{-4} \text{ Å year}^{-1}$$

Such a baseball would take over 10,000 years to travel the length of a carbon–carbon bond, 1.54 Å. This sort of motion is completely outside our experience with baseballs; thus we never regard baseballs as having wave properties.

8-5   THE UNCERTAINTY PRINCIPLE

One of the most important consequences of the dual nature of matter is the uncertainty principle, proposed in 1927 by Werner Heisenberg (1901–1976). This principle states that you cannot know *simultaneously* both the position and the momentum of any particle with absolute accuracy. The product of the uncertainty in position, Δx, and in momentum, $\Delta(mv)$, will be equal to or greater than Planck's constant divided by 4π:

$$[\Delta x][\Delta(mv_x)] \geq \frac{h}{4\pi} \tag{8-11}$$

We can understand this principle by considering how we determine the position of a particle. If the particle is large, we can touch it without disturbing it seriously. If the particle is small, a more delicate means of locating it is to shine a beam of light on it and observe the scattered rays. Yet light acts as if it were made of particles—photons—with energy proportional to frequency: $E = h\nu$. When we illuminate the object, we are pouring energy on it. If the object is large, it will become warmer; if the object is small

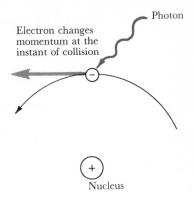

Electron changes
momentum at the
instant of collision

Photon

(−)

(+)
Nucleus

Figure 8-17

The position of an electron e^- at an instant of time should be deter-
minable by a "super microscope" using light of small wavelength,
λ (x rays or γ rays). However, photons of light of small λ have great
energy and therefore very large momentum. A collision of one of these
photons with an electron instantly changes the electronic momentum.
Thus, as the position is better resolved, the momentum becomes more
and more uncertain.

enough, it will be pushed away and its momentum will become uncertain.
The least interference that we can cause is to bounce a single photon off the
object and watch where the photon goes. Now we are caught in a dilemma.
The detail in an image of an object depends on the fineness of the wavelength of the light used to observe the object. (The shorter the wavelength,
the more detailed the image.) But if we want to avoid altering the momentum of the atom, we have to use a low-energy photon. However, the
wavelength of the low-energy photon would be so long that the position of
the atom would be unclear. Conversely, if we try to locate the atom accurately by using a short-wavelength photon, the energy of the photon sends
the atom ricocheting away with an uncertain momentum (Figure 8-17). We
can design an experiment to obtain an accurate value of either an atom's
momentum or its position, but the product of the errors in these quantities
is limited by equation 8-11.

Example 11

Suppose that we want to locate an electron whose velocity is 1.00×10^6 m
sec^{-1} by using a beam of green light whose frequency is 0.600×10^{15} sec^{-1}.
How does the energy of one photon of such light compare with the energy
of the electron to be located?

Solution

The energy of the electron is

$$E = \tfrac{1}{2}m_e v^2 = \frac{9.110 \times 10^{-31} \text{ kg} \times 1.00 \times 10^{12} \text{ m}^2 \text{ sec}^{-2}}{2}$$

$$= 4.56 \times 10^{-19} \text{ J}$$

But the energy of the photon is almost as large:

$$E_p = h\nu = 6.6262 \times 10^{-34} \text{ J sec} \times 0.600 \times 10^{15} \text{ sec}^{-1}$$

$$= 3.97 \times 10^{-19} \text{ J}$$

Finding the position and momentum of such an electron with green light is as questionable a procedure as finding the position and momentum of one billiard ball by striking it with another. In either case, you detect the particle at the price of disturbing its momentum. As a final difficulty, green light is a hopelessly coarse yardstick for finding objects of atomic dimensions. An atom is about 1 Å in radius, whereas the wavelength of green light is around 5000 Å. Shorter wavelengths make the energy quandary worse.

We do not see the uncertainty limitations in large objects because of the sizes of the masses and velocities involved. Compare the following two problems.

Example 12

An electron is moving with a velocity of 10^6 m sec^{-1}. Assume that we can measure its position to 0.01 Å, or 1% of a typical atomic radius. Compare the uncertainty in its momentum, p, with the momentum of the electron itself.

Solution

The uncertainty in position is $\Delta x \simeq 0.01$ Å $= 0.01 \times 10^{-10}$ m. The momentum of the electron is approximately

$$p = m_e v \simeq 10^{-30} \text{ kg} \times 10^6 \text{ m sec}^{-1} = 10^{-24} \text{ kg m sec}^{-1}$$

By the Heisenberg uncertainty principle, the uncertainty in the knowledge of the momentum is

$$\Delta p = \frac{h/4\pi}{\Delta x} \simeq \frac{0.5 \times 10^{-34} \text{ kg m}^2 \text{ sec}^{-1}}{0.01 \times 10^{-10} \text{ m}}$$

$$\simeq 0.5 \times 10^{-22} \text{ kg m sec}^{-1}$$

The *uncertainty* in the momentum of the electron is 50 times as great as the momentum itself!

Example 13

A baseball of mass 200 g is moving with a velocity of 30 m sec^{-1}. If we can locate the baseball with an error equal in magnitude to the wavelength of light used (e.g., 5000 Å), how will the uncertainty in momentum compare with the total momentum of the baseball?

Solution

The momentum, p, of the baseball is 6 kg m sec^{-1}, and $\Delta p = 1 \times 10^{-28}$ kg m sec^{-1}. The intrinsic uncertainty in the momentum is only one part in 10^{28}, far below any possibility of detection in an experiment.

8-6   WAVE EQUATIONS

In 1926, Erwin Schrödinger (1887–1961) proposed a general wave equation for a particle. The mathematics of the Schrödinger equation is beyond us, but the mode of attack, or the strategy of finding its solution, is not. If you can see how physicists go about solving the Schrödinger equation, even though you cannot solve it yourself, then quantization and quantum numbers may be a little less mysterious. This section is an attempt to explain the method of solving a differential equation of motion* of the type that we encounter in quantum mechanics. We shall explain the strategy with the simpler analogy of the equation of a vibrating string.

The de Broglie wave relationship and the Heisenberg uncertainty principle should prepare you for the two main features of quantum mechanics that contrast it with classical mechanics:

1. Information about a particle is obtained by solving an equation for a wave.

2. The information obtained about the particle is not its position; rather, it is the *probability* of finding the particle in a given region of space.

We can't say whether an electron is in a certain place around an atom, but we can measure the probability that it is there rather than somewhere else.

Wave equations are familiar in mechanics. For instance, the problem of the vibration of a violin string is solved in three steps:

1. Set up the equation of motion of a vibrating string. This equation will involve the displacement or amplitude of vibration, $A_{(x)}$, as a function of position along the string, x.

2. Solve the differential equation to obtain a general expression for amplitude. For a vibrating string with fixed ends, this general expression is a sine wave. As yet, there are no restrictions on wavelength or frequency of vibration.

3. Eliminate all solutions to the equation except those that leave the ends of the string stationary. This restriction on acceptable solutions of the wave equation is a boundary condition. Figure 8-15a shows solutions that fit this boundary condition of fixed ends of the string; Figure 8-15b shows solutions that fail. The only acceptable vibrations are those with $\lambda = 2a/n$, or $\bar{\nu} = n/2a$, in which $n = 1, 2, 3, 4, \ldots$. *The boundary conditions and not the wave equation are responsible for the quantization of the wavelengths of string vibration.*

---

*Equations of motion are always *differential equations* because they relate the change in one quantity to the change in another, such as change in position to change in time.

Exactly the same procedure is followed in quantum mechanics:

1. Set up a general wave equation for a particle. The Schrödinger equation is written in terms of the function $\psi_{(x,y,z)}$ (where ψ is the Greek letter psi), which is analogous to the amplitude, $A_{(x)}$, in our violin-string analogy. *The square of this amplitude, $|\psi|^2$, is the relative **probability density** of the particle at position (x, y, z).* That is, if a small element of volume, dv, is located at (x, y, z), the probability of finding an electron within that element of volume is $|\psi|^2 \, dv$.

2. Solve the Schrödinger equation to obtain the most general expression for $\psi_{(x,y,z)}$.

3. Apply the boundary conditions of the particular physical situation. If the particle is an electron in an atom, the boundary conditions are that $|\psi|^2$ must be *continuous, single-valued,* and *finite* everywhere. All these conditions are only common sense. First, probability functions do not fluctuate radically from one place to another; the probability of finding an electron a few thousandths of an angstrom from a given position will not be radically different from the probability at the original position. Second, the probability of finding an electron in a given place cannot have two different values simultaneously. Third, since the probability of finding an electron somewhere must be 100%, or 1.000, if the electron really exists, the probability at any one point cannot be infinite.

We now shall compare the wave equation for a vibrating string and the Schrödinger wave equation for a particle. In this text you will not be expected to do anything with either equation, but you should note the similarities between them.

■ *Vibrating string.* The amplitude of vibration at a distance x along the string is $A_{(x)}$. The differential equation of motion is

$$\frac{d^2 A_{(x)}}{dx^2} + 4\pi^2 \bar{\nu}^2 A_{(x)} = 0 \tag{8-12}$$

The general solution to this equation is a sine function

$$A_{(x)} = A_{\max} \sin\left(2\pi\bar{\nu}x + \alpha\right)$$

and the only acceptable solutions (Figure 8-15a) are those for which $\bar{\nu} = n/2a$, where $n = 1, 2, 3, 4, \ldots$, and for which the phase shift, α, is zero:

$$A_{(x)} = A_{\max} \sin n\left(\frac{\pi}{a}\right) x$$

■ *Schrödinger equation.* The square of the amplitude $|\psi_{(x,y,z)}|^2$ is the probability density of the particle at (x, y, z). The differential equation is

$$\frac{\partial^2 \psi}{\partial x^2} + \frac{\partial^2 \psi}{\partial y^2} + \frac{\partial^2 \psi}{\partial z^2} + \frac{8\pi^2 m_e}{h}(E - V_{(x,y,z)})\psi_{(x,y,z)} = 0 \qquad (8\text{-}13)$$

V is the potential energy function at (x, y, z), and m_e is the mass of the electron.

Although solving equation 8-13 is not a simple process, it is purely a mathematical operation; there is nothing in the least mysterious about it. The energy, E, is the variable that is restricted or quantized by the boundary conditions on $|\psi|^2$. Our next task is to determine what the possible energy states are.

8-7 THE HYDROGEN ATOM

The sine function that is the solution of the equation for the vibrating string is characterized by one integral quantum number: $n = 1, 2, 3, 4, \ldots$ The first few acceptable sine functions are

$$\left. \begin{aligned} A_{1(x)} &= A_0 \sin\left(\frac{\pi}{a}\right)x \\[2mm] A_{2(x)} &= A_0 \sin 2\left(\frac{\pi}{a}\right)x \\[2mm] A_{3(x)} &= A_0 \sin 3\left(\frac{\pi}{a}\right)x \\[2mm] A_{4(x)} &= A_0 \sin 4\left(\frac{\pi}{a}\right)x \end{aligned} \right\} \quad A_{n(x)} = A_0 \sin n\left(\frac{\pi}{a}\right)x \qquad (8\text{-}14)$$

These are the first four curves in Figure 8-15a.

An atom is three-dimensional, whereas the string has only length. The solutions of the Schrödinger equation for the hydrogen atom are characterized by three integer quantum numbers: n, l, and m. These arise when solving the equation for the wave function, ψ, which is analogous to the function $A_{n(x)}$ in the vibrating string analogy. In solving the Schrödinger equation, we divide it into three parts. The solution of the **radial** part describes how the wave function, ψ, varies with distance from the *center* of the atom. If we borrow the customary coordinate system of the earth, an **azimuthal** part produces a function that reveals how ψ varies with north or south latitude, or distance up or down from the equator of the atom. Finally, an **angular** part is a third function that suggests how the wave function varies with east–west longitude around the atom. The total wave function, ψ, is the product of these three functions. The wave functions that are solutions to the Schrödinger equation for the hydrogen atom are called **orbitals**.

In the process of separating the parts of the wave function, a constant, n, appears in the radial expression, another constant, l, occurs in the radial and azimuthal expressions, and m appears in the azimuthal and angular expressions. The boundary conditions that give physically sensible solutions to these three equations are that each function (radial, azimuthal, and angular) be continuous, single-valued, and finite at all points. These conditions will not be met unless n, l, and m are integers, l is zero or a positive integer less than n, and m has a value from $-l$ to $+l$. From a one-dimensional problem (the vibrating string) we obtained one quantum number. With a three-dimensional problem, we obtain three quantum numbers.

The **principal quantum number**, n, can be any positive integer: $n = 1, 2, 3, 4, 5, \ldots$. The **azimuthal quantum number**, l, can have any integral value from 0 to $n - 1$. The **magnetic quantum number**, m, can have any integral value from $-l$ to $+l$. The different quantum states that the electron can have are listed in Table 8-1. For one electron around an atomic nucleus, the energy depends only on n. Moreover, the energy expression is exactly the same as in the Bohr theory:

$$E_n = -\frac{Z^2 k}{n^2} \qquad k = \frac{2\pi^2 m_e e^4}{h^2}$$

For $Z = 1$ (the hydrogen atom), we have simply:

$$E_n = -\frac{k}{n^2}$$

where $k = 13.595$ eV or 1312 kJ mole^{-1}.

Quantum states, with $l = 0, 1, 2, 3, 4, 5, \ldots$, are called the $s, p, d, f, g, h, \ldots$ states, in an extension of the old spectroscopic notation (Figure 8-13). The wave functions corresponding to $s, p, d, \ldots$ states are called $s, p, d, \ldots$ orbitals. All of the l states for the same n have the *same* energy in the hydrogen atom; the energy-level diagram is as in Figure 8-10.

Example 14

An electron in atomic hydrogen has a principal quantum number of 5. What are the possible values of l for this electron? When $l = 3$, what are the possible values of m? What is the ionization energy (in electron volts) of this electron? What would it be in the same n state in He$^+$?

Solution

With $n = 5$, l may have a value of 4, 3, 2, 1, or 0. For $l = 3$, there are seven possible values of m: 3, 2, 1, 0, -1, -2, -3. The ionization energy of the electron depends only on n, according to:

$$IE = -E_n$$

$$E_n = -\frac{k}{n^2}$$

Table 8-1

Quantum States for the Hydrogen Atom through $n = 4$

n	l	m	s	Common name	Number of states
1	0	0	$\pm\frac{1}{2}$	$1s$	2
2	0	0	$\pm\frac{1}{2}$	$2s$	2
2	1	-1	$\pm\frac{1}{2}$		
		0	$\pm\frac{1}{2}$	$2p$	6
		$+1$	$\pm\frac{1}{2}$		
3	0	0	$\pm\frac{1}{2}$	$3s$	2
3	1	-1	$\pm\frac{1}{2}$		
		0	$\pm\frac{1}{2}$	$3p$	6
		$+1$	$\pm\frac{1}{2}$		
3	2	-2	$\pm\frac{1}{2}$		
		-1	$\pm\frac{1}{2}$		
		0	$\pm\frac{1}{2}$	$3d$	10
		$+1$	$\pm\frac{1}{2}$		
		$+2$	$\pm\frac{1}{2}$		
4	0	0	$\pm\frac{1}{2}$	$4s$	2
4	1	-1	$\pm\frac{1}{2}$		
		0	$\pm\frac{1}{2}$	$4p$	6
		$+1$	$\pm\frac{1}{2}$		
4	2	-2	$\pm\frac{1}{2}$		
		-1	$\pm\frac{1}{2}$		
		0	$\pm\frac{1}{2}$	$4d$	10
		$+1$	$\pm\frac{1}{2}$		
		$+2$	$\pm\frac{1}{2}$		
4	3	-3	$\pm\frac{1}{2}$		
		-2	$\pm\frac{1}{2}$		
		-1	$\pm\frac{1}{2}$		
		0	$\pm\frac{1}{2}$	$4f$	14
		$+1$	$\pm\frac{1}{2}$		
		$+2$	$\pm\frac{1}{2}$		
		$+3$	$\pm\frac{1}{2}$		
5	0	....		etc.	

$$IE = \frac{k}{n^2}$$

Since $k = 13.6$ eV, the IE of an electron with $n = 5$ is

$$IE = \frac{13.6 \text{ eV}}{5^2} = 0.544 \text{ eV}$$

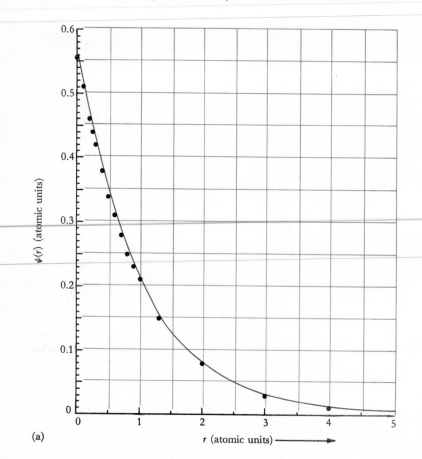

(a)

Figure 8-18

Graphs of (a) ψ and (b) $|\psi|^2$ for the $1s$ orbital of hydrogen, which has the equation $\psi_{1s}(r) = Ae^{-r}$. The distance r is measured in units of a_0, the Bohr radius ($a_0 = 0.529$ Å). Note that although the electron is most likely to be within 4 atomic units of the nucleus the probability curve never quite falls to zero, even at $r \to \infty$. In principle, the probability curve for the electron spreads over the entire universe. But the sphere around the nucleus that contains 99% of the probability within it has a radius of only 4.2 atomic units, or 2.2 Å.

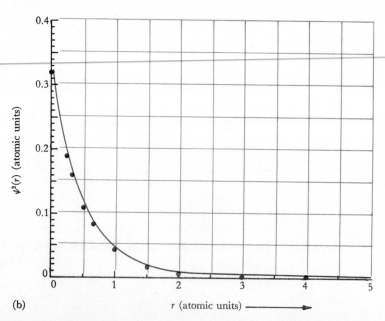

(b)

In general, for one-electron atomic species:

$$IE = -E_n$$

$$E_n = -\frac{Z^2 k}{n^2}$$

$$IE = \frac{Z^2 k}{n^2}$$

For He$^+$, $Z = 2$:

$$IE = \frac{2^2(13.6 \text{ eV})}{n^2}$$

For a He$^+$ electron with $n = 5$, we have

$$IE = \frac{4(13.6 \text{ eV})}{25} = 4 \times 0.544 \text{ eV} = 2.18 \text{ eV}$$

Each of the orbitals for the quantum states differentiated by n, l, and m in Table 8-1 corresponds to a different probability distribution function for the electron in space. The simplest such probability functions, for s orbitals ($l = 0$), are spherically symmetrical. The probability of finding the electron is the same in all directions but varies with distance from the nucleus. The dependence of ψ and of the probability density $|\psi|^2$ on the distance of the electron from the nucleus in the 1s orbital is plotted in Figure 8-18. You can see the spherical symmetry of this orbital more clearly in Figure 8-19. The quantity $|\psi|^2 \, dv$ can be thought of either as the probability of finding an electron in the volume element dv in one atom, or as the *average* electron density within the corresponding volume element in a great many different hydrogen atoms. The electron is no longer in orbit in the Bohr–Sommerfeld sense; rather, it is an electron probability cloud. Such probability density clouds are commonly used as pictorial representations of hydrogenlike **atomic orbitals**.

The 2s orbital is also spherically symmetrical, but its radial distribution function has a node, that is, zero probability, at $r = 2$ atomic units (1 atomic unit is $a_0 = 0.529$ Å). The probability density has a crest at 4 atomic units, which is the radius of the Bohr orbit for $n = 2$. There is a high probability of finding an electron in the 2s orbital closer to or farther from the nucleus than $r = 2$, but there is no probability of ever finding it in the spherical shell at a distance $r = 2$ from the nucleus (Figure 8-20). The 3s orbital has two such spherical nodes, and the 4s has three. However, these details are not as important in explaining bonding as are the general observations that s orbitals are spherically symmetrical and that they increase in size as n increases.

There are three 2p orbitals: 2p_x, 2p_y, 2p_z. Each orbital is cylindrically symmetrical with respect to rotation around one of the three principal axes

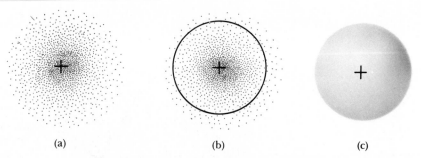

(a)   (b)   (c)

Figure 8-19   Three ways of representing the spherical electron probability density function of the $1s$ orbital of hydrogen. (a) $|\psi|^2$ represented by the density of stippling; (b) a black circle representing a cross section through the spherical shell that encloses 90% of the probability (radius 2.7 atomic units or 1.4 Å); (c) the 90% probability shell portrayed as a surface.

x, y, or z, as identified by the subscript. Each $2p$ orbital has two lobes of high electron density separated by a nodal plane of zero density (Figures 8-21 and 8-22). The sign of the wave function, ψ, is positive in one lobe and negative in the other. The $3p$, $4p$, and higher p orbitals have one, two, or more additional nodal shells around the nucleus (Figure 8-23); again, these details are of secondary importance. The significant facts are that the three p orbitals are mutually perpendicular, strongly directional, and of increasing size as n increases.

The five d orbitals first appear for $n = 3$. For $n = 3$, l can be 0, 1, or 2, thus s, p, and d orbitals are possible. The $3d$ orbitals are shown in Figure 8-24. Three of them, d_{xy}, d_{yz}, and d_{xz}, are identical in shape but different in orientation. Each has four lobes of electron density bisecting the angles between

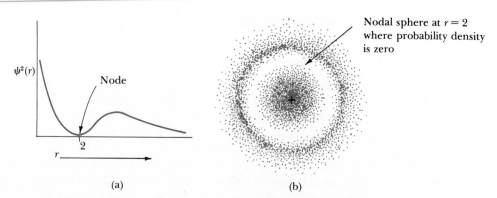

(a)   (b)

Figure 8-20   The $2s$ hydrogen orbital. (a) The graph of $|\psi|^2$ against r. (b) A cross section through the probability function plotted in three dimensions. Probability density is represented by stippling.

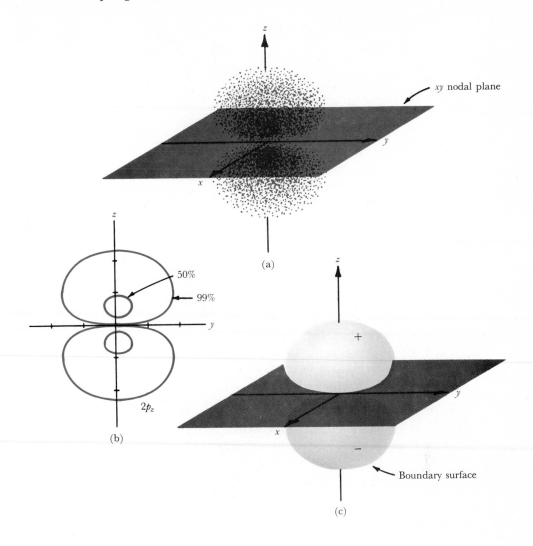

Figure 8-21   Three ways of representing the $2p_z$ atomic orbital of hydrogen. (a) $|\psi|^2$ represented by stippling. (b) Contour diagram of the $2p_z$ orbital. The contours represent lines of constant $|\psi|^2$ in the yz plane and have been chosen so that, in three dimensions, they enclose 50% or 99% of the total probability density. The $2p_z$ orbital is symmetrical around the z axis. (c) The 99% probability shell portrayed as a surface. The plus and minus signs on the two lobes represent the relative signs of ψ and should not be confused with electric charge. Note that there is no probability of finding the electron on the xy plane. Such a surface, which need not be planar, is called a *nodal surface.*

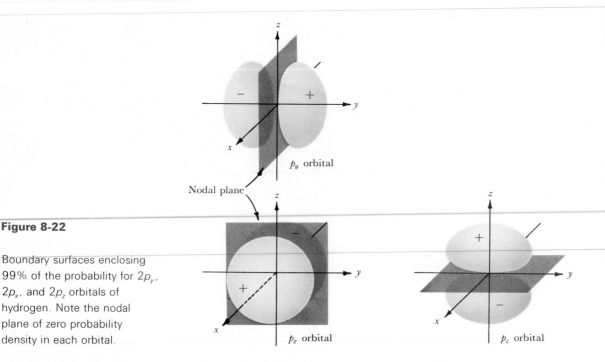

Figure 8-22

Boundary surfaces enclosing 99% of the probability for $2p_y$, $2p_x$, and $2p_z$ orbitals of hydrogen. Note the nodal plane of zero probability density in each orbital.

principal axes. The remaining two are somewhat unusual: The $d_{x^2-y^2}$ orbital has lobes of density along the x and y axes, and the d_{z^2} orbital has lobes along the z axis, with a small doughnut or ring in the xy plane. However, there is nothing sacrosanct about the z axis. The proper combination of wave functions of these five d orbitals will give us another set of five d orbitals in which the d_{z^2}-like orbital points along the x axis, or the y axis. We could even combine the wave functions to produce a set of orbitals, all of which were alike but differently oriented. However, the set that we have described, d_{xy}, d_{yz}, d_{xz}, $d_{x^2-y^2}$, and d_{z^2}, is convenient and is used conventionally in chemistry. The sign of the wave function, ψ, changes from lobe to lobe, as indicated in Figure 8-24.

The azimuthal quantum number l is related to the shape of the orbital, and is referred to as the **orbital-shape quantum number**: s orbitals with $l = 0$ are spherically symmetrical, p orbitals with $l = 1$ have plus and minus

Figure 8-23

Contour diagrams in the xz plane for hydrogen wave functions that show the 50% and 99% contours. The x and z axes are marked in intervals of 5 atomic units. All orbitals shown except the $3d_{xz}$ have rotational symmetry around the z axis. The $3p_z$ orbital differs from the $2p_z$ in having another nodal surface as a spherical shell around the nucleus at a distance of approximately 6 atomic units. But so far as chemical bonding is concerned, these inner details do not matter; the important difference between a $2p$ and a $3p$ orbital is size.

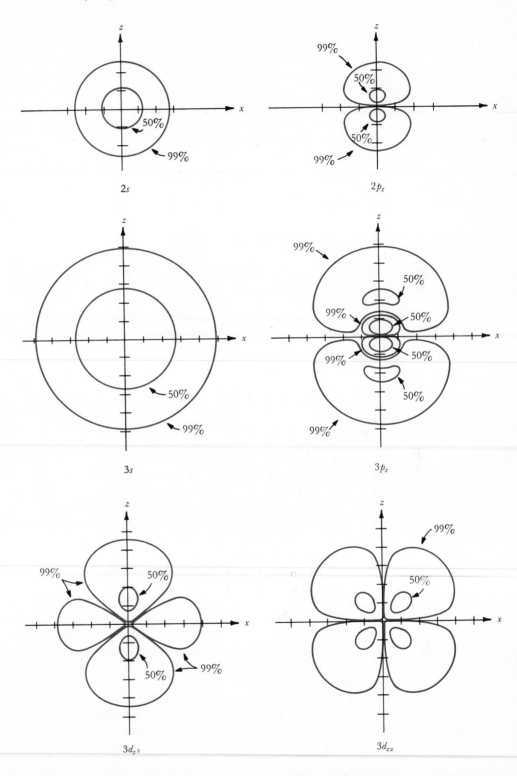

2s

2p_z

3s

3p_z

3d_{z^2}

3d_{xz}

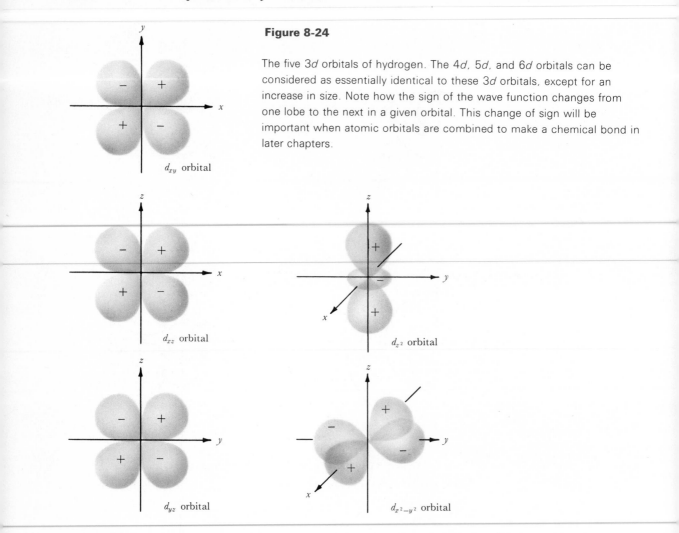

Figure 8-24

The five $3d$ orbitals of hydrogen. The $4d$, $5d$, and $6d$ orbitals can be considered as essentially identical to these $3d$ orbitals, except for an increase in size. Note how the sign of the wave function changes from one lobe to the next in a given orbital. This change of sign will be important when atomic orbitals are combined to make a chemical bond in later chapters.

extensions along one axis, and d orbitals with $l = 2$ have extensions along two mutually perpendicular directions (Figure 8-25). The third quantum number, m, describes the orientation of the orbital in space. It is sometimes called the magnetic quantum number because the usual way of distinguishing between orbitals with different spatial orientations is to place the atoms in a magnetic field and to note the differences in energy produced in the orbitals. We will use the more descriptive term, **orbital-orientation quantum number**.

    There is a fourth quantum number that has not been mentioned. Atomic spectra, and more direct experiments as well, indicate that an electron behaves as if it were spinning around an axis. Each electron has a choice of two **spin** states with **spin quantum numbers**, $s = +\frac{1}{2}$ or $-\frac{1}{2}$. A complete

Figure 8-25

Summary of the most important aspects of the hydrogen orbitals. (a) The principal quantum number, n, indicates the approximate relative size of the orbital. (b) The orbital-shape quantum number, l, indicates the shape or the degree of asymmetry of the orbital.

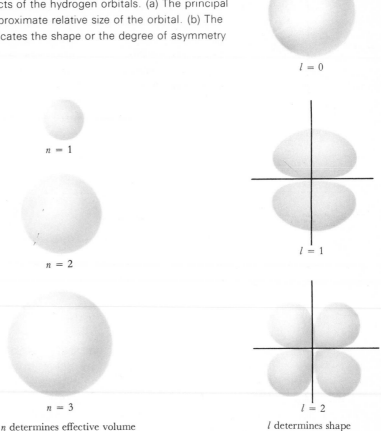

$l = 0$

$n = 1$

$l = 1$

$n = 2$

$n = 3$

$l = 2$

n determines effective volume

l determines shape

(a)

(b)

description of the state of an electron in a hydrogen atom requires the specification of all four quantum numbers: n, l, m, and s.

8-8 MANY-ELECTRON ATOMS

It is possible to set up the Schrödinger wave equation for lithium, which has a nucleus and three electrons, or uranium, which has a nucleus and 92 electrons. Unfortunately, we cannot solve the differential equations. There is little comfort in knowing that the structure of the uranium atom is calculable *in principle*, and that the fault lies with mathematics and not with physics. Physicists and physical chemists have developed many approximate methods that involve guesses and successive approximations to solutions of the Schrödinger equation. Electronic computers have been of immense value in such successive approximations. But the advantage of Schrödinger's theory of the hydrogen atom is that it gives us a clear *qualitative* picture of the electronic structure of many-electron atoms without such additional calculations. Bohr's theory was too simple and could not do this, even with Sommerfeld's help.

The extension of the hydrogen-atom picture to many-electron atoms is one of the most important steps in understanding chemistry, and we shall reserve it for the next chapter. We shall begin by assuming that electronic orbitals for other atoms are *similar* to the orbitals for hydrogen and that they can be described by the same four quantum numbers and have analogous probability distributions. If the energy levels deviate from the ones for hydrogen (which they do), then we shall have to provide a persuasive argument, in terms of the hydrogenlike orbitals, for these changes.

Summary

Rutherford's scattering experiments showed the atom to be composed of an extremely dense, positively charged nucleus surrounded by electrons. The nucleus is composed of protons and neutrons. A proton has one positive charge and a mass of 1.67×10^{-27} kg. A neutron is uncharged and has a mass of 1.67×10^{-27} kg.

Radio waves, infrared, visible, and ultraviolet light, x rays and γ rays are electromagnetic waves with different wavelengths. The speed of light, c, equal to 2.9979×10^{10} cm sec^{-1}, is related to its wavelength (λ) and frequency (ν) by $c = \nu\lambda$. The **wave number**, $\bar{\nu}$, is the reciprocal of the wavelength, $\bar{\nu} = 1/\lambda$. Hot objects radiate energy (**blackbody radiation**). Planck proposed that the energy of electromagnetic radiation is **quantized**. The energy of a **quantum** of electromagnetic radiation is proportional to its **frequency**, $E = h\nu$, in which h is **Planck's constant**, 6.6262×10^{-34} J sec. Electron ejection caused by light striking a metal surface is called the **photoelectric effect**. **Photon** is the name given to a quantum of light. The energy of a photon is equal to $h\nu$, in which ν is the frequency of the electromagnetic wave. The pattern of light absorption by an atom or molecule as a function of wavelength, frequency, or wave number is called an **absorption spectrum**. The related pattern of light emission from an atom or molecule is called an **emission spectrum**. The emission spectrum of atomic hydrogen is composed of several series of lines. The positions of these lines are given accurately by a single equation, the **Rydberg equation**,

$$\bar{\nu} = R_{\mathrm{H}} \times \left(\frac{1}{n_1^2} - \frac{1}{n_2^2} \right)$$

in which $\bar{\nu}$ is the wave number of a given line, R_{H} is the Rydberg constant, 109,677.581 cm^{-1}, and n_1 and n_2 are integers (n_2 is greater than n_1). The **Lyman series** is that group of lines with $n_1 = 1$ and $n_2 = 2, 3, 4, \ldots$. The

Balmer series has $n_1 = 2$ and $n_2 = 3, 4, 5, \ldots$, and the **Paschen series** has $n_1 = 3$ and $n_2 = 4, 5, 6, \ldots$.

Bohr pictured the hydrogen atom as containing an electron moving in a circular orbit around a central proton. He proposed that only certain orbits were allowed, corresponding to the following energies:

$$E = -\frac{k}{n^2}$$

in which E is the energy of an electron in the atom (relative to an ionized state, $H^+ + e^-$), k is a constant, equal to 13.595 eV atom^{-1} or 1312 kJ mole^{-1}, and n is a quantum number that can take only integer values from 1 to ∞. The radius of a Bohr orbit is $r = n^2 a_0$, where a_0 is called the **first Bohr radius**; $a_0 = 0.529$ Å. One atomic unit of length equals a_0. The **ground state** of atomic hydrogen is the lowest energy state, where $n = 1$. **Excited states** correspond to $n = 2, 3, 4, \ldots$. The energy levels in a general one-electron atomic species, such as He^+ and Li^{2+}, with atomic number Z, are given by

$$E = -\frac{Z^2 k}{n^2}$$

The wave nature of electrons was established when Davisson and Germer showed that metal foils diffract electrons in the same way that they diffract a beam of x rays. The wave–particle duality exhibited by electrons is present in all objects. For large objects (such as baseballs), particle behavior predominates to such an extent that wave properties are unimportant.

Heisenberg proposed that we cannot know both the position and the momentum of a particle with absolute accuracy. The product of the uncertainty in position, Δx, and momentum, $\Delta(mv)$, must be at least as large as $h/4\pi$:

$$[\Delta x][(\Delta m v_x)] \geq \frac{h}{4\pi}$$

The wave equation for a particle is called the **Schrödinger equation** The solutions to the Schrödinger equation, $\psi_{(x,y,z)}$, are called **wave functions** The square of the amplitude, $|\psi_{(x,y,z)}|^2$, is the relative **probability density** of the particle at position $_{(x,y,z)}$. A place where the amplitude of a wave is zero is called a **node**.

Solution of the Schrödinger equation for the hydrogen atom yields wave functions $\psi_{(x,y,z)}$ and discrete energy levels for the electron. The wave functions $\psi_{(x,y,z)}$ are called **orbitals**. An orbital is commonly represented as a probability density cloud, that is, a three-dimensional picture of $|\psi_{(x,y,z)}|^2$. Three quantum numbers are obtained from solving the Schrödinger equation: the **principal quantum number**, n, can be any positive integer ($n = 1, 2, 3, 4, \ldots$); the **azimuthal** (or orbital-shape) **quantum** number, l, can have any integral value from 0 to $n - 1$; the **magnetic** (or orbital-orientation) **quantum number,** m, takes integral values from $-l$ to $+l$. The energy levels depend only on n,

$$E = -\frac{k}{n^2}$$

Wave functions with $l = 0$ are called s orbitals; those with $l = 1$ are called p orbitals; those with $l = 2$ are called d orbitals; those with $l = 3, 4, 5, \ldots$, are called $f, g, h, \ldots$, orbitals. A fourth quantum number is needed to interpret atomic spectra. It is the **spin quantum number,** s, which can be $+\frac{1}{2}$ or $-\frac{1}{2}$.

Self-Study Questions

1. What are α, β, and γ rays? Which of them are composed of particles? Which are waves? Why is this an unfair question?
2. How are Thomson's and Rutherford's models of the atom different, and how does the scattering of α particles differentiate between them?
3. What is intolerable about Rutherford's atom, according to classical physics?
4. Which of the following is proportional to energy in electromagnetic radiation: speed, wave number, or wavelength?
5. If wavelengths are normally the measured quantities in spectroscopy, why are wave numbers preferable to frequencies when a quantity proportional to energy is desired?
6. What is the ultraviolet catastrophe, and how did Planck resolve it?
7. How did Planck's central assumption in explaining the ultraviolet catastrophe also account for the photoelectric effect?
8. What was the empirical formula for obtaining wave numbers of spectral lines in atomic hydrogen that was obtained by Balmer and others? How was it explained by Bohr? Can you think of any possible explanations for such a formula, other than ones of the type that Bohr proposed?
9. What was Bohr's basic assumption for obtaining the spectral formula from his model? What justification was there for that assumption? (Do not confuse Bohr's contribution with de Broglie's.)
10. What is quantization of energy? Tell how it arises in the following:
    a) The Bohr atom
    b) De Broglie's interpretation of the Bohr atom
    c) A vibrating violin string
    d) The Schrödinger equation for the hydrogen atom
    e) An organ pipe with closed ends
    f) An organ pipe with the top end open
11. Can you see any logical connection between the quantization of energy

in the hydrogen atom and Caruso shattering a wineglass with a high note?

12. How can the same atom of hydrogen, in quick succession, emit a photon in the Pfund, Brackett, Paschen, Balmer, and Lyman series? Can it emit them in the reverse order? Why, or why not?

13. How does the Bohr theory fail for the lithium atom? How did Sommerfeld try to overcome the difficulty? Where did his theory fail?

14. Why are energies of electrons in atoms always negative?

15. Why is wave behavior not seen in automatic rifle fire, although we can see it in a beam of neutrons?

16. What is the uncertainty principle? Why can we neglect it in everyday life?

17. What is a boundary condition in the solution of a wave equation? What is its physical meaning? What boundary conditions are imposed on the solution of the vibration of a violin string? What are the boundary conditions on the solution of the Schrödinger equation for an electron in a hydrogen atom?

18. What are the three quantum numbers that are found in the solution of the Schrödinger equation for the hydrogen atom? What values can each quantum number have, and what do they signify?

19. What is an atomic orbital? How does it differ from an orbit?

20. What is the difference between a probability density and a probability? Why is it incorrect to talk about a probability of an electron being at a particular point in space?

21. How many d levels are there in a quantum level? How do the shapes of the d orbitals, in their common chemical representation, differ from those of the p orbitals? How can the d orbitals be given different energies?

22. What is the spin quantum number? What numerical values can it have?

23. Consider two hydrogen atoms. The electron in the first hydrogen atom is in the $n = 1$ Bohr orbit. The electron in the second hydrogen atom is in the $n = 4$ Bohr orbit. (a) Which atom has the ground-state electronic configuration? (b) In which atom is the electron moving faster? (c) Which orbit has the larger radius? (d) Which atom has the lower potential energy? (e) Which atom has the higher ionization energy?

24. For each of the following statements, choose the possibility that most accurately completes the statement.

    a) Rutherford, Geiger, and Marsden performed experiments in which a beam of helium nuclei (α particles) was directed at a thin piece of gold foil. They found that the gold foil (1) severely deflected most of the particles of the beam directed at it; (2) deflected very few of the particles of the beam, and deflected these only very slightly; (3) deflected most of the particles of the beam, but deflected these only very slightly; (4) deflected very few of the particles of the beam, but deflected these severely.

b) From the results of the experiment, Rutherford concluded that (1) electrons are massive particles; (2) the positively charged parts of atoms are extremely small and extremely heavy particles; (3) the positively charged parts of atoms move about with a velocity approaching that of light; (4) the diameter of an electron is approximately equal to that of the nucleus.

c) Max Planck was led to his formulation of the quantum theory in attempting to explain why (1) electrons are emitted from a metal when light of sufficiently short wavelength falls upon it; (2) the thermal (or blackbody) radiation from a hot object contains a relatively large amount of ultraviolet light, contrary to the prediction of classical mechanics; (3) the thermal radiation from a hot object contains a relatively small amount of ultraviolet light, contrary to the prediction of classical mechanics; (4) the thermal radiation from a hot object can occur at all frequencies, contrary to the prediction of classical mechanics.

25. Which one of the following statements concerning the photoelectric effect is *not* true? (a) No electrons are emitted from the surface of a metal until the frequency of the light directed on the surface exceeds a certain "threshold" value. (b) Above the threshold frequency, the greater the intensity of the light, the greater will be the velocity of the emitted electrons. (c) Above the threshold frequency, the smaller the wavelength of the light, the greater will be the velocity of the emitted electrons. (d) Above the threshold frequency, the greater the intensity of the light, the greater will be the number of electrons emitted per second.

26. Which one of the following statements concerning the Bohr theory of the hydrogen atom is *not* true? (a) The theory successfully explained the observed emission and absorption spectra of the hydrogen atom. (b) The theory requires the energy of the electron in the hydrogen atom to be proportional to velocity. (c) The theory requires the electron in the hydrogen atom to have only certain discrete energy values. (d) The theory requires the distance of the electron from the nucleus in the hydrogen atom to have only certain discrete values.

27. Einstein interpreted the photoelectric effect in terms of which of the following ideas: (a) the particle nature of light; (b) the wave nature of light; (c) the wave nature of matter; (d) the uncertainty principle?

28. Which of the following best describes the emission spectrum of atomic hydrogen: (a) a continuous emission of light at all frequencies; (b) discrete series of lines, each line within a series equidistant from the next; (c) discrete lines occurring in pairs, each pair equidistant from the next pair; (d) only two lines observed over the entire spectrum; (e) discrete series of lines whose spacing within the series decreases at higher wave numbers?

29. Which of the following describes the process responsible for the emission spectrum of hydrogen, according to the Bohr theory?

a) Electrons are excited into higher-energy orbits and emit light when they fall back into lower-energy orbits.

b) The hydrogen atom emits light when electrons are excited into higher-energy orbits.

c) The hydrogen atom absorbs light when electrons are excited into higher-energy orbits.

d) Electrons are de-excited into lower-energy orbits and emit light when they return to higher-energy orbits.

30. In the Bohr theory of the atom, which of the following concepts was introduced arbitrarily by Bohr?

a) The electron is attracted to the nucleus by coulombic forces.

b) The electron moves in circular orbits about the nucleus.

c) The angular momentum of the electron is restricted to discrete values.

d) The kinetic energy of the electron is given by $\frac{1}{2}m_e v^2$.

e) The mass of the electron is restricted to discrete or quantized values.

31. Which of the following was not explained by the simple Bohr theory: (a) the ionization energy of hydrogen; (b) the details of atomic spectra of atoms with many electrons; (c) the locations of the lines in the hydrogen spectrum; (d) the spectra of hydrogenlike atoms such as He^+ and Li^{2+}; (e) the energy levels of the hydrogen atom?

32. Which of the following experiments most directly supports de Broglie's hypothesis of the wave nature of matter: (a) x-ray diffraction; (b) photoelectric effect; (c) α-particle scattering by metal foil; (d) the blackbody effect; (e) electron diffraction?

33. Which of the following aspects of the Bohr theory is not allowed by Heisenberg's uncertainty principle: (a) discrete atomic energy levels; (b) simple circular orbits; (c) quantum numbers; (d) electron orbitals; (e) electron waves? Why does the aspect that you chose clash with the uncertainty principle?

34. With which of the following is the quantum number m associated: (a) the spatial orientation of the orbital; (b) the shape of the orbital; (c) the energy of the orbital in the absence of a magnetic field; (d) the effective volume of the orbital?

35. The probability of finding a p-orbital electron at the nucleus of an atom is zero. A contradiction arises when the two lobes of a p orbital are described as touching each other. What is the contradiction? [From *J. Chem. Educ.* **38**, 20 (1961).]

Problems

Light and energy

1. Calculate the wavelength of a photon that has a frequency of 1.2×10^{15} Hz. What is the energy of the photon in joules per photon? What is the energy in kilojoules per mole? What do we usually call such radiation?

2. X rays typically have wavelengths of 1 Å to 10 Å. Calculate the energy in joules of photons with a 2-Å wavelength. Calculate the energy in kilojoules per mole of such 2-Å photons, and compare it with the bond energy of a carbon–carbon single bond, 347 kJ mole^{-1}. Would you expect x rays to be able to produce chemical reactions?

3. Calculate the energy of photons, in joules per photon and in kilojoules per mole, for 1000-kilocycle broadcast-band radio waves. (One kilocycle is a frequency of 10^3 Hz.) What is the wavelength of such photons? How does the energy compare with that for a carbon–carbon single bond? Would you expect radio waves to be able to produce chemical reactions?

4. The first ionization energy of Cs is 376 kJ mole^{-1}. Calculate the first ionization energy (in kilojoules and electron volts) for one atom of Cs. Calculate the wavelength of light that would be just sufficient to ionize a Cs atom.

5. In 1914, Moseley discovered that $\nu = c(Z - b)^2$, in which ν is the frequency of x rays emitted when an element is bombarded by a beam of electrons. Use the Bohr theory to explain the dependence of Z on the square root of ν in this expression.

6. What is the wavelength of photons that have an energy of 347 kJ mole^{-1}? What do we call such radiation? (See Figure 8-5a.)

7. Calculate the wavelength of a photon of visible light with a frequency of 0.66 × 10^{15} Hz. What is the energy of the photon in joules per photon? What is the wave number?

8. When a photon strikes a metal surface, it must have a certain minimum energy to eject an electron from the metal. This minimum or threshold energy is known as the *work function* of the metal. Any energy above this minimum in the original photon is translated into kinetic energy for the ejected electron. The threshold wavelength for photoelectric emission from Li, above which no electrons are emitted, is 5200 Å. Calculate the velocity of electrons emitted as the result of absorption of light at 3600 Å.

Bohr theory

9. How much energy (in electron volts per atom) is required to ionize a hydrogen atom in which the electron occupies the $n = 5$ Bohr orbit?

10. The Lyman series of lines results from transitions from more energetic orbits to which lower-energy Bohr orbit? A spectral line is found at 103,000 cm^{-1}. What is the quantum number of the initial orbit of the electron undergoing this transition?

11. Set up an expression for the wavelength of the radiation that will be emitted by a He$^+$ ion when it decays from an excited state having the principal quantum number $n = 4$ to a lower excited state having $n = 3$. Your expression should give the wavelength as a function of m_e, e, h, π, and c only. Calculate the numerical value of the wavelength of the emitted radiation.

12. Calculate the wave number of the photons emitted when a hydrogen atom decays from a state with $n = 3$ to one with $n = 2$. What is the name of the series containing such emission?

13. Calculate the energy (in electron volts per atom) released when a hydrogen atom decays from the state having the principal quantum number 4 to the state having the principal quantum number 3.

14. Calculate the frequency of the light emitted when a hydrogen atom decays as in Problem 13. Calculate the wave number of the light emitted in the decay of the hydrogen atom described.

15. If the energy associated with the first Bohr orbit is -13.60 eV atom^{-1}, what is the energy associated with the fourth Bohr orbit?

16. If the second Bohr orbit has a radius of 2.12 Å, what is the radius of the fourth Bohr orbit?

Electron orbitals

17. Draw spatial representations for the following orbitals and put in x, y, and z coordinates, if needed: $2p_z$; $3s$; $3d_{x^2-y^2}$; $3d_{xz}$; $n = 2$, $l = 1$.

18. Draw spatial representations for the following orbitals: $n = 2$, $l = 1$; $n = 1$, $l = 0$; $3d_{z^2}$.

19. An electron is in one of the $3d$ orbitals. What are the possible values of the orbital quantum numbers n, l, and m for the electron?

Quantum numbers

20. An electron in a hydrogen atom has a principal quantum number of 4. List the values of the second quantum number, l, that the electron can have.

21. The Balmer series for atomic hydrogen occurs in the visible region of the spectrum. Which series in the emission spectrum of Be^{3+} has its lowest-energy line closest to the first line in the hydrogen Balmer series? How many quantum-number combinations are there for each of the participating energy levels for this emission line in Be^{3+}?

22. If an electron has an orbital-shape quantum number of $l = 3$, what values of m can it have? What do we call such an $l = 3$ electron?

23. An electron is in a $4f$ orbital. What possible values for the quantum numbers n, l, m, and s can it have?

Uncertainty principle

24. Calculate the de Broglie wavelength of an average helium atom at 27°C (see Chapter 3). Assume that the position of an average He atom can be measured to 0.10 Å. Compare the uncertainty in the momentum of the He atom to its actual momentum.

25. Repeat Problem 24 for a Xe atom at 27°C.

Suggested Reading

A. W. Adamson, "Domain Representation of Orbitals," *J. Chem. Educ.* **42**, 141 (1965).

R. S. Berry, Advisory Council on College Chemistry Resource Paper on "Atomic Orbitals," *J. Chem. Educ.* **43**, 283 (1966).

J. B. Birks, Ed., *Rutherford at Manchester,* W. A. Benjamin, Menlo Park, Calif., 1963. A good account of what it was like to be a scientist in Britain in the early part of the twentieth century. Written by and about one of the sharpest minds, finest men, and best writers in physics.

I. Cohen and T. Bustard, "Atomic Orbitals: Limitations and Variations," *J. Chem. Educ.* **43**, 187 (1966).

U. Fano and L. Fano, *Physics of Atoms and Molecules,* University of Chicago Press, Chicago, 1972.

R. P. Feynman, R. B. Leighton, and M. Sands, *The Feynman Lectures on Physics: Quantum Mechanics,* Addison-Wesley, Reading, Mass., 1965. Exhilarating book. Ostensibly written for a beginning physics course, but excellent reading even when you cannot follow it all.

G. Gamow, *Mr. Tomkins in Wonderland,* Cambridge University Press, New York, 1939. What would our world be like if the speed of light were 10 miles per hour? If Planck's constant were 27 orders of magnitude larger? A collection of stories answering such questions, with each story based on a changed value for one important physical constant in quantum mechanics. The quantized world as seen through the eyes of a middle-class bank clerk with a taste for free public lectures. Highly recommended.

G. Gamow, *Thirty Years That Shook Physics: The Story of Quantum Theory,* Doubleday Anchor, New York, 1966.

W. Heisenberg, *The Physical Principles of Quantum Theory,* Dover, New York, 1930. An early statement by one of the pioneers in quantum mechanics.

C. N. Hinshelwood, *The Structure of Physical Chemistry,* Oxford University Press, New York, 1951.

R. M. Hochstrasser, *The Behavior of Electrons in Atoms,* W. A. Benjamin, Menlo Park, Calif., 1964. A supplement written for general chemistry students. Goes into more details than this chapter but at a similar level.

R. C. Johnson and R. R. Rettew, "Shapes of Atoms," *J. Chem. Educ.* **42**, 145 (1965).

E. A. Ogryzlo and G. B. Porter, "Contour Surfaces for Atomic and Molecular Orbitals," *J. Chem. Educ.* **40**, 256 (1963).

B. Perlmutter-Hayman, "The Graphical Representation of Hydrogen-Like Wave Functions," *J. Chem. Educ.* **46**, 428 (1969).

R. E. Powell, "The Five Equivalent *d* Orbitals," *J. Chem. Educ.* **45**, 1 (1968).

H. H. Sisler, *Electronic Structure, Properties and the Periodic Law,* Van Nostrand Reinhold, New York, 1963.

G. Thomson, *The Atom,* Oxford University Press, New York, 1962.

9

Electronic Structure and Atomic Properties

Key Concepts

9-1 Buildup of many-electron atoms. *Aufbau* process. Pauli exclusion principle and paired spins. Hund's rule. Effective nuclear charge. Orbital configuration and ionization energy. Valence electron and valence orbital. Representative elements, inner transition metals, transition metals, and noble gases. Electron affinity.

9-2 Types of bonding. Covalent bonds and ionic bonds. Bond energy, bond distance, and atomic radius.

9-3 Electronegativity. Partial ionic character.

I have been asked sometimes how one can be sure that elsewhere in the universe there may not be further elements, other than those in the periodic system. I have tried to answer by saying that it is like asking how one knows that elsewhere in the universe there may not be another whole number between 4 and 5. Unfortunately, some persons think that is a good question, too.

George Wald, 1964

We now know the wave functions and energy levels for a hydrogen atom. With this information and the *aufbau* (or buildup) process, we can go on to determine the electronic structures for atoms of all the elements. These structures lead directly to the periodic table of Figures 7-3 and 7-4. As we shall see, the structures explain the stability of eight-electron shells in noble gases and the trends in ionization energies and electron affinities of the elements.

9-1 BUILDUP OF MANY-ELECTRON ATOMS

Although we cannot solve the Schrödinger equation exactly for many-electron atoms, we can show that no radical new features are expected as the atomic number increases. There are the same quantum states, the same four quantum numbers (n, l, m, and s), and virtually the same electronic probability functions or electron-density clouds. The energies of the quantum levels are not identical for all elements; rather, they vary in a regular fashion from one element to the next.

In studying the electronic structure of a many-electron atom, we shall assume the existence of a nucleus and the required number of electrons. We shall assume that the possible electronic orbitals are hydrogenlike, if not identical to the hydrogen orbitals. Then we shall build the atom by adding electrons one at a time, placing each new electron in the lowest-energy orbital available. In this way we shall build a model of an atom in its ground state, or the state of lowest electronic energy. Wolfgang Pauli (1900–

1958) first suggested this treatment of many-electron atoms, and called it the *aufbau*, or buildup, process.

The *aufbau* process involves three principles:

1. No two electrons in the same atom can be in the same quantum state. This principle is known as the **Pauli exclusion principle**. It means that no two electrons can have the same n, l, m, and s values. Therefore, one atomic orbital, described by n, l, and m, can hold a maximum of two electrons: one of spin $+\frac{1}{2}$ and one of spin $-\frac{1}{2}$. We can represent an atomic orbital by a circle and an electron by an arrow:

When two electrons occupy one orbital with spins $+\frac{1}{2}$ and $-\frac{1}{2}$, we say that their spins are paired. A **paired spin** is represented as follows:

2. Orbitals are filled with electrons in order of increasing energies. The s orbital can hold a maximum of 2 electrons. The three p orbitals can hold a total of 6 electrons, the five d orbitals can hold 10, and the seven f orbitals can hold 14. We must decide on the order of increasing energies of the levels before we can begin the buildup process. For atoms with more than one electron, in the absence of an external electric or magnetic field, energy depends on n and l (the size and shape quantum numbers) and not on m, the orbital-orientation quantum number.

3. When electrons are added to orbitals of the same energy (such as the five $3d$ orbitals), one electron will enter each of the available orbitals before a second electron enters any one orbital. This follows **Hund's rule**, which states that in orbitals of identical energy electrons remain unpaired if possible. This behavior is understandable in terms of electron–electron repulsion. Two electrons, one in a p_x orbital and one in a p_y orbital, remain farther apart than two electrons paired in the same p_x orbital (Figure 8-22). A consequence of Hund's rule is that a **half-filled set** of orbitals (each orbital containing a single electron) is a particularly stable arrangement. The sixth electron in a set of five d orbitals is forced to pair with another electron in a previously occupied orbital. The mutual repulsion of negatively charged electrons means that less energy is required to remove this sixth electron than to remove one of the five in a set of five half-filled d orbitals. Similarly, the fourth electron in a set of three p orbitals is held less tightly than the third.

Relative Energies of Atomic Orbitals

The $3s$, $3p$, and $3d$ orbitals in the hydrogen atom have the same energy but differ in the closeness of approach of the electron to the nucleus (Figure 9-1).

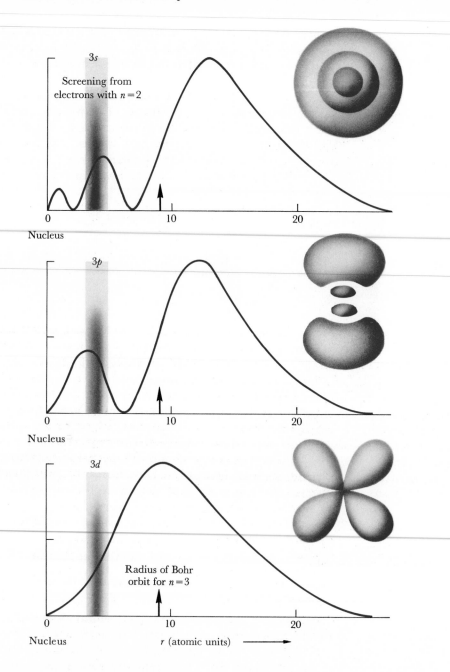

The energy of an electron in an orbital depends on the attraction exerted on it by the positively charged nucleus. Electrons with low principal quantum numbers will lie close to the nucleus and will screen some of this electrostatic attraction from electrons with higher principal quantum numbers. In the Li$^+$ ion, the **effective nuclear charge** beyond 1 or 2 atomic

Figure 9-1   Radial distribution functions for electrons in the 3s, 3p, and 3d atomic orbitals of hydrogen. These curves are obtained by spinning the orbital in all directions around the nucleus to smear out all details that depend on direction away from the nucleus, and then by measuring the smeared electron probability as a function of distance from the nucleus. The 3s orbital, which is already spherically symmetrical without the smearing operation, has a most probable radius at 13 atomic units and two minor peaks close to the nucleus. The 3p orbital has a maximum density near $r = 12$ atomic units, one spherical node at $r = 6$ atomic units and a density peak close to the nucleus. The 3d orbital has only one density peak, which occurs very close to the Bohr orbit radius of 9 atomic units. The shapes of the three orbitals before the spherical smearing process are to the right of each curve. An electron in the hydrogen atom with $n = 2$ will be in the neighborhood of $r = 4$ atomic units. The scale of distances changes in many-electron atoms, but relative distances in different orbitals in the atom are the same as in H. An electron in a 3s orbital is more stable than one in a 3p or 3d orbital because it has a greater probability of being inside the orbital of $n = 2$ electrons, in which it experiences a greater attraction from the nucleus. The 3p orbital is similarly more stable than the 3d.

units from the nucleus is not the true nuclear charge of $+3$, but a *net* charge of $+1$ produced by the nucleus plus the two 1s electrons. Similarly, the lone $n = 3$ electron in sodium experiences a net nuclear charge of approximately $+1$ rather than the full nuclear charge of $+11$.

If the net charge from the nucleus and the filled inner orbitals were concentrated at a point at the nucleus, then the energies of 3s, 3p, and 3d orbitals would be the same. But the screening electrons extend over an appreciable volume of space. The net attraction that an electron with a principal quantum number of 3 experiences depends on how close it comes to the nucleus, and whether it penetrates the lower screening electron clouds. As in Sommerfeld's elliptical-orbit model, the s orbital comes closer to the nucleus and is somewhat more stable than the p, and the p is more stable than the d. This is the reason for the variation of the l energy levels in the lithium energy-level diagram in Figure 8-13.

For a given value of the principal quantum number, n, the order of increasing energy is $s, p, d, f, g, \ldots$. It is less easy to decide whether and when the high l-value orbitals of one n overtake the low-l orbitals of the next: for example, whether a 4f orbital has a higher energy than a 5s, or a 3d a higher energy than a 4s. The question was originally settled empirically by choosing the order of overlap that accounted for the observed structure of the periodic table. The energies have since been calculated theoretically, and (fortunately for quantum mechanics) they agree with the observed order of levels. The sequence of energy levels is shown in Figure 9-2.

Orbital Configurations and First Ionization Energies

We shall build up the electronic structures of the atoms in the periodic table by adding electrons to the hydrogenlike orbitals in order of increasing

4. Write the ground-state electronic configurations of the following atoms or ions: (a) As; (b) Co^{2+}; (c) Cu; (d) S^{2-}; (e) Kr; (f) C; (g) W; (h) H^+; (i) H^-; (j) Cl^-.

5. Write the ground-state electronic configurations of the two atoms $^{18}_{8}O$ and $^{16}_{8}O$.

6. Several electronic configurations that may be correct for the nitrogen atom ($Z = 7$) are listed. For each configuration write one of the following words: *excited*, if the configuration represents a possible excited state of the nitrogen atom; *ground*, if the configuration represents the ground state of the nitrogen atom; *forbidden*, if the configuration in question cannot exist.

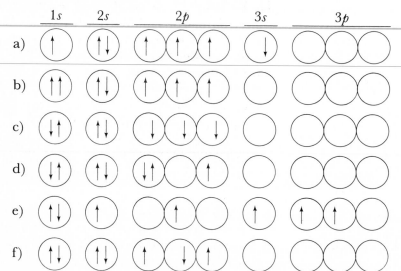

7. Which of the following configurations of electrons represent ground states, which represent excited states, and which are impossible? *Why* are the last unacceptable? What neutral atoms can have each permissible configuration?

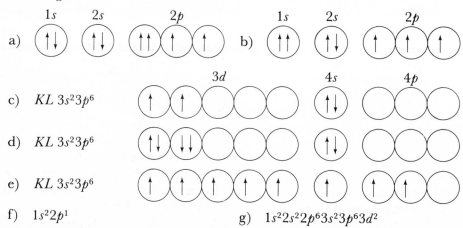

f) $1s^2 2p^1$

h) $1s^2 2s^2 2p^4$

g) $1s^2 2s^2 2p^6 3s^2 3p^6 3d^2$

i) $1s^2 2s^2 2p^6 2d^3 3s^2$

8. Ten atoms and electronic configurations are listed. For each, decide whether a neutral atom, a positive ion (cation), or a negative ion (anion) is represented. In addition, specify whether the electronic state represented is a ground state, an excited state, or impossible.

   a) $_3$Li $1s^2 2p^1$
   b) $_1$H $1s^2$
   c) $_{16}$S $1s^2 2s^2 2p^6 3s^2 3p^4$
   d) $_6$C $1s^2 2s^2 2p^1 2d^1$
   e) $_{10}$Ne $1s^2 2s^1 2p^7$
   f) $_7$N $1s^2 2s^1 2p^3$
   g) $_9$F $1s^2 2s^2 2p^5 3s^1$
   h) $_2$He $1p^1$
   i) $_{21}$Se $1s^2 2s^2 2p^6 3s^2 3p^6 3d^1 4s^2$
   j) $_8$O $1s^2 2s^2 2p^3$

9. Write ground-state electronic configurations for Li, Lu, La, and Lr and also for Li^+, Lu^{3+}, La^{3+}, and Lr^{3+}.

Atomic numbers

10. If the order of orbital energy levels were in strict numerical order ($1s$, $2s$, $2p$, $3s$, $3p$, $3d$, $4s$, $4p$, $4d$, $4f$, $5s$, $5p$, and so on) and if the stable elements that we call noble gases occurred when the last electron of a given n value was added, what atomic numbers would the noble gases have? Compare them with the true atomic numbers.

11. Suppose you discovered some material from another universe whose atomic "electrons" obeyed the following restrictions on quantum numbers:

$$n > 0$$
$$l = 0, 1, \ldots n - 1$$
$$m = +1 \text{ or } -1$$
$$s = +\tfrac{1}{2}$$

Assume that usual buildup rules still apply. What atomic numbers would the first two noble gases in that universe have?

Atomic properties

12. The ionization energies of polonium (Po) and astatine (At) are not given in Table 9-1 because these elements are not available in large quantities and no accurate IE measurements have been made. Using the information available in Table 9-1, estimate the values of IE for Po and At.

13. Write the electronic configuration ($1s^2 2s^2 \ldots$) for the following: F^-, Na^+, Ne, O^{2-}, and N^{3-}. What would you predict about the relative sizes of these species?

14. Which atom in each of the following pairs would you expect to have the larger electron affinity (EA): (a) Cu or Zn; (b) K or Ca; (c) S or Cl; (d) H or Li; (e) As or Ge?

15. The atoms of the yet-to-be-discovered "hypotransition" elements, starting at $Z = 121$, will have electrons in the $5g$ orbitals.

    a) How many elements will there be in the hypotransition metal series?
    b) Which electronic configurations in the series will have seven unpaired electrons?
    c) What is the maximum number of unpaired electrons an atom can have in the series? Will this be a new record for atoms in the periodic table?
    d) What is the IE of an electron in a $5g$ orbital of atomic hydrogen? Is this likely to be larger or smaller than the IE of a $5g$ electron in one of the hypotransition elements? Briefly explain your answer.
    e) In atomic hydrogen the $5s$, $5p$, $5d$, $5f$, and $5g$ orbitals all have the same energy. Will this be true for the hypotransition elements? If not, what will the energy order be? Explain briefly.

Electronegativity

16. Compute the electronegativity of a Cl atom. Assume $\chi_H = 2.20$ and take bond energies from Tables 12-2 and 12-4.

17. In Mulliken's first simple definition, the electronegativity of an element was proportional to the sum of its first ionization energy and its electron affinity. This relationship is not strictly true for the numerical values given in Table 9-1 because the ionization energies, electron affinities, and electronegativities in this table have been calculated by different people using different methods. Nevertheless, the proportionality is approximately valid. From the data in Table 9-1, plot a graph of the sum of ionization energy and electron affinity, against electronegativity, for the elements in the second and third periods of the table. (a) Draw the best straight line that you can through these data points and the origin. (b) Use this plot to estimate the electronegativity of Ne. If Ne—F bonds existed, would you expect them to be ionic or covalent? (c) Work backward from your plot to calculate the electron affinities of the fifth-period elements rubidium (Rb) through indium (In). Plot these values as a function of atomic number. In terms of electronic configurations of the atoms, explain the general trend of electron affinities across the transition metals in this period, and the striking behavior at Ag–Cd–In.

18. Calculate the electronegativity of H assuming the value 3.98 for χ_F. Bond energies are given in Tables 12-2 and 12-4. (The value you obtain will not agree exactly with that given in Table 9-1 because many χ differences [equation 9-1] were averaged to give the best values reported in the table.)

Suggested Reading

H. B. Gray, *Chemical Bonds,* W. A. Benjamin, Menlo Park, Calif., 1973.

R. L. Rich, *Periodic Correlations,* W. A. Benjamin, Menlo Park, Calif., 1965.

R. T. Sanderson, *Chemical Periodicity,* Van Nostrand Reinhold, New York, 1960.

H. H. Sisler, *Electronic Structure, Properties and the Periodic Law,* Van Nostrand Reinhold, New York, 1963.

10

Oxidation-Reduction and Chemical Properties

The electron has conquered physics, and
many worship the new idol rather blindly.

H. Poincaré (1907)

I n the preceding chapters we saw that the ionization energies, electron affinities, and electronegativities of atoms can be explained on the basis of atomic orbital electronic structures. Now we can proceed to relate electronic structure to the chemical properties of the elements and their compounds. We shall begin by discussing (and balancing equations for) reactions that involve loss and gain of electrons by reactants **(oxidation–reduction reactions)**. Then we shall systematically examine the properties of the elements and their compounds, with emphasis on the oxidation–reduction chemistry of the metallic elements.

10-1 OXIDATION NUMBERS

To make it easier to discuss oxidation–reduction chemistry, we assign an **oxidation number** to each atom in a molecule or complex ion, according to the following rules:

1. The oxidation number for an atom of any free (uncombined) element is zero; thus, the atoms in H_2, O_2, Fe, Cl_2, and Na have zero oxidation numbers.

2. The oxidation number for any simple one-atom ion is equal to its charge; thus the oxidation number of Na^+ is $+1$; of Ca^{2+}, $+2$; and of Cl^-, -1.

3. The oxidation number of hydrogen in any *nonionic* compound is $+1$. This rule applies to the great majority of hydrogen compounds, such as H_2O, NH_3, HCl, and CH_4. For the *ionic* metal hydrides, such as NaH, the oxidation number of hydrogen is -1.

4. The oxidation number of oxygen is -2 in all compounds in which oxygen does not form an O—O covalent bond. Thus, its oxidation number is -2 in H_2O, H_2SO_4, NO, CO_2, and CH_3OH, but in hydrogen peroxide, H_2O_2, it is -1. (Another exception to the rule that oxygen has an oxidation number of -2 is OF_2, in which O is $+2$ and F is -1.)

5. In combinations of nonmetals *not* involving hydrogen or oxygen, the nonmetal that is *more* electronegative is considered negative. Its oxidation number is given the same value as the charge on its most commonly encountered negative ion. In CCl_4, for instance, the oxidation number for chlorine is -1; for carbon it is $+4$. In CH_4, the oxidation number for hydrogen is $+1$; for carbon, it is -4. In SF_6, the oxidation number for fluorine is -1, and for sulfur, it is $+6$; but in CS_2, it is -2 for S and $+4$ for C. In molecules such as N_4S_4 in which the bonds are covalent (the combining atoms have the same or almost the same electronegativity), the concept of oxidation number loses usefulness.

6. The algebraic sum of the oxidation numbers of all atoms in the formula for a neutral compound must be zero. Hence, in NH_4Cl, the total oxidation number for the four hydrogen atoms is $4(+1) = +4$, and the oxidation number for Cl is -1, so the oxidation number for N must be -3.

7. The algebraic sum of oxidation numbers of all atoms in an ion must equal the charge on the ion. Thus, in NH_4^+, the oxidation number of N must be -3, so $-3 + 4 = +1$. In SO_4^{2-}, since the four oxygen atoms have a total oxidation number of -8, the oxidation number for sulfur must be $+6$ if the overall charge on the ion is -2.

8. In chemical reactions, the *total oxidation number is conserved*. It is this rule that makes oxidation numbers useful in modern chemistry. If the oxidation number of an atom increases during a chemical reaction, the atom is **oxidized**; if the number decreases, the atom is **reduced**. In a balanced chemical reaction, *oxidations and reductions must exactly compensate one another*.

Example 1

What is the oxidation number of each atom in NF_3?

Solution

Fluorine lies to the right of nitrogen in the periodic table. Therefore, it is considered negative. Fluorine is assigned an oxidation number of -1, corresponding to the F^- ion. Since there are three fluorines, each with an oxidation number of -1, nitrogen must have an oxidation number of $+3$ for the sum of all oxidation numbers to be zero.

Example 2

What is the oxidation number of Mo in the molybdate ion, MoO_4^{2-}?

Solution

Since the oxidation number of each oxygen is -2, Mo must have an oxidation number of $+6$ if the sum $1Mo(+6) + 4O(-2)$ is to be equal to -2.

Calculating Oxidation Numbers

From the preceding rules, we can calculate the oxidation numbers of the atoms in most molecules and complex ions. Certain oxidation numbers are characteristic of a given element, and these can be related to the position of the element in the periodic table. Figure 10-1 shows how oxidation numbers vary with atomic number. The *maximum* oxidation number generally increases across a period from $+1$ to $+7$.

■ *Representative metals.*   Metals in Groups I–III in the periodic table form ions with positive charges equal to the numbers of their respective groups; that is, their oxidation numbers are the same as their group numbers.

■ *Nonmetals.*   Nonmetals often assume either of two characteristic oxidation numbers. Their minimum oxidation number is usually $-(8 - gn)$, where gn is the number of the group in the periodic table; thus, each atom can combine with $8 - gn$ hydrogen atoms. For example, one sulfur atom (Group VI) combines with two hydrogen atoms, since sulfur has an oxidation number of -2. The maximum oxidation number of nonmetals is commonly $+gn$, especially in oxygen compounds. Examples are SO_3 and H_2SO_4, in which the oxidation number of sulfur is $+6$. Most nonmetals also exhibit intermediate oxidation numbers (see Table 10-1).

■ *Transition metals.*   Among transition metals oxidation numbers follow the trends illustrated in Table 10-2. Early members of the series of transition metals exhibit maximum oxidation numbers of increasing magnitude, up to $+7$ for manganese in MnO_4^-, which correspond to the group numbers. Thereafter, the maximum oxidation number usually falls again by *one* number for each step to the right across the second half of the transition metals.

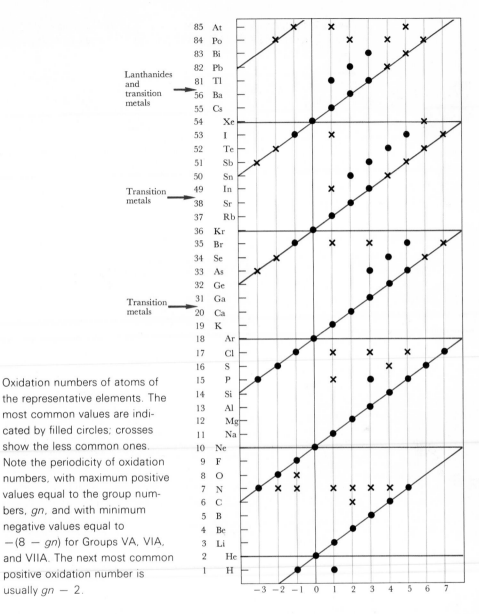

Figure 10-1

Oxidation numbers of atoms of the representative elements. The most common values are indicated by filled circles; crosses show the less common ones. Note the periodicity of oxidation numbers, with maximum positive values equal to the group numbers, gn, and with minimum negative values equal to $-(8 - gn)$ for Groups VA, VIA, and VIIA. The next most common positive oxidation number is usually $gn - 2$.

■ *Inner transition metals.* Lanthanide and actinide elements compose another type of transition series, in which adjacent elements have very similar properties. The oxidation number $+3$ is common to all lanthanides and actinides in their compounds. Other oxidation numbers are possible and in some cases more commonly encountered (e.g., Eu^{2+}, Ce^{4+}, and U^{6+}).

Table 10-1

Oxidation Numbers of Nonmetals

Element	Oxidation number	Representative compounds
F	-1	Fluorides: HF, Na^+F^-
O	-2	H_2O, OH^-, O^{2-}, SO_2
	-1	Peroxides: H_2O_2, O_2^{2-}
N	-3	NH_3, NH_4^+, N^{3-}
	$+5$	HNO_3, NO_3^-, N_2O_5
	All intermediate values	N_2H_4, NH_2OH, N_2O, NO, NO_2^-, NO_2
C	$+4$	CO_2, CCl_4, CF_4
	-4	CH_4
	Complicated by chain formation	C_2H_6, C_4H_{10}, C_2H_6O
Cl	-1	HCl, Cl^-
	$+7$	$HClO_4$, ClO_4^-
	Intermediate values	ClO^-, ClO_2^-, ClO_2, ClO_3^-
S	-2	H_2S, S^{2-}
	$+4$	H_2SO_3, SO_2, HSO_3^-, SO_3^{2-}
	$+6$	H_2SO_4, SO_3, SO_4^{2-}, SF_6
	Intermediate values	$S_2O_3^{2-}$, $S_2O_4^{2-}$, $S_5O_6^{2-}$
P	-3	PH_3, PH_4^+, P^{3-}
	$+5$	H_3PO_4, P_4O_{10}, PO_4^{3-}, PCl_5
	Intermediate values	H_3PO_3, H_3PO_2
Si	$+4$	SiO_2, SiO_4^{4-}
Br	-1	HBr, Br^-
	$+5$	$HBrO_3$, BrO_3^-, BrF_5
	Intermediate values	BrF, BrF_3
I	-1	HI, I^-
	$+5, +3, +1$	IO_3^-, ICl_4^-, ICl
	$+7$	HIO_4, H_5IO_6, IF_7
Se, Te	-2	H_2Se, H_2Te
	$+4$	SeO_2, TeO_2
	$+6$	H_2SeO_4, $Te(OH)_6$
As, Sb	-3	AsH_3, SbH_3
	$+3$	$AsCl_3$, $SbCl_3$
	$+5$	AsO_4^{3-}, $Sb(OH)_6^-$

10-2   OXIDATION–REDUCTION REACTIONS

An atom with a particular oxidation number is sometimes referred to as being "in the oxidation state" of that number; thus, in H_2O, H is in the $+1$ oxidation state and O is in the -2 oxidation state. Reactions in which the oxidation states of component atoms change are called **oxidation–**

Table 10-2

Oxidation Numbers of First-Row Transition Metals[a]

Oxidation number	IIIB	IVB	VB	VIB	VIIB	VIIIB			IB	IIB
7					MnO_4^-					
6				CrO_4^{2-}	$\underline{MnO_4^{2-}}$	FeO_4^{2-}				
5			VO_4^{3-}	$CrOCl_5^{2-}$	MnO_4^{3-}	*				
4		TiO_2	$\underline{VO^{2+}}$	*	$\underline{MnO_2}$	*	CoO_2	NiO_2		
3	Sc^{3+}	$\underline{Ti^{3+}}$	$\underline{V^{3+}}$	$\underline{Cr^{3+}}$	$\underline{Mn^{3+}}$	$\underline{Fe^{3+}}$	$\underline{Co^{3+}}$	Ni_2O_3	Cu^{3+}	
2	*	TiO	V^{2+}	$\underline{Cr^{2+}}$	Mn^{2+}	$\underline{Fe^{2+}}$	$\underline{Co^{2+}}$	Ni^{2+}	$\underline{Cu^{2+}}$	Zn^{2+}
1		*	*	*	$\underline{Mn(CN)_6^{5-}}$	*	*	$Ni_2(CN)_6^{4-}$	$\underline{Cu^+}$	
0		*	$V(CO)_6$	$Cr(CO)_6$	$Mn_2(CO)_{10}$	$Fe(CO)_5$	$Co_2(CO)_8$	$Ni(CO)_4$		

[a]Underlined species are those most commonly encountered under ordinary conditions in solids and in aqueous solutions. The asterisk indicates that oxidation numbers have been observed only in rare complex ions or unstable compounds.

reduction (redox) reactions. If an atom's oxidation number increases, the atom is oxidized; if its oxidation number decreases, it is reduced. Species containing an atom or atoms whose oxidation numbers increase are called **reducing agents** (or **reductants**); those containing an atom or atoms whose oxidation numbers decrease are called **oxidizing agents** (or **oxidants**). Some common oxidizing and reducing agents are listed in Table 10-3. The acceptance of electrons by a substance must cause a decrease in oxidation number, whereas the loss of electrons must involve an increase. It follows that an oxidation–reduction reaction may be thought of as one in which the reducing agent gives electrons to the oxidizing agent. Thus, in rusting, iron reduces oxygen, and oxygen oxidizes (the origin of the term) iron:

$$4Fe + 3O_2 \rightarrow 4Fe^{3+} + 6O^{2-} \quad \text{(in the form of } 2Fe_2O_3\text{)}$$
$$\underset{12e^-}{\underline{\qquad\qquad}}\uparrow$$

By donating electrons, copper reduces silver ions, and the silver ions oxidize the metallic copper in the reaction

$$Cu + 2Ag^+ \rightarrow Cu^{2+} + 2Ag$$
$$\underset{2e^-}{\underline{\qquad\qquad}}\uparrow$$

If an atom can have several oxidation states, in the intermediate oxidation states it can be either an oxidizing or a reducing agent. The ion Mn^{3+} can act as an oxidizing agent and be reduced to Mn^{2+}, or it can act as a reducing agent and be oxidized to Mn^{4+}. In fact, Mn^{3+} in solution is unstable and spontaneously **disproportionates** with self-oxidation–reduction to give the $+2$ and $+4$ oxidation states:

$$2Mn^{3+} + 2H_2O \rightarrow Mn^{2+} + MnO_2 + 4H^+$$

Table 10-3

Common Oxidizing and Reducing Agents

Oxidizing agents

1. Free (elemental) nonmetals become negative ions:

Fluorine	$F_2 + 2e^- \rightarrow 2F^-$
Oxygen	$O_2 + 4e^- \rightarrow 2O^{2-}$
Chlorine	$Cl_2 + 2e^- \rightarrow 2Cl^-$
Bromine	$Br_2 + 2e^- \rightarrow 2Br^-$
Iodine	$I_2 + 2e^- \rightarrow 2I^-$
Sulfur	$S + 2e^- \rightarrow S^{2-}$

2. Positive (usually metal) ions become neutral:

$$Ag^+ + e^- \rightarrow Ag$$
$$2H^+ + 2e^- \rightarrow H_2$$

3. Higher oxidation states become lower:

$$8H^+ + MnO_4^- + 5e^- \rightarrow Mn^{2+} + 4H_2O$$
$$Cu^{2+} + e^- \rightarrow Cu^+ \quad \text{(often written as } Cu^{2+}|Cu^+)$$
$$Fe^{3+} + e^- \rightarrow Fe^{2+} \quad \text{(or } Fe^{3+}|Fe^{2+})$$
$$Cr_2O_7^{2-}|Cr^{3+}$$
$$ClO_3^-|Cl^-$$
$$NO_3^-|(NO_2, NO, N_2O, NH_4^+, \text{etc.})$$
$$Ce^{4+}|Ce^{3+}$$

Reducing agents

1. Metals yield ions plus electrons:

$$Zn \rightarrow Zn^{2+} + 2e^-$$
$$Na \rightarrow Na^+ + e^-$$

All metals yielding their common ions may be included here.

2. Nonmetals combine with other nonmetals, such as O and F, which they take from compounds with metals:

$$C + [O^{2-}] \rightarrow CO + 2e^-$$

Here $[O^{2-}]$ represents oxygen in a -2 oxidation state in combination with a metal such as Fe in the following total equation:

$$3C + Fe_2O_3 \rightarrow 3CO + 2Fe$$

3. Lower oxidation states become higher:

$$Fe^{2+} \rightarrow Fe^{3+} + e^- \quad \text{(or } Fe^{2+}|Fe^{3+})$$
$$SO_3^{2-} + H_2O \rightarrow SO_4^{2-} + 2H^+ + 2e^- \quad \text{(or } SO_3^{2-}|SO_4^{2-})$$
$$NO + 2H_2O \rightarrow NO_3^- + 4H^+ + 3e^- \quad \text{(or } NO|NO_3^-)$$

Furthermore, when each oxidizing agent reacts it becomes a potential reducing agent, and vice versa. This process is similar to that described by the Brønsted–Lowry acid–base theory (Section 5-3), in which every acid, by giving up a proton, becomes a base, and every base, by accepting a proton, becomes an acid.

10-3 BALANCING OXIDATION–REDUCTION EQUATIONS

Let's look at a reaction involving $K_2Cr_2O_7$ and HI. If we assume that the reactants and products are known,* then the problem is how to find the mole ratios and balance the equation for the reaction. We start with the equation

$$K_2Cr_2O_7 + HI + HClO_4 \rightarrow KClO_4 + Cr(ClO_4)_3 + I_2 + H_2O \quad (10\text{-}1)$$

Two methods have been developed for balancing redox equations systematically. With the **oxidation-number method** we use the fact that the amount of oxidation must equal the amount of reduction in the total chemical reaction. With the **ion–electron method** we consider a redox reaction to be the formal sum of two half-reactions, one that donates electrons and one that accepts them.

Oxidation-Number Method

1. Identify the elements that change oxidation number during the reaction. Write the oxidation numbers of these atoms above the appropriate symbols on both sides of the equation. In equation 10-1, chromium (Cr) goes from $+6$ in $K_2Cr_2O_7$ to $+3$ in $Cr^{3+}(ClO_4^-)_3$. Imagine that each Cr atom accepts three electrons to change its oxidation state from $+6$ to $+3$. Iodine goes from -1 in HI to zero in I_2 and loses one electron per atom in the process.

2. Now choose enough of the reductant and oxidant so the electrons lost by one are used completely by the other. There must be three times as many I atoms involved as Cr, and since $K_2Cr_2O_7$ has two Cr atoms, the reaction requires six HI molecules:

$$\overset{+6}{K_2Cr_2O_7} + \overset{-1}{6HI} + HClO_4 \rightarrow KClO_4 + \overset{+3}{2Cr(ClO_4)_3} + \overset{0}{3I_2} + H_2O$$

---

*At this point you should not feel that you should be able to predict products of reactions. As you gain experience, especially in the laboratory, you will be able to make more and more predictions.

3. Balance the other metal ions that do not change oxidation number (K^+ in this case):

$$\overset{+6}{K_2Cr_2O_7} + \overset{-1}{6HI} + HClO_4 \rightarrow 2KClO_4 + \overset{+3}{2Cr(ClO_4)_3} + \overset{0}{3I_2} + H_2O$$

4. Balance the anions that do not change (ClO_4^- in this case):

$$K_2Cr_2O_7 + 6HI + 8HClO_4 \rightarrow 2KClO_4 + 2Cr(ClO_4)_3 + 3I_2 + H_2O$$

5. Balance the hydrogens, and make sure that oxygen is also balanced:

$$K_2Cr_2O_7 + 6HI + 8HClO_4 \rightarrow 2KClO_4 + 2Cr(ClO_4)_3 + 3I_2 + 7H_2O$$

The balancing process is thus completed. The sequence of balancing steps can be summarized as: oxidation numbers–cations–anions–hydrogens–oxygens. In what follows we shall balance the same equation by another method.

Ion–Electron (Half-Reaction) Method

It is often useful to pretend that oxidation and reduction are occurring separately, and then to combine enough of each half-reaction to cancel all the free electrons. Chemical reactions occurring at electrodes in batteries or electrolysis cells (Chapter 1, Section 1-7) are examples of half-reactions that actually occur. For example,

$$Cu^{2+} + 2e^- \rightarrow Cu \qquad \text{(cathode)}$$
$$2H_2O \rightarrow O_2 + 4H^+ + 4e^- \qquad \text{(anode)}$$

Redox reactions that occur in solution can be considered as the sum of two such half-reactions that proceed without the addition of an external driving force (the battery). In all electron-transfer reactions *the number of electrons donated by the reducing agent must equal the number of electrons accepted by the oxidizing agent.*

The $K_2Cr_2O_7$ reaction can be balanced by half-reactions as follows:

1. First, simplify the reaction by eliminating all **spectator ions** (those ions that do not really participate in the reaction), such as K^+ or ClO_4^-.

$$Cr_2O_7^{2-} + I^- + H^+ \rightarrow Cr^{3+} + I_2 + H_2O$$

2. Now construct two balanced half-reactions, one involving Cr and one involving I.

a. The unbalanced reactions are

$$Cr_2O_7^{2-} \rightarrow 2Cr^{3+}$$
$$2I^- \rightarrow I_2$$

b. Balance the atoms in each half-reaction by adding H^+ and H_2O if the reactions occur in an acid medium, or H_2O and OH^- if the reactions occur in a basic one:

$$Cr_2O_7^{2-} + 14H^+ \rightarrow 2Cr^{3+} + 7H_2O \qquad \text{(Cr, O, and H atoms balanced)}$$

$$2I^- \rightarrow I_2 \qquad \text{(no } H^+ \text{ or } OH^- \text{ on either side; no atoms needed)}$$

c. Balance the charge by adding electrons:

$$6e^- + Cr_2O_7^{2-} + 14H^+ \rightarrow 2Cr^{3+} + 7H_2O$$
$$2I^- \rightarrow I_2 + 2e^-$$

If the half-reaction is balanced properly, the number of electrons will indicate exactly the change in oxidation number. The two Cr require six electrons, and the two I^- produce two electrons.

3. Multiply the half-reactions by coefficients that make the number of electrons transferred in each half-reaction the same:

$$Cr_2O_7^{2-} + 14H^+ + 6e^- \rightarrow 2Cr^{3+} + 7H_2O$$
$$6I^- \rightarrow 3I_2 + 6e^-$$

4. Add the two half-reactions and cancel species that appear on both sides of the overall reaction:

$$Cr_2O_7^{2-} + 14H^+ + 6I^- \rightarrow 2Cr^{3+} + 3I_2 + 7H_2O$$

As a precaution, make sure that the *number of atoms* on both sides is the same, that the *charges* balance, and that there are *no net electrons left*. At this point the equation as it stands is balanced. However, for some applications it is useful to complete the equation by restoring the "uninvolved" species and by grouping ions to form known species:

$$Cr_2O_7^{2-} + 8H^+ + 6HI \rightarrow 2Cr^{3+} + 3I_2 + 7H_2O$$

$$K_2Cr_2O_7 + 8HClO_4 + 6HI \rightarrow 2Cr(ClO_4)_3 + 2KClO_4 + 3I_2 + 7H_2O$$

This process can be summarized as: half-reactions–whole reaction–uninvolved ions.

As a second example, let's balance the equation representing the reaction between potassium permanganate ($KMnO_4$) and ammonia (NH_3) that produces potassium nitrate (KNO_3), manganese dioxide (MnO_2), potassium hydroxide (KOH), and water.

■ *Oxidation-number method.*   The unbalanced reaction is

$$KMnO_4 + NH_3 \rightarrow KNO_3 + MnO_2 + KOH + H_2O$$

In this reaction manganese and nitrogen change oxidation number:

$$\begin{array}{cc} 7+ & 4+ \\ Mn \to Mn \end{array} \quad \text{(change of } -3)$$

$$\begin{array}{cc} 3- & 5+ \\ N \to N \end{array} \quad \text{(change of } +8)$$

To conserve overall oxidation numbers, we need eight manganese atoms for three nitrogen atoms:

$$\overset{+7}{8KMnO_4} + \overset{-3}{3NH_3} \to \overset{+5}{3KNO_3} + \overset{+4}{8MnO_2} + KOH + H_2O$$

Potassium (K^+) is the cation that does not change oxidation number; it now must be balanced:

$$\overset{+7}{8KMnO_4} + \overset{-3}{3NH_3} \to \overset{+5}{3KNO_3} + \overset{+4}{8MnO_2} + 5KOH + H_2O$$

The hydrogen atoms must be balanced:

$$8KMnO_4 + 3NH_3 \to 3KNO_3 + 8MnO_2 + 5KOH + 2H_2O$$

The oxygen atoms must balance; there are 32 on each side, and the process is complete.

■ *Half-reaction method.*   Begin by simplifying the reaction. The K^+ ion does not change, so it is omitted:

$$MnO_4^- + NH_3 \to NO_3^- + MnO_2 + OH^- + H_2O$$

In this reaction, MnO_4^- is reduced and NH_3 is oxidized:

$$MnO_4^- \to MnO_2$$
$$NH_3 \to NO_3^-$$

Since OH^- is involved in the reaction, H_2O and OH^- are used to balance the atoms in each half-reaction:

$$MnO_4^- + 2H_2O \to MnO_2 + 4OH^-$$
$$NH_3 + 9OH^- \to NO_3^- + 6H_2O$$

Electrons are added to balance the charge for each half-reaction:

$$3e^- + MnO_4^- + 2H_2O \to MnO_2 + 4OH^-$$
$$NH_3 + 9OH^- \to NO_3^- + 6H_2O + 8e^-$$

The half-reactions are multiplied by 8 and 3, respectively, and then added:

$$\require{cancel} \cancel{24e^-} + 8MnO_4^- + \overset{5}{\cancel{16H_2O}} \to 8MnO_2 + \cancel{32}OH^-$$

$$3NH_3 + \overset{2}{\cancel{27OH^-}} \to 3NO_3^- + \cancel{18}H_2O + \cancel{24e^-}$$

$$\overline{8MnO_4^- + 3NH_3 \to 8MnO_2 + 3NO_3^- + 5OH^- + 2H_2O}$$

Example 3

Balance the following equation involving the oxidation of sulfite (SO_3^{2-}) to sulfate (SO_4^{2-}) by chlorate (ClO_3^-). Use both oxidation-number and ion–electron methods.

$$ClO_3^- + SO_3^{2-} \rightarrow Cl^- + SO_4^{2-}$$

Solution

By the oxidation-number method,

$$\overset{\displaystyle \text{gains } 6e^-}{\underset{\displaystyle \text{loses } 2e^-}{\underset{+5 \quad\quad +4 \quad\quad\quad -1 \quad\quad +6}{ClO_3^- + SO_3^{2-} \rightarrow Cl^- + SO_4^{2-}}}}$$

Three moles of SO_3^{2-} are needed to balance the oxidation-number change for each ClO_3^-:

$$ClO_3^- + 3SO_3^{2-} \rightarrow Cl^- + 3SO_4^{2-}$$

Since 12 oxygens appear on each side, the equation is balanced.

In the ion–electron method the reaction is divided into halves:

$$ClO_3^- \rightarrow Cl^-$$
$$SO_3^{2-} \rightarrow SO_4^{2-}$$

Balance the oxygens by adding H^+ ions and H_2O:

$$6H^+ + ClO_3^- \rightarrow Cl^- + 3H_2O$$
$$H_2O + SO_3^{2-} \rightarrow SO_4^{2-} + 2H^+$$

Add electrons to each half-reaction to balance charge:

$$6e^- + 6H^+ + ClO_3^- \rightarrow Cl^- + 3H_2O$$
$$H_2O + SO_3^{2-} \rightarrow SO_4^{2-} + 2H^+ + 2e^-$$

Balance the electrons by multiplying the lower half-reaction by 3:

$$6e^- + 6H^+ + ClO_3^- \rightarrow Cl^- + 3H_2O$$
$$3H_2O + 3SO_3^{2-} \rightarrow 3SO_4^{2-} + 6H^+ + 6e^-$$

Add the two half-reactions to give a balanced equation:

$$\cancel{6e^-} + \cancel{6H^+} + ClO_3^- \rightarrow Cl^- + \cancel{3H_2O}$$
$$\underline{\cancel{3H_2O} + 3SO_3^{2-} \rightarrow 3SO_4^{2-} + \cancel{6H^+} + \cancel{6e^-}}$$
$$ClO_3^- + 3SO_3^{2-} \rightarrow Cl^- + 3SO_4^{2-}$$

10-4   REDOX TITRATIONS

One *equivalent* (equiv) of an acid or base in a neutralization reaction is the quantity of acid or base that will release or take up 1 mole of protons. In a similar way, 1 equiv of an oxidizing or reducing agent in a redox reaction is defined as the amount of compound that will produce 1 mole of oxidation-number change. In the reaction

$$Na \rightarrow Na^+ + e^-$$

the sodium undergoes a change in oxidation number of one unit, so the equivalent weight of sodium *in this reaction* is equal to its atomic weight. In the oxidation of a Group IIA metal,

$$Mg \rightarrow Mg^{2+} + 2e^-$$

each atom of magnesium changes by two oxidation-number units, and each mole of magnesium metal furnishes 2 equiv of reducing ability. Therefore, the equivalent weight of Mg in this reaction is half its atomic weight.

The equivalent weight of HCl in an acid–base neutralization reaction is equal to its molecular weight. The equivalent weight of HCl in a redox reaction depends on the change in oxidation number of chlorine during the reaction. If a chloride ion is oxidized to Cl_2,

$$\overset{-1}{Cl^-} \rightarrow \overset{0}{\tfrac{1}{2}Cl_2} + e^-$$

then there is one redox equivalent per HCl, and the equivalent weight and molecular weight of HCl are identical. But if the reaction is

$$\overset{-1}{Cl^-} + 3H_2O \rightarrow \overset{+5}{ClO_3^-} + 6H^+ + 6e^-$$

then each HCl furnishes 6 equiv of reducing power, and the equivalent weight is one-sixth the molecular weight.

In titrations using solutions of oxidants or reductants as reagents, it is convenient to use equivalents, for when all of an oxidizing agent in a sample has reacted with a solution's reducing agent from the burette, the number of equivalents of oxidant and reductant is the same. As with neutralization reactions, the *normality* of a solution is the *number of equivalents per liter of solution.*

Example 4

A 50.0-ml solution containing 1.00 g of $KMnO_4$ is used in titrating a reducing agent. During the reaction, MnO_4^- is reduced to Mn^{2+}. What is the molarity of the solution? What is the normality?

Solution   First, we find the molarity:

$$1.00 \text{ g } \cancel{KMnO_4} \times \frac{1 \text{ mole}}{158 \text{ g } \cancel{KMnO_4}} \times \frac{1}{0.0500 \text{ liter}} = 0.127 \text{ mole liter}^{-1}$$

$$= 0.127M$$

The reduction of MnO_4^- to Mn^{2+} is

$$\overset{+7}{MnO_4^-} + 8H^+ + 5e^- \rightarrow \overset{+2}{Mn^{2+}} + 4H_2O$$

Since Mn changes by five oxidation units in this reaction, the equivalent weight of $KMnO_4$ is one-fifth the molecular weight, and the normality is five times the molarity:

$$0.127 \text{ mole liter}^{-1} \times 5.00 \text{ equiv mole}^{-1} = 0.635 \text{ equiv liter}^{-1}$$
$$= 0.635N$$

Example 5

A 31.25-ml solution of $0.100M$ $Na_2C_2O_4$ (sodium oxalate) in acid is titrated with 17.38 ml of $KMnO_4$ solution of unknown strength. What is the normality of the $Na_2C_2O_4$ and of the $KMnO_4$, and the molarity of the $KMnO_4$?

Solution

The reaction is

$$\overset{+7}{2MnO_4^-} + \overset{+3}{5C_2O_4^{2-}} + 16H^+ \rightarrow \overset{+2}{2Mn^{2+}} + \overset{+4}{10CO_2} + 8H_2O$$

Manganese goes from $+7$ to $+2$, so each MnO_4^- provides 5 equiv of oxidizing power. Carbon goes from $+3$ to $+4$, so each $C_2O_4^{2-}$, with *two* carbon atoms, provides 2 equiv of reducing power. Another way of understanding this is to write the two half-reactions:

$$MnO_4^- + 8H^+ + 5e^- \rightarrow Mn^{2+} + 4H_2O$$
$$C_2O_4^{2-} \rightarrow 2CO_2 + 2e^-$$

For this reaction the $0.100M$ $Na_2C_2O_4$ solution is $0.200N$. The milliequivalents of oxidant and reductant are equal at neutralization. [One milliequivalent (meq) is 10^{-3} equiv.] So

$$\text{meq } Na_2C_2O_4 = \text{meq } KMnO_4$$
$$31.25 \text{ ml} \times \frac{0.200 \text{ meq}}{1 \text{ ml}} = 17.38 \text{ ml} \times \frac{x \text{ meq}}{1 \text{ ml}}$$

(Note that 1 meq ml^{-1} = 1 equiv liter^{-1}.)

$$\text{Normality of } KMnO_4 = x = 0.360N$$
$$\text{Molarity of } KMnO_4 = \frac{0.360 \text{ equiv liter}^{-1}}{5 \text{ equiv mole}^{-1}}$$
$$= 0.0720 \text{ mole liter}^{-1}$$
$$= 0.0720M$$

The importance of writing the equations or half-reactions when dealing with equivalents is illustrated by the fact that MnO_4^- can be reduced in various circumstances in the following ways:

$$MnO_4^- + e^- \rightarrow MnO_4^{2-}$$
$$MnO_4^- + 2H_2O + 3e^- \rightarrow MnO_2 + 4OH^-$$
$$MnO_4^- + 8H^+ + 4e^- \rightarrow Mn^{3+} + 4H_2O$$
$$MnO_4^- + 8H^+ + 5e^- \rightarrow Mn^{2+} + 4H_2O$$

The number of equivalent weights per mole of $KMnO_4$ in these examples is 1, 3, 4, and 5. The last reaction is the one most frequently encountered, but the others also occur. The normality of any $KMnO_4$ solution thus depends on how we use it.

10-5   OXIDATION AND REDUCTION POTENTIALS

The **oxidation potential**, $\mathscr{E}$, of a reaction is a measure of the tendency for a reaction of the type

Reduced substance $\rightarrow$ oxidized substance + electrons

to take place, in comparison with the reaction

$$\tfrac{1}{2}H_2(g) \rightarrow H^+ + e^-$$

which is *assigned* an oxidation potential of zero. If the oxidation potential of a reaction is positive, the reaction has a stronger tendency to occur than does the oxidation of H_2. This is true for sodium metal; its standard oxidation potential at $25°C$, $\mathscr{E}^0$, is

$$Na(s) \rightarrow Na^+ + e^-   \mathscr{E}^0 = +2.71 \text{ volts (V)}$$

The drive toward the oxidized state of Na in water is so strong that water itself is decomposed, and hydrogen ions are reduced to H_2 gas.

If the oxidation potential for a reaction is negative, the favored drive is toward the reduced rather than the oxidized state:

$$Ag(s) \rightarrow Ag^+ + e^-   \mathscr{E}^0 = -0.80 \text{ V}$$

Thus the reverse of this reaction will occur. We shall return to a systematic study of oxidation potentials in Chapter 19, where we will see how they are measured in electrolytic cells. At the moment, we want to use them only as measures of the relative tendency of elements to exist in different oxidation states in solution.

The qualification, "in solution," in the preceding sentence is an important one. The first ionization energy of sodium measures the tendency of a *gaseous atom* of Na to lose an electron and to form a *gaseous ion*. In contrast, the oxidation potential measures the tendency of *solid* Na to lose an electron and form a *hydrated* sodium ion in aqueous solution. This is a much more useful quantity in most chemical applications. Sometimes the result of oxidation of a metal in solution is not a hydrated cation but an oxide complex:

$$Mn(s) + 4H_2O \rightarrow MnO_4^- + 8H^+ + 7e^-   \mathscr{E}^0 = -0.771 \text{ V}$$

The oxidation potential of Mn to MnO_4^- is much more meaningful in solution chemistry than is the energy required to strip seven electrons from a Mn atom in the gas phase.

Although we find it convenient to work with oxidation potentials, the accepted international convention is to use **reduction potentials**. If the oxidation potential for the oxidation of metallic sodium to sodium ion is $+2.71$ V, the reduction potential for the reduction of a sodium ion to metallic sodium is -2.71 V.

Example 6

The oxidation potential of metallic calcium is $+2.76$ V. Write an equation for the oxidation of calcium in aqueous solution. What is the reduction potential of Ca^{2+}? Is Ca^{2+} a good oxidizing agent?

Solution

The equation is

$$Ca(s) \rightarrow Ca^{2+} + 2e^- \qquad \mathscr{E}^0 = +2.76 \text{ V}$$

The reduction potential refers to the reverse reaction

$$Ca^{2+} + 2e^- \rightarrow Ca(s) \qquad \mathscr{E}^0 = -2.76 \text{ V}$$

The very negative reduction potential (-2.76 V) shows that Ca^{2+} is definitely *not* a good oxidizing agent.

10-6   CHEMICAL PROPERTIES: THE *s*-ORBITAL METALS

Knowing the electronic structures of atoms, we can interpret the chemical properties of the metals in a reasonable way. Don't try to memorize all the facts given here; instead, try to pick out of the descriptive material those properties that show regular trends across the periodic table, and those that can be explained by electronic structure. Not every chemical property becomes absolutely clear once we know the electronic structure of an atom of an element, but much of what we observe *does* make sense now, and it is this sense that we shall look for in the mass of chemical data.

Group IA. Alkali Metals: Li, Na, K, Rb, and Cs

All the alkali metals have an s^1 outer electronic configuration. The electron is lost easily; these elements thus have low ionization energies and low electronegativities. Ionization energy and electronegativity decrease from Li to Cs as the distance between the outer shells and the nucleus increases.

These metals are the most reactive known, and never occur naturally in the metallic state. They occur in combination with oxygen, chlorine, or other elements, always in the $+1$ oxidation state. All their compounds are ionic, even the hydrides. Virtually any substance capable of being reduced will be reduced in the presence of an alkali metal. The oxidation potentials of the alkali metals, from Li to Cs, are

$$Li(s) \rightarrow Li^+ + e^- \qquad \mathscr{E}^0 = +3.05 \text{ V}$$
$$Na(s) \rightarrow Na^+ + e^- \qquad \mathscr{E}^0 = +2.71 \text{ V}$$
$$K(s) \rightarrow K^+ + e^- \qquad \mathscr{E}^0 = +2.92 \text{ V}$$
$$Rb(s) \rightarrow Rb^+ + e^- \qquad \mathscr{E}^0 = +2.92 \text{ V}$$
$$Cs(s) \rightarrow Cs^+ + e^- \qquad \mathscr{E}^0 = +2.92 \text{ V}$$

Each of these metals has a strong tendency to lose electrons and become oxidized in solution. In contrast, it is difficult to reduce their ions; potassium ions have a reduction potential of -2.92 V. Lithium loses electrons in solution more readily than Cs, in spite of the higher ionization energy of Li, because the small size of a Li^+ ion permits water molecules to approach the center of the ion more closely; this makes the hydrated ion quite stable.

Water attacks all the alkali metals, and the reaction with all these metals is violent and exothermic. A typical reaction is

$$Na(s) + H_2O \rightarrow Na^+ + OH^- + \tfrac{1}{2}H_2(g) \qquad \Delta H^0_{298} = -167 \text{ kJ}$$

The hydrogen gas evolved is ignited by the heat of the reaction and burns spontaneously in air. The alkali metals ordinarily are stored in kerosene or some other unreactive hydrocarbon.

Because the alkali metals are the strongest reducing agents known, the free metals cannot be prepared conveniently by reduction of their compounds with another substance:

$$Li^+ + \text{reducing substance} \rightarrow Li + \text{oxidized substance}$$

Instead, the metals are usually prepared by the electrolysis of their molten compounds.

The easily lost valence electron is responsible for the metallic properties of the alkali metals. With only one mobile electron per atom, metallic bonds are weak. The bonds become weaker with increasing atomic number as the valence electrons become more distant from the nucleus. The metals have low melting and boiling points, and are soft, malleable, and ductile. Lithium can be cut with a knife with difficulty, but Cs is as soft as cheese. Cesium has the lowest ionization energy of any element; its valence electron can be ejected most easily by light in a photoelectric cell. The photoelectric effect is used in photo cells and in television cameras such as the iconoscope, in which the optical image falling on cesium-coated cathodes is converted to electrical impulses.

Almost all the compounds of the alkali metals are soluble in water. The alkali metal ions in solution are colorless. Color is produced when an electron in an atom is excited from one energy level to another, and when the difference in energy of these levels is in the visible portion of the spectrum. The alkali metal ions have no free electrons to be excited by energies in the visible region. The oxides of the alkali metals are basic, and all react with water to form basic hydroxides that are soluble and completely dissociated.

Alkali metals have the interesting property of being soluble in liquid ammonia and forming intensely blue solutions that leave behind the original metal when the ammonia evaporates. The atoms dissociate into positive ions and electrons, and the electrons associate with the NH_3 solvent molecules. Such electrons are known as **solvated electrons**. The intense color has been shown to arise from the solvated electrons and not from the metal ions; the same color can be produced by introducing electrons into ammonia from a platinum electrode.

Group IIA. Alkaline Earth Metals: Be, Mg, Ca, Sr, and Ba

The chemistry of the alkaline earth metals is the chemistry of atoms with two easily lost electrons. All are typical metals and strong reducing agents (although not quite as strong as the alkali metals). The nuclear charge has increased by one from the alkali metals in a given period, but the screening of the nucleus by electrons in inner orbitals is similar for both groups, so the net nuclear charge is greater. Thus, the alkaline earth atoms are smaller and have higher first ionization energies than those of alkali metals in corresponding periods. Their oxidation potentials in aqueous solution are

$$Be(s) \rightarrow Be^{2+} + 2e^- \qquad \mathscr{E}^0 = +1.85 \text{ V}$$
$$Mg(s) \rightarrow Mg^{2+} + 2e^- \qquad \mathscr{E}^0 = +2.37 \text{ V}$$
$$Ca(s) \rightarrow Ca^{2+} + 2e^- \qquad \mathscr{E}^0 = +2.76 \text{ V}$$
$$Sr(s) \rightarrow Sr^{2+} + 2e^- \qquad \mathscr{E}^0 = +2.89 \text{ V}$$
$$Ba(s) \rightarrow Ba^{2+} + 2e^- \qquad \mathscr{E}^0 = +2.90 \text{ V}$$

They are more electronegative than are the alkali metals; nevertheless all their compounds, with the exception of some Be compounds, are ionic. Beryllium is the first example of the general observation that within a group elements with lower principal quantum number will be less metallic because their outer electrons are closer to the nucleus and are held more tightly. This behavior is reflected in the greater electronegativities of the smaller atoms within a group (Table 10-4). Beryllium has a lower oxidation potential, or lower tendency to lose an electron in solution, for the same reason that it has a higher first ionization energy than other elements in the group. It is true that Be, like Li, has a high hydration energy because of its small size. This leads us to expect a strong tendency to oxidize in aqueous solution and a large positive oxidation potential. However, Be has an extraordinarily high ionization energy and energy of vaporization (Table 10-4), and these two effects combine to dominate in the oxidation of Be to Be^{2+}, so the oxidation potential is somewhat lower than might be expected.

The second ionization energies of these metals are usually double their first ionization energies; thus, we might expect $+1$ ions to form and the $+1$ oxidation state to exist in solution. But this is not the case. The hydration of the doubly charged cation gives it enough extra stability to overcome the energy required to remove the second electron. Any solution of Ca^+ ions

Table 10-4

Properties of the Alkaline Earth Metals

Element	Be	Mg	Ca	Sr	Ba
Electronegativity	1.6	1.3	1.0	0.95	0.89
Metallic radius (Å)	0.89	1.36	1.74	1.91	1.98
Melting point (°C)	1278	651	842	769	725
Boiling point (°C)	2970	1107	1487	1384	1140
Heat of fusion (kJ mole^{-1})	11.7	9.2	9.2	9.2	7.5
Heat of vaporization (kJ mole^{-1})	295	129	150	139	151
MCl_2 (melting point, °C)	405	708	772	873	963
MCl_2 (boiling point, °C)	520	1412	1600	1250	1560
Equivalent conductivity of MCl_2 (ohm^{-1} mole^{-1})	0.086	29.0	52.0	—	—

would disproportionate spontaneously to Ca metal and Ca^{2+} ions:

$$2Ca^+ \rightarrow Ca(s) + Ca^{2+}$$

The solution chemistry of the alkaline earth metals is exclusively that of the +2 oxidation state.

The free metals do not occur in nature because they are too reactive. Beryllium and magnesium are found in complex silicate minerals such as beryl ($Be_3Al_2Si_6O_{18}$) and asbestos ($CaMg_3Si_4O_{12}$) (Chapter 14). Emerald is impure beryl, colored with a trace of chromium. Magnesium, calcium, strontium, and barium occur as the relatively insoluble carbonates, sulfates, and phosphates. Calcium and magnesium are much more common than the other elements in the group. Calcium carbonate, $CaCO_3$, is found as chalk, limestone, and marble, usually from deposits of shells and skeletons of marine organisms. Like the alkali metals, the pure alkaline earth metals are commonly prepared from molten compounds by electrolysis because of the difficulty of finding anything with a higher oxidation potential with which to reduce them chemically.

The pure metals have higher melting and boiling points than do alkali metals because they have two electrons per atom for forming metallic bonds. For the same reason, they are harder, although they still can be cut with a sharp steel knife. Beryllium and magnesium are the only elements in this group commonly used as structural metals; because of their lightness they are used pure or in alloys for aircraft and spacecraft, in which weight is an important factor.

Alkaline earth compounds are generally less soluble in water than the compounds of the alkali metals. The hydrides of Ca, Sr, and Ba (CaH_2, SrH_2, and BaH_2) are ionic and are white powders that release H_2 gas upon reaction with water:

$$CaH_2 + 2H_2O \rightarrow Ca^{2+} + 2OH^- + 2H_2$$

The oxides, all with the expected 1:1 atomic ratio (BeO, CaO, and so on), are hard, relatively insoluble in water, and basic with the exception of BeO. In water, the basic oxides form hydroxides, which also are only slightly soluble:

$$CaO(s) + H_2O \rightarrow Ca^{2+} + 2OH^- \rightarrow Ca(OH)_2(s)$$

$Ba(OH)_2$ is strongly basic; $Mg(OH)_2$ is weakly basic.

Beryllium is definitely the odd man out in Group IIA. Its oxide is **amphoteric**, showing both acidic and basic properties. It is virtually insoluble in water, but in strong acid it acts as if it were basic:

$$BeO + 2H^+ + 3H_2O \rightarrow Be(H_2O)_4^{2+}$$

and in strong base it acts as if it were acidic:

$$BeO + 2OH^- + H_2O \rightarrow Be(OH)_4^{2-}$$

In both cases, the cation is so small that only a coordination number of 4 is possible. The coordinating groups are arranged tetrahedrally around Be. The amphoteric behavior of BeO arises because Be is so small and electronegative; Be^{2+} attracts electrons from neighboring water molecules and makes it easier for them to lose a proton to the surroundings:

$$Be^{2+} + 4H_2O \rightarrow Be(OH)_4^{2-} + 4H^+$$

Beryllium shows many other signs of nonmetallic behavior in addition to the amphoterism of its oxide. Its melting and boiling points and heats of vaporization are unusually high, reversing the trends within the rest of Group IIA (Table 10-4). These facts suggest that the covalent bonds in Be, like those in diamond, persist in the liquid. Solid $BeCl_2$ is composed of covalently bonded chains that are held together only by weak intermolecular forces; $BeCl_2$ has the low melting point expected of a molecular, covalent compound instead of an ionic solid such as $CaCl_2$. Finally, liquid $BeCl_2$ does not conduct electricity, a fact that indicates the absence of ions.

10-7 THE FILLING OF THE d ORBITALS: TRANSITION METALS

The transition metals are hard metals with high melting and boiling points. The atoms tend to become smaller with increasing atomic number across a period because of the increased nuclear charge. Atoms in the second transition-metal series, Y to Cd, are larger than those in the first, Sc to Zn. But atoms in the third transition-metal series, Lu to Hg, are not as much larger than the atoms in the second series as would be expected. The reason is that the first *inner* transition-metal series, the lanthanides, is interposed after La. There is a steady decrease in size from La to Lu because of increasing nuclear charge, which produces the **lanthanide contraction**. Therefore, hafnium is not as large as it would have been had it followed directly after La. The nuclear charge in Zr is 18 greater than that in Ti, but that in Hf is 32 greater than that in Zr. The result is that the second- and third-series transition metals have not only the same outer electronic configurations in corresponding groups, but almost the same size as well. Thus, the second and third series are more similar in properties than either is to the first. Titanium resembles Zr and Hf less than Zr and Hf resemble one another. Vanadium is distinct from Nb and Ta, but the very names of Ta and Nb reflect how difficult it is to separate them. Tantalum and niobium were discovered in 1801 and 1802, but for nearly half a century many chemists thought that they were the same element. Because of the difficulty in isolating it, Ta was named after Tantalus, the Greek mythological figure who was doomed to an eternity of frustrating labor. Niobium, in turn, was named for Niobe, the daughter of Tantalus.

The Structure of Transition-Metal Ions

In the K^+ and Ca^{2+} ions, the $4s$ orbital is slightly more stable than the $3d$, and added electrons go into the $4s$ orbital. In contrast, at Sc^{3+}, the $3d$ orbital energy dips below the $4s$, and it remains there for all higher atomic numbers. The lone electron in Sc^{2+} is in a $3d$ orbital, not in the $4s$. This behavior is typical of all transition metals. The crossover of s- and d-orbital energies occurs at the beginning of a transition-metal series. Although the s orbitals fill first in Groups IA and IIA, it is the d orbitals that are occupied in transition-metal ions. The outer electronic configuration of Ti^{2+} is $3d^2$, not $4s^2$.

The lowest oxidation state in all the $3d$ transition metals, with the exception of Cu and a few rare compounds of other metals, is +2, with both s electrons lost. Other higher oxidation states occur with the loss of more electrons from the d orbitals, up to a maximum equal to the number of *unpaired* electrons in the d orbitals. This is why the maximum oxidation number increases from +3 in Sc to +7 in Mn (five d plus two s), and thereafter falls by one per group to +2 in Zn (loss of only the two s electrons). The most common oxidation states are +2 and +3. In the first half of the series, the maximum oxidation state for each element—Sc(III), Ti(IV), V(V), Cr(VI), and Mn(VII)—is also common (Table 10-2).

These generalizations are true for the first transition series. There are some higher oxidation states observed in the second and third series, as in RuO_4 and OsO_4. It is more important for you to know the behavior of the first transition series than to remember the exceptions in the heavier metals.

Oxidation Potentials

The oxidation potentials for the production of ions from neutral atoms are plotted in Figure 10-2 for the first transition series. The driving force for the production of ions decreases as the atomic number increases because the electrons are held more tightly. The +2-ion curve represents the removal of the two s electrons and the retention of the original d configuration. It is

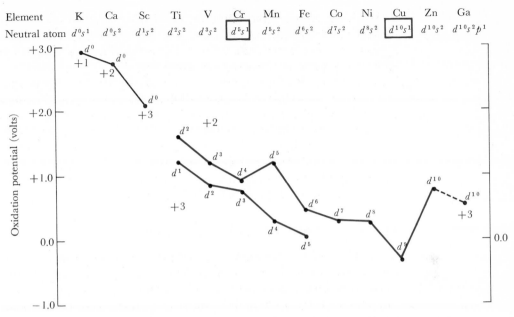

Figure 10-2   Oxidation potentials for the fourth-period metals, including the first transition-metal series. Potentials are for the production of simple cations in solution from the solid metals. The potentials for K, Ca, and Sc are given for the production of $+1$, $+2$, and $+3$ ions with the Ar noble-gas electronic structure. Transition-metal potentials are shown for $+2$ and $+3$ ions. By each point is given the outer electronic configuration of the ion.

particularly difficult to remove two electrons in Cu because only one is in the s orbital and the second would have to come from the filled d^{10} orbitals. In contrast, it is quite easy to remove the two s electrons in Zn and leave the stable, filled d^{10} orbitals untouched. This same effect is observed to a lesser extent in Cr and Mn, in which a half-filled, d^5 configuration is present instead of a filled d^{10}. The d electrons must be disturbed when a $+3$ ion is produced, so we do not have the local fluctuations observed in the $+2$ curve. The elements preceding the transition series become cations by losing all their electrons outside the inner noble-gas shells; the representative elements following a transition series achieve their maximum oxidation number by losing all their electrons outside the filled d^{10} shells.

Chemical Properties of Individual Groups: Sc and Ti Groups

The Sc–Y–Lu triad has the outer electronic configuration d^1s^2, and shows only $+3$ oxidation states. The properties of these elements are similar to those of Al in Group IIIA. All react with water, as does Al. But Sc_2O_3 is a basic oxide rather than amphoteric, like Al_2O_3, because Sc^{3+} is larger than Al^{3+}. The difference in behavior resembles that between CaO and BeO.

In the Ti–Zr–Hf triad, which has the d^2s^2 electronic configuration, Ti and Zr shows $+2$, $+3$, and $+4$ oxidation states, whereas Hf has only $+4$. This is an example of a general trend in the transition metals: The lower oxidation numbers are less important for the second and third transition series because the electrons are farther from the nucleus. If they lose some electrons, they are likely to lose all of them. The lower oxidation states of Ti are ionic, and the $+4$ state is more covalent and nonmetallic. Titanium (II) oxide, TiO, is basic and ionic; it has the NaCl crystal structure. In contrast, the dioxide, TiO_2, is a white, insoluble pigment that has both basic and acidic properties. The chlorides provide a particularly good illustration of this progression of properties. The dichloride, $TiCl_2$, is a strong reducing agent and oxidizes spontaneously in air. It is an ionic solid that decomposes at $475°C$ in a vacuum. Since it reduces water to H_2, there is no aqueous chemistry of Ti^{2+}. Titanium trichloride, $TiCl_3$, is another strong reducing agent and is an ionic solid that decomposes at $440°C$. In contrast, the tetrachloride, $TiCl_4$, is a stable liquid that freezes at $-25°C$ and boils at $+136°C$. It boils before $TiCl_3$ melts (decomposes) because it is a molecular compound with covalent bonds.

The Vanadium Group and the Colors of Ions and Complex Compounds

The chemistry of the V–Nb–Ta elements is similar to that of the previous triad: V and Ta have the d^3s^2 electronic configuration and Nb has the d^4s^1 configuration. Vanadium has oxidation states of $+2$, $+3$, $+4$, and $+5$, whereas only the $+5$ state is important in Nb and Ta (although some $+3$ and $+4$ compounds are known). Like Ti, Zr, and Hf, these metals react easily with N, C, and O at high temperatures, and it is difficult to prepare them by the high-temperature reduction processes used with Fe and other metals. At low temperatures an oxide coating protects them; consequently, the metals are more inert than their oxidation potentials would suggest. At the top of the group, V_2O_5 is amphoteric like TiO_2. It dissolves in both acids and bases to form complex and poorly characterized polymers. The $+4$ oxidation state of V is also on the borderline between ionic and covalent character; VCl_4 is a molecular liquid with a boiling point of $154°C$. In contrast, the V(III) compounds are ionic.

The vanadium ions provide good examples of the colors that are typical of transition-metal compounds. Vanadium(V) as VO_4^{3-} is colorless. In aqueous solution, the vanadyl ion, VO^{2+}, is deep blue, the V^{3+} ion is green, and the V^{2+} ion is violet. Only these colors are seen from the entire visible spectrum because the three solutions absorb orange light (ca. 610 nm), red light (ca. 680 nm), and yellow light (ca. 560 nm), respectively. The colors we see are the complementary colors to those absorbed (Table 20-2). Most electronic energy levels are so far apart that the radiation absorbed in an

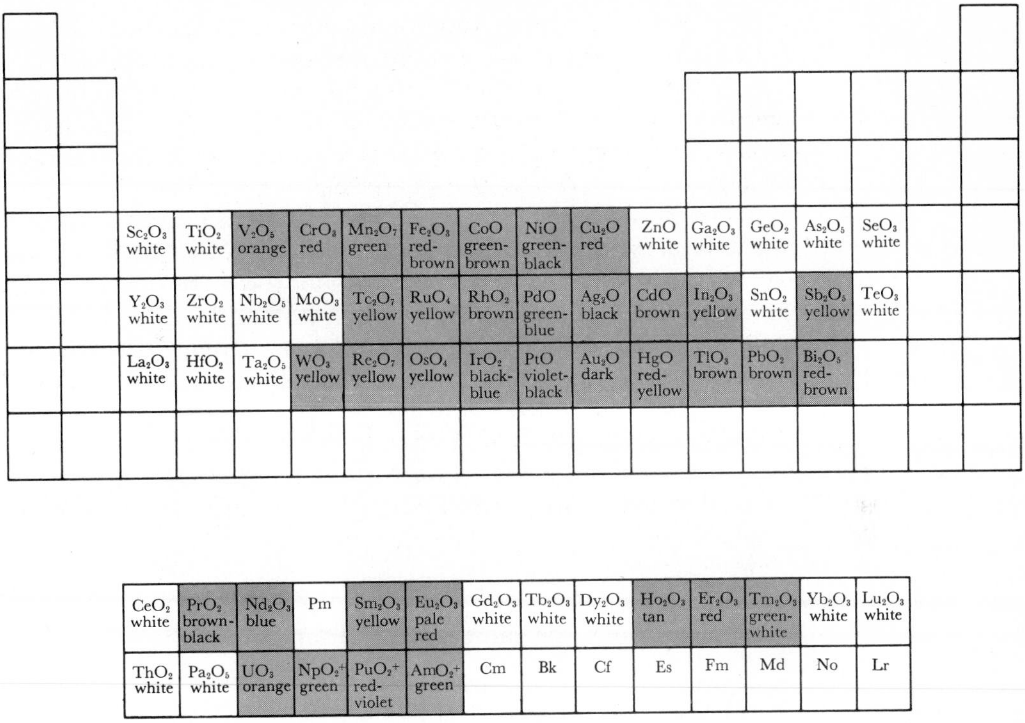

Sc_2O_3 white	TiO_2 white	V_2O_5 orange	CrO_3 red	Mn_2O_7 green	Fe_2O_3 red-brown	CoO green-brown	NiO green-black	Cu_2O red	ZnO white	Ga_2O_3 white	GeO_2 white	As_2O_5 white	SeO_3 white
Y_2O_3 white	ZrO_2 white	Nb_2O_5 white	MoO_3 white	Tc_2O_7 yellow	RuO_4 yellow	RhO_2 brown	PdO green-blue	Ag_2O black	CdO brown	In_2O_3 yellow	SnO_2 white	Sb_2O_5 yellow	TeO_3 white
La_2O_3 white	HfO_2 white	Ta_2O_5 white	WO_3 yellow	Re_2O_7 yellow	OsO_4 yellow	IrO_2 black-blue	PtO violet-black	Au_2O dark	HgO red-yellow	TlO_3 brown	PbO_2 brown	Bi_2O_5 red-brown	

CeO_2 white	PrO_2 brown-black	Nd_2O_3 blue	Pm	Sm_2O_3 yellow	Eu_2O_3 pale red	Gd_2O_3 white	Tb_2O_3 white	Dy_2O_3 white	Ho_2O_3 tan	Er_2O_3 red	Tm_2O_3 green-white	Yb_2O_3 white	Lu_2O_3 white
ThO_2 white	Pa_2O_5 white	UO_3 orange	NpO_2^+ green	PuO_2^+ red-violet	AmO_2^+ green	Cm	Bk	Cf	Es	Fm	Md	No	Lr

Figure 10-3   Oxides of many transition and inner transition metals are colored.

electronic excitation is in the ultraviolet. But in transition-metal complex ions and compounds, the d-orbital energy differences are so small that the frequencies of radiation required for transitions are in the visible region; hence, colors are produced (the theory of d-orbital energy differences, or splittings, in transition-metal complexes is given in Chapter 20). The smallest energy absorption, in the red region, would leave unabsorbed the wavelengths that produce a complementary blue-green color in the solution or compound. Larger and larger energy absorptions produce blue-green, blue, violet, purple, red-orange, and finally yellow; these are the colors complementary to red, orange, yellow, green, blue, and violet. Clearly, color is an approximate guide to the electronic energy differences in metal complex ions and compounds. The colors of several metal oxides are given in Figure 10-3.

The Chromium Group and the Chromate Ion

The elements Cr, Mo, and W have high melting and boiling points and are hard metals. They are relatively inert to corrosion because films of oxides formed on the surface adhere and protect the metal beneath. The thin layer of Cr_2O_3 on chromium metal makes chrome plating an efficient protection for the more easily attacked metals such as iron. Along with V, these three metals are used mainly as alloying agents in steels. Vanadium gives steels ductility, tensile strength, and shock resistance. Chromium makes stainless steels corrosion resistant, molybdenum acts as a toughening agent, and tungsten (W) is used in steel cutting tools that remain hard even when red hot.

Chromium(III) is the most prevalent oxidation state of chromium. Chromium(II) is a good reducing agent, and Cr(VI) is a good oxidizing agent. As we would expect, the acidity of the oxides varies with the oxidation state: CrO_3 is acidic, Cr_2O_3 is amphoteric, and CrO and $Cr(OH)_2$ are basic. A common anion is the yellow chromate ion, CrO_4^{2-}, which dimerizes in acid to form the orange dichromate ion:

$$2CrO_4^{2-} + 2H^+ \rightarrow Cr_2O_7^{2-} + H_2O$$

The dichromate ion is a powerful oxidizing agent, and the reaction by which dichromate ion is reduced to Cr^{3+} has a large positive reduction potential:

$$Cr_2O_7^{2-} + 14H^+ + 6e^- \rightarrow 2Cr^{3+} + 7H_2O \qquad \mathscr{E}^0 = +1.33 \text{ V}$$

We could express the same chemical fact by saying that the reaction by which Cr^{3+} is oxidized to dichromate ion has a low negative oxidation potential, and that the reaction tends to go in the other direction:

$$2Cr^{3+} + 7H_2O \rightarrow Cr_2O_7^{2-} + 14H^+ + 6e^- \qquad \mathscr{E}^0 = -1.33 \text{ V}$$

The Manganese Group and the Permanganate Ion

Of the Mn–Tc–Re triad, Mn is by far the most important. Rhenium was discovered in 1925, and Tc was the first element produced artificially. Technetium was discovered by Perrier and Segré, in 1937, in a sample of Mo that had been irradiated by deuterons ($_1^2H^+$ particles) in the Berkeley cyclotron by Ernest Lawrence (for whom, incidentally, element 103 was named). The new element was named technetium from the Greek *technetos,* "artificial."

The chief use of Mn metal is to make hard and tough manganese steels. Oxidation states from $+2$ to $+7$ are known; the two extremes are the most important. Unlike Ti^{2+}, V^{2+}, and Cr^{2+}, Mn^{2+} shows little tendency to go to higher oxidation states. It is strongly resistant to oxidation and is not a good reducing agent. Manganese(II) in water forms the pink $Mn(H_2O)_6^{2+}$ octahedral complex, and the $MnSO_4$ and $MnCl_2$ salts are also pink. The oxidation states Mn(III) through Mn(VI) are rare, except for the chief natural ore, MnO_2. Mn(VI) does exist as the manganate ion, MnO_4^{2-}. The Mn(VII) state is chiefly important for the deep purple *permanganate* ion, MnO_4^-. It is one of the most powerful common oxidizing agents, with a reduction potential of $+1.49$ V:

$$MnO_4^- + 8H^+ + 5e^- \rightarrow Mn^{2+} + 4H_2O \qquad \mathscr{E}^0 = +1.49 \text{ V}$$

Note again that a compound with a high positive reduction potential or a low negative oxidation potential will be a good oxidizing agent because the compound itself will have a strong tendency to go to the reduced form.

Solutions of the permanganate ion are used as disinfectants. (One of our black comedies on World War II has this bitter remark about the medical services available to enlisted men: "The first time you come in, they give you two aspirins; the second time they paint your gums purple. If you show up again, they arrest you for impersonating an officer." The purple, of course, is $KMnO_4$.) Manganese provides an excellent example of the dependence of chemical properties on oxidation state. Manganese(II) exists in solution as a cation and has a basic oxide, MnO, and hydroxide, $Mn(OH)_2$. At the other extreme, the $+6$ and $+7$ states exist as anions, MnO_4^{2-} and MnO_4^{-}, corresponding to the acidic oxides MnO_3 and Mn_2O_7.

The Iron Triad and the Platinum Metals

In Group VIII, the horizontal similarity between Fe, Co, and Ni is greater than that between these and the corresponding elements in the second and third transition series. These nine elements are usually separated into the iron triad, Fe–Co–Ni, and the light and heavy platinum triads, Ru–Rh–Pd and Os–Ir–Pt. Iron, cobalt, and nickel have the electronic configurations d^6s^2, d^7s^2, and d^8s^2, respectively. All show chiefly the $+2$ and $+3$ oxidation states. (The $+3$ state is very rare for Ni.) Iron is one of the most important structural metals. Many of the other transition metals are important chiefly as alloying agents with iron. Iron is found in three main oxide ores: FeO, Fe_2O_3, and the magnetic mixed oxide magnetite, Fe_3O_4 or $FeO \cdot Fe_2O_3$. Iron is produced by high-temperature reduction with the CO from coke in a blast furnace. The result is cast iron with 3–4% carbon. The open hearth and Bessemer processes are means of burning out most of this carbon with streams of oxygen to obtain steels with 0.1–1.5% carbon.

Iron is no more intrinsically reactive than the other transition metals that we have been discussing. Unfortunately, however, the iron oxides do not adhere to the surface of metallic iron. Rust (iron oxide) flakes off as it is formed and exposes fresh metal to attack. Chrome steel or stainless steel is more corrosion resistant, but the customary protection is an added surface layer such as chromium, tin, nickel, or paint. Ferrous or iron(II) compounds are usually green, and the hydrated ferric ion, $Fe(H_2O)_6^{3+}$, is a pale violet. Both $+2$ and $+3$ states form octahedral complexes with cyanide: $Fe(CN)_6^{4-}$ and $Fe(CN)_6^{3-}$. The traditional names for these anions are *ferrocyanide* and *ferricyanide*. In the modern systematic nomenclature, they are *hexacyanoferrate (II)* and *hexacyanoferrate(III)*. The nomenclature of complex ions is given in Chapter 20.

Cobalt in solution exists mainly as the $+2$ cation, since Co^{3+} is an excellent oxidizing agent and has a strong tendency to be reduced to Co^{2+}:

$$Co^{3+} + e^- \rightarrow Co^{2+} \qquad \mathscr{E}^0 = +1.84 \text{ V}$$

But in many octahedral complexes of Co(III) the ligands (the ions or molecules attached to Co) stabilize it against reduction. Nickel forms octahedral and square planar complexes in the Ni(II) state. Many Ni(II) salts, as well as the hydrated cation, are green. The square planar complexes are usually red or yellow.

Metals in the light and heavy platinum triads are relatively rare, and much work remains to be done on their reactions. All are relatively unreactive, and are found naturally as the pure metals. The oxidation states $+2$, $+3$, and $+4$ are most important, and the metals form octahedral or square planar complex ions in solution. Complex ions of Pt(IV) and Ir(III) are octahedral. Complexes of Pt(II) are square planar. The tetrachloroplatinate (II) ion, $PtCl_4^{2-}$, shows a strong tendency to bind to sulfur in proteins, and has been useful in preparing heavy-atom derivatives of proteins for x-ray crystallographic analysis.

The Coinage Metals

Copper, silver, and gold have the slightly irregular outer electronic configuration $d^{10}s^1$. They have lower melting and boiling points than the preceding transition metals and are moderately soft. These properties are part of a downward trend that began with Group VIB (Cr—Mo—W), which follows from the decreasing number of unpaired d electrons. The metals are excellent conductors of electricity and heat since their electronic arrangement makes the s electrons extremely mobile. They are malleable and ductile; they are inert and can be found naturally in the metallic state. Although rare enough to be prized, they are much less scarce than the platinum metals. Their relative abundance and their occurrence as uncombined metals made them the first metals to be collected and worked by man. The first metal to be reduced from its ore was probably copper. Metallurgy began when it was discovered that an alloy of copper with tin (a naturally occurring impurity) produced the much harder bronze. Copper artifacts have been unearthed in some of the earliest farming communities in the Middle East, dating from 7000–6000 B.C. Bronze was known in the Sumerian cities of Ur and Eridu from 3500 B.C., during the era that also saw the invention of writing.

These three metals have been the source of more strife and trouble than any other elements. Until a century ago, they were used mainly for their symbolic and decorative qualities. More recently, the physical properties of Ag and Au—electrical and thermal conductivity and corrosion resistance—have become so valuable that the metals no longer can be spared for their traditional coinage roles. Gold is now used for plating external surfaces of delicate components in satellites and space probes.

Copper, silver, and gold have little resemblance to the alkali metals, with which they are associated in the short form of the periodic table derived from Mendeleev's table (Figure 7-1). Copper shows mainly the $+2$ oxidation state in solution, and $+1$ to a lesser extent. For Ag, the reverse is true: The $+1$ state is common, and the $+2$ and $+3$ can be obtained only under extreme oxidizing conditions. Gold occurs in the $+3$ state and less frequently in the $+1$. The metals have low negative oxidation potentials, thereby indicating their inertness and reluctance to oxidize:

$$Cu \rightarrow Cu^{2+} + 2e^- \qquad \mathscr{E}^0 = -0.34 \text{ V}$$
$$Ag \rightarrow Ag^+ + e^- \qquad \mathscr{E}^0 = -0.80 \text{ V}$$
$$Au \rightarrow Au^{3+} + 3e^- \qquad \mathscr{E}^0 = -1.42 \text{ V}$$

Copper(I) is unstable in solution and disproportionates spontaneously to Cu and Cu^{2+}. However, it can be stabilized by complexes such as $CuCl_2^-$. Copper(I) exists as the solid and extremely insoluble Cu_2O and Cu_2S, which are the principal ores of copper. The chemistry of Cu(II) is similar to that of other transition metals in the $+2$ oxidation state. The hydrated Cu(II) ion has a characteristic blue color, and tetraamminecopper(II), $Cu(NH_3)_4^{2+}$, is an intense blue. The complex is square planar. Silver(I) forms complexes such as $AgCl_2^-$, $Ag(NH_3)_2^+$, and $Ag(S_2O_3)_2^{3-}$, and Au(III) forms the very stable $AuCl_4^-$ complex.

The Chemistry of Photography

All silver halides except AgF are sensitive to light and are the basis of the photographic process (Figure 10-4). In making photographic film, fine crystals of AgBr are spread in gelatin on a film backing. Light from the camera image interacts with the crystalline AgBr in a poorly understood process that appears to involve defects in the crystal structure, and makes the grains, or crystals, more sensitive to reduction. The sensitized AgBr is reduced in the developer by a mild organic reducing agent such as hydroquinone:

$$AgBr + e^- \text{ (reducing agent)} \rightarrow Ag + Br^-$$

Then the unsensitized AgBr grains are dissolved and washed away in a sodium thiosulfate solution, one of the few solutions in which silver halides are soluble:

$$AgBr + 2S_2O_3^{2-} \rightarrow Ag(S_2O_3)_2^{3-} + Br^-$$

(The old name for sodium thiosulfate was sodium hyposulfite; hence, the synonym "hypo" for fixer.)

The Low-Melting Transition Metals

The most distinguishing characteristic of Zn, Cd, and Hg is their weak coherence as metals. They are soft and have low melting and boiling points. Mercury is the only metal that is liquid at room temperature. Zinc and Cd resemble the alkaline earth metals in their chemical behavior. Mercury is more inert and resembles Cu, Ag, and Au. All three elements have a $+2$

Properties of Elements on the Metal–Nonmetal Border of the Periodic Table

$BeCl_2$ Covalent molecular chains held together in solid and liquid by weak intermolecular forces. Narrow liquid range; mp $400\,°C$, bp $520\,°C$. Forms a dimer, Be_2Cl_4	BCl_3 Molecular gas; mp $-107\,°C$, bp $13\,°C$. Hydrolyzes completely in solution	CCl_4 Molecular gas above $76.8\,°C$. Inert in water
BeF_2 Covalent molecular chains held together in solid by ionic forces, mp $800\,°C$. BeF_4^{2-} complex in solution	BF_3 Molecular gas; mp $-127\,°C$. BF_4^- complex in solution	CF_4 Molecular gas; mp $-184\,°C$. Inert in water
$Be(OH)_2$ Amphoteric hydroxide	$B(OH)_3$ Boric acid. Forms polymeric ions	CO_2 Gaseous, acidic oxide
$MgCl_2$ Ionic solid; mp $708\,°C$, bp $1412\,°C$	$AlCl_3$ Covalent network solid; sublimes $178\,°C$. Forms dimer, Al_2Cl_6	$SiCl_4$ Volatile molecular liquid; mp $-70\,°C$, bp $57.6\,°C$. Hydrolyzes completely in water
MgF_2 Ionic solid; mp $1266\,°C$, bp $2239\,°C$	AlF_3 Ionic solid; mp $1040\,°C$. AlF_6^{3-} complex in solution	SiF_4 Molecular gas; mp $-90\,°C$. SiF_6^{2-} complex in solution
$Mg(OH)_2$ Basic hydroxide	$Al(OH)_3$ Amphoteric hydroxide	SiO_2 Solid acidic oxide. Forms polymeric anions

But the chains break beyond $x = 6$, and even these low-molecular-weight silanes are explosively reactive with halogens and oxygen. Silicon can form another class of polymers, the **silicones**, in which the Si atoms are bridged with oxygen atoms:

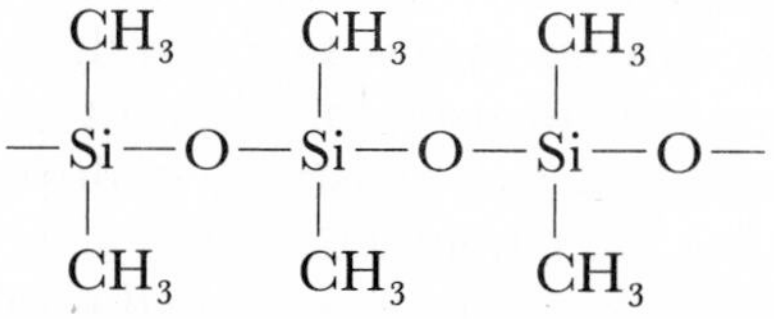

$$
\begin{array}{ccccccc}
& CH_3 & & CH_3 & & CH_3 & \\
& | & & | & & | & \\
-\,Si & - & O & -\,Si & - & O & -\,Si & - & O\,- \\
& | & & | & & | & \\
& CH_3 & & CH_3 & & CH_3 &
\end{array}
$$

These silicones are inert, water repellent, electrically insulating, and stable to heat. Silicon, despite the science fiction writers, is not a suitable alternative to carbon for life forms, at least under terrestrial conditions.

Germanium is a semimetal, and tin and lead are metals. Carbon and silicon show the $+4$ oxidation state in combination with oxygen-family elements and the halogens. For example, carbon has a $+4$ oxidation state in CCl_4, CO_2, and CS_2. Germanium and tin have both $+4$ and $+2$ states, and the chemistry of lead is almost wholly that of the $+2$ state.

The same behavior occurs in Group VA, but the break between metals and nonmetals is lower in the group. Nitrogen and phosphorus are nonmetals whose covalent chemistry and oxidation states are governed by the presence of five valence electrons: s^2p^3. Nitrogen and phosphorus most commonly have oxidation states -3, $+3$, or $+5$. Arsenic and antimony are semimetals with amphoteric oxides, and only Bi is metallic. For As and Sb, the $+3$ state is the most important. For Bi it is the only state, except under extraordinary conditions. Bismuth cannot lose all five valence electrons; the energy required is too high. However, it does lose the three $6p$ electrons to produce Bi^{3+}.

The trend in Group VIA is similar to that in the N group. Both oxygen and sulfur are nonmetals. Oxygen is strongly electronegative and has only the -2 oxidation state, except in OF_2 and the peroxides. Sulfur has the -2 state and several positive states as well, especially $+4$ and $+6$. Selenium and tellurium are semimetals, but have a chemistry that resembles that of sulfur. Polonium, a rare, radioactive element, has the electrical conductivity of a metal.

With Group VIIA, all metallic properties have been lost; thus the halogens are nonmetals. They are only one electron short of possessing a noble-gas electronic arrangement, and are reduced easily to anions with the s^2p^6 electronic configuration. Their reduction potentials are

$$F_2 + 2e^- \rightarrow 2F^- \qquad \mathscr{E}^0 = +2.86 \text{ V}$$
$$Cl_2 + 2e^- \rightarrow 2Cl^- \qquad \mathscr{E}^0 = +1.36 \text{ V}$$
$$Br_2 + 2e^- \rightarrow 2Br^- \qquad \mathscr{E}^0 = +1.09 \text{ V}$$
$$I_2 + 2e^- \rightarrow 2I^- \qquad \mathscr{E}^0 = +0.54 \text{ V}$$

For Cl, Br, and I, all odd oxidation numbers between -1 and $+7$ are known. But F, like O, is too electronegative to exhibit the positive states; it occurs only in the -1 state.

Summary

Atoms in complex ions and compounds are assigned **oxidation numbers** as an aid in keeping track of electrons during chemical reactions. Balancing a redox equation is equivalent to requiring that electrons be neither created nor destroyed. The maximum oxidation number exhibited by an atom of an element generally increases across a period. In the third period, for example, we have tne following: $Na^+(+1)$, $Mg^{2+}(+2)$, $Al^{3+}(+3)$, $SiCl_4$ $(+4)$, $PF_5(+5)$, $SF_6(+6)$, and $ClO_4^-(+7)$. The oxidation number of an atom is often called the **oxidation state** of that atom (or element) in a compound. Reactions in which the oxidation states of atoms change are called **oxidation–reduction (redox) reactions**. In such reactions the species whose oxidation number increases is called the **reducing agent (reductant)** and the species whose oxidation number decreases is called the **oxidizing agent (oxidant)**. In a redox reaction electrons are transferred from the reducing agent to the oxidizing agent. A species that undergoes self-oxidation–reduction is said to **disproportionate**. In a balanced redox equation the total number of electrons lost by the reducing agent is equal to the total number gained by the oxidizing agent. The **equivalent weight** of a redox agent is equal to its molecular weight divided by the oxidation-number change it undergoes in the reaction under consideration. The normality of a redox agent is the number of equivalents in 1 liter of solution. The normality therefore depends on the reaction that the redox agent undergoes.

The tendency of a substance to lose electrons in aqueous solution is known as its **oxidation potential**. The tendency for a substance to gain electrons in aqueous solution is called its **reduction potential**. The oxidation potentials of the alkali metal elements (Li, Na, K, Rb, Cs; all s^1 atoms) to form hydrated cations of $+1$ charge $[Li(s) \rightarrow Li^+ + e^-; \mathscr{E}^0 = +3.05 \text{ V}]$ are very large. The oxidation potentials of the alkaline earth metals (Be, Mg, Ca, Sr, Ba; all s^2) to form $+2$ hydrated cations $[Ca(s) \rightarrow Ca^{2+} + 2e^-; \mathscr{E}^0 = +2.76 \text{ V}]$ are also large. The reduction potentials of the halogens to form halide ions are all large ($F_2 + 2e^- \rightarrow 2F^-; \mathscr{E}^0 = +2.86 \text{ V}$).

The border between metals and nonmetals sweeps diagonally across the periodic table in an ill-defined band that goes approximately from Be and B to Po and At. The oxides of elements at or near the borderline often exhibit **amphoteric** (both acidic and basic) behavior. For example, BeO and Al_2O_3 are amphoteric. The transition metals undergo a smooth variation in melting point, boiling point, hardness, and common oxidation number, all of which can be related to the number of unpaired d electrons. Because of the **lanthanide contraction**, elements in the second and third transition series are more similar in size and in chemical properties than are those in the first and second periods.

Self-Study Questions

1. What is the oxidation number of Co in K_3CoF_6? In K_2CoI_4?
2. How are the suffixes *-ic* and *-ous* associated with oxidation state? Match the following formulas and names:
   a) Sulfuric and sulfurous acids: H_2SO_3 and H_2SO_4
   b) Nitric and nitrous acids: HNO_2 and HNO_3
3. Oxidation numbers are a convenient bookkeeping device for keeping track of electrons; they are useful even though in the actual reaction electrons are not totally removed from one atom and entirely given to another. The principle of conservation behind balancing redox equations is this: In a chemical reaction, electrons are neither created nor destroyed. How does this principle lead inevitably to Rule 8 (Section 10-1): In chemical reactions, the total oxidation number is conserved?
4. How does Rule 5, which deals with the relative oxidation numbers of nonmetals, follow from what you learned in Chapter 9 about ionization energies and electron affinities?
5. How are the most common oxidation states of the representative elements related to their group numbers?
6. What pattern of maximum oxidation numbers can be seen across a period in the transition metals?
7. Is a substance that gives up electrons in a reaction an oxidizing agent or a reducing agent? Is it oxidized or reduced? Does its oxidation number increase or decrease in the process?
8. Give the equivalent weight of sulfuric acid, H_2SO_4, in each of the following processes:
   a) An acid-base titration
   b) A redox reaction in which the sulfate ion goes to sulfite
   c) A redox reaction in which the sulfate ion goes to sulfide
9. What is the normality of a $0.1M$ solution of sulfuric acid in each of the three processes of Question 8?
10. The imaginary element turbidium (Tu) has the following oxidation potential:

$$Tu \rightarrow Tu^{3+} + 3e^- \qquad \mathscr{E}^0 = -3.00 \text{ V}$$

Is the turbidium a good oxidizing agent? A good reducing agent? Or is it neither? Is the Tu^{3+} ion a good oxidizing agent? A good reducing agent? Or is it neither?
11. What is the difference between oxidation potentials and reduction potentials?
12. Why are Li and Na not found free in nature, whereas Ag and Au are?
13. From which metal can electrons be ejected with light of a longer wavelength, Na or Cs? Why?

14. Which element is more metallic, Be or Ba? What evidence permits you to say this? How can you explain this in terms of electronic configuration?

15. Why is calcium harder than potassium?

16. Why is BeO amphoteric? What does this term mean in relation to chemical behavior?

17. What is the lanthanide contraction? Why does it occur? What detectable chemical effects does it produce?

18. If the first electrons after those in the Kr noble-gas shells go into the $5s$ orbital for Rb and Sr, why is the outer electronic configuration of Zr^{2+} $4d^2$, rather than $5s^2$ as in Sr?

19. How does the maximum oxidation number change with atomic number in the first transition-metal series, Sc to Zn?

20. In a transition metal with several oxidation states, which state is usually most metal-like in its compounds? Can you give an example?

21. As the principal quantum number of the element increases within a transition-metal group (vertical column), do the higher or lower oxidation states become more important? Can you give a reason for this behavior?

22. How is the color of a chemical related to electronic transitions between energy levels?

23. Why are transition-metal compounds more often colored than those of the representative elements?

24. How do the oxides of Cr, Ni, Cu, Al, and many other metals differ from iron oxide in a way that has great economic importance? Try to imagine a world in which iron oxide behaved the same way. What important chemical industry would be badly hurt if this were so?

25. Why do Zn, Cd, and Hg have such low melting points compared with Cr, Mo, and W?

26. Why do amphoteric oxides appear for those elements that lie on a diagonal across the periodic table (Be, Al, Ge, Sb) rather than in one group separating the metals from the nonmetals?

27. How does metallic character change within one group of the table?

28. Are the halogens good reducing agents or good oxidizing agents?

29. What aspect of electronic structure explains the high electrical conductivities of the coinage metals, Cu, Ag, and Au?

30. What are the principal oxidation states of nitrogen and phosphorus?

Problems

Oxidation numbers

1. Xenon forms several nonionic compounds with F and O. Give the oxidation number of the central Xe atom in each of the following compounds: XeO_4, XeF_2, XeO_3, XeF_4, and XeF_6.

2. Xenon can also form a number of ionic compounds, including $CsXeF_7$ and Cs_2XeF_8. What are the ions in each compound? What are the ionic charge and oxidation number of Xe in each ion containing Xe?

3. What is the oxidation number of the central atom in each of the following ions or molecules: $Co(CN)_6^{3-}$, $PtCl_6^{2-}$, CO_3^{2-}, SF_4?

4. What is the oxidation number of nitrogen in each of the following ions or molecules: NH_3, N_2H_4, NO, NO_2, NO_2^-, NO_3^-?

5. What is the oxidation number of platinum in the complex ion $PtCl_4^{2-}$?

6. Assign oxidation numbers to the atoms in the following chemical species: (a) gold, Au; (b) iodine, I_2; (c) barium chloride, $BaCl_2$; (d) ethane, C_2H_6; (e) stannous oxide, SnO; (f) stannic oxide, SnO_2; (g) nitrous oxide, N_2O; (h) phosphorus pentoxide, P_2O_5; (i) magnesium hydroxide, $Mg(OH)_2$; (j) sulfurous acid, H_2SO_3; (k) telluric acid, H_6TeO_6; (l) hypochlorous acid, $HClO$; (m) perchloric acid, $HClO_4$; (n) dichromate, $Cr_2O_7^{2-}$; (o) cyanide, CN^-.

7. What is the oxidation number of the underlined element in each ion or molecule: $\underline{V}O_2^+$, $P_2\underline{O}_7^{4-}$, $\underline{P}H_3$, $K\underline{N}O_2$, $H_2\underline{O}_2$, $Li\underline{H}$, $Mg_3\underline{N}_2$, $\underline{N}F_3$, $\underline{I}Cl_5$, $\underline{Ag}(NH_3)_2^+$?

Redox equations

8. When the equation

$$MnO_2 + I^- + H^+ \rightarrow$$
$$Mn^{2+} + I_2 + H_2O$$

is balanced, what is the net charge on each side of the equation? What element is oxidized? What element is reduced?

9. For each of the following reactions, list the substance reduced, the substance oxidized, the reducing agent, and the oxidizing agent.
 a) $6H^+ + 2MnO_4^- + 5SO_3^{2-} \rightarrow$
$$5SO_4^{2-} + 2Mn^{2+} + 3H_2O$$

b) $8H^+ + Cr_2O_7^{2-} + 6HI \rightarrow$
$$2Cr^{3+} + 3I_2 + 7H_2O$$
c) $3Cl_2 + 6OH^- \rightarrow$
$$ClO_3^- + 5Cl^- + 3H_2O$$

10. Phosphine, PH_3, is a colorless, highly toxic gas that smells like rotten fish and is produced in small amounts when animal and vegetable matter decay in moist situations such as damp graveyards. Traces of P_2H_4 are produced simultaneously and cause the PH_3 to ignite in air to give pale, flickering lights commonly called "corpse candles" or "will-o'-the-wisps." In the laboratory, phosphine can be prepared by adding water to calcium phosphide. Write a balanced equation for the reaction. Assign oxidation numbers to each of the elements present.

11. Balance the reaction

$$MnO_4^- + H^+ + H_2S \rightarrow$$
$$Mn^{2+} + H_2O + S$$

What is the oxidation number of Mn in MnO_4^-? What element is oxidized? What element is the oxidizing agent? Is this latter element oxidized or reduced?

12. Balance the following equations by the oxidation-number method:
 a) $H_2S + Cr_2O_7^{2-} \rightarrow$
$$S + Cr^{3+} \text{ (acidic solution)}$$
 b) $NH_3 + O_2 \rightarrow NO + H_2O$
 c) $NO_2^- + MnO_4^- \rightarrow$
$$NO_3^- + MnO_2 \text{ (basic solution)}$$
 d) $NH_3 + ClO^- \rightarrow$
$$N_2H_4 + Cl^- \text{ (basic solution)}$$
 e) $H_2 + OF_2 \rightarrow H_2O + HF$
 f) $MnO_2 + Al \rightarrow Al_2O_3 + Mn$

13. Balance the following equations by the half-reaction method:
 a) $MnO_2 + KOH + O_2 \rightarrow$
$$K_2MnO_4 + H_2O$$
 b) $CuCl_4^{2-} + Cu \rightarrow$
$$CuCl_2^- \text{ (acidic solution)}$$

c) $NO_3^- + Zn \rightarrow$
$$NH_4^+ + Zn^{2+} \text{ (acidic solution)}$$

d) $ClO_2 \rightarrow ClO_2^- + ClO_3^-$
$$\text{(basic solution)}$$

e) $Fe^{2+} + Cr_2O_7^{2-} \rightarrow$
$$Fe^{3+} + Cr^{3+} \text{ (acidic solution)}$$

f) $Cu + NO_3^- \rightarrow$
$$Cu^{2+} + NO \text{ (acidic solution)}$$

14. Balance the following equations by any method you choose:

a) $H_3PO_4 + CO_3^{2-} \rightarrow$
$$PO_4^{3-} + CO_2 + H_2O$$
$$\text{(neutral solution)}$$

b) $MnO_2 + SO_3^{2-} \rightarrow$
$$Mn(OH)_2 + SO_4^{2-}$$
$$\text{(basic solution)}$$

c) $H^+ + Cr_2O_7^{2-} + H_2SO_3 \rightarrow$
$$Cr^{3+} + HSO_4^- \text{ (acidic solution)}$$

d) $MnO_4^- + V^{2+} \rightarrow$
$$VO_2^+ + Mn^{2+} \text{ (acidic solution)}$$

e) $FeSO_4 + NaClO_2 \rightarrow$
$$Fe_2(SO_4)_3 + NaCl$$
$$\text{(in sulfuric acid solution)}$$

f) $KNO_2 + KMnO_4 \rightarrow$
$$KNO_3 + MnO_2$$
$$\text{(in KOH solution)}$$

g) $MnO_4^- + OH^- + I^- \rightarrow$
$$MnO_4^{2-} + IO_3^- + H_2O$$

h) $KMnO_4 + NH_3 \rightarrow$
$$KNO_3 + MnO_2 + KOH + H_2O$$

Molarity, normality, redox equivalents

15. Twelve grams of $KMnO_4$ were dissolved in sufficient water to make a liter of solution. Calculate the molarity of the solution. The solution was divided into parts and used in four different reactions. Reaction 1 was carried out in basic solution, thereby producing MnO_2 as a product. Reaction 2 was carried out in very strong base to give MnO_4^{2-}. Reaction 3 gave Mn^{3+} in acid solution, and reaction 4 yielded Mn^{2+}. Calculate the equivalent weight of $KMnO_4$ and the normality of the potassium permanganate solution for each reaction.

16. A solution of potassium dichromate, $K_2Cr_2O_7$, is made by adding enough water to 3.52 g of the salt to fill a 100-ml volumetric flask. (a) What is the molarity of the solution? (b) The solution will be used in a titration in which the dichromate ion is reduced to Cr^{3+}. What is the normality of the solution?

17. Find the equivalent weight of an oxidizing agent that oxidizes Fe^{2+} to Fe^{3+} if 0.664 g of the compound requires 23.5 ml of $0.540M$ Fe^{2+} solution.

18. (a) When H_3PO_4 reacts with NaOH to produce NaH_2PO_4, how many equivalents of H_3PO_4 are there per mole? (b) When H_3PO_4 is reduced to H_3PO_2, how many equivalents of H_3PO_4 are there per mole?

19. Exactly 6.40 g of gaseous SO_2 are absorbed in 95 ml of water. When absorption is complete, water is added so the final volume of solution is 100 ml. Calculate for each of the following reactions the volume of solution *of this composition* that contains 1 equiv of H_2SO_3: (a) neutralization to SO_3^{2-} with sodium hydroxide; (b) oxidation to SO_4^{2-}; (c) reduction to S^{2-}; (d) reduction to elemental sulfur.

20. An acidic solution is $0.10M$ in TiO^{2+}. What is its normality when reacted with (a) dilute NaOH to produce TiO_2; (b) dilute $FeSO_4$ to produce Ti^{3+}; (c) concentrated H_2SO_4 to produce Ti^{4+}?

21. A slightly acidic solution is $0.01M$ in Cl_2. What is its normality when reacted with (a) dilute $FeSO_4$ to produce Fe^{3+}; (b) dilute H_2O_2 to produce $HClO$?

22. If chlorine gas is bubbled through a basic solution of potassium iodide, the following reaction occurs:

$$KOH + Cl_2 + KI \rightarrow$$
$$KCl + KIO_3 + H_2O$$

Balance the equation. If chlorine gas is bubbled through 25 ml of a 0.10N solution of KI in aqueous KOH at STP, what volume of Cl_2 gas is required to react completely with the KI? (KI normality is based on its reducing action.)

23. Given the equation

$$H_2SO_4 + HI \rightarrow$$
$$H_2S + I_2 + H_2O$$

calculate the number of moles of sulfuric acid consumed by reaction with 25.00 ml of 0.100N HI. (Check first to see if the reaction is balanced.)

24. When H_2S reacts with $KMnO_4$ in acidic solution, the following reaction can be written:

$$MnO_4^- + H_2S + H^+ \rightarrow$$
$$Mn^{2+} + S + H_2O$$

Balance the equation. If a 0.05N solution of H_2S is used to titrate 50 ml of permanganate solution, 70 ml are required to reach the end point. What is the normality of the original permanganate solution? What is the molarity?

25. A 25.00-ml sample of an unknown copper solution is treated with excess potassium iodide in acidic solution, and the liberated iodine is titrated with 0.0250M sodium thiosulfate. If 12.50 ml of the

thiosulfate solution are required to reach the end point, what is the molarity of the unknown copper solution? The unbalanced reactions are

$$Cu^{2+} + I^- \rightarrow CuI + I_3^-$$
$$I_3^- + S_2O_3^{2-} \rightarrow I^- + S_4O_6^{2-}$$

26. In oxidizing H_2SO_3 to SO_4^{2-}, IO_3^- is reduced to I_2 in acidic solution. Write a balanced chemical equation for this redox reaction. What volume of 0.25N KIO_3 is required for the complete oxidation to sulfate of 125 ml of 0.10N H_2SO_3?

27. In acidic aqueous solution, permanganate ion will oxidize oxalic acid $(COOH)_2$ to carbon dioxide while being reduced to manganous ion, Mn^{2+}. (a) What are the oxidation numbers of C and Mn in the reactants and products? (None of the substances is a peroxide.) (b) Write a balanced equation for the reaction. (c) What are the molecular weights and equivalent weights of the reactants $KMnO_4$ and $(COOH)_2$? (d) How many moles of permanganate can be reduced by 0.01 mole of oxalic acid? (e) How many equivalents of oxalic acid can be oxidized by 0.04 equiv of potassium permanganate? (f) What is the normality of a solution that contains 15.8 g of potassium permanganate per liter? (g) What is the molarity of a solution that contains 4.5 g of oxalic acid per liter? (h) What weight of oxalic acid can be oxidized to carbon dioxide by 150.0 ml of a solution containing 6.25 g of potassium permanganate per liter?

Suggested Reading

M. J. Bigelow, *The Representative Elements,* Bogden and Quigley, New York, 1970.

J. L. Dye, "The Solvated Electron," *Scientific American,* February 1967.

J. L. Hall and D. A. Keyworth, *Brief Chemistry of the Elements,* W. A. Benjamin, Menlo Park, Calif., 1971.

K. J. Laidler and M. H. Ford-Smith, *The Chemical Elements,* Bogden and Quigley, New York, 1970.

E. M. Larsen, *Transitional Elements,* W. A. Benjamin, Menlo Park, Calif., 1965.

R. L. Rich, *Periodic Correlations,* W. A. Benjamin, Menlo Park, Calif., 1965.

R. T. Sanderson, *Chemical Periodicity,* Reinhold, New York, 1960.

H. H. Sisler, *Electronic Structure, Properties and the Periodic Law,* Reinhold, New York, 1963.

11

Lewis Structures and the VSEPR Method

Key Concepts

11-1 Lewis structures. Electron-pair sharing and electron-pair bonds. Lone pairs. Noble-gas configuration. Octet rule. Double bonds, triple bonds, and bond order. Open shells. Formal charges. Isoelectronic molecules. Lewis acids and Lewis bases. Resonance structures and resonance hybrids. Meaning of oxidation numbers.

11-2 Acidity of oxyacids. Amphoteric species and amphiprotic species. Acid strength and central-atom formal charge.

11-3 The valence-shell electron-pair repulsion (VSEPR) method. Electron-pair repulsion, steric number, and molecular geometry.

In this chapter we shall look at a simple method of representing covalent bonding using Lewis structures. We shall write Lewis structures for familiar molecules and ions and interpret them in terms of electron-pair sharing and the completion of noble-gas valence shells. After we explain oxidation number in terms of the inequality of electron-pair sharing by atoms with different electronegativities, we shall go on to correlate the acidities of certain molecules with the electronic structure of a central atom. And we shall conclude the chapter by showing how the valence-shell electron-pair repulsion (VSEPR) method can be used to predict molecular shape.

11-1   LEWIS STRUCTURES

Electron-dot formulas for chemical compounds were developed by G. N. Lewis, in 1916, as an attempt to comprehend covalent bonding. Our understanding of bonding rests on firmer ground now, but the dot formulas are still a convenient notation. Each valence electron (i.e., an electron in the outermost s and p orbitals) is represented by a dot placed beside the chemical symbol:

H ·   He:

In modern terminology, each of the four compass points of the symbol represents one of the s, p_x, p_y, and p_z atomic orbitals. For example, atoms of the second-period elements are written as

$$\text{Li}\cdot \quad \text{Be} \quad \cdot\text{B} \quad \cdot\dot{\text{C}}\cdot \quad \cdot\dot{\text{N}}\cdot \quad :\dot{\text{O}}\cdot \quad :\ddot{\text{F}}: \quad :\ddot{\text{Ne}}:$$

The loss and gain of electrons in the formation of ions can be illustrated by the formation of sodium chloride from sodium and chlorine atoms:

$$\text{Na}\cdot + :\dot{\text{Cl}}: \rightarrow \text{Na}^+:\ddot{\text{Cl}}:^-$$

Each ion in sodium chloride has the outer electronic configuration of a noble gas: The sodium ion has the configuration of Ne, and the chloride ion has the configuration of Ar. This transfer of an electron occurs because Cl is more electronegative than Na (3.16 for Cl versus 0.93 for Na). What happens in HI, in which the electronegativities are nearly equal (2.20 and 2.66, respectively)?

According to Lewis' theory of covalence, each atom completes a noble-gas configuration, not by transferring but by *sharing* an electron:

$$\text{H}\cdot + \cdot\ddot{\text{I}}: \rightarrow \text{H}:\ddot{\text{I}}:$$

The H atom now has two electrons in its outer valence orbital, as in He, and I has eight electrons, as in Xe. Lewis set forth this principle: *Atoms form bonds by losing, gaining, or sharing enough electrons to achieve the outer electronic configurations of noble gases.* The type of bond, ionic or covalent, depends on whether electrons are transferred or shared. The combining capacities of atoms are a consequence of the proportions in which they must associate to achieve noble-gas configurations. Lewis' theory explains bond type and the pattern of connections of atoms within a molecule. However, it is not able to explain the geometry of molecules.

The Lewis theory made bonding between like atoms, as it occurs in H_2, F_2, or N_2, understandable for the first time. Two hydrogen atoms share their electrons to provide each atom with the He closed-shell structure:

$$\text{H}\cdot + \text{H}\cdot \rightarrow \text{H}:\text{H} \quad \text{or} \quad \text{H}-\text{H}$$

A straight-line bond symbol is often used, as here, in the special sense of a symbol for a Lewis electron-pair bond. Two fluorine atoms share one pair of electrons; thus, each F atom has the Ne structure:

$$:\ddot{\text{F}}\cdot + :\ddot{\text{F}}\cdot \rightarrow :\ddot{\text{F}}:\ddot{\text{F}}: \quad \text{or} \quad :\ddot{\text{F}}-\ddot{\text{F}}:$$

The unshared pairs of electrons on F are called **lone pairs**; we now would interpret them as spin-paired electrons in atomic orbitals that are not involved in bonding. The bond energy, the energy required to break the diatomic molecule into two infinitely separated atoms, is 432 kJ mole^{-1}

for H_2 and only 139 kJ mole^{-1} for F_2. Part of this relative instability of the F_2 molecule may arise from electrostatic repulsion between lone pairs of electrons on the two F atoms.

  The H_2 and F_2 molecules are representative of many molecules in which electron-pair bonds are formed such that each atom achieves a closed-shell configuration. Hydrogen needs two electrons to achieve a closed shell, filling the $1s$ valence orbital. Each atom in the second period requires eight electrons (an **octet**) to achieve a closed shell, because the $2s$ and $2p$ orbitals can accommodate a total of eight electrons ($2s^2 2p^6$). This requirement is known as the **octet rule**. In the example of the F_2 molecule, after bonding each F atom has eight electrons associated with it.

Multiple Bonds

If we try to construct O_2 as we did F_2, we end with unpaired electrons and only seven electrons in the neighborhood of each O atom:

$$:\ddot{O}\cdot \; + \; :\ddot{O}\cdot \; \rightarrow \; :\ddot{O}:\ddot{O}: \quad \text{or} \quad :\dot{\ddot{O}}-\ddot{O}:$$

This defect can be eliminated by assuming that the oxygen atoms share *two* pairs of electrons (without regard to the geometry of the process):

$$:\ddot{O}::\ddot{O}: \quad \text{or} \quad :\ddot{O}=\ddot{O}:$$

Thus, there is a **double bond** between two oxygen atoms. A **triple bond** must be assumed in N_2 to give each nitrogen atom a noble-gas configuration:

$$:N:::N: \quad \text{or} \quad :N\equiv N:$$

This concept of multiple bonds is not all imagination; bond energies and bond lengths both support the idea of a single bond in F_2, a double bond in O_2, and a triple bond in N_2:

	N_2	O_2	F_2
Bond energy (kJ mole^{-1})	942	494	139
Bond length (Å)	1.10	1.21	1.42

A molecule such as N_2, with a triple bond between two atoms, is said to have a **bond order** of three. (The bond order is the number of electron-pair bonds.) The oxygen molecule has a bond order of two, and the F_2 molecule has a bond order of one. The higher the bond order, the more tightly held the atoms, the greater the bond energy, and the shorter the bond.

  An interesting problem arises in the Lewis structural formulation of the common air-pollutant molecule, nitric oxide (NO). A closed-shell configuration cannot be constructed for NO because it has an odd number of valence electrons. Nitric oxide has 11 valence electrons, 5 valence electrons originally associated with the nitrogen atom and 6 electrons originally associated with the oxygen atom. Thus either N or O will "own" only

7 electrons in the NO molecule. We choose N because it is less electro-negative than O. Therefore the best structure for NO is

$$:\overset{\cdot}{N}=\underset{\cdot\cdot}{O}:$$

Lewis structures for molecules such as NO, which have an odd number of electrons, cannot have closed shells associated with each atom. At least one atom, nitrogen in the NO example, is left with an **open shell**.

Formal Charges

Carbon dioxide is easy to represent using Lewis structures, but carbon monoxide raises a problem. Each O needs two electrons to achieve the stable eight-electron (octet) structure; thus it should share two electron pairs with C. Yet the carbon atom needs four electrons and should share four pairs. The only satisfactory Lewis structure for CO is obtained by letting three pairs be shared, and by distributing the other four valence electrons in such a way as to complete an eight-electron shell around each atom.

Carbon dioxide: $:\overset{\cdot\cdot}{O}::C::\overset{\cdot\cdot}{O}:$ or $:\overset{\cdot\cdot}{O}=C=\overset{\cdot\cdot}{O}:$

Carbon monoxide: $:C:::O:$ or $:\overset{\ominus}{C}\equiv\overset{\oplus}{O}:$

Carbon monoxide is isoelectronic with N_2, so we might expect the hypothe-sis of a triple bond in CO to be completely satisfactory. (We can imagine some hypothetical "Schrödinger's Demon" making a molecule of carbon monoxide from a molecule of N_2 by removing a proton from one nitrogen nucleus and adding it to the other.) Nevertheless, the triple-bond hypothesis for CO does present a problem. If we assume that each shared electron pair is shared equally between atoms, then carbon has three of the six electrons from the triple bond, plus the two in the lone pair. It has five electrons but a nuclear charge that will counterbalance only four of them. Similarly, oxygen has five valence electrons but a nuclear charge designed for six. Therefore, carbon has a **formal charge** of -1, and oxygen has a formal charge of $+1$. This statement does not mean that these charges are fully present; it indicates only that the demands of bonding lead to a non-uniform distribution of charge. The formal charges in the case of CO are indicated by $C^{\ominus}$ and $O^{\oplus}$ (in the Lewis structure shown).

When calculating the formal charge on the atoms in a molecule, we assign to each atom one electron for each covalent electron-pair bond that it makes, plus all of its lone-pair electrons. The formal charge on the atom is then the charge that it would have if it were an isolated ion with the same number of valence electrons:

$$\text{Formal charge} = Z - (N_{\text{bonds}} + N_{\text{nonbonding}})$$

Here Z is the atomic number, N_{bonds} is the number of covalent electron-pair bonds that the atom makes with other atoms, and $N_{\text{nonbonding}}$ is the total number of electrons the atom has that are not involved in covalent bonds. You should verify for yourself that in every uncharged molecule discussed so far in this section (except CO) the formal charge on each atom is zero.

Example 1

Write an acceptable Lewis structure for the cyanide molecule (CN).

Solution

A closed-shell configuration cannot be constructed for CN because it has an odd number of valence electrons. The CN molecule has nine valence electrons, four valence electrons associated with the carbon atom and five valence electrons associated with the nitrogen atom. Thus either C or N will "own" only seven electrons in the CN molecule. We choose C because it is less electronegative than N. Therefore the better Lewis structure for CN is

$$\cdot C \equiv N :$$

The alternative structure,

$$: \overset{\ominus}{C} \equiv \overset{\oplus}{N} \cdot$$

results in formal charges on both C and N atoms. *Such a formal-charge separation should be avoided, if possible, for bonds involving atoms of about the same electronegativity.* Because of the octet rule, a formal-charge separation for CO could not be avoided. For CN, however, it can, and $\cdot C \equiv N :$ is clearly the better Lewis structure for the molecule.

Lewis structures for molecules such as CN and NO, which have an odd number of electrons, cannot have closed shells associated with each atom. At least one atom, carbon in the CN example, is left with an open shell. As a consequence of this open-shell structure in CN, two molecules of CN combine to form the dimer $(CN)_2$, which is called *cyanogen*. The driving force for this reaction is readily seen to be the formation of a new carbon–carbon bond without any significant weakening of the carbon–nitrogen triple bond:

$$\cdot C \equiv N : + \cdot C \equiv N : \rightarrow : N \equiv C - C \equiv N :$$

The odd electron in CN is unpaired. The presence of one or more unpaired electrons in an atom or molecule gives rise to a physical property known as *paramagnetism,* which we shall discuss in the next chapter. Experiments show that the paramagnetism of the CN molecule is in accord with the presence of one unpaired electron, as the Lewis structure predicts. However, not all paramagnetic molecules are easily explained by Lewis structures. A multiple-bonded molecule with particularly vexing magnetic properties (for Lewis' structural theory) is O_2, which is known to have two unpaired electrons in its ground state and to be paramagnetic. An unusual structure such as

$$:\overset{\cdot}{O}\equiv\overset{\cdot}{O}: \quad \text{or} \quad :\overset{\cdot\cdot}{O}-\overset{\cdot\cdot}{O}:$$

would be required to explain this magnetic behavior. However, the observed bond length and bond energy of O_2 are completely consistent with the simple double-bond structure,

$$:\overset{\cdot\cdot}{O}=\overset{\cdot\cdot}{O}:$$

as we have already shown. We will see in Chapter 12 that the molecular orbital theory provides a satisfactory explanation of both the paramagnetism and the bond properties of the oxygen molecule.

Example 2

Write an acceptable Lewis structure for $MgCl_2$.

Solution

A magnesium atom has a low electronegativity and atomic chlorine has a high electronegativity; thus magnesium chloride requires a structure showing ionic bonds. One electron is transferred from the magnesium atom to each of the chlorine atoms in $MgCl_2$. The correct structure indicates the charges:

$$[:\overset{\cdot\cdot}{\underset{\cdot\cdot}{Cl}}:^-][Mg^{2+}][:\overset{\cdot\cdot}{\underset{\cdot\cdot}{Cl}}:^-]$$

Some Polyatomic Molecules

The Lewis structures of methane (CH_4), ammonia (NH_3), and water (H_2O) are

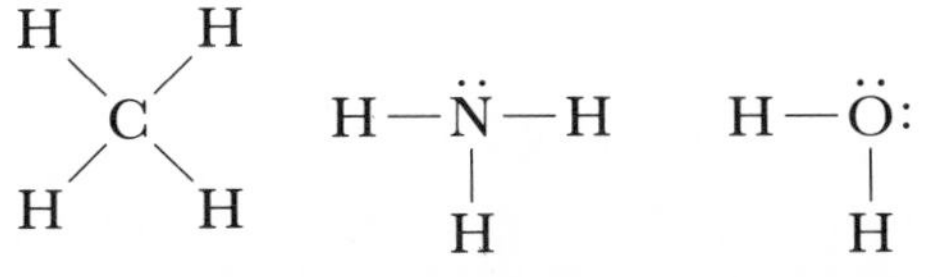

$$
\begin{array}{ccc}
\overset{\displaystyle H\qquad H}{\underset{\displaystyle H\qquad H}{\diagdown \diagup \atop \diagup \diagdown}}\!\!C & H-\overset{\cdot\cdot}{N}-H & H-\overset{\cdot\cdot}{O}: \\
 & \mid & \mid \\
 & H & H
\end{array}
$$

methane   ammonia   water

These three molecules are isoelectronic; they have the same number of electrons. The eight valence electrons around the central atom illustrate the octet rule. In CH_4 all eight electrons are involved in bond pairs, whereas the other molecules contain lone electron pairs. Ammonia has three bond pairs and one lone pair, and water has two bond pairs and two lone pairs.

The usefulness of the octet rule is that it allows us to predict which molecules will be stable under ordinary conditions of temperature and pressure. For example, all the following carbon hydrides are known:

$$CH \qquad CH_2 \qquad CH_3 \qquad CH_4$$

However, only CH_4 obeys the octet rule and is stable under ordinary conditions.

The simple hydrocarbons ethylene and acetylene illustrate multiple bonding; ethane is shown for comparison:

The lone-electron-pair notation is useful when atoms other than C and H are involved:

methyl alcohol   acetic acid   ethylamine

The Ammonium Chloride Molecule

The ammonium chloride molecule, NH_4Cl, contains NH_4^+ and Cl^- ions. The Lewis structures for the nitrogen atom and for each hydrogen atom are

$$:\overset{\displaystyle \cdot}{\underset{\displaystyle \cdot}{N}}\cdot \qquad H\cdot$$

However, the NH_4^+ ion has one positive charge, which means that it has lost one of its nine electrons. Since all the hydrogen atoms in the ion are equivalent, we give the nitrogen atom the positive charge,

$$\cdot\overset{\displaystyle \cdot}{\underset{\displaystyle \cdot}{N}}\cdot\, +$$

and write

$$
\left[\begin{array}{c} H \\ | \\ H-\overset{\oplus}{N}-H \\ | \\ H \end{array} \right]^{+}
$$

Thus the correct Lewis structure for the NH_4^+ ion has four single bonds and no unshared electrons. Notice that the ammonium ion is isoelectronic with methane. A formal charge of $+1$ is assigned to the nitrogen atom.

Now, we assign the electron that was removed from NH_4 (to give NH_4^+) to the chlorine atom, to give the ion-pair structure for the ammonium chloride molecule:

$$
\left[\begin{array}{c} H \\ | \\ H-\overset{\oplus}{N}-H \\ | \\ H \end{array} \right]^{+} \qquad [:\overset{..}{\underset{..}{Cl}}:]^{-}
$$

In summary, the nitrogen–hydrogen bonds in NH_4^+ are covalent but have some ionic character, whereas the NH_4^+ ion is attached to Cl^- by an ionic bond.

Lewis Acids and Bases

The BF_3 molecule is represented by the Lewis structure

$$
:\overset{..}{\underset{..}{F}}-B-\overset{..}{\underset{..}{F}}: \\
| \\
:\overset{..}{F}:
$$

It is unusual because it does not have four electron pairs around the B atom. It reacts with ammonia to form the addition compound BF_3NH_3 as follows:

$$
\begin{array}{ccc}
:\overset{..}{F}: \quad H & & :\overset{..}{F}: \quad H \\
| \quad\quad | & & | \quad\quad | \\
:\overset{..}{\underset{..}{F}}-B + :N-H \rightarrow & & :\overset{..}{\underset{..}{F}}-\overset{\ominus}{B}-\overset{\oplus}{N}-H \\
| \quad\quad | & & | \quad\quad | \\
:\overset{..}{F}: \quad H & & :\overset{..}{F}: \quad H
\end{array}
$$

In this compound, nitrogen, with a lone pair, donates both the electrons of the covalent bond. Such a donor–acceptor bond is sometimes called a **coordinate covalent bond**; however, the distinction is pointless, because once the bond is formed it is like any other covalent bond.

The compound BF_3NH_3 is isoelectronic with CF_3CH_3, and differs from it only by the formal charges that are assigned to the central atoms. The charge is zero for each atom in the carbon compound. But if you work out the formal assignment of charge for the boron compound, you can see that B has a formal charge of -1, and N, of $+1$. Since formal charges arise from the way in which electrons are distributed in a molecule or ion, the total formal charge of all the atoms must be equal to the total charge on the ion, or zero for a neutral molecule.

A compound such as BF_3, which can accept an electron pair, is called a **Lewis acid**, and an electron-pair donor is a **Lewis base**. This terminology, like that of Brønsted, is an extension of the simple Arrhenius acid–base theory. By the Arrhenius theory, an acid is a substance that produces hydrogen ions or protons in aqueous solution, and a base is a substance that produces hydroxide ions. Brønsted's terminology is more general: An acid is any substance that can donate protons, and a base is a substance that can accept them. To illustrate the differences in the three definitions, consider the neutralization of HCl and NaOH,

$$HCl + NaOH \rightarrow H_2O + NaCl$$

In terms of the species present in aqueous solution, the reaction should be written as

$$H_3O^+ + Cl^- + Na^+ + OH^- \rightarrow Na^+ + Cl^- + 2H_2O$$

To Arrhenius, HCl is the acid, and NaOH is the base. To Brønsted, H_3O^+ is the acid, and the hydroxide ion (OH^-) is the base since it is the species that combines with the proton. To Lewis, the proton is the acid, because it will combine with the lone pair on the hydroxide ion; the hydroxide ion is the electron-pair donor and hence the base:

$$H^+ + :\overset{..}{\underset{..}{O}}-H^- \rightarrow H-\overset{..}{\underset{..}{O}}-H$$

The Brønsted and Lewis theories are both applicable to nonaqueous solutions, whereas Arrhenius' theory is not. Both Brønsted's and Lewis' theories will be useful later. The general Lewis definition of acids and bases is helpful because it includes compounds that do not contain hydrogen, and which we might not recognize as having properties of acids with Arrhenius' theory. For example, because it is an electron acceptor, BF_3 will often catalyze organic reactions that are also catalyzed by protons.

Bonding to Heavier Atoms

The octet rule is extremely valuable as a guide in writing Lewis structural formulas. For second-row nonmetallic elements (B, C, N, O, F), exceptions to the rule are very rare. It is easy to explain why this is so. Atoms of the second-row elements have stable $2s$ and $2p$ orbitals, and the "magic number" of 8 corresponds to the closed valence-orbital configuration $2s^2 2p^6$. Adding more electrons to such a configuration is impossible, because the next atomic orbital available to a second-row element is the high-energy $3s$ orbital.

Beyond the second row in the periodic table the octet rule is not obeyed with such satisfying regularity. However, it remains a useful rule, as illustrated by molecules such as PH_3, PF_3, H_2S, and SF_2.

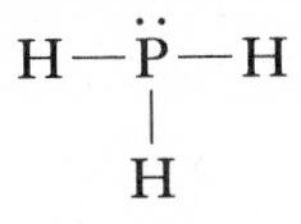

$$H—\overset{\cdot\cdot}{P}—H \qquad :\overset{\cdot\cdot}{F}—\overset{\cdot\cdot}{P}—\overset{\cdot\cdot}{F}: \qquad H—\overset{\cdot\cdot}{\underset{\cdot\cdot}{S}}: \qquad :\overset{\cdot\cdot}{F}—\overset{\cdot\cdot}{\underset{\cdot\cdot}{S}}:$$

$$\qquad\quad H \qquad\qquad\quad :\overset{\cdot\cdot}{\underset{\cdot\cdot}{F}}: \qquad\qquad H \qquad\qquad :\overset{\cdot\cdot}{\underset{\cdot\cdot}{F}}:$$

| phosphine | phosphorus trifluoride | hydrogen sulfide | sulfur difluoride |

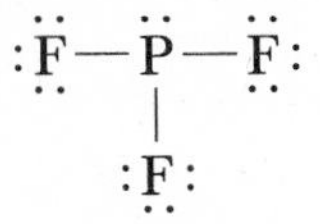

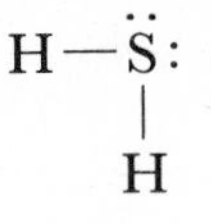

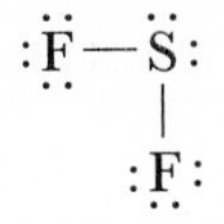

Atoms of the heavier elements often do not obey the octet rule. Some of them show a surprising ability to bind more atoms (or associate with more electron pairs) than would be predicted from the octet rule. For example, phosphorus and sulfur form the compounds PF_5 and SF_6, respectively. Lewis structures for these compounds use all the valence electrons of the heavy element in bonding:

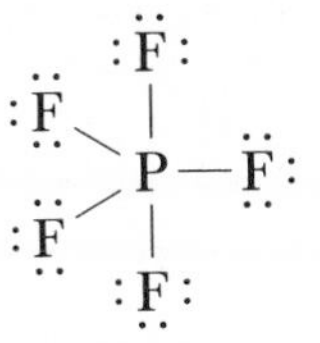

phosphorus pentafluoride

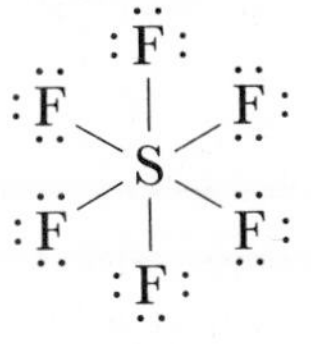

sulfur hexafluoride

That phosphorus shares 10 electrons and sulfur shares 12 electrons obviously violates the octet rule. The theory of atomic structure helps us see why the violation has occurred. The noble gas in the third row with phosphorus and sulfur is argon. The argon electronic structure fills the $3s$ and $3p$ orbitals, but leaves the five $3d$ orbitals vacant. If some of these $3d$ orbitals are used for electron-pair sharing, extra bonds are possible. The atomic theory thus provides an explanation of the enhanced bonding versatility of elements in the third row and beyond.

Perhaps the most important consequence of the use of d orbitals is the formation of an important series of oxyacids. The best-known examples are phosphoric acid (H_3PO_4), sulfuric acid (H_2SO_4), and perchloric acid ($HClO_4$). It is possible to write a Lewis structure for sulfuric acid that obeys the octet rule:

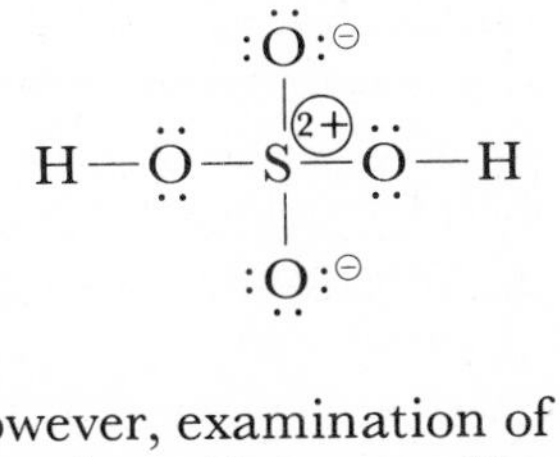

However, examination of this structure reveals that a formal charge of $+2$ is on the sulfur atom. Development of a large positive formal charge on an electronegative nonmetal atom is not very reasonable. The formal charge can be removed if we write two sulfur–oxygen double bonds, thereby allowing the sulfur atom to share 12 electrons:

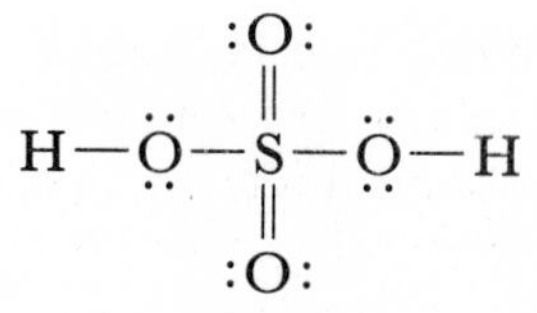

sulfuric acid

Similar Lewis structures can be written for other oxyacids:

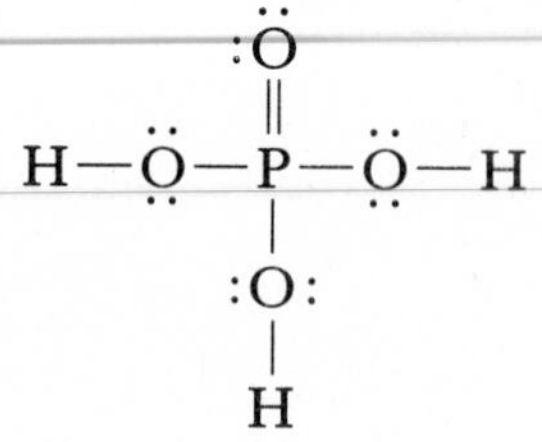

phosphoric acid 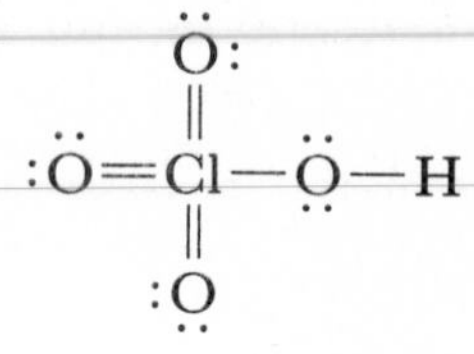   perchloric acid

Resonance Structures

There are molecules and ions for which more than one satisfactory Lewis structure can be drawn. For example, the nitrite ion, NO_2^-, can be formulated as either

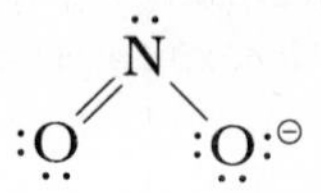

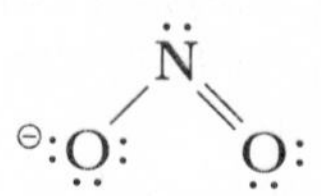

In either case the octet rule is satisfied. If either of these structures were the "correct" one, the ion would have two distinguishable nitrogen–oxygen bonds, one single and one double. Double bonds are shorter than single bonds, but structural studies of NO_2^- show that the two nitrogen–oxygen bonds are indistinguishable.

Consideration of NO_2^- and many other molecules and ions shows that our simple scheme for counting electrons and assigning them to the valence shells of atoms as bonds or unshared pairs is not entirely satisfactory. Fortunately, the simple model may be altered fairly easily to fit many of the awkward cases. The problem with NO_2^- is that the ion is actually more symmetrical than either one of the Lewis electronic structures that we wrote. However, if we took photographs of the two structures shown previously and superimposed the pictures, we would obtain a new structure having the same symmetry as the molecule. The photographic superimposition method is the same as writing a structure such as

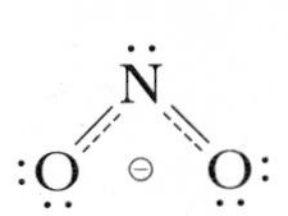

This structure would imply, "NO_2^- is a symmetrical ion, having partial double-bond character in each of the nitrogen–oxygen bonds." For some purposes the structure tells us enough. However, keeping track of the electrons in such a structure requires some rather special notation. What we actually do most of the time in such situations is to write two or more Lewis structures, which we call **resonance structures**, and connect them with a symbol that means: "Superimpose these structures to get a reasonable representation of the molecule." Applied to NO_2^- the structures are

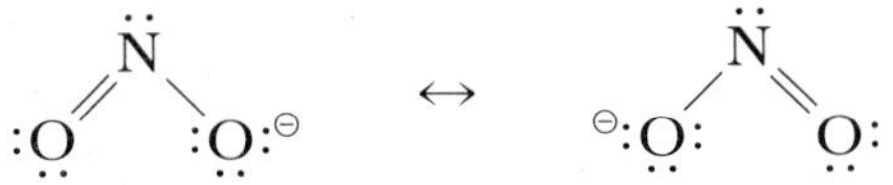

The double-headed arrow is the symbol for superimposition. It should not be confused with the symbol consisting of two arrows pointing in opposite directions, $\rightleftarrows$, which indicates that a reversible chemical reaction is occurring. The double-headed arrow does not imply that a molecule or ion flips back and forth between two structures. Instead, it tells us that the electronic formula of NO_2^- is a hybrid of the two resonance structures. When two or more resonance structures can be drawn for a molecule or ion, the electronic formula for the species is considered to be a **resonance hybrid** of these structures.

When we consider the benzene molecule, C_6H_6, which has six carbon atoms arranged in a ring, we can draw two equally satisfactory structures:

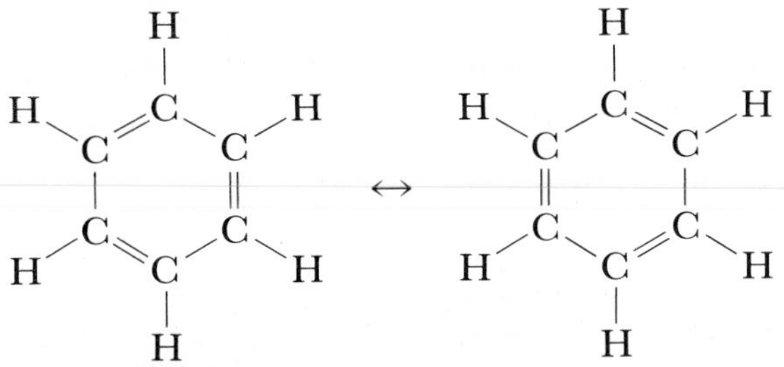

Both resonance structures show the ring to be composed of alternating single and double bonds. However, structural studies reveal that all the carbon–carbon bond distances are equal, as would be expected for a resonance hybrid of the two structures. The full symmetry of the molecule can also be indicated by a single structure with a special dashed-line notation:

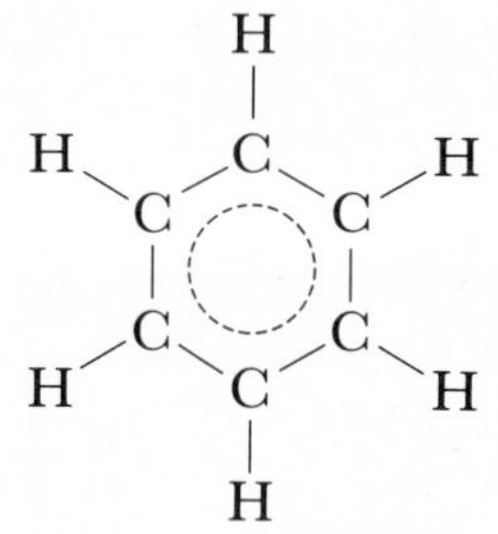

Resonance structures are required in many cases other than those in which they are demanded by symmetry. For example, compare two well-known anions, nitrate (NO_3^-) and nitroamide (^-O_2NNH). Because nitrate has three equivalent nitrogen–oxygen bonds, we write a set of three equivalent resonance structures, and say that the correct formula is a resonance hybrid:

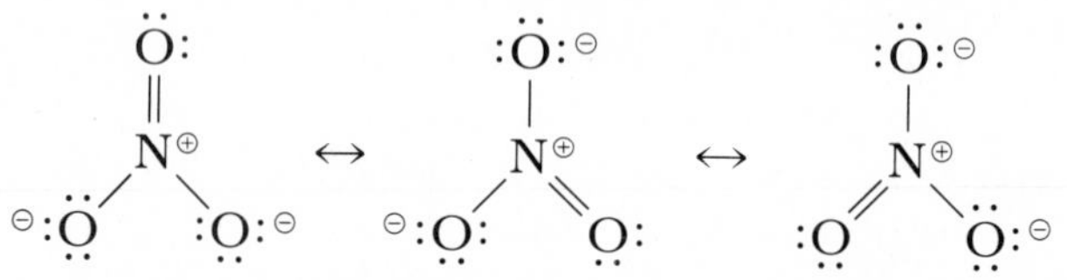

For the nitroamide ion we can write two equivalent structures (I and II), plus a third (III) that is not equivalent to the other two:

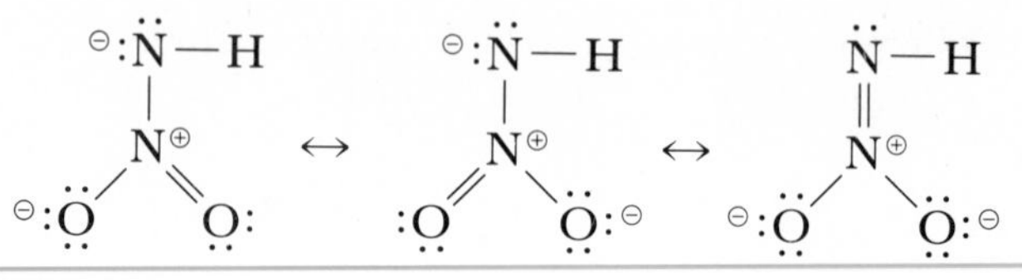

Common sense tells us that all three structures should contribute to our description of the ion. Since the structures are not equivalent, the resonance symbol no longer means, "Mix these structures equally in your thinking." It merely means, "Mix them." Therefore no quantitative implications are intended by the double-headed arrow. When we become semiquantitative in our description we state that structure III "contributes" more to the resonance hybrid of the nitroamide ion than either of the equivalent structures I and II because III places both formal negative charges on the oxygen atoms.

The last of the polyatomic molecules we will discuss is the anion obtained by removing the two protons in sulfuric acid, the sulfate ion, SO_4^{2-}. As in the case of H_2SO_4, an octet-rule structure with only single bonds can be written by assigning three lone pairs to each oxygen atom:

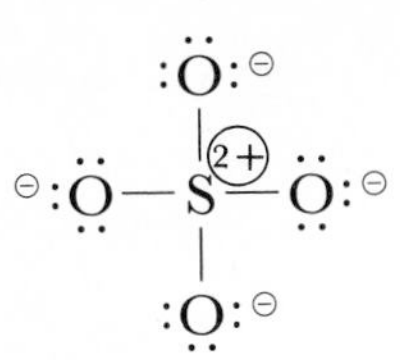

$$
\begin{array}{c}
:\!\overset{..}{O}\!:^{\ominus} \\
| \\
^{\ominus}:\!\overset{..}{O}\!\!-\!\!\overset{(2+)}{S}\!\!-\!\!\overset{..}{O}\!:^{\ominus} \\
| \\
:\!\overset{..}{O}\!:^{\ominus}
\end{array}
$$

However, if we consider the large positive formal charge on the sulfur atom we conclude that this is not a particularly appropriate structure. A much better representation of the bonding in SO_4^{2-} removes the $+2$ formal charge on the central sulfur atom by forming two sulfur–oxygen double bonds.

There are six equivalent structures with two sulfur–oxygen double bonds and two sulfur–oxygen single bonds. Thus we represent the bonding in SO_4^{2-} as a resonance hybrid of the following six equivalent structures:

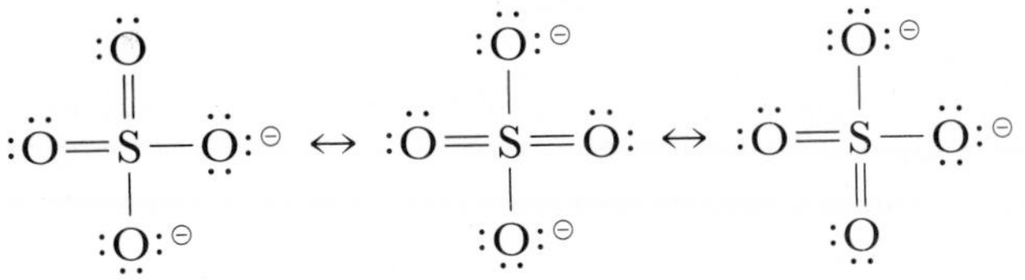

I   II   III

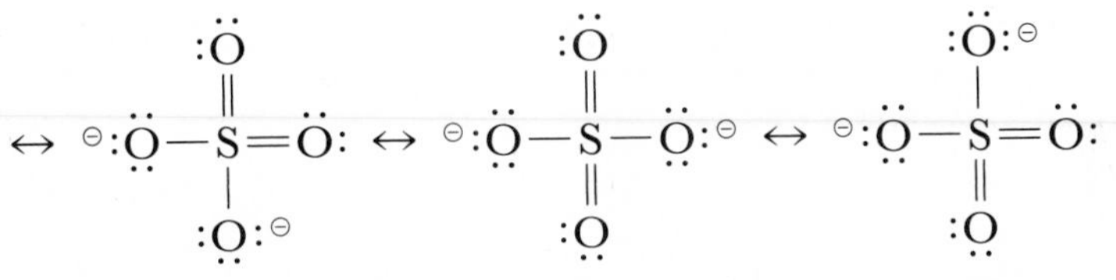

IV   V   VI

The resonance hybrid of the six equivalent structures (I–VI) of SO_4^{2-} would have an average sulfur–oxygen bond order of $1\frac{1}{2}$. In accord with this model of partial double-bond character is the fact that the observed sulfur–oxygen bond length in SO_4^{2-} (1.49 Å) is 0.21 Å shorter than the standard sulfur–oxygen single-bond length of 1.70 Å, which is obtained by adding the atomic radii of sulfur (1.04 Å) and oxygen (0.66 Å) (see Figure 9-5).

Example 3

Write Lewis structures for SO_2 and SO_3. Can you write a Lewis structure for SO_3 in which all three sulfur–oxygen bonds are equivalent? What is the formal charge on S then?

Solution   An SO_2 structure with equivalent bonds is

$$:\ddot{O}\!=\!\ddot{S}\!=\!\ddot{O}:$$

This SO_2 structure is satisfactory in that both sulfur–oxygen bonds are equivalent and there is no formal charge on the sulfur atom; however, it places 10 valence electrons around the S atom. But we already have seen in SO_4^{2-} that as many as 12 valence electrons can be placed around a central sulfur atom. If we insist on no more than 8 electrons around each atom, we cannot draw a single Lewis structure with equivalent bonds for SO_2. However, we can draw two Lewis structures for SO_2 in which one sulfur–oxygen bond is single and one is double, and in which there is a formal-charge separation. An adequate representation of SO_2 would be a resonance hybrid of these two structures:

$$^{\ominus}:\ddot{O}\!-\!\ddot{\overset{\oplus}{S}}\!=\!\ddot{O}: \leftrightarrow :\ddot{O}\!=\!\ddot{\overset{\oplus}{S}}\!-\!\ddot{O}:^{\ominus}$$

If we had to choose a single structure to represent SO_2, it would be $:O\!=\!S\!=\!O:$, because the sulfur–oxygen bonds are equivalent and there is no formal-charge separation. A better representation would be a resonance hybrid of this structure and the two others,

$$:\ddot{O}\!=\!\ddot{S}\!=\!\ddot{O}: \leftrightarrow ^{\ominus}:\ddot{O}\!-\!\ddot{\overset{\oplus}{S}}\!=\!\ddot{O}: \leftrightarrow :\ddot{O}\!=\!\ddot{\overset{\oplus}{S}}\!-\!\ddot{O}:^{\ominus}$$

This is because the observed sulfur–oxygen bond length in SO_2 is a bit longer than expected for a double bond.

The SO_3 molecule involves more difficulties. There are two Lewis structures in which all three sulfur–oxygen bonds are equivalent:

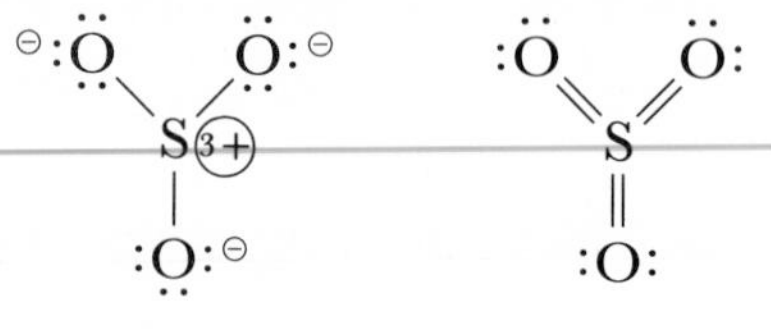

The first structure places a formal charge of $+3$ on S and has only three electron pairs around the central atom. This structure may be rejected. The second structure avoids formal charges, but at the expense of surrounding the S atom with six electron pairs. Three equivalent resonance structures can be drawn with an octet of electrons around S, but they all give S a $+2$ formal charge:

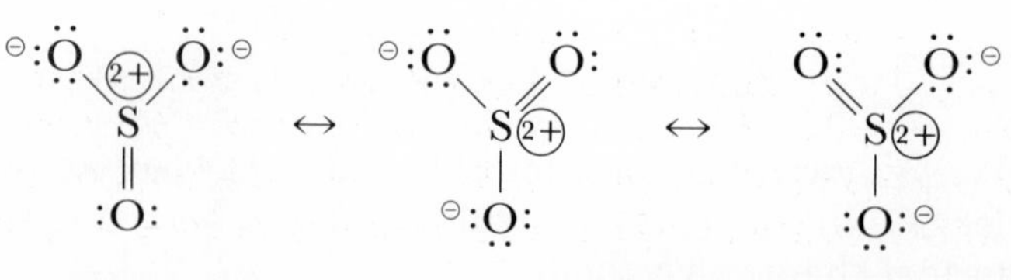

Three additional equivalent resonance structures place a $+1$ formal charge on S:

$$:\ddot{O} \diagdown \quad \diagup \ddot{O}: \qquad :\ddot{O} \diagdown \quad \diagup \ddot{O}:^{\ominus} \qquad :\ddot{O}^{\ominus} \diagdown \quad \diagup \ddot{O}:$$

$$\overset{\oplus}{S} \quad \longleftrightarrow \quad \overset{\oplus}{S} \quad \longleftrightarrow \quad \overset{\oplus}{S}$$

$$:\underset{\cdot\cdot}{O}:^{\ominus} \qquad\qquad :\underset{\cdot\cdot}{O}: \qquad\qquad :\underset{\cdot\cdot}{O}:$$

The x-ray and spectroscopic data available reveal that all three bonds in SO_3 are identical, and that their length is shorter than that expected for a single bond but longer than for a double bond. Therefore, we cannot describe the actual molecule accurately with any one Lewis structure, and our failure illustrates the inadequacy of such a simple bond model. We can compromise and say that the structure of the SO_3 molecule is a resonance hybrid of the all-double-bonded structure with a little of the other six structures as well.

The Meaning of Oxidation Numbers

Chlorine is found in a series of oxyanions, ClO^-, ClO_2^-, ClO_3^-, and ClO_4^-, that illustrate its entire range of positive oxidation states. The chloride ion has the Ar noble-gas structure with four pairs of valence electrons. The four oxyanions can be thought of as the products of the reaction of this Cl^- ion as a Lewis base with one to four oxygen atoms, which act as electron-pair acceptors and Lewis acids. Following are the four reactions and the oxidation number of Cl in each oxyanion:

$$:\ddot{Cl}:^- \quad + \ddot{O}: \rightarrow :\ddot{Cl}:\ddot{O}:^- \qquad +1$$

$$:\ddot{Cl}:\ddot{O}:^- \quad + \ddot{O}: \rightarrow :\ddot{O}:\ddot{Cl}:\ddot{O}:^- \qquad +3$$

$$:\ddot{O}:\ddot{Cl}:\ddot{O}:^- + \ddot{O}: \rightarrow :\ddot{O}:\underset{:\underset{\cdot\cdot}{O}:}{\overset{:\ddot{O}:}{\ddot{Cl}}}:\ddot{O}:^- \qquad +5$$

$$:\underset{:\underset{\cdot\cdot}{O}:}{\overset{:\ddot{O}:}{\ddot{O}:\ddot{Cl}:\ddot{O}:}}^- + \ddot{O}: \rightarrow :\underset{:\underset{\cdot\cdot}{O}:}{\overset{:\ddot{O}:}{\ddot{O}:\ddot{Cl}:\ddot{O}:}}^- \qquad +7$$

There are no chlorine oxyanions with more than four oxygens because there are no more valence electron pairs on Cl^-.

The formal charge on each atom is determined by assigning one electron in a bond to each participating atom. In contrast, the oxidation number is found by assigning *both* electrons in a bond to the more electronegative of the two atoms. (This is the meaning of Rule 5 in Section 10-1.) Thus the oxidation number is the charge that the atom would have if it were an isolated ion with the assigned number of electrons:

$$\text{Oxidation number} = Z - (N_{\text{assigned}} + N_{\text{nonbonding}})$$

Here Z is the atomic number, N_{assigned} is the total number of electrons in bonds between the given atom and atoms that are *less* electronegative than it is, and $N_{\text{nonbonding}}$ is the total number of electrons possessed by the atom that are *not* involved in covalent bonds. In calculating oxidation numbers, we always pretend that both electrons in a bond belong to the more electronegative of two bonded atoms. Fluorine, the most electronegative element, always has an oxidation number of -1. Oxygen always has an oxidation number of -2, except in peroxides and compounds of fluorine. In the Cl oxyanions, since Cl has an electronegativity of 3.16, and O, of 3.44, both electrons in a bond are assigned to oxygen. In Cl^-, chlorine has a net charge of -1 and an oxidation number of -1. In ClO^-, chlorine has six assigned electrons and an oxidation number of $+1$, since the bonding electron pair has been assigned to oxygen because of oxygen's greater electronegativity. In ClO_4^-, all four electron pairs have been "abducted" by the more electronegative oxygen atoms, and Cl has an oxidation number of $+7$, *as if* it really had lost all seven of its valence electrons. Oxidation numbers are couched in the language of electron loss and gain. What they really measure, however, is the extent of combination of an atom with other atoms more electronegative than itself.

Example 4

What are the formal charges on Cl in the Lewis structures given previously for ClO^-, ClO_2^-, ClO_3^-, and ClO_4^-? Can better Lewis structures be written for these ions? What are they? Do these new structures change the oxidation numbers of Cl in each case?

Solution

The formal charges on Cl are as follows:

$:\!\ddot{Cl}\!-\!\ddot{O}\!:^{\ominus}$ (0 on Cl)

$:\!\ddot{Cl}^{\oplus}\!-\!\ddot{O}\!:^{\ominus}$ with $:\!\ddot{O}\!:^{\ominus}$ above (+1 on Cl)

$^{\ominus}:\!\ddot{O}\!-\!\overset{2+}{Cl}\!-\!\ddot{O}\!:^{\ominus}$ with $:\!\ddot{O}\!:^{\ominus}$ above (+2 on Cl)

$^{\ominus}:\!\ddot{O}\!-\!\overset{3+}{Cl}\!-\!\ddot{O}\!:^{\ominus}$ with $:\!\ddot{O}\!:^{\ominus}$ above and $:\!\ddot{O}\!:_{\ominus}$ below (+3 on Cl)

In every ion but ClO^-, a formal positive charge is present on Cl. Better

Lewis structures can be written for each ion, in which the formal charge is reduced to zero on Cl. These structures involve Cl=O: bonds, and will result in more than eight electrons around the Cl; this larger number of electrons is acceptable for a third-period atom (see the previous discussion of SO_4^{2-}). Thus we have

$$
\overset{\overset{\cdot\cdot}{O}:}{\underset{}{\|}} \qquad :\overset{\cdot\cdot}{O}:^{\ominus} \\
:\overset{\cdot\cdot}{\underset{\cdot\cdot}{Cl}}-\overset{\cdot\cdot}{\underset{\cdot\cdot}{O}}:^{\ominus} \leftrightarrow :\overset{\cdot\cdot}{\underset{\cdot\cdot}{Cl}}=\overset{\cdot\cdot}{\underset{\cdot\cdot}{O}}:
$$

two equivalent resonance structures, each with zero formal charge on Cl; both chlorine–oxygen bonds are equivalent. Next, we have

$$
\overset{\overset{\cdot\cdot}{O}:}{\|} \qquad :\overset{\cdot\cdot}{O}:^{\ominus} \qquad \overset{\overset{\cdot\cdot}{O}:}{\|} \\
^{\ominus}:\overset{\cdot\cdot}{\underset{\cdot\cdot}{O}}-\overset{\cdot\cdot}{\underset{\cdot\cdot}{Cl}}=\overset{\cdot\cdot}{\underset{\cdot\cdot}{O}}: \leftrightarrow :\overset{\cdot\cdot}{\underset{\cdot\cdot}{O}}=\overset{\cdot\cdot}{\underset{\cdot\cdot}{Cl}}=\overset{\cdot\cdot}{\underset{\cdot\cdot}{O}}: \leftrightarrow :\overset{\cdot\cdot}{\underset{\cdot\cdot}{O}}=\overset{\cdot\cdot}{\underset{\cdot\cdot}{Cl}}-\overset{\cdot\cdot}{\underset{\cdot\cdot}{O}}:^{\ominus}
$$

three equivalent resonance structures, each with zero formal charge on Cl; all three chlorine–oxygen bonds are equivalent. And,

$$
\overset{\overset{\cdot\cdot}{O}:}{\|} \qquad :\overset{\cdot\cdot}{O}:^{\ominus} \qquad \overset{\overset{\cdot\cdot}{O}:}{\|} \qquad \overset{\overset{\cdot\cdot}{O}:}{\|} \\
^{\ominus}:\overset{\cdot\cdot}{\underset{\cdot\cdot}{O}}-\underset{\underset{:O}{\|}}{Cl}=\overset{\cdot\cdot}{\underset{\cdot\cdot}{O}}: \leftrightarrow :\overset{\cdot\cdot}{\underset{\cdot\cdot}{O}}=\underset{\underset{:O}{\|}}{Cl}=\overset{\cdot\cdot}{\underset{\cdot\cdot}{O}}: \leftrightarrow :\overset{\cdot\cdot}{\underset{\cdot\cdot}{O}}=\underset{\underset{:O}{\|}}{Cl}-\overset{\cdot\cdot}{\underset{\cdot\cdot}{O}}:^{\ominus} \leftrightarrow :\overset{\cdot\cdot}{\underset{\cdot\cdot}{O}}=\underset{\underset{:O:_{\ominus}}{|}}{Cl}=\overset{\cdot\cdot}{O}:
$$

four equivalent resonance structures, each with zero formal charge on Cl; all four chlorine–oxygen bonds are equivalent.

The better Lewis structures do *not* change the oxidation-number assignments. In each, *all* bonding electrons are assigned to the more electronegative atom, which is oxygen. So, for perchlorate (ClO_4^-), for example, we make the following comparison:

$$
:\overset{\cdot\cdot}{O}:^{\ominus} \qquad\qquad\qquad \overset{\cdot\cdot}{O}: \\
^{\ominus}:\overset{\cdot\cdot}{\underset{\cdot\cdot}{O}}-\underset{\underset{:O:^{\ominus}}{|}}{\overset{|}{Cl}}\text{③⁺}\overset{\cdot\cdot}{\underset{\cdot\cdot}{O}}:^{\ominus} \qquad\qquad ^{\ominus}:\overset{\cdot\cdot}{\underset{\cdot\cdot}{O}}-\underset{\underset{:O}{\|}}{\overset{\|}{Cl}}=\overset{\cdot\cdot}{O}:
$$

(at right: (and three other resonance structures))

Lewis structure with large formal-charge separation (poor); formal charge on Cl = +3; oxidation number of Cl obtained by assigning all four bonding pairs to O, as if the bonds were all ionic; oxidation number of Cl = +7

Lewis structure with zero formal charge on Cl (better); oxidation number of Cl obtained by assigning all seven bonding pairs to O, as if the bonds were all ionic; oxidation number of Cl = +7

11-2   ACIDITY OF OXYACIDS

The simple Lewis model of bonding, with all its defects, does give us a physical understanding of the relative acidity of compounds that contain a central atom bonded to oxygen atoms alone, or to the oxygen atoms in hydroxide ion and water. Oxides of nonmetals dissolve in water to form acids. For example,

$$SO_3 + H_2O \rightarrow H_2SO_4$$

(Sulfuric acid is not prepared this way commercially, however.) In such oxyacids the protons are bound to oxygen:

$$
\begin{array}{c}
\ddot{\text{O}}: \\
\parallel \\
:\ddot{\text{O}}\!=\!\text{S}\!-\!\ddot{\text{O}}\!-\!\text{H} \\
| \\
:\text{O}: \\
| \\
\text{H}
\end{array}
$$

Consider a series of compounds containing hydroxide groups bound to positive ions ranging from Na^+ to Cl^{7+}: $NaOH$, $Mg(OH)_2$, $Al(OH)_3$, $Si(OH)_4$, $OP(OH)_3$ or H_3PO_4, $O_2S(OH)_2$ or H_2SO_4, and O_3ClOH or $HClO_4$. Now consider the bonds in the structure

$$M\overset{1}{-}O\overset{2}{-}H$$

in which M is the central atom. If the bond breaks at position 1 to form M^+ and OH^-, the compound is basic. If it breaks at position 2 to form $M-O^-$ and H^+, the compound is acidic. For the preceding series of compounds, Na^+OH^- is an ionic compound that is very soluble in water; thus, OH^- and Na^+ are separated easily. The smaller, more highly charged Mg^{2+} ion binds more tightly to OH^-, thereby making magnesium hydroxide, $Mg(OH)_2$, less soluble than $NaOH$ and a weaker base. Aluminum hydroxide is virtually insoluble in water, but loses OH^- to strong acids and H^+ to strong bases:

$$Al(OH)_3 + 3H^+ \rightarrow Al^{3+} + 3H_2O$$
$$Al(OH)_3 + OH^- \rightarrow AlO(OH)_2^- + H_2O$$

As we have seen, a compound such as aluminum hydroxide, which is able to react either as a base with an acid, or as an acid with a base, is called **amphoteric**. Hydrogen species such as HCO_3^- that can either lose a proton (giving CO_3^{2-}) or gain a proton (forming H_2CO_3) are called **amphiprotic**.

The exact formula of the ion represented by $AlO(OH)_2^-$ is uncertain. Recent work has shown that the hydrated aluminum ion is $Al(H_2O)_6^{3+}$. Removal of three protons from this would give neutral, insoluble $Al(OH)_3(H_2O)_3$, which could then lose another proton and form

$Al(OH)_4(H_2O)_2^-$. This formula probably most nearly represents the actual situation.

Silicic acid, $Si(OH)_4$, easily gives up water molecules to form SiO_2. It is a weak acid and reacts with NaOH. However, it does not react with HCl; thus, the compound is not amphoteric:

$$Si(OH)_4 + 2NaOH \rightarrow Na_2SiO_3 + 3H_2O$$
$$Si(OH)_4 + HCl \rightarrow \text{no reaction}$$

The compounds H_3PO_4, H_2SO_4, and $HClO_4$ become progressively more acidic; $HClO_4$ is the strongest oxyacid known. It appears that the oxygen–hydrogen bonds in this series become easier to break as the oxidation number of the central atom increases. However, HNO_3, in which N is in the $+5$ oxidation state, is a much stronger acid than $Te(OH)_6$, in which Te is in the $+6$ oxidation state. A better correlation can be made between acidity of oxyacids and the formal charge on the central atom, provided all Lewis structures are written to be consistent with the octet rule. With this provision, the formal charges on the central atoms in several oxyacids are

Formal charge: N = +1   Formal charge: S = +2

Formal charge: Cl = 0   Formal charge: Cl = +1   Formal charge: Cl = +3

In general, acidity increases with increasing formal charge on the central atom. The higher the formal positive charge on a central atom, the more the central atom will attract electrons from attached oxygen atoms. This results in a weakening of the oxygen–hydrogen bond, thereby allowing easier removal of the H^+ and therefore an increase in acid strength. The effect of formal charge on acidity for similar species is shown by the data in Table 11-1; a small increase in negative charge on an atom to which a proton is attached enormously decreases the acidity of the species.

There is another, simpler correlation between the acidity of oxyacids and their structures. Acidity increases as the number of oxygen atoms

Table 11-1

Effect of Formal Charge on Acid Strength

Substance	Formal charge on O, N, S	pK
H_3O^+	+1	−1.75
H_2O	0	+15.7
OH^-	−1	25
NH_4^+	+1	9.25
NH_3	0	35
H_2S	0	7.04
HS^-	−1	11.96

without hydrogen atoms attached increases. Table 11-2 shows such a correlation in which the pK's of oxyacids are clearly functions of n in the formula $XO_n(OH)_m$ and are much less influenced by the value of m.

The correlations of acidity with structure are approximate but have led to interesting discoveries. For instance, the pK of H_3PO_3 is 1.8 for the first hydrogen ionization. A chemist would instinctively write the structure as

$$
\begin{array}{c}
H \\
| \\
:\!\ddot{O}\!: \\
| \\
H\!-\!\ddot{O}\!-\!P\!-\!\ddot{O}\!-\!H
\end{array}
$$

for this acid, which would suggest a pK of about 7 to 9 (see the "very weak" group in Table 11-2). However, the pK_1 of 1.8 suggests that phosphorus has one oxygen bound to it, which is not bound to a hydrogen atom. Structural studies show that only two protons can be ionized and that the third proton is attached to phosphorus. Thus, the structure of H_3PO_3 should be written as

$$
\begin{array}{c}
:\!\ddot{O} \\
\| \\
H\!-\!\ddot{O}\!-\!P\!-\!\ddot{O}\!-\!H \\
| \\
H
\end{array}
$$

This structure is clearly consistent with the correlation in Table 11-2.

Example 5

Would you expect perchloric acid, $HClO_4$, to be stronger than nitric acid, HNO_3? Why?

Table 11-2

pK Values for Inorganic Oxyacids[a]

X(OH)$_m$ (very weak)		XO(OH)$_m$ (weak)		XO$_2$(OH)$_m$ (strong)		XO$_3$(OH)$_m$ (very strong)	
Cl(OH)	7.5	ClO(OH)	2.0	ClO$_2$(OH)	(−3)	ClO$_3$OH	(−8)
Br(OH)	8.7	NO(OH)	3.4	NO$_2$(OH)	−1.4	MnO$_3$(OH)	(−8)
I(OH)	10.6	IO(OH)	1.6	IO$_2$(OH)	0.8		
B(OH)$_3$	9.2	SO(OH)$_2$	1.8	SO$_2$(OH)$_2$	(−3)		
Sb(OH)$_3$	11.0	SeO(OH)$_2$	2.5	SeO$_2$(OH)$_2$	(−3)		
Si(OH)$_4$	9.7	TeO(OH)$_2$	2.5				
Ge(OH)$_4$	8.6	CO(OH)$_2$	6.4				
Te(OH)$_6$	7.7	PO(OH)$_3$	2.1				
As(OH)$_3$	9.2	AsO(OH)$_3$	2.3				
		HPO(OH)$_2$	1.8				
		H$_2$PO(OH)	2.0				

[a]Values given within parentheses are estimated values.

Solution

The Lewis structure of $HClO_4$ that places exactly eight electrons around Cl is

The formal charge on Cl is +3. The related structure for HNO_3,

places a +1 formal charge on N. Thus we would expect $HClO_4$ to be a stronger acid than HNO_3 (which it is; see Table 11-2). Since both acids have the same number of —OH groups (one each), the oxidation-number method also predicts the relative acidities correctly. The oxidation number of Cl in $HClO_4$ is +7, which is greater than the +5 for N in HNO_3. Finally, the number of oxygen atoms without hydrogen atoms attached to the central atom is greater for $HClO_4$ than for HNO_3 ($n = 3$ for $HClO_4$; $n = 2$ for HNO_3), again suggesting that $HClO_4$ is the stronger acid.

11-3   THE VSEPR METHOD
AND MOLECULAR GEOMETRY

Up to this point we have been concerned mainly with identifying the bonds and lone pairs in the Lewis structures of molecules and ions. But we can say much more about the structure of a molecule than is implied by its Lewis diagram. Molecules and complex ions have shapes (geometries), and it is this aspect of structure that we shall discuss next.

Molecules have all sorts of geometrical structures. For example, the ground-state molecular geometry of CO_2 is linear, whereas that of OCl_2 is angular (bent); BF_3 is trigonal planar, NF_3 is trigonal pyramidal, and ClF_3

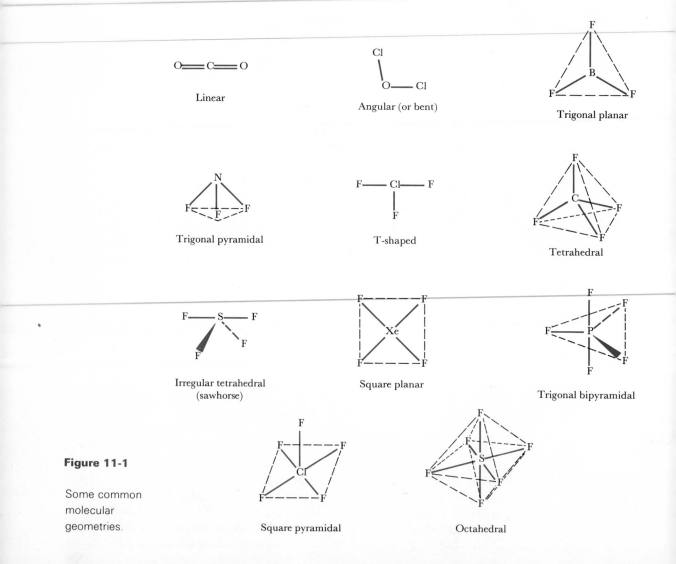

Figure 11-1

Some common molecular geometries.

is T-shaped; CF_4 is tetrahedral, SF_4 is irregular tetrahedral (sawhorse), and ClF_4^- is square planar; PF_5 is trigonal bipyramidal but ClF_5 is square pyramidal; and finally, SF_6 is octahedral. These geometrical structures are shown in Figure 11-1.

How are we to understand all these geometries? Is there any simple way to predict them (and predict others as well)? The answer to the last question, surprisingly, is that there is—it is a method that was described by N. V. Sidgwick and H. M. Powell in 1940, and extended by R. J. Gillespie and R. S. Nyholm in 1957. This approach is called the **valence-shell electron-pair repulsion (VSEPR) method**; it states that the bonding electron pairs and lone electron pairs of an atom will adopt a spatial arrangement that minimizes electron-pair repulsion around that atom.

Before applying the VSEPR method to molecular systems, let us first consider the most favorable arrangement for n electron pairs circumscribed on a sphere. For values of n from 2 through 6, the predicted arrangement is shown in Figure 11-2. The predicted arrangement for 7 through 12 electron pairs will be discussed later in this section.

To apply the VSEPR method to molecules we simply count the number of lone electron pairs and the number of atoms around the central atom in a polyatomic molecule. We will call the total number of lone pairs of electrons and attached atoms the **steric number**, SN. If there are no lone pairs of electrons around the central atom (A) and the SN arises only from attached atoms (X), the observed molecular geometry agrees with that shown in Figure 11-2. In each of the examples shown in Table 11-3, the predicted geometry involves separating the bonding electron pairs as much as possible, to minimize electron-pair repulsion. It should be emphasized that multiple bonds between atoms do not alter the predictions of molecular

Table 11-3

Predicted Geometries for Various Molecules

Molecule	Steric number	Predicted geometry		Example
AX_2	2	Linear	$180°$	CO_2
AX_3	3	Trigonal planar	$120°$	BF_3
AX_4	4	Tetrahedral	$109.5°$	CF_4
AX_5	5	Trigonal bipyramidal	$90°$ $120°$	PF_5
AX_6	6	Octahedral	$90°$ $90°$	SF_6

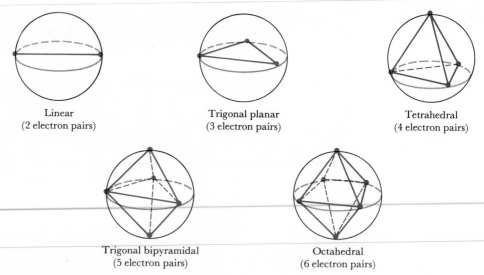

Linear
(2 electron pairs)

Trigonal planar
(3 electron pairs)

Tetrahedral
(4 electron pairs)

Trigonal bipyramidal
(5 electron pairs)

Octahedral
(6 electron pairs)

Figure 11-2   The arrangement of electron pairs on the surface of a sphere so that the distance between electron pairs is maximized.

geometry. For example, beryllium dihydride and carbon dioxide each have $SN = 2$ and are predicted to be linear:

$$H—Be—H \qquad :\ddot{O}=C=\ddot{O}:$$

However, if one of the attached atoms is replaced by a lone pair of electrons, the molecular geometry does change. By *molecular geometry* we mean the positions of the atoms that we can determine experimentally, but not the placement of lone pairs of electrons, which can only be inferred. Thus for AX_3 and AX_2E where X represents an attached atom and E a lone pair of electrons, we obtain the following structures:

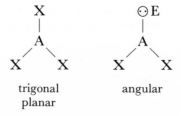

trigonal
planar

angular

We can predict not only that an AX_2E molecule will be angular but also that the XAX angle will be *less* than 120°. This prediction results from the postulate that lone electron pairs are closer to the central atom and therefore have a repulsive effect larger than that of bonding electron pairs.

We can apply this same postulate to predict the relative bond angles in the isoelectronic sequence CH_4, $:NH_3$, and $H_2\ddot{O}:$. Each of these molecules has $SN = 4$. However, in view of the lone-pair/bond-pair repulsions in NH_3 and H_2O, we predict the bond angles in these two molecules to be less than the tetrahedral angle (109.5°). The experimental bond angles agree with these predictions.

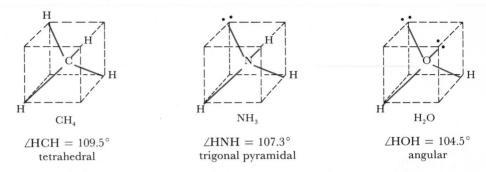

CH_4	NH_3	H_2O
∠HCH = 109.5°	∠HNH = 107.3°	∠HOH = 104.5°
tetrahedral	trigonal pyramidal	angular

Examples of predicted molecular shapes from $SN = 2$ through $SN = 6$ are shown in Figure 11-3.

The structure of a molecule that contains different atoms attached to a central atom exhibits distortions from the idealized structures shown in Figure 11-3. Thus CH_3Cl has HCH angles of 110.5° and ClCH angles of 108.5°, compared to the perfect tetrahedron for which all bond angles are 109.5°. Further examples of these distortions from an idealized geometry are provided by ethylene and formaldehyde. In both molecules the carbon atoms have $SN = 3$, for which the idealized bond angles are 120°. The observed structures are

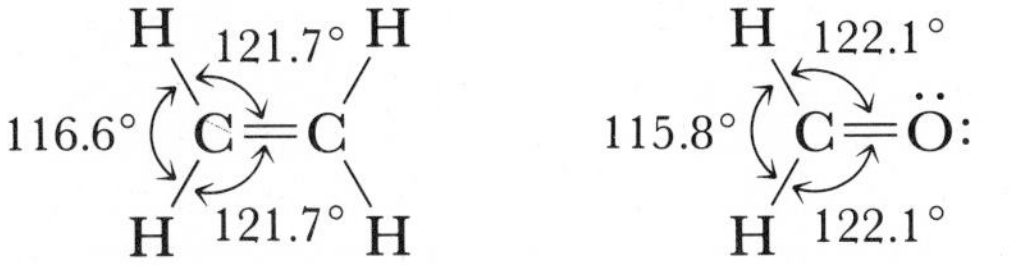

$$\begin{array}{cc}
\text{H} \quad 121.7° \quad \text{H} & \text{H} \quad 122.1° \\
116.6° \left(\text{C} = \text{C} \right. & 115.8° \left(\text{C} = \ddot{\text{O}}: \right. \\
\text{H} \quad 121.7° \quad \text{H} & \text{H} \quad 122.1°
\end{array}$$

Here one can consider that the two pairs of electrons in a double bond exert more of an electrostatic repulsion than the electron pairs in the single bonds, pushing the C—H bonds slightly together.

The placement of electron pairs for $SN = 5$ deserves special comment. The spatial arrangement for five electron pairs is a trigonal bipyramid, in which there are three equatorial and two axial positions:

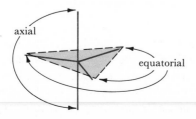

SN	Number of lone pairs	Molecular shape	Example
2	0	 linear	BeH_2, CO_2
3	0	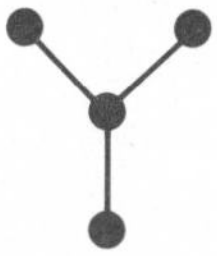 trigonal planar	SO_3, BF_3
3	1	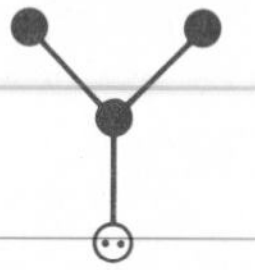 angular	SO_2, O_3
4	0	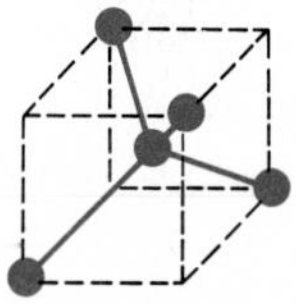 tetrahedral	CH_4, CF_4, SO_4^{2-}
4	1	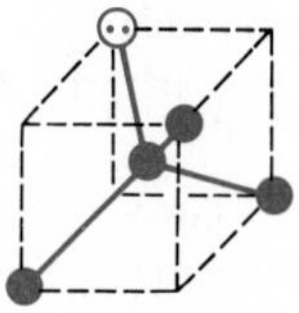 trigonal pyramidal	NH_3, PF_3, $AsCl_3$
4	2	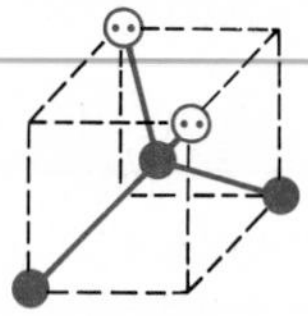 angular	H_2O, H_2S, SF_2
5	0	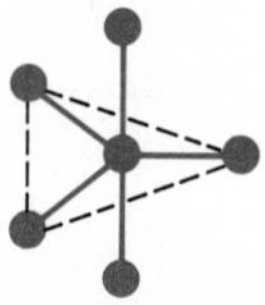 trigonal bipyramidal	PF_5, PCl_5, AsF_5

SN	Number of lone pairs	Molecular shape	Example
5	1	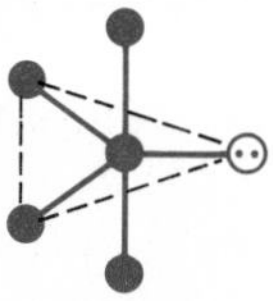 sawhorse	SF_4
5	2	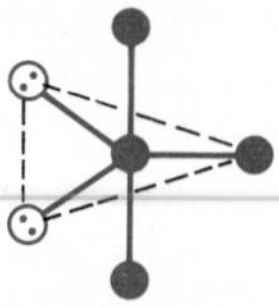 T-shaped	ClF_3
5	3	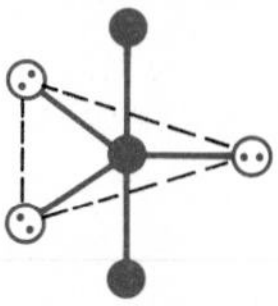 linear	XeF_2, I_3^-, IF_2^-
6	0	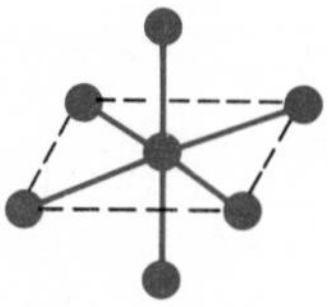 octahedral	SF_6, PF_6^-, SiF_6^{2-}
6	1	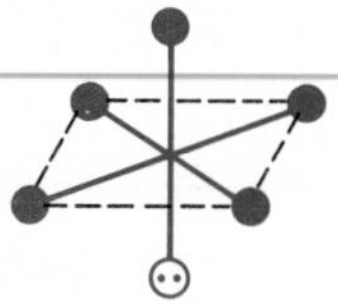 square pyramidal	IF_5, BrF_3
6	2	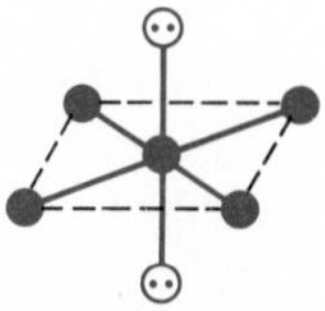 square planar	XeF_4, IF_4^-

Figure 11-3   Molecular shapes predicted by the VSEPR method.

The shape of a molecule such as PF_5 must be trigonal bipyramidal because there are no lone pairs surrounding the central phosphorus atom. However, it is not necessary that the axial bond lengths be equal to the equatorial bond lengths. Since each axial bond experiences three 90° repulsions compared to only two 90° repulsions for the equatorial positions, we predict that the axial bonds will be longer than the equatorial bonds (the 90° interactions are much larger than 120° or 180° interactions, because electron-pair repulsions decrease very rapidly as the distance between pairs increases). Experimentally, the axial P—F bond lengths are 1.577 Å, compared to 1.534 Å for the equatorial bond lengths.

Next we consider the SF_4 molecule, which has four bond pairs and one lone pair surrounding the sulfur atom. Where should we place the lone electron pair in SF_4? According to the VSEPR method, the most prohibitive repulsion is lone pair/lone pair, followed in order by lone pair/bond pair and bond pair/bond pair. Therefore, in SF_4 the "worst" repulsion is lone pair/bond pair because there is only one lone pair. If we place the lone pair in one of the equatorial orbitals, it will repel only two bonded pairs at a 90° angle, whereas in an axial orbital it will repel three pairs at 90°:

 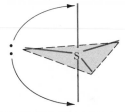

three 90° interactions   two 90° interactions
(axial placement of lone pair)   (equatorial placement of lone pair)

Therefore the VSEPR choice is equatorial placement of the lone electron pair, which results in a smaller number of 90° interactions. Accordingly, we also place the second (e.g., ClF_3) and third (e.g., I_3^-) lone pairs into equatorial orbitals, and predict the shapes shown in Figure 11-3.

The equatorial position of the lone pairs in SF_4 and ClF_3 should result in distortions of the idealized 90° and 120° angles between bonds. Indeed, the experimental geometries indicate the expected distortions:

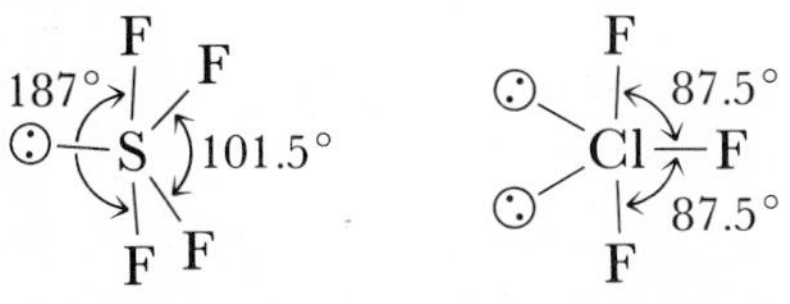

As our last example in the discussion of $SN = 5$, we consider those molecules in which the attached atoms are not all identical. Examples of such molecules are CH_3PF_4 and SOF_4. In each of these molecules, the less electronegative groups occupy the equatorial positions and cause distortions from the idealized 90° and 120° angles similar to those caused by lone electron pairs. Thus we find the following structures:

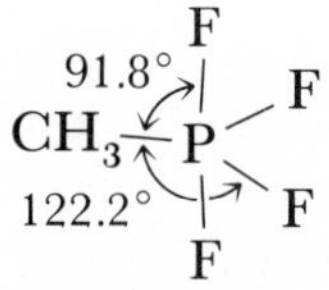

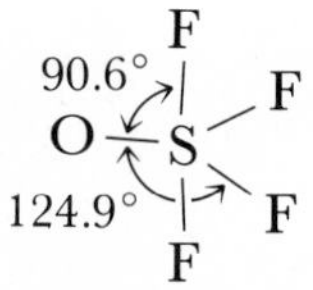

$$\underset{122.2°}{\overset{91.8°}{CH_3 \rightarrow P}} \begin{array}{c} F \\ \diagup F \\ \diagdown F \end{array} \qquad \underset{124.9°}{\overset{90.6°}{O \rightarrow S}} \begin{array}{c} F \\ \diagup F \\ \diagdown F \end{array}$$

The VSEPR method is easy to use and gives the correct molecular shape for a remarkably large number of molecules. All the predicted shapes in Figure 11-3 are in agreement with experimentally determined molecular structures.

The VSEPR method can be summarized by three simple rules:

1. The electron pairs around a central atom will adopt a spatial arrangement that minimizes electron-pair repulsion.
2. The most prohibitive repulsion is lone pair/lone pair, followed in order by lone pair/bond pair and bond pair/bond pair.
3. Among several structures involving 90° interactions, the most favored structure is the one that results in a smaller number of 90° lone-pair interactions.

Steric Numbers Greater Than Six

The idealized geometries for 7 through 12 electron pairs are shown in Figure 11-4. For any real molecular system there are several different geometries that have nearly the same energy. Thus, the IF_7 molecule ($SN = 7$) has a pentagonal bipyramidal structure rather than the monocapped octahedral structure shown in Figure 11-4a.

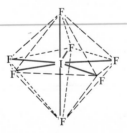

pentagonal bipyramidal structure
of IF_7

Another molecule with $SN = 7$ is XeF_6. Its molecular geometry is predicted to be *irregular octahedral* because of the presence of the lone pair on the xenon atom. The precise structure has not been determined experimentally because the geometry is not static—an intramolecular rearrangement process rapidly interconverts fluorine positions. In agreement

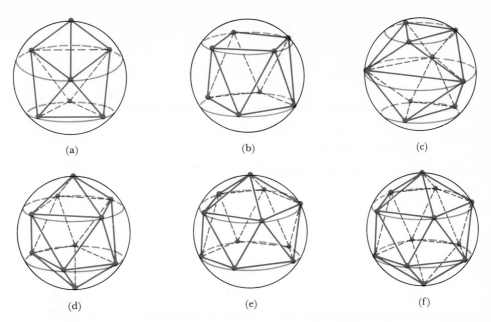

Figure 11-4 Arrangements of 7 through 12 electron pairs on a sphere that maximize their distance apart: (a) 7 pairs—monocapped octahedron; (b) 8 pairs—square antiprism; (c) 9 pairs—tricapped trigonal prism; (d) 10 pairs—bicapped square antiprism; (e) 11 pairs—monocapped pentagonal antiprism; (f) 12 pairs—icosahedron.

with the VSEPR prediction, however, the structure is definitely not regular octahedral.

Examples of species with SN equal to 8 and 9 are TaF_8^{3-} and ReH_9^{2-}. These complex ions have the predicted square antiprismatic and tricapped trigonal prismatic structures, respectively. Steric numbers higher than 9 are rare and will not be discussed here.

Exceptions to the VSEPR Rules

So far we have applied the VSEPR method to predict the molecular shapes of molecules that have all their electrons in pairs. If a molecule has one electron that is unpaired, we cannot predict molecular shape so simply. Consider the isoelectronic series BeH_3^{2-}, BH_3^-, CH_3, and NH_3^+, all of which have the Lewis electron-dot structure

$$H-\overset{\displaystyle .}{A}-H$$
$$\underset{\displaystyle H}{|}$$

Experimental evidence indicates that NH_3^+ is strictly trigonal planar, whereas CH_3 probably is trigonal planar but may be slightly pyramidal. No experimental evidence exists for BeH_3^{2-} and BH_3^-, but accurate quantum-

Table 11-4

Bond Angles in Group VA and Group VIA Hydrides

Hydride	Bond angle
NH_3	107.3°
PH_3	93.3°
AsH_3	91.8°
H_2O	104.5°
H_2S	92.2°
H_2Se	91.0°

mechanical calculations predict the degree of distortion from trigonal planar to trigonal pyramidal structures to follow the order $NH_3^+ < CH_3 < BH_3^- < BeH_3^{2-}$.

Other molecules that are not readily understood in terms of the VSEPR rules are the third-, fourth-, and subsequent-row compounds of Groups VA and VIA. The geometries of the hydrides of these compounds are compared to those of the second-row elements in Table 11-4. It is clear that only the second-row compounds exhibit bond angles near the expected tetrahedral angle. The subsequent-row compounds all exhibit bond angles near 90°. A weak explanation of this phenomenon is that because of the larger size of the central atom, there is less bond pair/bond pair repulsion in these compounds, so the bond angles are considerably smaller than the tetrahedral angle (see Chapter 13, Example 1).

Other exceptions to the VSEPR rules are provided by the ions $TeCl_6^{2-}$, $TeBr_6^{2-}$, and $SbBr_6^{3-}$, which are found to be regular octahedral, even though they are all of the AX_6E type and should be distorted just as predicted for XeF_6. This is explained after the fact by noting that the Br and Cl atoms are fairly bulky and consequently that the atom–atom repulsions perhaps dominate over the lone pair/bond pair repulsions.

Don't be discouraged by the fact that there are a few exceptions to the VSEPR rules. In most cases they work fine, which is more than can be said for the great majority of chemical "theories"!

Example 6

Predict the molecular geometry of SO_2F_2 using the VSEPR method.

Solution   The Lewis structure of SO_2F_2 is

The *SN* is 4; the structure is tetrahedral:

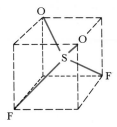

Example 7

Predict the shape of ClO_2^-.

Solution   The Lewis structure of ClO_2^- is

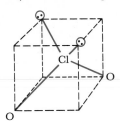

The *SN* is 4. The two O atoms plus the two lone pairs are arranged tetrahedrally around Cl, giving the ClO_2^- ion an *angular* shape:

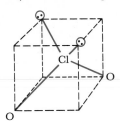

Summary

Lewis structures are the electronic formulas of molecules and complex ions
that use lines to indicate **shared electron pairs** (bonds) between atoms and
two dots on each atom to indicate a **lone pair** of electrons. For molecules and

complex ions containing only first- and second-period atoms, the best Lewis structures are always ones in which each atom is surrounded by the same number of electrons as in the next noble-gas atom. For H, this is two electrons (one electron pair as in He); for the second-period atoms of non-metallic elements (B, C, N, O, F), this is eight electrons (four electron pairs as in Ne). Because exactly eight electrons complete a closed $2s^2 2p^6$ shell, the rule in drawing Lewis structures is to surround each second-period atom with an octet of electrons. This is known as the **octet rule**

A bond in which two electron pairs are shared between atoms is called a **double bond**; one that involves three shared electron pairs is called a **triple bond**. The **bond order** is the number of shared electron pairs between two atoms. Bond orders of single, double, and triple bonds are 1, 2, and 3, respectively. As the bond order increases between any two atoms, the bond strength increases and the bond length decreases. If an unpaired (or odd) electron is left on an atom in a Lewis structure, the molecule or ion has an **open shell**.

If an atom in a Lewis structure "owns" more or fewer electrons than it does as a free neutral atom, then it possesses a **formal charge**. For the purpose of determining formal charge, each shared electron pair counts as one "owned" electron, and each lone pair counts as two. Thus in the ammonium ion, NH_4^+, N has a formal charge of $+1$ because it owns only four electrons (and it owns five valence electrons in atomic nitrogen, $2s^2 2p^3$). **Isoelectronic** species possess the same number of electrons; CH_4 and NH_4^+ are isoelectronic.

A **Lewis acid** is an electron-pair acceptor (an example is BF_3); a Lewis base is an electron-pair donor (an example is $:NH_3$).

If two or more acceptable Lewis structures can be written for a molecule or ion, together they describe the electronic structure of the species. The individual Lewis structures are called **resonance structures** and the "mixed" structure they represent by the notation of a double-headed arrow is called a **resonance hybrid**.

If atoms of different electronegativities are bonded together in a Lewis structure, the **oxidation numbers** are obtained by arbitrarily assigning all bonding electron pairs to the more electronegative partner, as if the structure were ionic. For example, the Lewis structure for CO_2 is

$$:\ddot{O}=C=\ddot{O}:$$

Assigning all the bonding pairs to the oxygens gives

$$:\ddot{\underset{..}{O}}:^{2-} \qquad C^{4+} \qquad :\ddot{\underset{..}{O}}:^{2-}$$

The oxidation number of carbon is $+4$, and the oxidation number of oxygen is -2 in CO_2.

An **oxyacid** contains the unit $M—O—H$; that is, it has a hydrogen atom that can be lost as H^+ bonded to an oxygen atom, which in turn is bonded to a so-called **central atom**. A molecule or ion of the $M-O-H$ type that can act both as a base (donating OH^- to an acid) and as an acid (donating H^+ to a base) is **amphoteric**. A species such as HSO_4^- that can either lose or gain a proton (H^+) is **amphiprotic**:

$$SO_4^{2-} \xleftarrow{-H^+} HSO_4^- \xrightarrow{+H^+} H_2SO_4$$

The strengths of oxyacids increase as the formal charge on the central atom increases, provided those charges are determined consistently from Lewis structures in which the octet rule is obeyed. An alternative procedure is to count the number of oxygens attached to M but *not* bound to hydrogens. That is, for the general oxyacid formula $MO_n(OH)_m$, acid strength increases as n increases. Nitric acid, HNO_3, or $NO_2(OH)$, is weaker than perchloric acid, $HClO_4$, or $ClO_3(OH)$.

The **valence-shell electron-pair repulsion (VSEPR) method** predicts the **geometries (shapes)** of molecules and complex ions. The main VSEPR rule is that the atoms and lone pairs around a central atom will adopt a spatial arrangement that minimizes **electron-pair repulsion**. The **steric number (SN)** is the total number of atoms and lone pairs that are attached to a given central atom. Arrangements that minimize electron-pair repulsions are as follows: $SN = 2$, linear; $SN = 3$, trigonal planar; $SN = 4$, tetrahedral; $SN = 5$, trigonal bipyramidal; and $SN = 6$, octahedral (see Figure 11-2).

Self-Study Questions

1. What experimental evidence is there that O_2 has a bond order of 2, and that N_2 has a bond order of 3?
2. How are N_2, O_2, and F_2 represented by Lewis structures?
3. Which molecule has the greatest bond energy: O_2, F_2, or N_2?
4. Is the HI bond covalent, or ionic?
5. Is the NaCl bond covalent, or ionic?
6. What is meant by an open shell? Name a molecule that has one.
7. How is formal charge defined?
8. Which has a higher $C-C$ bond order, ethylene or acetylene?
9. What is meant by the statement that the ammonium ion and methane are isoelectronic? If they are, then why do they not have similar chemical properties?
10. What combination of boron and hydrogen would be isoelectronic with methane and the ammonium ion? Can you think of any good reasons why such a boron compound might or might not exist?
11. Is the addition compound of BF_3 and NH_3 isoelectronic with any organic compound?
12. How are Lewis acids and bases different from Brønsted acids and bases, and from Arrhenius acids and bases?

13. In the water solution of ammonia,

$$NH_3 + H_2O \rightarrow NH_4^+ + OH^-$$

which of the species on the left (or the species that can be derived from them) is the acid and which is the base, according to the definitions of Arrhenius, Brønsted, and Lewis?

14. Why can sulfur bond to six fluorines (in SF_6), whereas oxygen bonds to a maximum of two (in OF_2)?

15. What are resonance structures?

16. What is a resonance hybrid?

17. Why are three resonance structures required to describe the bonding in the nitrate ion, NO_3^-?

18. What is the difference between oxidation number and formal charge? How is each one calculated for an atom in a molecule?

19. Which acid is stronger, $Te(OH)_6$ or H_3PO_4? Why?

20. How does the acidity of an oxyacid depend on the formal charge on the central atom?

21. How does acidity depend on the number of oxygens (without hydrogens) attached to the central atom of an oxyacid?

22. Which is the stronger acid, perchloric ($HClO_4$) or nitric (HNO_3)? Why?

23. What is the VSEPR method? What are the rules used in applying it to predict molecular geometry?

24. What is the steric number (SN)? What is the steric number of CO_2? Of SF_4? Of PF_5? Of SF_6?

25. How will six electron pairs arrange themselves to minimize repulsions?

26. What is the geometry of SF_4? Of ClF_3?

27. How does the VSEPR method explain the geometry of XeF_6?

28. Why is the geometry of IF_7 an exception to the VSEPR prediction? What geometry is it predicted to have?

29. What is the geometry of $TeCl_6^{2-}$? Does this agree with the VSEPR prediction? Suggest a reason for the discrepancy.

30. Which are greater, repulsions between two lone pairs or repulsions between two bond pairs? How does this order of repulsions affect the bond angle in H_2O?

Problems

Lewis structures

1. Write Lewis structures for the atoms or ions Na, C, Si, Cl, Ca^{2+}, K^+, Ar, Cl^-, and S^{2-}.

2. What are the Lewis structures for the diatomic molecules or ions O_2, CO, Li_2^+, and CN^-?

3. What are the Lewis structures for Cl_2, N_2, NO, and HCl? Which of these molecules would you expect to be paramagnetic, and why?

4. Write Lewis structures for $BaCl_2$, PH_3, NH_4Cl, HOCl (no H—Cl bond), H_2O, H_2O_2, and NO_2^-.

5. Write Lewis structures for $CaCl_2$, SiH_4, CS_2 (no S—S bond), ClO_2 (chlorine dioxide), ClO_2^- (chlorite ion), and N_2O.

6. Write Lewis structures for BrO_4^-, SiH_4, PCl_4^+, CH_2Cl_2, and BF_4^-.

7. Xenon forms a number of interesting molecules and ions with fluorine and oxygen. Write a Lewis structure for each of the following: XeO_4, XeO_3, XeF_8^{2-}, XeF_6, XeF_4, XeF_2, and XeF^+. Show the placement of formal charges in the Lewis structures. Avoid structures with formal-charge separation, if possible. Using expected trends in effective atomic radii, predict whether the Xe—F bond length in XeF_4 will be longer or shorter than the I—F bond length in the related ion IF_4^-.

8. Nitrogen forms a trifluoride, NF_3, but NF_5 does not exist. For phosphorus both PF_3 and PF_5 are known. Write Lewis structures for NF_3, PF_3, and PF_5. Present possible explanations for the fact that PF_5 is stable, whereas NF_5 is not. In the light of your explanations, which of the following molecules would you expect to be nonexistent: OF_2, OF_4, OF_6, SF_2, SF_4, SF_6? Write Lewis structures and appropriate comments to support your case.

9. Write Lewis structures for CO_2 and SO_2. Are the C—O bonds primarily ionic, or primarily covalent? What shape do you expect for each molecule?

10. The acetylene molecule, HCCH, is linear. Write a Lewis structure for acetylene. Do you expect the C—C bond to be longer in C_2H_2 than in C_2H_4? Compare the energies of the C—C bonds in C_2H_4 and C_2H_2.

11. Iodine forms several oxyanions of the type IO_x^{n-}. Write Lewis structures for IO_3^-, IO_4^-, and IO_6^{5-}. Predict the relative I—O bond lengths in these oxyanions.

12. Write a Lewis structure for S_2. Do you expect the bond energy of S_2 to be larger, or smaller, than that of Cl_2?

13. Write Lewis structures for BF_3 and NO_3^-. Do these molecular species have anything in common?

14. Write Lewis structures for CN^- and CO. The C—O bond length is shorter in CO than it is in CO_2. Explain.

15. Write Lewis structures for the following molecules and ions. Show resonance structures if appropriate. Also indicate in each case whether there are formal charges on one or more atoms.
    a) FBr   j) SO_3 s) XeO_3
    b) S_2 k) CO_3^{2-} t) SO_2
    c) Cl_2 l) CF_4 u) SF_6
    d) P_2 m) $SiBr_4$ v) Na_2O
    e) NCO^- n) BF_4^- w) ClO_2
    f) CNO^- o) NCl_3 x) N_2F_2
    g) $BeCl_2$ p) PF_3 y) CsF
    h) CS_2 q) CH_3^- z) SrO
    i) BF_3 r) SF_2

16. For each of the following descriptions give the Lewis structure of a known chemical example:
    a) A diatomic molecule with one unpaired electron
    b) A triatomic molecule with two double bonds
    c) A diatomic molecule with formal-charge separation
    d) A diatomic molecule with partial ionic character
    e) An alkaline-earth oxide
    f) A molecule or ion with two equivalent resonance structures
    g) A molecule or ion with three equivalent resonance structures

17. Predict relative N—O bond lengths for NO_2^+, NO_2, and NO_2^-.

18. Borazine has the formula $(BHNH)_3$. The combination of a boron atom and a nitrogen atom is isoelectronic with two carbon atoms and has the same sum of atomic weights. Formulate an acceptable structure for borazine. Can you draw more than one equivalent structure?

Formal charge

19. For the following species, assign formal charges to each of the atoms and indicate the net charge on each species:

$$[:C\equiv N:] \qquad [:\ddot{S}=C=\ddot{N}:]$$
$$[:\dot{N}=N=\ddot{N}:] \qquad [:\ddot{S}=C=\ddot{O}:]$$

20. In diazomethane (H_2CNN), one nitrogen atom is attached directly to the carbon atom, and the second nitrogen atom is attached to the first. Draw Lewis structures for this molecule in which (a) the two N atoms are joined by a triple bond and (b) the middle N atom forms two double bonds to C and to N. When correctly drawn, each C and N atom should have eight electrons in its valence shell. What is the formal charge on each atom in each of the two structures?

Lewis acids and bases

21. For each of the following reactions, indicate which molecule or ion is the Lewis acid and which is the Lewis base:
a) $Ag^+ + 2NH_3 \rightarrow Ag(NH_3)_2^+$
b) $C_2H_3O_2^- + HF \rightarrow HC_2H_3O_2 + F^-$

22. Trimethyl phosphine, $(CH_3)_3P$, reacts with oxygen atoms to give trimethylphosphine oxide, $(CH_3)_3PO$. Is the trimethyl phosphine a Lewis acid or Lewis base in this reaction?

Resonance structures

23. When acetic acid is ionized in solution, the two carbon–oxygen bonds have the same length. Write two resonance structures for the acetate ion that account for this fact.

24. The electronic structure of the thiocyanate ion, NCS^-, can be represented as a hybrid of two resonance structures. Write these two structures and give the carbon–nitrogen and carbon–sulfur bond orders for each structure.

Oxyacids

25. Periodic acid, $(HO)_5IO$, is an oxyacid. Write a Lewis structure for periodic acid. Explain why the I—O bond length (1.78 Å) is shorter than the I—OH bond lengths (1.89 Å).

26. Predict whether $(HO)_5IO$ will be a stronger or weaker acid than $Te(OH)_6$.

VSEPR method

27. Predict the molecular shape of each of the following molecules:
a) CS_2 f) SiH_4 j) ClO_2
b) SO_3 g) SF_2 k) IF_5
c) ICl_3 h) SeF_6 l) OF_2
d) BF_3 i) PF_3 m) H_2Te
e) CBr_4

28. What is the geometry around the central atom in each of the following species: CH_4, BF_3, NF_3, ICl_4^-, H_2O?

29. What is the geometry around the central atom in each of the following molecules and ions: BrO_3^-, $CHCl_3$, ClO_4^-, H_2S?

30. Give examples of ions or molecules that have the following structures:

a) $[AB_3]^{2-}$ planar
b) $[AB_3]$ planar
c) $[AB_3]$ pyramidal
d) $[AB_3]^-$ pyramidal
e) $[AB_4]^-$ tetrahedral
f) $[AB_4]^{2-}$ tetrahedral
g) $[AB_2]$ linear
h) $[AB_2]$ bent

31. What is the geometry around each carbon atom in acetic acid,

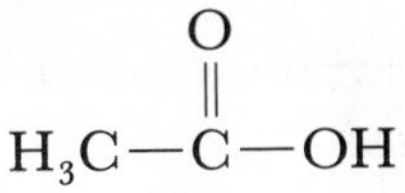

$$\underset{\displaystyle H_3C-\overset{\displaystyle \overset{O}{\parallel}}{C}-OH}{}$$

Which carbon–oxygen bond will be longer?

Suggested Reading

H. A. Bent, "Isoelectronic Systems," *J. Chem. Educ.* **43,** 170 (1966).

H. J. Eméleus and A. G. Sharpe, *Modern Aspects of Inorganic Chemistry,* Halsted Press, Div. of John Wiley & Sons, Inc., New York, 1973, 4th ed.

R. J. Gillespie, *Molecular Geometry,* Van Nostrand Reinhold, London, 1972.

B. M. Gimarc, "The Shapes of Simple Polyatomic Molecules and Ions," *J. Amer. Chem. Soc.* **92,** 266 (1970).

H. B. Gray, *Chemical Bonds,* W. A. Benjamin, Menlo Park, Calif., 1973.

J. F. Liebman, "Comments and a Warning about Isoelectronic Systems," *J. Chem. Educ.* **48,** 188 (1971).

K. M. Mackay and R. A. Mackay, *Introduction to Modern Inorganic Chemistry,* Intertext Books, London, 1972, 2nd ed.

L. Pauling, *The Chemical Bond,* Cornell University Press, Ithaca, N.Y., 1967, 3rd ed.

G. C. Pimentel and R. D. Spratley, *Understanding Chemistry,* Holden-Day, San Francisco, 1971.

G. E. Ryschkewitsch, *Chemical Bonding and the Geometry of Molecules,* Van Nostrand Reinhold, New York, 1972, 2nd ed.

R. T. Sanderson, *Chemical Bonds and Bond Energy,* Academic Press, New York, 1976, 2nd ed.

12

Diatomic Molecules

The language of chemistry is filled with Lewis line and dot structural formulas. If we know the Lewis structure for a molecule, we can say something about the stability, bond orders, bond energies, and bond lengths of that molecule. And, if we use the VSEPR method, we often can predict the geometry of the molecule. In this chapter we shall see that it is possible to go even further in formulating the electronic structures of molecules, by considering the actual shapes and energies of the valence orbitals that are involved in chemical bond formation. This more advanced method of analysis is known as the **molecular orbital theory**.

12-1 MOLECULAR ORBITALS

In Chapter 9, we used atomic orbitals to explain the properties of atoms. From the Schrödinger equation (Chapter 8) we obtained a set of wave functions, $\psi_{(x,y,z)}$, of such a nature that $|\psi_{(x,y,z)}|^2$ at any point is the electron probability density at that point. If the electron is in the quantum state described by n, l, and m, the probability of finding the electron within a small element of volume dv at (x, y, z) is

$$|\psi_{n,l,m(x,y,z)}|^2 \, dv$$

In building a picture of many-electron atoms, we envisioned the probability functions or orbitals as if they had a shadowy existence of their own and then we filled these orbitals by adding electrons like peas dropped into cups. To build a picture of a molecule, we need to find a set of **molecular orbitals** for a given arrangement of atoms and then fill the orbitals with the available electrons, no more than two to an orbital as before. But before we do this, let's look at what happens when two hydrogen atoms come together to make a molecule.

Bonding in the H₂ Molecule

If two hydrogen atoms are far apart, they have no effect on one another. But as they are brought closer together, they begin to exert an effect. The two nuclei, having the same positive charge, repel one another, and the two electron clouds also repel one another. However, most important of all is the attraction between the nucleus of one atom and the electron cloud of the other atom. As the atoms approach, the electron clouds are pulled toward the region between the nuclei (Figure 12-1d). The combination of two nuclei and two electrons is more stable (has lower energy) than two isolated nuclei, each with its electron. The closer the nuclei come, the more electron density is attracted between them, the lower the energy falls, and the more stable the assembly (which can now be called a molecule) becomes. However, there is a limit to this process. If the nuclei are too close, the repulsion between them begins to dominate. Beyond this limit, the nuclear repulsion is greater than the nucleus–electron cloud attraction.

The potential energy of an H₂ molecule as a function of the distance between the two H atoms is shown in Figure 12-2. There is an intermediate *equilibrium* distance at which the attractive and repulsive forces are balanced. Pull the atoms apart, and attractive forces pull them back again. Push them together, and repulsive forces push back. The two atoms act very much as if they were tied together by a spring. This condition of balance, or equilibrium distance, is what we normally mean when we speak of the **bond length** (Figures 12-1e and 12-2).

The attraction that makes the molecule stable is the attraction of the nuclei for the electron density concentrated between them. We can think of this concentration as an **overlap** of the $1s$ atomic orbitals. If, for convenience, we represent the atomic wave function ψ_{1s} simply by the symbol $1s$, then electron density in the atom is represented by $[1s]^2$. [Square brackets

Figure 12-1   Bonding in the H₂ molecule. (a) Probability density in the $1s$ atomic orbital of hydrogen. ▶ (b) The spherical surface that encloses 99% of the probability density. (c) Two hydrogen atoms sufficiently far apart will exert no effect on one another. (d) As the atoms are brought together, each electron cloud begins to respond to the attraction of the nucleus of the other atom. The electron clouds become distorted, and electron density increases in the region between the nuclei. (e) At closer proximity, repulsion between the nuclei becomes significant. The equilibrium bond distance in the H₂ molecule is the point of balance between this attraction and repulsion.

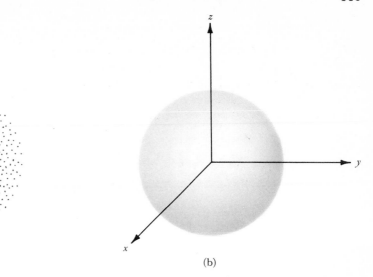

(a)

(b)

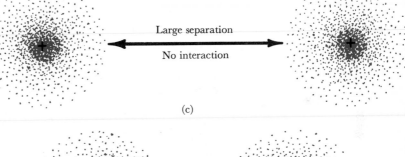

Large separation

No interaction

(c)

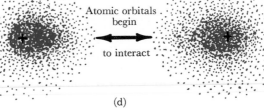

Atomic orbitals
begin

to interact

(d)

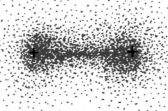

(e)

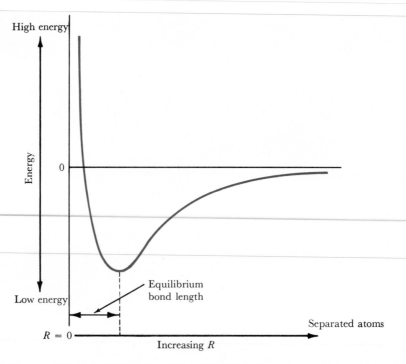

Figure 12-2   Potential-energy curve for the H_2 molecule. As atoms come closer together and the distance between nuclei (R) decreases, the potential energy decreases because of electron cloud—nucleus attraction, and then increases because of nucleus—nucleus repulsion. The equilibrium bond length is the bond length at the minimum value of the potential energy (where attraction and repulsion are balanced and the molecule is stable).

are used here to avoid confusion with the notation for electronic configuration in atoms: $(1s)^2(2s)^1$, and so on.] We can construct a molecular orbital (which we shall sometimes abbreviate MO) by adding the two atomic wave functions from atoms a and b to produce the molecular wave function $1s_a + 1s_b$. The electron probability density in such a molecular state is given by the square of the molecular wave function: $[1s_a + 1s_b]^2$. As you can see in Figure 12-3a, such a combination of atomic orbitals produces the pileup of electron density that we have been using to explain bonding. In the hydrogen molecule, this molecular orbital is filled with the two electrons having opposite spins (paired), and a single covalent bond is formed. This type of molecular orbital is a **bonding orbital**.

There is more than one way of combining two atomic wave functions, $1s_a$ and $1s_b$. What if they were subtracted instead of added? Expressed dif-

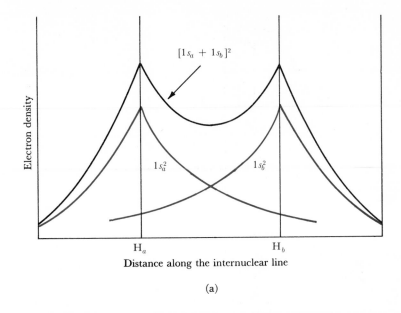

Distance along the internuclear line

(a)

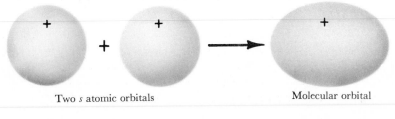

Two *s* atomic orbitals   Molecular orbital

(b)

Figure 12-3   Molecular orbitals are obtained by taking linear combinations (sums and differences) of atomic orbitals. If the hydrogen $1s$ wave function, ψ_{1s}, is represented simply by $1s$, the electron density is given by $[1s]^2$. Similarly, the electron density in the combined molecular orbital is given by $[1s_a + 1s_b]^2$, in which $1s_a$ and $1s_b$ are the wave functions of the individual atoms, H_a and H_b. (a) A plot of electron density in atomic orbitals (color) and the molecular orbital (black). (b) A conventional representation of the combination of two atomic orbitals to make a molecular orbital. Plus signs indicate signs of wave functions, not charge.

ferently, what if the atomic wave functions were combined with opposite sign, or were out of phase? The results are compared in Figure 12-4. The first drawing (Figure 12-4a) shows the addition of atomic wave functions to make the molecular orbital $[1s_a + 1s_b]^2$. The second drawing (Figure 12-4b) shows the subtraction of one from the other to make the molecular orbital $[1s_a - 1s_b]^2$. The wave function, $1s_a - 1s_b$, changes sign halfway from one

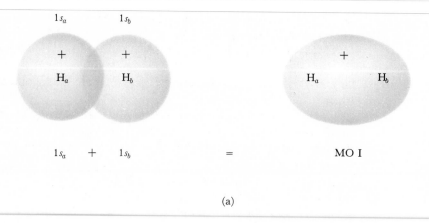

$$1s_a \quad + \quad 1s_b \quad\quad\quad = \quad\quad\quad \text{MO I}$$

(a)

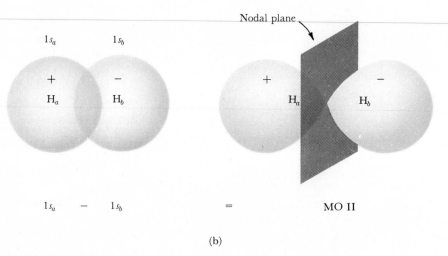

$$1s_a \quad - \quad 1s_b \quad\quad\quad = \quad\quad\quad \text{MO II}$$

(b)

Figure 12-4 Two atomic $1s$ orbitals give rise to two molecular orbitals (MO). (a) If the two atomic wave functions are added, or combined with the same sign, the resulting molecular orbital has high electron density between the nuclei. Electrons in such an orbital hold the molecule together, and it is called a *bonding* orbital. (b) If the two atomic functions are subtracted, or combined with opposite signs, the electron density in the molecular orbital is concentrated away from the internuclear region. There is zero probability of finding the electron in a nodal plane halfway between the nuclei. Electrons in such a molecular orbital pull the molecule apart, so it is called an *antibonding* orbital.

nucleus to the other, so its square falls to zero at the nodal plane. If electrons are in this molecular orbital in the molecule, there is *no* probability of finding them on a plane halfway between the nuclei. In fact, most of the electron density is concentrated outside the two nuclei. Rather than being pulled together, the nuclei are pulled apart. This type of molecular orbital is an **antibonding orbital**.

The potential energies of the bonding and antibonding orbitals are shown in Figure 12-5a. The closer the nuclei come in the antibonding state, the more they are held back by the drag of their electron clouds and the higher the energy of the molecule is. At every point, the energy of the molecule is greater than that of two isolated atoms. The energies of the two molecular orbitals at the equilibrium bonding distance are plotted in Figure 12-5b and compared with the energy of the electrons in $1s$ orbitals of isolated atoms.

In summary, the two atomic $1s$ orbitals can be combined in two different ways to produce two molecular orbitals, one bonding and one antibonding. The bonding orbital concentrates electron density between the nuclei; the antibonding orbital concentrates it outside the two nuclei and has no density at all on a plane halfway between them. Both these molecular orbitals are symmetrical with respect to rotation around the line joining the nuclei; that is, when the orbital is spun around this line, neither the appearance of the electron-density cloud nor the sign of the wave function composing it is altered. Orbitals with such symmetry are called **sigma** (σ) orbitals. The bonding orbital is given the superscript b and the superscript * is given to the antibonding orbital. (The types of molecular orbitals are described by the symbols sigma (σ), pi (π), delta (δ), . . . , by analogy with $s, p, d, . . .$ for atomic orbitals.)

The Pauli Buildup Process in Molecules

Now we can use an *aufbau* process to explain the occurrence or nonoccurrence of the molecules H_2^+, H_2, He_2^+, and He_2. The hydrogen molecule–ion, H_2^+, has two nuclei but only one electron. By Pauli's reasoning, this electron will be in the lowest-energy molecular orbital, which Figure 12-5b indicates is the bonding σ^b orbital. The H_2^+ molecule–ion should be weakly stable.

The hydrogen molecule, H_2, has two nuclei and two electrons. Both electrons can be accommodated in the σ^b orbital if their spins are paired, and a covalent electron-pair bond is created. The bond energy (the energy needed to pull the atoms apart) should be substantially larger than that of the hydrogen molecule–ion.

The helium molecule–ion, He_2^+, has two helium nuclei and three electrons. Although the energies of the helium orbitals, atomic and molecular, are different from those of hydrogen because of the different nuclear charge, the relative arrangement of atomic and molecular energy levels is similar. We can use Figure 12-5b for He as well as for H if we make the proper adjustments to the energy scale on the left.

The first two electrons in He_2^+ pair their spins and fill the σ^b bonding orbital. But what happens to the third electron? By the Pauli exclusion principle, it cannot occupy the σ^b state, so it must go into the next lowest

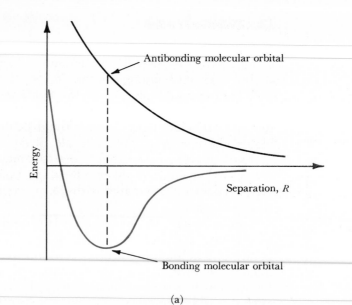

Antibonding molecular orbital

Energy

Separation, R

Bonding molecular orbital

(a)

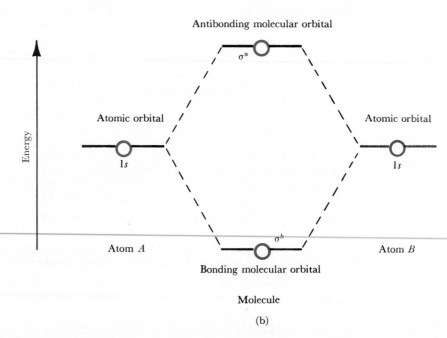

Antibonding molecular orbital

Energy

Atomic orbital

$1s$

σ^*

Atomic orbital

$1s$

Atom A

σ^b

Atom B

Bonding molecular orbital

Molecule

(b)

Figure 12-5   (a) The energy of a molecule with electrons in the bonding orbital falls to a minimum at the observed interatomic distance. The energy of a molecule with electrons in the anti-bonding orbital is always greater than the energy of completely separated atoms; it increases steadily as the atoms are brought closer together. (b) The two lowest molecular orbitals for the hydrogen molecule, and the atomic $1s$ orbitals from which they came. The symbol σ (sigma) indicates that the orbital is symmetrical around the line between the nuclei, and that the orbital could be spun around that line as an axis without changing the orbital's appearance. The superscript b indicates bonding character, and * indicates antibonding character.

Table 12-1

Comparison of Predicted and Observed Bonding in Simple Diatomic Molecules

| | Molecular-orbital-theory predictions | | | | Experimental observations | |
Molecule	Molecular orbital configuration	Bonding electrons	Antibonding electrons	Net bonding electrons	Bond length (Å)	Bond energy (kJ mole⁻¹)
H_2^+	$(\sigma^b)^1$	1	0	1	1.06	255
H_2	$(\sigma^b)^2$	2	0	2	0.74	432
He_2^+	$(\sigma^b)^2(\sigma^*)^1$	2	1	1	1.08	322
He_2	$(\sigma^b)^2(\sigma^*)^2$	2	2	0	none	none

energy level, which is the antibonding σ^* orbital. This third electron is pushed away from the region between the nuclei by the presence of the first two and is forced into the region outside the two nuclei. This electron is a disruptive influence; it pulls the nuclei apart. The molecule would be more stable if the third electron were not there. The electron effectively counteracts the contribution of one of the bonding electrons, thereby leaving a **net bonding action** of one electron, or half a covalent bond. The bond energy of He_2^+ should be less than that of H_2.

In He_2, the fourth electron also must go into the antibonding orbital. Now there are two bonding electrons and two antibonding electrons. The molecule is no more stable than the isolated atoms and falls apart. We should not expect to find an He_2 molecule.

Enough of theory for a moment. What actually happens? Table 12-1 lists the observed bond energies and bond lengths for H_2^+, H_2, and He_2^+; as predicted, He_2 does not exist. Moreover, the measured bond energies are consistent with the number of net bonding electrons given by molecular orbital theory. Bond lengths, too, are consistent. The more bonding electrons, the tighter the interaction and the shorter the bond length.

Example 1

What is the electronic configuration of H_2^-? How many bonding electrons are there? How many antibonding electrons? What is the net number of bonds? Will the bond length of H_2^- be longer or shorter than that of H_2? What molecule–ion is H_2^- isoelectronic with?

Solution

The molecule–ion H_2^- has three electrons. The ground state of H_2^- is therefore $(\sigma^b)^2(\sigma^*)^1$. There are two bonding electrons and one antibonding electron; there is a net of one bonding electron, or half a bond. The bond length of H_2^- will be greater than that of H_2, because there is less net bonding in H_2^- (one half-bond versus one full bond). Finally, H_2^- is isoelectronic with He_2^+ (each has three electrons).

Thus far, molecular orbital theory explains the data well. How can we extend this explanation to more complicated molecules? The process that we shall use to explain first the diatomic molecules of heavier atoms and then more complicated molecules can be summarized as follows:

1. Combine atomic orbitals in a suitable way to obtain a set of molecular orbitals. The total number of molecular orbitals obtained will always be equal to the number of atomic orbitals that we began with.

2. Decide the order of energies of these molecular orbitals.

3. Feed all the electrons in the molecule into these molecular orbitals. Start from the lowest and work up; place no more than two electrons in any one orbital.

4. Examine the filled bonding and antibonding orbitals to determine the net number of bonding electrons. (Some antibonding orbitals will have lower energy than other bonding orbitals and will be filled before these bonding orbitals are. The criterion for a bonding orbital is not that it have a low energy, but that it have a *minimum* in energy, as in Figure 12-5a, at some interatomic distance.) Two net bonding electrons correspond to what we have called a single bond in the Lewis model.

12-2   DIATOMIC MOLECULES WITH ONE TYPE OF ATOM

The O_2, N_2, and Cl_2 molecules, with only one type of atom, are called **homonuclear** molecules. In contrast, HCl, CO, and HI are **heteronuclear**. We want to extend the simple molecular orbital treatment of H_2 and He_2 to homonuclear diatomic molecules of elements in the second period of the periodic table. Some of these molecules, such as N_2, O_2, and F_2, are stable at STP. Others, such as C_2 and Li_2, are found only at high temperatures. Some do not exist at all. What are the predictions of molecular orbital theory?

The first step in the treatment is to construct molecular orbitals. The atomic orbitals available are the $2s$ and three $2p$ orbitals from each of the two atoms. Their energies are diagrammed in Figure 12-6, and the molecular orbitals that result from their combination are shown in Figure 12-7.

The two $2s$ atomic orbitals can be combined into a bonding σ_s^b and an antibonding σ_s^* orbital in the same manner as for the $1s$. If the line joining the nuclei is the z axis, then there are two kinds of $2p$ orbitals: the $2p_z$ orbital, which is parallel to the internuclear axis, and the $2p_x$ and $2p_y$ orbitals, which are perpendicular to it. The two $2p_z$ orbitals from the two atoms can be combined with the same signs in the internuclear region, thereby pro-

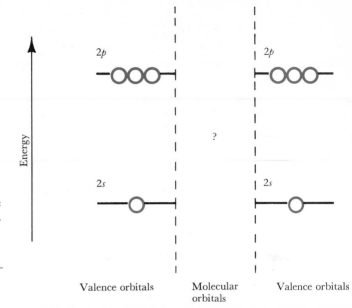

Figure 12-6   The 2s and 2p atomic orbital energies of elements in the second period of the periodic table, before combination into molecular orbitals.

ducing a concentration of electron density between the nuclei. They can also be combined with opposing signs, thus producing a deficiency of electron density between the nuclei. These molecular orbitals are labeled σ_z^b and σ_z^*, and are at the left and right center in Figure 12-7. They are still σ orbitals because they are rotationally symmetrical around the z axis.

The $2p_x$ orbitals on the two atoms can be combined as sums (lower left, Figure 12-7) or differences (lower right). The first combination, $[2p_x + 2p_x]^2$, produces a molecular orbital that looks like an exaggerated version of the original $2p$ orbitals. Maximum electron density occurs in two watermelon-shaped lobes, above and below a nodal plane that was also the nodal plane of the original atomic orbitals. The wave function itself has opposite signs in the two lobes. The other combination, $[2p_x - 2p_x]^2$, leads to a molecular orbital with a second nodal plane, and four lobes whose electron densities lie mostly outside the internuclear region (lower right, Figure 12-7). The two-lobed orbital is bonding; the four-lobed orbital is antibonding. They are called π **orbitals** and have a different kind of symmetry around the z axis. If either orbital is rotated 180° around the z axis, the electron-density cloud has the same appearance, but the signs of the wave function in the different lobes are reversed. The two orbitals are labeled π_x^b and π_x^*. A corresponding pair, π_y^b and π_y^*, results from the two $2p_y$ atomic orbitals.

From eight atomic orbitals we have obtained eight molecular orbitals, four of them bonding (σ_s^b, σ_z^b, π_x^b, π_y^b) and the other four antibonding (σ_s^*, σ_z^*, π_x^*, π_y^*). What is the order of energy of these orbitals?

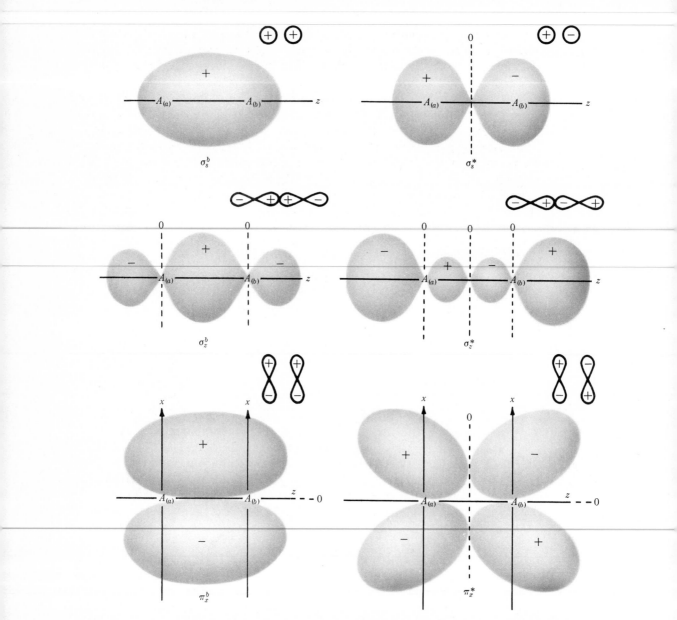

Figure 12-7 The six different kinds of molecular orbitals formed from the s, p_x, p_y, and p_z orbitals of two similar atoms in a diatomic molecule. The line drawn through the two nuclei is chosen as the z axis. The symbol π indicates that if the molecular orbital is rotated 180° around the axis the electron distribution is unchanged. The only effect is to reverse the signs of the parts of the wave function. Plus and minus signs represent only the signs on the wave function, and not electric charge. The atomic orbitals from which these are obtained are shown, with their appropriate signs, at the upper right of each molecular orbital. The atomic orbitals used are s (top row), p_z (middle row), and p_x (bottom row), which is equivalent to p_y. Bonding orbitals are in the left column; antibonding orbitals are in the right one. Dashed lines are nodal planes of zero electron density.

The orbitals derived from the s atomic orbitals will have lower energy than those from the p orbitals. Moreover, of two orbitals derived from the same atomic orbitals, the bonding orbital will lie lower than the antibonding orbital. Therefore, the first two most stable levels are the σ_s^b and σ_s^*. The most stable of the bonding orbitals obtained from $2p$ orbitals are π_x^b and π_y^b rather than σ_z^b, which is contrary to earlier ideas and to the diagrams in many older texts. This order of levels has been found from recent, careful, spectroscopic and magnetic studies of B_2 and N_2^+. It is reasonable, because electrons added to π_x^b are farther removed in space from those in filled σ_s^b and σ_s^* than they would be if they were in the σ_z^b orbital. (There is a crossover of energy levels at O, so in O_2 and F_2 the σ_z^b is more stable than π_x^b and π_y^b. This really does not matter in our discussion since all three levels are filled in O_2 and F_2 anyway.) The π_x^b and π_y^b orbitals have the same energy, and are said to be **degenerate** energy levels. Above these orbitals lies the σ_z^b, then the two antibonding π_x^* and π_y^*, and last of all the antibonding σ_z^*. The complete energy diagram of the molecular orbitals from $2s$ and $2p$ atomic orbitals appears in Figure 12-8.

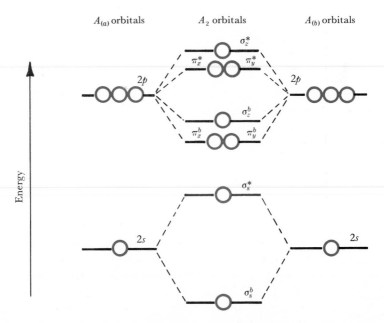

Figure 12-8   Energy levels for the molecular orbitals shown in Figure 12-7. Among the orbitals that come from either s or p atomic orbitals, bonding molecular orbitals are more stable than antibonding orbitals. The π_x^b and π_y^b orbitals are more stable than the σ_z^b because they permit the electrons to remain farther away from the filled σ_s^b orbital.

Paramagnetism and Unpaired Electrons

Substances whose molecules and ions have electrons with unpaired spins tend to be drawn into magnetic fields. The magnetic field aligns the spins and magnetizes the substance. Many substances, called **paramagnetic** substances, lose their magnetism when removed from the magnetic field. These materials have unpaired electrons, and the strength of the attraction by a magnetic field can be used to determine how many such unpaired electrons there are per mole of substance. If a molecule has no unpaired electrons, it is **diamagnetic** and is slightly repelled by a magnetic field because of the small opposing magnetic moments induced in it by the field. The number of unpaired electrons in a molecule of a substance can be determined with a magnetic balance as shown schematically in Figure 12-9. In the next section we shall see examples of both paramagnetic and diamagnetic molecules.

Figure 12-9

(a) The presence or absence of unpaired electron spins can be determined by a magnetic or Gouy balance. (b) A diamagnetic substance, with no unpaired electrons, is slightly repelled by a magnetic field. (c) A paramagnetic substance, with unpaired electron spins, is attracted into the magnetic field.

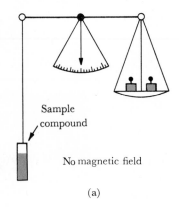

Sample compound

No magnetic field

(a)

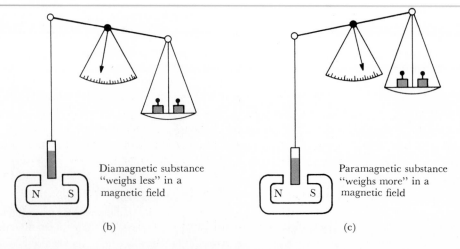

Diamagnetic substance "weighs less" in a magnetic field

(b)

Paramagnetic substance "weighs more" in a magnetic field

(c)

Buildup of Diatomic Molecules

We are now ready to feed electrons into molecular orbitals, two electrons to an orbital, and to build the diatomic molecules from Li_2 through Ne_2. There will always be four electrons from the lower $n = 1$ atomic orbitals. In the diatomic molecule, two of these inner-shell electrons will be in the σ_{1s}^b bonding molecular orbital and two will be in the σ_{1s}^* antibonding orbital. However, it makes no difference to the net bonding whether we think of them as in $1s$ atomic orbitals or in the molecular orbitals obtained from $1s$ orbitals. The bonding properties of the molecule arise only from the outer shell of $n = 2$ electrons, and we need to consider only the molecular orbitals derived from $2s$ and $2p$ atomic orbitals.

■ *Lithium.* The Li atom has one valence electron, so the Li_2 molecule has two potential bonding electrons. These are paired in the lowest available molecular orbital, σ_s^b. Therefore, the Li_2 molecule contains a single covalent bond. The bond length is longer than in H_2, 2.67 Å compared to 0.74 Å, because the larger $n = 2$ orbitals are involved, rather than the $n = 1$. For the same reason the bond is weaker: 110 kJ mole^{-1} rather than 432 kJ mole^{-1} as in H_2. The nuclei are farther apart, the electron cloud is spread over a greater volume, and the overall attractive forces are weaker.

■ *Beryllium.* There are four valence electrons available in the Be_2 molecule. Two are paired in the bonding σ_s^b molecular orbital, and two are paired in the antibonding σ_s^*. This configuration gives no net bonds, which is consistent with the absence of Be_2 from the family of stable second-row diatomic molecules.

■ *Boron.* The two additional valence electrons of B_2 go into the next lowest unfilled molecular orbitals, π_x^b and π_y^b. By Hund's rule, electron–electron repulsion ensures that one electron occupies each orbital rather than both being spin-paired in one. Whether or not the electrons are paired, the effect of two bonding electrons is a single covalent bond. The electronic configuration for B_2 is

$$KK(\sigma_s^b)^2(\sigma_s^*)^2(\pi_x^b)^1(\pi_y^b)^1$$

The symbol KK represents the four electrons in the inner $n = 1$ shells that have no effect on bonding. The experimental bond length in B_2, 1.59 Å, is less than that in Li_2, 2.67 Å. The bond energy is greater; it is 274 kJ mole^{-1} rather than 110 kJ mole^{-1}. Both effects arise from the greater positive charge on the B nucleus and the tightness with which the electrons are held. Perhaps the most satisfying test of the molecular orbital theory is the finding of two unpaired electrons in B_2 from magnetic measurements. This is a direct confirmation of the order of σ_z^b and π_x^b orbital energies in Figure 12-8; if the order were reversed, both electrons would be paired in σ_z^b and the molecule would have no unpaired spins. (As a matter of historical fact, the unpaired electrons in B_2 were not predicted in advance. The existence of the unpaired electrons compelled scientists to revise their original order of orbital energies to that of Figure 12-8.)

■ *Carbon.* The two additional electrons in carbon, C_2, complete the π_x^b and π_y^b molecular orbitals. There are four net bonding electrons, and hence two covalent bonds in Lewis' terminology. There should be no unpaired spins in the ground electronic state. True to predictions, the bond energy of C_2 is twice that of B_2 (603 kJ mole^{-1} and 274 kJ mole^{-1}), and the bond length is less (1.24 Å and 1.59 Å). Moreover, C_2 is not paramagnetic.

■ *Nitrogen.* With nitrogen, all the bonding orbitals in Figure 12-8 are filled. The N_2 molecule has the electronic configuration

$$KK(\sigma_s^b)^2(\sigma_s^*)^2(\pi_x^b)^2(\pi_y^b)^2(\sigma_z^b)^2 \qquad \text{or} \qquad KK(\sigma_s^b)^2(\sigma_s^*)^2(\pi_{x,y}^b)^4(\sigma_z^b)^2$$

There are six net bonding electrons, so the N_2 molecule contains a triple bond. Since there are no unpaired electrons, no paramagnetism is expected.

Nitrogen has the greatest bond energy and the shortest bond length of any element in the second period, 942 kJ mole^{-1} and 1.10 Å. The increase in bond energy with theoretical bond order (single, double, or triple bonds), shown in Figure 12-10, is remarkably constant. As predicted, N_2 is diamagnetic.

We now can interpret the Lewis structure of N_2,

$$\overset{\cdot\cdot}{N} \equiv \overset{\cdot\cdot}{N}$$

The three bonds involve the π_x^b, π_y^b, and σ_z^b orbitals. The two lone pairs correspond, at least formally, to the self-canceling pair of orbitals: $(\sigma_s^b)^2(\sigma_s^*)^2$.

We did not try to write a Lewis structure for C_2. Nothing in Lewis' covalence theory suggests that it should exist. Now, by analogy with N_2, we would write the C_2 molecule as

$$\overset{\cdot\cdot}{C} = \overset{\cdot\cdot}{C}$$

in which the bonds correspond to the filled π_x^b and π_y^b orbitals, and the lone pairs are as in N_2. But C_2, like BF_3, is electron-deficient; there are only six valence electrons around each carbon atom. We might expect that C_2 would accept electron pairs from a donor in the way that BF_3 accepts them from NH_3 to make the addition compound BF_3NH_3. But C_2 also has a lone pair and can play the role of donor as well. The C_2 molecule is found only at high temperatures. At lower temperatures, each atom in C_2 accepts electrons from one new C and donates electrons to another. The result is a network in which each C is covalently linked to at least three others (graphite) or, alternatively, to four (diamond).

■ *Oxygen.* In oxygen, the next 2 electrons must go into the two antibonding orbitals π_x^* and π_y^*, one in each by Hund's rule. Of the 12 valence

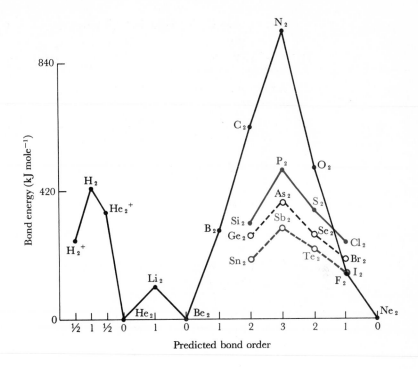

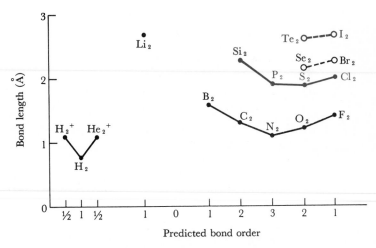

Figure 12-10

Plot of bond energies and bond lengths against predicted bond order for homonuclear diatomic molecules. Bond energies increase with increasing bond order, and bond lengths decrease.

electrons in O_2, a total of 8 are in bonding orbitals and 4 are in antibonding orbitals. There are 4 *net* bonding electrons, thus the molecule has a double bond. The two additional electrons, which go into antibonding orbitals, cancel the effects of two of the electrons in the orbitals that gave N_2 a triple bond. Both bond length and bond energy agree well with theory (Figure 12-10). The electronic configuration for O_2 is

$$KK(\sigma_s^b)^2(\sigma_s^*)^2(\sigma_z^b)^2(\pi_{x,y}^b)^4(\pi_{x,y}^*)^2$$

Notice that the relative order of energy levels σ_z^b and $\pi_{x,y}^b$ has changed, as was mentioned previously.

Molecular orbital theory explains why O_2 is paramagnetic, indicating two unpaired electrons, whereas the Lewis theory fails. The Lewis structure for O_2 has no unpaired electrons:

$$\ddot{\text{:O}}\!=\!\ddot{\text{O}}\text{:}$$

The only possible Lewis structures with a double bond and two unpaired electrons violate the symmetry of the molecule by making the oxygen atoms different, and make it appear that the unpaired electrons are both associated with a particular atom:

$$\cdot\dot{\ddot{\text{O}}}\!=\!\ddot{\text{O}}\text{:}\qquad\text{:}\ddot{\text{O}}\!=\!\dot{\ddot{\text{O}}}\cdot$$

You can partially redeem the Lewis structures by saying that these two structures are the two resonance structures for O_2, and that the true structure is unrepresentable but has the character of both resonance structures in equal amounts. But this treatment hardly seems worth the effort. It is easier to abandon Lewis structures and to think in molecular orbital terms.

■ *Fluorine.* In F_2, all of the molecular orbitals in Figure 12-8 are occupied except the highest one. The molecule has one net covalent bond from its two net bonding electrons, and the electronic structure is

$$KK(\sigma_s^b)^2(\sigma_s^*)^2(\sigma_z^b)^2(\pi_{x,y}^b)^4(\pi_{x,y}^*)^4$$

Bond energy and bond length are as expected for a single bond, and the F_2 molecule is diamagnetic.

■ *Neon.* The Ne_2 molecule would have all the molecular orbitals in the center of Figure 12-8 filled, and an equal number of bonding and antibonding electrons. There would be no net bonding electrons and no reason for the atoms to remain together. As predicted, there is no Ne_2 molecule.

■ *Later periods in the table.* Experimental data on several diatomic molecules and molecule–ions are given in Table 12-2. Some of these data have been plotted in Figure 12-10 as well. The trends for the nonmetals are regular and understandable in terms of larger orbitals (in which $n = 3$, 4, and 5) and weaker forces holding the electrons. The unexpected weakness of the F_2 bond is odd. The lone electron pairs in F_2 are considerably closer together than they are in the larger halogens, and we think that such close lone-pair repulsion may be at least part of the reason for the weak F_2 bond.

Example 2

Write the molecular orbital electronic structure of the molecule–ion O_2^-. What is the bond order, and how many unpaired electrons are there?

Table 12-2

Bond Properties of Some Homonuclear Diatomic Molecules and Ions

Molecule	Bond length (Å)	Bond energy (kJ mole^{-1})
Ag_2	—	161.9
As_2	2.288	382.0
Au_2	2.472	225.5
B_2	1.589	274.0
Bi_2	—	195.0
Br_2	2.2809	190.1
C_2	1.2425	602.5
Cl_2	1.988	239.2
Cl_2^+	1.8917	415.0
Cs_2	—	43.5
Cu_2	2.2195	197.9
F_2	1.417	138.9
Ge_2	—	272
H_2	0.74116	432.0
H_2^+	1.06	255.5
He_2^+	1.080	322.2
I_2	2.6666	148.7
K_2	3.923	49.4
Li_2	2.672	110.0
N_2	1.0976	941.7
N_2^+	1.116	842.2
Na_2	3.078	72.4
O_2	1.20741	493.5
O_2^+	1.1227	—
O_2^-	1.26	392.9
O_2^{2-}	1.49	—
P_2	1.8937	477.0
Pb_2	—	96.2
Rb_2	—	47.3
S_2	1.889	421.3
Sb_2	2.21	298.3
Se_2	2.1663	324.7
Si_2	2.246	314
Sn_2	—	192
Te_2	2.5574	260.7

Solution   The ion has 6 valence electrons from each oxygen atom plus 1 extra for the -1 charge, or 13 valence electrons. Filling orbitals in Figure 12-8 from the bottom, we find an electronic structure of

$$KK(\sigma_s^b)^2(\sigma_s^*)^2(\sigma_z^b)^2(\pi_{x,y}^b)^4(\pi_{x,y}^*)^3$$

There are three net bonding electrons, hence a bond order of $1\frac{1}{2}$. The molecule has one unpaired electron.

Example 3

With what neutral molecule is O_2^{2-} isoelectronic? Explain the fact that the O—O bond lengths increase in the following order:

$$O_2^+ < O_2 < O_2^- < O_2^{2-}$$

Solution   The molecular orbital configuration of O_2^{2-} is $KK(\sigma_s^b)^2(\sigma_s^*)^2(\sigma_z^b)^2(\pi_{x,y}^b)^4$ $(\pi_{x,y}^*)^4$; O_2^{2-} is isoelectronic with F_2 and has two net bonding electrons, corresponding to a single bond (bond order of 1). As electrons are removed from the antibonding $\pi_{x,y}^*$ level, the bond order increases as follows: O_2^{2-}, 1; O_2^-, $1\frac{1}{2}$; O_2, 2; O_2^+, $2\frac{1}{2}$. As the bond order increases, the bond length shortens. Thus we would predict the O—O bond length to increase in the order observed.

12-3 DIATOMIC MOLECULES WITH DIFFERENT ATOMS

Keeping the methods used for homonuclear diatomic molecules in mind, let's examine the molecular orbital treatment of heteronuclear molecules, those with two different atoms.

Hydrogen Fluoride and Potassium Fluoride

When we carry out the mathematical operations that are behind the expression, "combining two atomic orbitals to produce an antibonding and a bonding molecular orbital," we find that the two atomic orbitals should be reasonably close in energy. In the H_2 molecule, each of the two molecular orbitals has a 50% contribution from each of the two hydrogen $1s$ atomic orbitals. At the other extreme, if in a molecule of the type AB we combined an orbital from A having extremely high energy with one from B of quite low energy, we would find at the end of the mathematical analysis that the antibonding molecular orbital was almost pure A, and the bonding orbital was almost pure B. Then a pair of electrons in this "bonding" orbital would not be in a true covalent bonding orbital at all. It would be a lone electron pair in a B orbital. The interaction of these two atomic orbitals would be negligible. We shall see, for the HF molecule, what this means in terms of partial ionic character in a bond.

In HF, the energies of the hydrogen $1s$ and fluorine $1s$ atomic orbitals are so different that there is effectively no interaction between them. The fluorine $2s$ orbital has too low an energy as well. Only the $2p$ orbitals are close enough in energy to the hydrogen $1s$ to effect an appreciable combination into molecular orbitals. Moreover, the $2p_x$ and $2p_y$ orbitals have the wrong symmetry for combining with hydrogen $1s$, as Figure 12-11 shows. The total overlap of either p with the $1s$ is zero if proper account is taken of the signs of the wave functions. The molecular orbitals in HF are obtained by a combination of hydrogen $1s$ and fluorine $2p_z$ atomic orbitals. This combination produces two orbitals with σ symmetry, one bonding (σ^b) and one antibonding (σ^*).

The energy levels for HF appear in Figure 12-12. The π_x and π_y orbitals are essentially lone-pair orbitals on fluorine and might just as well be designated $2p_x$ and $2p_y$. A third lone pair occupies the fluorine $2s$ orbital. There are eight valence electrons in HF; seven from F and one from H. These electrons fill all the HF orbitals except the highest antibonding σ^*. In the HF molecule this assignment produces one covalent bond and three lone pairs on F. The Lewis structure is

$$\mathrm{H} - \ddot{\underset{..}{\mathrm{F}}}:$$

and it is accurate.

The energies of the $1s$ atomic orbital at the left of Figure 12-12 and the $2p$ orbitals at the right are obtainable from the first ionization energies of H and F. If 1310 kJ mole^{-1} are required to remove the electron from H, the energy of the electron before removal is -1310 kJ mole^{-1}. Similarly, the first ionization energy of F is 1682 kJ mole^{-1}, so the energy of the $2p$ levels is -1682 kJ mole^{-1}. The two atomic levels differ by 372 kJ mole^{-1}. The σ^b molecular orbital is closer in energy to the fluorine $2p$ than to the hydrogen $1s$. This means that there will be more of a fluorine $2p$ character to the σ^b orbital. The covalent bond is not perfectly symmetrical; there is a small inequality in charge distribution and a partial ionic character to the bond. Electrons in the σ^b orbital will have a greater probability of being near the F atom. A small charge displacement is represented by a lowercase delta, δ. We can show the partial ionic character of the HF molecule by $\mathrm{H}^{\delta+} \mathrm{F}^{\delta-}$.

Imagine what would happen to the H—F bond if the hydrogen $1s$ atomic orbital energy were to fall slowly. The energy separations between the σ^b molecular orbital and the two atomic orbitals from which it came would equalize; σ^b would assume an equal contribution from each. The charge displacement would diminish, and the bond would approach the perfectly symmetrical covalent bond of F_2 or H_2. This is more nearly the situation for HCl, in which the first ionization energies of H and Cl are close: 1310 kJ mole^{-1} and 1255 kJ mole^{-1}. In HCl, HBr, and HI the bonds are much more covalent and the charge separation in the molecule much less than in HF.

In the HCl example, the numbers just given make it appear that the electrons would be more attracted to H than to Cl since the first ionization energy of H (1310 kJ mole^{-1}) is larger than that of Cl (1255 kJ mole^{-1}). But ionization energies are only part of the story; relative electron affinities

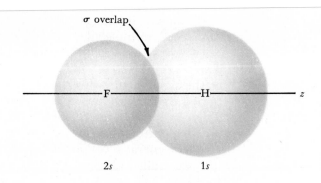

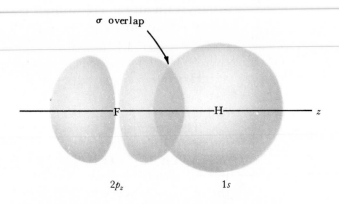

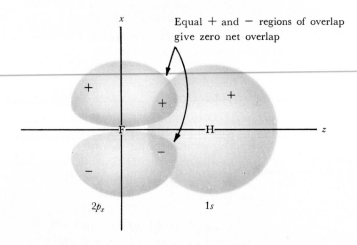

Figure 12-11   Overlap of the hydrogen $1s$ atomic orbital with the valence orbitals of fluorine. The net overlap of a $2p_x$ or $2p_y$ orbital of fluorine with the hydrogen $1s$ orbital is zero, and these two p orbitals cannot be used in forming molecular orbitals.

must also be considered. The electron affinity of Cl (356 kJ mole^{-1}) is so much larger than that of H (67 kJ mole^{-1}) that the prediction based on ionization energies alone is reversed. The combination of ionization energy and electron affinity—the electronegativity of each atom—is the true deciding factor in determining charge distribution in the bond.

Now imagine instead that the H orbital at the left of Figure 12-12 is raised from its present position of 1310 kJ mole^{-1} toward an eventual limit of zero energy. As this happens, the σ^b molecular orbital becomes even more like the original $2p_z$. The limit of this trend is for the hydrogen $1s$ orbital to go to zero energy (which means complete dissociation of the electron) and for the σ^b containing the two bonding electrons to become the $2p_z$ of F

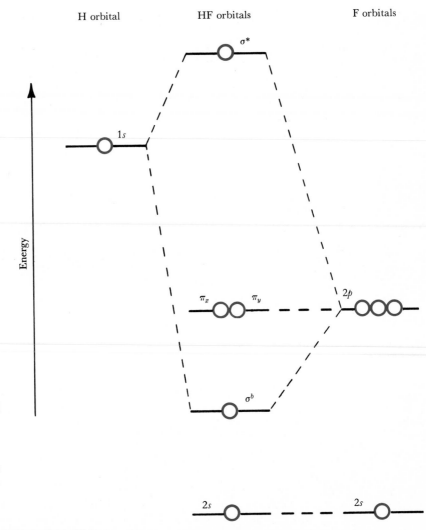

Figure 12-12

Relative energies of atomic and molecular orbitals in HF. The energy of an electron in the hydrogen atom $1s$ orbital is -1310 kJ mole^{-1} (the first ionization energy of H is $+1310$ kJ mole^{-1}), and the energy in the $2p$ orbitals in F is -1682 kJ mole^{-1} (first ionization energy of $+1682$ kJ mole^{-1}).

(which means the formation of a F^- anion). This behavior is approached in KF. Here the first ionization energy of K is only 418 kJ mole^{-1}, and the energy of the K $3s$ level is -418 kJ mole^{-1}.

Dipole Moments

A heteronuclear diatomic molecule such as HF possesses an **electric dipole moment** caused by the separation of positive and negative charges. If a positive and a negative charge of magnitude q are separated by a distance r, the dipole moment, μ (the Greek letter mu), is

$$\mu = qr$$

The measured experimental dipole moment of HF is 1.82 debye units. [One debye (D) unit is 10^{-20} esu (or electrostatic unit) m. Since the charge on an electron in the electrostatic system of units is 4.80×10^{-10} esu, two unit charges of opposite sign separated by 1 Å will have a dipole moment of $4.80 \times 10^{-10} \times 10^{-10}$ esu m $= 4.80 \times 10^{-20}$ esu m $= 4.80$ D.] If H really had a full $+1$ charge and F had a full -1 charge, and if these charges were separated by the true bond length of 0.92 Å, the dipole moment of HF would be calculated as 4.4 D, from the formula just given. The separated partial charges on the HF molecule are given by the ratio of true or experimental dipole moment (μ_e) to calculated dipole moment (μ_c): $1.82/4.4 = 0.41$. We say that the HF bond has 41% ionic character.

The **percent ionic characters** of several other diatomic molecules are listed in Table 12-3. The HCl bond has only 17% ionic character, and the KF bond is 83% ionic by the dipole-moment criterion.

This treatment of HF indicates that no bond is purely ionic or purely covalent. These are not two different mechanisms of bonding but only two extreme examples of a continuous range of polarity. What matters in molecular orbital theory is the degree of match or mismatch of energy levels from the two atoms. This match or mismatch is related to the electronegativities of the atoms.

Example 4

The dipole moment of the diatomic KBr molecule is 10.41 D, and that of KCl is 10.27 D. Which bond has greater ionic character?

Solution

The bond length of KBr is 2.82 Å; that of KCl is 2.67 Å (Table 12-4). The ionic structures have the following calculated dipole moments, μ_c:

$$\mu_c \text{ (KBr)} = 4.80(2.82) = 13.5 \text{ D}$$
$$\mu_c \text{ (KCl)} = 4.80(2.67) = 12.8 \text{ D}$$

Using the formula $\mu_e/\mu_c \times 100\%$, we find the percent ionic character in each case to be

Table 12-3

Percent Ionic Character of Bonds, from Dipole Moments

Molecule	Bond length, r (Å)	Calculated dipole moment, μ_c	Experimental dipole moment, μ_e	Percent ionic character, $\mu_e/\mu_c \times 100$
H_2	0.74	3.6	0.00	0
F_2	1.42	6.82	0.00	0
HI	1.60	7.68	0.38	5
BrCl	2.14	10.3	0.57	5
ICl	2.32	11.1	0.65	6
FCl	1.63	7.82	0.88	11
HBr	1.41	6.77	0.79	12
FBr	1.76	8.45	1.29	15
HCl	1.27	6.10	1.07	17
HF	0.92	4.4	1.82	41
KI	3.05	14.6	9.24	63
LiH	1.60	7.68	5.88	77
KF	2.17	10.4	8.60	83

KBr: $10.41/13.5 \times 100\% = 77\%$

KCl: $10.27/12.8 \times 100\% = 80\%$

The KCl bond is more ionic than the KBr bond, which is consistent with the fact that Cl is more electronegative than Br.

A General *AB* Type Diatomic Molecule

The treatment of heteronuclear diatomic molecules of the general *AB* type is similar to that of homonuclear molecules. The energy-level diagram is similar, except that the atomic levels of the more electronegative atom are lower than those of the more electropositive atom (Figure 12-13). Therefore, bonding orbitals have more of the character of the electronegative atom, and antibonding orbitals, of the electropositive atom. The molecular orbitals are skewed toward one or the other atom, as shown in Figure 12-14.

Filling of orbitals with electrons occurs exactly as before. The BN molecule is isoelectronic with C_2, except that the $\pi_{x,y}^b$ and σ_z^b levels are so close together that the energy required to promote one electron to the σ_z^b orbital can be provided by the energy gained in unpairing two electrons. The electronic configuration of BN is

$$KK(\sigma_s^b)^2(\sigma_s^*)^2(\pi_{x,y}^b)^3(\sigma_3^b)^1$$

The BN bond energy of 385 kJ mole^{-1} is suspiciously low in comparison

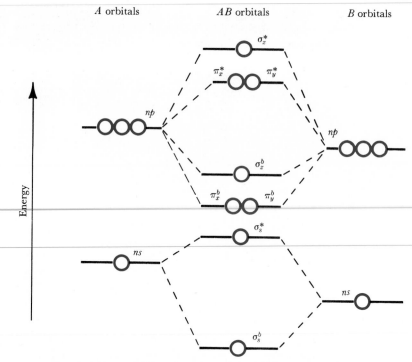

A orbitals   AB orbitals   B orbitals

Figure 12-13   Energy levels for a general *AB* molecule, in which *B* is more electronegative than *A*. Compare with the homonuclear *AA* molecular levels in Figure 12-8. As atom *B* becomes more electronegative, its atomic energy levels decrease in energy and the bonding molecular orbitals assume more *B*-atom character.

with 603 kJ mole^{-1} for C_2. Further experimental work is necessary to verify the BN bond energy.

The species BO, CN, and CO$^+$ have 9 valence electrons. From molecular orbital theory we can predict a bond order of $2\frac{1}{2}$ for them. The ions and molecules NO$^+$, CO, and CN$^-$ have 10 valence electrons and are isoelectronic with N_2. Nitric oxide, NO, has 11 electrons and is one of the few common gases with an odd number of electrons. The electronic configuration of NO is

$$KK(\sigma_s^b)^2(\sigma_s^*)^2(\pi_{x,y}^b)^4(\sigma_z^b)^2(\pi_{x,y}^*)^1$$

It has a bond order of $2\frac{1}{2}$, and both its bond energy and its bond length are intermediate between those of N_2 and O_2. Data on other *AB* diatomic molecules are in Table 12-4.

Example 5

What is the electronic configuration of CF? What is its bond order? Does it have unpaired electrons?

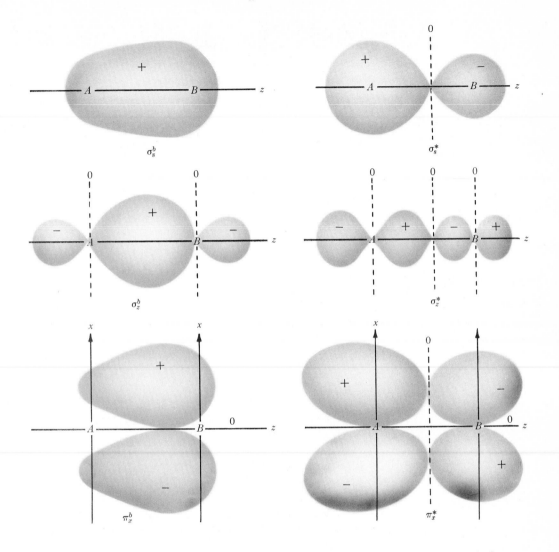

Figure 12-14   Molecular orbitals in an AB molecule, in which B is more electronegative than A. Compare with Figure 12-7. Note the increased electron probability near the more electronegative atom in bonding orbitals, and the opposite trend in antibonding orbitals.

Solution   CF has 11 valence electrons (it is isoelectronic with NO). Therefore its electronic configuration is $KK(\sigma_s^b)^2(\sigma_s^*)^2(\pi_{x,y}^b)^4(\sigma_z^b)^2(\pi_{x,y}^*)^1$, with a bond order of $2\frac{1}{2}$. It has one unpaired electron.

Table 12-4

Bond Properties of Some Heteronuclear Diatomic Molecules and Ions

Molecule	Bond length (Å)	Bond energy (kJ mole^{-1})
AsN	1.620	481
AsO	1.623	473
BF	1.262	548
BH	1.2325	293
BN	1.281	385
BO	1.2043	800.0
BaO	1.940	545.6
BeF	1.3614	568.6
BeH	1.297	222
BeO	1.3308	443.9
BrCl	2.138	218
BrF	1.7555	230
CF	1.2718	443
CH	1.1202	335
CN	1.1719	787
CN$^+$	1.1727	—
CN$^-$	1.14	—
CO	1.1283	1070.3
CO$^+$	1.1152	805.0
CP	1.5583	510.9
CS	1.5349	726.3
CSe	1.66	577
CaO	1.822	382.1
ClF	1.6281	252.3
CsBr	3.072	382.8
CsCl	2.9062	425.5
CsF	2.345	510
CsH	2.494	176
CsI	3.315	315.5
GeO	1.650	657
HBr	1.4145	361.9
HBr$^+$	1.459	—
HCl	1.2744	427.6
HCl$^+$	1.3153	453.1
HF	0.91680	565.3
HI	1.6090	295.0
HS	1.3503	340.6
IBr	2.485	175.3
ICl	2.32070	207.6
IF	1.908	191.2
KBr	2.8207	382.4
KCl	2.6666	421.8
KF	2.1715	497.5

Table 12-4 (*Continued*)

Bond Properties of Some Heteronuclear Diatomic Molecules and Ions

Molecule	Bond length (Å)	Bond energy (kJ mole^{-1})
KH	2.244	180
KI	3.0478	323.0
LiBr	2.1704	423
LiCl	2.018	473.8
LiF	1.5639	568.2
LiH	1.5953	234
LiI	2.3919	339
MgO	1.749	339
NH	1.045	356
NH+	1.081	—
NO	1.1508	678
NO+	1.0619	—
NP	1.4910	—
NS	1.495	481
NS+	1.25	—
NaBr	2.502	368
NaCl	2.3606	412
NaF	1.9260	476.6
NaH	1.8873	197
NaI	2.7115	289
NaK	—	59.8
NaRb	—	57.7
OH	0.9706	424.7
OH+	1.0289	422.6
PH	1.4328	—
PN	1.4869	730.5
PO	1.473	519
RbBr	2.9448	380.3
RbCl	2.7868	430.1
RbF	2.2704	500.0
RbH	2.367	163
RbI	3.1769	325.1
SO	1.4810	517.4
SbO	1.848	310
SiF	1.6008	541.8
SiH	1.5201	310
SiN	1.575	435
SiO	1.5097	764.8
SiS	1.929	619
SnH	1.785	310
SnO	1.838	529.3
SnS	2.209	461.5
SrO	1.9199	415.0

Summary

A **molecular orbital** is a wave function for an electron in a molecule. It is usually an additive or subtractive combination of atomic orbitals on the atomic centers that are bound together in the molecule. Orbitals on different atoms often overlap, particularly in the region between the nuclei. Such overlap is called **orbital overlap**

The wave function of a **bonding orbital** is additive in the overlap region. As a result, electron density is concentrated between the nuclei, and a bonding orbital is of lower energy than its atomic-orbital components. The energy of an electron in a bonding orbital goes through a minimum as the internuclear separation decreases. Placement of an electron in a bonding orbital enhances bond strength and stabilizes a molecule.

The wave function of an **antibonding orbital** is subtractive in the overlap region; such an orbital does *not* concentrate electron density between the nuclei. Rather, it forces electron density outside this region. An antibonding orbital has a **nodal plane**, or a plane at which the molecular wave function (and the electron density) is zero. This plane lies between the nuclei and is perpendicular to the internuclear axis. The energy of an electron in an antibonding orbital is greater than that of an electron in either of its atomic-orbital components; this energy does not go through a minimum, but continuously increases as the internuclear separation decreases. Placement of an electron in an antibonding orbital decreases bond strength and decreases the stability of a molecule.

Wave functions that are completely symmetric about the internuclear axis are called **sigma** (σ) orbitals. Those whose wave functions change sign on a 180° rotation about the internuclear axis are called **pi** (π) orbitals. Both σ and π orbitals can be bonding (σ^b or π^b), or antibonding (σ^* or π^*), as shown:

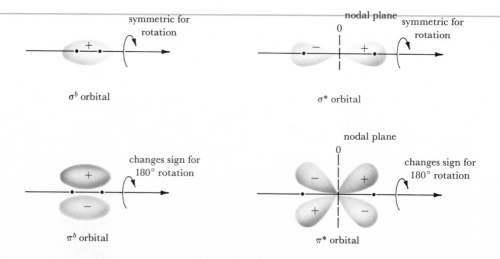

symmetric for rotation

σ^b orbital

nodal plane   symmetric for rotation

σ^* orbital

changes sign for 180° rotation

π^b orbital

nodal plane   changes sign for 180° rotation

π^* orbital

A **homonuclear** diatomic molecule is one in which the two atoms are the same; a **heteronuclear** diatomic molecule contains two different atoms. The molecular orbitals of a homonuclear diatomic molecule with ns and np valence orbitals increase in energy according to $\sigma_s^b < \sigma_s^* < \pi_{x,y}^b < \sigma_z^b < \pi_{x,y}^* < \sigma_z^*$ (except for the molecules O_2 and F_2, for which σ_z^b is below $\pi_{x,y}^b$). The π_x^b and π_y^b orbitals have the same energy; orbitals with the same energy are said to be **degenerate**. The π_x^b and π_y^b orbitals correspond to a **degenerate energy level** that is denoted $\pi_{x,y}^b$; $\pi_{x,y}^*$ also is a degenerate energy level composed of π_x^* and π_y^* orbitals.

To build up the electronic structures of homonuclear diatomic molecules, valence electrons are fed into the molecular orbitals σ_s^b through σ_z^*, in order of increasing energy. Thus Li_2 has the molecular orbital configuration $KK(\sigma_s^b)^2$ (one σ bond) and the N_2 molecule has the configuration $KK(\sigma_s^b)^2(\sigma_s^*)^2(\pi_{x,y}^b)^4(\sigma_z^b)^2$, with three net bonds (one σ, two π). The net number of bonding electrons divided by 2 gives the conventional bond order: Li_2 has a bond order of 1, N_2 has a bond order of 3. As the bond order in a given homonuclear diatomic system increases, bond length decreases and bond energy increases.

Molecules with unpaired electrons (such as B_2 and O_2) are **paramagnetic**. Molecules with all their electrons paired (such as Li_2 and N_2) are **diamagnetic**.

In a heteronuclear diatomic molecule, AB, where B is more electronegative than A, the bonding molecular orbitals have more B atomic-orbital character, and the antibonding molecular orbitals are more closely associated with A. Where the difference in electronegativity is extremely large, such as in KF, the valence electrons become localized on the more electronegative atom (F in this case) and the concept of a covalent bonding orbital loses significance. In such a situation an ionic structure K^+F^- is appropriate. Most heteronuclear diatomic molecules are somewhere between ion pairs and covalently bound atoms; that is, they possess some degree of ionic character, $A^{\delta+}B^{\delta-}$.

A diatomic molecule with charge separation possesses an **electric dipole moment** (μ):

$$\mu = qr$$

If q is taken as 4.80 debyes $\overset{\circ}{A}^{-1}$, and r (distance between positive and negative charges) is taken in $\overset{\circ}{A}$, then the dipole moment, μ, is in debye units (10^{-20} esu m = 1 debye, D). That is, the dipole moment of a molecule in which unit positive and negative charges are separated by 1 $\overset{\circ}{A}$ is 4.80 D.

The **percent ionic character** of a molecule may be evaluated by dividing the experimentally observed dipole moment (μ_e) by the calculated dipole moment (μ_c) for an ionic structure:

Percent ionic character $= \mu_e/\mu_c \times 100\%$

For example, HF has $\mu_e = 1.82$ D, and $\mu_c = 4.4$ D (4.80 D $\overset{\circ}{A}^{-1} \times 0.92 \overset{\circ}{A} = 4.4$ D); therefore HF has $1.82/4.4 \times 100\% = 41\%$ ionic character.

Self-Study Questions

1. Why does the potential-energy curve in Figure 12-2 have a minimum?

2. Figure 12-3a shows profiles through the two atomic electron-density clouds, $[1s_a]^2$ and $[1s_b]^2$, and the profile through the bonding molecular orbital formed from $[1s_a + 1s_b]^2$. What would profiles through the two atomic wave functions, $1s_a$ and $1s_b$, look like? Draw the profile through the antibonding molecular wave function, $1s_a - 1s_b$, and through the resulting electron-density distribution in the antibonding molecular orbital, $[1s_a - 1s_b]^2$.

3. What do the symmetry symbols for molecular orbitals, σ and π, indicate?

4. Why does molecular orbital theory predict that He_2 should not exist, whereas He_2^+ should, under the right conditions?

5. What is the equivalent of a Lewis covalent bond in molecular orbital theory?

6. What is wrong with the general statement that bonding orbitals have low energies and antibonding orbitals have high energies? What feature of a bonding orbital makes it a bonding orbital?

7. What is the order of increasing energy among the molecular orbitals formed from the $2s$ and $2p$ atomic orbitals in diatomic molecules? What experimental evidence is there that this order is correct?

8. What are homonuclear and heteronuclear diatomic molecules?

9. What do the small positive and negative signs in the lobes of the orbitals in Figure 12-7 signify? What do the small line drawings at the upper right of each orbital drawing represent?

10. Why is the bond energy in Li_2 less than in H_2?

11. Each boron atom has three valence electrons. Why is the B_2 molecule not held together by a triple bond as N_2 is?

12. What is the ground-state configuration of the C_2 molecule?

13. Why is O_2 paramagnetic, whereas N_2 is not? Support your argument with the ground-state electronic configurations of the two molecules.

14. Describe the electronic structure of the diatomic molecule O_2 by using molecular orbitals. Is O_2 paramagnetic, from predictions based on molecular orbital theory, and does this prediction agree with the possible predictions based on the Lewis structure? Which molecule would you expect to have the greater bond energy, O_2 or NO?

15. The molecular orbital description of the B_2 molecule can be written $KK(\sigma_s^b)^2(\sigma_s^*)^2(\pi_{x,y}^b)^2$. What does the KK symbol mean? In the same notation, what is the molecular orbital description of F_2?

16. What are the molecular orbital descriptions of Li_2 and Be_2? Which molecule should not exist, and why?

17. What happens to the molecular orbitals when two atomic orbitals with radically different energies on two different atoms are combined? If we combine two atomic orbitals with drastically different energies, and

place two electrons in the lower of the two resulting molecular orbitals, what will be the nature of the electrons? Will they be bonding electrons?

18. Why can't the $2p_x$ and $1s$ atomic orbitals in Figure 12-11 be combined to produce two molecular orbitals?

19. In Figure 12-12, is the σ^b molecular orbital more like the hydrogen $1s$ or the fluorine $2p$ atomic orbital? Which atomic orbital makes more of a contribution to the σ^* molecular orbital?

20. What experimental data give us the relative positions of the hydrogen $1s$ and fluorine $2p$ atomic orbitals in Figure 12-12?

21. If the hydrogen $1s$ and fluorine $2p$ atomic orbitals were by some process made equal in energy, what effect would this have on the character of the bond in HF?

22. What is a purely ionic bond, in the language of molecular orbital theory and Figure 12-12?

23. How can dipole moments give us an estimate of the ionic character of a bond? How ionic is the HF bond?

24. Why do the orbital drawings for heteronuclear diatomic molecules in Figure 12-14 differ from those for homonuclear diatomic molecules in Figure 12-7?

25. Which of the following molecules are paramagnetic: CO, Cl_2, NO, N_2?

Problems

MO configuration and bond properties

1. What is the molecular orbital configuration of the diatomic ion Li_2^+?

2. What is the bond order of Li_2^+? Is the Li_2^+ ion paramagnetic?

3. Describe the electronic structure of the diatomic molecule NO by using molecular orbitals. From the molecular orbital diagram, would you expect the molecule to be paramagnetic? Does your answer agree with the predictions that you can make from the Lewis structure? Would you predict the bond energy of NO to be greater than, equal to, or less than that of the ion NO^+?

4. Using molecular orbitals, describe the electronic structures of the peroxide ion, O_2^{2-}, and the superoxide ion, O_2^-. Are these ions diamagnetic, or paramagnetic? How does the strength of the oxygen–oxygen bond in each of these ions compare to that in O_2?

5. What is the molecular orbital configuration of the P_2 molecule? How many unpaired electrons does it have? What is its bond order? Would you expect its bond energy to be greater, or less, than that of S_2? Why? Which should have the longer bond length, P_2 or S_2?

6. Write the molecular orbital configurations for the ground states of NF, NF^+, and NF^-. Which of these species are paramagnetic? How many unpaired electrons does each have? Predict the bond orders and relative bond lengths for each species.

7. Which would you expect to have the greater bond energy, NF or NF^+?

8. What is the bond order of Cl_2^+? Should its bond energy be larger, or smaller, than that of Cl_2? Why? Is Cl_2^+ paramagnetic?

9. What is the molecular orbital configuration of Br_2^-? Is Br_2^- paramagnetic?

10. Write the molecular orbital configuration of the SO molecule. How many unpaired electrons does it have? What is its bond order?

11. How many unpaired electrons does PO have? Would you expect the bond energy of PO to be greater, or smaller, than that of SO? Why? Should the bond length of PO be greater, or smaller, than that of SO?

12. Why is the bond energy of B_2 greater than that of F_2? Can you explain the fact that the F_2 bond distance is *shorter* than the B_2 bond distance?

13. The ground state of H_2 has the molecular orbital configuration $(\sigma^b)^2$. In addition to the ground state there are excited states possessing the following configurations:

a) σ^b σ^* b) σ^b σ^*

Predict which of these states is higher in energy and which is lower. Explain your reasoning. Would you expect the lower excited state of H_2 to be paramagnetic or diamagnetic?

14. The neutral, diatomic OH molecule has been observed in outer space. Formulate its electronic structure in terms of molecular orbital theory using only the $2p$ oxygen and the $1s$ hydrogen orbitals. What type of molecular orbital contains the unpaired electron? Is this orbital associated with both oxygen and hydrogen atoms, or is it localized on a single atom? If it is localized, which atom is it on?

15. What type of molecular orbital contains the unpaired electron in HF^+? Is it localized on H, or on F? What is the molecular configuration of HF^-? What is the bond order of HF^-? Of HF^+? Which should have the larger bond energy, HF^+ or HF^-? Should HF^- be paramagnetic?

Dipole moments and ionic character

16. Which of the following molecules will be expected to have a dipole moment: H_2, O_2, HF, HI, I_2? Assuming that the molecules with dipole moments are completely ionic, calculate their dipole moments. (Needed data are in Tables 12-2 and 12-4.)

17. The measured dipole moment of carbon monoxide (CO) is 0.112 D. What is the partial ionic character of the C—O bond?

18. Which substance has bonds of greater ionic character, KI or BaO? What is the percent ionic character of each? (Dipole-moment data are available in the *CRC Handbook of Chemistry and Physics*.)

19. Compare the percent ionic character of the bonds in HCl, CsCl, and TlCl. Can you interpret your results in terms of the periodic table? (The Tl—Cl bond length is 3.2 Å. Measured dipole moment for TlCl molecules in the gas phase is 4.44 D; for CsCl, 10.42 D.)

20. Predict the order of increasing ionic character of the following molecules: ClF, BrF, IF. Which should have the larger dipole moment, BrF or IF?

21. Which molecule has greater ionic character, LiH or CsH? Which should have the larger dipole moment?

Suggested Reading

C. A. Coulson, *Valence,* Oxford, New York, 1961, 2nd ed.

H. B. Gray, *Chemical Bonds,* W. A. Benjamin, Menlo Park, Calif., 1973.

H. B. Gray, *Electrons and Chemical Bonding,* W. A. Benjamin, Menlo Park, Calif., 1965.

R. C. Johnson and R. R. Rettew, "Shapes of Atoms," *J. Chem. Educ.* **42**, 145 (1965).

E. A. Ogryzlo and G. B. Porter, "Contour Surfaces for Atomic and Molecular Orbitals," *J. Chem. Educ.* **40**, 256 (1963).

G. C. Pimentel and R. D. Spratley, *Chemical Bonding Clarified through Quantum Mechanics,* Holden-Day, San Francisco, 1969.

G. E. Ryschkewitsch, *Chemical Bonding and the Geometry of Molecules,* Van Nostrand Reinhold, New York, 1972, 2nd ed.

A. C. Wahl, "Chemistry by Computer," *Scientific American,* April 1970.

13

Polyatomic Molecules

In the space of one hundred and seventy-six years the Lower Mississippi has shortened itself two hundred and forty-two miles. That is an average of a trifle over one mile and a third per year. Therefore, any calm person, who is not blind or idiotic, can see that in the Old Oölitic Silurian Period, just a million years ago next November, the Lower Mississippi River was upwards of one million three hundred thousand miles long, and stuck out over the Gulf of Mexico like a fishing-rod. And by the same token any person can see that seven hundred and forty-two years from now the Lower Mississippi will be only a mile and three quarters long, and Cairo and New Orleans will have joined their streets together and be plodding comfortably along under a single mayor and a mutual board of aldermen. There is something fascinating about science. One gets such wholesale returns of conjecture out of such a trifling investment of fact.

M. Twain
The Atlantic Monthly **36, 193 (1875)**

The molecular orbital method that we applied to diatomic molecules provides a logical starting point for understanding polyatomic systems. A general method for constructing molecular wave functions for polyatomic molecules is to use atomic orbitals in linear combinations. Electrons in these molecular orbitals are not localized between two atoms of a polyatomic molecule; rather, they are **delocalized** among several atoms. This model is conceptually very different from the Lewis picture, in which two electrons between two atoms are equivalent to one chemical bond.

An alternative method for dealing with complex molecules is to use **localized** two-atom molecular orbitals. We shall emphasize the localized-bond theory in this chapter, as it provides a simple framework for the discussion of many ground-state properties, particularly molecular geometry. The delocalized-molecular-orbital theory is useful in discussing π bonding in molecules such as benzene, where two or more resonance structures are needed in Lewis formulations. Thus we shall treat the π bonding in benzene as an example of the application of the delocalized-molecular-orbital theory.

13-1 LOCALIZED MOLECULAR ORBITALS FOR BeH₂, BH₃, AND CH₄

Beryllium hydride, BeH_2, has a linear structure. To make localized bonding molecular orbitals, we first form two equivalent valence orbitals centered on Be, which are directed at the two hydrogen atoms, H_a and H_b, respectively. This is accomplished by **hybridizing** or mixing the $2s$ and $2p$ orbitals to give two equivalent sp hybrid orbitals, as shown in Figure 13-1. One hybrid orbital, sp_a, is directed at H_a and strongly overlaps the $1s_a$ orbital. The other hybrid orbital, sp_b, is directed at H_b and strongly overlaps the $1s_b$ orbital. In this scheme the two bonding molecular orbitals in BeH_2 are formed by making two equivalent linear combinations, each of which is localized between two atoms:

$$sp_a + 1s_a$$
$$sp_b + 1s_b$$

The two localized molecular orbitals are shown in Figure 13-2. The four valence electrons participate in two localized-electron-pair bonds, as in the Lewis structure for BeH_2. Each of the *linear sp* hybrid orbitals has half s and half p character, and the two sp orbitals are sufficient to attach two hydrogen atoms to the central beryllium atom in BeH_2.

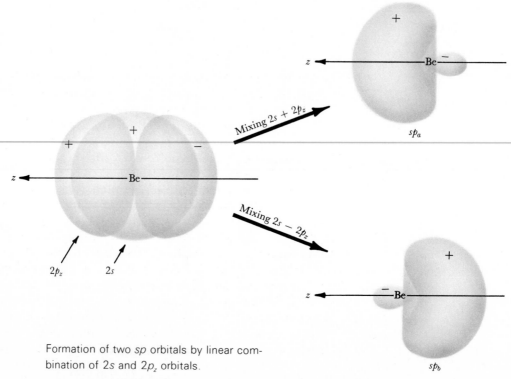

Figure 13-1   Formation of two *sp* orbitals by linear combination of $2s$ and $2p_z$ orbitals.

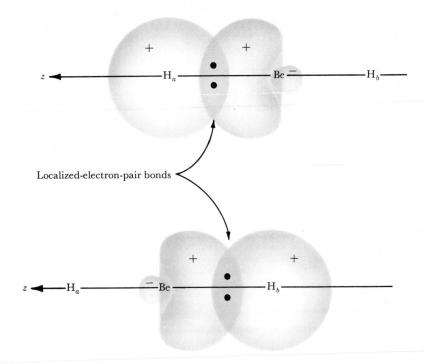

Localized-electron-pair bonds

Figure 13-2

Localized-electron-pair bonds for BeH$_2$ that are formed from two equivalent *sp* hybrid orbitals centered at the Be nucleus. Each Be *sp* orbital forms a localized bonding molecular orbital with a hydrogen 1*s* orbital.

Now consider the BH$_3$ molecule (which is observed in a mass spectrometer as a fragment of the B$_2$H$_6$ molecule; see Section 13-2) in which three hydrogen atoms are bonded to the central boron atom. According to the localized-molecular-orbital method this bonding is accomplished by hybridizing the 2*s* and 2*p* orbitals of a boron atom to give three equivalent *sp^2* hybrid orbitals (Figure 13-3). Each *sp^2* hybrid orbital has one-third *s* and two-thirds *p* character. Since any two *p* orbitals lie in the same plane and the *s* orbital is nondirectional, the *sp^2* hybrid orbitals lie in a plane. The three *sp^2* hybrid orbitals form three equivalent, localized bonding orbitals with the three hydrogen 1*s* orbitals. Each of the *sp^2* + 1*s* bonding orbitals is occupied by an electron pair, as illustrated in Figure 13-4. Using hybrid orbital theory, we predict the structure of BH$_3$ to be trigonal planar. The angle between the H—B—H internuclear lines, which is called the H—B—H bond angle, is expected to be 120°.

Methane, CH$_4$, has four equivalent atoms attached to a central atom. All the carbon valence orbitals are needed to attach the four hydrogen atoms. Thus by hybridizing the 2*s* and the three 2*p* orbitals we obtain four equivalent *sp^3* hybrid orbitals (Figure 13-5). Each *sp^3* hybrid orbital

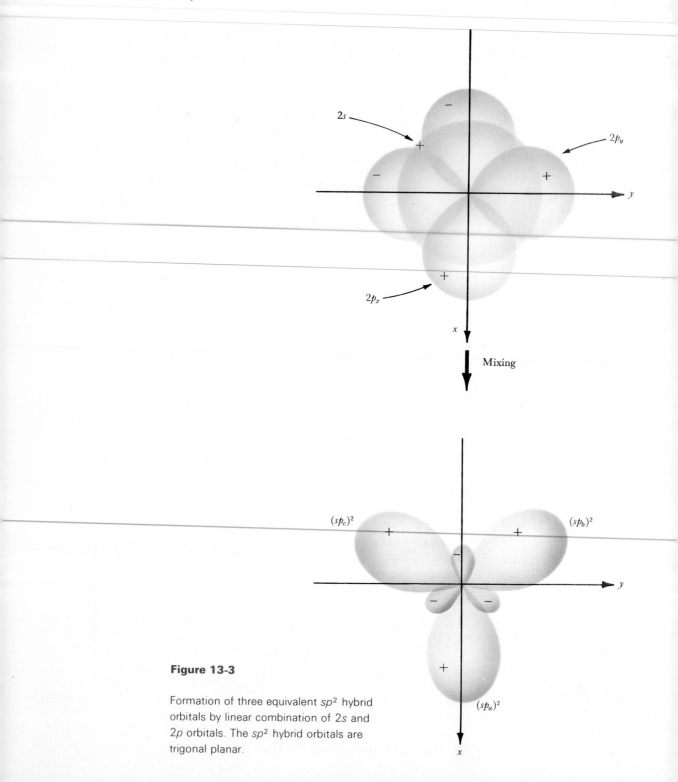

Figure 13-3

Formation of three equivalent sp^2 hybrid orbitals by linear combination of $2s$ and $2p$ orbitals. The sp^2 hybrid orbitals are trigonal planar.

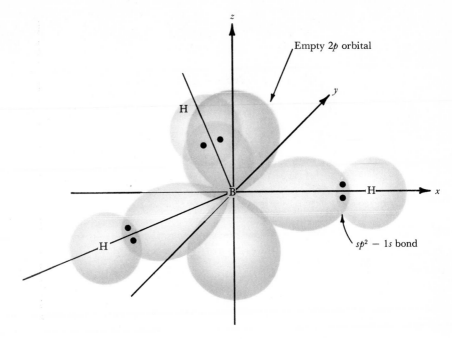

Figure 13-4   Localized-electron-pair bonds in BH$_3$.

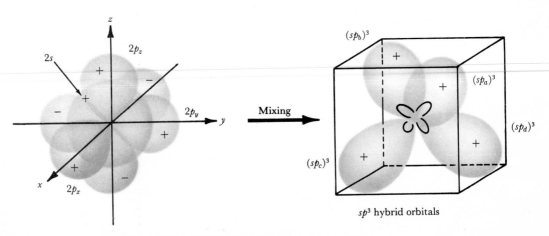

Figure 13-5   Formation of four equivalent, tetrahedral sp^3 hybrid orbitals.

Table 13-1

Localized-Orbital Theory and Molecular Geometry

Molecule	Groups attached to central atom	Hybrid orbitals appropriate for central atom	Molecular geometry
BeH_2	2	sp	Linear (angle H—Be—H = 180°)
BH_3	3	sp^2	Trigonal planar (angle H—B—H = 120°)
CH_4	4	sp^3	Tetrahedral (angle H—C—H = 109.5°)

has one fourth s and three fourths p character. The four sp^3 orbitals are directed toward the corners of a regular tetrahedron; thus the sp^3 orbitals are called **tetrahedral hybrids**. Four localized bonding orbitals can be made by combining each hydrogen $1s$ orbital with an sp^3 hybrid orbital. The best overlap between the sp^3 orbitals and the $1s$ orbitals is obtained by placing the four hydrogen atoms at the corners of a regular tetrahedron, as indicated in Figure 13-6. There are eight valence electrons (four from the carbon atom and one from each of the four hydrogen atoms) to distribute in the four localized bonding orbitals. These eight electrons account for the four equivalent, localized-electron-pair bonds shown in Figure 13-6.

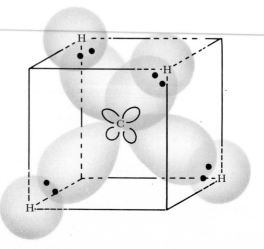

Figure 13-6   Localized-electron-pair structure for CH_4.

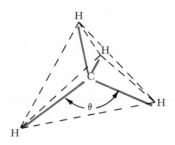
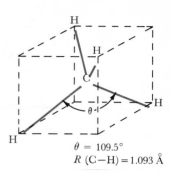

$\theta = 109.5°$
$R\ (C—H) = 1.093\ Å$

Figure 13-7   Tetrahedral molecular structure of CH_4.

The structure of CH_4 has been determined by several experimental methods. All data reveal that the structure of CH_4 is tetrahedral (Figure 13-7), which is completely consistent with localized-bond-orbital theory. The H—C—H bond angle is 109.5° and the C—H bond length is 1.093 Å. A summary of localized-bond-orbital theory for BeH_2, BH_3, and CH_4 is given in Table 13-1.

13-2 HYDROGEN IN BRIDGE BONDS

Of the three molecules discussed in the preceding section, only CH_4 has a closed-valence-shell bonding configuration. At normal temperatures and pressures both BeH_2 and BH_3 use their empty valence orbitals to form larger molecular aggregates. Beryllium hydride is a solid in which hydrogen atoms share electrons with adjacent beryllium atoms in **bridge bonds**, which may be represented as

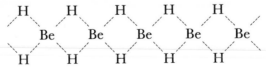

In a sense each Be shares eight electrons in the solid, thereby achieving a closed valence shell.

Under ordinary conditions the compound of empirical formula BH_3 has the molecular formula B_2H_6 and is called *diborane*. The experimentally determined structure of B_2H_6 reveals two types of bonds associated with the hydrogen atoms, as shown in Figure 13-8. Two BH_2 units are held together through two B—H—B bridge, or *three-center*, bonds. In the B_2H_6 structure the regular (or terminal) B—H bond length is shorter than the B---H distance in the bridge bonds.

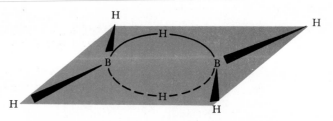

Figure 13-8   Bridging bonds in diborane, B_2H_6. The arc from B through H to B represents a three-center electron-pair bond. The B—H distance in the bridge bonds is 1.334 ± 0.027 Å, as compared to the terminal B—H bond length of 1.187 ± 0.030 Å.

One way to formulate the electronic structure of B_2H_6 employing localized molecular orbitals is shown in Figure 13-9. Each boron atom uses two sp^3 hybrid orbitals to attach the two terminal hydrogen atoms. Each of the remaining two sp^3 orbitals forms a three-center bonding orbital with a hydrogen $1s$ orbital and an sp^3 orbital on the other boron atom. In this model the bridging hydrogen atoms are positioned above and below the plane of the connected BH_2 fragments, as is observed experimentally.

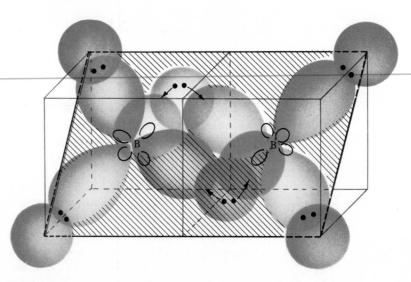

Figure 13-9   A localized-orbital model of the three-center electron-pair bonds in diborane. The $1s$ orbitals of bridging hydrogen atoms overlap with the sp^3 hybrid orbitals on each B atom.

13-3   LOCALIZED-MOLECULAR-ORBITAL THEORY FOR MOLECULES WITH LONE ELECTRON PAIRS

The ammonia molecule, NH_3, is similar to CH_4 in that it has four valence-electron pairs associated with the central atom:

$$
\begin{array}{c}
H \\
| \\
H - \underset{\displaystyle ..}{N} - H
\end{array}
$$

However, in NH_3 not all these electron pairs are equivalent. As we can see from its Lewis structure, NH_3 has three $N—H$ single bonds and one lone electron pair. We know that the three hydrogen atoms in NH_3 are equivalent. One simple formulation of the bonding in NH_3 involves three localized-electron-pair bonds between the nitrogen $2p$ and the hydrogen $1s$ orbitals (Figure 13-10). In this model the lone electron pair is in the nitrogen $2s$ orbital.

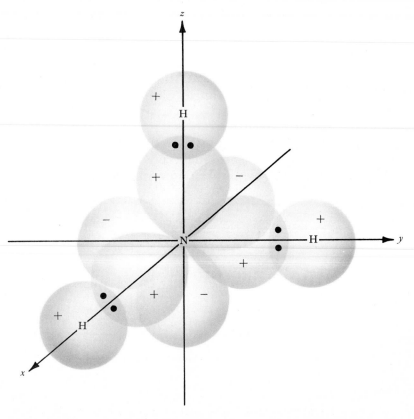

Figure 13-10   Simple picture of the bonding in NH_3 that uses only the nitrogen $2p$ orbitals.

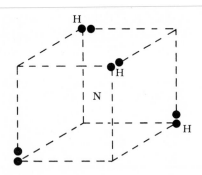

Figure 13-11

Tetrahedral placement of four electron pairs in NH_3 to minimize the electron—electron repulsions. This is the placement prescribed by the VSEPR method (see Chapter 11).

Another simple bonding scheme for NH_3 minimizes the repulsions of the four valence-electron pairs by placing them in a tetrahedral arrangement (Figure 13-11). This bonding scheme is the one that is consistent with the VSEPR method that we studied in Chapter 11. To accommodate the electrons in a tetrahedral arrangement, we use four equivalent sp^3 hybrid orbitals on the nitrogen atom. In this model the nitrogen $2s$ orbital is involved in the N—H bonds. Each of the three N—H bonding orbitals is constructed from one nitrogen sp^3 hybrid orbital and one hydrogen $1s$ orbital. The lone electron pair is assigned to the remaining sp^3 hybrid orbital. This model of the electronic structure of NH_3 is shown in Figure 13-12.

The two models of the electronic structure of NH_3 predict different H—N—H bond angles, although both predictions give the same general molecular shape. By the molecular shape we mean the positions of the atoms, which we can determine experimentally, but not the placement of lone-pair electrons, which we can only infer. From both the $(2p + 1s)$ and $(sp^3 + 1s)$ bonding schemes we predict that the molecular shape of NH_3 is trigonal pyramidal. However, from the $(2p + 1s)$ scheme, we predict an

Figure 13-12

Localized-bond-orbital structure for NH_3 that uses sp^3 hybrid orbitals for nitrogen.

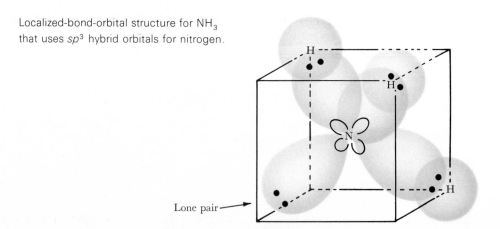

Lone pair

Figure 13-13

A view of the trigonal pyramidal molecular structure of NH_3. The H—N—H bond angle, θ, is 107°. The observed bond angle is very near the 109.5° value predicted by the sp^3 bonding model of Figure 13-12.

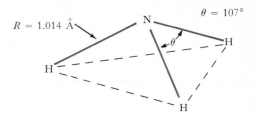

H—N—H bond angle of 90° (the angle between the p orbitals), whereas from the $(sp^3 + 1s)$ scheme, we predict the tetrahedral angle of 109.5° (the angle between the sp^3 hybrid orbitals).

The experimentally determined molecular structure of NH_3 is shown in Figure 13-13. Ammonia is trigonal pyramidal and the observed H—N—H bond angle is 107°, which is much closer to the tetrahedral angle predicted from the $(sp^3 + 1s)$ model, minimizing valence-electron-pair repulsions. Thus we see once again that repulsions of valence-electron pairs play an important role in determining molecular geometry. The N—H bond length is 1.014 Å, which is slightly shorter than the C—H bond length of 1.093 Å in CH_4. This shorter N—H bond length is consistent with the fact that the nitrogen atom is smaller than the carbon atom.

Example 1

Formulate the $(2p + 1s)$ and the $(sp^3 + 1s)$ localized-bonding models for H_2O. How do the predictions of these models compare with the observed H—O—H bond angle of 105°? The H—S—H angle in hydrogen sulfide, H_2S, is 92°. Suggest a reason why the bond angle in H_2S is much closer to the 90° prediction of one of the models for H_2O.

Solution

We use an approach similar to that discussed for NH_3. The Lewis structure of H_2O is

$$H—\overset{..}{\underset{|}{O}}:$$
$$H$$

with two H—O single bonds and two lone electron pairs. Since atomic oxygen has the valence electronic configuration $2s^2 2p^4$, we can construct two localized-electron-pair bonds, using $(2p + 1s)$ bonding orbitals (Figure 13-14). In this scheme the lone electron pairs are in the $(2s)^2$ and $(2p_y)^2$ orbitals. If we consider the repulsions of the four electron pairs to be of primary importance, the $2s$ and the three $2p$ orbitals of oxygen must be hybridized to make four sp^3 orbitals. Two $(sp^3 + 1s)$ bonding orbitals are formed between the oxygen atom and the two hydrogen atoms. The two

Figure 13-14   Simple picture of the bonding in H_2O. Only the oxygen $2p$ orbitals are used, thus an H—O—H bond angle of 90° is predicted.

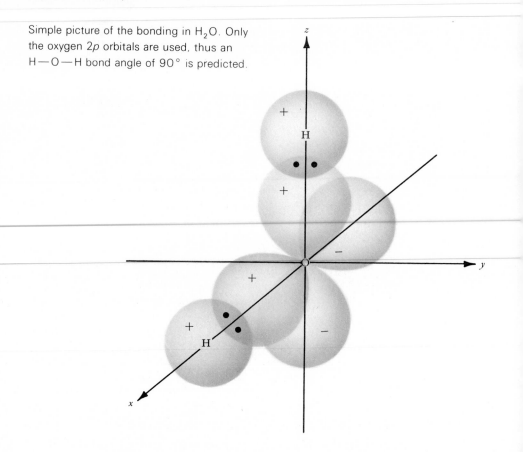

Figure 13-15

Localized-bond structure for H_2O with sp^3 orbitals for oxygen. Using this model we predict an H—O—H bond angle of 109.5°.

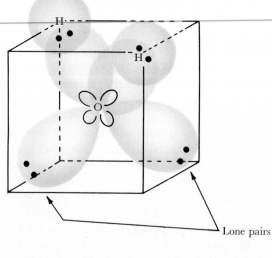

Lone pairs

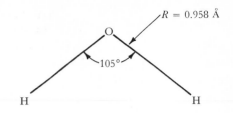

Figure 13-16   Angular molecular structure of H_2O.

lone electron pairs are equivalent and occupy the remaining two sp^3 orbitals. This model of the electronic structure of H_2O is shown in Figure 13-15.

    From both electronic structural descriptions we predict correctly that the shape of the H_2O molecule is angular. With the $(2p + 1s)$ bonding scheme we predict an H—O—H bond angle of 90°, whereas with the $(sp^3 + 1s)$ bonding scheme we predict an angle of 109.5°. The actual molecular geometry of H_2O is illustrated in Figure 13-16. The observed H—O—H bond angle of 105° again is much closer to the angle predicted from the $(sp^3 + 1s)$ model, which minimizes the repulsions of the four electron pairs.

    As the central atom in a molecule becomes larger the electrons in valence orbitals, on the average, are farther from each other. Consequently, interelectronic repulsions play a proportionately smaller role in determining molecular shapes. For example, a sulfur atom is effectively larger than an oxygen atom, and from atomic spectra we know that interelectronic repulsions in the sulfur valence orbitals are substantially smaller than in the oxygen valence orbitals. This is probably the reason why the H—S—H bond angle in hydrogen sulfide (H_2S) is 92°, which is much closer to the bond angle that is predicted from the $(3p + 1s)$ bonding model (Figure 13-17). Apparently the repulsion of the two bonding electron pairs in H_2S is much smaller than the repulsion of the two bonding pairs in H_2O.

Example 2

Formulate a localized-bond model for the methyl anion, CH_3^-, using hybrid orbitals for the central carbon atom. What molecular geometry should CH_3^- have?

Solution

The Lewis structure for CH_3^-,

$$H-\overset{..\ominus}{C}-H$$
$$|$$
$$H$$

shows that there are three bonding electron pairs and one lone pair asso-

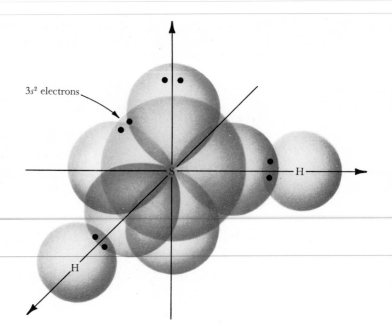

Figure 13-17

Simple $3p + 1s$ localized-bonding structure for H_2S.

ciated with the central carbon (CH_3^- is isoelectronic with NH_3). Thus we may use three sp^3 hybrid orbitals to attach the three hydrogens ($sp^3 + 1s$ bonds) and use the remaining sp^3 orbital for the lone pair:

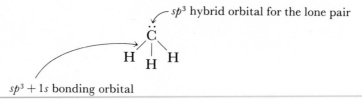

We predict a trigonal pyramidal geometry for CH_3^-, with $H-C-H$ angles of a little less than 109.5°.

13-4   SINGLE AND MULTIPLE BONDS IN CARBON COMPOUNDS

The tetrahedral sp^3 hybrid atomic orbitals explain the bonding in methane quite well. They also explain the structures of ethane, C_2H_6, and many other organic compounds in which carbon atoms are linked in chains by single bonds. In ethane, each of the carbon atoms has three hydrogen atoms linked to it by covalent bonds that use three of the four sp^3 hybrid orbitals. The fourth sp^3 orbital on each carbon atom links the carbon atoms together in a covalent bond. In forming the bond, the two sp^3 atomic orbitals combine to produce a stable bonding molecular orbital and an unstable antibonding orbital. The bonding orbital, which is symmetrical around the C—C axis and is thus a σ^b orbital, is filled by two spin-paired electrons, and the bond is complete.

The disposition of sp^3 orbitals is shown in Figure 13-18a; the measured bond lengths and bond angles are shown in Figure 13-18b. Propane (CH$_3$—CH$_2$—CH$_3$), butane (CH$_3$—CH$_2$—CH$_2$—CH$_3$), and all of the large array of continuous-chain and branched-chain hydrocarbons, including the components of kerosene, gasoline, and paraffin wax, can be constructed from tetrahedrally hybridized orbitals of carbon atoms that combine with one another and with hydrogen atomic orbitals. Such hydrocarbons are said to be **saturated** because each carbon atom has used its four valence orbitals to attach four atoms through σ bonds.

In **unsaturated** organic molecules at least some of the carbon atoms use only two or three valence orbitals to attach other atoms through σ bonds. One of the simplest examples is ethylene, C$_2$H$_4$, whose Lewis structure assumes a double bond between carbon atoms:

$$
\begin{array}{ccc}
\text{H} & & \text{H} \\
| & & | \\
\text{C} & = & \text{C} \\
| & & | \\
\text{H} & & \text{H}
\end{array}
$$

In the best model of this molecule, each carbon atom uses trigonal, sp^2 hybridization. Two of the three equivalent hybrid orbitals on each carbon

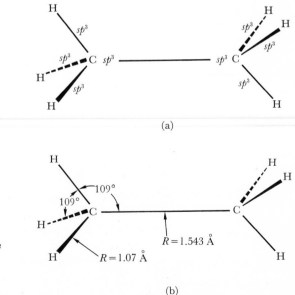

(a)

(b)

Figure 13-18   Bonding in ethane, C$_2$H$_6$. (a) The atomic orbitals contributed by carbon. (b) Observed structural parameters in ethane.

atom are used to bond to two hydrogen atoms; the third orbital from each carbon atom participates in a single C—C bond with σ symmetry (Figure 13-19a). The second bond of the double bond arises from a combination of the two $2p$ atomic orbitals that were not involved in hybridization, producing a π molecular orbital (Figure 13-19b). There are 12 valence electrons in ethylene: 4 each from the two carbon atoms, and 1 each from the four hydrogen atoms. Eight of these electrons are used in the four electron-pair bonds to H, and 2 more in the σ C—C bond. The last 2 valence electrons occupy the π C—C bond and complete the expected double bond.

In this model of ethylene, all six atoms must lie in a plane, because if one —CH$_2$ group were to be twisted relative to the other around the C—C line, the overlap between $2p$ orbitals in the π^b molecular orbital would be weakened, and the bond would be reduced to something approaching a single σ^b bond. The bond energy of the single C—C bond in ethane is

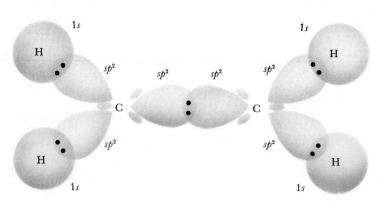

5 σ-bonding pairs = 10 electrons

(a)

Figure 13-19

Bonding in ethylene, C$_2$H$_4$. (a) The σ-bonding structure. (b) the π-orbital contribution to the double bond.

(b)

347 kJ mole^{-1}, and that of the double bond in ethylene is 523 kJ mole^{-1}. The energy required to twist the ethylene molecule by 90° should be the difference between these two numbers, or 176 kJ mole^{-1}. This is a formidable amount of energy, so the ethylene molecule should be planar.

X-ray analysis shows that ethylene *is* planar, and that its bond angles in the plane agree closely with the 120° predicted from sp^2 hybridization: 117° for each H—C—H angle and 121.5° for each H—C—C angle. The molecular structure of C_2H_4 is in close agreement with the molecular orbital picture, and we have a good example of the structure of a double bond.

There is another way in which the double bond in ethylene could be explained by molecular orbitals: tetrahedral sp^3 hybridization. In this model, two of the four sp^3 orbitals on one carbon atom overlap with two similar orbitals on the other carbon atom. The two carbon tetrahedra share an edge in the manner illustrated earlier for B_2H_6 (Figure 13-9). However, the total overlap of atomic orbitals is less in this model than in the sp^2 hybridization model, which means that the bond is not as strong. In addition, the tetrahedral model with two bent bonds predicts that the H—C—H angle is close to the tetrahedral value of 109° rather than the sp^2 value of 120°. The observed value of 117° is a strong argument for the model of the double bond in Figure 13-19 rather than the bent-bond sp^3 hybrid model.

In C_2H_2, there is only one hydrogen atom attached to each carbon atom. We can construct a localized-bonding model for C_2H_2 as follows. The s and one p orbital on each carbon atom combine to produce two sp hybrid orbitals that are oriented 180° to each other. One sp orbital is used for the σ bond to a hydrogen atom and the other is used for the σ C—C single bond. Each C has two unused $2p$ orbitals. These combine to form two π^b molecular orbitals. This model is shown in Figure 13-20.

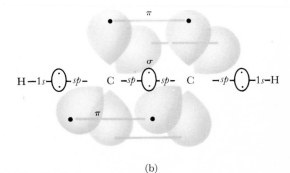

(a)

(b)

Figure 13-20　Bonding in acetylene, C_2H_2. (a) The Lewis structure. (b) The σ bonding from sp carbon orbitals and the two π bonds from carbon p orbitals.

Table 13-2

Effect of Bond Order on Bond Length and Bond Energy

Molecule	C—C bond order	C—C bond length (Å)	$H_nC—CH_n$ energy (kJ mole^{-1})
C_2H_6	1	1.54	347
C_2H_4	2	1.35	523
C_2H_2	3	1.21	962

In summary, according to molecular orbital theory, the double bond in C_2H_4 consists of one σ bond and one π bond. The triple bond in C_2H_2 is made up of one σ bond and two π bonds. The relationship of bond order, bond distance, and bond energy is clearly illustrated by experimental data for these three compounds. As the C—C bond order increases, the bond length decreases, and the energy required to break the bond increases, as shown in Table 13-2.

Example 3

Formulate a localized-bond-orbital model for propylene, $CH_3CH{=}CH_2$.

Solution

The two carbons involved in the double bond use sp^2 hybrids, leaving one $2p$ orbital left over for each atom. These extra $2p$ orbitals are used to form a π bond. The carbon of the methyl group (CH_3) is sp^3 hybridized, as it is bonded to four groups (one carbon and three hydrogens). The following localized-bond-orbital model describes the bonding in propylene:

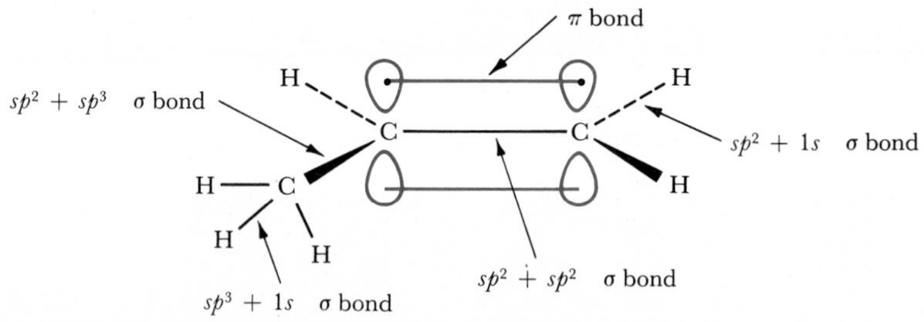

The localized-bond-orbital model predicts that the

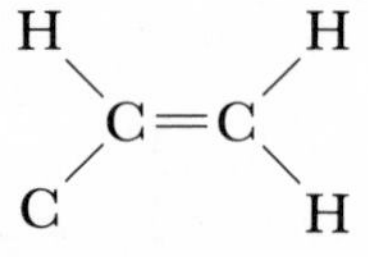

$$
\begin{array}{ccc}
H & & H \\
\diagdown & & \diagup \\
& C{=}C & \\
\diagup & & \diagdown \\
C & & H
\end{array}
$$

framework will be planar, in agreement with experimental observations. The structure about the methyl carbon,

$$\begin{array}{c} H \\ | \\ H-C- \\ | \\ H \end{array}$$

is tetrahedral, also as predicted.

Example 4

Formulate a localized-bond-orbital model for acetaldehyde, CH_3CHO. Is the $\overset{H}{\underset{C}{\diagdown}} C=O$ unit planar? What should the $H-C=O$ bond angle be?

Solution

The Lewis structure for acetaldehyde shows that the central carbon is attached to three groups (CH_3, H, and O): $\overset{H}{\underset{H_3C}{\diagdown}} C=\ddot{O}:$

Thus the central carbon should use three sp^2 orbitals to attach these groups. The $2p$ orbital left over on the central carbon is used to form a π bond with an oxygen $2p$ orbital. The carbon of the methyl group is sp^3 hybridized. The following localized-bond-orbital model is appropriate for acetaldehyde:

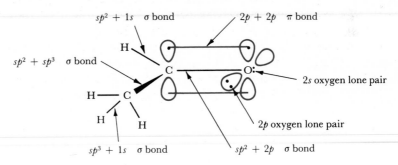

The localized bond model shows that the $\overset{H}{\underset{C}{\diagdown}} C=O$ unit is planar. (The sp^2 orbitals about the central carbon are trigonal planar; therefore the three atoms attached to the central carbon must be in the same plane.) The $H-C=O$ bond angle is predicted to be $120°$, which is the angle between sp^2 hybrid orbitals.

13-5   BENZENE AND DELOCALIZED ORBITALS

With benzene, the localized-molecular-orbital treatment fails. Just as we could not produce a satisfactory Lewis structure for O_2, so we cannot produce a satisfactory localized-molecular-orbital structure for C_6H_6.

Benzene has the planar hexagonal skeleton shown in Figure 13-21. Each carbon atom in the hexagon is attached to one hydrogen atom and two other carbon atoms with bond angles of 120°. The bond angles alone suggest sp^2 hybridization. If we use sp^2 hybridization for carbon, we form the σ bonding network shown in Figure 13-22. Each carbon atom is joined to one H atom and two other C atoms by single covalent bonds.

This cannot be the whole story, because the observed C—C bond length in benzene, 1.390 Å, is too short for a single bond (1.54 Å). Each carbon atom would have an unused $2p$ orbital perpendicular to the plane of the hexagonal ring (Figure 13-23). There are 30 valence electrons in benzene: 4 each from the six carbon atoms, and 1 each from the six hydrogen atoms. Twelve are used in the six C—H σ single bonds, and 12 more in the six C—C σ single bonds. Six electrons remain, along with six p atomic orbitals. It would be logical to use the orbitals in three pairs for making three more covalent bonds. But how are these pairs to be chosen?

Kekulé proposed that the pairs be formed between adjacent carbon atoms around the ring (Figure 13-24). If this were so, then the C—C bond lengths around the benzene ring should alternate between a single-bond length of 1.54 Å and 1.35 Å, the double-bond length in ethylene. However, x-ray analysis shows that all six bonds are equal. Three other structures, with different combinations of three covalent bonds between p orbitals, were proposed by Dewar (Figure 13-24). Each of them by itself is even less

Figure 13-21   The skeleton of the benzene molecule. The C—C bond length has been determined by x-ray crystal structure analysis.

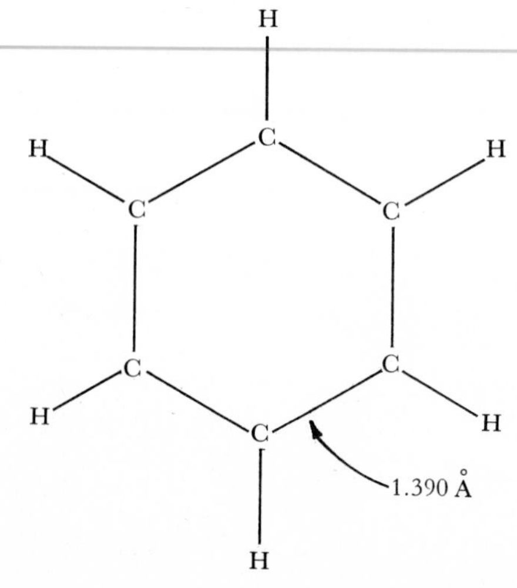

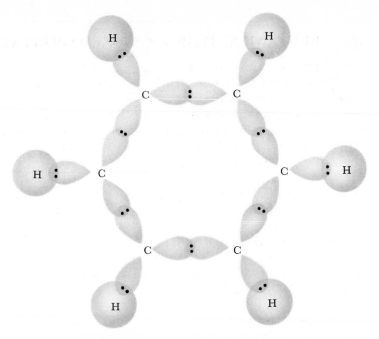

Figure 13-22   The pattern of σ bonding in benzene. These six C—C bonds and six C—H bonds use 24 atomic orbitals and 24 electrons. Six 2p carbon atomic orbitals and 6 valence electrons remain unused.

satisfactory than a Kekulé structure. It is impossible to draw any one bond structure that explains the benzene molecule. The fault lies in our notion that a bond is something that is formed in a private sort of way between two atoms in a molecule, with the other atoms uninvolved.

In a strict sense, every molecular wave function should include atomic orbitals from all of the atoms of a molecule. Usually, all atoms except two make negligible contributions to a given wave function, and we can consider that wave function as an accurate description of the bond between

Figure 13-23   The unused 2p atomic orbitals in benzene. These orbitals combine to make molecular orbitals that can accommodate the six valence electrons unused in the σ bonding.

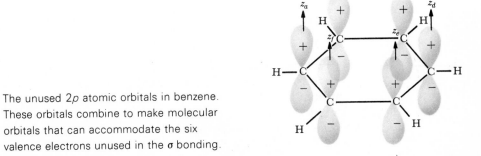

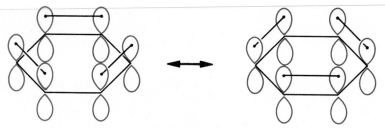

The two Kekulé structures

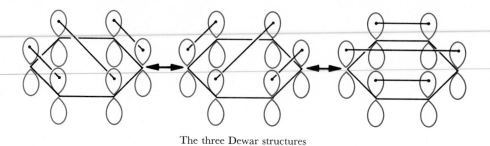

The three Dewar structures

Figure 13-24   The six p orbitals can be combined in pairs to make covalent bonds in several ways.

the two atoms. But anomalies such as benzene occur often enough to remind us of the flaws in our assumption about localized bonds.

Using symmetry arguments beyond the scope of this text, we can construct a set of six full-molecule wave functions and orbitals from the six atomic $2p$ orbitals. Let the six orbitals be labeled $z_a, z_b, z_c, z_d, z_e,$ and z_f, as in Figure 13-23. Let the sign of each z orbital be plus or minus, depending on whether the plus lobe of the p wave function is up or down in Figure 13-23. Then the six full-molecule wave functions are given by

$$\pi_1^b = z_a + z_b + z_c + z_d + z_e + z_f$$
$$\pi_2^b = 2z_a + z_b - z_c - 2z_d - z_e + z_f$$
$$\pi_3^b = z_b + z_c - z_e - z_f$$
$$\pi_1^* = 2z_a - z_b - z_c + 2z_d - z_e - z_f$$
$$\pi_2^* = z_b - z_c + z_e - z_f$$
$$\pi_3^* = z_a - z_b + z_c - z_d + z_e - z_f$$

The squares of these functions are the electron-density distributions. These six orbitals are pictured in Figure 13-25. Three are bonding, and three are antibonding. Their energies are shown in Figure 13-26. Note how these orbitals illustrate the rule that, in general, the more nodes an orbital has the higher its energy will be. You can test the validity of this rule with the orbitals for homonuclear and heteronuclear diatomic molecules discussed in Chapter 12, and even with the hydrogen atomic wave functions.

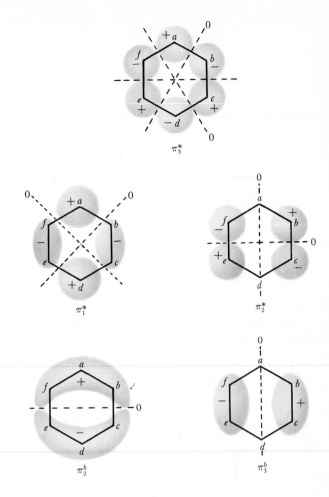

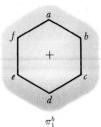

Figure 13-25

The six full-molecule delocalized molecular orbitals obtained from the six $2p$ orbitals of benzene. The dashed lines indicate nodes of zero electron density, and the plus and minus signs mark the relative signs of the wave functions on either side of a nodal surface. The greater the number of nodal surfaces, the higher the energy of the wave function. All six of these orbitals have a nodal plane in the plane of the paper; for example, the π_1^b orbital has a region of high probability density above the plane of the benzene ring and another region of high density below the plane, in which the wave function itself has the opposite sign. The π_3^* has 12 lobes of high density: 6 above the plane of the ring as shown here, and another 6 below the ring with opposite signs for the wave functions.

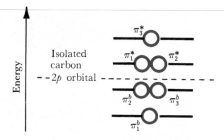

Figure 13-26

Energy-level diagram for the six benzene delocalized orbitals diagrammed in Figure 13-25. A generally valid principle in quantum mechanics is that the more space there is available in which a particle can move the lower and more closely spaced its energy levels will be. As an extreme example, this is why we notice the quantization of energy of an electron in a hydrogen atom, but do not notice the quantization of energy of a baseball in Yankee Stadium. The mass of the baseball is so great, and the volume in which it can move is so large, that its quantized energy levels are too close together to detect. In addition to the mass difference, the baseball example is a case of extreme delocalization. Even at the molecular level, the more room in which the electrons have to move (the more they are delocalized), the more stable the molecule will be, if other factors are neglected. The delocalization stability of each of the π_1^b, π_2^b, and π_3^b orbitals is given by its distance below the horizontal dashed line representing the stability of electrons in the isolated 2p orbitals.

The six unused valence electrons in benzene occupy the three bonding orbitals in Figure 13-26. No one electron pair belongs to any pair of C atoms; thus, the six electrons are delocalized. Each carbon–carbon bond consists of one full σ bond and half a π bond. The C—C bond length, 1.390 Å, is intermediate between single- and double-bond lengths.

Benzene is actually more stable than might be expected for a molecule with six C—C single bonds, six C—H single bonds, and three C—C π bonds. This added stability results because the electrons in the three π bonds are delocalized over all six carbon atoms. Orbital π_1^b in Figure 13-25 is symmetrical with respect to all six carbon atoms. Orbitals π_2^b and π_3^b look unsymmetrical, but the combination of the two is symmetrical. There is nothing special about atoms a and d; we could have written π_2^b and π_3^b so atoms f and c appeared to be the "axis" of the molecule. If we did not allow the delocalization of electrons in C_6H_6, the bonding would be as in one of the Kekulé or Dewar structures shown in Figure 13-25 or 13-27. Instead, the bonding structure in benzene can be represented best as in the bottom drawing of Figure 13-27. As we can calculate from experimental data, the molecule is 167 kJ mole^{-1} more stable than expected from the sum of the bond energies of six C—H, three C—C, and three C=C bonds.

Another way of treating the symmetrical benzene molecule is by using the idea of resonance mentioned previously with Lewis structures (Chapter

Kekulé structures

Figure 13-27

The two Kekulé structures for benzene, the three Dewar structures, and a diagrammatic representation of the delocalized electrons in the benzene ring. The Kekulé and Dewar structures sometimes are called resonance structures for benzene. The implication intended by this rather unfortunate terminology is not that the bonds flip back and forth or resonate from one structure to the other, but only that the true structure, which cannot be represented by localized bonds, has something of the character of each of these structures.

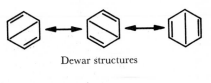

Dewar structures

Simple molecular orbital structure

11). In this treatment, we state that although the benzene molecule cannot be represented accurately by any one localized-bond model the real molecule has some of the character of both Kekulé structures and all three Dewar structures in Figure 13-27. We can write the complete wave function of benzene as a linear combination of the wave functions of the two Kekulé structures (K_1 and K_2), and the three Dewar structures (D_1, D_2, and D_3)

$$\psi = uK_1 + vK_2 + wD_1 + xD_2 + yD_3$$

There are ways of calculating the coefficients u through y, and the result is that the best approximation to the real benzene molecule occurs if we assume that each Kekulé structure contributes 39% to the real molecule and each Dewar structure contributes 7%. These five models are called the *resonance structures for benzene,* and the extra 167 kJ mole^{-1} of stability is called the *resonance stabilization energy.* Again, beware of the terminology, and *do not* think of the benzene molecule as resonating or flipping back and forth from one resonance structure to another.

Example 5

Addition of an electron to benzene gives the benzene radical anion, $C_6H_6^-$. Which π orbital does the extra electron enter? What is the average C—C π bond order in $C_6H_6^-$?

Solution

The extra electron must be placed in the lowest unoccupied π molecular orbital of C_6H_6, which is one of the degenerate π_1^*, π_2^* set. Thus the π orbital configuration of $C_6H_6^-$ is $(\pi_1^b)^2(\pi_2^b, \pi_3^b)^4(\pi_1^*)^1$. Since π_1^* is antibonding, there are now 3 minus $\frac{1}{2}$, or $2\frac{1}{2}$ net π bonds for the entire C_6 framework. Each of the six C—C bonds has a π bond order of $\frac{5}{12}$ in $C_6H_6^-$ ($2\frac{1}{2}$ divided by 6), as compared to $\frac{6}{12}$ (or $\frac{1}{2}$) in C_6H_6.

13-6 POLAR AND NONPOLAR POLYATOMIC MOLECULES

In Chapter 12 we discussed the electronic structure of HCl and pointed out that heteronuclear diatomic molecules are polar, whereas homonuclear diatomic molecules are nonpolar. A nonpolar molecule has zero (or nearly zero) dipole moment. For polyatomic molecules there are numerous instances in which individual bonds are polar although the molecule as a whole is nonpolar. An example is CCl_4. The molecular structure of CCl_4 is shown in Figure 13-28a. Since chlorine is more electronegative than carbon, the bonding electron pairs are pulled toward the chlorine atoms. Thus each C—Cl unit has a small bond dipole. The bond dipoles can be resolved into equal and opposite CCl_2 dipoles, as shown in Figure 13-28b. The symmetrical (tetrahedral) molecular shape of CCl_4 results in a zero dipole moment; thus CCl_4 is nonpolar.

An example of a polar polyatomic molecule is CH_3Cl. Since carbon and hydrogen have electronegativities that are nearly the same, the contribution of the three C—H bonds to the net dipole moment is negligible. The electronegativity difference between carbon and chlorine is large, however, and it is this highly polar bond and the lone-pair electrons on chlorine that account for most of the dipole moment of CH_3Cl ($\mu = 1.87$ D):

$$:\ddot{C}l:$$

resultant dipole moment

Another example of a polar polyatomic molecule is H_2O. Because oxygen is more electronegative than hydrogen, the electron pairs in the two O—H bonds are pulled more toward the oxygen atom. In addition, the oxygen atom has two lone pairs of electrons. The result is a separation of charge in the H_2O molecule, in which the oxygen atom is relatively negative and the hydrogen atoms are relatively positive:

$$\delta-O^{\delta-}$$
$$\delta+H \qquad H^{\delta+}$$

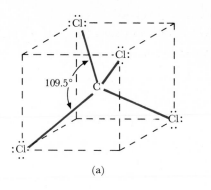

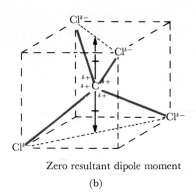

Zero resultant dipole moment

(b)

Figure 13-28 (a) Tetrahedral structure of CCl_4. (b) Canceling bond dipoles in CCl_4.

Because of the angular shape of H_2O, the H—O bonds and lone-pair contributions combine, as shown in Figure 13-29, to give the dipole moment of 1.85 D.

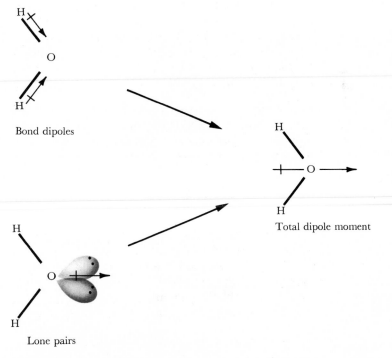

Bond dipoles

Lone pairs

Total dipole moment

Figure 13-29 Contributions to the 1.85 D dipole moment of H_2O.

Example 6

Which should have the larger dipole moment, $CHBr_3$ or CBr_4?

Solution   By virtue of the molecule's tetrahedral geometry, the four $C^{\delta+}$—$Br^{\delta-}$ bond dipoles of CBr_4 exactly cancel (as in CCl_4), yielding a zero dipole moment. The three $C^{\delta+}$—$Br^{\delta-}$ bond dipoles in $CHBr_3$, however, do not cancel:

$$
\begin{array}{c}
H \\
| \\
{}^{\delta+}Br{-}\overset{\displaystyle {}^{\delta+}C^{\delta+}}{\underset{\displaystyle Br^{\delta-}}{\big|}}{-}Br^{\delta-}
\end{array}
$$

$CHBr_3$ has a larger dipole moment than CBr_4. $CHBr_3$ is polar, whereas CBr_4 is nonpolar.

Example 7

Which should have the larger dipole moment, ammonia (NH_3) or arsine (AsH_3)?

Solution   Both NH_3 and AsH_3 have trigonal pyramidal geometries with bond dipoles:

$$
{}^{\delta+}H\diagup\overset{{}^{\delta-}\ddot{N}{}^{\delta-}}{\underset{H^{\delta+}}{\big|}}\diagdown H^{\delta+} \qquad\qquad {}^{\delta+}H\diagup\overset{{}^{\delta-}\ddot{As}{}^{\delta-}}{\underset{{}^{\delta+}H}{\big|}}\diagdown H^{\delta+}
$$

Since N is more electronegative than As, the individual $H^{\delta+}$—$N^{\delta-}$ bond dipoles should be larger than the $H^{\delta+}$—$As^{\delta-}$ ones. Therefore the resultant dipole moment for NH_3 should be larger than that of AsH_3. This prediction agrees with the known values for these two molecules (NH_3, 1.47 D; AsH_3, 0.16 D).

Let's now consider ethylene and the various isomers* of dichloroethylene. As a result of their symmetrical structures, ethylene and *trans*-1,2-dichloroethylene are nonpolar molecules:

$$
\begin{array}{cc}
\underset{H}{\overset{H}{\diagdown}}C=C\underset{H}{\overset{H}{\diagup}} & \qquad \underset{Cl}{\overset{H}{\diagdown}}C=C\underset{H}{\overset{Cl}{\diagup}}
\end{array}
$$

$$\mu = 0.00\ D \qquad\qquad\qquad \mu = 0.00\ D$$
$$\text{ethylene} \qquad\qquad trans\text{-1,2-dichloroethylene}$$

*Isomers have the same molecular formula but different molecular structures.

However, two other isomers of dichloroethylene are polar molecules:

$\mu = 1.90$ D
cis-1,2-dichloroethylene

$\mu = 1.34$ D
1,1-dichloroethylene

The relationship between dipole moment and molecular geometry can be illustrated further by a comparison of the three isomers of dichlorobenzene:

$\mu = 0.00$ D
*para*dichlorobenzene

$\mu = 1.72$ D
*meta*dichlorobenzene

$\mu = 2.50$ D
*ortho*dichlorobenzene

It is clear that the para isomer must have a zero dipole moment by reason of its symmetry. Consideration of the individual C—Cl bond dipoles also leads one to conclude that the ortho isomer should have a larger dipole moment than the meta isomer, as observed.

In summary, polar molecules have bond dipoles that add to give a resultant nonzero dipole moment. Nonpolar molecules have either pure covalent bonds (equal sharing) or bond dipoles that are placed in a geometric arrangement in which they cancel each other. Table 13-3 gives the molecular shapes and the polarities of several representative polyatomic molecules.

13-7 MOLECULAR SPECTROSCOPY

In addition to electronic energy levels, molecules possess energy levels associated with rotational (Figure 13-30) and vibrational (Figure 13-31) motion. Generally, any linear polyatomic molecule can rotate around the three mutually perpendicular axes through the center of gravity of the molecule, as shown in Figure 13-30. For a linear molecule (a diatomic molecule must be linear), one of these axes lies on the line between the

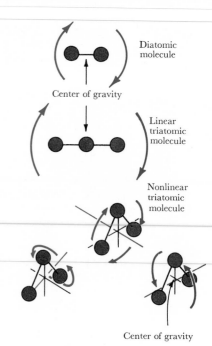

Diatomic
molecule

Center of gravity

Linear
triatomic
molecule

Nonlinear
triatomic
molecule

Center of gravity

Figure 13-30   Rotational motion in molecules.

centers of the atoms; thus there are only two ways of rotation. Figure 13-31 shows the modes of vibration of diatomic, linear triatomic, and nonlinear triatomic molecules. Often it is helpful in discussing vibrations in molecules to treat the bonds between the atoms as springs, as shown.

A reasonable approximation for the total molecular energy is

$$E_{\text{total}} = E_{\text{electronic}} + E_{\text{vibrational}} + E_{\text{rotational}}$$

Figure 13-32 shows a generalized energy-level diagram for a molecule. Two electronic levels, E_1 and E_2, are shown with their vibrational and rotational levels. Separations between electronic energy levels are usually much larger than those between vibrational levels, which in turn are much larger than those between rotational levels. **Electronic transitions** correspond to absorption of radiation in the visible and ultraviolet portion of the spectrum; **vibrational transitions** correspond to absorption in the near-infrared and infrared regions; and **rotational transitions** correspond to absorption in the far-infrared to microwave regions.

Table 13-4 shows the range of electromagnetic radiation, gives energies in some commonly used units, and indicates the kind of spectroscopy that is used in each range. The quantized properties of molecular energy levels have been utilized in modern spectroscopic techniques to identify molecules and to assign molecular structures. For example, the study of rota-

Table 13-3

Molecular Shapes and Dipole Moments of Selected Polyatomic Molecules

Molecule	Shape	Dipole moment, μ (D)	Classification
CS_2	Linear	0	Nonpolar
HCN	Linear	2.98	Polar
COS	Linear	0.712	Polar
BF_3	Trigonal planar	0	Nonpolar
SO_3	Trigonal planar	0	Nonpolar
CF_4	Tetrahedral	0	Nonpolar
CCl_4	Tetrahedral	0	Nonpolar
NH_3	Trigonal pyramidal	1.47	Polar
PF_3	Trigonal pyramidal	1.03	Polar
AsF_3	Trigonal pyramidal	2.59	Polar
H_2O	Angular	1.85	Polar
H_2S	Angular	0.97	Polar
SO_2	Angular	1.63	Polar
NO_2	Angular	0.316	Polar
O_3	Angular	0.53	Polar

tional transitions by far-infrared and microwave spectroscopy yields extremely precise data on bond angles and bond lengths.

In **infrared spectroscopy** a beam of infrared radiation, whose wavelength, λ, varies from 2.5 μm to 15 μm* (wave number, $\bar{\nu}$, varies from 4000 cm^{-1} to 667 cm^{-1}) is passed through a sample of a compound (Figure 13-33). Often, the sample is first compressed under pressure into a thin wafer and inserted into infrared-transparent sodium chloride sample holders. (Sodium chloride must be used because quartz or glass is opaque to infrared light.) The sodium chloride prism position and the slit widths determine the wavelength of the radiation reaching the detector. The absorption of radiation at different wave numbers corresponds to the excitation of molecules from a low (usually the lowest) vibrational energy level to the next higher vibrational energy level.

Absorbed radiation is identified by its wavelength, its frequency, or its wave number. Radiation absorption is detected electronically and recorded in some suitable form as a graphic trace. A strong absorption throughout a narrow range of frequencies causes a sharp "peak" or "line" in the recorded spectrum. Absorption peaks are not always narrow and sharp because each vibrational energy level has superimposed upon it an array of

*Wavelength is often given in Angstroms (Å), micrometers (μm), or nanometers (nm); 1 μm $= 10^3$ nm $= 10^4$ Å $= 10^{-4}$ cm.

Table 13-4

Electromagnetic Radiation and Spectroscopy

Frequency (Hz)	10^6	10^7	10^8	10^9	10^{10}	10^{11}
Wave number (cm^{-1})	3.3×10^{-5}	3.3×10^{-4}	0.0033	0.0333	0.333	3.33
Wavelength	300 m	30 m	3 m	30 cm	3 cm	0.3 cm
Energy	0.0004 J mole^{-1}	0.004 J mole^{-1}	0.04 J mole^{-1}	0.4 J mole^{-1}	4 J mole^{-1}	40 J mole^{-1}
Name	Radio (Long-wave)	(Short-wave)	(Television and FM) (UHF)	Microwave →		Infrared → (Far IR)
Source and detector	(Not used)	Vacuum tubes, wires, antenna, coil		Klystron, guide, cavity		Hot wire, etc.

Frequency (Hz)	10^{12}	10^{13}	10^{14}	10^{15}	10^{16}	10^{17}
Wave number (cm^{-1})	33	333	3333	33333	3.3×10^5	3.3×10^6
Wavelength	300 μm	30 μm	3 μm / 30,000 Å	300 nm / 3000 Å	30 nm / 300 Å	3 nm / 30 Å
Energy	0.4 kJ mole^{-1} / 0.16 kT	4 kJ mole^{-1} / 1.6 kT	40 kJ mole^{-1} / 16 kT	400 kJ mole^{-1} / 4 eV	4000 kJ mole^{-1} / 40 eV	400 eV
Name	Infrared → (Near IR)		Visible → (Red → blue)	Ultraviolet → (Near UV)		(Far UV)
Source and detector	Lamp, prism, grating: phototube or photographic plate			Lamp, etc., grating, phototube		

Frequency (Hz)	10^{18}	10^{19}	10^{20}
Wave number (cm^{-1})	3.3×10^7	3.3×10^8	3.3×10^9
Wavelength	3 Å	0.3 Å	0.03 Å
Energy	4 keV	40 keV	400 keV / 0.4 MeV
Name	x rays →		γ rays →
Source and detector	x-ray tube, photographic plate		Nuclear reaction, counter

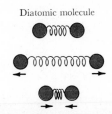

Diatomic molecule

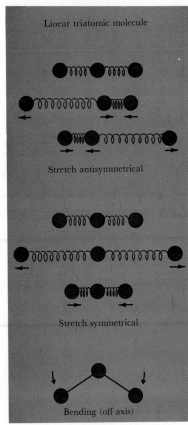

Linear triatomic molecule

Stretch antisymmetrical

Stretch symmetrical

Bending (off axis)

Nonlinear triatomic molecule

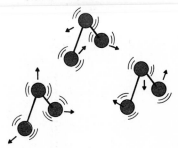

Figure 13-31   Vibrational motion in molecules.

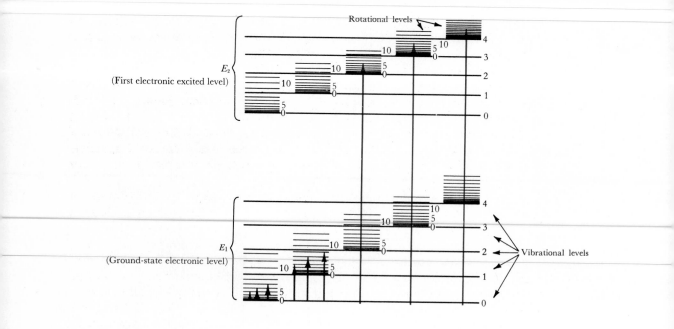

Figure 13-32   Generalized energy-level diagram for a molecule.

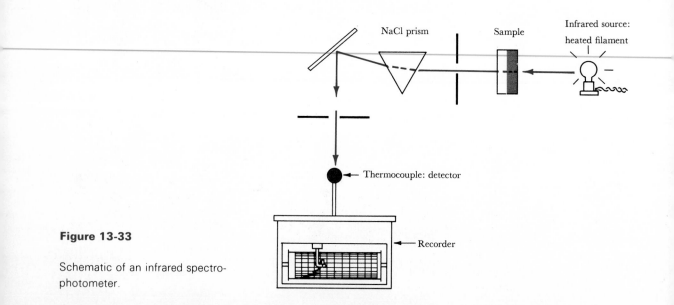

Figure 13-33

Schematic of an infrared spectrophotometer.

rotational energy levels (Figure 13-32); therefore a particular vibrational transition is really the superposition of transitions from many vibrational–rotational levels.

Different absorbed radiation frequencies correspond to different intramolecular excitations. For example, stretching vibrations for $C=O$ bonds occur in the region below 6 μm (1780 cm^{-1} to 1850 cm^{-1}) and for $C-H$ bonds in the region between 3 μm and 4 μm (2800 cm^{-1} to 3000 cm^{-1}). Absorption peaks in other regions of the infrared (IR) spectrum correspond to energy changes in other bonds or to complex intramolecular vibrations. Table 13-5 lists the characteristic IR absorption wave numbers of certain bonds.

Infrared spectroscopy is extremely useful in the identification of unknown materials because each chemical compound has a unique IR spectrum. Figure 13-34 shows the IR spectra of tetrachloroethylene and cyclohexene.

Not all molecules respond to infrared radiation. In particular, molecules with certain elements of symmetry, such as homonuclear diatomic molecules, do not absorb infrared radiation. In more complex molecules not all modes of vibration respond to infrared radiation. For example, symmetrical molecules such as ethylene, $H_2C=CH_2$, do not exhibit all

Table 13-5

Positions of Some Characteristic Infrared Bands of Molecular Groups

Bond	Group	Description	$\bar{\nu}$ (cm^{-1})	λ (μm)
C—H	CH_2, CH_3	Stretching	3000–2900	3.3–3.4
C—H	$\equiv C-H$	Stretching	3300	3.0
C—H	Aromatic (benzene)	Stretching	3030	3.3
C—H	$-CH_2-$	Bending	1465	6.8
C—H	$-CH=CH-$ (trans)	Out-of-plane bending	970–960	10.3–10.4
		In-plane bending	1310–1295	7.6–7.7
C—H	$-CH=CH-$ (cis)	Out-of-plane bending	~690	14.5
O—C	$>C=O$	Stretching	1850–1700	5.4–5.9
O—H	$-O-H^a$ (alcohols)	Stretching	3650–3590	2.7–2.8
O—H	Hydrogen bonded (alcohols)	Stretching	3400–3200	3.0–3.1

^{a}Refers to an O—H group that is not hydrogen bonded to a neighboring molecule. Hydrogen bonding is discussed in Chapter 14.

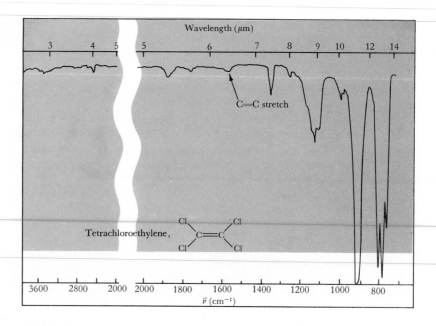

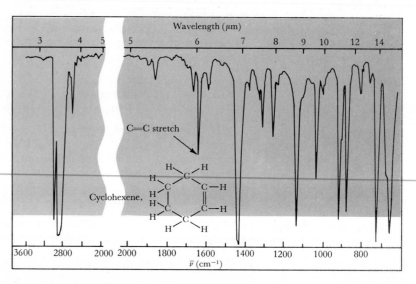

Figure 13-34   Infrared spectra of tetrachloroethylene and cyclohexene.

their vibrations in infrared spectra. To assist in examining the vibrations of these molecules, **Raman spectroscopy** often can be used. The Raman spectrum results from the irradiation of molecules with light (usually in the visible region) of a known wavelength. A laser beam is commonly used to irradiate the sample in Raman spectrometers (Figure 13-35). Radiation

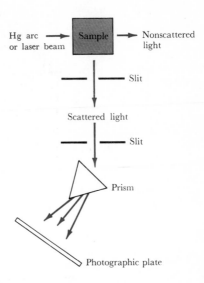

Figure 13-35   Schematic of a Raman spectrometer.

absorption is not noted directly. When irradiated with light of high energy, the molecules may add to, or extract from, the incident light small amounts of energy corresponding to the energy of some particular molecular vibration. The incident light is said to be *scattered*, rather than absorbed, and may be observed as light with a wavelength different from that of the incident light. Figure 13-36 shows the Raman spectra of tetrachloroethylene and cyclohexene.

As stated previously, electronic transitions correspond to the absorption of larger amounts of energy than are absorbed in vibrational or rotational transitions. Electronic excitations are usually associated with absorption of visible and ultraviolet light. Just as vibrational absorption "bands" are widened by a superposition of many vibrational–rotational transitions, absorption spectra recorded in the visible–ultraviolet region exhibit broad bands, rather than sharp peaks, because of the superposition of many vibrational–electronic transitions (Figure 13-37). Absorption bands are characterized by a maximum at a particular wavelength, λ_{max}.

A schematic diagram of a visible and ultraviolet spectrophotometer is shown in Figure 13-38. A prism, or diffraction grating, and the width of the slit regulate the spread of wavelengths through the sample. The sample in solution is compared to a reference blank containing pure solvent. The light incident on the sample is denoted by I_0 and the transmitted light is denoted by I. The absorption of light can be expressed quantitatively by the relationship known as *Beer's law*, $A = \epsilon c l$, in which A is the absorbance ($\log I_0/I$), c is the concentration of the substance (in moles liter^{-1}), and l is the length (in cm) of the light path through the substance. The **molar extinction coefficient**, ϵ, is characteristic of the absorbing sample. Values for molar extinction vary considerably from compound to compound and from

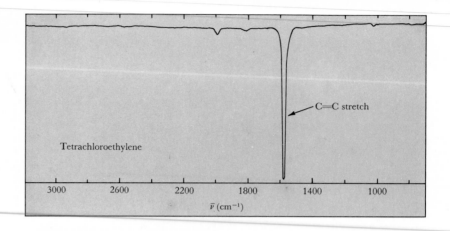

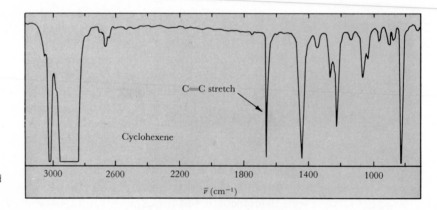

Figure 13-36

Raman spectra of
tetrachloroethylene and
cyclohexene.

peak to peak in the absorption spectrum of the same compound. The extent to which an electronic transition is "allowed" or "forbidden" by quantum mechanical selection rules is reflected in the value of ϵ of the absorption peak accompanying that transition. Thus the experimentally determined ϵ value can often be extremely useful in assigning a given absorption peak to a particular type of electronic transition. If ϵ is in the range of 10^3–10^5, the transition meets all the requirements of the quantum mechanical selection rules.

The excitation of an electron in the π bonding orbital to the π anti-bonding orbital of C_2H_4 gives rise to an absorption band with a maximum at 171 nm (58,500 cm^{-1}). This $\pi^b \rightarrow \pi^*$ transition is allowed, as the ϵ value is approximately 10^4. Unsaturated hydrocarbons such as ethylene absorb light at longer wavelengths (lower energies) than do saturated hydrocarbons. For example, ethane, a saturated hydrocarbon, does not begin to absorb strongly before 160 nm. This fact suggests that the separation be-

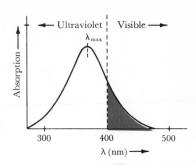

Figure 13-37   A typical electronic absorption band with λ_{max} in the near-ultraviolet region.

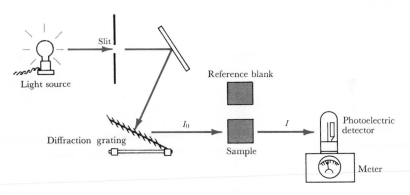

Figure 13-38   Schematic diagram of a visible and ultraviolet spectrophotometer.

tween σ bonding and σ antibonding orbitals in hydrocarbons is larger than the separation between π bonding and π antibonding orbitals. For this reason it is common to ignore the σ^* levels but to include both π^b and π^* orbitals in a molecular orbital formulation of unsaturated hydrocarbons. Spatial representations and relative energies of the π^b and π^* orbitals of C_2H_4 are shown in Figure 13-39.

If the π-electron system of an unsaturated hydrocarbon is more extensive than that of ethylene, the energy separation between the highest occupied π^b orbital and the lowest unoccupied π^* level becomes smaller, and energy absorption occurs at longer wavelengths. Such extensive π-electron systems are found in **conjugated polyenes**, compounds in which conventional structural formulas show alternate single and double bonds:

$$-\overset{\displaystyle |}{C}=\overset{\displaystyle |}{\underset{\displaystyle |}{C}}-\overset{\displaystyle |}{\underset{\displaystyle |}{C}}=\overset{\displaystyle |}{\underset{\displaystyle |}{C}}-\overset{\displaystyle |}{\underset{\displaystyle |}{C}}=\overset{\displaystyle |}{\underset{\displaystyle |}{C}}-$$

Polyenes having 10 or more conjugated double bonds absorb visible light; hence they are colored. The pigments responsible for light perception in

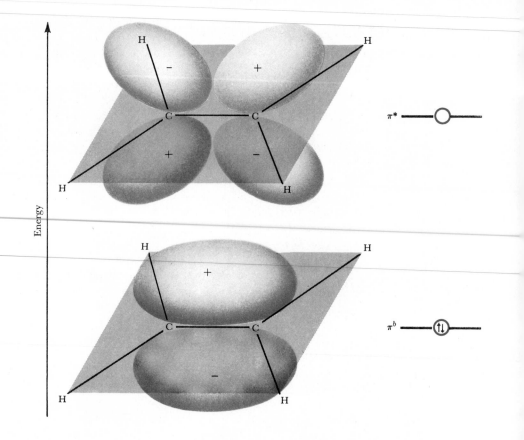

Figure 13-39   The bonding and antibonding π molecular orbitals in C_2H_4. The π electronic structure of the ground state is $(\pi^b)^2$. An absorption band peaking at 58,500 cm^{-1} in the spectrum of ethylene is due to the electronic transition $\pi^b \rightarrow \pi^*$

the human eye contain long, conjugated polyene chains, as do some vegetable pigments such as carotene, the colored substance in carrots.

   An important example of a forbidden transition is the excitation of a nonbonding $2p$ oxygen electron in molecules containing the carbonyl group (C=O) to the π^* orbital. This commonly is called an $n \rightarrow \pi^*$ transition. The selection rules are only approximate when they describe a transition as forbidden because in actual practice an absorption band generally can be observed. Even so, the intensity of the band will be diminished significantly if it corresponds to a forbidden transition; in this situation ϵ is usually in the range 10^{-2}–10^2.

Summary

The electronic structures of polyatomic molecules may be described by forming **localized** molecular orbitals between every pair of adjacent atoms in a molecule. It is often convenient to use **hybrid orbitals** to explain bonding between a centrally located atom, such as the carbon in CH_4, and the outer attached atoms (the four hydrogens in CH_4) in a localized bonding scheme. If four groups are to be attached, four equivalent sp^3 (tetrahedral) hybrids are used; for three groups, three equivalent sp^2 (trigonal planar) hybrids are used; for two groups, two equivalent sp (linear) hybrids are used. For example, each C—H bond in CH_4 may be described as an electron pair in a localized bonding molecular orbital built from a carbon sp^3 hybrid and a hydrogen $1s$ ($sp^3 + 1s$).

There are some molecules that contain one or more **bridge bonds**; an example is diborane, B_2H_6, in which two three-center, two-electron B—H—B bonds are believed to be present.

Molecules such as NH_3 and H_2O may be formulated in terms of localized molecular orbitals. In NH_3, for example, there are three bonding pairs in localized ($sp^3 + 1s$) orbitals, and the lone pair is in the remaining sp^3 hybrid orbital.

In **saturated** hydrocarbons each carbon uses four sp^3 hybrids to attach four groups in a tetrahedral geometry; **unsaturated** hydrocarbons are those in which at least two carbons are sp^2 or sp hybridized, so that they have not reached their full bonding capacity (of four groups).

Molecules such as benzene (C_6H_6), in which there are many connected unsaturated carbon centers, are best described by formulating **delocalized** π orbitals that are built from the $2p$ orbitals on the attached sp^2-hybridized carbon centers. Electrons in these delocalized π molecular orbitals are not restricted to conventional two-atom bonds; rather, they are spread over the entire molecule.

Polyatomic molecules may be polar or nonpolar. It is convenient to think of each bond in the molecule as having a dipole moment; these individual **bond dipoles** may add to give the molecule a net dipole moment, or they may subtract giving no net dipole moment. Whether they add or subtract depends on the molecular geometry; in H_2O, which is angular, each $H^{\delta+}O^{\delta-}$ bond dipole adds to give a resultant dipole moment. Water is therefore a polar molecule. In linear CO_2, however, the individual $C^{\delta+}O^{\delta-}$ bond dipoles cancel each other, yielding a zero resultant dipole moment. Therefore, CO_2 is an example of a **symmetric nonpolar molecule**.

Molecules have **electronic energy levels**, **vibrational energy levels**, and **rotational energy levels**. Transitions between rotational levels fall in the microwave region of the spectrum; those between vibrational energy levels fall in the infrared region; and those between electronic energy levels fall in the visible and ultraviolet regions. **Infrared spectroscopy** and **Raman** spectroscopy are used to observe vibrational transitions in molecules. Absorption of visible and ultraviolet light by molecules occurs because of electronic transitions The intensity of this absorption as a function of the wavelength of light is known as an absorption spectrum

Self-Study Questions

1. What are localized and delocalized molecular orbitals?
2. How are sp hybrid orbitals used to describe the bonding in BeH_2?
3. What are bridge bonds? Why are they needed in describing the bonding in diborane, B_2H_6?
4. What happens to the third p orbital in sp^2 hybridization?
5. Why is sp^2 trigonal hybridization wrong for the ammonia molecule?
6. What is the localized $(sp^3 + 1s)$ bond theory, and how can it be used to predict the structures of H_2O and NH_3? Why is the theory less accurate with H_2S than with H_2O?
7. How is the second bond of the $C{=}C$ double bond in ethylene formed?
8. Why does the presence of the double bond in ethylene keep the molecule planar? What would happen if the two ends of the ethylene molecule were twisted around the $C{=}C$ bond axis?
9. Why is the sp^3 hybridization model for ethylene less correct than the sp^2 hybridization model?
10. What experimental evidence do we have that a Kekulé structure for bonding in benzene is wrong?
11. What is the relationship between the number of nodes in the wave functions for benzene and their energies?
12. Why is CCl_4 nonpolar, whereas CH_3Cl is polar?
13. What is a bond dipole?
14. Which isomer of 1,2-dichloroethylene is polar?
15. Why is *para*dichlorobenzene nonpolar?
16. What type of transition is observed in infrared spectroscopy?
17. What type of transition is observed in Raman spectroscopy?
18. What type of transition gives rise to absorption in the visible and ultraviolet regions of the spectrum?
19. Why do saturated hydrocarbons absorb at higher energies than does ethylene?
20. What is the nature of the forbidden electronic transition in H_2CO?

Problems

Hybridization and molecular properties

1. What is the hybridization of the Ge valence orbitals in GeH_4?

2. Indicate an appropriate hybridization for the central-atom valence orbitals and predict the molecular shape and polarity for each of the following molecules: (a) CS_2; (b) CBr_4; (c) PF_3; (d) H_2Te; (e) SiH_4; (f) SF_2; (g) BF_3; (h) OF_2.

3. Formulate the hybridization of the central atom and predict the molecular geometry of CH_3^+.

4. What is the geometry of bonding around each carbon atom in acetic acid,

$$H_3C-\overset{\overset{\textstyle O}{\|}}{C}-OH$$

What is the hybridization of the H_3C- carbon? Of the $-\overset{\overset{\textstyle O}{\|}}{C}-$ carbon? Which of the carbon–oxygen bonds in the acetic acid molecule is longer?

5. Formulate the hybridization of the central carbon atom in acetone,

$$CH_3-\overset{\overset{\textstyle O}{\|}}{C}-CH_3$$

6. What is the hybridization of the central carbon atom in acetonitrile, $CH_3C\equiv N$?

Localized molecular orbitals

7. For the molecules and ions CO_2, NO_2^+, NO_2, NO_2^-, and SO_2:
   a) Represent their electronic structures by drawing localized molecular orbital structures.
   b) Predict their molecular shapes and indicate which neutral molecules are polar.
   c) Predict the number of unpaired electrons for each molecule.
   d) Predict the bond angles in NO_2^+ and NO_2^-.
   e) Predict relative N—O bond lengths for NO_2^+, NO_2, and NO_2^-.

8. Draw a localized-molecular-orbital structure for N_3. How many unpaired electrons does it have? Is the molecule polar? Predict the relative N—N bond lengths in N_2 and N_3.

9. Formulate a localized-bond model for H_3O^+.

10. Borazine has the formula $(BHNH)_3$. What do you think the molecular structure of borazine is? How would you formulate the electronic structure of borazine in terms of localized molecular orbitals?

11. Formulate a localized-orbital model for both the σ and π bonds in the ozone molecule, O_3. What is the hybridization of the central oxygen atom? How many π bonds are there between each pair of oxygen atoms?

12. Draw localized-orbital representations including π bonds for NO_3^-, NCO^-, and CNO^-.

13. The allyl cation has the formula $C_3H_5^+$. What structure does it have? Formulate a localized-orbital model of the electronic structure of $C_3H_5{}^+$. Repeat the problem for the allyl radical, C_3H_5, and the allyl anion, $C_3H_5^-$.

14. Formulate a localized-bond-orbital model for methyl fluoride, CH_3F. What H—C—F angle do you predict?

15. Formulate a localized-bond-orbital model for the amide ion, NH_2^-. Is NH_2^- linear, or angular?

16. Formulate a localized-bond-orbital model for hydrazine, N_2H_4. Is N_2H_4 planar? Is there a double, or a single, bond between the two nitrogen atoms?

Dipole moments

17. Would you expect COS to have a higher dipole moment than CS_2, or a lower one? Explain.

18. Which should have the larger dipole moment, CS_2 or H_2S? Why?

19. The dipole moment of H_2O is 1.85 D whereas that of F_2O is only 0.297 D, although the bond angles in the two molecules are nearly the same. Explain why the dipole moment of F_2O is so much smaller than that of H_2O.

20. Attempt an explanation of the following dipole moments in terms of a localized-orbital model of electronic structure: NH_3, $\mu = 1.47$ D; PH_3, $\mu = 0.55$ D; NF_3, $\mu = 0.23$ D.

21. Why should the dipole moment of H_2Se be lower than that of H_2O?

Ionization energies

22. Which molecule would you expect to have the higher ionization energy, ethylene or ethane? Why?

23. Predict the relative ionization energies of ethylene and acetylene.

Electronic spectroscopy

24. Formaldehyde (H_2CO) exhibits a strong electronic absorption band in the ultraviolet region, which may be assigned to a $\pi^b \rightarrow \pi^*$ transition (see the discussion of ethylene in Section 13-6). In addition, a weaker, longer-wavelength peak is observed in the spectra of H_2CO, and all organic compounds containing a carbonyl (C=O) group, in the region 270–300 nm. Formulate the molecular orbitals for H_2CO and suggest a possible assignment for the long-wavelength peak.

25. Would you expect benzene to absorb light of lower, or higher, energy than ethylene does? Why?

26. One-electron oxidation of benzene gives the benzene radical cation, $C_6H_6^+$. What is the π-orbital configuration of $C_6H_6^+$? Would you expect $C_6H_6^+$ to absorb light due to electronic transitions at lower, or higher, energy than does benzene itself? Why?

Suggested Reading

E. Cartmell and G. W. A. Fowles, *Valency and Molecular Structure*, Butterworths, London, 1977, 4th ed.

C. A. Coulson, *Valence*, Oxford, New York, 1961, 2nd ed.

H. B. Gray, *Chemical Bonds*, W. A. Benjamin, Menlo Park, Calif., 1973.

H. B. Gray, *Electrons and Chemical Bonding,* W. A. Benjamin, Menlo Park, Calif., 1965.

M. Karplus and R. N. Porter, *Atoms and Molecules: An Introduction for Students of Physical Chemistry,* W. A. Benjamin, Menlo Park, Calif., 1970.

L. Pauling, *The Chemical Bond,* Cornell University Press, Ithaca, N.Y., 1967, 3rd ed.

G. C. Pimentel and R. D. Spratley, *Chemical Bonding Clarified through Quantum Mechanics,* Holden-Day, San Francisco, 1969.

14

Bonding in Solids and Liquids

Imagine two hundred brilliant violin players playing the same piece with perfectly tuned instruments, but commencing at different places selected at random. The effect would not be pleasing, and even the finest ear could not recognize what was being played. Such music is made for us by the molecules of gases, liquids, and ordinary solids. . . . A crystal, on the other hand, corresponds with the orchestra led by a vigorous conductor when all eyes intently follow his nod, and all hands follow the exact beat. . . . To me, the music of physical law sounds forth in no other department in such full and rich accord as in crystal physics.

W. Voigt

Now that we know how bonding occurs between small numbers of atoms, we can look at bonding in solids and liquids. A simple but quite useful theory of the electrical properties of solids considers the entire solid body as one large molecule and uses delocalized molecular orbitals that extend over the entire solid. This is the **band theory** of metals and insulators.

How are compounds that definitely exist as molecules held together in a solid? Why aren't Br_2, I_2, and all organic substances gases at room temperature? What is the force that keeps the hydrocarbon molecules of gasoline in the liquid state? Why is sugar crystalline if no covalent or ionic bonds hold one molecule to another? Molecular solids become comprehensible as soon as we recognize the contributions of the weak forces known as van der Waals attraction and hydrogen bonding.

Five types of chemical bonding—covalent bonding in nonmetals, metallic bonding, ionic attraction, hydrogen bonding, and van der Waals attraction—are sufficient to explain atomic and molecular interactions in solids, liquids, and gases. Each type contributes a different kind of stability to the atoms it binds. In what follows, we shall examine how the types differ.

Figure 14-1   The structure of solid argon. Each sphere represents an individual Ar atom, in cubic close packing (see Figure 14-7) with 3.8 Å between the atomic centers.

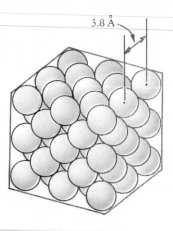

3.8 Å

14-1   ELEMENTAL SOLIDS

Solids that are built by weak attractive interactions between individual molecules are called **molecular solids**. At very low temperatures the noble gases (He, Ne, Ar, Kr, Xe, Rn) exist as molecular solids that are held together by weak interatomic forces. For example, Ar freezes at $-189°C$ to make the close-packed structure shown in Figure 14-1. Examples of elemental substances that crystallize to give molecular solids include the halogens; for example, Br_2 freezes at $-7°C$ to build the structure shown in Figure 14-2.

Group VIA atoms such as oxygen and sulfur (s^2p^4) have two vacancies in their valence shells and hence form two electron-pair bonds per atom. Under normal conditions of temperature and pressure, the most stable

Figure 14-2   The structure of crystalline bromine, Br_2. Spheres with solid outlines represent one layer of packed molecules, and those with dashed outlines represent the layer beneath. The molecules have been shrunk for clarity in this drawing; they are actually in close contact within a layer, and the layers are packed against one another.

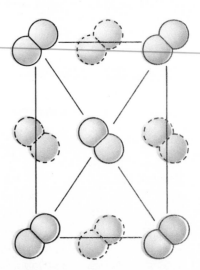

Figure 14-3   Ring molecule of eight sulfur atoms as found in solid sulfur.

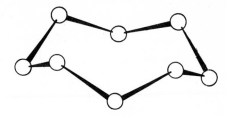

form of elemental oxygen is the diatomic molecule, whereas sulfur exists as a solid, and its two principal allotropes* consist of discrete S_8 rings (Figure 14-3). There are two other allotropes of sulfur, one of which has S_6 rings; the other contains helical chains of S atoms.

Atoms of the Group VA elements have three vacancies in their valence electronic configuration ($s^2 p^3$) and thus are expected to form three electron-pair bonds per atom. The most stable form of elemental nitrogen is the diatomic molecule, whereas the allotrope white phosphorus exists as a solid containing discrete tetrahedral P_4 units (Figure 14-4).

We turn next to the Group IVA elements carbon and silicon, which have the valence electronic configuration $s^2 p^2$ with only two unpaired electrons. We might expect only *two* electron-pair bonds per atom as in the diatomic molecules

$$:C{=}C: \qquad :Si{=}Si:$$

However, the C_2 molecule is orbital rich and electron deficient, since it has not achieved an octet around each atom. Each carbon atom prefers to form four electron-pair bonds, as illustrated by the two common allotropes,

*Allotropes of an element possess different interatomic structures and have different physical and chemical properties.

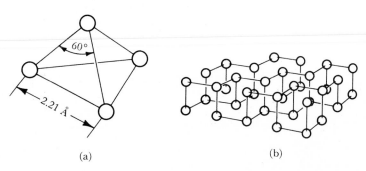

(a)   (b)

Figure 14-4   Structures of two allotropes of solid phosphorus. (a) White phosphorus consists of discrete P_4 molecules. (b) Black phosphorus, which is more stable, has an infinite network structure.

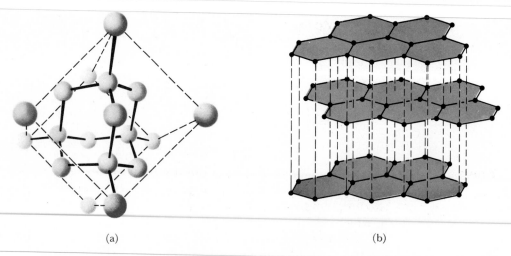

(a)   (b)

Figure 14-5   Crystalline carbon. (a) Diamond structure. The coordination number of carbon in diamond
                   is 4. Each atom is surrounded tetrahedrally by four equidistant atoms. The C—C bond
                   distance is 1.54 Å. (b) Graphite structure. This is the more stable structure of carbon.
                   Strong carbon—carbon bonding occurs within a layer, and weaker bonding occurs be-
                   tween layers.

diamond and graphite (Figure 14-5). Similarly, Si_2 is electron deficient, and
does not exist as an individual molecule in solid silicon. Rather, the structure of solid silicon is analogous to that of diamond (Figure 14-5a).

Diamond and graphite are called **nonmetallic network solids**, because
they consist of infinite arrays of bonded atoms; no discrete molecules can
be distinguished. Thus any given piece of network solid may be considered a giant, covalently bonded molecule. Network solids generally are
poor conductors of heat and electricity. Strong covalent bonds among
neighboring atoms throughout the structure give these solids strength and
high melting temperatures. Diamond sublimes (does not melt but volatil-
izes directly to a gas) at 3500°C and above. Some of the hardest substances
known are nonmetallic network solids.

The only Group IIIA element with nonmetallic properties is boron
($2s^2 2p^1$). There are three principal allotropic forms of elemental boron, and
all have network structures based on a B_{12} unit. Each B_{12} unit is icosahedral
(Figure 14-6). In the three allotropes the B_{12} icosahedra are linked together

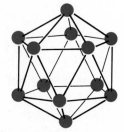

Figure 14-6   Structure of the B_{12} icosahedral unit. In the different crystalline
                   forms (allotropes) of boron, these units are linked together in
                   various ways.

in different ways, but in general the bonds between individual icosahedra are weaker than those within any one icosahedron, where each boron atom is bonded to five others.

Atoms of the metallic elements generally have fewer valence electrons than the number of available orbitals; that is, they are electron deficient. Consequently these atoms tend to share their electron density with several other atoms to achieve a maximum bonding capacity. In most metals at least eight "nearest-neighbor" atoms surround a particular atom in one of three common structures shown in Figure 14-7. In both hexagonal close packing and cubic close packing, each sphere touches 12 other spheres, 6 in a plane, 3 above, and 3 below. It has been shown through x-ray analysis that two-thirds of all metals crystallize in one of these two structures. A majority of the other one-third crystallize as body-centered cubes, in which each atom has only eight nearest neighbors.

Lithium and sodium, which have the s^1 valence electronic configuration, adopt the body-centered cubic structure (Figure 14-7a). Solid beryllium and magnesium (s^2 atoms) both crystallize in the hexagonal close-packed structure (Figure 14-7b). The crystal structure of aluminum ($s^2 p^1$) is cubic close packed (Figure 14-7c).

In the periodic table shown in Figure 14-8, the elemental solids are classified as metallic, network nonmetallic, or molecular. In Table 14-1 the correlation between coordination number and structure in elemental solids is presented. The majority of elements crystallize in metallic structures in which each atom has a high coordination number. Included as metals are elements such as tin and bismuth, which crystallize in structures with relatively low atomic coordination numbers but which still have strong metallic properties. The light blue area of the periodic table includes elements

Table 14-1

The Correlation between Coordination Number and Structure in Elemental Solids

Bonding coordination number	Type of solid structure
0	Atomic solids, low melting and boiling points
1	Diatomic molecular solids, low melting and boiling points
2	Rings or chains. Solids with packed ring molecules are less metallic than those with packed chains
3	P_4 tetrahedra or sheets. Solids with packed tetrahedral molecules are less metallic than those with packed sheets
4	Three-dimensional nonmetallic networks
5	B sheet curved in on itself in a B_{12} icosahedron
6 or more	Packed metallic solids

Figure 14-7

Common structures of metals. (a) Body-centered
cubic (e.g., Na, V, and Ba); (b) hexagonal close
packing (e.g., Mg, Ir, and Cd); (c) cubic close
packing, or face-centered cubic (e.g., Al, Cu,
and Au).

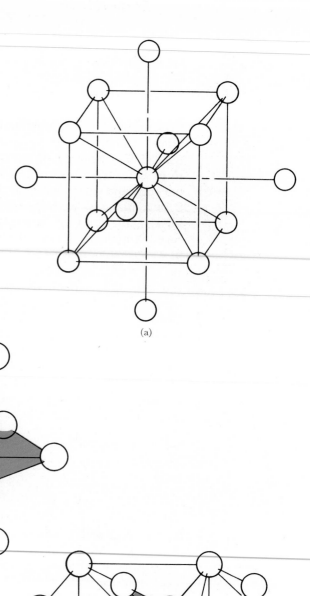

(a)

(b)

(c)

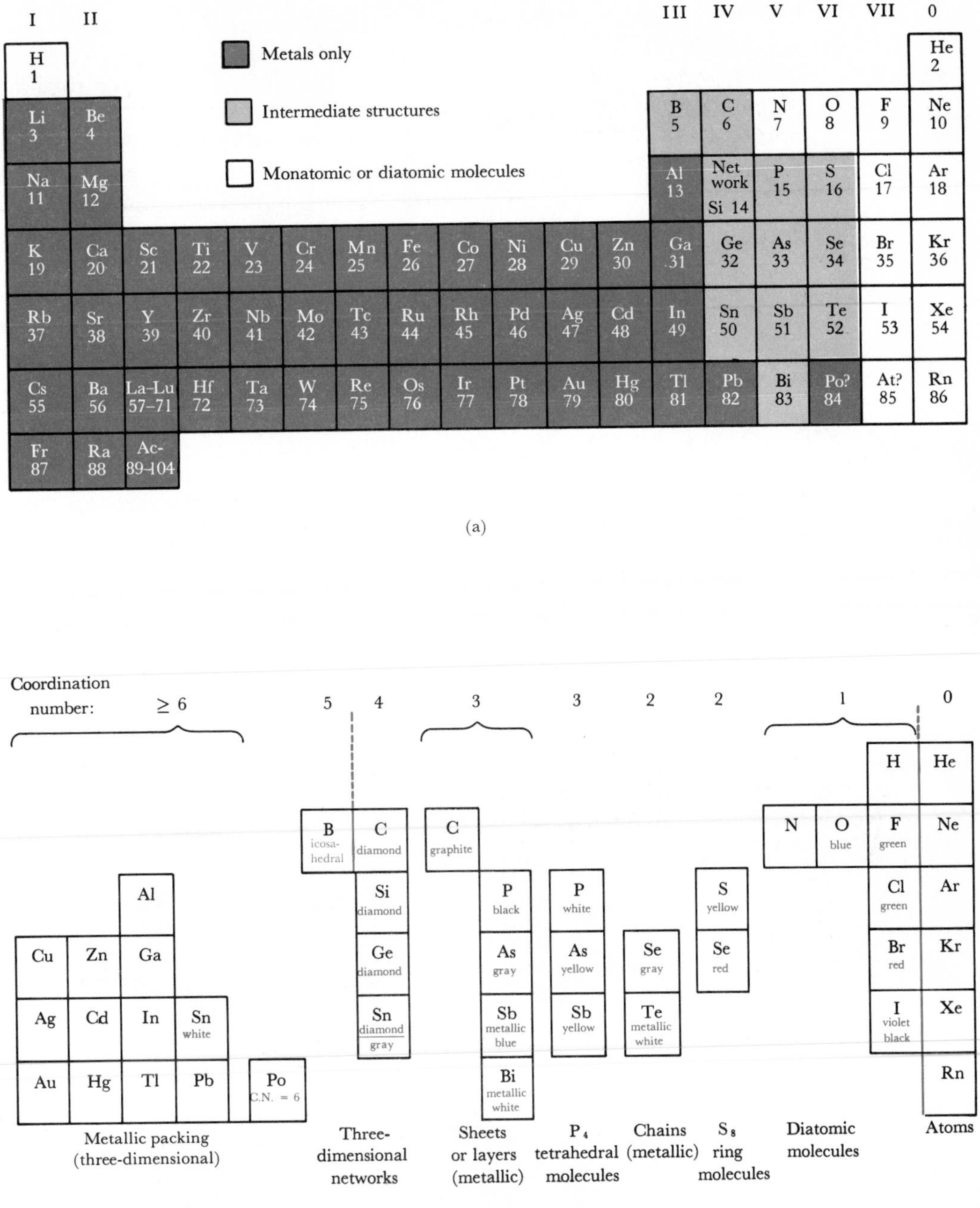

(a)

(b)

Figure 14-8 Solid structures of the elements. (a) General trends in bonding in solid elements. (b) The types of bonding in the variable zone of the table. The structures of the nonmetals are determined by their coordination numbers, which are 8 minus the group number except when multiple bonds are used as in graphite, N_2, and O_2 (see Table 14-1).

that have borderline properties. Although germanium crystallizes in a diamondlike structure in which the coordination number of each atom is only 4, some of its properties resemble those of metals. This similarity to metals indicates that the valence electrons in germanium are not held as tightly as would be expected in a true nonmetallic network solid. Arsenic, antimony, and selenium exist as either molecular or metallic solids, although the so-called metallic structures have relatively low atomic coordination numbers. We know that tellurium crystallizes in a metallic structure, and it seems reasonable to predict that it may also exist as a molecular solid. From its position in the periodic table we predict intermediate properties for astatine, which has not been studied in detail.

14-2   IONIC SOLIDS

Ionic solids consist of infinite arrays of positive and negative ions that are held together by electrostatic forces. These forces are the same as those that hold a molecule of NaCl together in the vapor phase. In solid NaCl the Na^+ and Cl^- ions are arranged to maximize the electrostatic attraction, as shown in Figure 14-9. The coordination number of each Na^+ ion is 6, and

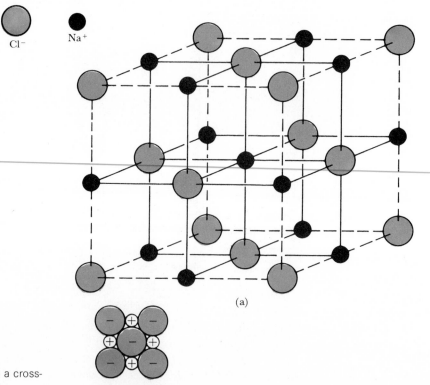

(a)

(b)

Figure 14-9

(a) The ionic NaCl structure; (b) a cross-sectional view.

each Cl^- ion similarly is surrounded by six Na^+ ions. Because ionic bonds are very strong, much energy is required to break down the structure in solid-to-liquid or liquid-to-gas transitions. Thus ionic compounds have high melting and boiling temperatures.

The crystal structures of several typical ionic solids are shown in Figure 14-10. Cesium chloride crystallizes in a structure in which the cation and

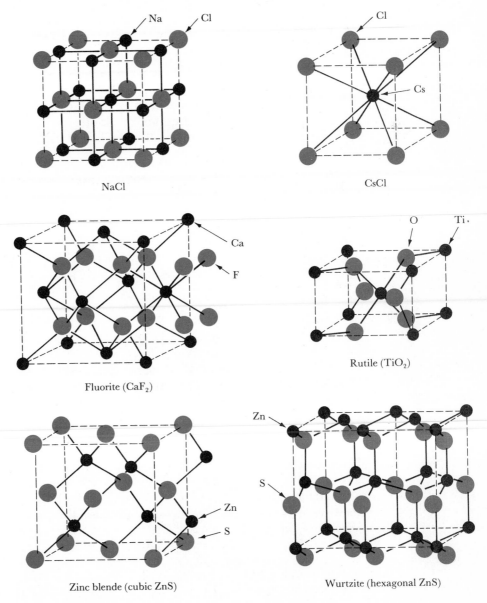

NaCl

CsCl

Fluorite (CaF_2)

Rutile (TiO_2)

Zinc blende (cubic ZnS)

Wurtzite (hexagonal ZnS)

Figure 14-10   Some of the common structural types found for ionic substances.

anion each have a coordination number of 8. Zinc sulfide crystallizes in two distinct structures—the so-called zinc blende and wurtzite structures—in which the cation and anion each have a coordination number of 4. Calcium fluoride crystallizes in the fluorite structure. The coordination numbers are 8 for the cation (eight fluorides surround each calcium) and 4 for the anion. One of the crystalline forms of titanium dioxide is rutile, in which the coordination numbers are 6 for the cation and 3 for the anion.

14-3   MOLECULAR SOLIDS AND LIQUIDS

Molecules such as H_2, N_2, O_2, and F_2 form molecular solids because all the valence orbitals are either used for *intramolecular* bonding or occupied with nonbonding electrons. Thus any intermolecular bonding that holds molecules together in the solid must be weak compared with the strength of the intramolecular bonding in the molecules. The weak forces that contribute to intermolecular bonding are called **van der Waals forces**.

Van der Waals Forces

There are two principal van der Waals forces. The most important force at short range is the repulsion between electrons in the filled orbitals of atoms on neighboring molecules. This electron-pair repulsion is illustrated in Figure 14-11. The analytical expression commonly used to describe the energy resulting from this interaction is

$$\text{van der Waals repulsion energy} = be^{-ar} \qquad (14\text{-}1)$$

in which b and a are constants for two interacting atoms. Notice that this repulsion term is very small at large values of the interatomic distance, r.

The second van der Waals force is the attraction that results when electrons in the occupied orbitals of the interacting atoms synchronize their motion to avoid one another as much as possible. For example, as shown in Figure 14-12, electrons in orbitals of atoms belonging to interacting molecules can synchronize their motion to produce an instantaneous dipole–induced dipole attraction. If at any instant the atom on the left in Figure 14-12 has more of its electron density at the left (as shown) then the atom becomes a tiny dipole with a negative left side and a positive right side. This positive side attracts electrons of the atom on the right in the figure and changes this atom into a dipole with a similar orientation. Therefore these two atoms attract each other because the positive end of the left atom and the negative end of the right atom are close. Similarly, fluctuation in electron density of the right atom induces a temporary dipole, or asymmetry of electron density, in the left atom. The electron densities fluctuate continually, yet the net effect is an extremely small but important attraction between atoms. The energy resulting from this attractive force

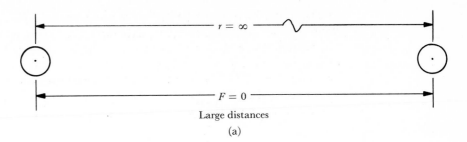

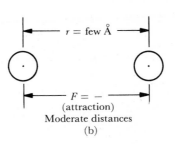

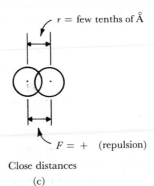

Figure 14-11

Repulsion of electrons in filled orbitals. (a) At very large distances two atoms or molecules behave as neutral species and neither repel nor attract one another. The force between them, *F*, is zero. (b) At moderate distances two atoms or molecules have not yet come close enough for repulsion to be appreciable. However, they do attract one another (see Figure 14-12) because of deformations of their electron densities. (c) At close range, when the electron density around one atom or molecule is large in the same region of space as the electron density around the other atom or molecule (i.e., when the filled orbitals overlap), coulomblike repulsion dominates and the two molecules repel one another.

is known as the **London energy**, after Fritz London, who derived the quantum mechanical theory for this attraction in 1930. The London energy varies inversely with the sixth power of the separation between atoms:

$$\text{London energy} = -\frac{d}{r^6} \tag{14-2}$$

in which d is a constant and r is the distance between atoms. This "inverse sixth" attractive energy decreases rapidly with increasing r, but not nearly as rapidly as the van der Waals repulsion energy. Thus at longer distances the London attraction is more important than the van der Waals repulsion; consequently, a small net attraction results.

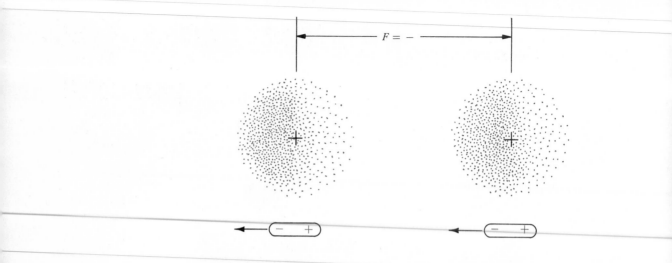

Instantaneous polarization of an atom leaves, for a moment, more electron density on the left than on the right, thus creating an "instantaneous dipole"

This "instantaneous dipole" can polarize another atom by attracting more electron density to the left, thereby creating an "induced dipole"

Figure 14-12   Schematic illustration of the instantaneous dipole—induced dipole interaction that gives rise to a weak attraction. For the brief instant that this figure describes, there is an attractive force, F, between the instantaneous dipole and the induced dipole. The effect is reciprocal; each atom induces a polarization in the other.

The total potential energy (PE) of van der Waals interactions is the sum of the attractive energy of equation 14-2 and the repulsive energy of equation 14-1:

$$PE = be^{-ar} - \frac{d}{r^6} \tag{14-3}$$

The total van der Waals potential energy can be compared quantitatively with ordinary covalent bond energies by examining systems for which the curves of potential energy versus interatomic distance r are known accurately. We can calculate values for the constants a, b, and d from experimental data on the deviation of real gases from ideal gas behavior. Some of these values for interactions of noble gases are listed in Table 14-2.

The potential-energy curve for van der Waals interactions between helium atoms is illustrated in Figure 14-13. At separations of more than 3.5 Å, the second term in equation 14-3 predominates. As the atoms move closer, they attract each other more, and the energy of the system decreases. However, at distances less than 3 Å the strong electron-pair repulsion overwhelms the London attraction, and the potential-energy curve in Figure 14-13 rises. A balance between attraction and repulsion exists at a

Table 14-2
Van der Waals Energy Parameters

Interaction pair	a (a.u.)$^{-1a}$	b (kJ mole^{-1})	d [kJ mole^{-1} (a.u.)^6]
He—He	2.10	17.1×10^3	6.3×10^3
He—Ne	2.27	86.6×10^3	12.1×10^3
He—Ar	2.01	125.5×10^3	40.6×10^3
He—Kr	1.85	68.6×10^3	57.3×10^3
He—Xe	1.83	111.3×10^3	89.1×10^3
Ne—Ne	2.44	438.5×10^3	23.8×10^3
Ne—Ar	2.18	635.1×10^3	80.3×10^3
Ne—Kr	2.02	346.4×10^3	111.7×10^3
Ne—Xe	2.00	561.5×10^3	173.6×10^3
Ar—Ar	1.95	918.4×10^3	270.3×10^3
Ar—Kr	1.76	501.2×10^3	377.0×10^3
Ar—Xe	1.74	813.4×10^3	582.8×10^3
Kr—Kr	1.61	272.8×10^3	524.7×10^3
Kr—Xe	1.58	443.5×10^3	813.4×10^3
Xe—Xe	1.55	718.8×10^3	1259.8×10^3

[a] 1 a.u. = 1 atomic unit = 0.529 Å. The value of r in equation 14-3 must be expressed in atomic units as well.

separation of 3 Å and the He—He "molecule" is 76.1 J mole^{-1} more stable than two isolated atoms.

Figure 14-13 also shows the marked contrast between van der Waals attraction and covalent bonding. In the H_2 molecule strong electron–proton attractions in the bonding molecular orbital cause the potential energy to decrease as the H atoms approach one another, and it is proton–proton repulsion that makes the energy increase sharply if the atoms are pushed too closely together. This proton–proton repulsion operates at smaller distances than the electronic repulsion between the two He atoms. The H—H bond length in the H_2 molecule is 0.74 Å, whereas the equilibrium distance of van der Waals–bonded He atoms is 3 Å. Moreover, a covalent bond is much stronger than a weak van der Waals interaction. Only 76.1 J mole^{-1} are required to separate helium atoms at their equilibrium distance, but 431,000 J mole^{-1} are needed to break the covalent bond in H_2.

Molecular solids, in which only van der Waals intermolecular bonding exists, generally melt at low temperatures. This is because relatively little energy of thermal motion is needed to overcome the energy of van der Waals bonding. The liquid and solid phases of helium, which result from weak van der Waals "bonds," exist only at temperatures below 4.6 K. Even at temperatures near absolute zero, solid helium can be produced only at high pressures (29.6 atm at 1.76 K).

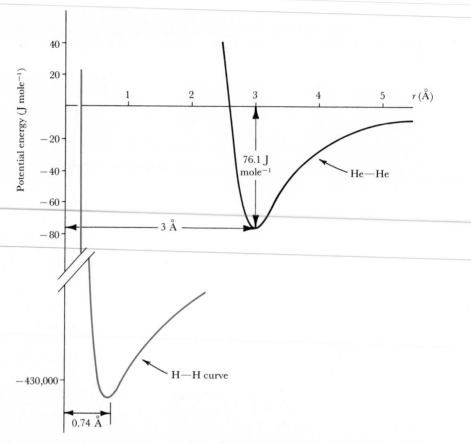

Figure 14-13   A comparison of the potential-energy curves for van der Waals attraction between two He atoms (black curve) and covalent bonding between two H atoms (colored curve). Notice that the energy scale is in joules rather than kilojoules per mole. The covalent bond is more than 5000 times as stable as the van der Waals bond.

Van der Waals bonds in molecular solids and liquids generally are stronger with increasing size of the atoms and molecules involved. For example, as the atomic number of the noble gases increases, the strength of the van der Waals bonding also increases, as shown by the argon–argon potential-energy curve in Figure 14-14. The attraction between the heavier atoms is stronger, presumably because the outer electrons are held more loosely, and larger instantaneous dipoles and induced dipoles are possible. Because of this stronger van der Waals bonding, solid argon melts at $-184°C$, or 89 K, which is considerably higher than the melting point for solid helium.

An example of the effect of molecular size on melting and boiling temperatures is provided by a series of continuous-chain alkanes, with formulas C_nH_{2n+2}, depicted in Figure 14-15 for $n = 1$ through 20. Part of

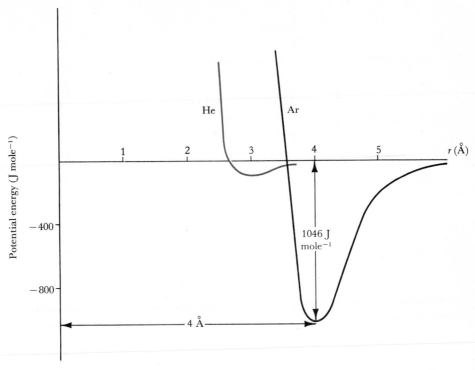

Figure 14-14   A comparison of the potential-energy curves for van der Waals attraction between two Ar atoms (black curve) and two He atoms (colored curve). The larger Ar atoms are more tightly held, although the bond energy is still one four-hundredth that of a H—H bond.

the increase in melting and boiling temperatures with increasing molecular size and weight arises from the greater energy needed to move a heavy molecule. However, another important factor is the large surface area of a molecule such as eicosane ($C_{20}H_{42}$), compared with methane, and the greater stability that eicosane can therefore gain from intermolecular van der Waals attractions. The mass effect is similar for both melting and boiling temperatures. However, molecular surface area affects the boiling temperatures more because molecules in the liquid phase are still close enough to exert van der Waals attractions. In fact, without these attractions, which are broken during vaporization, the liquid state could not exist.

Polar Molecules and Hydrogen Bonds

Polar molecules are stabilized in a molecular solid by the attractive interaction of oppositely charged ends of the molecules (Figure 14-16). This is

Figure 14-15

Melting and boiling temperatures of the continuous-chain hydrocarbons as a function of the length of the carbon chain. More energy is required to separate two molecules of eicosane (20 carbons) than two molecules of ethane (2 carbons) because of the more numerous van der Waals interactions between the two larger molecules.

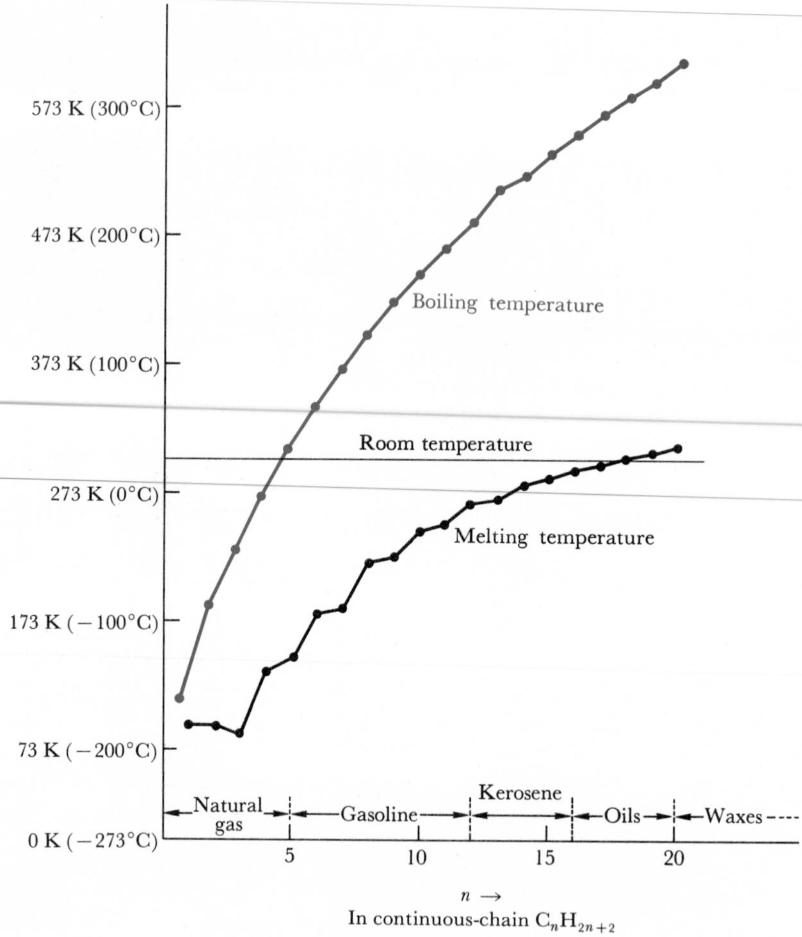

573 K (300°C)

473 K (200°C)

Boiling temperature

373 K (100°C)

Room temperature

273 K (0°C)

Melting temperature

173 K (−100°C)

73 K (−200°C)

Kerosene

← Natural gas → ← Gasoline → ← → ← Oils → ← Waxes ----

0 K (−273°C)

5   10   15   20

$n \rightarrow$

In continuous-chain $C_n H_{2n+2}$

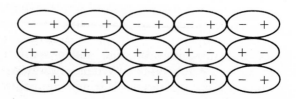

Figure 14-16   Diagrammatic representation of the packing of polar molecules into a crystalline solid. Packing occurs so partial charges of opposite sign are in close proximity.

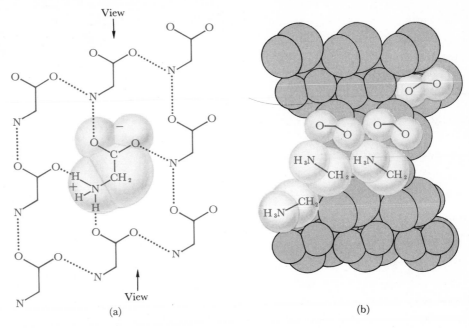

Figure 14-17   The bonding in the sheet structure of solid glycine, $^+H_3N-CH_2-COO^-$. (a) Molecules in a layer are packed tightly and are held together by van der Waals attractions and by hydrogen bonds (dotted). (b) The layers are stacked on top of one another and held together by van der Waals attractions. With this perspective the layers are on edge, and in a horizontal position. The view of the layers in (b) is marked by arrows in (a).

called **dipole–dipole interaction**. A particularly important kind of polar interaction is the **hydrogen bond**. This is a bond, primarily electrostatic, between a positively charged hydrogen atom and a small, electronegative atom, usually N, O, or F. For example, glycine molecules are held in a sheet structure by van der Waals forces and hydrogen bonds (Figure 14-17). Ice provides another example of the importance of hydrogen bonding in building intermolecular structures. As shown in Figure 14-18, each oxygen atom of a polar H_2O molecule is tetrahedrally coordinated to four other oxygen atoms in a structure that resembles the diamond structure. Each oxygen atom is bound to its four neighbor oxygen atoms by hydrogen bonds. In two of these hydrogen bonds the central H_2O molecule supplies the hydrogen atoms; in the other two bonds the hydrogen atoms come from neighboring H_2O molecules. Such bonds are weak compared with covalent bonds. A typical covalent bond energy is about 400 kJ mole^{-1}, whereas a hydrogen bond between H and O is approximately 20 kJ mole^{-1}. But hydrogen bonds are important for the same reason that van der Waals bonds are: They may be weak but there are many of them.

Figure 14-18   In crystalline ice each oxygen atom is hydrogen bonded to two others by means of its own hydrogen atoms, and bonded to two more oxygen atoms by means of their hydrogen atoms. The coordination is tetrahedral and the structure is similar to that of diamond.

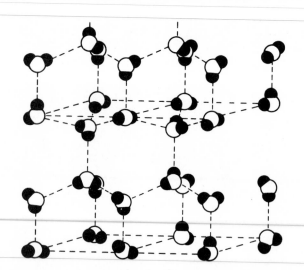

Hydrogen bonding in water is responsible for many of its most important properties. Because of hydrogen bonds in both solid and liquid phases the melting and boiling temperatures of water are unexpectedly high when compared with those of H_2S, H_2Se, and H_2Te, which are hydrogen compounds of elements also in Group VIA of the periodic table. Solid and liquid ammonia and hydrogen fluoride show anomalous behavior similar to that of water and for the same reason (Figure 14-19). However, hydrogen bonding is less pronounced in ammonia than it is in water for two reasons: N is less electronegative than O, and NH_3 has only one lone pair of electrons to attract the H from a neighboring molecule. In contrast, HF is not as well hydrogen-bonded as H_2O, despite the greater electronegativity of F and the presence of three lone pairs. This is because HF has only one hydrogen atom to use in making such bonds.

Since hydrogen bonding causes an open network structure in ice (Figure 14-18), ice is less dense than water at the melting temperature. Upon melting, part of this open-cage structure collapses, and the liquid is more compact than the solid. The measured heat of fusion of ice is only 5.9 kJ mole^{-1}, whereas the energy of its hydrogen bonds is 20 kJ mole^{-1}. This indicates that only about 30% of the hydrogen bonds of ice are broken when it melts. Water is not composed of isolated, unbonded molecules of H_2O; rather, it has regions or clusters of hydrogen-bonded molecules. That is, part of the hydrogen-bonded structure of the solid persists in the liquid. As the temperature is increased these clusters break up, and the volume continues to shrink. If the temperature is increased still more, the expected thermal expansion begins to take place. Consequently, liquid water has a minimum molar volume, or a maximum density, at 4°C.

Because the hydrogen-bonded H_2O clusters are broken slowly as heat is added, water has a higher heat capacity than any other common liquid

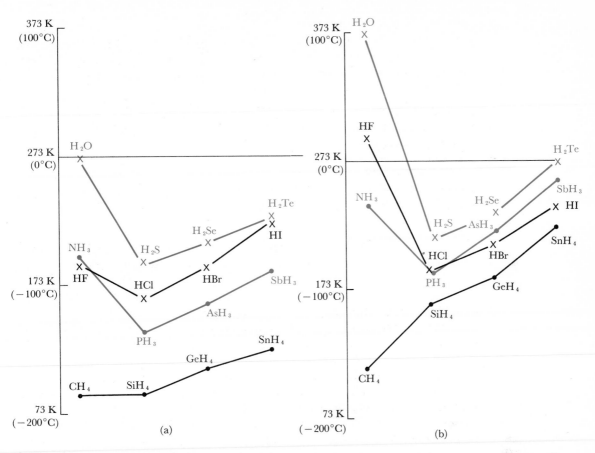

Figure 14-19   (a) Melting temperatures and (b) boiling temperatures for binary hydrogen compounds of some elements. Generally, melting and boiling temperatures increase with molecular weight within a group. The anomalous compounds HF, H_2O, and NH_3 have hydrogen bonds between molecules in both solid and liquid states.

except ammonia. Water also has an unusually high heat of fusion and heat of vaporization. All three of these properties give water the capability to act as a large thermostat, which confines the temperature on the earth within moderate limits. Ice absorbs a large amount of heat when it melts, and water absorbs more heat per unit of temperature increase than almost any other substance. Correspondingly, as water cools, it gives off more heat to its surroundings than other substances.

Coastal regions never experience the extremes of heat and cold that are typical of continental regions like the American Great Plains and the steppes of Central Asia and Siberia. It is unlikely that life could evolve and develop to a high level on planets where the extremes of temperature were not moderated by a high-specific-heat liquid such as H_2O.

Hydrogen bonds are even more important to life than the water structure suggests. They are one of the principal means of holding protein molecules together, as we shall see in Chapter 21. Without such bonds between carbonyl oxygen atoms and amine hydrogen atoms, no polypeptide chain would fold properly to make its protein.

Polar Molecules as Solvents

The polar nature of liquid water makes it an excellent solvent for ionic solids such as NaCl. Water can dissolve NaCl and separate the oppositely charged Na^+ and Cl^- ions because the energy required to separate the ions is provided by the formation of hydrated ions (Figure 14-20). Each Na^+ ion in solution still has an octahedron of negative charges around it, but instead of being Cl^- ions they are the negative poles of the oxygen atoms of the water molecules. The Cl^- ions also are hydrated, but it is the positive ends (H) of the water molecules that approach the Cl^- ions. A nonpolar solvent such as gasoline, a liquid composed of hydrocarbon molecules, cannot form such *ion–dipole bonds* with Na^+ and Cl^-. Consequently NaCl and other salts are insoluble in gasoline.

Polar solvents dissolve polar molecular solids because of dipole–dipole interactions. The energy released by the formation of dipole–dipole bonds between a polar solvent and solute molecules is sufficient to break the intermolecular forces in the molecular solids (Figure 14-21). For example, ice is soluble in liquid ammonia but not in benzene because NH_3 is a polar molecule, whereas C_6H_6 is nonpolar.

Figure 14-20   In solution, the hydrated Na^+ ion is surrounded by an octahedron of negative charges, but these negative charges are from the dipolar solvent molecule, H_2O, instead of from Cl^-.

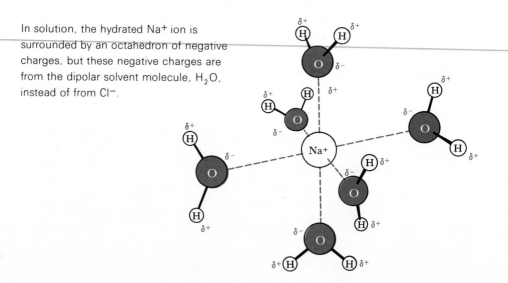

Energy is required to break up a solid molecular lattice

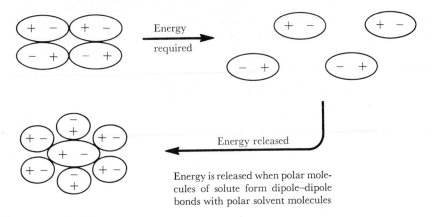

Energy required

Energy released

Energy is released when polar molecules of solute form dipole–dipole bonds with polar solvent molecules

Figure 14-21   When a crystalline solid composed of polar molecules dissolves, stability is lost when oppositely charged ends of neighboring molecules are removed. This loss is compensated by the stability produced by solvating the polar molecules in solution. A solvent that cannot provide such stabilization cannot dissolve the solid.

14-4   METALS

The high thermal and electrical conductivities of metals suggest that the valence electrons are relatively free to move through the crystal structure. Figure 14-22 illustrates one model in which the electrons form a sea of negative charges that holds the atoms tightly together. The circled positive charges represent the positively charged ions remaining when valence electrons are stripped away, thereby leaving the nuclei and the filled electron

Figure 14-22   Cross section of a crystal structure of a metal with the sea of electrons. Each circled positive charge represents the nucleus and filled, non-valence electron shells of a metal atom. The shaded area surrounding the positive metal ions represents the mobile sea of electrons.

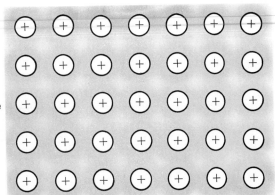

shells. Since metals generally have high melting temperatures and high densities, especially in comparison with molecular solids, the "electron sea" must strongly bind the positive ions in the crystal.

The simple electron-sea model for metallic bonding is also consistent with two other commonly observed properties of metals: malleability and ductility. A malleable material can be hammered easily into sheets; a ductile material can be drawn into thin wires. For metals to be shaped and drawn without fracturing, the atoms in the planes of the crystal structure must be displaced easily with respect to one another. This displacement does not result in the development of strong repulsive forces in metals because the mobile sea of electrons provides a constant buffer, or shield, between the positive ions. This situation is in direct contrast to that of ionic crystals, in which the binding forces are due almost entirely to electrostatic attractions between oppositely charged ions. In an ionic crystal, valence electrons are bound firmly to the atomic nuclei. Displacement of layers of ions in such a crystal brings ions of like charge together and causes strong repulsions that can lead to crystal fracture (Figure 14-23).

Electronic Bands in Metals

The molecular orbital theory provides an adequate model for metallic bonding. According to this model, the entire block of metal is considered to be a giant molecule. All the atomic orbitals of a particular type in the crystal interact to form a set of delocalized orbitals that extend throughout the entire block. For a particular crystal, assume that the number of valence orbitals is on the order of 10^{23}. To visualize the interaction of such a large number of valence orbitals, consider the hypothetical sequence of linear lithium molecules Li_2, Li_3, and Li_4 in which the important valence orbitals are $2s$ orbitals. Figure 14-24 shows the buildup of molecular orbitals for these three molecules. Notice that because of the delocalization of the molecular orbitals, none of the electrons is required to reside in an antibonding orbital. The spacing between the orbitals also becomes smaller. In the limit of 10^{23} equivalent atoms, the combination of atomic orbitals produces a band of closely spaced energy levels.

Figure 14-25 illustrates the three bands of energy levels formed by the $1s$, $2s$, and $2p$ orbitals of the simplest metal, lithium. The $1s$ molecular orbitals are filled completely because the $1s$ atomic orbitals in isolated lithium atoms are filled. Thus the $1s$ electrons make no contribution to bonding. They are part of the positive ion cores and can be eliminated from the discussion. Atomic lithium has one valence electron in a $2s$ orbital. If there are 10^{23} atoms in a lithium crystal, the 10^{23} $2s$ orbitals interact to form a band of 10^{23} delocalized orbitals. As usual, each of these orbitals can accommodate two electrons, so the capacity of the band is 2×10^{23} electrons. Lithium metal has enough electrons to fill only the lower half of the $2s$ band, as illustrated in Figure 14-25.

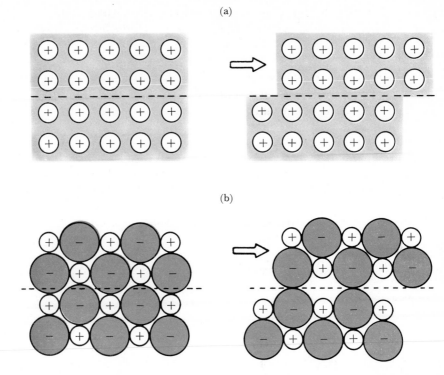

Figure 14-23   (a) Shift of metallic crystal along a plane results in no strong repulsive forces. (b) Shift of an ionic crystal along a plane results in strong repulsive forces and crystal distortion.

The presence of a partially filled band of delocalized orbitals accounts for bonding and electrical conduction in metals. Electrons in the lower filled orbital band move throughout the crystal in a random fashion such that their motion results in no *net* separation of electrons and positive ions in the metal. For a metal to conduct an electric current, electrons must be excited to unfilled delocalized orbitals in such a way that their movement in one direction is not exactly canceled by electrons moving in the opposite direction. Such concerted electron movement occurs only when an electric potential difference is applied between two regions of a metal. Then electrons are excited to the unfilled delocalized molecular orbitals that are part of the same band (the 2s band for lithium) and just slightly higher in energy. Therefore we can expect a metal to conduct electricity. Conduction is restricted by the frequent collisions of electrons with the positive ions, which have kinetic energy and thus vibrate randomly within their crystal sites. As the temperature increases, vibration of the positive ions increases, and collisions with the conduction electrons are more frequent. Therefore electrical conductivity in metals decreases as the temperature increases.

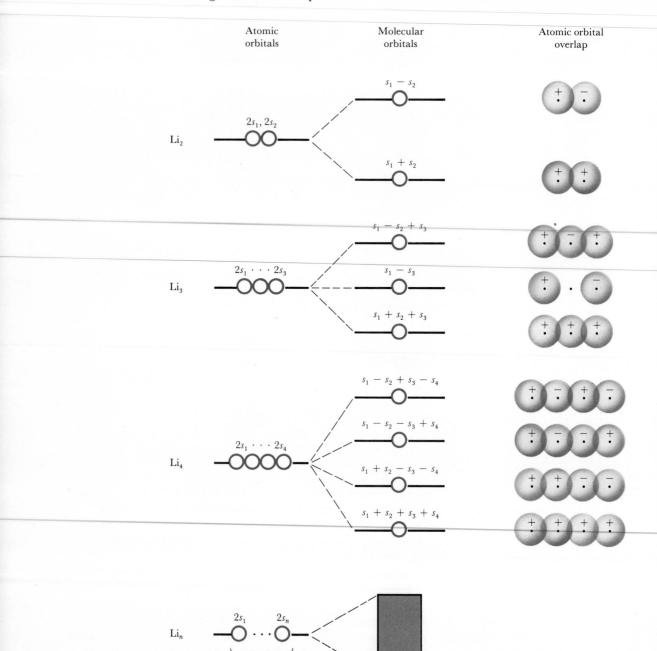

Figure 14-24   Molecular orbital development of the band theory of metals.

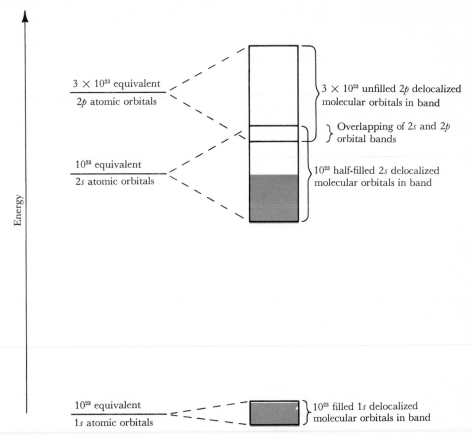

3×10^{23} equivalent
2p atomic orbitals

3×10^{23} unfilled 2p delocalized molecular orbitals in band

} Overlapping of 2s and 2p
} orbital bands

10^{23} equivalent
2s atomic orbitals

10^{23} half-filled 2s delocalized molecular orbitals in band

10^{23} equivalent
1s atomic orbitals

10^{23} filled 1s delocalized molecular orbitals in band

Figure 14-25   Delocalized-molecular-orbital bands in lithium. The original 2s and 2p atomic orbitals are so close in energy that the molecular orbital bands overlap. Lithium has one electron in every 2s atomic orbital, hence only half as many electrons as can be accommodated in the 2s atomic orbitals or in the delocalized-molecular-orbital band. There are unfilled energy states an infinitesimal distance above the highest-energy filled state, so an infinitesimal energy is required to excite an electron and send it moving through the metal. Thus lithium is a conductor.

Beryllium is a more complicated example of a metal than lithium is. An isolated beryllium atom has exactly enough electrons to fill its 2s orbital. Accordingly, beryllium metal has enough electrons to fill its 2s delocalized band. If the 2p band did not overlap the 2s (Figure 14-26), beryllium would not conduct well because an energy equal to the gap between bands would be required before electrons could move through the solid. However, the two bands do overlap and beryllium has unoccupied delocalized orbitals that are an infinitesimal distance above the most energetic filled orbitals. Consequently, beryllium is a metallic conductor.

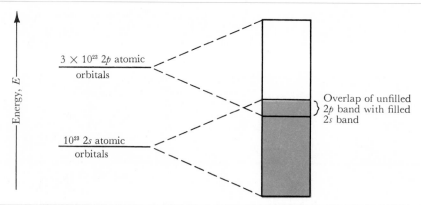

Figure 14-26   Band-filling diagram for beryllium. A Be atom has enough electrons (two) to fill its $2s$ orbital, so Be metal has enough electrons to fill its $2s$ delocalized-molecular-orbital band. If the $2s$ and $2p$ bands did not overlap, Be would be an insulator because an appreciable amount of energy would be required to make electrons flow in the solid. But with the band overlap shown here, an infinitesimal amount of energy excites electrons to the $2p$ band orbitals and electrons flow.

14-5   NONMETALLIC NETWORK SOLIDS

Nonmetallic network materials such as carbon or silicon are **insulators**, that is, they do not conduct electrical current. There is no simple way to apply the molecular orbital model to a discussion of the bonding in nonmetallic network solids. Suffice it to say that in nonmetallic network solids it is usually possible to "count" electrons in the Lewis electron-dot sense and show that the octet is achieved. This is because the atoms in nonmetallic network solids usually have at least as many valence electrons as the number of valence orbitals. Consequently, low coordination numbers are preferred, and simple electron-pair bonds can be formed between each atom and its nearest neighbors. Because of the low coordination numbers, the potential energy is not constant throughout the crystal; rather, it is greatly lowered in the internuclear region and the electrons are not free to move throughout the crystal as they are in the case of the metals.

For example, in diamond each carbon atom has a coordination number of 4. The hybrid orbital model adequately describes the bonding by assigning to each carbon atom four localized tetrahedral sp^3 hybrid orbitals (Figure 14-27). The four valence electrons in each carbon atom are sufficient to fill these bonding orbitals. Thus all electrons in diamond are used for bonding, leaving none to move freely to conduct electricity.

The effect of coordination number on the electronic bands of a solid can be illustrated with respect to carbon. Calculations have shown that the electronic bands for carbon would be delocalized *if* carbon crystallized in a

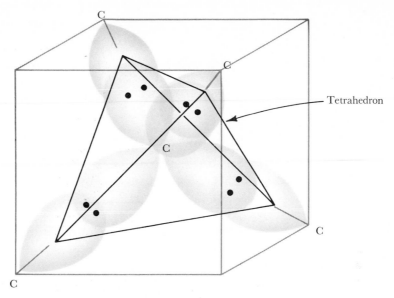

Figure 14-27 Schematic representation of the overlap of the four sp^3 hybrid orbitals of a carbon atom with similar orbitals from four other carbon atoms in diamond crystals.

structure with a high coordination number such as is found for the metals (Figure 14-28a). Consequently carbon would be an electrical conductor. Experimentally it is found that carbon has a coordination number of 4, and this causes the extended bands shown in Figure 14-28a to split into a filled band corresponding to bonding orbitals, and an unfilled band of anti-bonding orbitals. A "forbidden zone" or "band gap" separates the bands (Figure 14-28b).

For an insulator to conduct, it needs sufficient energy to excite the electrons in the filled band to move across the band gap into the unfilled molecular orbitals. This energy is the activation energy of the conduction process. Only high temperatures or extremely strong electrical fields will provide enough energy to an appreciable number of electrons for conduction to occur. In diamond the band gap (between the top of the filled or valence band and the bottom of the unfilled or conduction band) is 5.2 eV, or 502 kJ mole^{-1}.

Semiconductors

The borderline between metallic and nonmetallic network structures of elements in the periodic table is not sharp (Figure 14-8). This is shown by the fact that several elemental solids have properties that are intermediate

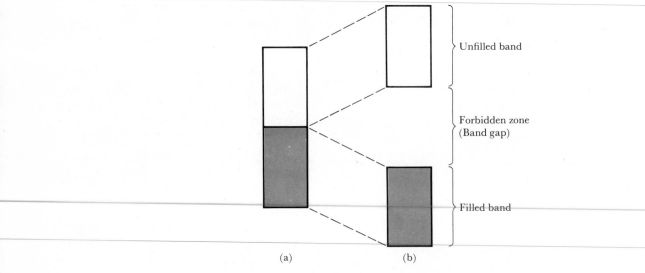

Figure 14-28   Two calculated band structures of crystalline carbon. The structure in (a) is based on the assumption that carbon crystallizes in the body-centered cubic structure with each carbon atom surrounded by eight other carbon atoms; the structure in (b) is based on the assumption that carbon crystallizes in the known diamond structure wherein each carbon atom is surrounded by four other carbon atoms. In (a) carbon would behave as an electrical conductor; in (b) carbon would behave as an insulator, which is observed experimentally for diamond.

between conductors and insulators. Silicon, germanium, and α-gray tin all have the diamond structure. However, the band gap between filled and empty bands for these solids is much smaller than it is for carbon. The gap for silicon is only 105 kJ mole^{-1}. (As we have seen, it is 502 kJ mole^{-1} for carbon.) For germanium it is 59 kJ mole^{-1}, and for α-gray tin it is 7.5 kJ mole^{-1}. The metalloids silicon and germanium are called **semiconductors**.

A semiconductor can carry a current if the relatively small energy required to excite electrons from the lower filled valence band to the upper empty conduction band is provided. Since the number of excited electrons increases as the temperature increases, the conductance of the semiconductor increases with temperature. This behavior is exactly the opposite of that of metals.

Conduction in materials such as silicon and germanium can be enhanced by adding small amounts of certain impurities. The band gap in silicon can be narrowed effectively if impurities such as boron or phosphorus are added to silicon crystals. Small amounts of boron or phosphorus (a few parts per million) can be incorporated into the silicon structure when the crystal is grown. Phosphorus has five valence electrons and thus has an extra free electron even after four electrons have been used in the covalent bonds of the silicon structure. This fifth electron can be moved away from a phosphorus atom by an electric field; hence we say phosphorus is an electron donor. Only 1.05 kJ mole^{-1} is required to free the donated electrons, thereby making a conductor out of silicon to which a small amount of phosphorus has been added. The opposite effect occurs if boron is added to silicon. Atomic boron has one too few electrons for complete covalent bonding. Thus for each boron atom in the silicon crystal there is a single vacancy in a bonding orbital. It is possible to excite the valence electrons of silicon into these vacant orbitals in the boron atoms, thereby causing the electrons to move through the crystal. To accomplish this conduction an electron from a silicon neighbor drops into the empty boron orbital. Then an electron that is two atoms away can fill the silicon atom's newly created vacancy. The result is a cascade effect, whereby an electron from each of a row of atoms moves one place to the neighboring atom. Physicists prefer to describe this phenomenon as a hole moving in the opposite direction. No matter which description is used, it is a fact that less energy is required to make a material such as silicon conduct if the crystal contains small amounts of either an electron donor such as phosphorus or an electron acceptor such as boron.

14-6   THE FRAMEWORK OF THE PLANET: SILICATE MINERALS

The core of our planet is believed to be mainly iron and nickel, with a radius of approximately 2200 miles. This core gives earth its magnetic field, which the moon and our neighbor planets, Mars and Venus, apparently lack. The earth's core is at a high pressure and temperature, and is probably fluid. An old theory of the origin of our planet presumed that it formed when hot gases collected and cooled. By this theory, the core is a relic of the first hot period; it has not solidified because of the insulating effect of the outer layers.

The current view is that the earth grew by cold accretion of solid debris and dust. After a certain critical mass was reached, heat from the interior could not be lost to the surroundings as rapidly as it was generated by natural radioactivity and pressure, and the center of the planet liquefied. This could occur only in a planet above a critical size, which presumably explains the absence of a molten core and a strong magnetic field in Mars and the moon. The phenomenon is similar to that of critical mass in uranium fission. Below a certain mass of ^{235}U the metal loses too many neutrons to the outside to sustain a chain reaction, so nuclear explosion occurs only in pieces of ^{235}U above this critical size.

For 1800 miles above the core of the earth extends the *mantle,* a layer

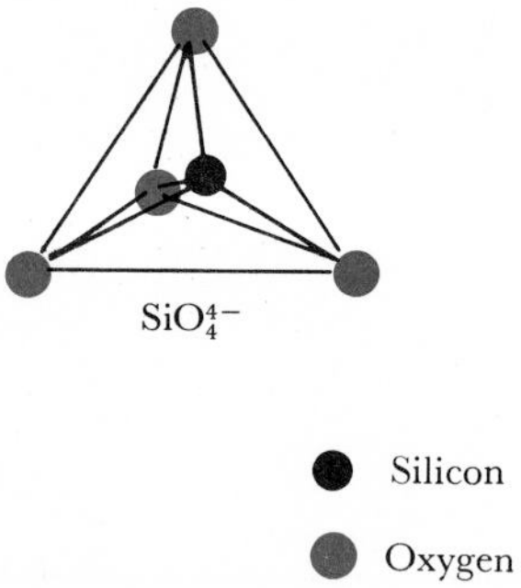

Figure 14-29   The SiO_4^{4-} tetrahedron, which is the building block of most silicate minerals. The Si atom (black) is covalently bonded to four oxygen atoms at the corners of a tetrahedron (color). The black lines between oxygen atoms are included only to give form to the tetrahedron.

probably composed of a dense silicate material similar to basalt. The upper 20 miles under the continents, or as little as three miles under the ocean beds, is the *crust,* the only part about which we have any real chemical knowledge. On top of this crust—itself only 1.5% of the volume of the planet—is spread the thin layer containing virtually all the matter with which we are concerned.

The crust is 48% oxygen by weight, in the form of the various silicate minerals. It is also 26% silicon, 8% aluminum, 5% iron, and 2% to 5% calcium, sodium, potassium, and magnesium. Remarkably enough, these silicate minerals are more than 90% oxygen by volume. Most of the crust is some form of the mixture of silicate minerals known as *granite.*

The basic building block in silicates is the orthosilicate ion, SiO_4^{4-}, shown in Figure 14-29. Each silicon atom is covalently bonded to four oxygen atoms at the corners of a tetrahedron. The SiO_4^{4-} anion occurs in simple minerals such as zircon ($ZrSiO_4$), garnet, and topaz. Two tetrahedra can share a corner oxygen atom to form a discrete $Si_2O_7^{6-}$ anion, or three tetrahedra can form a ring, shown in Figure 14-30. Benitoite, $BaTiSi_3O_9$, is the best known example of this uncommon kind of silicate. Beryl, $Be_3Al_2Si_6O_{18}$, a common source of beryllium, has anions composed of rings of six tetrahedra with six shared oxygen atoms.

Chain Structures

All the silicates mentioned so far are made from discrete anions. A second class is made of endless strands or chains of linked tetrahedra. Some minerals have single silicate strands with the formula $(SiO_3)_n^{2n-}$. A form of asbestos has the double-stranded structure shown in Figure 14-31. The double-stranded chains are held together by electrostatic forces between themselves and the Na^+, Fe^{2+}, and Fe^{3+} cations packed around them. The chains can be pulled apart with much less effort than is required to snap the covalent bonds within a chain. Therefore, asbestos has a stringy, fibrous texture. Aluminum ions can replace as many as one quarter of the silicon ions in the tetrahedra. However, each replacement requires one more positive charge

Figure 14-30   A ring of three tetrahedra, with three oxygen atoms shared between pairs of tetrahedra, has the formula $Si_3O_9^{6-}$. This structure occurs as the anion in soft, crumbly rocks such as benitoite, $BaTiSi_3O_9$.

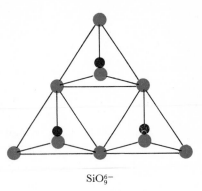

$$SiO_9^{6-}$$

from another cation (such as K^+) to balance the charge on the silicate oxygen atoms. The physical properties of the silicate minerals are influenced strongly by how many Al^{3+} ions replace Si^{4+} ions, and by how many extra cations are therefore needed to balance the charge.

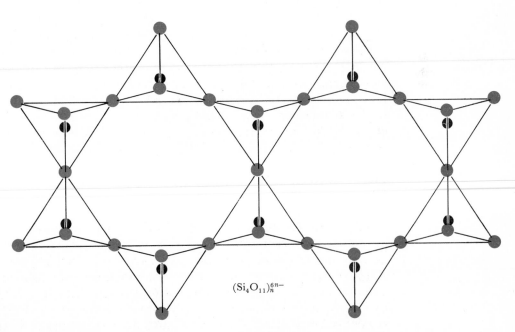

$$(Si_4O_{11})_n^{6n-}$$

Figure 14-31   Long double-stranded chains of silicate tetrahedra are in fibrous minerals such as asbestos.

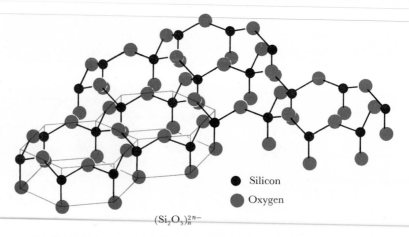

$(Si_2O_5)_n^{2n-}$

Figure 14-32   In talc, mica, and the clay minerals, silicate tetrahedra each share three of their corner oxygen atoms to make endless sheets. All of the unshared oxygen atoms point down in this drawing on the same side of the sheet.

Sheet Structures

Continuous broadening of double-stranded silicate chains produces planar sheets of silicate structures (Figure 14-32). Talc, or soapstone, has this structure, in which none of the Si^{4+} is replaced with Al^{3+}. Therefore, no additional cations between the sheets are required to balance charges. The silicate sheets in talc are held together primarily by van der Waals forces. Because of these weak forces the layers slide past one another relatively easily, and produce the slippery feel that is characteristic of talcum powder.

Mica resembles talc, but one quarter of the Si^{4+} ions in the tetrahedra are replaced by Al^{3+} ions. Thus an additional positive charge is required for each replacement to balance charges. Mica has the layer structure shown in Figure 14-33. The layers of cations (Al^{3+} serves as a cation between layers as well as a substitute in the silicate tetrahedra) hold the silicate sheets together electrostatically with much greater strength than in talc. Thus mica is not slippery to the touch and is not a good lubricant. However, it cleaves easily, thereby splitting into sheets parallel to the silicate layers. Little effort is required to flake off a chip of mica, but much more strength is needed to bend the flake across the middle and break it.

The clay minerals are silicates with sheet structures such as in mica. These layer structures have enormous "inner surfaces," and can often absorb large amounts of water and other substances between the silicate layers. This is why clay soils are such useful growth media for plants. It is also why clays are used as beds for metal catalysts. The common catalyst platinum black is finely divided platinum metal obtained by precipitation from solution. The catalytic activity of platinum black is enhanced by the large amount of exposed metal surface. The same effect can be achieved

Figure 14-33

In a form of mica called muscovite, $K_2Al_4Si_6Al_2O_{20}(OH)_4$, anionic sheets of silicate tetrahedra (Figure 14-32) alternate with layers of potassium ions and aluminum ions sandwiched between hydroxide ions. This layer structure gives mica its flaky cleavage properties.

by precipitating a metal to be used as a catalyst (Pt, Ni, or Co) onto clays. The metal atoms coat the interior walls of the silicate sheets, and the clay structure prevents the metal from consolidating into a useless mass. J. D. Bernal has suggested that the first catalyzed reactions in the early stages of the evolution of life, before biological catalysts (enzymes) existed, may have occurred on the surfaces of clay minerals.

Three-Dimensional Networks

The three-dimensional silicate networks, in which all four oxygen atoms of SiO_4^{4-} are shared with other Si^{4+} ions, are typified by quartz, $(SiO_2)_n$, shown in Figure 14-34. In quartz, all the tetrahedral structures have Si^{4+} ions, but in other network minerals, up to half the Si^{4+} ions can be replaced with Al^{3+} ions. These minerals include the feldspars, for which a typical empirical formula is $KAlSi_3O_8$. Feldspars are nearly as hard as quartz. Basalt is a compact mineral related to feldspar. Granite, the chief component of the earth's crust, is a mixture of crystallites of mica, feldspar, and quartz.

Glasses are amorphous, disordered, noncrystalline aggregates with linked silicate chains of the sort depicted in Figure 14-35. Common soda-lime glass is made with sand (SiO_2), limestone ($CaCO_3$), and sodium carbonate (Na_2CO_3) or sodium sulfate (Na_2SO_4), which are melted together and allowed to cool. Other glasses with special properties are made by using other metal carbonates and oxides. Pyrex glass has boron as well as silicon and some aluminum in its silicate framework. Glasses are not true solids, but are extremely viscous liquids. If you examine the panes of glass in a very old New England home, you can sometimes see that the bottom of the

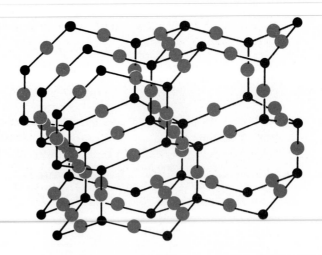

● Silicon atom—each attached to 4 oxygen atoms

● Oxygen atom—each attached to 2 silicon atoms

Figure 14-34   The three-dimensional network of silicate tetrahedra in quartz, SiO_2.

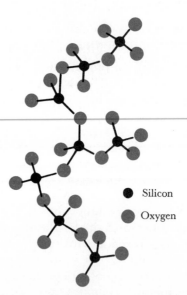

● Silicon

● Oxygen

Figure 14-35   Glasses are amorphous, disordered chains of silicate tetrahedra that are fused with metal oxides or carbonates such as Na_2CO_3 or $CaCO_3$.

pane is slightly thicker than the top because of two centuries of slow, viscous flow of the glass.

Summary

The types of bonding in solids that we have seen are summarized in Table 14-3. **Ionic** or **electrostatic bonds** and electron-pair **covalent bonds** both have bond energies of approximately 400 kJ mole^{-1}. **Metallic bonds** are variable, but are of comparable strength. **Hydrogen bonds** are much weaker: The bond between O and H is about 20 kJ mole^{-1}. **Van der Waals attractions** are weaker yet; from a few tenths to 2 kJ mole^{-1}. Hydrogen

Table 14-3

Types of Bonding in Solids

	Molecular	Nonmetallic network	Metallic	Ionic
Structural unit:	Molecule	Atom	Atom	Ion
Principal bonding between units:	Weak van der Waals and, in polar molecules, a stronger dipole–dipole bond	Strong covalent bonds	Delocalized electron sea through a system of metallic positive ion centers	Strong ionic bonding (electrostatic)
Properties:	Soft Low melting point Insulator	Hard High melting point Insulator or semiconductor	Wide range of hardness Wide range of melting points Conductor	Hard High melting point Insulator
Usually occurs in:	Nonmetals at right of periodic table and compounds predominantly composed of nonmetals	Nonmetals in center of periodic table	Metals in left half of periodic table	Compounds of metals and nonmetals
Examples	O_2, C_6H_6, H_2O, Br_2	Diamond, Si, ZnS, SiO_2	Na, Zn, Au, brass, bronze	KI, Na_2CO_3, LiH

bonds and van der Waals attractions assume a greater importance than their strength would suggest because of the large number of such bonds that can form.

Hydrogen bonding results from the attractive interaction between a positive hydrogen atom and an electronegative atom so small that the proton can approach it closely. Oxygen and fluorine can participate in such hydrogen bonds, as can nitrogen to a lesser degree, but chlorine is usually too large. Hydrogen bonds are responsible for many of the familiar properties of water and ice.

Both **metals** and **covalent network solids** can be interpreted by using delocalized molecular orbitals, in which the "molecule" extends over the entire piece of matter under study. This **band theory** accounts for many of the observed properties of **conductors, semiconductors,** and **insulators**.

The basic subunit of the **silicates**, a SiO_4^{4-} tetrahedron, can be organized into rings, chains, sheets, and three-dimensional networks. Aluminum can replace some silicon, but other cations must be added to balance the charge, thereby increasing the electrostatic contribution to holding the solid together. The silicates illustrate four of the five types of bonding that have been discussed in this chapter: covalent bonding between Si and O in the tetrahedra, van der Waals forces between silicate sheets in talc, ionic attractions between charged sheets and chains, and hydrogen bonds between water molecules and the silicate oxygen atoms in clays. If we include Ni catalysts prepared on a clay support, the fifth type of bonding (metallic) is represented as well.

Self-Study Questions

1. What types of forces hold molecules together in crystals and liquids?
2. Why do the halogen molecules form molecular solids?
3. Why do carbon and silicon form network solids?
4. What is the main structural unit in elemental boron?
5. Why are nonmetallic network solids usually quite hard?
6. What physical effect is responsible for the attraction in van der Waals interactions? What is responsible for the repulsion in such interactions? Compare the origin of attraction and repulsion in van der Waals interactions with that in ionic and covalent bonds.
7. Which molecule would you predict to have the higher boiling point, H_2 or CH_4? Why?
8. If van der Waals bonds are extremely weak, why are they discussed at all?
9. What effect do hydrogen bonds have on the boiling temperatures of liquids?

10. How do hydrogen bonds participate in the structure of ice? What effects do they have on its properties?

11. How do we know that some hydrogen bonding in water persists in the liquid phase?

12. Why are ionic solids soluble in polar solvents? Why are they insoluble in nonpolar solvents?

13. What is the coordination number of a metallic atom in a cubic close-packed structure? A hexagonal close-packed structure? A body-centered cubic structure?

14. Why are metals found on the left side of the periodic table?

15. In the delocalized-molecular-orbital theory of metals, in what sense do we say that the entire piece of metal is a large molecule?

16. Where does the "band" of energy levels come from in the delocalized-molecular-orbital theory of electronic structure of metals?

17. Why would beryllium be an insulator if the $2s$ and $2p$ molecular orbital bands did not overlap?

18. What is the structural difference between metals, semiconductors, and insulators?

19. What effect do small amounts of boron or phosphorus have on the conducting properties of silicon?

20. What is the difference between white phosphorus and black phosphorus?

21. Why is diamond an insulator? What properties would carbon have if it were to crystallize in a body-centered cubic structure?

22. Which has the largest gap between the valence band and the conduction band: an insulator, a metal, or a semiconductor? Which has the smallest?

23. What is the basic building block of silicate minerals?

24. How does mica differ structurally from talc?

25. Why are clays useful in industrial catalysis?

26. How do glasses differ from quartz?

Problems

Van der Waals interactions

1. Explain the trend in the melting temperatures of the following tetrahedral molecules: CF_4, 90 K; CCl_4, 250 K; CBr_4, 350 K; CI_4, 440 K.

2. Make a sketch of the way in which the repulsion part of the van der Waals interaction (equation 14-1) varies with distance r between atomic centers. Make a similar sketch for the attraction terms, equation 14-2. Add these two curves in an approximate way and satisfy yourself that a potential-energy curve such as Figure 14-13 is the result.

3. Construct the potential-energy curve for the van der Waals interaction between two Kr atoms. How strong is the van der Waals bond? Estimate the Kr—Kr bond distance in solid krypton.

4. The molecule RbBr is held together primarily by an ionic bond. The distance between Rb^+ and Br^- in the molecule is 2.945 Å. The closed electron shells of Rb^+ and Br^- both have the configuration of the noble gas Kr. From the energy curve constructed for Problem 3, estimate the van der Waals energy between Rb^+ and Br^-, assuming that the energy is the same as for a pair of Kr atoms separated by a distance of 2.945 Å. Is the repulsive part, or the attractive part, of the interaction dominant? How important is the van der Waals energy compared to the overall bond energy of 377 kJ mole^{-1} in RbBr? Examine the Kr—Kr van der Waals energy for distances of 2 Å and 1 Å and then explain what prevents Rb^+ and Br^- ions from approaching each other too closely in an ionic solid.

Solid structures

5. The internuclear distances in gas-phase ionic molecules are considerably smaller than those in the corresponding crystals. For example, the internuclear distance in NaCl(g) is 2.36 Å, whereas in NaCl(s) the shortest internuclear distance is 2.81 Å. Examine the NaCl(s) structure in Figure 14-9 and explain why this should be so.

6. Solid carbon dioxide behaves as a molecular solid (it is easily compressible and sublimes at 195 K), whereas solid silicon dioxide (quartz, Figure 14-34) is a nonmetallic network solid (it is very hard and has a melting point of 1883 K). Rationalize this difference in behavior in terms of the relative σ and π bonding in the molecular species CO_2 and SiO_2.

7. Compare the structures for diamond and graphite shown in Figure 14-5.
   a) What type of model (metallic, nonmetallic covalent, or van der Waals) best describes the bonding within a layer of the graphite structure?
   b) What type of model best describes the bonding between layers in the graphite structure?
   c) Explain why graphite, unlike diamond, is very soft, whereas like diamond it has a very high melting point.
   d) Draw a Lewis structure for a fragment of one graphite layer.
   e) Graphite is a relatively good electrical conductor. Use the Lewis structure obtained in part d to rationalize the conductivity of graphite. Which type of electrons (σ or π) do you suspect are mobile and therefore able to conduct electrical current?

8. Replacement of the zinc and sulfur atoms by carbon atoms in the wurtzite structure of zinc sulfide (Figure 14-10) results in the diamond structure (Figure 14-5). Optical and electrical measurements on samples of ZnS show that it has a band gap of approximately 3.6 eV. Discuss the nature of ZnS in terms of the three models: nonmetallic network, ionic, and metallic, all of which have been applied to describe this solid.

9. Provide a structural explanation for the fact that quartz is hard, asbestos fibrous and stringy, and mica platelike.

10. What type of solid will BF_3 and NF_3 molecules build? What kinds of intermolecular interactions are likely to be important in each case?

11. Which should have the higher melting temperature, BF_3 or NF_3?

Band gaps

12. It requires 5.2 eV, or 502 kJ mole^{-1}, to excite electrons in a diamond crystal from the valence band to the conduction band. What frequency of light is needed to bring about this excitation? What wavelength? What wave number? What part of the electromagnetic spectrum does this correspond to?

13. Using data given in this chapter, repeat Problem 12 for the semiconductors silicon and germanium.

Suggested Reading

L. Brewer, "Bonding and Structures of Transition Metals," *Science* **161,** 115 (1968).

E. Cartmell and G. W. A. Fowles, *Valency and Molecular Structure,* Butterworths, London, 1977, 4th ed.

F. A. Cotton and G. Wilkinson, *Advanced Inorganic Chemistry,* Wiley-Interscience, New York, 1972, 3rd ed.

A. H. Cottrell, "The Nature of Metals," *Scientific American,* September, 1967.

W. E. Dasent, *Inorganic Energetics,* Penguin, Middlesex, England, 1970.

W. A. Deer, R. A. Howie, and J. Zussman, *Rock Forming Minerals,* Vols. 1–5, Wiley, New York, 1962.

H. B. Gray, *Chemical Bonds,* W. A. Benjamin, Menlo Park, Calif. 1973.

T. L. Hill, *Matter and Equilibrium,* W. A. Benjamin, Menlo Park, Calif., 1966.

J. E. Huheey, *Inorganic Chemistry: Principles of Structure and Reactivity,* Harper & Row, New York, 1978, 2nd ed.

W. J. Moore, *Seven Solid States,* W. A. Benjamin, Menlo Park, Calif., 1967.

N. Mott, "The Solid State," *Scientific American,* September, 1967.

C. S. G. Phillips and R. J. P. Williams, *Inorganic Chemistry,* Vol. 1, Clarendon Press, Oxford, England, 1965.

G. C. Pimentel and R. D. Spratley, *Understanding Chemistry,* Holden-Day, San Francisco, 1971.

K. F. Purcell and J. C. Kotz, *Inorganic Chemistry,* W. B. Saunders, Philadelphia, 1977.

H. Reiss, "Chemical Properties of Materials," *Scientific American,* September, 1967.

A. F. Wells, *Structural Inorganic Chemistry,* 3rd ed. Oxford University Press, New York, 1962.

15

Energy and Enthalpy in Chemical Systems

Heat and cold are nature's two hands by which she chiefly worketh.

Francis Bacon (1627)

An old motto from the time of World War II (and probably earlier) is, "The difficult we do at once; the impossible takes a little longer." In this chapter and the next we shall discover what is *possible* in chemical reactions. This does not mean that everything that is possible by the laws of thermodynamics will take place in a short time. When the chemical thermodynamicist says that a reaction is spontaneous, he makes no predictions whatsoever about elapsed time; he only says that, given *enough* time, the reaction can happen. To the thermodynamicist, the explosion produced by dropping sodium in water and the weathering away of the entire North American continent are both spontaneous processes.

To the chemist, it is important to know whether a reaction is spontaneous in the thermodynamic sense. If it is slow but spontaneous, then some means, such as catalysis, may be found to hasten the process. If the reaction is not spontaneous, the search for an accelerator is doomed at the start; another means must be devised to force the desired reaction to occur.

By what criterion does a chemist say that a reaction is spontaneous? In Chapter 4 we discussed the ideas of spontaneity and equilibrium, but we took the numerical values of equilibrium constants on faith. Now we shall see how these constants can be related to other measurable properties of a reaction. Most spontaneous reactions release heat. Explosions and other combustions are familiar examples. Is it a valid generalization to say that *all* spontaneous reactions release heat? Why do some reactions go to completion so thoroughly that essentially no reactants are left, whereas others appear to halt when a mixture of reactants and products is present? Can we predict in advance that a given reaction will behave in either of these two ways? What effect does the amount of a reactant or a product have on the spontaneity of a reaction?

These are some of the questions that we will answer in the course of this chapter and the next one. However, you should not forget that thermodynamics only describes what *can* happen (or better, what is not forbidden). Making it happen, and making it happen in a reasonable time, is the task of the research chemist.

15-1   WORK, HEAT, AND CALORIC

One of Lavoisier's great contributions to chemistry was to undercut the phlogiston theory, as we have seen in Chapter 6. He demonstrated that combustion was a combination with oxygen and not a loss of phlogiston. He was less perceptive in his ideas about the source of the heat, which is so prominent a feature of combustion. Lavoisier coined the term *caloric,* in 1789, for what he regarded as the "imponderable matter of heat." Heat was considered to be a fluid, probably weightless, that surrounded the atoms of substances and could be drained away in reactions that produced heat.

Dalton conceived of each atom as existing in an "atmosphere" of heat. In 1808, he wrote:

"The most probable opinion concerning the nature of caloric is that of its being an elastic fluid of great subtility, the particles of which repel one another, but are attracted by all other bodies."

According to this generally accepted idea of Dalton's time, a gas is heated when it is compressed because the particles of caloric, repelling one another as they do, are squeezed out of the gas. Heat of friction develops when the frictional motion strips caloric away from its atoms. The caloric theory of heat was accepted by most scientists for the first half of the nineteenth century.

The Cannons of Bavaria

In 1798, Benjamin Thompson (Count Rumford) conducted some experiments with friction that, if they had been appreciated fully, would have done away with caloric as Lavoisier did away with phlogiston. Thompson was superintending the boring of cannons at the military arsenal in Munich. The process involved cast metal cannon blanks and drill bits that were turned by horses. Thompson was impressed by the considerable heat evolved during the drilling. He tried boring the cannons under water and determined that the same length of time was always required to bring a given amount of water to the boiling point. He also observed that the generation of heat apparently could be continued indefinitely. He interpreted what he saw correctly; the work provided by the horses was being converted into heat. Thompson wrote:

"It is hardly necessary to add that anything which any insulated body or system of bodies can continually be furnished without limitation cannot possibly be a material substance, and it appears to me to be extremely difficult if not quite impossible to form any distinct ideas of anything capable of being excited and communicated in the manner the *Heat* was excited and communicated in these experiments except it be *Motion.*"

His experiments failed to convince others. Those who believed in the caloric theory were ready with the explanation that the friction of the drill bit rubbed caloric away from the metal atoms and brought it to the surface. They failed to appreciate the significance of Count Rumford's ability to continue to produce heat indefinitely. According to the caloric theory, after the supply of caloric had been rubbed away from the metal, further boring should not produce heat. Unfortunately, scientists were not accustomed to thinking of heat in quantitative terms, just as they were not accustomed to thinking of matter in quantitative terms before Lavoisier's proposal. Rumford's work had little impact.

Blood, Sweat, and Gears

The men who finally convinced scientists *that heat and work were equivalent, and that both were forms of energy,* were Julius Mayer (1814–1878) and Hermann von Helmholtz (1821–1894), both German physicians, and James Joule (1818–1889). In 1840, Mayer signed onto a ship bound for Java as ship's doctor. He noted that the blood from the veins of the Javanese and from his own ship's crew was a brighter red than that he had seen from patients in Germany. He interpreted this correctly as indicating that more oxygen remained in the veins of inhabitants from the tropics than in the veins of people from cold climates, because less combustion of foods was required to maintain a constant body temperature in the tropics. This train of thought led him to the further conclusion that the heat of combustion of foods was used both to maintain body temperature and to carry out the work done by an individual. Heat could be converted into work, and both were forms of the same thing, energy. On his return to Germany he tried to calculate the conversion factor between heat and work by using stirring devices for water and expanding gases into chambers. The experiments were difficult to perform accurately, because the temperature increases were fractions of a degree. Nevertheless, he obtained an approximate value for the mechanical equivalent of heat and submitted an account of his work to the *Annalen der Physik*. The *Annalen der Physik* rejected his paper as unfit for publication. He reworked it and submitted it to the *Annalen der Chemie und Pharmacie* instead. It was published in 1842, and aroused no comment whatever. Like Newlands with his classification of the elements, Mayer had expected controversy, but encountered only indifference.

At the same time, Joule, in England, was doing virtually the same experiments and meeting the same indifference and disbelief. Joule was a student of Dalton and the son of a Lancashire brewer. At the age of 19, he began building electric motors and generators with the intention of converting the brewery from steam power to electricity. These attempts were abortive, but Joule became interested in the relationship among the work of cranking the dynamo, the electricity generated, and the heat produced by electricity. Later he dropped the electricity from the sequence and studied the heat produced by stirring water mechanically with paddles driven by a falling weight (Figure 15-1). Like Mayer, Joule found the experiments difficult because of the small temperature changes produced. In spite of this, he obtained a conversion factor that, expressed in calories, is 42.4 kg cm cal^{-1}, within 1% of the currently accepted value of 42.67 kg cm cal^{-1}. That is, a 1-kg weight, falling through a distance of 42.67 cm, can do enough work (by turning a stirring paddle, for example) to add 1 cal of heat to the water. If the experiment is performed with an insulated 1-liter container of water, then, since the heat capacity of water is 1 cal deg^{-1} g^{-1}, the temperature will increase only by one-thousandth of a degree. It was a remarkable achievement of Joule's to come so close to the best modern value with home-made and home-calibrated thermometers.

The connection between heat and work is much more straightforward in SI units. If a mass of m kilograms is lifted a distance h against the acceleration of gravity, $g = 9.806$ m sec^{-2}, the work done is

$$w = mgh \text{ joules}$$

From this and Joule's experiment we can find the conversion between calories and joules. For $m = 1.000$ kg and $h = 0.4267$ m:

$$w = 1.000 \text{ cal} = 1.000 \times 9.806 \times 0.4267 \text{ kg m}^2 \text{ sec}^{-2}$$
$$1.000 \text{ cal} = 4.184 \text{ J}$$

In 1843, Joule submitted his results to the British Association. They were received with disbelief and general silence. A year later a paper on the subject was rejected by the Royal Society. In 1845, Joule again presented his ideas on the equivalence of work and heat to the British Association. He suggested that the water at the bottom of Niagara Falls should be 0.2°F warmer than the water at the top because of the energy gained in the fall. He also proposed the idea of an absolute zero of temperature, based on the thermal expansion of gases, at -480°F (-284°C).

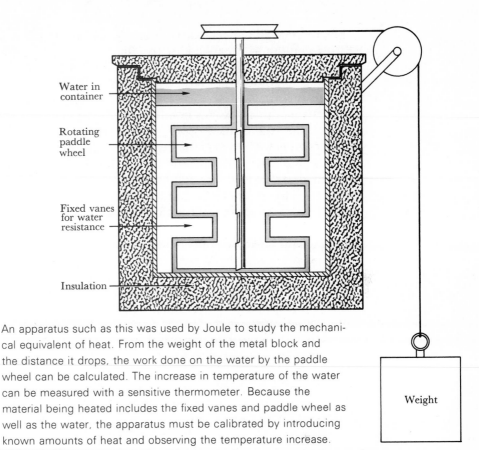

Figure 15-1 An apparatus such as this was used by Joule to study the mechanical equivalent of heat. From the weight of the metal block and the distance it drops, the work done on the water by the paddle wheel can be calculated. The increase in temperature of the water can be measured with a sensitive thermometer. Because the material being heated includes the fixed vanes and paddle wheel as well as the water, the apparatus must be calibrated by introducing known amounts of heat and observing the temperature increase.

No one listened. He tried again in 1847, and in Joule's own words, written in 1885:

"The communication would have passed without comment if a young man had not risen in the section, and by his intelligent observations created a lively interest in the new theory. The young man was William Thomson, who had two years previously passed the University of Cambridge with the highest honour, and is now probably the foremost scientific authority of the age."

Thomson, later to become Lord Kelvin, was 26 years old at the time. Neither he nor Faraday, who was also at the meeting, was convinced by Joule's case, which depended on temperature increases of hundredths of a degree; but at long last, Joule had forced his peers to discuss his ideas. Thomson wrote later that, two weeks after the 1847 meeting, he was walking from Chamonix to begin a tour of Mont Blanc when . . .

"whom should I meet walking up but Joule, with a long thermometer in his hand, and a carriage with a lady in it not far off. He told me that he had been married since we parted at Oxford! and he was going to try for elevation of temperatures in waterfalls."

In 1849, a paper by Joule entitled "On the Mechanical Equivalent of Heat" was communicated to the Royal Society by Faraday, and it appeared in their *Philosophical Transactions* the next year.

Mayer suffered the same pangs that Newlands did; he saw what he regarded as his own ideas being acclaimed by others but attributed to Joule. Mayer's despondency led him to attempt suicide in 1850, and he was committed to a mental asylum for two years thereafter. He continued to receive little credit or attention until late in his life, when John Tyndall, in England, and Rudolf Clausius and Hermann von Helmholtz, in Germany, made a concerted effort to secure proper recognition for Mayer.

The man who finally convinced scientists of the validity of the equivalence of heat and work was Helmholtz. In 1847, he submitted a paper to the *Annalen der Physik* that outlined the principles of the *conservation of energy* and the *equivalence of heat and work* in more general terms than either Mayer or Joule had done. The paper was rejected. Helmholtz presented the paper at a meeting in Berlin and had it published privately.

Helmholtz' analysis of heat, work, and energy convinced Faraday and Thomson. Joule's experiments gradually began to be accepted. Ultimately, the German physicist Rudolf Clausius (1822–1888) stated, in 1850, the first law of thermodynamics as it usually is given today:

Heat and work are both forms of energy. In any process, energy can be changed from one form to another (including heat and work), but it is never created or destroyed.

Helmholtz' conservation of energy joined Lavoisier's conservation of mass as one of the great generalizations of science.

15-2   THE FIRST LAW OF THERMODYNAMICS

Thermodynamicists talk continually about **thermodynamic systems** and their **surroundings**; so will we. We shall look at the work that a system does on its surroundings, or the work that the surroundings do on the system. We shall note the loss or gain of heat of a system to or from its surroundings. What is a thermodynamic system?

A thermodynamic system is any part of the universe that we want to focus attention upon, and its surroundings are that part of the universe with which it can exchange energy, heat, or work. A suitable system could be a balloon full of gas, or a flask with reacting chemicals, or a locomotive engine, or just the cylinders and pistons of the engine. If we are looking at the energy balance on our planet, then the earth itself would be a thermodynamic system, and the sun would be part of its surroundings. An **isolated system** is one that does not exchange energy, heat, or work with its surroundings. So far as thermodynamics is concerned, it has no surroundings. A **closed system** has walls that permit heat and energy, but not matter, to flow in and out. An **open system** can exchange both matter and energy with its surroundings. The word *system* is a pointing finger on an old-fashioned signboard; it calls attention to whatever region of space we want to examine.

Example 1

What kind of thermodynamic systems are the following: (a) a beaker half full of chemicals; (b) the same chemicals in a screw-top jar with the lid on; (c) the same chemicals in a sealed vacuum bottle; (d) a human being; (e) a tin can of beans; (f) the planet earth; (g) the solar system; (h) the Andromeda galaxy; (i) the entire material universe?

Solution

(a) Open; (b) closed; (c) isolated, if the bottle is a good one; (d) open (how could someone turn you into a closed system, and what would happen to you then?); (e) closed; (f) open (the earth receives energy from the sun, and exchanges astronauts and other forms of matter with its neighbors); (g) an open system (think of starlight and comets); (h) very nearly isolated; (i) isolated, by definition.

It is often easiest to think of an ideal gas in some type of enclosure as a typical thermodynamic system. Many of the thermodynamic properties common to all systems are comprehended most readily in such a simple system. When we heat a gas, it expands unless it is constrained. As it expands, it pushes against the pressure of the atmosphere and therefore does work against this pressure. We say that heat, q, has been added to the gas from its surroundings, and that the gas has done work on the surroundings. If we add heat to the gas but constrain it so it cannot expand, the temperature and pressure increase as given by the ideal gas law, derived in Chapter 3,

$$PV = nRT \tag{3-8}$$

Heat again has been added to the gas, but no work has been done by the gas. If the gas is initially at a high pressure, we can allow it to expand without heating it. In this case, the gas does work against its environment or surroundings, without having heat added to it. However, the gas at the conclusion of the expansion is cooler than it was initially.

A variation of this experiment is shown in Figure 15-2. In this experiment, the way in which the expanding gas can be made to do work is

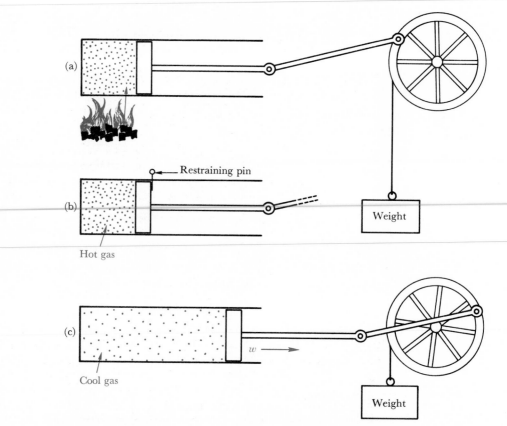

Figure 15-2   If the gas in the cylinder is heated (a), but the gas is prevented from expanding (b), its temperature increases. Conversely, if the gas is allowed to expand (c), it can do mechanical work, and its final temperature will not be so high. In the first example, heat is converted into internal energy; in the second example, it is converted into work.

somewhat more obvious. Heat, q, can be added to the gas with or without its doing work, and work, w, can be obtained from the gas if heat is added and sometimes even if it is not.

How is work measured in an expanding gas? Work is defined in physics as the product of the force against which motion takes place times the distance moved. Endless motion of an object produces no work if there is no resisting force to the motion. Moreover, no matter how large the resisting force to the motion of an object might be, no work is done unless the object moves against that force. For an infinitesimal movement, ds, against an opposing force, F, the infinitesimal amount of work done is $dw = F\,ds$. If the

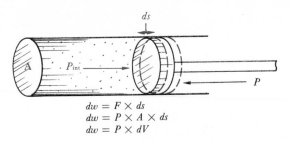

$$dw = F \times ds$$
$$dw = P \times A \times ds$$
$$dw = P \times dV$$

Figure 15-3

The work done by the gas in the cylinder when it moves the piston an infinitesimal distance, ds, against an opposing pressure, P, is: $dw = F\,ds = PA\,ds = P\,dV$. ($A$ is the area over which the pressure is exerted.) The pressure in the work expression is the external pressure against which the motion takes place, not the internal pressure (P_{int}) of the expanding gas. However, for an expansion to occur, P_{int} must be greater than the external pressure.

object moves through a finite distance, s, against a constant force, F, the work done is $w = F\,s$, as we saw in Chapter 2.

Let's suppose that the gas is enclosed in a cylinder with a piston (Figure 15-3) and that the pressure inside the cylinder, P_{int}, is greater than the constant atmospheric pressure outside, P. As the gas expands and moves the piston an infinitesimal distance, ds, the force against the piston from the outside remains constant and equal to the product of the pressure, P, and the area, A, over which this pressure is exerted. The work done, as shown in the figure, is the product of the volume increase times the external pressure against which the expansion takes place: $dw = P\,dV$. For an expansion such as this one, in which the resisting pressure remains constant, the work done in a measurable volume change of ΔV is $w = P\,\Delta V$. These relationships, although derived here only for gas expanding in a cylinder, are generally true in gas expansions. This kind of work is commonly called *expansion work*, or *PV work*. Other kinds of work are possible. We can do gravitational work by lifting a weight to a position where it has greater potential energy and can fall to its original position. We can do electrical work by moving charged ions or other objects in an electrical potential field. We can do magnetic work by pulling a compass needle away from the direction to which it points if undisturbed. All of these types of work are included in the generalization known as the **first law of thermodynamics**.

In a thermodynamic system, heat can go in or out, and work can be done on or by the system. The first law states that in all of these processes *energy in the system is neither created nor destroyed.* The energy of the system is not necessarily constant; it can rise or fall, depending on what we do to or with the system. But the *change* in energy of the system is equal to the *net* heat added to the system less the *net* work done by the system on its surroundings:

$$\Delta E = q - w \tag{15-1}$$

A Different View of the First Law

Another way of looking at the first law of thermodynamics is especially meaningful for chemists. In this view we think of equation 15-1 as no more than a definition of a bookkeeping function known as the **internal energy, E.** Recall from the discussion of Figure 15-1 that we can heat a gas and do work with it. We can also reverse the process. We can do work on a gas by compressing it, and we can drain away the heat produced. We can heat a gas without letting it do work, in which case the temperature increases. Conversely, we can let a gas at high pressure expand and do work without being heated, but we will find that the gas cools in the process. With the right conditions, q and w can be manipulated independently. It's easier to keep track of what's happening if we define change in the internal energy, ΔE, as the *difference* between the heat added and the work done, as in equation 15-1. If heat is added and an exactly equivalent amount of work is done, the internal energy of the system is unchanged. If we heat the gas but constrain it so it cannot expand and do work, the internal energy increases by an amount equal to the heat added. Finally, if we use the gas to do work without adding heat, the internal energy decreases by an amount equal to the work done. Our common-sense observations in this paragraph about when the gas is heated and cooled suggest that internal energy and temperature should be related.

Thus far, we have done nothing remarkable. From this viewpoint, equation 15-1 is not the first law; it is only the definition of a bookkeeping device or a fudge factor. The first law is the statement that *this new bookkeeping function is a state function.*

State Functions

State functions are extremely important in thermodynamics, especially to chemists. A **state function** is a property whose value is determined completely by the state of a system at a given instant; it is not dependent on the past history of the system.

To illustrate what this means, imagine the following Cold War scenario. Suppose that a river flows from West Germany into East Germany at Einstadt and reemerges to West Germany many miles further at Ausdorf (Figure 15-4). West German observers are not allowed behind the Iron Curtain. The East German authorities are suspected of building an atomic power station by the banks of the river and of using the river water as a coolant and working medium for their steam turbines. Can the presence of the reactor be detected by the West German observers?

First, let's suppose that the heat from the reactor converts river water to steam, which then drives steam turbines and generates electricity. The river water is now the thermodynamic system. The water is heated and vaporized by the heat supplied by the atomic pile; then it is cooled as it expands in the turbine and does work to turn the turbine rotor. The East

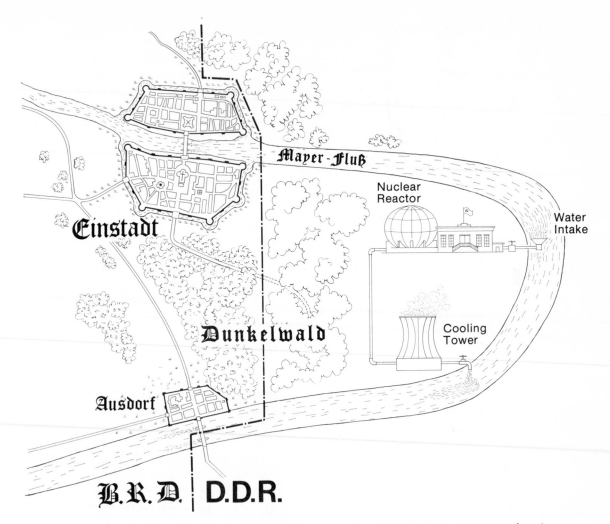

Figure 15-4   *State functions* of the river water (or any other thermodynamic system) are functions whose value depends only on the present state of the water and not upon its history. A scientist at Einstadt and Ausdorf cannot tell, by measurements on the water as it flows into or out of East Germany, whether the nuclear reactor is operating or not, as long as the effluent water is carefully recooled. For analogies with the first law of thermodynamics, see text.

German power authorities suspect that their West German colleagues are investigating, so they take care to cool the water to the same temperature as it had at the intake channel before they dump it back in the river.

Next, suppose that May Day is a holiday at the reactor, and that on May 1 of every year the engineers shut down the reactor and close the intake channel at the river. Can the West German observers detect the presence of the power station by taking measurements on the river at Einstadt and Ausdorf?

Unfortunately for the peace of mind of West German intelligence, they cannot. What physical measurements might be made on the thermodynamic system (the river)? They could measure the temperature of the water, its density, viscosity, molar volume, electrical conductivity, ratio of ^{16}O to ^{18}O, melting and boiling points, chemical purity, and many other properties. All these properties are state functions. The temperature of the water depends on its present state and not on its past history. (That is, the present temperature may be a consequence of what happened in the past, but we do not need to know that history to measure the temperature.) The change in temperature of the water as it flows from Einstadt to Ausdorf can be determined by measuring the temperature at the two towns:

$$\Delta T = T_{Aus} - T_{Ein} = T_2 - T_1$$

Similarly, the change in density, viscosity, or any other state function is obtained by taking the difference between the density, or viscosity, or other property at Einstadt and at Ausdorf. We do not need to know what happened to the river in East Germany.

All of this may seem trivial until you realize that we cannot do the same thing with heat or work. There is no such thing as a "heat content" that we can measure at Einstadt and Ausdorf to indicate how much heat has been added to the water during its route through East Germany. Similarly, there is no property known as "work content" that can be measured at the two towns to determine how much work the East Germans obtained from the water. So long as the power authorities are careful to make the temperature of the exit water from the reactor match the temperature of the intake water, there will be no difference in the water at Ausdorf when the reactor is running, or on May Day when it is not. The existence or nonexistence of the power station will be a mystery to the West German observers, at least from their measurements on the river water.

If we observe that the change in water temperature at Einstadt and Ausdorf is the same on May Day as on any other day, then we can say that whatever amount of heat was added to the water in the reactor must have been balanced precisely by the work done in the turbines (or by the cooling done before the water was dumped back into the river). Conversely, every bit of work obtained in the turbines must have been compensated by heat from the reactor or else the water would have been cooler at Ausdorf. No information about the water at the two towns will suggest the amount of heat, q, added to the water or the amount of work done, w. Neither heat nor work is a state function, *but their difference is.* If the quantity $q - w$ is not kept constant, differences will appear in measurable properties of the water at Ausdorf. The most obvious such property is temperature, but molar volume, density, electrical conductivity, and other properties will change as well. Turning this statement around, if we specify the state of the water at Einstadt and Ausdorf, then we have specified the quantity $q - w$ during the transit to East Germany, even though we have not specified q and w individually. Their difference is the *change in a state function, E.*

For an ideal gas, this internal energy, E, is the same function as the mean molar kinetic energy, E_k, that we encountered in Chapter 3. The internal energy is directly proportional to temperature:

$$E = \tfrac{3}{2}PV = \tfrac{3}{2}RT \qquad \text{(for 1 mole of ideal monatomic gas)} \qquad (3\text{-}26)$$

For nonideal gases, E will be approximately proportional to temperature; and for substances in general, an increase in temperature accompanies an increase in internal energy.

State functions are useful to chemists precisely because they do not depend on the history of a chemical system. Energy is a state function. So are pressure, temperature, volume, and all the other quantities that we commonly think of as properties of a substance. The very term *property* suggests something that a substance has, independent of any factors other than its present condition. We never speak of the work that a substance has, and should not speak of the heat that it possesses. If the final state of a system is identified by a subscript 2, and the initial state is given the subscript 1, then equation 15-1 becomes the first law of thermodynamics if we expand it to

$$\Delta E = E_2 - E_1 = q - w \qquad (15\text{-}2)$$

The state-function properties are implicit in the middle term.

Example 2

A traveler goes from Pasadena, California, to Aspen, Colorado. Which of the following are state functions of the trip, and which are not: (a) distance traveled; (b) latitude change; (c) altitude change; (d) gasoline consumed; (e) elapsed time; (f) longitude change; (g) work done to move the vehicle; (h) oxygen consumed by passengers and vehicle?

Solution

Only b, c, and f are state functions, independent of the way the trip was carried out.

15-3 ENERGY, ENTHALPY, AND HEAT CAPACITY

When chemists first began to study heats of reaction systematically they discovered that a particularly convenient type of reaction was one constrained to a fixed volume in a **bomb calorimeter** (Figure 2-4). This is a sturdy steel container with a tight lid, immersed in a water bath and provided with electrical leads to detonate the reaction inside. The heat evolved in such a reaction at constant volume is measured by the increase in temperature of the water bath.

If the chemical system inside the metal container is not allowed to change in volume, then it cannot do PV work. If no other types of work are involved, the heat liberated by the reaction is equal to the decrease in internal energy:

$$\Delta E = q_v \qquad \text{(at constant volume)} \tag{15-3}$$

In the absence of any work effects, the gain or loss of heat by the contents of the container is a direct measure of the increase or decrease of internal energy of the reacting substances. If the reaction releases heat, it is called an **exothermic** reaction; if it absorbs heat, it is **endothermic**.

Most reactions take place at constant pressure rather than at constant volume, and it would be useful to have a thermodynamic function that behaves at constant pressure as E does at constant volume; that is, a measure of the heat of the reaction under those conditions. Such a function is the **enthalpy**, H, defined by

$$H \equiv E + PV \tag{15-4}$$

At constant pressure, the change in enthalpy of a system is

$$\Delta H = \Delta E + P\,\Delta V \tag{15-5}$$

From the first law, we can write this as

$$\Delta H = q - w + P\,\Delta V$$

If we rule out electrical, gravitational, magnetic, and all kinds of work other than PV work, then $w = P\Delta V$ and the last two terms cancel in the preceding equation. We then have the statement that the heat of a reaction at constant pressure is equal to the change in enthalpy of the system:

$$\Delta H = q_P \tag{15-6}$$

(The subscript P indicates that the transfer of heat takes place at constant pressure.) At constant pressure, enthalpy *increases* in an endothermic reaction as heat flows into the system, and enthalpy *decreases* in an exothermic process as heat flows out of the system.

All the functions on the right of the equivalence sign in equation 15-4 are state functions, so H is a state function as well. The change in enthalpy of a system depends on the enthalpy of the system before and after a process and not at all on the path by which the system went from the initial state to the final state; thus

$$\Delta H = H_2 - H_1 \tag{15-7}$$

In summary, internal energy, E, and enthalpy, H, are complementary functions because they both measure the transfer of heat in a process under certain conditions: E at constant volume, and H at constant pressure. Both E and H are state functions, so the change in E, for example, is the same whether the process is carried out at absolutely constant volume, or whether the volume is merely brought back to its original value at the end. Similarly, the enthalpy change during a reaction is the same if the reaction is carried out at constant pressure, or if the pressure is allowed to vary but is brought back to its starting value at the end. Hence the expressions *constant volume* or *constant pressure* in connection with state functions need only mean "no *net* change in volume or pressure." This is an advantage of state functions.

The heat that must be added to a specified amount of any substance in order to increase the temperature by one kelvin is defined as the **heat capacity** of the substance, C. Heat capacities are expressed in units of joules per mole per kelvin. Since the heat capacity is the rate of addition of heat with temperature change, it can be written as*

$$C = \frac{dq}{dT} \tag{15-8}$$

Under constant-pressure conditions (or at least with no net change in pressure), transfer of heat is measured by change in enthalpy, so the heat capacity at constant pressure is the rate of change of enthalpy with temperature:

$$C_P = \left(\frac{dH}{dT}\right)_P \tag{15-9}$$

---

*For those for whom differential notation is unfamiliar, the *average* heat capacity when a quantity of heat, Δq, is added over a temperature change, ΔT, is

$$C_{av} = \frac{\Delta q}{\Delta T}$$

The heat capacity at a specific temperature, T, is the limit of this average as the size of the temperature interval ΔT goes to zero:

$$C_T = \lim_{\Delta T \to 0} \left(\frac{\Delta q}{\Delta T}\right) = \frac{dq}{dT}$$

Differential notation may be easier to understand with speed, v, which is the the rate of change of position, x, with time, t. Here v_{av} corresponds to an average speed over a time interval Δt:

$$v_{av} = \frac{\Delta x}{\Delta t}$$

whereas v_t corresponds to an instantaneous speed at time t:

$$v_t = \lim_{\Delta t \to 0} \left(\frac{\Delta x}{\Delta t}\right) = \frac{dx}{dt}$$

Energy plays the corresponding role under constant volume conditions, and the heat capacity at constant volume is

$$C_V = \left(\frac{dE}{dT}\right)_V \tag{15-10}$$

As we saw in Chapter 3, the energy per mole of an ideal monatomic gas is $E = \frac{3}{2}RT$. Hence the heat capacity at constant volume is $C_V = \frac{3}{2}R = \frac{3}{2} \times 8.314$ J mole^{-1} = 12.47 J mole^{-1}. It is a test of the validity of ideal gas theory that the measured heat capacity of monatomic gases such as helium or neon at room temperature is extremely close to this figure. Moreover, the heat capacity at constant pressure is expected to be

$$C_P = \frac{5}{2}R \tag{15-11}$$

since $H = E + PV = E + RT$ for 1 mole of an ideal gas. The fact that C_P for real monatomic gases is very close to $\frac{5}{2}R$ or 20.79 J mole^{-1}, and approaches this value most closely under conditions of low pressure or high temperature, gives us confidence in the essential correctness of the kinetic molecular theory of gases outlined in Chapter 3.

15-4   THE FIRST LAW AND CHEMICAL REACTIONS

Equation 15-7 is the most important single consequence of the first law of thermodynamics for chemistry. It tells us that the heat of a reaction carried out at constant pressure is a state function. The heat of reaction is the difference between the enthalpy of the products and the enthalpy of the reactants. It is the same whether the actual reaction occurs in one step, or in half a dozen intermediate steps. This principle of the additivity of heats of reaction was introduced in Chapter 2 without proof, but now we can see why it is so. In our example from Section 2-6 on the synthesis of diamond, the heat of preparation of diamond from methane is the same, regardless of whether diamond is made directly from methane or the methane is oxidized to CO_2, which is then used to make diamond:

$$
\begin{array}{lll}
CH_4(g) + 2O_2(g) \rightarrow CO_2(g) + 2H_2O(l) & \Delta H = -890 \text{ kJ} & (2\text{-}12) \\
\underline{CO_2(g) \rightarrow C(di) + O_2(g)} & \underline{\Delta H = +395 \text{ kJ}} & (2\text{-}14) \\
CH_4(g) + O_2(g) \rightarrow C(di) + 2H_2O(l) & \Delta H = -495 \text{ kJ} & (2\text{-}15)
\end{array}
$$

Because enthalpy, H, is a state function, *heats of reactions are additive* in the same way that the reactions to which they pertain are additive. This statement is sometimes called Hess' law, although it is really only a consequence of the first law of thermodynamics.

Figure 15-5

The heat of preparation of diamond from methane is the same whether the reaction occurs in one step or whether CO_2 is made from methane, and then diamond is made from CO_2. Such a statement can be made only because the heat of a reaction at constant pressure and temperature is equal to the change in enthalpy, H, and enthalpy is a state function.

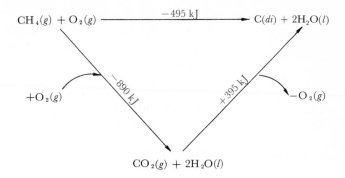

The independence of enthalpy change of the path of a reaction can be diagrammed for the diamond synthesis reactions by a cycle, as in Figure 15-5. The first law states that either way around the cycle (the one- or two-step path) leads to the same ΔH and heat of reaction. The enthalpy changes also can be represented on an energy-level diagram (Figure 15-6). Note that the absolute numerical value of the enthalpy is not defined, but only the changes in going from one state of reactants or products to another. In the past, every time that we drew an energy-level diagram we were unconsciously using the state function properties of energy. Neither heat nor work can be represented on such a diagram (except for the special cases in which one or the other of these is equal to a state function).

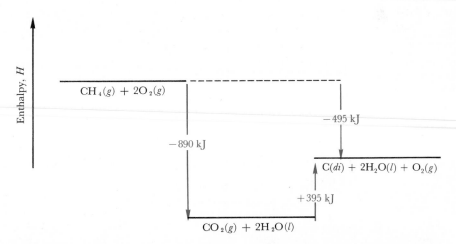

Figure 15-6

For the same reasons that the reactions of Figure 15-5 can be drawn as a cycle, the enthalpies can be represented in an energy-level diagram such as this one. The enthalpy change in going from one state to another depends only on the levels of the two states in this diagram and not on the manner of going from one state to the other.

All of the earlier discussion of Section 2-6 is valid because H is a state function. It is unnecessary to tabulate the heats of all reactions; we need to list only those from which all other reactions can be obtained by a proper combination of reactions. The reactions chosen are the reactions for the formation of compounds from their elements in standard states. The **standard state** of a gas at a chosen temperature is 1 atm partial pressure; that of a liquid or solid is the pure liquid or solid at 1 atm external pressure. The chosen temperature is usually 298 K for most thermodynamic tabulations. Standard heats of formation for many substances are listed in Appendix 3.

Example 3

What is the standard heat of the reaction for the formation of anhydrous crystalline copper sulfate from its elements in their standard states?

Solution   The reaction is

$$Cu(s) + S(s) + 2O_2(g) \rightarrow CuSO_4(s)$$

This is the reaction for which the heat of formation is tabulated in Appendix 3. For this reaction, $\Delta H^0_{298} = -769.9$ kJ mole^{-1}.

Example 4

The standard heat of formation of gaseous B_5H_9 is given in Appendix 3 as $+62.8$ kJ mole^{-1}. Of what reaction is this the heat of reaction?

Solution   The reaction is that of synthesis of B_5H_9 from elemental solid boron and hydrogen gas at 1 atm and 298 K:

$$5B(s) + 4\tfrac{1}{2}H_2(g) \rightarrow B_5H_9(g) \qquad \Delta H^0_{298} = +62.8 \text{ kJ}$$

When a heat term is written after a reaction, as in the preceding equation, the units are understood to be kilojoules *per stoichiometric unit* of reaction as written. The heat of the reaction here is 62.8 kJ per mole of B_5H_9, but only 12.6 kJ per mole of boron used, or 62.8 kJ for every $4\tfrac{1}{2}$ moles of hydrogen gas used. Heats of formation are always tabulated per mole of the compound formed.

Example 5

The compound B_5H_9 ignites spontaneously in air with a green flash to produce B_2O_3 and water. What is the heat of the reaction under standard conditions?

Solution

The reaction in unbalanced form is

$$B_5H_9(g) + O_2(g) \rightarrow B_2O_3(s) + H_2O(l)$$

Two moles of B_5H_9 are needed for every 5 moles of B_2O_3 to account for the boron. Hence, the 18 hydrogen atoms must appear as 9 water molecules. The balanced reaction is

$$2B_5H_9(g) + 12O_2(g) \rightarrow 5B_2O_3(s) + 9H_2O(l)$$

The tabulated standard heats of formation of reactants and products are

Substance	ΔH^0_{298}
$B_5H_9(g)$	$+62.8$ kJ mole^{-1}
$O_2(g)$	0.0 kJ mole^{-1}
$B_2O_3(s)$	-1264 kJ mole^{-1}
$H_2O(l)$	-285.8 kJ mole^{-1}

The three reactions of formation that when added produce the desired reaction are

$$
\begin{array}{ll}
& \Delta H^0_{298} \\
2B_5H_9(g) \rightarrow 10B(s) + 9H_2(g) & -2 \times (+62.8 \text{ kJ}) = -126 \text{ kJ} \\
10B(s) + 7\frac{1}{2}O_2(g) \rightarrow 5B_2O_3(s) & +5 \times (-1264 \text{ kJ}) = -6320 \text{ kJ} \\
9H_2(g) + 4\frac{1}{2}O_2(g) \rightarrow 9H_2O(l) & +9 \times (-285.8 \text{ kJ}) = -2572 \text{ kJ} \\
\hline
2B_5H_9(g) + 12O_2(g) \rightarrow 5B_2O_3(s) + 9H_2O(l) & \Delta H^0_{298} = -9018 \text{ kJ}
\end{array}
$$

Boron hydrides were once considered as rocket fuels because of their extremely high heats of combustion.

We can take a shortcut with these tables of enthalpies or heats of formation by treating the numbers as though they were absolute enthalpies of the compounds, rather than enthalpies of formation from elements. The result is the same, since reactants and products must be composed of the same number and same kind of atoms. Then for the reaction

$$2B_5H_9(g) + 12O_2(g) \rightarrow 5B_2O_3(s) + 9H_2O(l)$$

the enthalpy of reaction is nine times the standard enthalpy of liquid water, plus five times the standard enthalpy of solid B_2O_3, less two times the standard enthalpy of gaseous B_5H_9. The standard enthalpy of elemental O_2 is zero.

Example 6

The compound B_5H_9 can be prepared from diborane, B_2H_6, which reacts at the proper temperature to give B_5H_9 and H_2. Is the reaction exothermic or endothermic? What is the heat of reaction at 298 K per mole of diborane consumed?

Solution   The reaction is exothermic; $\Delta H^0_{298} = -5.9$ kJ mole^{-1} of B_2H_6.

15-5   BOND ENERGIES

In the language of localized bond models, we say that a molecule of methane, CH_4, is held together by four equivalent C—H single bonds. If this idea is valid, the heat of decomposition of methane to isolated carbon and hydrogen atoms should be four times the bond energy of a C—H bond. (Although we shall work consistently with enthalpies, we shall adopt the common but loose terminology and refer to our results as bond energies rather than bond enthalpies. The difference is small and is within the limits of accuracy of the bond-energy approach itself.)

The heat of formation of methane is

$$C(gr) + 2H_2(g) \rightarrow CH_4(g) \qquad \Delta H^0_{298} = -74.85 \text{ kJ}$$

But to calculate bond energies we need the decomposition to atomic gaseous carbon and hydrogen, not solid graphite and diatomic H_2. The atomization reactions are

$$C(gr) \rightarrow C(g) \qquad \Delta H^0_{298} = +718.38 \text{ kJ}$$
$$\tfrac{1}{2}H_2(g) \rightarrow H(g) \qquad \Delta H^0_{298} = +217.94 \text{ kJ}$$

These are the standard heats of formation of the gaseous atoms from the elements in their standard states, and are tabulated in Appendix 3 along with the other heats of formation.

The desired reaction for decomposing methane into isolated atoms can be constructed from the preceding reactions:

	ΔH^0_{298}
$CH_4(g) \rightarrow C(gr) + 2H_2(g)$	$-1 \times (-74.85 \text{ kJ}) = + 74.85$ kJ
$C(gr) \rightarrow C(g)$	$+1 \times (+718.38 \text{ kJ}) = + 718.38$ kJ
$2H_2(g) \rightarrow 4H(g)$	$+4 \times (+217.94 \text{ kJ}) = + 871.76$ kJ
$CH_4(g) \rightarrow C(g) + 4H(g)$	$\Delta H^0_{298} = +1665.0$ kJ

If this is the heat needed to break four C—H bonds, then the bond energy (strictly speaking, the bond enthalpy) of one C—H bond is one-fourth this figure. The bond energy of a C—H bond in methane is 416 kJ mole^{-1} of bonds.

Bond Energy of a C—C Single Bond

From the heat of formation of ethane, C_2H_6, we can obtain a value for the bond energy of a carbon–carbon single bond. From Appendix 3,

$$\Delta H^0_{298}$$

$$
\begin{array}{ll}
C_2H_6(g) \rightarrow 2C(gr) + 3H_2(g) & -1 \times (-84.7 \text{ kJ}) = + 84.7 \text{ kJ} \\
2C(gr) \rightarrow 2C(g) & +2 \times (+718.4 \text{ kJ}) = +1436.8 \text{ kJ} \\
\underline{3H_2(g) \rightarrow 6H(g)} & \underline{+6 \times (+217.9 \text{ kJ}) = +1307.4 \text{ kJ}} \\
C_2H_6(g) \rightarrow 2C(g) + 6H(g) & \Delta H^0_{298} = +2829 \text{ kJ}
\end{array}
$$

In the localized bond model of ethane, the molecule has six C—H bonds and one C—C bond:

$$
\begin{array}{cc}
H & H \\
| & | \\
H-C-C-H \\
| & | \\
H & H
\end{array}
$$

If the value of 416 kJ mole^{-1} is accepted for the C—H bonds in methane, the six C—H bonds in ethane will account for 2496 kJ mole^{-1}. The remaining 333 kJ must be the bond energy of a mole of C—C single bonds.

We can test the validity of this entire approach by calculating the expected heat of formation of propane, C_3H_8, from graphite and hydrogen gas. In the localized bond model, propane has two C—C bonds and eight C—H bonds:

$$
\begin{array}{ccc}
H & H & H \\
| & | & | \\
H-C-C-C-H \\
| & | & | \\
H & H & H
\end{array}
$$

The heat of formation is calculated as follows:

$$\Delta H^0_{298}$$

$$
\begin{array}{ll}
3C(g) + 8H(g) \rightarrow C_3H_8(g) & 2 \times (-333 \text{ kJ}) + 8 \times (-416 \text{ kJ}) \\
& = -3994 \text{ kJ} \\
3C(gr) \rightarrow 3C(g) & 3 \times (+718.4 \text{ kJ}) = +2155 \text{ kJ} \\
\underline{4H_2(g) \rightarrow 8H(g)} & \underline{8 \times (+217.9 \text{ kJ}) = +1743 \text{ kJ}} \\
3C(gr) + 4H_2(g) \rightarrow C_3H_8(g) & \Delta H^0_{298} = -3994 + 3898 = -96 \text{ kJ}
\end{array}
$$

The observed heat of formation of propane is -104 kJ mole^{-1}, which gives you some idea of the degree of accuracy of bond-energy calculations. We are unfortunately in the difficult position of wanting the small difference between two large numbers. Errors and approximations in the data and in the assumption of localized bonds contribute to an error of 8 kJ mole^{-1}.

The calculations for propane can be diagrammed on an energy scale, as in Figure 15-7. The addition of 3898 kJ to atomize graphite and dis-

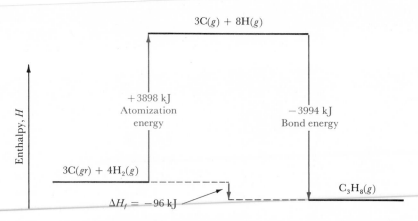

Figure 15-7   An energy-level diagram representing the formation of propane, C_3H_8, from graphite and hydrogen gas. Bond-energy calculations involve a hypothetical intermediate high-energy state in which all atoms are isolated from one another in the gas phase. The difference between the energy required to turn reactants into gaseous atoms and that given off when products are made from these same atoms is the heat of formation.

sociate hydrogen gas represents a move from a lower to a higher energy level, that of separated atoms. The combination of those atoms into a molecule of C_3H_8 then represents a drop in energy of 3994 kJ, to a still lower state. The difference between initial and final energy levels, of graphite and H_2 as reactants and C_3H_8 as product, is the calculated heat of formation, $\Delta H_{298} = -96$ kJ.

Tabulation of Bond Energies

Now we can proceed to calculate the bond energies of bonds of all types.

Example 7

From the data for ethene in Appendix 3, calculate the bond energy of a C=C double bond. The bond structure of ethene is

$$\begin{array}{ccc} H & & H \\ \diagdown & & \diagup \\ & C = C & \\ \diagup & & \diagdown \\ H & & H \end{array}$$

Solution   The bond energy is 592 kJ mole^{-1}.

Example 8

From the data for water in Appendix 3, calculate the bond energy of an O—H bond.

Solution

The bond energy is 463 kJ mole^{-1}. Note that you need the heat of atomization of oxygen and that you must use the heat of formation of water vapor, not of liquid water.

The most useful bond energies are obtained not from the heats of formation of individual compounds like methane or ethane, but by averaging the values obtained from entire classes of compounds, such as the hydrocarbons for C—H and C—C bond energies. These adjusted best values for several types of bonds are given in Table 15-1. Note that the adjusted value for a C—H bond differs by 3 kJ mole^{-1} from the value obtained from methane alone. Errors of 5 or 10 kJ mole^{-1} are considered acceptable in bond-energy calculations.

Table 15-1

Approximate Molar Heats of Atomization and Bond Energies[a] at 298 K

	Molar heat of atomization (kJ mole^{-1})	Bond energy[b] (kJ mole^{-1})								
		H—	C—	C=	C≡	N—	N=	N≡	O—	O=
H	217.9	436	413			391			463	
C	718.4	413	348	615	812	292	615	891	351	728
N	472.6	391	292	615	891	161	418	946		
O	247.5	463	351	728					139	485
F	76.6	563	441			270			185	
Si	368.4	295	290						369	
P	314.5	320								
S	222.8	339	259	477						
Cl	121.4	432	328			200			203	
Br	111.8	366	276							
I	106.6	299	240							

[a]Values from L. Pauling, *The Nature of the Chemical Bond,* Cornell University Press, Ithaca, N.Y., 1960, 3rd ed. See also T. L. Cottrell, *The Strengths of Chemical Bonds,* Butterworths, London, 1958, 2nd ed.

[b]This is an example of loose but convenient terminology. These are actually bond enthalpies at 298 K, in the sense that they were obtained from enthalpies of formation.

Applications of Bond-Energy Calculations

Using the best mean values for bond energies given in Table 15-1, we can calculate the expected heat of formation of a molecule from its elements. This is useful in cases where the method works; but paradoxically it is even more useful in cases where the method does not work, or where the calculated heat of formation differs sharply from the experimentally measured value. When this occurs, it is a sign that our simple picture of bonds between pairs of atoms is not good enough. Bond-energy calculations can be a tool for learning about bonding in molecules.

As an example, small hydrocarbon molecules are known in which the carbon atoms are in a ring, and each carbon has two hydrogens bonded to it. The simplest of these cyclic hydrocarbons have the structures:

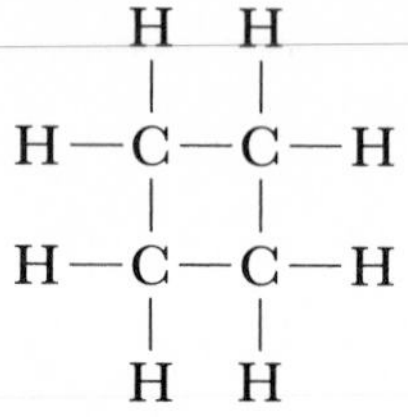

cyclobutane

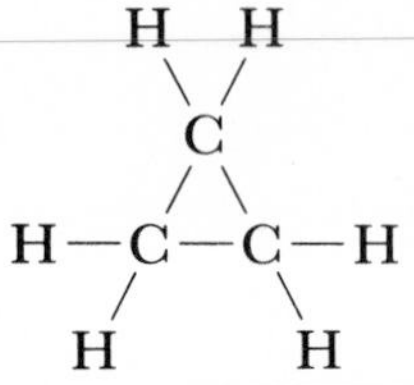

cyclopropane

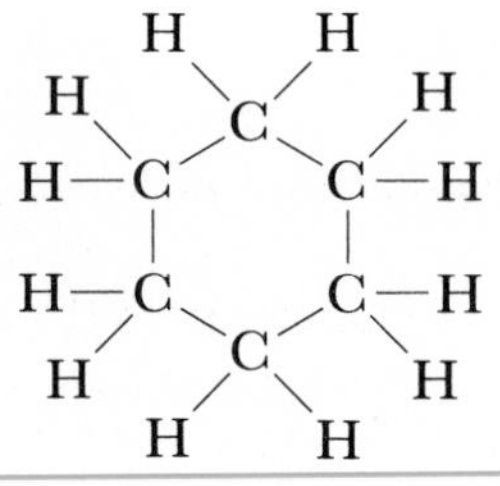

cyclohexane

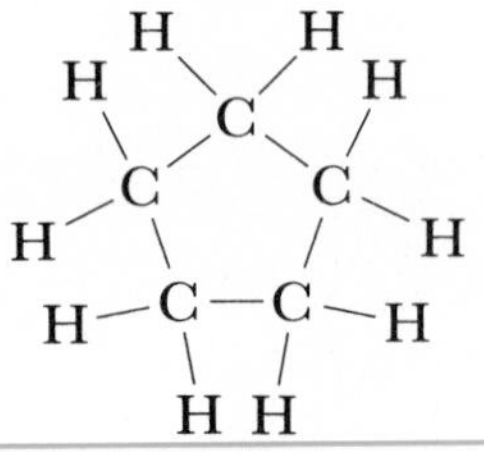

cyclopentane

Let's compare the expected heats of formation of these molecules with the observed values. In Table 15-2, line A gives the total bond energy of each molecule, or the energy given off when the molecule is made from isolated carbon and hydrogen atoms. Line B gives the energy needed to make the requisite number of C and H atoms from graphite and H_2 gas. The difference, in line C, is the calculated heat of formation, and the measured value is in line D. Agreement is quite good for cyclohexane, but becomes progressively worse as the molecules become smaller. The cyclopropane molecule is expected to be 60 kJ mole^{-1} more stable than its component elements, but is actually 53 kJ mole^{-1} less stable. The molecule is 113 kJ mole^{-1} less stable than it is expected to be on the basis of three

Table 15-2

Strain Energy in Cyclic Hydrocarbon Molecules (kilojoules per mole)

		Cyclopropane, C_3H_6	Cyclobutane, C_4H_8	Cyclopentane, C_5H_{10}	Cyclohexane, C_6H_{12}
	No. C—C bonds	3	4	5	6
	No. C—H bonds	6	8	10	12
	C—C bond energy	1044	1392	1740	2088
	C—H bond energy	2478	3304	4130	4956
A.	Total bond energy	3522	4696	5870	7044
	No. C atoms	3	4	5	6
	No. H atoms	6	8	10	12
	C vaporization energy	2155	2874	3592	4310
	H_2 dissociation energy	1307	1743	2179	2615
B.	Total atomization energy	3462	4617	5771	6925
C.	ΔH, calculated (B − A)	−60	−79	−99	−119
D.	ΔH, measured	+53	+27	−77	−123
E.	Instability of molecule (D − C)	+113	+106	+22	−4

C—C bonds and six C—H bonds. These energy levels are diagrammed in Figure 15-8.

The reason for the unexpected instability of the cyclopropane molecule is that it is highly strained. The three-carbon ring must have bond angles of 60° rather than the preferred tetrahedral 109.5°. In terms of molecular orbitals, the overlap between atomic sp^3 orbitals on the carbon atoms is poor, so the bonds are weak. Cyclobutane has somewhat less strain or bond distortion, and cyclopentane has even less. Table 15-2 shows that cyclohexane is essentially an unstrained molecule. If you try to build models using tetrahedrally bonded carbon atoms, cyclopropane and cyclobutane cannot be built without bending the bonds, cyclopentane requires only a warping of bond angles from 109.5° to the 108° of a pentagon, and the cyclohexane molecule is loose and free to adopt more than one configuration.

The Heat of Formation of Benzene

The bond-energy method is a conspicuous failure again in predicting the heat of formation of benzene. As before, this failure suggests a great deal

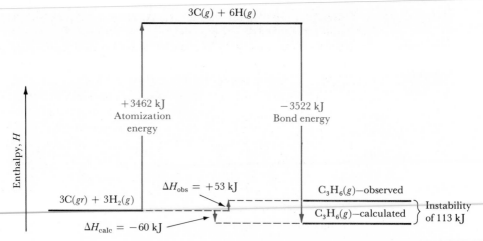

Figure 15-8 Energy-level diagram for the formation of cyclopropane from graphite and hydrogen gas. Bond strain makes the product molecule 113 kJ mole⁻¹ less stable than would be expected from simple bond-energy calculations.

about the benzene molecule. Let us assume that benzene has one of the structures proposed for it by the organic chemist Kekulé.

From this model, benzene would contain six C—H single bonds, three C—C single bonds, and three C=C double bonds. Hence per mole of benzene the total bond energy is (using values from Table 15-1)

Six C—H bonds:	6×413 kJ	$= 2478$ kJ
Three C—C bonds:	3×348 kJ	$= 1044$ kJ
Three C=C bonds:	3×615 kJ	$= 1845$ kJ
Total bond energy:		5367 kJ

The heat of formation reaction then is constructed:

$$\Delta H^0_{298}$$

$6C(g) + 6H(g) \rightarrow C_6H_6(g)$	$-(\text{bond energy}) = -5367$ kJ
$6C(gr) \rightarrow 6C(g)$	$6 \times (718 \text{ kJ}) = +4308$ kJ
$3H_2(g) \rightarrow 6H(g)$	$6 \times (218 \text{ kJ}) = +1308$ kJ
$6C(gr) + 3H_2(g) \rightarrow C_6H_6(g)$	$\Delta H^0_{298} = +\ 249$ kJ

There is a flaw somewhere, because the standard heat of formation of gaseous benzene, as measured in the laboratory, is not 249 kJ mole^{-1} of benzene, but only 83 kJ mole^{-1}. The benzene molecule is more stable by 166 kJ mole^{-1} than predicted for a molecule with the Kekulé structure. These energy levels are shown in Figure 15-9.

The flaw lies in the assumption of the Kekulé structure. In Chapter 13 we discovered that the Kekulé structure failed to explain the six equal bond lengths between carbon atoms in the benzene ring, but that a delocalized-molecular-orbital theory could satisfactorily account for bonding. In Chapter 21, we shall look at the large class of aromatic compounds—compounds with just such delocalized electrons. In general, delocalization makes the molecule more stable by lowering the energy of the delocalized electrons. Bond energies provide a way to calculate this stabilization from measurements of heats of formation of aromatic compounds.

Example 9

Calculate the standard heat of formation of carbon dioxide, $O=C=O$. Assume the presence of two double $C=O$ bonds. Compare your value with the measured value in Appendix 3. Do your figures predict delocalization in CO_2?

Solution   The calculated heat of formation is -243 kJ mole^{-1}. Yes; the stabilization energy is 151 kJ mole^{-1}.

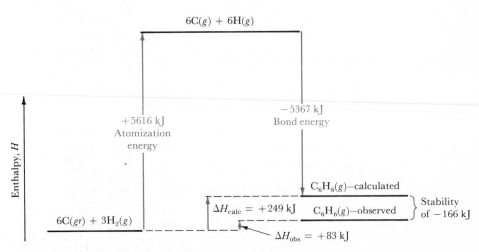

Figure 15-9   Energy-level diagram for formation of benzene. The expected heat of formation, ΔH_{calc}, is $+249$ kJ, but the observed value, ΔH_{obs}, is only $+83$ kJ. Hence the benzene molecule is 166 kJ more stable than the bond structure suggested by Kekulé would predict.

Summary

This chapter has provided the foundation for the treatment of energy in chemical reactions that was introduced in Chapter 2. There are many ways in which the first law of thermodynamics can be viewed, but we have confined our discussion to the aspects that are most useful to a chemist interested in heats of reaction and bond energies.

Heat and work are alternative forms of energy and, within limits, they can be interconverted. In any such process, energy is conserved. The difference between the heat fed into a system and the work done by the system on its surroundings is defined as the change in **internal energy**, E:

$$\Delta E = q - w$$

A **state function** is a function for which we can calculate the change during any process, knowing only the starting and ending states, without information about how the process was carried out. In short, state functions depend only on the state of a system and not on its past history. Neither q nor w by itself is a state function, but the **first law of thermodynamics** asserts that the *difference* between heat added and work done is a state function. If the initial state is represented by the subscript 1 and the final state is represented by 2, then the first law is

$$\Delta E = E_2 - E_1 = q - w$$

The internal energy then is a property that a substance exhibits under a certain set of conditions. From intuition we know that internal energy is somehow correlated with temperature, and the simple kinetic molecular theory of gases introduced in Chapter 3 tells us that for ideal, monatomic gases, $E = \frac{3}{2}RT$.

The work done when a gas increases in volume by dV against an opposing pressure of P is $dw = P\, dV$. If the volume is kept constant, $w = 0$, and the internal energy increase is simply equal to the amount of heat fed in: $\Delta E_V = q_V$. The heat of a reaction carried out at constant volume is measured by the change in internal energy.

Most reactions are carried out at constant pressure, not constant volume. If the **enthalpy**, H, is defined by $H = E + PV$, then the heat of a reaction carried out at constant pressure is ΔH, or $\Delta H_P = q_P$. Because E, P, and V are state functions, H also is a state function. This has important consequences for heats of reaction. Because only the initial and final states of a reaction are important for ΔH, the term *constant pressure* need only mean that the pressure at the end of the reaction is brought back to its original value. Furthermore, the heat of a reaction that can be written as a sum of several reactions is found by summing the individual heats of those reactions. Heats of reaction are additive because H is a state function.

This additivity results in a great simplification in data handling. Instead of tabulating heats of all possible reactions, we need list only those for the formation of each molecule from its elements in agreed-upon standard states. The heat of any other reaction can be found by combining heats of formation in the same way that the formation reactions are combined to make the reaction in question.

The heat required to cause a 1 K temperature increase per mole of a substance is the **molar heat capacity**, C. At constant volume, $C_V = \frac{3}{2}R$ for an ideal monatomic gas, and at constant pressure, $C_P = \frac{5}{2}R$. The extent to which real monatomic gases approach these relationships is a measure of the validity of the ideal gas model.

The simplest picture of chemical bonding imagines that a molecule is held together by individual two-electron bonds connecting pairs of atoms. For the great majority of molecules, standard bond energy values can be set up that reproduce the experimental heats of formation of molecules within 5 or 10 kJ. For some molecules, however, the bond-energy calculations appear to fail. The molecules are more stable, or less stable, than calculations using simple bond models predict. This is a sign that the simple bond model is wrong. Molecules with strained geometries may be less stable than predicted, and molecules with delocalization of electrons may be more stable.

Self-Study Questions

1. What is the difference between a spontaneous process and a rapid process? Give an example of a reaction that is either spontaneous or rapid, but not both.

2. Give an example of the conversion of (a) heat into work; (b) kinetic energy into heat; (c) work into heat; (d) work into kinetic energy; (e) potential energy into work; (f) work into potential energy; (g) kinetic energy into work; (h) heat into kinetic energy.

3. Characterize the following thermodynamic systems as open, closed, or isolated: (a) an astronaut in the briefing room at the Houston space center; (b) the same astronaut in a space capsule without a window; (c) the same astronaut in Skylab, with a hydroponic garden; (d) Dracula in his Transylvanian castle; (e) Dracula sealed in a lead coffin.

4. What will happen to any living organism if steps are taken to turn it into an isolated thermodynamic system?

5. How is work related to pressure, volume, and temperature in the expansion of a gas?

6. In what way is a state function different from any other type of function? Are pressure and volume state functions? Why are state functions particularly convenient to work with?

7. A careless driver keeps putting dents into his Porsche, but he has a skilled repairman who smooths them out with creative bodywork almost as fast as they occur. On the first of each month, his friendly banker, who really owns the car, tallies the net number of dents present in the car body, as an estimate of the market value of the vehicle. Show that this market value is a state function, whereas the number of dents incurred in a given month and the amount of bodywork done are not. Why would the number of dents received become a state function if the auto mechanic went on vacation for a month?

8. In the analogy in Question 7, which quantities correspond to work, heat, and energy? Why is a repairman on vacation like a closed steel container?

9. What is the first law of thermodynamics, and how do state functions enter it?

10. When we define the molar energy of a monatomic ideal gas as the sum of the kinetic energies of the individual molecules, or as Avogadro's number times the average molecular energy as in Chapter 3, unconsciously we are saying that E is a state function. Why is this so?

11. Under what conditions will the following functions be state functions: (a) internal energy; (b) heat; (c) volume; (d) work; (e) enthalpy?

12. What is Hess' law? How is it only a special case of the first law of thermodynamics?

13. Suppose that you take two identical clock springs, leave one slack, wind the other tightly and tie it with catgut (acid-impervious), and then dissolve each spring in a beaker of acid. What do you think happens to the work that you exerted to wind the second spring?

14. What is heat capacity? For an ideal monatomic gas, what is the difference between heat capacity at constant volume and heat capacity at constant pressure? Why, from a molecular viewpoint, would you expect C_P to be greater than C_V? What heat-compensating abilities does the gas have at constant pressure that it does not have at constant volume?

15. Why does the first law of thermodynamics permit us to add heats of reaction along with the reactions themselves, as we did in Chapter 2?

16. Why does the first law of thermodynamics make it unnecessary for us to tabulate the heat of each and every chemical reaction?

17. What is a standard heat of formation? Formation from what? And why "standard"? What is the heat of combustion of a substance? Can you think of two substances for which a heat of formation is also a heat of combustion?

18. How does an energy-level diagram such as Figure 15-6 imply that the energy function plotted is a state function?

19. What important assumption lies behind a table of bond energies such as Table 15-1? Why is this assumption valid for methane, ethane, and propane, but not for benzene or cyclopropane?

20. Suppose that you wanted to calculate an experimental C—O single-bond energy in methanol, CH_3—OH, but had *only* the information contained in Appendix 3. How would you go about it?

21. What effects do strain and electron delocalization have on the stability of a molecule? How can bond-energy calculations reveal these two effects?

22. What would be the effect on the petroleum industry if work were a state function?

Problems

First law

1. When a bottle containing copper shot is shaken, its temperature increases. Explain this phenomenon in terms of the first law of thermodynamics.

2. What interconversion of quantities measured in joules is involved when a moving automobile skids to a braking stop? What happens to the kinetic energy of motion that the vehicle once had?

Heat of formation

3. The standard heat of formation of liquid water is -285.8 kJ mole^{-1}. (a) Write a balanced equation for this reaction. Is heat absorbed or given off? (b) What is the heat of combustion of hydrogen gas per mole?

4. (a) What is the standard heat of formation of liquid benzene, from Appendix 3? (b) Write a balanced equation for this reaction. Is heat emitted or absorbed? (c) What is the heat of combustion of liquid benzene?

Heat of combustion

5. The heat of combustion of a mole of liquid acetaldehyde, CH_3CHO, to carbon dioxide and liquid water is -1164 kJ. (a) Write a balanced equation for the combustion reaction. (b) How much heat is evolved per mole of acetaldehyde burned? Per mole of liquid water produced? Per mole of oxygen used? (c) How much heat is released when a gram of acetaldehyde is burned? (d) Use the data you have obtained plus data on carbon dioxide and water from Appendix 3 to calculate the standard heat of formation of acetaldehyde. Compare your answer with that tabulated in Appendix 3.

6. The heat of combustion of solid urea, $(NH_2)_2CO$, to CO_2, N_2, and liquid water is -632.2 kJ per mole of urea burned. (a) Write a balanced equation for the reaction. (b) How much heat is given off per mole of oxygen used? (c) Use the data you have obtained plus data for CO_2 and water from Appendix 3 to calculate the standard heat of formation of urea. Compare your figure with the value tabulated in Appendix 3.

7. (a) Write a balanced equation for the combustion of liquid methanol, CH_3OH, in an ample supply of oxygen to produce liquid water. (b) From data

in Appendix 3, calculate the heat given off during this reaction.

8. (a) Calculate the heat of combustion of ethanol,

$$CH_3CH_2OH(l) + \tfrac{7}{2}O_2(g) \rightarrow$$
$$2CO_2(g) + 3H_2O(l)$$

(b) Do the same for glucose, $C_6H_{12}O_6(s)$. (c) How do the heats of combustion of ethanol and glucose compare on a joule-per-gram basis? By this calculation, which is a better energy source, gin or candy? (Assume gin to be 45% ethanol by weight, and candy 100% glucose.) (d) Many microorganisms, including yeast, obtain their energy by fermenting glucose to ethanol, which they give off as a waste product:

$$C_6H_{12}O_6(s) \rightarrow$$
$$2C_2H_5OH(l) + 2CO_2(g)$$

We use the ethanol in wine, and even use the carbon dioxide in champagne and other sparkling wines. Calculate the energy that the yeast obtains per mole of glucose. What can you deduce from this calculation and part b about the advantages of combustion of glucose with O_2, over simple fermentation without O_2?

Fuel combustion

9. (a) How much heat is released when a mole of gasoline, C_8H_{18}, is burned in the open air? Write a balanced equation for the reaction that produces liquid water. (b) How much heat is obtained if the gasoline is burned with a restricted oxygen supply, so CO is produced instead of CO_2? (c) How much heat results if this CO is oxidized to CO_2? (d) How do your answers to parts b and c compare with that to part a? What principle does this represent?

Heat of condensation

10. When an inch of rain falls on New York City, it results in a rainfall of 19.7×10^9 liters over the city's 300-square-mile area. (a) Assume a density of liquid water of 1.00 g cm^{-3}, how much heat is released when this quantity of water condenses from vapor in the rain clouds? (Think of condensation as a chemical reaction: $H_2O(g) \rightarrow H_2O(l)$, and use thermodynamic data from Appendix 3.) (b) A ton of TNT releases around 4000 kJ of energy. How many megatons of TNT (1 megaton $= 10^6$ tons) would be needed for an explosion that released as much energy as the inch of rain in part a?

Heat of reaction

11. Use the following heats of reaction to calculate the heat of formation of NO. Compare your answer with that given in Appendix 3.

$$4NH_3 + 5O_2 \rightarrow 4NO + 6H_2O(l)$$
$$\Delta H = -1170 \text{ kJ}$$
$$4NH_3 + 3O_2 \rightarrow 2N_2 + 6H_2O(l)$$
$$\Delta H = -1530 \text{ kJ}$$

12. From data in Appendix 3, calculate the heat given off when limestone is dissolved by acid:

$$CaCO_3(s) + 2H^+ \rightarrow$$
$$Ca^{2+} + CO_2(g) + H_2O(l)$$

Rocket fuels

13. Calculate the heat of formation of liquid hydrazine, N_2H_4, from the following data:

$$2NH_3 + 3N_2O \rightarrow 4N_2 + 3H_2O(l)$$
$$\Delta H = -1010 \text{ kJ}$$
$$N_2O + 3H_2 \rightarrow N_2H_4(l) + H_2O(l)$$
$$\Delta H = -317 \text{ kJ}$$

$$2NH_3 + \tfrac{1}{2}O_2 \rightarrow N_2H_4(l) + H_2O(l)$$
$$\Delta H = -143 \text{ kJ}$$
$$H_2 + \tfrac{1}{2}O_2 \rightarrow H_2O(l) \quad \Delta H = -286 \text{ kJ}$$

Write the balanced equation for the combustion of N_2H_4 in oxygen to form nitrogen gas and liquid water. What is the heat of combustion of hydrazine?

14. (a) In liquid-fuel rockets, such as the lunar module of the Apollo moon missions, the fuel is liquid hydrazine, N_2H_4, and the oxidant is N_2O_4. Write a balanced equation for the reaction of these two substances to form liquid water and N_2 gas. (b) How much heat is given off in this reaction per mole of hydrazine? (c) Would more, or less, heat be given off if the oxidant were O_2 instead of dinitrogen tetroxide? How much?

Industrial fuel gas

15. In the manufacture of water gas, a commercial heating gas, steam is passed through hot coke, and the following reaction occurs:

$$C + H_2O(g) \rightarrow CO + H_2$$

(a) What is the standard enthalpy of this reaction per mole of carbon in the coke? How much heat is stored per mole of carbon? (b) Write the reactions that occur when water gas (the mixture of CO and H_2) is burned in air. How much energy is released when water gas containing a mole each of CO and H_2 is burned to CO_2 and liquid water? (c) How much heat is given off when 100 liters of water gas (measured at 1 atm pressure and 298 K) are burned?

16. Another way of storing the energy of coke in a combustible gas is by passing dry air (20% O_2 and 80% N_2 by volume or moles) through hot coke so the following reaction occurs:

$$C + \tfrac{1}{2}O_2 \rightarrow CO$$

(a) If 100 liters of air are passed through the furnace, how many liters of O_2 are used up? How many liters of CO are formed? How many liters of N_2 go through the furnace unchanged? What volume of CO gas is obtained from 100 liters of air? (b) How much heat is given off when 100 liters of CO gas (1 atm and 298 K) are burned?

Bond energies

17. The heat of combustion of gaseous dimethyl ether, CH_3—O—CH_3, to carbon dioxide and liquid water is -1461 kJ mole^{-1} of ether. (a) Calculate the standard heat of formation of dimethyl ether and compare your value with that tabulated in Appendix 3. (b) Use the table of bond energies to calculate the standard heat of formation of dimethyl ether. Illustrate your calculation with an energy-level diagram similar to Figure 15-7, labeling all energy levels and energy transitions. How does your answer compare with that of part a?

18. Use bond energies to calculate the heat of formation of acetaldehyde vapor, $CH_3CHO(g)$, at 25°C from graphite and O_2 and H_2 gases. Illustrate your calculations with an energy-level diagram. Compare your answer with the thermodynamic value in Appendix 3. How good is the bond-energy method, according to this calculation?

19. From bond energies and atomic heats of formation, calculate the standard heat of formation of the following gases: (a) C_2H_6; (b) CH_3SH; (c) CH_3NH_2. How do these heats compare with the measured values in Appendix 3?

20. The following are some of the gas reactions that take place in our atmosphere.

Use bond-energy information to calculate the enthalpy for each reaction:
a) $N + O_2 \rightarrow NO + O$
b) $NO_2 + O \rightarrow NO + O_2$
c) $N + N \rightarrow N_2$
d) $O + O_3 \rightarrow 2O_2$

Bond energies and molecular structure

21. (a) From data in Appendix 3, calculate the heat of isomerization of liquid ethanol to dimethyl ether:
$$CH_3-CH_2-OH(l) \rightarrow$$
$$CH_3-O-CH_3(g)$$

(b) Calculate the corresponding heat of isomerization beginning with ethanol vapor. Explain the difference between this value and the answer to part a. (c) Calculate the heat of the reaction in part b from bond energies. How different are the bond energy and thermodynamic values? (d) Account for the heat of isomerization of ethanol vapor in terms of the bonding in each molecule.

22. (a) Assume that the bond structure of carbon monoxide is $C\!=\!O$; of carbon dioxide, $O\!=\!C\!=\!O$. Calculate the standard enthalpy of the following reaction from bond-energy tables:

$$CO(g) + \tfrac{1}{2}O_2 \rightarrow CO_2(g)$$

Compare this enthalpy with the thermodynamically measured value. How big is the error, and how good do your assumptions about the bond structures of these two molecules appear to be? (b) Select two other chemical reactions from the data in Appendix 3, one involving CO but not CO_2, and the other involving CO_2 but not CO. Calculate the enthalpies of these two reactions from bond energies, compare them with thermodynamic values, and decide which bond assumption of part a is worse, that of $C\!=\!O$ or that of $O\!=\!C\!=\!O$.

23. The heat of combustion of gaseous isoprene, $CH_2\!=\!CH\!-\!C(CH_3)\!=\!CH_2$ or C_5H_8, to carbon dioxide gas and liquid water is -3186 kJ mole^{-1}. Calculate the heat of formation, and, by comparison with a bond-energy calculation, estimate the resonance energy of isoprene. Can you draw several possible resonance structures?

Postscript: Count Rumford versus the World

In *Order and Chaos,* S. W. Angrist and L. G. Hepler remark:

❝It is well established that people with widely differing backgrounds and positions have made significant contributions to the theory of heat and energy. Consider the following list of contributors to the science and practice of heat: (1) a spy for the British government in the employ of General Gage, who was British Commandant in Boston at the time of the American Revolution, (2) the Secretary of the Province of Georgia in the British Foreign Office in 1779, (3) The Undersecretary of State for the Northern Department in the British Foreign Office in 1780, (4) a lieutenant colonel in the King's American Dragoons, (5) a Knight in the court of George III, (6) a British spy in the court of the Elector of Bavaria, (7) the founder of the Munich Military Workhouse, (8) the designer of Munich's English

Gardens, (9) a lieutenant general in the service of the Elector of Bavaria, (10) a member of the Polish Order of St. Stanislaus with the rank of White Eagle, (11) a Count of the Holy Roman Empire, (12) the founder of the Royal Institution, (13) a foreign associate of the French Academy of Sciences, and (14) Lavoisier's widow's second husband.**"***

As the authors point out, all these people are really only one individual, Benjamin Thompson, later Count Rumford.

To this list they could have added: the inventor of the combustion calorimeter, the comparative photometer with the International Standard Candle, the kitchen range or cookstove, the double boiler, the baking oven, the portable stove and army field kitchen, the drip coffee maker, the modern steam heating system, the smoke shelf and damper system now used in all fireplaces, an improved oil study lamp of unprecedented illumination, a naval signaling system used by Great Britain, and an improved ballistic pendulum for measuring the force of gunpowder; the discoverer of convection currents in gases and liquids, the maximum density of water at 4°C, and the superior absorption and emission of radiation by black instead of polished objects; one of the earliest investigators of the tensile strength of fibers and the insulating properties of cloth; the founder of one of the earliest public schools, and of the first international scientific medal and prize (still awarded); and the intended first head of West Point (declined by prearrangement for political reasons). The list is still incomplete. Thompson was a practical genius and inventor in the same league with Thomas Edison. He revolutionized nutrition in Europe in the late 1700s in the same way that Edison revolutionized life, a century later, with the practical use of electricity. He was certainly a more prolific inventor than Franklin, and probably a better scientist. Why, then, is he virtually unknown except to historians of science and students of thermodynamics?

The reason lies largely in the personality of the man. Thompson was ambitious and utterly without scruples or principles. He toadied to his superiors, was caustic and treacherous to his peers, and tyrannical to his subordinates. No one could work with him, and he made a host of enemies wherever he went. He was, in short, an intolerable genius.

Thompson was born in Woburn, Massachusetts, in 1753. He was a member of a large farm family. He appears to have been a compulsive organizer and student. Notebooks from his youth give a daily schedule of subjects to be studied ("Munday—Anatomy, Tewsday—Anatomy, Wednesday—Institutes of Physick, Thursday—Surgery, Fryday—Chimistry with the Materia Medica, Saturday—Physick $\frac{1}{2}$ and Surgery $\frac{1}{2}$") as well as hourly schedules for each day. He was first apprenticed to a dry-goods dealer, then to a local doctor. Neither apprenticeship was satisfactory, and he became a schoolteacher in Concord, New Hampshire (originally called Rumford, N.H.). Here Thompson made his first—and in many respects typical—step

*S. W. Angrist and L. G. Hepler, *Order and Chaos: Laws of Energy and Entropy*, Basic Books, New York, 1967, p. 9.

upward in life. In 1772, he married the young widow of a wealthy New Hampshire landowner.

Thompson's natural autocratic leanings and the connections of his wife and her late husband led him to become a favorite of the British Royal Governor of New Hampshire. He became an informer and spy for the British, who needed information on caches of arms and supplies that the Colonial Militia and Minuteman groups were secreting about the New England countryside. He was suspected of being an informer, and the New Hampshire Committee of Public Safety called him before them to answer charges that he was "unfriendly to the cause of freedom." Nothing could be proved against him. But a week before Christmas, 1774, Thompson learned that a group of "patriots" was coming for him that evening with tar and feathers. He left his wife, baby daughter, and elderly father-in-law to face the mob alone and rode for Boston. He never came back.

Thompson continued to spy for the British in Massachusetts, and had another brush with the Committee of Public Safety in that state. Again, he was too clever for anything to be proved against him, but thereafter he was watched carefully and lost his advantage as a spy. When the British army was forced out of Boston in March, 1776, Thompson went with them. He soon arrived in London, where he found employment, first as an expert on the Revolutionary War (the equivalent of our "Kremlinologists"), then in several governmental posts. After seven years, during which time he made several important inventions, was suspected of slipping British naval intelligence to the French in the La Motte case, and made innumerable personal enemies, he felt obliged to seek employment elsewhere. He soon appeared in Munich as a colonel and military advisor to the Elector Karl Theodore of Bavaria. (He sent several military intelligence reports on the state of the army of his new employer back to England in cipher.)

The Bavarian army was in wretched shape. It had no discipline, training, decent equipment, supply procedures, or morale, and was ridden with graft, corruption, and inefficiency. Thompson was given the responsibility for whipping it into a decent fighting force. His situation in Munich was similar to Lavoisier's in the *Ferme Générale,* or Tax Farm. The businessmen of the *Ferme Générale* contracted with the French Crown to deliver a certain amount in tax revenues to the treasury each year. Any taxes that they collected over and above this amount, they could keep. Colonel Thompson was given a fixed sum of money each year to run the Bavarian army. If the operation of the army became more effective and Thompson simultaneously found ways to cut expenses, the money saved was his own. It paid both Lavoisier and Thompson to carry out their duties in the most efficient possible manner. Thompson's experiments with clothing, nutrition, cannon boring and the Munich military workhouses were all part of his plan to make the Bavarian army efficient. When conservative manufacturers refused to weave cloth and construct equipment to his specifications, he used the army to round up the thousands of street beggars of Munich in one night's sweep and set up the military workhouses as his factories. He gave

Figure 15-10   Count Rumford supervising a public lecture at his Royal Institution in London, in 1802. Rumford is the hook-nosed figure smiling benignly at the upper right. The lecturer is Thomas Young, a Professor of Natural Philosophy at the Royal Institution, and his assistant with the bellows and an evil leer is the young Humphrey Davy. The "victim" of the demonstration is Sir John Hippisley, manager of the Royal Institution. Davy worked extensively with the physiological effects of various gases. He had almost killed himself inhaling methane two years before, and caused a sensation at a lecture in 1801 by giving laughing gas (nitrous oxide) to volunteers from the audience. James Gillray, the artist, was the Herblock or Mauldin of his era and was famous for his devastating political cartoons. He considered these Royal Institution lectures a sham because, although intended as an education for working people, they had become the fashionable entertainment of the wealthy, as caricatured here. Davy and Michael Faraday continued a tradition of public lectures which has been maintained to the present day. One of us [R.E.D.] gave a Royal Institution lecture in 1970 in what was recognizably the same lecture hall as shown here in 1802. (Photograph of the original etching courtesy The Fisher Collection.)

each worker room and board, and set up free schools for their children (until they were old enough to work). He built the famous Munich "English Gardens" as demonstration gardens for his innovations in agriculture and nutrition. His "Rumford soup," developed for the workhouses, was an attempt to provide a complete food at the lowest possible cost. He introduced potatoes to Bavaria, although he had to smuggle the first ones into his kitchens by stealth because the Bavarians considered them unfit to eat. He propagandized coffee as a stimulating substitute for alcohol and invented the drip percolator to make it popular. Soldiers in European armies of the time found food where they could get it, and cooked it themselves over open fires in camp. Thompson first designed a collapsible one-man field stove and then conceived the idea of a traveling field kitchen to cook for the army. His cannon-boring experiments, virtually the only achievement for which he is still remembered, were only an incident in a colorful career in Munich.

From a colonel in the Bavarian army, Thompson rose to be minister of war, minister of police, major general, chamberlain of the Bavarian Court, and state councillor. He held all these offices simultaneously and was the second most powerful man in Bavaria, after the elector himself. His ultimate title was that of count of the Holy Roman Empire. Thompson chose as his title the original name of Concord, New Hampshire, and, after 1792, insisted on being addressed as "Count Rumford" rather than as Benjamin Thompson. The choice of "Rumford" may have been a belated acknowledgment of his wife and child, whom he had deserted 18 years earlier, or it might have derived from his pretentions that he had come from a wealthy landowning family in the Colonies.

By 1795, his intensive work had begun to damage his health, and his many enemies in the Court of Bavaria were becoming too powerful. He left Munich and returned in triumph to London. He was given almost overwhelming adulation, by both governmental figures and the general public, as a great philanthropist, philosopher, and benefactor. No matter what his difficulties in getting along with people and his personal defects might have been, Rumford's improvements in housing, lighting, clothing, and nutrition had made a real difference to the average citizen of Europe of the time. The Royal Institution of Great Britain, now a respected research laboratory, was initially created as a showcase for Thompson's inventions and innovations. He brought in a young country boy by the name of Humphry Davy (the Sir Humphry Davy of the Dalton postscript to Chapter 6) to assist in giving public demonstrations and lectures (Figure 15-10). Typically, Thompson conceived of the Royal Institution as a place where the uninformed would come to ask Count Rumford how their lives should be run. The fact that he was so often right made little difference to those who were put off by his arrogance. Within two years, he was forced out of active control of the Royal Institution, although the Institution went on to be the brilliant showcase and laboratory for Humphry Davy, Michael Faraday, and a continuing procession of noted scientists.

At this point in his career, Rumford was only 49 years old. We must leave him, except to remark that he began a new career in France by marrying the widow of Lavoisier. His first wife had conveniently died by this time. The new marriage was notoriously stormy and lasted only two years. But by the end of this time, he was as firmly established in French affairs as he had been in those of Munich and London.

Rumford died suddenly in 1814. He managed his death as efficiently as he had his life. In a curious will, he left all his possessions to Harvard University, which still looks after his grave in France. Upon his death, he sank into obscurity as rapidly as he had risen from it. He was not remembered, as were Lavoisier, or Dalton, or Franklin. People who tried to live and work with him found it so difficult to give him credit for his real achievements that when his life was over they simply forgot him as fast as possible. He had proclaimed so often in life what a great man he was, that people were content to let the issue rest after his death.

In a funeral eulogy before the French Academy, the naturalist Baron Cuvier summarized the flaws in this remarkable man:

> He considered the Chinese government as the nearest to perfection, because in delivering up the people to the absolute power of men of knowledge alone, and in raising each of these in the hierarchy according to the degree of his knowledge, it made in some measure so many millions of hands the passive organs of the will of a few good heads. An empire such as he conceived would not have been more difficult for him to manage than his barracks and poorhouses. . . . The world requires a little more freedom and is so constituted that a certain height of perfection often appears to it a defect, when the person does not take as much pains to conceal his knowledge as he has taken to acquire it.

Suggested Reading

S. C. Brown, *Count Rumford, Physicist Extraordinary,* Anchor, New York, 1962. An entertaining biography of one of the greatest practical geniuses and blackguards in the annals of science.

I. M. Klotz and R. M. Rosenberg, *Introduction to Chemical Thermodynamics,* W. A. Benjamin, Menlo Park, Calif., 1972, 2nd ed.

B. H. Mahan, *Elementary Chemical Thermodynamics,* W. A. Benjamin, Menlo Park, Calif., 1963. A good introduction to classical thermodynamics (that is, without the statistical interpretation of entropy). Moderate use of calculus, which is mostly explained as it is introduced.

L. Nash, *Introduction to Chemical Thermodynamics,* Addison-Wesley, Reading, Mass., 1963.

J. Waser, *Basic Chemical Thermodynamics,* W. A. Benjamin, Menlo Park, Calif., 1966. Both this and Nash's book are at a level similar to Mahan and this chapter. Highly recommended.

16

Entropy,
Free Energy,
and
Chemical Reactions

Maximum disorder was our equilibrium.

T. E. Lawrence, *Seven Pillars of Wisdom* **(1926)**

The previous chapter answered a fundamental question: "If a chemical reaction takes place by itself, will it give off or absorb heat?" This is a useful question for builders of heat engines and designers of reaction vessels. But an even more fundamental question is the following: "If left to itself, will a given reaction take place at all without outside interference?" This question of **spontaneity** is the subject of this chapter.

16-1   SPONTANEITY, REVERSIBILITY, AND EQUILIBRIUM

If we place equivalent amounts of hydrogen and oxygen gases in a container and apply a flame or a platinum catalyst, there will be a violent explosion. The H_2 and O_2 will disappear, and water vapor will form in their place. Similarly, a mixture of H_2 and Cl_2, if triggered by light, will explode and produce HCl gas. In contrast, a mixture of H_2 and N_2 gases will react much less violently, and the final product will be a mixture of H_2, N_2, and NH_3 gases.

The water and HCl reactions are good illustrations of highly spontaneous processes. As we saw in Chapter 4, a **spontaneous process** is one that has enough impetus to proceed on its own without further input from the rest of the universe.* We shall learn later in this chapter how that impetus is measured. For the moment we can say that the reaction to produce HCl from H_2 and Cl_2,

$$H_2 + Cl_2 \rightarrow 2HCl$$

has a far greater tendency to occur than the reverse process, dissociation of HCl,

$$2HCl \rightarrow H_2 + Cl_2$$

The reaction synthesizing ammonia from H_2 and N_2 also has a somewhat greater initial tendency to occur than the decomposition reaction. Reaction 16-1 has more of a drive than reaction 16-2:

$$3H_2 + N_2 \rightarrow 2NH_3 \qquad \text{(synthesis)} \tag{16-1}$$

$$2NH_3 \rightarrow 3H_2 + N_2 \qquad \text{(decomposition)} \tag{16-2}$$

But as more NH_3 accumulates, and as less N_2 and H_2 are left, synthesis becomes slower and decomposition accelerates. Ammonia decomposes more rapidly as more of it is present to decompose. At some concentration of H_2, N_2, and NH_3, the two reactions proceed at the same rate. Ammonia is produced exactly as fast as it is broken down. Although synthesis and decomposition still occur at the molecular level, we see no net change in the composition of the gas mixture. The gas gives the appearance of having ceased to change. This condition of balance between two opposing reactions is called **chemical equilibrium**.

A reaction that is at equilibrium is a **reversible reaction**. To understand what this means, let's examine our equilibrium mixture of H_2, N_2, and NH_3. An increase in pressure favors reaction 16-1 over 16-2 because 16-1 leads to a smaller number of moles (or molecules) of gas and relieves the stress on the system caused by the pressure increase. Similarly, a decrease in pressure favors the decomposition of ammonia to produce more moles of gas. Both of these are applications of Le Chatelier's principle: When a system at equilibrium is subjected to a stress of any kind, the system shifts toward a new equilibrium condition in such a way as to relieve that stress. The synthesis of ammonia from its elements is an exothermic process; the ΔH^0_{298} of reaction 16-1 is -92.38 kJ per mole of reaction, as written, or -46.19 kJ mole^{-1} of ammonia. If the temperature of the container is increased, reaction 16-1 will be hindered, and reaction 16-2 will be favored

---

*In a strict sense, spontaneity has nothing to do with time. A thermodynamically spontaneous reaction is one that will occur on its own, even if it requires virtually forever to do so. The role of a catalyst is to bring about in a short time that which would occur anyway, but only over a longer time interval. Thermodynamics answers the question, "Will it occur, eventually?" To answer the question, "How soon will it occur?" we must turn to kinetics (Chapter 22).

because it absorbs heat and partially counteracts the temperature increase. If more ammonia is added to the container from an outside source, reaction 16-1 will be hindered again and 16-2 will be favored because it relieves the stress of the added ammonia. Le Chatelier's principle is useful in predicting qualitatively what an equilibrium system will do when acted upon by an outside influence.

If the N_2–H_2–NH_3 system is truly at equilibrium, the changes in pressure, temperature, or concentration of one component required to alter the relative rates of the two reactions are infinitesimally small. Just as the lightest weight can tip a balance in mechanical equilibrium, so the smallest change can affect a system in chemical equilibrium. This is why the term *reversible* is applied to such situations. A fingertip touch cannot halt a falling boulder, and an infinitesimal change in pressure, temperature, concentration, or any other variable cannot halt the explosion of H_2 and Cl_2 or the less spectacular reaction of N_2 and H_2 before equilibrium is reached. Such chemical systems are not at equilibrium; their processes are irreversible.

In summary, an equilibrium process is reversible, and a nonequilibrium or spontaneous process is irreversible. We shall want to know how to calculate the equilibrium conditions for a system of chemical substances, for two main reasons: because it is often useful to know the relative amounts of reactants and products at equilibrium, and because the distance of a given chemical situation from equilibrium is a measure of how strong the drive is in a direction toward equilibrium.

16-2 HEAT, ENERGY, AND MOLECULAR MOTION

When an object is heated, its molecular motion increases. Heat is not a fluid that can be forced away from the atoms by friction. Instead, it is an expression of the state of motion of the molecules and atoms, which are made to move more rapidly by the mechanical forces of friction. These conclusions were suggested by experiments that demonstrated the mechanical equivalence of work and heat, and were made palatable by the kinetic theory of gases and its extension to the molecular theory of liquids and solids.

In previous chapters we talked about two types of energy that an object could have: kinetic energy and potential energy. Kinetic energy is possessed by a moving body and is represented by $E_k = \frac{1}{2}mv^2$. Potential energy is possessed by a body because of where that body is. If a mass can perform work by moving from point A to point B in space, we say that the body has a greater gravitational potential energy at A than at B. If we want, we can talk about a gravitational potential field in which the body moves, but we are only rephrasing the observation, not explaining it. The idea of a gravitational field *comes from* the observation that work can be done when the body moves from one place to another. Similarly, if a positive or negative charge can be made to do work as it moves from point A to point B, we say that the charge has a greater electrostatic potential energy at A than it has at B. Again, we can describe (not explain) the observation by talking about an electrostatic field.

Now we have a third kind of energy to deal with: the energy possessed by a body because its atoms and molecules are in a state of motion, even though the body might be stationary. This molecular motion is heat, and it is measured by the temperature of the object. The temperature scale is based on the expansion behavior of an ideal gas, as we noted in Chapter 3. Heat is measured in the same units as work and energy. The amount of heat required for a mole of a substance to experience a temperature increase of 1 K is called the **heat capacity** of the substance and is measured in $J K^{-1} mole^{-1}$ (joules per kelvin per mole).

If this third kind of energy did not exist, the first law of thermodynamics would be

$$\Delta E = E_2 - E_1 = -w$$

and would read as follows: The change in the internal energy of a system balances any work that the system does on its surroundings. Or, since $-w$ represents work done on the system by its surroundings, the equation also would state that the increase in internal energy of the system equals the work done on it from the outside. This work could be used to accelerate objects in the system and give them greater kinetic energy, or it could be used to lift them and give them greater potential energy.

The full statement of the first law,

$$\Delta E = E_2 - E_1 = q - w$$

is as follows: The change in the internal energy of a system is the sum of the work done on it by the environment and the increase in random motion given its molecules by the environment. This increase in molecular motion is described as a flow of heat.

It is always possible to change work into heat. Friction is often used as an example because it is so simple. A block, moving as a unit with a large velocity, but with its molecules in relatively slow random motion, comes to a stop on a surface because of friction. After it stops, it has no velocity of motion as a unit. However, its molecules, and the molecules of the surface on which it had been sliding, are moving with greater individual speed than before. If the object is partly a gas, this motion may be straight-line motion throughout the container. If it is a solid, the motion will be vibration of atoms and molecules around average positions in the crystal. In either case, large-scale motion has been converted to microscopic motion.

This process is not completely reversible. The example of a skidding automobile in Section 3-6 is an illustration of this fact. Generally, we cannot take random molecular motion and convert it to coordinated motion of the entire object with 100% efficiency. The expression of our inability to do this is the **second law of thermodynamics**. Two slightly different versions of the second law were proposed in the middle of the nineteenth century. One version, by William Thomson, says this: *One cannot convert a quantity of heat completely into work without wasting some of this heat at a lower temperature.* The other version, by Rudolf Clausius, says this: *One cannot transfer heat from a cold object to a hot object without using work to make the transfer.* Both statements are *summaries of experience,* and are "statements of impotence." They are statements about the *limitations* on what we can do in the real world. Either can be shown to follow if the other is assumed first.

16-3 ENTROPY AND DISORDER

Either form of the second law of thermodynamics leads, with the help of some calculus that we shall not reproduce here, to a new state function, S. This new state function is called the **entropy** of the system. We shall not use the most general expression for entropy, but one special case is useful. If an amount of heat, q, is added to a system *in a reversible manner* at a temperature T, then the entropy of the system increases by

$$\Delta S = \frac{q}{T} \tag{16-3}$$

If the heat is added *irreversibly,* the entropy increases by more than q/T:

$$\Delta S > \frac{q_{\text{irr}}}{T} \tag{16-4}$$

The quantity q/T is a lower limit to the entropy increase that is applicable only when the heat transfer is reversible, that is, when the object being heated is in thermal equilibrium with the object donating the heat. Heat flows from one body to another because they are *not* in equilibrium, and it would take an infinite time for reversible heat flow to take place. Truly reversible processes are idealizations for real, irreversible processes. What we should say is that in any real (irreversible) process, the entropy increase will be greater than q/T; but the more slowly and carefully we carry out the heat transfer, the less ΔS will exceed q/T.

The entropy as derived from the second law has no obvious molecular interpretation. But Ludwig Boltzmann (1844–1906), an Austrian physicist, showed in 1877 that entropy has a fundamental molecular significance: It is a measure of the **disorder** of a system. Boltzmann proposed that entropy, S, is related to the number of different microscopic ways of obtaining a specified macroscopically definable and observable situation. If the number of equivalent ways of constructing a situation is W, then the entropy is proportional to the logarithm of W:

$$S = k \ln W \tag{16-5}$$

The proportionality constant, k, is the gas constant per molecule, or

$$k = \frac{R}{N} \tag{16-6}$$

where R is the familiar ideal gas constant and N is Avogadro's number. Appropriately enough, we now call k **Boltzmann's constant**.

The important physical quantity in the equation is the number of ways of obtaining a given state, W. There is only one way of putting together a perfect crystal, provided of course that the molecules are indistinguishable from one another and are motionless, packed against their neighbors (which means at absolute zero temperature). For a perfect crystal with motionless molecules at 0 K, $W = 1$ and $S = k \ln 1 = 0$. In contrast, there are many equivalent ways of building a liter of a certain gas at a given temperature and pressure. The individual positions of molecules in a gas do not have to be specified, nor do their individual speeds. A gas will match the given specifications if the total number of its molecules of each kind and its total energy per mole are correct; all gases that satisfy these conditions will appear alike to an outside observer. It follows that for a gas, W is large, so $\ln W$ is a positive number, and $S = k \ln W$ is greater than zero. Of course, even a perfect crystal will have some entropy if it is warmed above 0 K, since the individual molecules will begin to vibrate about their equilibrium positions in the crystal lattice, and there will be several ways of constructing a vibrating-molecule crystal, all of which will look the same to the external observer. The entropy of a crystal of rock salt at room temperature, however, will be far less than that of a comparable quantity of gas.

Boltzmann made the crucial connection between thermodynamic *entropy* and *disorder*. Any situation that is so definite that it can be put together only in one or a small number of ways is recognized by our minds as orderly. Any situation that could be reproduced in thousands or in millions of different but entirely equivalent ways is disorderly. Boltzmann's law tells us that the most perfect, orderly object conceivable in the universe would be a perfect crystal at absolute zero. Anything else—a crystal at any temperature above 0 K, a liquid, a gas, or a mixture of substances—is more disordered and hence has a positive entropy. The higher the entropy, S, the greater the disorder.

When we combine Boltzmann's ideas about entropy with thermodynamics, we arrive at one of the most important principles of science: *In any real, spontaneous processes, including chemical reactions, the disorder of the universe always increases.* In any isolated system, in which the total energy cannot change, a spontaneous reaction is one in which entropy (and disorder) increases. No process that produces order, or lowers the entropy, can occur without outside help. If we supply enough energy, we can make a reaction occur even though entropy decreases in the process. But if we do not provide enough energy, a reaction leading to increased order will not take place.

Life in a Nine-Point Universe

What do we mean when we say that entropy, S, can be calculated from the expression $S = k \ln W$, in which W is the number of equivalent ways the molecules can be arranged to give the same observable result? Why should a gas inevitably have a higher entropy than a crystal of the same substance? It is hard to answer such questions in our own universe without getting bogged down in mathematics. But it is much easier in an imaginary universe with only four atoms in it, and only nine places where the atoms can be.

Imagine that the nine places in our mini-universe are arranged in the 3×3 grid shown in Figure 16-1a. All four atoms placed in a close-packed square will constitute a crystal in our imaginary space, and any other arrangement of the four atoms will be called a gas. Examples of crystals and gases are given in Figure 16-1b. If we examine every possible arrangement of four atoms in a nine-point universe, how many of these arrangements will lead to crystals and how many to gases?

First of all, how many total arrangements are there for both gases *and* crystals? The first atom can go to any one of nine places in the universe. The second atom has only eight places left open, the third has seven places, and the final atom has only six unoccupied places. The total number of ways of placing four atoms in the nine locations appears to be $9 \times 8 \times 7 \times 6 = 3024$ ways.

This is not quite correct, because we have overcounted. The answer of 3024 would be correct if the atoms had names or labels, and if arrangements of the type shown in Figure 16-1c really were different. But atoms have no labels. If by some miraculous process we could photograph the atoms at a chosen instant and study them we could tell the difference between the four arrangements in Figure 16-1b because the atoms occupy different places. However, we could see absolutely no difference among the arrangements in Figure 16-1c, because one atom is just like every other atom of the same type. The most that we could say is that atoms were present at positions 3, 4, 7, and 8 of our mini-universe (Figure 16-1d).

How can we correct for this overcounting? How do we "remove the labels" from the atoms? As a correction factor, how many different label shufflings can be made for each arrangement of four atoms? Label a could be given to any one of the four atoms, label b to any of the remaining three, label c to two, and label d then has to go to the last atom. There are $4 \times 3 \times 2 \times 1 = 24$ meaningless permutations of labels for every really different arrangement of atoms. We have overcounted by a multiplicative factor of 24. Hence the 3024 ways of arranging atoms must be divided by 24 to remove the labels on the atoms. The number of different ways of arranging four *indistinguishable* atoms among nine locations is

$$W = \frac{9 \times 8 \times 7 \times 6}{4 \times 3 \times 2 \times 1} = \frac{3024}{24} = 126$$

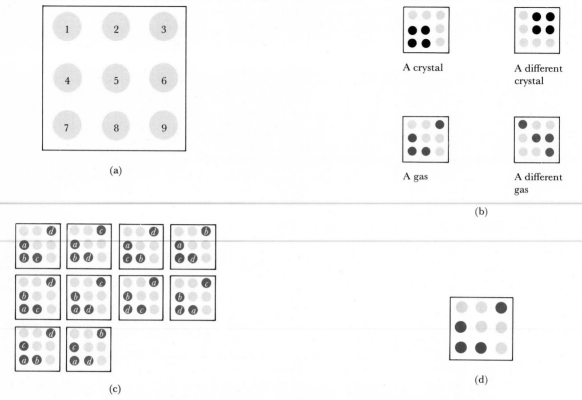

Figure 16-1   A mini-universe with only nine "locations" or places where an object can be, and only four objects to fill them. (a) The "universe." (b) Different arrangements of atoms leading to different gases and crystals. (c) If atoms had labels, all these patterns and others like them would represent different arrangements of atoms. (d) But since atoms do not have names or labels, and all look alike, all the pictures in (c) correspond to a single atomic arrangement: four atoms at positions 3, 4, 7, and 8. From Dickerson and Geis, *Chemistry, Matter, and the Universe.*

If you are skeptical about the logic of the derivation just given, you may check all 126 arrangements in Figure 16-2. Of the 126 possible arrangements, only four are crystals (shown in black), and the other 122 are gases. Even in such a tiny and restricted universe, a gas is far more likely to result from a random arrangement of atoms than is a crystal. This is true because the specifications for a crystal are so much more restrictive:

Crystal: Four adjacent atoms in a square

Gas: Four atoms in any arrangement *except* those that lead to a crystal

For crystals, $W_c = 4$, and for gases, $W_g = 122$, so a gas is more than 30 times as likely to occur randomly as is a crystal.

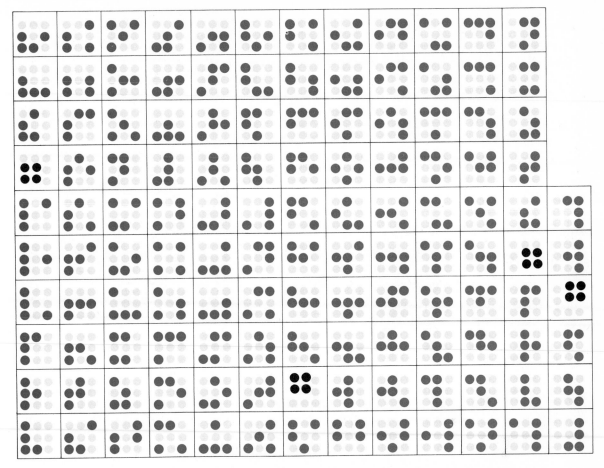

Figure 16-2　　Out of the 126 possible arrangements of four atoms in our nine-point universe, only the four shown in black are "crystals." The others, shown with dark-colored dots, are "gases." From Dickerson and Geis, *Chemistry, Matter, and the Universe.*

If we were to repeat the experiment with four atoms in a 4 × 4 "universe," we would find that

$$W = \frac{16 \times 15 \times 14 \times 13}{4 \times 3 \times 2 \times 1} = 1820$$

different arrangements would be possible, of which only 9 would be crystalline and 1811 would be gases (Figure 16-3). In that universe, a gas is more than 200 times as probable as a crystal.

When we jump to the real world, with many times Avogadro's number of atoms and many, many places for atoms to be, W becomes astronomical. It is then easier to use a logarithmic representation to keep down powers

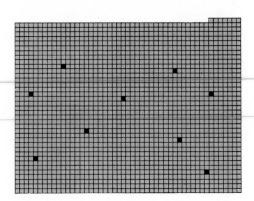

Figure 16-3   With four atoms in a 16-point universe, there are 9 random arrangements leading to crystals, and 1811 yielding gases. From Dickerson and Geis, *Chemistry, Matter, and the Universe.*

of 10, reducing 10 and 10,000,000 and 100,000,000,000,000 (or 10^1, 10^7, and 10^{14}) to a more manageable 1, 7, and 14. The entropy, S, is simply the number of ways of obtaining a given state of matter, expressed logarithmically instead of linearly:

$$S = k \ln W = 2.303\, k \log_{10} W \qquad (16\text{-}7)$$

16-4 ENTROPY AND CHEMICAL INTUITION

Boltzmann gave us a very precise interpretation of entropy in terms of order and disorder at the molecular level. In Appendix 3, along with standard heats of formation, there are tabulated standard entropies of substances, S^0_{298}. These values were *not* obtained from Boltzmann's $S = k \ln W$ expression. They are the results of thermal measurements of heat capacities of solids, liquids, and gases; of heats of fusion and heats of vaporization, from room temperature down to within extrapolation distance of absolute zero. (One learns in advanced courses how to calculate values for S from such purely thermal data.) These tabulated values of S^0_{298} are sometimes called **third-law entropies** because the logic of their calculation from thermal data is incomplete without the assumption of the **third law of thermodynamics**: *The entropy of a perfect crystal at absolute zero is zero.* The third law is obvious, of course, if you accept Boltzmann's statistical interpretation of S.

The beauty of third-law entropies is that, although they were not derived from Boltzmann's statistical interpretation, they agree with it completely. Let's look at matter through Boltzmann's eyes, and interpret these measured third-law entropies in terms of order and disorder. We can see several clear-cut trends, and they all become obvious if we replace the word *entropy* with *disorder*.

1. The entropy (or disorder) increases whenever a liquid or a solid is converted to a gas. Examples from Appendix 3 are

Substance:	$Na(s)$	$Br_2(l)$	$I_2(s)$	$H_2O(l)$	$CH_3OH(l)$
S^0 (solid or liquid):	51.0	152.3	117.0	69.9	127.0
S^0 (gas):	153.6	245.4	260.6	188.7	236.0

The units of entropy here and elsewhere are $J K^{-1}$ mole^{-1}, or entropy units (e.u.) mole^{-1}, an entropy unit being defined as a joule per kelvin. In each case, the third-law entropy of the condensed phase is around 100 e.u. mole^{-1} less than that of the same substance as a gas, because gases are inherently more disordered.

2. Entropy increases when a solid or a liquid is dissolved in water or other solvent:

Substance:	$CH_3OH(l)$	$HCOOH(l)$	$NaCl(s)$
S^0 (solid or liquid):	127.0	129	72.4
S^0 (after solution in H_2O):	132.3	164	115.4

The entropy of methanol, CH_3OH, rises only slightly, for a mole of methanol molecules interspersed among water molecules is only slightly more disordered than a mole of pure liquid methanol. Formic acid, $HCOOH$, undergoes a larger entropy increase when it is dissolved, because the molecules partially dissociate into protons and formate ions, $HCOO^-$, producing two entities where only one was present before. The crystal lattice of sodium chloride breaks up completely into hydrated Na^+ and Cl^- ions, leading to an increase in disorder even though some water molecules are tied up by hydrating the ions. Notice that the entropy figure for NaCl was obtained from Appendix 3 by adding the entropies of the two ions:

$$60.2 \ (Na^+) + 55.2 \ (Cl^-) = 115.4 \text{ e.u. mole}^{-1}$$

3. Entropy falls when a gas is dissolved in water or other solvent:

Substance:	$CH_3OH(g)$	$HCOOH(g)$	$HCl(g)$
S^0 (gas):	236	251	186.7
S^0 (after solution in H_2O):	132.3	164	55.2

Dissolving a gas in water is somewhat like condensing it to a liquid, in terms of closeness of neighbor molecule contacts. As before, the entropy of a dissolved ionic compound is found by adding the entropies of its hydrated ions.

4. Entropy rises with increasing mass, other things being equal:

Substance:	F_2	Cl_2	Br_2	I_2		O	O_2	O_3
S^0 (gas):	203	223	245	261		161	205	238

This is an important principle, but one that requires quantum mechanics for an explanation. Simple quantum theory tells us that as the mass and size of an atom or molecule increase, the spacings between its energy levels decrease. A large, massive object with a certain total energy hence has more quantum states available to it, so W is larger and $S = k \ln W$ is greater. This is illustrated in Figure 16-4 with four molecules having a total of six units of energy, for light molecules with widely spaced energy levels (Figure 16-4a) and for heavier molecules with closely spaced energy levels (Figure 16-4b).

5. Entropy is lower in covalently bonded solids, with strong, directional bonds, than in solids with partial metallic character:

C (diamond): 2.44	C (graphite): 5.69
Sn (gray, diamond): 44.8	Sn (white, metallic): 51.5

Both diamond and gray tin have a tetrahedrally bonded, three-dimensional covalently bonded structure. Both graphite in its packed two-dimensional sheets and white tin with its metallic packing of atoms are less orderly than the two diamond structures, so their entropy is greater.

6. In general, entropy rises with increasing softness and with weakness of bonds between atoms:

Figure 16-4

Entropy increases with mass. A heavier molecule has more closely spaced energy levels, and there are more ways of obtaining a mole of such molecules having a given total energy. In this example, four molecules together have six units of energy, and we can represent the energy of each molecule by locating it on a diagram of possible molecular energy levels. (a) Light molecule with widely spaced energy levels; $W = 2$. (b) Heavier molecule with more closely spaced energy levels; $W = 9$. From Dickerson and Geis, *Chemistry, Matter, and the Universe.*

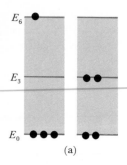

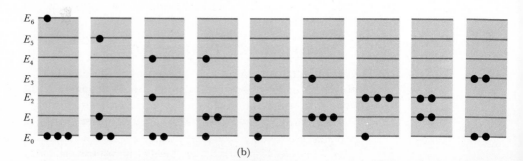

Substance:	C(di)	Be(s)	SiO$_2$(s)	Pb(s)	Hg(l)	Hg(g)
S^0:	2.44	9.54	41.8	64.9	77.4	174.9
State:	diamond	hard metal	quartz	soft metal	liquid	gas

7. Entropy increases with chemical complexity. This principle holds for ionic salt crystals with increasing numbers of ions per mole:

Substance:	NaCl	MgCl$_2$	AlCl$_3$
$S^0(s)$:	72.4	89.5	167

It also holds for crystals such as CuSO$_4 \cdot n$H$_2$O with increasing numbers of water molecules of hydration:

n:	0	1	3	5
S^0:	113	150	225	305

and for organic compounds with larger carbon frameworks:

Substance:	CH$_4$	C$_2$H$_6$	C$_3$H$_8$	C$_4$H$_{10}$
$S^0(g)$:	186	230	270	310

The molar entropies for pure elements in different physical states are plotted in Figure 16-5. All metallic solids have entropies below 80 e.u. mole^{-1}, monatomic gases lie between 130 and 180 e.u. mole^{-1}, and diatomic and polyatomic gases have higher entropies yet. Third-law entropies, although they come purely from thermal measurements, tell us something about molecular structure if we know how to interpret them. Although entropy originated as a rather abstract concept involving heat, it has a clear molecular interpretation. Entropy is a direct and quantitative measure of disorder.

16-5 FREE ENERGY AND SPONTANEITY IN CHEMICAL REACTIONS

What determines whether or not a chemical reaction is spontaneous? What measurable or calculable properties for the system of H$_2$, Cl$_2$, and HCl indicate that the reaction of H$_2$ and Cl$_2$ is explosively spontaneous under conditions for which the decomposition of HCl to H$_2$ and Cl$_2$ is scarcely detectable at all? Marcellin Berthelot and Julius Thomsen,* French and

---

*This Berthelot is not the Berthollet you already have encountered, nor is this Thomsen any of the Thomsons or Thompsons of past reference. For clarity, here is a brief glossary of these gentlemen:

Claude Berthollet (1748–1822). French chemist, opponent of definite proportions for compounds.

Marcellin Berthelot (1827–1907). French thermodynamicist.

Benjamin Thompson (1753–1814). American adventurer and spy, Bavarian munitions maker, founder of Royal Institution, London. (continued on p. 615)

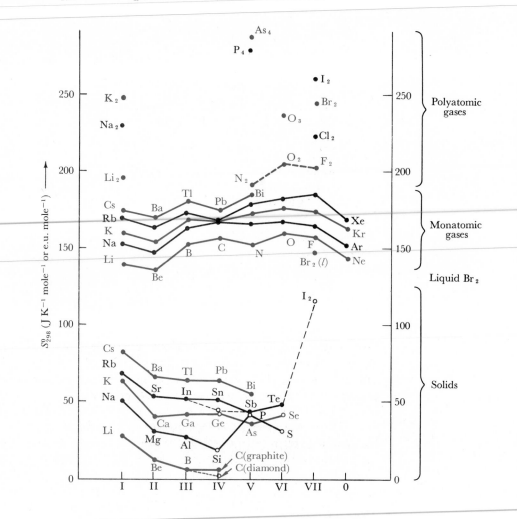

Figure 16-5   Third-law entropies in joules per kelvin per mole or entropy units per mole for various elements as solids, liquids, and monatomic and polyatomic gases. Polyatomic gases have higher entropies than monatomic gases because the mass of the molecular unit is greater. All monatomic gases have approximately the same entropy per mole, with a gradual increase in entropy with mass. Solids with stronger bonds have lower entropies. Filled circles in solids indicate metallic structures; open circles, nonmetallic structures. The two structures for carbon are graphite (filled) and diamond (open). The two structures for tin are metallic white tin (filled) and gray tin with the diamond structure (open). The entropy of the molecular solid of I_2 molecules is similar to those of crystals of other polyatomic small molecules such as ICN (129), glycine (109), oxalic acid (120), and urea (105).

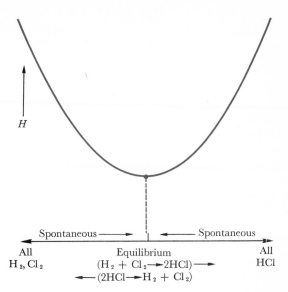

Figure 16-6

If the principle of Berthelot and Thomsen were correct and all spontaneous reactions liberated heat, then enthalpy, H, would be a chemical potential function that is minimized at equilibrium. This is not so; we can find spontaneous processes that absorb heat. The most obvious is the evaporation of a liquid.

Danish thermodynamicists, proposed the wrong answer, in 1878, in the form of their *principle of Berthelot and Thomsen:* Every chemical change accomplished without the intervention of an external energy tends toward the production of the body or the system of bodies that sets free the most heat. In other words, all spontaneous reactions are exothermic.

If the principle of Berthelot and Thomsen were correct, and if the enthalpy of a system of reacting chemicals *decreased* during any spontaneous process, equilibrium would occur at the minimum of enthalpy since any spontaneous process is moving toward equilibrium. The plot of enthalpy, H, against the extent of reaction on the horizontal axis would look something like Figure 16-6.

Unfortunately for the principle, we can easily find exceptions: reactions that are spontaneous but absorb heat. One of these is the vaporization of water or any other substance at a partial pressure less than its vapor pressure. When a pan of water evaporates, heat is absorbed:

$$H_2O(l) \rightarrow H_2O(g) \qquad \Delta H^0_{298} = +44.0 \text{ kJ mole}^{-1}$$

If the principle of Berthelot and Thomsen were true, all gases would con-

---

William Thomson (1824–1907). British thermodynamicist, later Lord Kelvin.

Julius Thomsen (1826–1909). Danish thermodynamicist.

J. J. Thomson (1856–1940). British physicist, discoverer of electron, Nobel Prize in 1906.

G. P. Thomson (1892–1976). British physicist, Nobel Prize in 1937 for electron diffraction. Son of J. J. Thomson.

dense spontaneously to liquids, and all liquids would solidify to solids since by doing so they would give off enthalpy.

Like the vaporization of water, the solution of ammonium chloride crystals in water is both spontaneous and endothermic:

$$NH_4Cl(s) + H_2O \rightarrow NH_4(aq)^+ + Cl(aq)^- \qquad \Delta H^0_{298} = +15.1 \text{ kJ}$$

Adding water to solid NH_4Cl makes a beaker cold enough to freeze pure water on the outside of the beaker. Yet we do not see dilute solutions of ammonium chloride separate spontaneously into crystals and pure water simply because heat would be released in the process.

As a last example, dinitrogen pentoxide is an unstable solid that reacts, sometimes explosively, to produce NO_2 and O_2:

$$N_2O_5(s) \rightarrow 2NO_2(g) + \tfrac{1}{2}O_2(g) \qquad \Delta H^0_{298} = +109.5 \text{ kJ}$$

However, a large quantity of heat is absorbed in this decomposition.

There are many other examples of spontaneous processes that absorb heat. We cannot find the point of equilibrium by minimizing H. Enthalpy is not a measure of the tendency of a reaction to proceed spontaneously.

The three preceding reactions occur despite their requirement of heat because their products are much more disordered than their reactants. Water vapor is more disordered and has a higher entropy than liquid water. Hydrated NH_4^+ and Cl^- ions have a higher entropy than crystalline NH_4Cl. Gaseous NO_2 and O_2 are more disordered and have a higher entropy than solid N_2O_5. A chemical system seeks not only the state of lowest energy or enthalpy, but also the state of greatest disorder, probability, or entropy. A new state function can be defined, the **free energy**, **G**:

$$G \equiv H - TS \tag{16-8}$$

We can show very simply that for reactions at constant pressure and temperature *any reaction is spontaneous whose free energy decreases.* If we look at the total free energy, G, of a collection of compounds in a beaker at constant temperature, the change in total free energy brought about by chemical reaction is related to the changes in enthalpy and entropy by

$$\Delta G = \Delta H - T \Delta S \tag{16-9}$$

But $H = E + PV$, and at constant pressure

$$\Delta H = \Delta E + P \Delta V$$

and by the first law of thermodynamics

$$\Delta E = q - w$$

Let us assume that no electrical or other work is done, aside from PV work as the reactants and products expand and contract. Then $w = P \Delta V$. And if we substitute into the preceding equations in reverse order,

$$\Delta E = q - P \Delta V$$
$$\Delta H = q - P \Delta V + P \Delta V = q$$

This last equation states that the enthalpy change equals the heat of reaction at constant pressure, which we already have encountered. Substituting q for ΔH into equation 16-9 yields

$$\Delta G = q - T\,\Delta S \tag{16-10}$$

For a reversible reaction, we mentioned previously (equation 16-3) that $\Delta S = q/T$. Therefore, $q = T\,\Delta S$, and

$$\Delta G = 0 \tag{16-11}$$

for a reversible reaction carried out at constant pressure and temperature. What happens in an irreversible reaction? Does the free energy increase, decrease, or do both, depending on the circumstances?

We can answer this question by using equation 16-4. In any real, irreversible reaction, the entropy change is *greater* than q/T. Thus $\Delta S > q/T$, or $T\,\Delta S > q$. Therefore, by equation 16-10, $\Delta G < 0$ for such a reaction.

Both results can be summarized as follows: *In any spontaneous reaction at constant pressure and temperature, the free energy, G, always decreases. When the reaction system reaches equilibrium, G is at a minimum, and ΔG equals zero. This behavior of G is represented in Figure 16-7.*

It is difficult to overestimate the importance of these results to chemists. Free energy is now seen to be the touchstone by which we can determine in advance whether a given reaction will proceed spontaneously, remain at equilibrium, or occur spontaneously in the reverse direction.

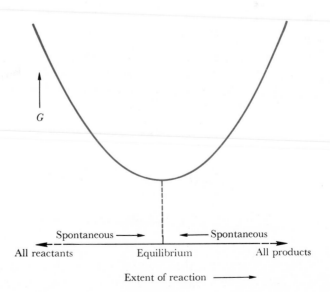

Figure 16-7   The true chemical potential function under conditions of constant pressure and temperature is the free energy, *G.* In all spontaneous reactions, the free energy decreases, and at equilibrium the free energy change of the reaction in either direction is zero.

One last item of thermodynamic ingenuity needs to be mentioned. Since G, like the functions H, T, and S of which it is composed, is a state function, it does not matter if the pressure and temperature change during a reaction, so long as they are brought back to the starting pressure and temperature at the conclusion of the reaction. The preceding comments about G and equilibrium, although derived for unchanging pressure and unchanging temperature, apply equally well to a high-temperature explosion, provided that the reaction vessel and its contents are returned to 298 K and 1 atm at the conclusion of the reaction.*

Free Energy Changes When External Work Is Done

In a reversible process in which no external work is involved, as we have just seen, the free energy of a reacting system does not change: $\Delta G = 0$. What happens if, during the reversible process, the system does electrical, magnetic, or gravitational work on its surroundings? We shall need the answer to this question in Chapter 19, where we treat electrochemical cells.

If PV work is not the only kind of work involved, then

$$w = P\,\Delta V + w_{ext} \qquad (16\text{-}12)$$

in which w_{ext} represents all other kinds of work. Hence,

$$\Delta E = q - P\,\Delta V - w_{ext}$$

and the enthalpy and free energy changes are derived as in the preceding pages:

$$\Delta H = q - w_{ext}$$
$$\Delta G = q - T\,\Delta S - w_{ext} = -w_{ext}$$

This final result is the object of our derivation:

$$\Delta G = -w_{ext} \qquad (16\text{-}13)$$

When a chemical system does work on its surroundings in a reversible manner, the decrease in free energy of the system exactly balances the work done *other than* pressure–volume work. In an electrochemical cell, the work done by the cell is a measure of the decrease of free energy within the cell. Conversely, if a potential is applied across the terminals of an electrolysis cell of the type discussed in Section 1-7, the electrical work done on the cell (measured by methods that we shall examine in Chapter 19) is identical to the increase in free energy of the chemicals in the cell. When water is dissociated electrolytically by passing a current through it, the electrical work required is stored as the increase in free energy of hydrogen

---

*In this chapter we deal with reactions at 298 K only. Free energies are sensitive to temperature changes; we will see examples in the next chapter.

and oxygen gas from the free energy of liquid water:

$$H_2O(l) \rightarrow H_2(g) + \tfrac{1}{2}O_2(g) \qquad \Delta G^0 = +237.2 \text{ kJ}$$

This free energy can be recovered as heat when hydrogen and oxygen gases are burned. Alternatively, if the proper apparatus is used, the free energy can be converted to work again. (A fuel cell such as the one for generating electricity in lunar space capsules uses this $H_2 + \tfrac{1}{2}O_2$ reaction. If the gases are simply burned, part of the free energy is converted to heat and, by the second law, is not recoverable as work. The trick in efficient utilization of energy is to avoid turning it into heat at any step in the process. This is the secret of the efficiency of metabolic processes [Chapter 21].)

Calculations with Standard Free Energies

Standard free energies of formation of compounds from elements in their standard states are tabulated in Appendix 3. The standard state for a gas, pure liquid, or pure solid is the same as with enthalpies: gas at 1 atm partial pressure, pure liquid, pure solid—usually at 298 K. The standard state for a solute in solution is a concentration of 1 mole per liter of solution, or a $1M$ solution. The standard state of a solution component for tabulating enthalpy was not this $1M$ solution, but was a solution so dilute that adding more solvent had no additional heat effect. However, since enthalpy does not change very much with concentration (unlike free energy, as we shall see in Section 16-6) we can also use the tabulated enthalpy values as if they were for $1M$ solutions.

Let's return to our chemical examples and interpret them in terms of free energy. The explosive reaction of H_2 with Cl_2 has the following molar free energies, enthalpies, and entropies of reaction:

	H_2	+	Cl_2	$\rightarrow$	2 HCl
ΔH^0(kJ mole^{-1}):	0.0		0.0		-92.31
ΔG^0(kJ mole^{-1}):	0.0		0.0		-95.27
ΔS^0(J K^{-1} mole^{-1}):	130.6		223.0		186.7

The reaction liberates 92.31 kJ of heat *per mole* of HCl produced, and the free energy decreases by more than this amount: 95.27 kJ mole^{-1} HCl. Where does this extra impetus for reaction come from?

For the reaction as written, which produces 2 moles of HCl, $\Delta H^0 = -184.6$ kJ and $\Delta G^0 = -190.5$ kJ. The entropy of the reaction is $\Delta S^0 = 2 \times (186.7) - 130.6 - 223.0 = +19.8$ e.u. Thus, $T\Delta S^0$ is 298 K $\times$ 19.8 e.u. $= 5900$ J or 5.90 kJ. Equation 16-9 is verified:

$$\Delta G = \Delta H - T\Delta S$$
$$-190.5 = -184.6 - 5.9$$
$$-190.5 = -190.5$$

Two moles of HCl are slightly more disordered and have a slightly higher entropy than 1 mole each of H_2 and Cl_2 gases. Most of the drive behind the reaction, as expressed by the free energy, comes from the liberation of heat, but 3% of it originates because the products have a higher entropy than the reactants.

A standard free energy of -191 kJ indicates a tremendous impetus toward reaction—the impetus that often accompanies an explosion. Now let's look at a gentler reaction, that of making ammonia from hydrogen and nitrogen.

Example 1

What are the changes in standard free energy, enthalpy, and entropy for the reaction $3H_2 + N_2 \rightarrow 2NH_3$? Do the heat and the disorder factors promote or oppose this reaction in the direction written?

Solution

From the data in Appendix 3,

$$\Delta G^0 = 2(-16.64) - 3(0.0) - (0.0) = -33.28 \text{ kJ}$$
$$\Delta H^0 = 2(-46.19) - 3(0.0) - (0.0) = -92.38 \text{ kJ}$$
$$\Delta S^0 = 2(192.5) - 3(130.59) - (191.49) = -198.26 \text{ e.u.}$$

and at 298 K, $T \Delta S^0 = -59.08$ kJ.

Checking equation 16-9, we find that

$$\Delta G = \Delta H - T \Delta S$$
$$-33.28 = -92.38 + 59.08$$

The reaction is favored by the liberation of 92.38 kJ of heat, but opposed by 59.08 kJ because the products are so much more ordered and have 198.3 e.u. lower entropy than the reactants. Another way of considering the reaction is to say that, of the 92.38 kJ of heat liberated, 59.08 kJ were required to pay for the creation of an ordered system, and only 33.28 kJ remain to drive the reaction.

The drive toward the synthesis of ammonia is much smaller than the drive toward the synthesis of HCl, primarily because of the entropy factor. Is it ever possible for the entropy term to overwhelm the heat term and send the reaction in the direction opposite to that indicated by enthalpy alone? Yes, it is, and these are precisely the instances in which the principle of Berthelot and Thomsen fails.

Example 2

Calculate the free energy, enthalpy, and entropy changes for the vaporization of liquid water. Check equation 16-9 with your results. Which term is responsible for the evaporation's taking place?

This is a trick question. The *standard* values are $\Delta G^0 = +8.60$ kJ, $\Delta H^0 = +44.01$ kJ, $\Delta S^0 = 118.8$ e.u., $T \Delta S^0 = +35.41$ kJ. It still looks as if water should not evaporate. But all that this means is that if water vapor at a partial pressure of 1 atm is present at 298 K, it will condense spontaneously. If the partial pressure of the water is only a few millimeters, it will evaporate spontaneously instead. The dependence of ΔG on concentration is the topic of the next section.

Example 3

Calculate the free energy, enthalpy, and entropy changes for the reaction $NH_4Cl(s) \rightarrow NH_4(aq)^+ + Cl(aq)^-$. (The solvent, water, is assumed to be present in equal concentrations before and after reaction, and is omitted.) Do the heat and entropy terms promote or oppose the solution of ammonium chloride?

Solution $\Delta G^0 = -6.77$ kJ, $\Delta H^0 = +15.14$ kJ, $\Delta S^0 = +73.4$ e.u.; $T \Delta S^0 = +21.9$ kJ. The enthalpy change opposes the reaction by 15.14 kJ, but the entropy change favors it by 21.9 kJ. The net drive in free energy is 6.77 kJ.

Example 4

Calculate the free energy, enthalpy, and entropy changes for the decomposition of N_2O_5,

$$N_2O_5(s) \rightarrow 2NO_2(g) + \tfrac{1}{2}O_2(g)$$

Solution

$$\Delta G^0 = 2(+51.84) + \tfrac{1}{2}(0.0) \quad - (134) \quad = \quad 30.3 \text{ kJ}$$
$$\Delta H^0 = 2(+33.85) + \tfrac{1}{2}(0.0) \quad - (-41.8) = +109.5 \text{ kJ}$$
$$\Delta S^0 = 2(+240.5) \quad + \tfrac{1}{2}(205.0) - (113) \quad = +470 \text{ e.u.}$$

$$T \Delta S^0 = \frac{298 \times (470 \text{ J K}^{-1})}{1000 \text{ J kJ}^{-1}} = +140 \text{ kJ}$$

$$\Delta G^0 = \quad \Delta H^0 \quad - T \Delta S^0$$
$$-30.3 = +109.5 - 140$$
$$-30.3 \simeq -30.5$$

The disadvantage of having to absorb about 110 kJ of heat on decomposition is more than compensated by the greater entropy of the gaseous products, and the reaction proceeds with a standard driving force of about 30 kJ of free energy.

In summary, the drive of a reaction, carried out at constant pressure and temperature, is measured by its free energy change. If the free energy change is negative, the reaction is spontaneous; if the free energy change is positive, the reaction is spontaneous in the reverse direction; if the free energy change is zero, reactants and products are at equilibrium. The free energy change has two components: $\Delta G = \Delta H - T \Delta S$. A large decrease in enthalpy, meaning the emission of heat, favors a reaction. But there is a second factor as well. A large increase in entropy when reactants form products also favors reaction. The entropy term at normal temperatures is generally small, so ΔG and ΔH have the same sign. In such cases, spontaneous reactions *are* exothermic. Yet there are other instances in which the entropy and enthalpy terms work against one another, and even in which the entropy term dominates. This is especially true in reactions in which solids or liquids change to gases or solutions as products.

Thus far, we have used only standard concentrations, meaning 1 atm partial pressure for gases, pure liquids, and pure solids for condensed phases, and $1M$ solutions for solutes. How does free energy change with changes in concentration?

16-6   FREE ENERGY AND CONCENTRATION

To this point, we have managed to avoid calculus. We shall need it briefly to derive the equation for the dependence of free energy on concentration, but shall not use it thereafter. Recall the basic definition of free energy,

$$G = H - TS = E + PV - TS$$

Instead of looking at small but finite changes (ΔG, ΔE, ΔS) at constant temperature, as in equation 16-9, let's examine infinitesimally small changes (dG, dE, and dS) under the most general experimental conditions. Then the equation for the basic definition of free energy becomes

$$dG = dE + P\,dV + V\,dP - T\,dS - S\,dT$$

The first law of thermodynamics, expressed in the form of infinitesimally small changes, is

$$dE = dq - dw \tag{16-14}$$

and the free energy expression is

$$dG = dq - dw + P\,dV + V\,dP - T\,dS - S\,dT$$

We can simplify this expression considerably. If the reaction takes place at constant temperature, then $S\,dT = 0$ since dT, the change in temperature, is zero. If the reaction is reversible, $dq = T\,dS$, and if only PV or expansion work is permitted, $dw = P\,dV$. All terms on the right except one cancel, and

$$dG = V\,dP \tag{16-15}$$

For 1 mole of an ideal gas, $V = RT/P$, and

$$dG = RT\frac{dP}{P} = RT\,d\ln P \tag{16-16}$$

Our last use of calculus is the integration (summing all the infinitesimal changes) of this equation:

$$G_2 = G_1 + RT\ln\frac{p_2}{p_1} \tag{16-17}$$

This equation means that if we know the molar free energy of an ideal gas at partial pressure p_1 to be G_1, the molar free energy at some other partial pressure p_2 is G_2. Although we derived this for a reversible change of conditions from p_1 to p_2, once we have it we can use it for irreversible changes as well, since by the state function properties of G it is irrelevant how we go from state 1 to state 2.

Now let us make state 1 our chosen standard state of 1 atm partial pressure and make state 2 any state at all. The more general form of equation 16-17 then is

$$G = G^0 + RT\ln\left(\frac{p}{p^0}\right) \tag{16-18}$$

in which $p^0 = 1$ atm. For the free energy of the compound in the standard state, G^0, we can use instead the free energy of formation from Appendix 3, ΔG^0 at 298 K. This is legitimate because, in any chemical reaction, matter is neither created nor destroyed, and both reactants and products must be made from the *same type and quantity of elements* in their standard states. When we adapt equation 16-18 to ΔG for whole reactions, the contributions of the elements will cancel in reactants and products.

We can now calculate how the free energy of ammonia depends on its partial pressure in a mixture of gases (or more precisely, the free energy of formation of ammonia from its elements in their standard states). Because the standard free energy of formation of ammonia at 298 K is -16.64 kJ mole^{-1},

$$G_{NH_3} = -16.64 + RT\ln\left(\frac{p_{NH_3}}{1\text{ atm}}\right)$$

$$G_{NH_3} = -16.64 + 2.303\,RT\log_{10}\left(\frac{p_{NH_3}}{1\text{ atm}}\right)$$

Since $R = 8.314$ J deg^{-1} mole^{-1} and $T = 298$ K, $2.303\,RT = 5706$ J mole^{-1} or 5.706 kJ mole^{-1} (this is a handy number to remember). Thus

$$G_{NH_3} = -16.64 + 5.706\log_{10}\left(\frac{p_{NH_3}}{1\text{ atm}}\right)$$

The molar free energy of ammonia is plotted against partial pressure in Figure 16-8. (This is actually the molar free energy of formation of ammonia at a specified pressure, from H_2 and N_2 in their standard states.) Note that the plot is a straight line because pressure was plotted on a logarithmic scale. The ratio of partial pressure to partial pressure in the standard state is usually abbreviated as the **activity**, a:

$$a = \frac{p}{p^0}$$

Since the standard state is 1 atm, the activity is numerically equal to the partial pressure of the gas, but activity has no units. This makes activity easier to use; we do not find ourselves wondering how, in the expression "$\ln p$," we can take the logarithm of a number with units attached to it, and then having to remember that the proper form of the logarithm is "$\ln (p/1 \text{ atm})$."

In summary, for ammonia we can write the general expression for the relationship between free energy and pressure as

$$G_{NH_3} = G^0_{NH_3} + RT \ln a_{NH_3}$$

We can do the same for all the reactants and products, j, in a chemical process:

$$G_j = G^0_j + RT \ln a_j = G^0_j + 2.303 \, RT \log_{10} a_j \tag{16-19}$$

Figure 16-8

The free energy of a gas depends on its partial pressure according to the expression $G = G^0 + RT \ln (p/p^0)$. This plot is of the free energy of ammonia, as given by $G_{NH_3} = -16.64 + 5.706 \log_{10} (p_{NH_3}/1 \text{ atm})$. Pressure, p, is plotted on a logarithmic scale.

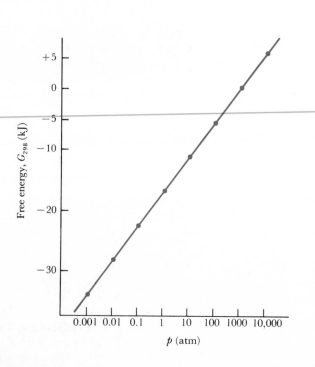

Let's apply this to the synthesis of ammonia and see what we can learn. In writing the equation as

$$N_2 + 3H_2 \rightleftarrows 2NH_3$$

we have changed from the single arrow to a double arrow because we now are considering both the forward and the reverse reactions. For each of the reactants and products, we can write

$$G_{N_2} = G^0_{N_2} + RT \ln a_{N_2}$$
$$G_{H_2} = G^0_{H_2} + RT \ln a_{H_2}$$
$$G_{NH_3} = G^0_{NH_3} + RT \ln a_{NH_3}$$

The total free energy of the chemical reaction is

$$\Delta G = 2G_{NH_3} - G_{N_2} - 3G_{H_2}$$
$$\Delta G = 2G^0_{NH_3} - G^0_{N_2} - 3G^0_{H_2} + 2RT \ln a_{NH_3} - RT \ln a_{N_2} - 3RT \ln a_{H_2}$$

The first three terms on the right are the *standard* free energy of reaction that we have already used, and are combined as ΔG^0. Then if we take the coefficients in front of the RT terms inside the logarithms as exponents, we can write

$$\Delta G = \Delta G^0 + RT \ln a^2_{NH_3} - RT \ln a_{N_2} - RT \ln a^3_{H_2}$$

Finally,

$$\Delta G = \Delta G^0 + RT \ln \left(\frac{a^2_{NH_3}}{a_{N_2} a^3_{H_2}} \right) \tag{16-20}$$

Take a close look at the ratio in parentheses. It is the ratio of activities of products to reactants, and in fact is simply the reaction quotient, Q of Chapter 4, with concentrations expressed in terms of activities rather than moles per liter:

$$Q \equiv \left(\frac{a^2_{NH_3}}{a_{N_2} a^3_{H_2}} \right) \tag{16-21}$$

This is a more general treatment than we saw previously, and will lead to a more general concept of equilibrium. This reaction quotient, Q, can be calculated for any given set of experimental conditions from the partial pressures of reactants and products.

The true free energy change in the ammonia reaction is the combination of the standard free energy change (for which $Q = 1$ and $\ln Q = 0$) and the reaction quotient term describing the actual experimental conditions. With numbers inserted, equation 16-20 becomes

$$\Delta G = -33.30 + 5.706 \log_{10} Q \quad \text{(units of kJ)} \tag{16-22}$$

Note that since $\Delta G^0 = 2G^0_{NH_3} - G^0_{N_2} - 3G^0_{H_2} = 2 \Delta G^0_{NH_3}$, the free energy value in equation 16-22 is twice the value in Appendix 3 for the free energy of formation of 1 mole of NH_3 gas.

Table 16-1 lists the results of applying equation 16-22 to 11 different sets of starting conditions for ammonia synthesis. These are not successive points in the same reaction from the same starting conditions; they are each separate conditions. If the concentrations of reactants and products were as shown in each experiment, what would be the free energy change in the ammonia synthesis reaction? These are the ΔG values at the right of the table. We shall defer until Chapter 17 the question of how to follow a given reaction from start to finish.

In experiment A, nitrogen and hydrogen gases are at 1 atm partial pressure, and there is no ammonia present. Hence, the drive for the production of ammonia is infinitely strong. But as soon as the ammonia concentration is even 10^{-3} atm, the free energy change has increased from $-\infty$ to -67.51 kJ, and the drive to produce more ammonia has slackened (experiment B). If we were to set up experiment C, with N_2 and H_2 at 1 atm partial pressure and NH_3 at 0.1 atm, we would find that the free energy is -44.69 kJ. The more products and the less reactants there are, the smaller is the drive to produce more products. At uniform concentrations of 1 atm, the standard free energy results (experiment D). In experiment E, the excess of ammonia makes the drive smaller than in the standard state. In experiments F, G, and H, the drive toward products is stopped entirely. This is the

Table 16-1

Free Energy of Synthesis of Ammonia at 298 K

$$N_2(g) + 3H_2(g) \rightleftarrows 2NH_3(g)$$

$$\Delta G = -33.30 + 5.706 \log_{10} Q$$

$$Q = \frac{a^2_{NH_3}}{a_{N_2} a^3_{H_2}}$$

Experiment	a_{N_2}	a_{H_2}	a_{NH_3}	Q	$5.706 \log_{10} Q$ (kJ)	ΔG (kJ)
A	1	1	0	0	$-\infty$	$-\infty$
B	1	1	0.001	10^{-6}	-34.24	-67.51
C	1	1	0.1	10^{-2}	-11.64	-44.69
D	1	1	1	1	0	-33.30
E	1	1	100	10^4	$+22.83$	-10.44
F	1	1	825	6.8×10^5	$+33.30$	0
G	1.47	0.01	1	6.8×10^5	$+33.30$	0
H	0.01	0.1	2.61	6.8×10^5	$+33.30$	0
I	0.01	0.1	26.1	6.8×10^7	$+44.77$	$+11.50$
J	0.01	0.01	100	10^{12}	$+68.48$	$+35.21$
K	0	1	1	∞	$+\infty$	$+\infty$

equilibrium state. At equilibrium

$$\Delta G = 0 \quad \text{and} \quad \Delta G^0 = -RT \ln Q$$

As we saw in Chapter 4, the reaction quotient at equilibrium is the **equilibrium constant**, K_{eq}. We can calculate the value of this equilibrium constant from the standard free energy of the reaction:

$$K_{eq} = \frac{a_{NH_3}^2}{a_{N_2} a_{H_2}^3} = e^{-(\Delta G^0/RT)} = 10^{-[(-33.30)/2.303\,RT]} = 6.8 \times 10^5$$

If the reaction quotient is smaller than the equilibrium constant, products will be formed spontaneously. Whenever conditions are such that the reaction quotient is greater than the equilibrium constant, K_{eq}, the reaction will be spontaneous in the reverse direction (experiments, I, J, and K in Table 16-1). When the reaction quotient equals the equilibrium constant, $Q = K_{eq}$, the forward and reverse reactions proceed at the same rate, and the reacting system of chemicals is at equilibrium.

General Expressions

So far, we have stated the free energy derivations in terms of the ammonia reaction. Let's generalize to a reaction in which r moles of compound R combine with s moles of S to produce t moles of T and u moles of U:

$$r\,R + s\,S = t\,T + u\,U \tag{16-23}$$

The free energy change for the reaction, ΔG, is

$$\Delta G = t\,G_T + u\,G_U - r\,G_R - s\,G_S \tag{16-24}$$

Each free energy can be expressed in terms of activity:

$$\begin{aligned}
G_R &= G_R^0 + RT \ln a_R \\
G_S &= G_S^0 + RT \ln a_S \\
G_T &= G_T^0 + RT \ln a_T \\
G_U &= G_U^0 + RT \ln a_U
\end{aligned} \tag{16-25}$$

These can be substituted into equation 16-24,

$$\Delta G = \Delta G^0 + RT \ln Q \tag{16-26}$$

in which

$$\Delta G^0 = t\,G_T^0 + u\,G_U^0 - r\,G_R^0 - s\,G_S^0 \tag{16-27}$$

and

$$Q = \frac{a_T^t a_U^u}{a_R^r a_S^s} \tag{16-28}$$

Equation 16-26 gives the free energy change for the reaction under any conditions. For the special case of equilibrium, $\Delta G = 0$, and

$$Q = K_{eq} \qquad \text{(the equilibrium constant)} \qquad \text{(16-29)}$$

$$-\Delta G^0 = RT \ln K_{eq} \qquad \text{(16-30)}$$

$$K_{eq} = e^{[-\Delta G^0/RT]} \qquad \text{(16-31)}$$

Notice that the form of the reaction quotient, Q, and the equilibrium constant, K_{eq}, depends only on the overall stoichiometry of the reaction, and not on any particular reaction mechanism. It isn't necessary to know in molecular detail how a reaction takes place to write the equilibrium-constant expression. We first mentioned this labor-saving fact in Chapter 4; now we have given it a thermodynamic proof.

Example 5

Calculate the equilibrium constant at 298 K for the reaction $H_2(g) + Cl_2(g) = 2HCl(g)$, and compare it with the equilibrium constant for the ammonia synthesis.

Solution

We found earlier that the free energy of this reaction is -95.27 kJ mole^{-1} of HCl, or -190.5 kJ for the reaction as written to produce 2 moles of HCl. Thus

$$K_{eq} = e^{[-(-190.5)/RT]} = 10^{+(190.5/5.706)} = 10^{33.4} = 2.5 \times 10^{33}$$

The K_{eq} for the ammonia synthesis is only 6.8×10^5. The HCl reaction is different; equilibrium lies far on the side of much HCl and few reactants. The expression for the amounts of H_2, Cl_2, and HCl at equilibrium is

$$\frac{a_{HCl}^2}{a_{H_2}a_{Cl_2}} = K_{eq} = 2.5 \times 10^{33}$$

If pure HCl gas is enclosed in a container at 1 atm, enough HCl can dissociate spontaneously to yield partial pressures of H_2 and Cl_2 that are approximately 2×10^{-17} atm, hardly a significant amount of either reactant. In Chapter 17 we shall see how to make such statements for reactions that are less completely skewed in one direction than the HCl reaction is.

Summary

In this chapter we have answered an important chemical question that has been deferred since we first began thinking about energy in chemical reactions in Chapter 2: By what criterion can we decide whether a given chemical reaction is spontaneous, and how can we find the conditions of equilibrium? The criterion is **free energy**, G (at least for reactions run at constant temperature and pressure), and this goal was reached via another new function, **entropy**, S.

Entropy is a measure of the disorder of a system. It can be calculated, in principle and sometimes in fact, from the number of different microscopic ways of building the same observable situation. Third-law entropies, obtained from purely thermal measurements, agree well with what we would expect for different substances from Boltzmann's statistical explanation of entropy.

The spontaneity of a reaction at constant pressure and temperature is measured by its **change in free energy**, ΔG, per unit of reaction. For a reaction in which no work other than pressure–volume work is done, if ΔG is negative, the reaction is spontaneous. If ΔG is positive, the reaction is spontaneous in the reverse direction. If ΔG is zero, the reaction is at equilibrium. Expressed another way, the free energy is the chemical potential function that we minimize to find the point of chemical equilibrium.

If electrical work, or some form of work other than pressure–volume work, is involved in a chemical process accomplished in a reversible manner, then the free energy change during the reaction is not zero. Instead, the free energy of the reacting system decreases by an amount equal to the work done on the surroundings: $\Delta G = -w_{\text{ext}}$.

The free energy of a reaction is the net result of two effects, heat and disorder: $\Delta G = \Delta H - T \Delta S$. A reaction is favored if it releases heat (ΔH is negative) and if the products are more disordered than the reactants ($-T \Delta S$ negative, or ΔS positive). Usually, but not always, ΔH is the dominant term on the right side of the equation.

The free energy of a gas varies with its partial pressure by the relationship $G_2 = G_1 + RT \ln (p_2/p_1)$. The **activity** of the gas, a, is the ratio of its pressure to that in a standard state of 1 atm pressure. Thus the free energy at any pressure is given by $G = G^0 + RT \ln a$.

The free energy change of a gas reaction varies with the partial pressures of its components according to the expression $\Delta G = \Delta G^0 + RT \ln Q$. The quantity Q is the reaction quotient, first defined in Chapter 4. For the special case of equilibrium, the free energy of reaction is zero, $\Delta G = 0$, and the reaction quotient becomes the equilibrium constant, $Q = K_{\text{eq}}$. Then the *standard* free energy of reaction and the equilibrium constant are related by

$$\Delta G^0 = -RT \ln K_{\text{eq}} \qquad \text{or} \qquad K_{\text{eq}} = e^{-(\Delta G^0/RT)}$$

For any given set of experimental conditions, if the reaction quotient is smaller than the equilibrium constant, the reaction is spontaneous in the forward direction. If Q is greater than K_{eq}, the reverse reaction is spontaneous. At equilibrium, $Q = K_{\text{eq}}$.

Self-Study Questions

1. Interpret the skidding automobile example of Section 3-6 in terms of the first and second laws of thermodynamics. How does the first law enter? How does reversibility or irreversibility enter, and how is it tied to the second law?

2. Explain how heat and kinetic energy represent motion of different orders of size. Which corresponds to motion at the molecular level?

3. Which energy conversion is subject to severe limitations: converting large-scale motion to molecular motion, or converting molecular motion to large-scale motion? In what way does the second law of thermodynamics refer to this dilemma? (Which is easier: mixing a bag of red beans with a bag of white beans, or unmixing them?)

4. How is entropy related to heat and temperature, according to classical thermodynamics? How does the relationship differ, depending on whether the process is carried out reversibly or irreversibly?

5. To what did Boltzmann relate entropy? Why is it convenient for his definition of S to be in logarithmic form?

6. Why is the entropy of a solid less than that of a gas of the same substance?

7. Why, according to Boltzmann, will the entropy of a perfect crystal be zero only at absolute zero temperature?

8. Why should dissolving a solid or liquid in water lead to an increase in entropy, whereas dissolving a gas causes an entropy decrease?

9. Why should entropy increase with mass?

10. Why should the entropy of a cross-linked polymer be less than that of the polymer subunits before they have been polymerized?

11. Does an aqueous solution of calcium ions have a larger entropy before or after hydration by water molecules? Why, then, are the ions hydrated?

12. Which has the higher entropy: a mole of ice or a mole of liquid water? Why, then, does water freeze?

13. Which has the higher enthalpy: a mole of water or a mole of water vapor? Why, then, does water evaporate?

14. What is wrong with the statement, "In all spontaneous processes, the system moves toward a state of lowest energy"? What is the correct statement?

15. What is wrong with the statement, "In all spontaneous processes, the entropy rises"? For what kinds of thermodynamic system is this statement true?

16. How do enthalpy change and entropy change contribute to the total free energy change, at constant temperature?

17. How does your answer to Questions 14–16 lead to the conclusion that spontaneous reactions are those for which the free energy of reaction decreases?

18. How is electrical or other work related to free energy changes in a reaction carried out reversibly at constant pressure and temperature?

19. Appendix 3 tabulates ΔG^0, ΔH^0, and S^0 at 298 K for chemical substances. Show that some of this information is redundant, and could be dispensed with. Why, then, is there duplication of information in these tables? What could you delete from the tables, and how would you calculate it from the data that were left?

20. How does free energy depend on concentrations of reactants and products?

21. What is the thermodynamic activity of a substance? How are the activity and partial pressure of an ideal gas related?

22. What is the relationship between the reaction quotient and the equilibrium constant for a reaction?

23. How can you tell from the numerical value of the reaction quotient whether a given reaction is at equilibrium, spontaneous in the forward direction, or spontaneous in reverse?

Problems

Disorder and entropy

1. In each of the following pairs, which item will have the greater entropy: (a) a packaged deck of cards or the same cards spread out on a table; (b) an assembled automobile or the unassembled parts needed to make the automobile; (c) carbon dioxide, water, nitrogen compounds and minerals, or the tree that grows from them?

2. In each of the following pairs, which state has the higher entropy: (a) a mole of liquid water, or a mole of water vapor at 1 atm pressure and 25°C; (b) a mole of dry ice or a mole of carbon dioxide gas at 1 atm and 195 K; (c) five dimes on a tabletop showing four heads and one tail, or showing three heads and two tails; (d) 100 g of liquid H_2O and 100 g of liquid D_2O (D is the symbol for deuterium) in separate beakers, or the 200 g mixture of the two; (e) a mole of gaseous CO_2, or a mole of CO_2 in the form of carbonated water?

Entropy changes

3. Will the entropy change in each of the following processes be positive or negative? Will the disorder in each process increase or decrease?
   a) 1 mole of solid methanol → 1 mole of gaseous methanol
   b) 1 mole of solid methanol → 1 mole of liquid methanol
   c) $\frac{1}{2}$ mole of gaseous O_2 + 2 moles of solid Na → 1 mole of solid Na_2O
   d) 1 mole of solid XeO_4 → 1 mole of gaseous Xe + 2 moles of gaseous O_2
   Rank these four processes in order of increasing ΔS.

4. In 1884, Frederick Trouton discovered that for many liquids the heat of vaporization is directly proportional to the normal boiling point, or that the ratio of heat of vaporization to boiling point is constant:

$$\frac{\Delta H_{vap}}{T_b} = 88 \text{ J K}^{-1} \text{ mole}^{-1}$$

We now would explain Trouton's rule by saying that the molar entropy of vaporization of many liquids is approximately the same, or that the disorder produced by evaporation is comparable for many liquids. But HF is odd in that its molar entropy of vaporization is higher, 105 J K^{-1} mole^{-1}. Why is this so? (*Clue:* The molar entropy of HF gas is not different enough from that of other gases to account for the discrepancy.)

5. Predict the sign of the entropy change in each of the following reactions:

   a) $2CO(g) + O_2(g) \rightarrow 2CO_2(g)$
   b) $Mg(s) + Cl_2(g) \rightarrow MgCl_2(s)$
   c) $Al(s) \rightarrow Al(l)$
   d) $I_2(s) \rightarrow I_2(g)$
   e) $CH_4(g) + 2O_2(g) \rightarrow$
   $$CO_2(g) + 2H_2O(l)$$

   In each case give the molecular reason for your answer.

6. Calculate the entropy changes of the following chemical reactions, all at 298 K:

   a) $Ba(s) + \frac{1}{2}O_2(g) \rightarrow BaO(s)$
   b) $BaCO_3(s) \rightarrow BaO(s) + CO_2(g)$
   c) $Br_2(g) \rightarrow 2Br(g)$
   d) $H_2(g) + Br_2(l) \rightarrow 2HBr(g)$

   Explain the sign of each of the entropy changes by comparing qualitatively the freedom of motion, or molecular disorder, in reactants and products.

7. Explain the entropy increase in the reaction

   $$Br_2(l) + Cl_2(g) \rightarrow 2BrCl(g)$$

8. Explain the entropy increase in the reaction

   $$Br_2(g) + Cl_2(g) \rightarrow 2BrCl(g)$$

   (This is not a repetition of Problem 7.)

Entropy and free energy

9. For the reaction (at 298 K and 1 atm),

   $$2Ag(s) + Br_2(l) \rightarrow 2AgBr(s)$$

   $\Delta H^0 = -199.2$ kJ mole^{-1} and $\Delta G^0 = -191.6$ kJ mole^{-1}.
   a) Calculate the standard entropy change, ΔS^0, for the reaction.
   b) What does this value of ΔS^0 tell you about the relative degree of order in reactants and products?

10. The reaction

    $$2H_2(g) + O_2(g) \rightarrow 2H_2O(g)$$

    proceeds spontaneously even though there is an increase in order within the system. How can this be?

11. Calculate the change in entropy, ΔS^0, for the reaction

    $$S(s, rhombic) + O_2(g) \rightarrow SO_2(g)$$

    given data in Appendix 3.

12. Using data in Appendix 3, calculate ΔS^0 in two different ways for the reaction

    $$\tfrac{1}{2}H_2(g) + \tfrac{1}{2}Cl_2(g) \rightarrow HCl(g)$$

Contributions of *H* and *S* to *G*

13. Calculate the standard free energy change, ΔG^0, for the reaction

    $$Fe_2O_3(s) + 3C(gr) \rightarrow 2Fe(s) + 3CO(g)$$

    Is this reaction spontaneous at 25°C? Calculate the standard enthalpy change, ΔH^0, and entropy change, ΔS^0, of the reaction at 298 K, and verify that

    $$\Delta G^0 = \Delta H^0 - T\,\Delta S^0$$

    Do the enthalpy change and entropy change work for, or against, spontaneity in the reaction? Which factor predominates?

14. Calculate ΔG^0, ΔH^0, and ΔS^0 for the reaction (at 298 K),

$$2Ag(s) + Hg_2Cl_2(s) \rightarrow \\ 2AgCl(s) + 2Hg(l)$$

Show that $\Delta G^0 = \Delta H^0 - T \Delta S^0$. Is the reaction spontaneous? Do the enthalpy change and entropy change each work for, or against, spontaneity for the reaction? Which factor predominates? Explain the entropy effect on molecular grounds.

Free energy and K_{eq}

15. What is the standard free energy change for synthesis of ammonia at 298 K? The reaction is

$$3H_2(g) + N_2(g) \rightarrow 2NH_3(g)$$

Calculate the equilibrium constant, K_{eq}, and write the expression for K_{eq} in terms of concentrations of reactants and products present at equilibrium.

16. Give the standard free energy change at 25°C for the reaction

$$Cl_2(g) + I_2(s) \rightarrow 2ICl(g)$$

Will the reaction as written be spontaneous under standard conditions? What are the relative contributions of enthalpy and entropy to the spontaneity of the reaction? Which effect predominates? Calculate K_{eq} for this reaction.

Molar entropy

17. Arrange the following substances in order of increasing molar entropy:

$N_2O_4(s)$, $Na(s)$, $NaCl(s)$, $Br_2(l)$, $Br_2(g)$.

18. Calculate the change in free energy for the reaction that converts diamond to graphite. In view of your answer, why don't diamonds in diamond rings spontaneously turn into lumps of graphite?

Free energy and spontaneity

19. Consider the reaction
$$CH_4(g) + 2O_2(g) \rightarrow \\ CO_2(g) + 2H_2O(l)$$

(a) According to the calculated ΔG^0, is this reaction thermodynamically spontaneous? (b) How do you account for the fact that methane and oxygen gases can remain mixed for long periods of time without detectable reaction?

20. Calculate ΔG^0 at 25°C for the following reactions, using data in Appendix 3:
a)
$$2NaF(s) + Cl_2(g) \rightarrow 2NaCl(s) + F_2(g)$$
b)
$$PbO_2(s) + 2Zn(s) \rightarrow Pb(s) + 2ZnO(s)$$
In view of your answers, comment on (a) the likelihood of obtaining fluorine gas by treating NaF with chlorine gas, and (b) the use of zinc to reduce PbO_2 to lead.

21. The standard free energy change of the reaction

$$2C(gr) + H_2(g) \rightarrow HC\equiv CH(g)$$

is $+209$ kJ mole^{-1} of acetylene, C_2H_2. (a) Is this a practical route for the synthesis of acetylene at room temperature? What is the equilibrium constant for this reaction? Does it favor reactants, or products? (b) Calculate the free energy change for the reaction

$$2CH_4(g) + \tfrac{3}{2}O_2(g) \rightarrow \\ HC\equiv CH(g) + 3H_2O(g)$$

Would this be a better approach for synthesis of acetylene? Why, or why not?

22. Calculate the free energy change associated with the combustion of liquid methanol:

$$CH_3OH(l) + \tfrac{3}{2}O_2(g) \rightarrow$$
$$CO_2(g) + 2H_2O(l)$$

(a) Is this reaction spontaneous under standard conditions? What is the equilibrium constant at 298 K? Does it favor reactants, or products? (b) What effect would an increase in pressure have on the spontaneity of the reaction? (c) What effect would an increase in temperature have on the spontaneity of the reaction?

Change of phase

23. If 333.5 J of heat are required to melt 1 g of ice at 0°C, what is the molar heat of fusion of ice at this temperature? What is the entropy change when 1 g of ice melts at 0°C? What is the free energy change for the process?

24. (a) Consider the evaporation of water as a chemical process, and calculate ΔH^0, ΔS^0, and ΔG^0 for the process under standard conditions at 25°C. (b) Recalling that

$$\Delta H^0 - T\,\Delta S^0 = \Delta G^0$$

$$= -RT \ln \frac{a(g)}{a(l)}$$

where $a(g)$ and $a(l)$ are the activities of H$_2$O gas and liquid, derive an expression for the vapor pressure of water as a function of temperature. (c) Calculate the equilibrium vapor pressure of water at 50°C and at 100°C. (Enthalpy and entropy are essentially constant at these temperatures.)

25. One mole of benzene, C_6H_6, is vaporized at its boiling point under a constant pressure of 1 atm. The heat of vaporization measured in a calorimeter at constant pressure is 30.54 kJ mole^{-1}. The boiling point of benzene at 1 atm is 80°C. (a) Calculate ΔH^0, ΔG^0, and ΔS^0 for this process. (b) Calculate ΔE^0 for the process, assuming that benzene vapor is an ideal gas. (c) Compare your answers to the standard heat and free energy of vaporization obtained from Appendix 3. Why are the values different? Can you explain the difference using Le Chatelier's principle?

Reversible and irreversible processes

26. A reaction, A → B, was carried out at 25°C in such a way that no useful work was done. In this process, 41.84 kJ mole^{-1} of heat were evolved. In a second experiment, the same overall reaction was carried out in such a way that the maximum amount of useful work was done. In this process, only 1.673 kJ mole^{-1} of heat were evolved. For each of the two reactions, calculate q, w, ΔE^0, ΔH^0, ΔS^0, and ΔG^0.

Suggested Readings

All of the readings for Chapter 15 are useful, plus:

H. A. Bent, *The Second Law: An Introduction to Classical and Statistical Thermodynamics*, Oxford University Press, New York, 1965.

R. E. Dickerson, *Molecular Thermodynamics*, W. A. Benjamin, Menlo Park, Calif., 1969. Somewhat more advanced than Mahan, Nash, or Waser, but with a strong emphasis on the statistical interpretation of entropy.

17

Free Energy and Equilibrium

I look upon amity and enmity as proper-
ties of intelligent beings, and I have not
yet found it explained by any, how these
appetites can be ascribed to bodies
inanimate and devoid of knowledge or of
so much as sense.

Robert Boyle (1661)

In this chapter we shall provide a theoretical basis for some ideas about spontaneity and equilibrium that were presented only as observed facts in Chapter 4. When we first encountered equilibrium constants, they were purely experimental numbers, which were convenient for predicting how a chemical reaction would behave after enough time had elapsed. There was no way to calculate them from other measurable properties of the chemical substances. Nor could we prove that the equilibrium constant, K_{eq}, which varies with temperature but is independent of concentrations, *should* exist for a reaction. We tried to explain K_{eq} by mass-action arguments, with the warning that the arguments held only for the simplest, one-step reactions. All we really could say was that observation has shown us this is how nature behaves.

The thermodynamics discussed in the preceding chapter placed K_{eq} on a much firmer footing. We can now prove that a K_{eq} exists for every reaction from the equation

$$\Delta G^0 = -RT \ln K_{eq}$$

or

$$K_{eq} = e^{-(\Delta G^0/RT)}$$

in which ΔG^0 is the free energy of the reaction at a specified temperature when all reactants and products are in their standard states. Since specifying standard states determines the concentrations ($1M$ for solutions, 1 atm partial pressure for gases and so on), the standard free energy, ΔG^0, is independent of the actual concentrations in any real experiment. Therefore, K_{eq} *also is independent of concentrations.* However, since the standard free energy is defined for a given temperature, and will have a different numerical value at different temperatures, the equilibrium constant also will be a function of temperature.

In this chapter we shall look more closely at the relationships between ΔG, K_{eq}, and T. In the process we shall review the most important properties of equilibrium and the equilibrium constant, which were introduced empirically in Chapter 4, but now we can give more satisfactory explanations of these properties.

17-1 THE PROPERTIES OF EQUILIBRIUM

Keep the following fundamental features of equilibrium in mind as you study this chapter:

1. Equilibrium is a dynamic rather than a static process. Consider the reaction for the dissociation of phosphorus pentachloride vapor that starts with pure PCl_5:

Dissociation $\rightarrow$
$$PCl_5(g) \rightleftarrows PCl_3(g) + Cl_2(g) \qquad (17\text{-}1)$$
$\leftarrow$ Recombination

If the reactants and products are confined in a closed vessel, eventually no more chlorine gas will appear to form. When the reaction is monitored by measuring the total pressure with a manometer, we discover that the pressure increases more slowly with time and eventually reaches a constant, maximum value. Dissociation of PCl_5 has not stopped; rather, dissociation and recombination are occurring at equal rates so there is no *net* change. You can prove this by performing an experiment at equilibrium conditions, using chlorine gas composed of the pure ^{37}Cl isotope instead of the natural 3:1 mixture of ^{35}Cl to ^{37}Cl. Even though equilibrium is maintained and no reaction seems to be occurring, you will soon find that more than the normal amount of ^{37}Cl isotope is present both in PCl_5 and in PCl_3. Some of the "labeled" Cl_2 has combined with PCl_3 that has the normal Cl isotope ratio to make ^{37}Cl-rich PCl_5; then some of this PCl_5 has dissociated to produce PCl_3 that is also enriched in ^{37}Cl. The reversible nature of equilibrium is symbolized in chemical equations by using double horizontal arrows pointing in both directions, or by using an equals sign rather than a single arrow to the right.

2. There is a spontaneous tendency toward equilibrium. This statement implies nothing about the *rate* at which a reaction may reach equilibrium; it deals only with the *drive toward* equilibrium. This chemical drive is measured as the free energy of the reaction. The **free energy change** represents the amount of energy that is available either to do work or to serve as the driving force in a chemical reaction. A hydrostatic analogue to free energy of reaction is shown in Figure 17-1, in which the drop in water level between reservoir and outflow tank plays the role of the free energy change, ΔG.

3. The driving force toward equilibrium diminishes as equilibrium is approached. Thus the appearance of products actually decreases the forward impetus of the reaction, making the free energy change, ΔG, less negative. Each successive increment of reaction gives rise to a smaller free energy change and hence can do less work.

4. Equilibrium is reached when ΔG becomes zero. The hydrostatic analogue of chemical equilibrium is shown in Figure 17-1c. Chemical equilibrium is a balance between two effects, heat and entropy or disorder. At equilibrium

$$\Delta G = 0 = \Delta H - T \, \Delta S$$
$$\Delta H = T \, \Delta S$$

Reaction in one direction is favored if it releases heat and leads to a lower enthalpy. Reaction in the other direction is favored if it leads to less ordered substances with higher entropy. At equilibrium these two effects balance each other.

5. The equilibrium position is the same at a given temperature, no matter from which direction it is approached. Thus in the dissociation of N_2O_4,

$$N_2O_4(g) \rightleftarrows 2NO_2(g) \tag{17-2}$$

the *relative* concentrations of N_2O_4 and NO_2 at equilibrium (assuming constant pressure and temperature) are independent of whether the initial material was almost pure N_2O_4 or almost pure NO_2. In the dissociation of PCl_5 (equation 17-1) the equilibrium point will be the same if you begin with pure PCl_5 or with *equimolar amounts* of PCl_3 and Cl_2, since PCl_3 and Cl_2 combine in equimolar amounts. If you begin with twice the number of moles of PCl_3 as Cl_2, the equilibrium concentrations will differ from those achieved by starting with pure PCl_5, but the equilibrium constant will be the same.

6. Equilibrium for a reaction is always characterized by an equilibrium constant, K_{eq}, expressed in terms of the activities, a, of reactants and products:

$$K_{eq} = \frac{a_{PCl_3} a_{Cl_2}}{a_{PCl_5}} \qquad \text{(for reaction 17-1)} \tag{17-3}$$

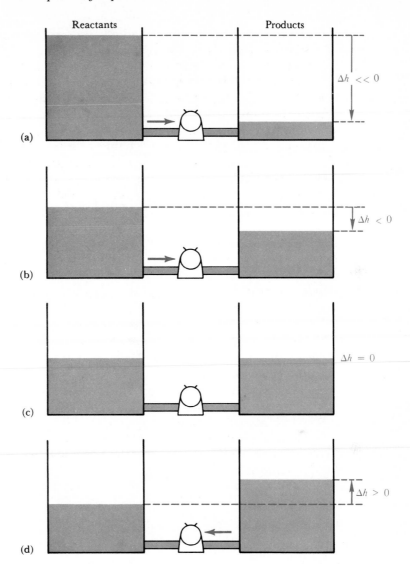

Figure 17-1 Hydrostatic analogy to the free energy of a reaction. Imagine that two water reservoirs are connected by a pipe that leads through a turbine for generating electricity. The amount of work that each cubic meter of water can do as it passes through the turbine depends on the hydrostatic head, or the distance from reactant water level to product water level, Δh. (a) At the start, a maximum amount of work is obtained per unit of water because Δh is large and negative. But every unit of water run through the turbine decreases the difference in water level and therefore decreases the amount of work that the next unit of water will do. (b) At this stage in the process, each cubic meter of water yields less than a third as much work as at the start. (c) When the water levels in both reservoirs are equal, no more work can be obtained. The system is at equilibrium, and $\Delta h = 0$. (d) This last state will never be attained from the first state spontaneously, because Δh can never spontaneously increase in the positive direction. If the experiment is begun as in (d), water will flow spontaneously in the reverse direction.

Any amounts of PCl_5, PCl_3, and Cl_2 that satisfy the equilibrium-constant expression will lead to an equilibrium situation. For ideal systems, the activity of each substance is the unitless ratio of its actual concentration to the concentration in its standard state. For nonideal systems, the activity is at least related to concentration, and can be thought of as an "effective concentration."

7. The equilibrium constant for a chemical reaction can be calculated from the standard free energy change of reaction, ΔG^0:

$$\Delta G^0 = -RT \ln K_{eq} \qquad \text{or} \qquad K_{eq} = e^{-(\Delta G^0/RT)} \qquad (17\text{-}4)$$

Since, in the standard free energy change for reaction, ΔG^0, we assume that the partial pressures of all components are 1 atm, the standard free energy change is not a function of pressure. Any pressure variations in reactants or products are expressed by the logarithmic term in the general free energy equation

$$\Delta G = \Delta G^0 + RT \ln Q \qquad (17\text{-}5)$$

Because ΔG^0 is independent of pressure, the equilibrium constant, K_{eq}, also is independent of pressure. No matter how much we compress a mixture of PCl_5, PCl_3, and Cl_2 gases, the *ratio* of their activities at equilibrium (equation 17-3) will not change. When the gases are compressed, the equilibrium conditions will shift so that the equilibrium constant is the *same* after the shift as before.

The standard free energy change *does* vary with temperature. In all of our free energy calculations, we have assumed that the temperature is 298 K, often omitting the subscript 298 from the term ΔG^0_{298}. But the free energy of a gas reaction with all gases at 1 atm partial pressure, for example, generally will not be the same at 1000 K as at 298 K. Hence, the equilibrium constant, K_{eq}, also will vary with temperature. In what follows we shall see some examples of this behavior.

Stoichiometry and the Equilibrium Constant

The expression for the equilibrium constant depends on how the equation for the reaction is written. For example, the reaction for the production of ammonia can be written in terms of producing 1 mole of ammonia:

$$(1) \quad \tfrac{1}{2}N_2(g) + \tfrac{3}{2}H_2(g) \rightleftarrows NH_3(g) \qquad K_1 = \frac{a_{NH_3}}{a_{N_2}^{1/2} a_{H_2}^{3/2}}$$

or of using 1 mole of nitrogen or 1 mole of hydrogen:

$$(2) \quad N_2(g) + 3H_2(g) \rightleftarrows 2NH_3(g) \qquad K_2 = \frac{a_{NH_3}^2}{a_{N_2} a_{H_2}^3}$$

$$(3) \quad \tfrac{1}{3}N_2(g) + H_2(g) \rightleftarrows \tfrac{2}{3}NH_3(g) \qquad K_3 = \frac{a_{NH_3}^{2/3}}{a_{N_2}^{1/3} a_{H_2}}$$

In each case the exponent of the activity, a, in the equilibrium-constant expression is the coefficient of that substance in the chemical equation.

Reaction 2 is twice reaction 1 and three times reaction 3; thus equilibrium constant K_2 is the square of K_1 and the cube of K_3. In general, multiplication of the reaction equation by any number, n, raises the corresponding equilibrium constant to the nth power.

The reverse of reaction 2 is

$$(4) \quad 2NH_3(g) \rightleftarrows N_2(g) + 3H_2(g) \qquad K_4 = \frac{a_{N_2} a_{H_2}^3}{a_{NH_3}^2} = \frac{1}{K_2}$$

Thus the equilibrium constant for the reverse reaction is the inverse of the equilibrium constant for the forward reaction.

If we add two reactions to give a third reaction, the equilibrium constant of the third reaction is the *product* of the equilibrium constants of the first two:

$$(1) \quad C(s) + O_2(g) \rightleftarrows CO_2(g) \qquad\qquad K_1 = \frac{a_{CO_2}}{a_C a_{O_2}}$$

$$(2) \quad H_2(g) + CO_2(g) \rightleftarrows H_2O(g) + CO(g) \qquad K_2 = \frac{a_{H_2O} a_{CO}}{a_{H_2} a_{CO_2}}$$

$$(3) \quad H_2(g) + C(s) + O_2(g) \rightleftarrows H_2O(g) + CO(g) \qquad K_3 = \frac{a_{H_2O} a_{CO}}{a_{H_2} a_C a_{O_2}}$$

$$K_3 = K_1 K_2$$

If $(1) + (2) = (3)$, then $K_1 K_2 = K_3$. This multiplicative property of equilibrium constants corresponds to the principle of additivity of free energies, introduced in Chapter 2 for enthalpies as Hess' law.

Standard States and Activities

We can extend the concept of activity, and therefore of free energy changes and equilibrium constants, to solids, liquids, and components of solutions, if we define the activity of any substance as the ratio of the concentration of that substance to its concentration in a chosen standard state. When calculating equilibrium constants, this standard state obviously must be the same as the standard state for which the thermodynamic data are tabulated if we are to calculate K_{eq} from such data. The standard states used for calculating the free energy values in Appendix 3 are listed in Table 17-1.

For gas reactions the activities of reactants and products are unitless quantities that are numerically equal to the partial pressures in atmospheres. It is common to use equilibrium constants calculated from the partial pressures of the reactants and products. For example,

$$K_p = \frac{p_{PCl_3} p_{Cl_2}}{p_{PCl_5}}$$

We shall use this terminology occasionally since it is so common. However,

Table 17-1

Standard States and Activities in Free Energy Tabulations and Equilibrium-Constant Calculations

Substance	Standard state	Activity, a
Gas, pure or in a mixture	1 atm partial pressure of gas in question	Numerically equal to the partial pressure: $a_j = |p_j|$
Pure liquid or solid	Pure liquid or solid	Unit activity: $a_j = 1$
Solvent in a dilute solution	Pure solvent	Activity nearly equal to 1: $a_j \simeq 1$
Solute	$1M$ solution of the solute	Numerically equal to the molar concentration of solute: $a_j = |c_j|$

remember that, in the strict sense, these "partial pressures" are really the activities of the individual gases; they are the ratios of partial pressures under experimental conditions to the partial pressures in the standard state of 1 atm. When activities are used, the equilibrium constant, K_{eq}, is a unitless number, as it must be if $\ln K_{eq}$ is to have any meaning.

A pure solid or liquid component in a reaction acts like an infinite reservoir of material, and the amount of solid or liquid does not affect the equilibrium as long as some solid or liquid is present. In the decomposition of limestone to lime and carbon dioxide,

$$CaCO_3(s) \rightleftarrows CaO(s) + CO_2(g)$$

the free energy of the reaction depends only on the partial pressure of the carbon dioxide above the solids, and not on the amount of limestone or lime that is present. The choice of the standard state for solids as the pure solid itself makes the activities unity and eliminates them from the equilibrium constant and reaction quotient:

$$\Delta G = \Delta G^0 + RT \ln \left(\frac{1 \times a_{CO_2}}{1} \right) = \Delta G^0 + RT \ln a_{CO_2}$$
$$K_{eq} = e^{-(\Delta G^0/RT)} = a_{CO_2}$$

At equilibrium, the partial pressure of CO_2 in a container with CaO and $CaCO_3$ is constant for a given temperature, exactly as the equations predict. In all our applications a solvent in a *dilute* solution can be considered an inexhaustible source of pure solvent material. Thus the activity of a solvent also is unity and does not enter into the equilibrium constant.

17-2 REACTIONS INVOLVING GASES

Let's apply the previous conclusions to reactions in which all reactants and products are gases. We shall do this in a series of examples, each of which will illustrate a new idea.

Experimental Measurement of Equilibrium Constants

Example 1

Ammonia, nitrogen, and hydrogen are at equilibrium in a steel tank at 298 K. An analysis of the contents of the tank shows the following partial pressures of the three gases:

$$p_{N_2} = 0.080 \text{ atm}$$
$$p_{H_2} = 0.050 \text{ atm}$$
$$p_{NH_3} = 2.60 \text{ atm}$$

What is the equilibrium constant, K_{eq}, for the reaction

$$N_2(g) + 3H_2(g) \rightleftharpoons 2NH_3(g) \tag{17-6}$$

Solution

The equilibrium constant is calculated to be

$$K_{eq} = \frac{a_{NH_3}^2}{a_{N_2} a_{H_2}^3} = \frac{(2.60)^2}{(0.080)(0.050)^3} = 6.8 \times 10^5$$

Example 2

The preparation of sulfur trioxide from sulfur dioxide is an important step in making sulfuric acid. A mixture of SO_2 and O_2 gases is passed slowly through a tube containing a platinum catalyst heated to 1000 K. The outflow gases are analyzed and the following partial pressures are found:

$$p_{SO_2} = 0.559 \text{ atm}$$
$$p_{O_2} = 0.101 \text{ atm}$$
$$p_{SO_3} = 0.331 \text{ atm}$$

What is the equilibrium constant, K_{eq}, for the reaction

$$2SO_2(g) + O_2(g) \rightleftharpoons 2SO_3(g) \tag{17-7}$$

Solution

The answer is

$$K_{eq} = \frac{a_{SO_3}^2}{a_{SO_2}^2 a_{O_2}} = 3.47$$

Calculation of Equilibrium Constants

Now we can calculate equilibrium constants from thermodynamic data, which we could not do in Chapter 4. Given either the equilibrium constant or the standard free energy of a reaction, we can calculate the other quantity.

Example 3

Calculate the equilibrium constant for reaction 17-6 from standard free energies of formation.

Solution

$$\Delta G^0 = 2(-16.64) - (0.0) - 3(0.0) = -33.28 \text{ kJ}$$

$$K_{eq} = e^{-(\Delta G^0/RT)} = 10^{-(-33.28/5.706)} = 10^{+5.832} = 6.8 \times 10^5$$

Don't attach undue importance to the fact that this is exactly the same value that we obtained previously from partial pressures. Equilibrium constants rarely are known to better than 5% accuracy.

Example 4

Calculate the dissociation constant for acetic acid from the free energy tables in Appendix 3.

Solution

The reaction is

$$CH_3COOH(aq) \rightarrow H^+(aq) + CH_3COO^-(aq)$$

and the free energy of dissociation is

$$\Delta G^0 = 0.0 + (-372.5) - (-399.6) = +27.1 \text{ kJ}$$

The dissociation constant is found from

$$K = e^{-(\Delta G^0/RT)} = 10^{-(27.1/5.706)} = 10^{-4.75}$$
$$pK = 4.75 \quad \text{and} \quad K = 1.76 \times 10^{-5}$$

You can check this result against Table 5-3. Notice that if the dissociation reaction had been written

$$CH_3COOH(aq) + H_2O \rightarrow CH_3COO^-(aq) + H_3O^+(aq)$$

the result using data in Appendix 3 would have been the same.

Example 5

From the equilibrium constant that you calculated previously for the sulfur trioxide reaction, find the standard free energy of reaction 17-7 at 1000 K.

Solution $\Delta G^0_{1000} = -10.3$ kJ, or -5.15 kJ mole^{-1} of SO_3

The Partial Pressure of One Component

We can calculate the partial pressure of one component of a system at equilibrium if we know K_{eq} and the partial pressures of the other components of the system.

Example 6

In another experiment with the ammonia reaction in a steel tank, the partial pressures of ammonia and hydrogen at 298 K were found to be

$$p_{NH_3} = 1.53 \text{ atm}$$
$$p_{H_2} = 0.50 \text{ atm}$$

No nitrogen could be detected. What must the partial pressure of N_2 have been if the contents of the tank were at equilibrium?

Solution   Let y be the unknown partial pressure of nitrogen. Then

$$K_{eq} = \frac{(1.53)^2}{y(0.50)^3} = 6.8 \times 10^5$$

$$y = \frac{(1.53)^2}{0.125 \times 6.8 \times 10^5} = 2.8 \times 10^{-5} \text{ atm}$$

Because we have calculated the value of the equilibrium constant for this reaction from other experiments with more nearly equal amounts of reactants and products, it is possible to calculate how much nitrogen is present even if we cannot measure it.

Example 7

In the SO_3 equilibrium at 1000 K (equation 17-7), what would the partial pressure of oxygen gas have to be to give us equal amounts of SO_2 and SO_3?

Solution $p_{O_2} = 0.288$ atm

Alteration of Stoichiometry

Example 8

What is the equilibrium constant for the reaction

$$2NH_3(g) \rightleftarrows N_2(g) + 3H_2(g) \qquad (17\text{-}8)$$

Solution This reaction is the reverse of reaction 17-6, so the equilibrium constant is the reciprocal of the forward reaction:

$$K_{eq} = \frac{a_{N_2}a_{H_2}^3}{a_{NH_3}^2} = \frac{1}{6.8 \times 10^5} = 1.5 \times 10^{-6}$$

Example 9

What is the equilibrium constant for the reaction

$$NH_3(g) = \tfrac{1}{2}N_2(g) + \tfrac{3}{2}H_2(g) \qquad (17\text{-}9)$$

Solution $K_{eq} = 1.2 \times 10^{-3}$

Extent of Reaction

Often it is useful to be able to calculate the extent of reaction at equilibrium, expressed as the fraction or percentage of the pure starting material that has reacted. For example, we can express the progress of the decomposition of ammonia into nitrogen and hydrogen in terms of a *percent dissociation* of pure ammonia. Pure ammonia at a total pressure of 1 atm will dissociate spontaneously, to a small extent, to H_2 and N_2. If the pressure is kept constant at 1 atm and the temperature is 298 K, what fraction of the ammonia is dissociated?

The reaction is shown in equation 17-8 and at the top of Table 17-2. Since the equilibrium-constant expression is in terms of partial pressures, our first task is to express concentrations in partial pressures also. We then can set up the equilibrium-constant expression, use the known value of the equilibrium constant, and solve for the percent dissociation.

It is helpful to make a table such as Table 17-2. Assume that pure NH_3 was present at the beginning of the experiment and that no N_2 or H_2 was present. This description need not correspond to any real experimental situation; it is merely a framework for calculating concentrations at equilibrium. Since 2 moles of ammonia are involved in the reaction as written, assume that $2n$ moles of ammonia are present at the start. Now assume that the system comes to equilibrium at some later time when a

Table 17-2

Dissociation of Ammonia at 298 K

	$2NH_3(g) \rightleftarrows N_2(g) + 3H_2(g)$			Total
Start:	$2n$	0	0	$2n$
Equilibrium:	$2n(1 - \alpha)$	$n\alpha$	$3n\alpha$	$2n(1 + \alpha)$
Mole fraction:	$\dfrac{(1 - \alpha)}{1 + \alpha}$	$\dfrac{\alpha}{2(1 + \alpha)}$	$\dfrac{3\alpha}{2(1 + \alpha)}$	1
Partial pressure:	$\dfrac{(1 - \alpha)P}{1 + \alpha}$	$\dfrac{\alpha P}{2(1 + \alpha)}$	$\dfrac{3\alpha P}{2(1 + \alpha)}$	P

$$K_{eq} = \frac{a_{N_2} a_{H_2}^3}{a_{NH_3}^2} = \frac{\alpha P}{2(1 + \alpha)} \cdot \frac{27\alpha^3 P^3}{8(1 + \alpha)^3} \cdot \frac{(1 + \alpha)^2}{(1 - \alpha)^2 P^2} = \frac{27}{16} \frac{\alpha^4 P^2}{(1 - \alpha^2)^2}$$

$$\Delta G^0 = (0.0) + 3(0.0) - 2(-16.63) = +33.27 \text{ kJ}$$

$$K_{eq} = 10^{-(33.27/5.831)} = 10^{-5.832} = 1.5 \times 10^{-6}$$

$$K_{eq} = \frac{27}{16} \frac{\alpha^4 P^2}{(1 - \alpha^2)^2} = 1.5 \times 10^{-6}$$

$$\frac{\alpha^2 P}{1 - \alpha^2} = 0.94 \times 10^{-3}$$

fraction, α, of the ammonia has dissociated. Then, of the original $2n$ moles of ammonia, $2n\alpha$ moles will have reacted and $2n(1 - \alpha)$ moles will be left.

For every 2 moles of ammonia that decompose, 1 mole of nitrogen and 3 moles of hydrogen are formed. At equilibrium, the $2n\alpha$ moles of dissociating ammonia produce $n\alpha$ moles of nitrogen and $3n\alpha$ moles of hydrogen, as shown on the line of Table 17-2 marked "Equilibrium." The total number of moles of the three types of molecules has increased as a result of the dissociation. There now are $2n(1 - \alpha) + n\alpha + 3n\alpha = 2n(1 + \alpha)$ moles of gas, whereas prior to dissociation there were only $2n$ moles. The total number of moles is recorded at the right of the table.

The mole fraction of each component in the reaction mixture is found by dividing the number of moles of each component by the total number of moles. At this stage, the advantage of using *fractional* dissociation, α, is obvious, because the number of moles, n, drops out of the problem. In the third line of the table, headed "Mole fraction," the mole fraction of each component at equilibrium is given in terms of the degree of dissociation of ammonia.

In the next line of the table, each mole fraction is multiplied by the total pressure to yield the partial pressure of that component. These partial pressures are then substituted into the equilibrium-constant expression, assuming the activities of the three gases to be numerically equal to their partial pressures. The result, after simplifying as much as possible, is an expression for the equilibrium constant in terms of the degree of dissociation:

$$K_{eq} = \frac{27\alpha^4 P^2}{16(1 - \alpha^2)^2}$$

From the free energy change for the reaction, $+33.28$ kJ, we can use equation 17-4 to calculate the numerical value of the equilibrium constant:

$$K_{eq} = 1.5 \times 10^{-6}$$

Using this value, we have an expression,

$$\frac{27\alpha^4 P^2}{16(1 - \alpha^2)^2} = 1.5 \times 10^{-6}$$

which after combining numerical constants and taking the square root of both sides of the equation becomes

$$\frac{\alpha^2 P}{1 - \alpha^2} = 0.94 \times 10^{-3}$$

This expression can be solved exactly for α^2:

$$\alpha^2 = \frac{0.00094}{P + 0.00094}$$

Since $P = 1$ atm, we can neglect the 0.00094 term in the denominator and solve for $\alpha = 0.0307$. The answer to our original question is that at 298 K and 1 atm pressure, ammonia is approximately 3% dissociated.

Example 10

A molecule of N_2O_4 gas dissociates spontaneously into two molecules of NO_2. Derive an expression for the degree of dissociation, α, as a function of the total pressure on the system at 298 K. What fraction of the N_2O_4 will have dissociated at 1 atm pressure?

Solution

The derivation is shown in Table 17-3, to which you can refer if you have trouble. The expression for K_{eq} in the table again can be solved exactly for α^2:

$$\alpha^2 = \frac{K_{eq}}{4P + K_{eq}} = \frac{1}{(4P/K_{eq}) + 1} \tag{17-10}$$

and at $P = 1$ atm and $K_{eq} = 0.114$, $\alpha = 0.167$. At 1 atm and 298 K, one N_2O_4 molecule in six is dissociated. Note that in this problem, K_{eq} was large enough so that it could not be neglected in the denominator in comparison with $4P$ when $P = 1$ atm.

Table 17-3

Dissociation of N_2O_4 at 298 K

	$N_2O_4(g)$ $\rightleftarrows$	$2NO_2(g)$	Total
Start:	n	0	n
Equilibrium:	$n(1 - \alpha)$	$2n\alpha$	$n(1 + \alpha)$
Mole fraction:	$\dfrac{1 - \alpha}{1 + \alpha}$	$\dfrac{2\alpha}{1 + \alpha}$	1
Partial pressure:	$\dfrac{1 - \alpha}{1 + \alpha} \cdot P$	$\dfrac{2\alpha}{1 + \alpha} \cdot P$	P

$$K_{eq} = \frac{a_{NO_2}^2}{a_{N_2O_4}} = \frac{4\alpha^2 P^2}{(1 + \alpha)^2} \cdot \frac{(1 + \alpha)}{(1 - \alpha)P} = \frac{4\alpha^2 P}{1 - \alpha^2}$$

$$\Delta G^0 = 2(51.84 \text{ kJ}) - (98.29 \text{ kJ}) = +5.39 \text{ kJ}$$

$$K_{eq} = e^{-(5.39/RT)} = 10^{-(5.39/5.706)} = 10^{-0.945} = 0.114$$

$$\frac{4\alpha^2 P}{1 - \alpha^2} = 0.114 \quad \text{or} \quad \frac{\alpha^2 P}{1 - \alpha^2} = 0.0285$$

What will be the effect on the dissociation of N_2O_4 when we increase the pressure? Since reassociation decreases the total number of moles, Le Chatelier's principle predicts that an increase in pressure will favor re-association. In Figure 17-2a the degree of dissociation, α, obtained from equation 17-10, is plotted against the total pressure. You can see that Le Chatelier's principle is borne out. Above 1000 atm pressure the gas mixture is nearly all N_2O_4; below 0.001 atm, it is almost entirely NO_2. In Figure 17-2a, the pressure was plotted as a logarithmic function, that is, in increasing powers of 10. The change in α with changing pressure is displayed clearly. In contrast, in Figure 17-2b, which has pressure plotted on a linear scale, the curve flattens over half the α range on the vertical axis. For any α above 0.5, Figure 17-2b gives no clear description of what is happening. Any other choice of a linear plotting range for pressure (0 to 150 atm, 0 to 1.5 atm, or 0 to 0.15 atm) would have similar drawbacks. We shall find that logarithmic plots are used widely in equilibrium problems.

These two examples, the ammonia reaction and the dissociation of N_2O_4, produced the same mathematical expression for the degree of dissociation. With ammonia, there was 3% dissociation; with N_2O_4, there was 16.7% dissociation. These relative values could have been predicted from the standard free energy changes for the reactions: $+16.64 \text{ kJ mole}^{-1}$ for NH_3 and $+5.39 \text{ kJ mole}^{-1}$ for N_2O_4. The standard free energy change for dissociation is higher for NH_3 than for N_2O_4; thus the drive toward dissociation is less and the equilibrium point is closer to the side of no dissociation.

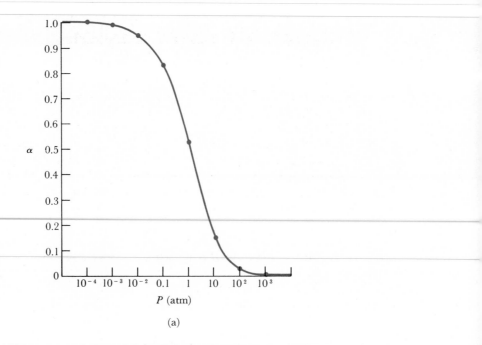

(a)

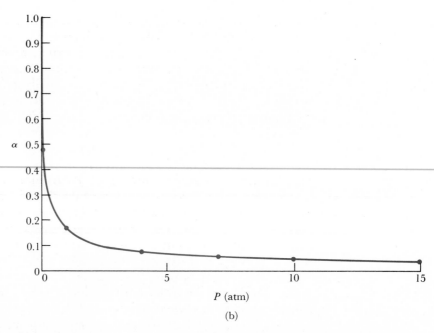

(b)

Figure 17-2   Plots of the degree of dissociation of N_2O_4, α, as a function of pressure at 298 K. The function plotted is equation 17-10. (a) Plot of α against $\log_{10} P$ showing the full detail across the entire range of α. (b) A less informative plot of α against P that obscures the behavior of the curve above $\alpha = 0.5$.

17-3 LE CHATELIER'S PRINCIPLE

Le Chatelier's principle states that when a stress is applied to a system at equilibrium the equilibrium conditions shift in such a way as to relieve the stress. In the two dissociation reactions we have examined so far (those of N_2O_4 and NH_3) the pressure dependence has appeared in the numerator of the equilibrium-constant expression. Since K_{eq} is independent of pressure, α must decrease as P increases if the quantity on the right side of the equilibrium-constant expansion is to remain unchanged. Le Chatelier's principle predicts that at higher pressures the equilibrium conditions will shift in the direction that produces the smaller number of moles of gas, which is in agreement with the equilibrium-constant derivations.

What would this principle predict about equilibrium at higher pressures for the water-gas reaction,

$$H_2(g) + CO_2(g) \rightleftarrows H_2O(g) + CO(g) \tag{17-11}$$

If you set up a table such as Table 17-2 for this reaction, and let α be the fraction of H_2 that has reacted, you will find that the equilibrium constant is

$$K_{eq} = \frac{a_{H_2O}a_{CO}}{a_{H_2}a_{CO_2}} = \frac{\alpha^2}{(1 - \alpha)^2} \tag{17-12}$$

and that pressure does not appear in the equilibrium-constant expression. This is because the same number of moles of gas are found on each side of the equation, and all of the pressure terms cancel. For the same reason, Le Chatelier's principle predicts that the water-gas equilibrium will be insensitive to changes in pressure.

The Effect of Temperature

What does Le Chatelier's principle predict about the effect of temperature on the equilibrium constant? The temperature of a reacting system is increased by adding heat. The stress of the added heat can be relieved if the equilibrium conditions shift in the direction that absorbs heat. If a reaction is endothermic, its equilibrium constant will increase with temperature; if it is exothermic, its equilibrium constant will decrease.

The SO_3 dissociation,

$$2SO_3(g) \rightleftarrows 2SO_2(g) + O_2(g)$$

is highly endothermic. At 298 K,

$$\Delta G^0 = 2(-300.4) - 2(-370.4) = +140.0 \text{ kJ per 2 moles } SO_3$$
$$\Delta H^0 = 2(-296.9) - 2(-395.2) = +196.6 \text{ kJ per 2 moles } SO_3$$
$$\Delta S^0 = 2(248.5) + (205.0) - 2(256.2) = +189.6 \text{ J K}^{-1} \text{ per 2 moles}$$
$$SO_3 \text{ dissociated}$$

Since so much heat is absorbed in the dissociation, this dissociation is strongly favored by higher temperatures. Thus, a difference in 1100 K results in a change in K_{eq} of 27 orders of magnitude (Table 17-4).

17-4 THE ANATOMY OF A REACTION

Le Chatelier's principle gives an indication of how a reaction will proceed; however, it does not explain why, except in the most intuitive way. Why does the position of equilibrium change with temperature? Why does the driving impetus of the SO_3 dissociation reaction increase so sharply with temperature? To answer these questions we must look at the behavior of the free energy, enthalpy, and entropy of reaction as the temperature changes.

At a given temperature the standard free energy, enthalpy, and entropy of reaction are related by the expression

$$\Delta G^0 = \Delta H^0 - T\,\Delta S^0 \tag{17-13}$$

The standard free energy changes for the SO_3 dissociation reaction at various temperatures, as calculated from experimental dissociation constants, are listed in Table 17-4. As the temperature is increased, the standard free energy change for the reaction becomes more negative, the equilibrium constant becomes larger, and the reaction must proceed farther to the right before equilibrium is attained. The heats and entropies of reaction can both be deduced from these data. From equation 17-13 and the use of calculus,

Table 17-4

Variation of K_{eq} with Temperature for the Reaction
$2SO_3(g) \rightleftarrows 2SO_2(g) + O_2(g)$

Temperature (K)	ΔG^0 (kJ per 2 moles SO_3)	K_{eq}
298	140.0	2.82×10^{-25}
400	120.7	1.78×10^{-16}
500	101.5	2.51×10^{-11}
600	82.1	1.94×10^{-8}
700	63.6	1.82×10^{-5}
800	44.4	1.29×10^{-3}
900	27.7	0.0248
1000	11.1	0.264
1100	−5.9	1.89
1200	−23.0	10.0
1300	−40.2	40.8
1400	−57.0	132

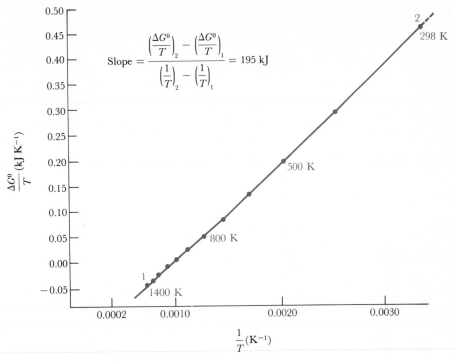

Figure 17-3

A Gibbs–Helmholtz plot for the dissociation of SO_3 to SO_2 and O_2. The slope at any point of the curve (which here appears to be nearly a straight line) gives the heat of the reaction at that point. The enthalpy of the SO_3 dissociation does not vary significantly with temperature from 298 K to 1400 K. The ΔG^0 data are from Table 17-4.

we can derive a relationship known as the Gibbs–Helmholtz equation. This equation need not concern us except for its prediction that if we plot $\Delta G^0/T$ against $1/T$ the slope of the curve at any point is ΔH^0 at that temperature.

A plot of the data in Table 17-4 appears in Figure 17-3. The curve is nearly a straight line, which indicates that the standard heat of the SO_3 dissociation reaction does not change very much between 298 K and 1400 K. The average slope over this entire temperature range yields an average enthalpy of reaction of $+195$ kJ, whereas the measured value at one extreme end of the range, at 298 K, is $+196.6$ kJ. To a close approximation, we can consider the heat of the reaction to be constant for all temperatures.

The heat of reaction, ΔH^0, and the free energy change, ΔG^0, are plotted against temperature in Figure 17-4. The difference between them, $\Delta H^0 - \Delta G^0$, is $T\,\Delta S^0$. If both ΔH^0 and ΔG^0 are approximated by straight lines, then $T\,\Delta S^0$ is proportional to T; hence ΔS^0 also is approximately inde-

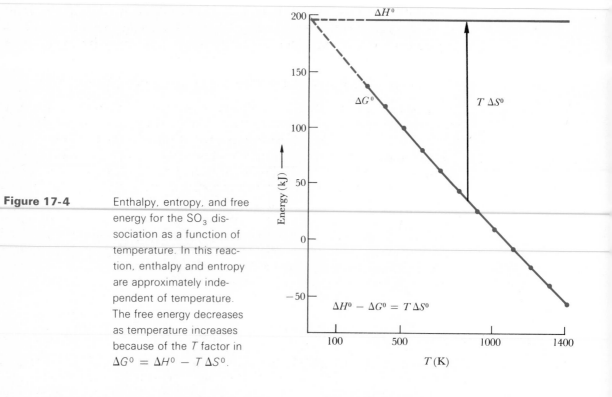

Figure 17-4   Enthalpy, entropy, and free energy for the SO_3 dissociation as a function of temperature. In this reaction, enthalpy and entropy are approximately independent of temperature. The free energy decreases as temperature increases because of the T factor in $\Delta G^0 = \Delta H^0 - T\,\Delta S^0$.

pendent of temperature. If we extrapolate the ΔH^0 and ΔG^0 lines to absolute zero, we see that they intersect. At 0 K, $T\,\Delta S^0 = 0$, and $\Delta H^0 = \Delta G^0$.

Now we can explain what happens in this reaction as the temperature changes. When two molecules of SO_3 dissociate,

$$2SO_3(g) \rightleftarrows 2SO_2(g) + O_2(g) \tag{17-14}$$

two S—O bonds are broken and one O—O bond is made. The enthalpy required to accomplish this, 195 kJ for 2 moles of SO_3, is so large that the minor fluctuations arising with temperature can be ignored, and ΔH^0 is approximately constant. This enthalpy factor by itself would keep all of the SO_3 undissociated. Berthelot and Thomsen would have been satisfied with this conclusion (Section 16-5).

However, there is a second factor to consider, entropy. Two molecules of SO_2 and one molecule of O_2 are more disordered than two molecules of SO_3. The relative molar entropies are exactly what you might predict for the three molecules based on the complexity of each molecule. From Appendix 3,

Molecule:	O_2	SO_2	SO_3
Molar entropy, S^0 ($J\,K^{-1}\,mole^{-1}$):	205.03	248.5	256.2

However, the entropies per mole are numerically so similar for the three kinds of molecules that the deciding factor is the change in the number of molecules during the reaction. Two molecules of reactant (SO_3) yield three molecules of product ($2SO_2$, O_2), and the overall entropy change is

$$\Delta S^0 = 2(248.5) + (205.03) - 2(256.2) = +189.6 \, J \, K^{-1}$$

Entropy, like enthalpy, is not very sensitive to temperature. The disorder produced per unit of reaction at 298 K is approximately the same as that produced at 1400 K. Nevertheless, the effect of this entropy change is more marked at high temperatures because the entropy term is multiplied by absolute temperature, T. The higher the temperature, the more of an effect a given increase in disorder has on the reaction. The impetus toward reaction, or the standard free energy, is the combination of the heat effect and the entropy effect. As you can see from Figure 17-4, enthalpy continues to oppose the dissociation of SO_3 to approximately the same extent at all temperatures. But entropy provides a steadily increasing driving force in favor of dissociation. Thus, at 298 K, the dissociation constant is only 2.82×10^{-25}, whereas at 1400 K it is 132.

Example 11

What would be the value of the dissociation constant, K_{eq}, if the reactants and products in the SO_3 dissociation were of equal entropy? This also would be the dissociation constant at absolute zero if the dashed-line extrapolations in Figure 17-4 were valid. (They nearly are.)

Solution $K_{eq} = 6 \times 10^{-35}$

Thus far our conclusions have been derived from a special case, the dissociation of SO_3. Yet it is generally true that at low temperatures the enthalpy or heat is more important in determining which way a chemical reaction will go, and at high temperatures the entropy or disorder is more important.

Summary

The **free energy change** of a reaction, ΔG, is related to the ratio of concentrations of products to reactants known as the reaction quotient, Q, by the expression

$$\Delta G = \Delta G^0 + RT \ln Q$$

The term ΔG^0 is the free energy change of the reaction when all substances are in their standard states, so that $Q = 1$.

At equilibrium, $\Delta G = 0$, and therefore

$$\Delta G^0 = -RT \ln K_{eq}$$

This important expression relates the equilibrium constant of a reaction to its standard free energy change. The equilibrium constant, K_{eq}, is just the reaction quotient under conditions of equilibrium.

The reaction quotient and equilibrium-constant expressions have the form introduced in Chapters 4 and 5—the concentrations of products divided by concentrations of reactants, each separately raised to a power given by the coefficient multiplying that reactant or product in the balanced chemical equation. But now it is more convenient to generalize from any one concentration unit to the **activity**, defined as the ratio of the concentration under given conditions to the concentration in the standard state. Pure solids or liquids, or other components already in their standard states, therefore have activities of 1, and can be neglected in equilibrium-constant calculations.

For gases, the activity is a unitless number whose numerical value is equal to the partial pressure in atmospheres, so it is easiest to work with partial pressures in setting up an equilibrium-constant expression. Beginning with a hypothetical state of pure reactants (no reaction) or pure products (complete reaction), we express the actual degree of reaction in terms of a fractional variable, α. Then, in a table, we can calculate mole fractions of reactants and products in terms of this degree-of-reaction variable, α, and continue to set up expressions for partial pressures and finally for K_{eq}, all in terms of α. This last expression can then be solved for α, using the known value of the equilibrium constant.

The framework of gas-phase equilibrium calculations that we have just set up permits us to verify Le Chatelier's principle quantitatively. An increase in temperature favors the heat-absorbing direction of a reaction. An increase in total pressure will not change K_{eq}, nor the ratio of partial pressures, but it will change the mole fractions of the various components if reactants and products contain different numbers of moles of gases. Hence α will change even though K_{eq} does not. Equilibrium will shift so as to decrease the total number of moles of gas.

Self-Study Questions

1. How is the equilibrium constant for a reaction related to thermodynamic quantities? How does this relationship prove that K_{eq} is independent of the concentrations of reactants and products?

2. What can you say about the standard free energy change for a reaction if the reaction is spontaneous when its reactants and products are in their standard states? What can you say about the equilibrium constant for the reaction, and how does it compare with the reaction quotient?

3. If the standard free energy change is negative, must the reaction be spontaneous under all experimental conditions?

4. If the heat that can be obtained by burning gasoline is more or less constant and independent of burning conditions, why does the useful work that can be obtained per gram of fuel decrease steadily as equilibrium is approached? (Look again at the hydrostatic analogue in Figure 17-1). Why is this falloff of useful work per gram of little practical concern in an automobile engine?

5. In Chapter 4, the equilibrium constants were expressed in terms of partial pressures (gases) or moles per liter (gases or solutes). In this chapter, activities have been used instead. What is the relationship between the two usages, and why are activities useful? Once we have connected K_{eq} with ΔG^0, why is it illogical for us to talk about a K_{eq} that has units or dimensions?

6. How does the use of activities in equilibrium expressions make the treatment of condensed phases simpler?

7. Why are Tables 17-2 and 17-3, which help in setting up the equilibrium-constant expressions, more complex than tables with a similar goal in the text of Section 4-4? Does this give you a clue as to why we moved as rapidly as possible from gases to solutions in Chapters 4 and 5?

8. If hydrogen and iodine react at 300°C, the free energy change for the reaction depends on the concentrations or partial pressures of H_2, I_2, and HI. In contrast, if the reaction is run at room temperature, only the concentrations of H_2 and HI have an effect on the free energy change for the reaction. Why is this so?

9. What effect will an increase in pressure have on the PCl_5 reaction of equation 17-1?

10. What effect will an increase in temperature have on the PCl_5 reaction in equation 17-1?

11. What do you think would happen to the equilibrium of equation 17-1 if you introduced (a) nitrogen gas at constant pressure, (b) nitrogen gas at constant volume, or (c) chlorine gas at constant pressure?

12. How do bond breaking and the creating of disorder influence the equilibrium conditions for a chemical reaction? If bond energies were the only important factor, what would be the equilibrium constant for the dissociation of hydrogen gas molecules into atoms? If entropy were the only operating factor, what would be the equilibrium constant for the hydrogen dissociation? Using your answers to these questions, and the relationship between G, H, and S, explain why dissociation of hydrogen gas should be more pronounced at higher temperatures.

Problems

Calculation of Δ

1. A system consisting of nitrogen, hydrogen, and ammonia—all gases—is allowed to come to equilibrium at a total pressure of 5 atm. The partial pressures of the three gases are measured to be $p_{N_2} = 1$ atm, $p_{H_2} = 2$ atm, and $p_{NH_3} = 2$ atm. What is K_{eq} for the reaction

$$N_2(g) + 3H_2(g) \rightleftarrows 2NH_3(g)$$

What is K_{eq} for the reaction

$$NH_3(g) \rightleftarrows \tfrac{1}{2}N_2(g) + 1\tfrac{1}{2}H_2(g)$$

2. In an experiment at an elevated temperature, a mixture of SO_2, O_2, and SO_3 is allowed to equilibrate by the reaction

$$SO_2(g) + \tfrac{1}{2}O_2(g) \rightleftarrows SO_3(g)$$

The partial pressures are then measured as $p_{SO_2} = 0.0526$ atm, $p_{O_2} = 0.0263$ atm, and $p_{SO_3} = 1.053$ atm. Calculate the equilibrium constant, K_{eq}, for the reaction as written.

3. For the reaction of nitric acid with hydrogen sulfide at $25\,°C$,

$$2H^+(aq) + 2NO_3^-(aq) + 3H_2S(aq) \rightleftarrows$$
$$2NO(g) + 4H_2O(l) + 3S(s)$$

K_{eq} is 1×10^{81}. Given that the dissociation constant of H_2S is 1×10^{-19} and the solubility product of CdS is 8×10^{-27}, calculate the equilibrium constant of the reaction

$$3CdS(s) + 8H^+(aq) + 2NO_3^-(aq) \rightleftarrows$$
$$2NO(g) + 4H_2O(l)$$
$$+ 3Cd^{2+}(aq) + 3S(s)$$

Do you think CdS will dissolve in aqueous nitric acid? Calculate ΔG for the reaction.

4. Phosphonium chloride, PH_4Cl, which is unstable even at low temperatures, dis-

sociates into PH_3 and HCl. If 1 mole of PH_4Cl partially decomposes in a 5-liter flask, determine the equilibrium constant for the reaction

$$PH_4Cl(s) \rightleftarrows PH_3(g) + HCl(g)$$

if the partial pressure of PH_3 is 5.0 atm.

Form of K_{eq}

5. Steam reacts with iron at $500\,°C$ to produce hydrogen gas and Fe_3O_4. Write the equilibrium constant for the reaction in terms of quantities (activities) of reactants and products present.

Le Chatelier's principle

6. The density of ice at 273 K is 0.917 g cm^{-3}, and the density of liquid water is 1.00 g cm^{-3} at the same temperature. What will be the effect of a pressure increase on the freezing point of water?

7. Is the reaction

$$2NO(g) + O_2(g) \rightleftarrows 2NO_2(g)$$

endothermic or exothermic when run under standard conditions? If the temperature increases, will K_{eq} increase, decrease, or remain unchanged? If the pressure is increased, will K_{eq} increase, decrease, or remain unchanged? If a catalyst is added, will K_{eq} increase, decrease, or remain unchanged?

8. At $25\,°C$ and 20 atm, the reaction

$$N_2(g) + 3H_2(g) \rightleftarrows 2NH_3(g)$$

has a ΔH of -92.5 kJ. If the temperature is increased to $300\,°C$ while the pressure is held at 20 atm, will more or less ammonia be present at equilibrium? If the pressure is increased to 30 atm while the temperature remains at 25°C, will more or less ammonia be present than in the initial conditions? If half the ammonia is removed and the system is allowed to come to equilibrium again, will the amount of nitrogen gas present increase or decrease? What will be the effect on the original equilibrium mixture if a catalyst for ammonia synthesis is added?

9. At 298 K, what is the standard enthalpy change of the reaction

$$2NO_2(g) \rightleftarrows N_2O_4(g)$$

Calculate the equilibrium constant for the reaction. How is the extent of conversion of NO_2 to N_2O_4 affected by increasing the temperature? By increasing the pressure? By increasing the volume at constant temperature? Solid Na_2O reacts with NO_2 to produce $NaNO_3$, but it will not react with N_2O_4. What is the effect on the extent of conversion of NO_2 to N_2O_4 if Na_2O is added? What is the effect on K_{eq}?

10. At a certain temperature, T, and a total pressure of 2.00 atm, phosphorus pentachloride is 75.0% dissociated:

$$PCl_5(g) \rightleftarrows PCl_3(g) + Cl_2(g)$$

What is K_{eq} at this temperature? What is K_{eq} at 298 K? Do you think that temperature T is greater or less than 298 K? On what evidence do you base your decision?

Calculation of K_{eq} $G°$

11. When 1 mole of gaseous HI is sealed in a 1-liter flask at 225°C, it decomposes to form 0.182 mole each of hydrogen and iodine:

$$2HI(g) \rightleftarrows H_2(g) + I_2(g)$$

What is the value of the equilibrium constant K_{eq} at 225°C? Calculate the standard free energy change, $\Delta G°$, for the reaction at 225°C. From the data in Appendix 3, calculate the standard free energy change and the equilibrium constant at 25°C. Why the difference in values between 25°C and 225°C?

12. Metacresol (which we shall represent by HCre) is a weak organic acid with $K_{eq} = 1.0 \times 10^{-10}$. Write the equilibrium-constant expression for this process. What is the concentration of the Cre$^-$ ion in a 1.00M solution of metacresol in water? What is the pH of this solution? What is the standard free energy change of the metacresol dissociation reaction in aqueous solution?

13. Hydrazine is a weak base that dissociates in water according to the equation

$$N_2H_4 + H_2O \rightleftarrows N_2H_5^+ + OH^-$$

The equilibrium constant for this dissociation at 25°C is 2.0×10^{-6} and the standard free energy of undissociated hydrazine in aqueous solution is 127.9 kJ mole^{-1}. Recalling that the hydrated H$^+$ ion is assigned (by convention) a standard free energy of 0.00 kJ mole^{-1}, calculate the standard free energy of the hydrazinium ion.

Ionic reactions

14. From data in Appendix 3, calculate the standard free energy of the reaction

$$CH_3COOH(aq) \rightleftarrows H^+(aq) + CH_3COO^-(aq)$$

Use this value to calculate K_{eq} for the reaction. Compare your value with that in Table 5-3.

15. Use the data in Appendix 3 to calculate the dissociation constant for formic

acid in aqueous solution. Compare your result with the value in Table 5-3. What is the pH of a 0.075M solution of formic acid?

Gibbs—Helmholtz plot

16. The solubility of potassium chloride in water is 347 g liter^{-1} at 20°C, and 802 g liter^{-1} at 100°C. Calculate K_{sp} for KCl at each temperature. Using a Gibbs–Helmholtz plot such as Figure 17-3, calculate the heat of solution of KCl. Is the dissolving process exothermic or endothermic?

Perturbation of equilibrium

17. At 25°C, ΔH^0 is 92.6 kJ for the reaction

$$PCl_5(g) \rightleftarrows PCl_3(g) + Cl_2(g)$$

An equilibrium mixture of $PCl_5(g)$, $PCl_3(g)$, and $Cl_2(g)$ is subjected to various operations (a–e). For each operation determine whether the position of equilibrium shifts to the right, shifts to the left, or remains the same. Also determine whether the equilibrium constant increases, decreases, or remains the same.
a) A catalyst is added.
b) The volume of the container is decreased.
c) $Cl_2(g)$ is added.
d) The temperature is increased.
e) An inert ideal gas is added to the container.

Degree of dissociation

18. Derive an expression for the degree of dissociation, α, of $PCl_5(g)$ as a function of the total pressure of the system. What fraction of $PCl_5(g)$ will have dissociated at 1 atm and 25°C?

19. If 0.50 g of N_2O_4 is allowed to evaporate into a 2-liter flask at 25°C, the reaction $N_2O_4(g) \rightleftarrows 2NO_2(g)$ takes place. Calculate the partial pressure exerted by N_2O_4; K_{eq} for the reaction at 25°C is 0.114.

Suggested Reading

A. J. Bard, *Chemical Equilibrium,* Harper and Row, New York, 1966.

R. E. Dickerson, *Molecular Thermodynamics,* W. A. Benjamin, Menlo Park, Calif., 1969. Chapter 5, "Thermodynamics of Phase Changes and Chemical Reactions," has a more detailed treatment of thermodynamics and equilibrium.

W. Kauzmann, *Thermodynamics and Statistics, with Applications to Gases,* W. A. Benjamin, Menlo Park, Calif., 1967. Chapter 5 has a good treatment of thermodynamics and equilibrium constants.

I. M. Klotz and R. M. Rosenberg, *Introduction to Chemical Thermodynamics,* W. A. Benjamin, Menlo Park, Calif., 1972, 2nd ed.

B. H. Mahan, *Elementary Chemical Thermodynamics,* W. A. Benjamin, Menlo Park, Calif., 1964. Chapter 3, "The Second Law," has a simpler treatment of thermodynamics and equilibrium than Dickerson gives, but no applications to real problems.

L. Nash, *ChemThermo: A Statistical Approach to Classical Chemical Thermodynamics,* Addison-Wesley, Reading, Mass., 1972.

M. J. Sienko, *Chemistry Problems,* W. A. Benjamin, Menlo Park, Calif., 1972, 2nd ed. Contains several hundred problems on equilibrium, with answers. Emphasis is on practical problem solving.

18

Equilibria Involving Liquids and Solids

A liquid is a strange substance. The principles that govern the behavior of solids and gases are much better understood than those that govern the behavior of liquids. The marvel is not that liquids behave as they do, but that they exist at all. In theory, it might seem more reasonable for a crystalline solid to melt to a fluid having molecules initially touching one another, and for further heating to cause the molecules to move faster and farther apart until something like a gas is produced, without any sharp transition in fluid properties along the way. This theoretical possibility is diagrammed in Figure 18-1 in a plot of volume per mole against temperature.

Real substances actually do behave this way above what is called their **critical pressure**, P_c, which is 218 atm for H_2O and 72 atm for CO_2. But at lower pressures their behavior is more like that shown in Figure 18-2. The molar volume usually increases slightly upon melting (point e to point d), and then makes a sudden jump at the boiling point, T_b, where the liquid changes to a gas (point b to point c). The molecules that were in contact in the liquid are pulled away from one another, and thereafter touch only when they collide and rebound.

If you scan the thermodynamic tables in Appendix 3, you'll notice that the designation for liquid (l) rarely appears—illustrating how unusual liquids really are. Most substances at 298 K are either crystalline solids or gases. This relative scarcity of liquids in nature is obscured by the overwhelming abundance of one particular liquid, H_2O, which is doubly nontypical because it expands rather than contracting when it solidifies.

This chapter will be a brief introduction to the properties of solids, liquids, and gases, the thermodynamics of transitions from one phase to another, and the conditions for equilibrium between phases.

18-1 MELTING, EVAPORATION, AND SUBLIMATION

Figure 18-1 describes the behavior of substances at pressures above the critical pressure, P_c. The crystalline solid expands slightly upon heating below its melting temperature, T_m, and then melts to a fluid. Since molecules are in contact both before and after melting, melting takes place with a relatively small increase in molar volume. Above T_m the molar volume increases at first slowly and then more rapidly, until the substance approaches the $P\overline{V} = RT$ behavior of an ideal gas. (The bar over the V indicates a volume per mole, or V/n.)

Below the critical pressure we see quite different behavior (Figure 18-2). As before, the crystalline solid may expand slightly with heating. At its melting point its molecules slide past one another, destroying the geometrical order that had been present in the crystal, but they remain in contact. This is a definition of a liquid: a fluid whose molecules are free to move past one another while remaining in sliding contact. Most substances increase slightly in molar volume upon melting, since the molecules of the liquid are not packed as efficiently or as tightly as they were in the solid. Water is unusual because its molar volume decreases when ice melts. The H_2O molecules in ice are held apart by hydrogen bonds in an open cage structure that begins to collapse upon melting. At first the molar volume of liquid water continues to drop with increasing temperature, as more and more of the hydrogen-bonded framework remaining from the ice phase

Figure 18-1

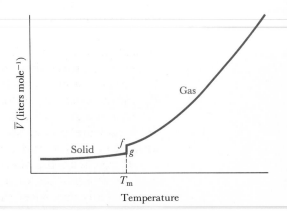

A plot of volume per mole, $\overline{V}$, against temperature for a hypothetical substance that passes directly from a solid to a gas without going through a liquid phase. Real substances behave this way above the critical pressure. During melting, the molar volume increases from point g to point f.

Figure 18-2   A plot of molar volume, $\overline{V}$, against temperature for a substance that melts to a liquid and then evaporates to a gas, with an increase in volume at each phase transition.

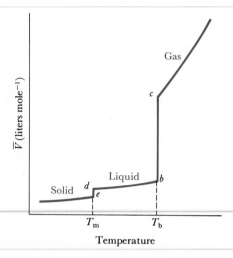

disintegrates, but normal thermal expansion finally predominates. Hence water has a minimum molar volume, and maximum density, at 4°C.

Heat must be supplied to melt a crystalline solid, and the order of the crystal lattice must be destroyed. Hence both enthalpy and entropy increase during melting. The enthalpy change is called the **heat of fusion**, ΔH^0_{fus}. For ice at 273 K,

$$\Delta H^0_{\text{fus}} = +6.01 \text{ kJ mole}^{-1}$$
$$\Delta S^0_{\text{fus}} = +22.2 \text{ e.u. mole}^{-1}$$

(Recall that 1 e.u. or entropy unit $= 1 \text{ J K}^{-1}$.) These are the heat and entropy required to break up the ordered structure of crystalline ice and send the molecules sliding past one another. Compare these values with the corresponding heat and entropy of vaporization, when the molecules of liquid are actually pulled away from one another to create a gas. For water at 298 K,

$$\Delta H^0_{\text{vap}} = +44.01 \text{ kJ mole}^{-1}$$
$$\Delta S^0_{\text{vap}} = +118.8 \text{ e.u. mole}^{-1}$$

More energy is required, and more disorder is created, in evaporation than in melting.

The heat and entropy of vaporization just given can be obtained by thinking of the evaporation of water as a two-component chemical "reaction":

$$H_2O(l) \rightleftarrows H_2O(g)$$

The heat of vaporization per mole is the difference between the standard heats of formation of liquid and gaseous H_2O in Appendix 3:

$$\Delta H^0_{\text{vap}} = \Delta H^0_{(g)} - \Delta H^0_{(l)} = -241.83 - (-285.84)$$
$$= +44.01 \text{ kJ mole}^{-1}$$

Strictly speaking, this is the heat required to convert 1 mole of liquid water to 1 mole of water vapor at 1 atm partial pressure and a temperature of 298 K. But examples in earlier chapters will have prepared you to discover that the heat of this reaction, like most others, is relatively insensitive to temperature. Approximately 45 kJ of heat are required to pull the molecules in 1 mole of water away from one another, no matter what the temperature is.

The entropy of vaporization at 298 K can be found from Appendix 3 in a similar way:

$$\Delta S^0_{vap} = S^0_{(g)} - S^0_{(l)} = 188.72 - 69.94 = 118.78 \text{ e.u. mole}^{-1}$$

As with enthalpy, this number is not very sensitive to temperature; disorder is disorder no matter at what temperature it is introduced. (But as we shall see, the *effect* of this disorder on a chemical reaction depends very much on the temperature.)

Heats and entropies of vaporization of several common liquids are compared in Table 18-1. Notice first of all that the entropies of vaporization are approximately the same for all liquids. The disorder introduced by taking 6.022×10^{23} molecules packed against one another in a liquid, and separating them in a gas, is relatively independent of the nature of the molecules. This generalization is known as **Trouton's rule**, after the man who discovered it empirically in the nineteenth century. The highest molar entropies of vaporization, by 10 or 20 e.u., are those of methanol, ethanol, and water. These larger entropies arise because the molecules are held together in the liquid by polar attractions and hydrogen bonding. The extra degree of liquid order means that the disordering required to produce a gas is slightly greater. Since more energy is needed to pull these interacting molecules of liquid apart, the heat of vaporization is slightly greater also. The heat of vaporization of mercury is large because the individual mercury

Table 18-1

Boiling Points and Heats and Entropies of Vaporization for Representative Liquids[a]

Molecule	T_b (K)	ΔH^0_{vap} (kJ mole^{-1})	ΔS^0_{vap} (e.u. mole^{-1})	$\Delta H^0_{vap}/\Delta S^0_{vap}$ (K)
BCl_3	285.6	23.0	80.8	285
Br_2	331.9	30.7	92.3	333
CH_3OH	338.1	37.4	110.9	337
C_2H_5OH	351.7	42.2	121.3	348
C_6H_6	353.3	33.9	96.2	352
H_2O	373.15	44.2	118.8	372
Hg	629.7	61.0	97.5	626

[a]Data from Appendix 3.

atoms are heavy and hard to move about, but this does not mean that the entropy of vaporization is correspondingly larger.

At sufficiently low pressure a solid also can go directly to the gas phase; this process is called **sublimation**. Sublimation is the normal behavior of solid carbon dioxide at 1 atm pressure, which is why it is commonly called "dry ice." Water ice normally melts to a liquid, but on a cold winter morning with dry air, a snowbank can sublime directly to water vapor without going through a liquid stage. Since enthalpies and entropies are state functions, the heat or entropy of sublimation must be the sum of the heats or entropies of fusion and vaporization at the same temperature. For water, making the approximation that ΔH and ΔS are the same at 273 K as they are at 298 K, we have

$$\Delta H^0_{sub} = \Delta H^0_{fus} + \Delta H^0_{vap} = 6.01 + 44.01 = 50.02 \text{ kJ mole}^{-1}$$
$$\Delta S^0_{sub} = \Delta S^0_{fus} + \Delta S^0_{vap} = 22.2 + 118.8 = 141.0 \text{ e.u. mole}^{-1}$$

18-2 FREE ENERGY OF VAPORIZATION AND VAPOR PRESSURE

The free energy of vaporization is found from

$$\Delta G^0_{vap} = \Delta H^0_{vap} - T \Delta S^0_{vap} \qquad \text{(at constant } T\text{)} \qquad (18\text{-}1)$$

In contrast to ΔH^0_{vap} and ΔS^0_{vap}, which are nearly constant with temperature, ΔG^0_{vap} is very temperature-sensitive because T occurs explicitly in equation 18-1. If we assume, for illustrative purposes, that the enthalpy and entropy changes are constant, then ΔH, ΔS, and ΔG can be represented for H_2O as in Figure 18-3. At high temperatures, $T \Delta S^0$ is larger than ΔH^0, the free energy of vaporization is negative, and evaporation of water to vapor at 1 atm partial pressure is thermodynamically spontaneous. At low temperatures, ΔH^0 is greater than $T \Delta S^0$, so ΔG^0 is positive, and it is condensation that is spontaneous. At some intermediate temperature, the enthalpy and entropy effects exactly cancel, ΔG^0 is zero, and liquid water is in equilibrium with water vapor at 1 atm partial pressure. This is defined as the **normal or sea-level boiling point** of the liquid, T_b. For water this temperature is 100°C or 373.15 K. At lower atmospheric pressures at high altitudes, water will boil below 100°C.

Since ΔG must be zero at the boiling point, at which liquid and gas are in equilibrium, then $\Delta H = T \Delta S$, and in principle the boiling point can be found from

$$T_b = \frac{\Delta H^0_{vap}}{\Delta S^0_{vap}} \qquad (18\text{-}2)$$

Equation 18-2 is strictly true only for enthalpy and entropy values at the boiling point, and these values are not readily available. But because neither H nor S is very temperature-sensitive, we can obtain approximate

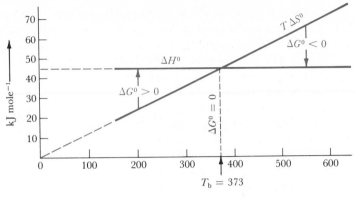

Figure 18-3   Contributions of enthalpy, ΔH^0, and entropy in the form of $T\ \Delta S^0$, to the free energy change of vaporization of liquid water. Enthalpy dominates below 373 K, and the liquid is the thermodynamically stable form relative to vapor at 1 atm partial pressure. Entropy dominates above 373 K and vapor at 1 atm pressure is the stable form. At the boiling point of 373 K, ΔG^0 is zero, and liquid and vapor are in equilibrium.

boiling points from $\Delta H^0_{\mathrm{vap}}$ and $\Delta S^0_{\mathrm{vap}}$ at 298 K as calculated from Appendix 3. These approximate boiling points are given in the last column of Table 18-1, and are within a few degrees of the true T_{b}. The discrepancies arise not because of any approximation in equation 18-2, but because $\Delta H^0_{\mathrm{vap}}$ and $\Delta S^0_{\mathrm{vap}}$ at 298 K are not quite identical with their values at the boiling point.

At temperatures below the boiling point, liquid water is not in equilibrium with water vapor at 1 atm partial pressure; this does *not* mean, however, that water vapor cannot exist below the boiling point. Rather, the lowered temperature favors condensation, so a new equilibrium is established with vapor at a lower partial pressure. If we again regard evaporation as a chemical reaction, we have a general expression for free energy (recall equation 16-26):

$$\Delta G = \Delta G^0 + RT \ln \frac{a_{(g)}}{a_{(l)}} \tag{18-3}$$

Since the standard state of the liquid is just the pure liquid itself, $a_{(l)} = 1$. In turn, $a_{(g)}$ is a unitless number that is numerically equal to the partial pressure of water vapor above the liquid. It is customary to use the symbol for partial pressure, $p_{(g)}$, instead of $a_{(g)}$, even though it is mathematically proper to take the logarithm only of a unitless number. Then

$$\Delta G = \Delta G^0 + RT \ln p_{(g)} \tag{18-4}$$

The partial pressure of vapor that is in equilibrium with a liquid at a specified temperature is defined as the **equilibrium vapor pressure**, p_{v}. Since

$\Delta G = 0$ at equilibrium, equation 18-4 can be simplified:

$$\Delta G^0 = -RT \ln p_v \qquad \text{or} \qquad p_v = e^{-\Delta G^0/RT} \qquad (18\text{-}5)$$

By the definition of the normal boiling point, $p_v = 1$ atm for water at 373 K. Let's see what it is at room temperature, 298 K.

Example 1

Calculate ΔG^0_{vap} and the vapor pressure of water at 298 K.

Solution   From ΔH^0_{vap} and ΔS^0_{vap} found earlier,

$$\Delta G^0_{vap} = +44{,}010 - 298(+118.8) = 8608 \text{ J mole}^{-1}$$

Free energies from Appendix 3 yield the same answer:

$$\Delta G^0_{vap} = -228.59 - (-237.19) = 8.60 \text{ kJ mole}^{-1}$$

The equilibrium vapor pressure then is

$$p_v = e^{-8600/8.314 \times 298} = e^{-3.47} = 10^{-3.47/2.303} = 0.0311 \text{ atm}$$

Before we can calculate p_v at any other temperature, we need to know how ΔG^0_{vap} changes with temperature. We could use the Gibbs–Helmholtz equation mentioned in Section 17-4, but this is a clumsy procedure. A surprisingly good approximation is to assume that enthalpy and entropy are effectively constant with temperature, so that free energy has a temperature dependence of the form $a - bT$. The values of p_v obtained from this oversimplification are plotted in Figure 18-4 as four circled points, beside a smooth curve representing the true vapor pressure of water as a function of temperature. Near the boiling point the curve rises so steeply that the approximate p_v is badly in error, by 0.11 atm, even though T_b is wrong by less than three degrees.

Figure 18-4 can be regarded as a plot of either the partial pressure of vapor in equilibrium with liquid as a function of temperature, or the way that the boiling point of water changes with pressure, since the boiling point is defined as the temperature at which p_v equals the external pressure. At temperatures below T_b, molecules can evaporate from the surface of the liquid, but any bubbles of vapor that form in the interior will be flattened by the external pressure on the top of the liquid. However, at T_b, the pressure inside these bubbles becomes as great as the surrounding pressure. Bubbles of vapor begin to form in the interior of the liquid, and rise to the surface in the agitated motion that we call "boiling."

At high altitudes, where the atmospheric pressure is less than 1 standard atm, the boiling point of water is lower. The U.S. Weather Bureau rule of thumb for pressure changes with altitude is "an inch of mercury per 1000 feet." (Standard atmospheric pressure is 29.9 in. of mercury.)

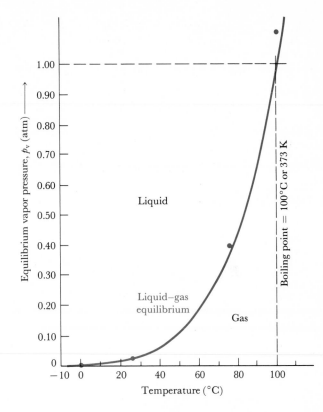

Figure 18-4   A plot of equilibrium vapor pressure of water against temperature. At the boiling point, $p_v = 1$ atm. The four solid points are results of an approximate calculation of p_v, assuming that enthalpy and entropy of vaporization are temperature-independent.

Example 2

Use the foregoing approximation to calculate the boiling point of water at 12,000 ft in the Colorado Rockies.

Solution

The pressure is lower than standard by 12 in. of mercury, or $12/29.9 = 0.40$ atm. If $P = 1.00 - 0.40 = 0.60$ atm, then from Figure 18-4, $T_b = 86°C$.

18-3   THE CRITICAL POINT

The vapor pressure curve of Figure 18-4 does not rise forever; it comes to an abrupt halt at the critical point, which for water is at $P_c = 218$ atm and $T_c = 374°C$. As this point is approached, the molar volume of the gas phase decreases because of increasing pressure, and the molar volume of the liquid, which being condensed is less susceptible to pressure changes, increases because of increasing temperature. As the liquid–vapor equilibrium curve is followed past the top of Figure 18-4, the liquid phase expands and the gas phase contracts. The critical point is reached when the molar volumes of liquid and gas become the same, so the distinction between the two separate phases vanishes. At the critical point, H_2O in its single phase has a molar volume of 57 cm^3 mole^{-1}, to be contrasted with 18 cm^3 mole^{-1} for liquid water (density 1 g cm^{-3}) and 24,450 cm^3 mole^{-1} for the gas at 298 K. Hence the critical point is reached not so much by expanding the liquid, as by compressing the gas until it matches the liquid.

Another view of the critical point is provided by Figure 18-5, a plot of volume versus pressure for CO_2 at a series of temperatures. Along the constant-temperature line or isotherm for 30.4°C, starting at point *a*, the volume of gaseous CO_2 decreases as the pressure is increased to 71.8 atm at point *b*. This is the vapor pressure of CO_2 that is in equilibrium with liquid at 30.4°C, so further attempts at compression only cause more vapor to liquefy without change in overall pressure, until point *c* is reached. At this point, all the CO_2 has been liquefied, and the curve rises steeply to point *d* and beyond, since the liquid is less compressible than the gas. A large pressure increase produces only a small decrease in volume.

The dome-shaped region outlined by the dashed line represents conditions of gas–liquid equilibrium. At the left end of a horizontal tie line such as *c*–*b* is the molar volume of the liquid that is in equilibrium with gas of molar volume given at the right end of the same tie line, for a specified temperature and pressure. You can see that as the temperature increases, the dome contracts and the molar volumes of the two phases approach one another. At the top of the dome, these molar volumes of liquid and gas become identical and the distinction between phases vanishes. This is the

Figure 18-5   A plot of molar volume of carbon dioxide as a function of pressure, for a series of constant temperatures. The dome-shaped central region (dashed line) is one of two-phase liquid–vapor equilibrium. Liquid conditions are to the left of the dome, and vapor conditions are to the right. Points *a* through *g* are discussed in the text.

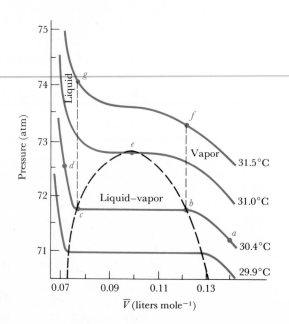

critical point, *e*. At still higher temperatures, and at larger molar volumes off to the right of Figure 18-5, the *P* versus *V* isotherms approach the *PV* = constant behavior expected of an ideal gas, as in Figure 3-3a.

The paradoxical nature of the critical point can be illustrated by the path marked *c–b–f–g–c*, partly outlined by dotted lines, in Figure 18-5. Suppose that we begin with liquid CO_2 in a cylinder with an observation window, at the temperature, pressure, and volume represented by point *c*, and then increase the volume slowly at constant temperature toward point *b*. Some of the liquid will begin to evaporate, and two phases will be visible in the cylinder, with a liquid meniscus between them. When point *b* is reached, the last of the liquid will just have evaporated, and only CO_2 gas will remain. Now suppose that this gas is heated at constant volume from 30.4°C to 31.5°C, taking the system from point *b* to point *f*, compressed isothermally at 31.5°C from point *f* to point *g*, and finally cooled at constant volume to point *c*. At no time along this path would two phases be present together in the cylinder. Starting with a pure gas at *b*, one would end with a pure liquid at *c*. Then where along the pathway *b–f–g–c* did the gas cease being a gas and become a liquid? This is only a word game, conditioned by our experience that if we go directly from *b* to *c* along the 30.4°C isotherm, separation into two phases does take place along the way. But it is possible to go from *b* to *c* by a route that achieves only a gradually diminishing separation between molecules in a single phase.

Critical-point constants for several substances are compared in Table 18-2. Notice that for no substance is the normal earth atmospheric pressure greater than P_c. Hence melting to a liquid, followed by evaporation to a

Table 18-2

Critical-Point Constants for Some Common Substances

Substance	P_c (atm)	T_c (K)	$\overline{V}_c$ (cm^3 mole^{-1})	T_b (K, at 1 atm)
He	2.34	5.3	62	4.5
H_2	13	33.2	65	21
Ne	26.9	44		27
N_2	34	126	90	77
O_2	50.5	154	75	90
CH_4	46.5	191		109
CO_2	72.8	304	95	195 (sublimes)
HCl	83.2	325		188
H_2S	90.4	373		212
NH_3	113	405	72	240
H_2O	218	647	57	373
Hg	1036	1735	40	630

gas, is for us the "normal" way for matter to behave when heated. Room temperature, 298 K, is greater than T_c for the substances from helium through methane in the table, meaning that none of these gases can be liquefied with a phase change by increasing pressure, without first cooling them below their respective critical temperatures.

18-4   PHASE DIAGRAMS

All the behavior described in the preceding two sections can be represented on a phase diagram—a plot of pressure versus temperature, indicating conditions at which the solid, liquid, or gas is the thermodynamically stable form, and at which two or even all three phases are in equilibrium. The phase diagram for CO_2 in Figure 18-6 is typical for substances that expand upon melting, the most common behavior. The vapor pressure curve that we have been examining now extends from the **triple point**, where solid, liquid, and gas are in equilibrium, to the critical point. Liquid and gas are in equilibrium along this line. Liquid is the stable phase above the curve, and gas is the stable phase below.

Solid and gas are in equilibrium along the curve from the origin in Figure 18-6 to the triple point. This curve can be regarded as indicating the vapor pressure above solid CO_2 as a function of temperature, or the change in sublimation temperature, T_s, with pressure. The straight line up from the triple point marks the conditions of equilibrium between solid and liquid, showing how the melting point changes with pressure.

The positive slope of all three of these curves is predictable by Le Chatelier's principle. An increase in pressure favors the more dense phase. The molar volume of the liquid is greater than that of the solid, and that of the

Figure 18-6   A pressure—temperature phase diagram for carbon dioxide. The areas are regions of single phases, and the boundary lines are conditions of two-phase equilibrium. Three phases coexist at the triple point. P_c and T_c are coordinates of the critical point, and P_3 and T_3 are coordinates of the triple point; T_s is the sublimation temperature at 1 atm pressure.

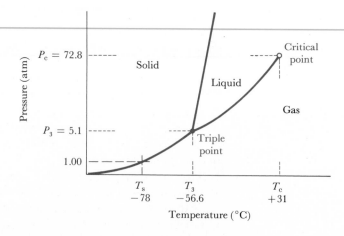

gas is greater than either. Hence the temperatures of sublimation, T_s, melting, T_m, and boiling, T_b, all increase as pressure increases, since more energy is required to bring molecules into the less dense phase.

At only one set of conditions, the triple point, can solid, liquid, and gaseous CO_2 coexist in equilibrium. The *triple point* for CO_2 occurs at 5.1 atm pressure and $-56.6°C$. We ordinarily think of solid CO_2 as subliming rather than melting, but this is only because atmospheric pressure is less than the triple-point pressure. If solid CO_2 under 1 atm pressure is warmed slowly, its vapor pressure rises along the solid–gas curve in Figure 18-6 until it reaches atmospheric pressure at $-78°C$. At temperatures higher than this sublimation point the solid is no longer stable at 1 atm external pressure.

Above 5.1 atm pressure, in contrast, CO_2 exhibits the melting and vaporization behaviors that are familiar from water and other substances that show a liquid phase. If we extend a horizontal line across Figure 18-6 at 6 atm pressure, for example, the intersection with the solid–liquid line is the melting point of the solid, and that with the liquid–gas curve is the boiling point of the liquid. On a planet where the atmospheric pressure was more than 5.1 standard earth atmospheres, the inhabitants might go swimming in lakes of liquid carbon dioxide. At pressure higher than 72.8 atm, the distinction between liquid and gas disappears, and only one phase transition is seen, that from solid to fluid, whatever you choose to call it.

Water is unusual because its solid phase is less dense than the liquid. Ice floats in water, whereas almost every other solid sinks in its liquid. Only a few metal alloys also expand when they freeze; they are used to cast type for printing, a sharp-edged replica of the mold being desirable. Water vapor is less dense (has greater molar volume) than either liquid or solid, but the liquid is denser than ice and hence is favored at higher pressures. This means that the liquid state becomes easier to attain at elevated pressures, or that the melting point falls as P increases. Hence the solid–liquid curve leans to the left, as in Figure 18-7. Ice melts under pressure, whereas the liquid state of almost every other substance solidifies upon compression. (What would be the effect on the climate of our planet if the ice of the oceans sank to the bottom rather than floating?)

The triple point for water, where gas, liquid, and solid can coexist at equilibrium, is at $0.0098°C$ and 0.0060 atm pressure. Because normal atmospheric pressure is greater than this, we are accustomed to seeing ice melt to liquid water rather than subliming as solid CO_2 does. The intersection of the $P = 1$ atm line with the solid–liquid equilibrium curve gives the melting point of $0°C$ for ice, and the intersection with the liquid–gas curve gives the boiling point of $100°C$ for water. The distinction between liquid and gas vanishes for water only above the critical pressure of 218 atm.

Similar phase diagrams can be made for all substances that exist as solid, liquid, and gas. The triple points of several substances are compared in Table 18-3, and phase diagrams can be constructed for them with appro-

Figure 18-7

A phase diagram for H_2O. Subscripts c and 3 denote critical and triple points as in Figure 18-6. T_m and T_b are the melting and boiling temperatures at 1 atm pressure. Because ice contracts when it melts, the solid—liquid equilibrium line slopes to the left, unlike that for carbon dioxide and most other substances.

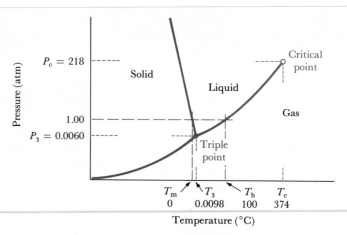

priate changes of scale. Figure 18-6 for CO_2 is the model for these phase diagrams because of the positive slope of the melting-point line. Notice that conditions where only one phase is present—solid, liquid, or gas—appear as *areas* on the phase diagram. Within these areas, pressure and temperature can be varied independently. Conditions under which two phases are in equilibrium appear as lines or *curves* on the phase diagram. Only one variable—pressure or temperature—can be changed independently while a two-phase equilibrium is maintained. With liquid and gas, for example, if the temperature is specified, then the vapor pressure of the gas in equilibrium with the liquid is likewise fixed. Alternatively, the boiling point of a liquid is determined once the ambient atmospheric pressure is specified. The unique set of conditions under which all three phases can coexist at equilibrium is located by a *point* on the phase diagram: the triple point.

Table 18-3

Triple Points of Some Common Substances

Substance	P_3 (atm)	T_3 (°C)
O_2	0.0026	−218
N_2	0.127	−210
Ar	0.674	−190
CO_2	5.11	−56.6
H_2O	0.0060	+0.0098

There are no degrees of freedom at the triple point, as neither P nor T can be varied without causing at least one of the three phases to disappear.

The brilliant American mathematician and physicist J. Willard Gibbs summarized this behavior in his **phase rule**: If f represents the number of degrees of freedom, or of independently variable parameters such as pressure and temperature, and if p represents the number of phases present, then

$$f = 3 - p \tag{18-6}$$

This version of the phase rule is for a one-component system or a pure substance. The rule becomes more useful when several substances are present, as with multicomponent solutions. For these, the phase rule in its most general form is

$$f = 2 + c - p \tag{18-7}$$

where c represents the total number of component substances in the system. Our examples of pure water and of CO_2 have been one-component systems, for which $c = 1$ and equation 18-6 results.

There is an interesting sidelight for the geometrically minded. For polyhedral solids such as cubes, octaheda, tetrahedra, and even those with more than one kind of face, if F is the total number of faces on the solid, V is the number of vertices, and E the number of edges, then

$$F = 2 + E - V \tag{18-8}$$

Gibbs the mathematician certainly knew this rule, and he was clever enough to think about phase equilibria geometrically and see that the same expression could be applied to solutions in equilibrium.

18-5 SOLUTIONS AND RAOULT'S LAW

The equilibrium vapor pressure measures the tendency of molecules to escape from the liquid into the gas. As we have seen, this tendency is greater at high temperatures and smaller at low temperatures. It is also smaller when there are fewer molecules of liquid available to escape: In a solution in which half the molecules are water and half are some other substance, the vapor pressure of water will be cut in half. In general, if X_A is the mole fraction of component A in a solution and if $p_A^\bullet$ is the vapor pressure of pure substance A at that temperature, then the vapor pressure contribution from the solution is

$$p_A = X_A p_A^\bullet \tag{18-9}$$

(The black dot superscript means "pure component.") This is **Raoult's law**, and solutions that obey this expression for all of their components are termed **ideal solutions**.

Figure 18-8

Partial pressures of benzene, p_B, and toluene, p_T, and the total vapor pressure, P_{total}, for a solution of the two liquids. They form an ideal solution, with Raoult's law obeyed for both components. The pressures are each a linear function of mole fraction.

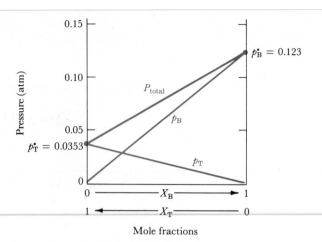

Of course, if the other substance in the solution is volatile as well, then it makes a contribution to the vapor above the solution:

$$p_B = X_B p_B^\bullet \tag{18-10}$$

The total pressure above the solution is the sum of the partial pressures of all components:

$$P_{total} = p_A + p_B = X_A p_A^\bullet + X_B p_B^\bullet \tag{18-11}$$

Since $X_A + X_B = 1$, we can write

$$P_{total} = X_A p_A^\bullet + (1 - X_A)p_B^\bullet = p_B^\bullet + (p_A^\bullet - p_B^\bullet)X_A \tag{18-12}$$

and the total pressure is a linear function of the mole fraction of either component.

This ideal solution behavior is followed by liquids with closely similar properties, in which the interactions between unlike and like molecules are virtually identical. An example is benzene and toluene, for which the two partial pressures and the total pressure are plotted in Figure 18-8. Other solutions can deviate from Raoult's law. If unlike molecules interact with one another less strongly than like molecules do, then the vapor pressure contributions will be greater than predicted by Raoult's law. Replacing the A molecules around a given molecule A with molecules of type B will increase the chance of that molecule escaping into the gas phase. Instead of Raoult's law, we would have

$$p_A > X_A p_A^\bullet \quad \text{and} \quad p_B > X_B p_B^\bullet \tag{18-13}$$

This is known as **positive deviation** from ideal behavior, and is observed in solutions of acetone and carbon disulfide, as shown in Figure 18-9.

For some combinations of substances, the attractions between unlike molecules are stronger than the attractions between molecules of the same kind. Water and ethanol exemplify this behavior. Adding ethanol to water

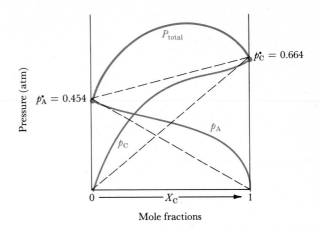

Figure 18-9

Partial pressures and total pressure for a solution of acetone (A) and carbon disulfide (C), which shows positive deviation from ideality. Dashed lines indicate ideal, Raoult's law behavior.

actually decreases the chance of an H_2O molecule's escaping, and lowers the vapor pressure more than the simple dilution and Raoult's law predict. Then, for both components,

$$p_A < X_A p_A^* \quad \text{and} \quad p_B < X_B p_B^* \tag{18-14}$$

Because the vapor pressures are lowered, this behavior is termed **negative deviation**. An example for acetone and chloroform is shown in Figure 18-10.

Although positive and negative deviations from Raoult's law are important for real solutions, just as deviations from the ideal gas law are important for real gases, we will be concerned chiefly with ideal solution behavior, and with situations where Raoult's law is at least approximately obeyed.

Example 3

The vapor pressures of water and ethanol at 25°C are 0.0311 atm and 0.0713 atm respectively. What is the vapor pressure of a solution that is 25 mole percent alcohol, and what are the contributions of each of the components? Assume ideal solution behavior.

Solution

The individual contributions are as follows:

Ethanol: $p = 0.25 \times 0.0713$ atm $= 0.0178$ atm
Water: $p = 0.75 \times 0.0311$ atm $= 0.0233$ atm

The total vapor pressure above the solution is the sum of these, or 0.0411 atm.

18-6 COLLIGATIVE PROPERTIES

Four properties of dilute solutions of a nonvolatile solute in a volatile solvent have traditionally been known as the **colligative properties**:

Figure 18-10   Partial pressures and total pressure for a solution of acetone (A) and chloroform (C), which shows negative deviation from ideality. Dashed lines indicate ideal behavior.

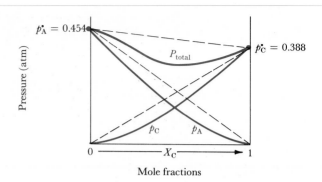

1. Lowering of vapor pressure of the solvent by the solute
2. Elevation of the boiling point of the solvent by the solute
3. Lowering of the freezing point of the solvent by the solute
4. The phenomenon of osmotic pressure

These are called *colligative* (meaning "collective") properties because they depend on how many molecules or ions of solute are present, and not on what the particles are (so long as they are not volatile and appear only in the liquid phase). Colligative properties were important to early chemists because they provided information about the number of particles of solute present, and hence about molecular weights and degrees of ionization in solution. The colligative properties were valuable to Arrhenius because he could show that more particles were present in solution than there were molecules of solute, and hence that the solute molecules were breaking apart into ions. Today the colligative properties are most useful in determining molecular weights of unknown materials, as we shall see.

Vapor Pressure Lowering

When a pure liquid, B, is in equilibrium with its vapor, the free energy of liquid and gaseous B must be the same. Evaporation and condensation occur at the same rate. If a small amount of nonvolatile solute, A, is added to the liquid, the free energy or escaping tendency of B in the solution is lowered, since a certain fraction of the solution molecules approaching the liquid–gas interface will then be A, not B. However, the reverse tendency, the condensation of vapor to liquid, is unaffected, because no type A molecules are present in the gas to interfere. At constant temperature, the frequency with which a molecule in the solution approaches the surface with enough kinetic energy to escape to the gas phase is the same in both pure B and in solution, assuming ideal solution behavior (Figure 18-11). But solute A is assumed to be nonvolatile. Hence not all molecules approaching with the right energy to escape can in fact do so. If 1% of the molecules in solution are A, then the vapor pressure of B will be only 99% as great as it was in pure B. This is just Raoult's law:

$$p_B = X_B p_B^{\cdot} \tag{18-10}$$

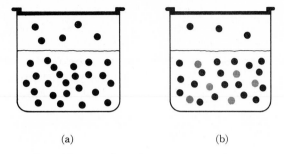

(a)   (b)

Figure 18-11   Vapor pressure lowering. (a) Equilibrium exists between a liquid and its vapor when the molar free energy or escaping tendency is the same in both phases. (b) Adding solute (colored dots) decreases the concentration and hence the escaping tendency of the solvent and causes the system to shift toward more condensation of gas. Le Chatelier's principle can be invoked here. When solvent is diluted by solute, the equilibrium position shifts in a direction that will tend to restore the solvent concentration to its previous value.

The lowering of vapor pressure of solvent B produced by solute A is

$$\Delta p = p_B^* - p_B = (1 - X_B)p_B^* = X_A p_B^* \qquad (18\text{-}15)$$

The change in vapor pressure is proportional to the amount of A present, but not to the particular nature of the A molecules.

The effect of nonvolatile solute on the vapor pressure curve of a liquid is shown in Figure 18-12. At every temperature, the vapor pressure is reduced to a fraction, X_B, of its original value, so the vapor pressure curve in the solution (dashed line) follows below that of the pure solvent (solid line).

Boiling Point Elevation

Figure 18-12 also shows a second colligative property: boiling point elevation. Suppose p_B^* is equal to the ambient atmospheric pressure, so T_b^* is the

Figure 18-12   Effect of solute on the liquid–vapor equilibrium curve of the solvent. At each temperature, the vapor pressure is less by a factor of X_B. The boiling point of the solution, T_b, is higher than that of the pure solvent, T_b^*.

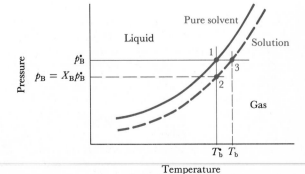

boiling point of pure liquid B (point 1). Addition of enough solute, A, to drop the mole fraction of B down from 1 to X_B means that the vapor pressure drops from p_B^* to $p_B = X_B p_B^*$ (point 2). This is less than atmospheric pressure, so the solution no longer boils. To make the solution boil again, we must increase the temperature along the dashed vapor pressure curve from point 2 to point 3, where the vapor pressure once again is equal to atmospheric pressure.

For dilute solutions, where the amount of added solute, A, is small, the change in boiling point, ΔT_b, is directly proportional to the *molality* of A, m_A, or the moles of A per kilogram of solvent:

$$\Delta T_b = T_b - T_b^* = k_b m_A \tag{18-16}$$

The molal boiling point constant for the solvent B, k_b, depends only on properties of the solvent such as its normal boiling point, T_b^*, its molecular weight, M_b, and the heat of vaporization of B, ΔH_{vap}. It does not depend in any way on the properties of the solute A that is causing the change in boiling point:

$$k_b = \frac{RT_b^{*2}M_b}{1000\,\Delta H_{vap}} \tag{18-17}$$

Boiling point constants for several common solvents are given in Table 18-4.

Example 4

What is the boiling point of 3.00 g of glucose, $C_6H_{12}O_6$, dissolved in 100 cm³ of water?

Table 18-4

Freezing Point and Boiling Point Constants for Common Solvents[a]

Solvent	T_m (°C)	k_f (deg mole⁻¹ kg⁻¹)	T_b (°C)	k_b (deg mole⁻¹ kg⁻¹)	M (g)
Water (H_2O)	0	1.86	100	+0.52	18.0
Carbon tetrachloride (CCl_4)	−22.99	~30.00	76.5	5.03	154.0
Chloroform ($CHCl_3$)	−63.5	4.70	61.7	3.63	119.5
Benzene (C_6H_6)	5.5	5.12	80.1	2.53	78.0
Carbon disulfide (CS_2)	−111.5	3.83	46.2	2.34	76.0
Ether ($C_4H_{10}O$)	−116.2	1.79	34.5	2.02	74.0
Camphor ($C_{10}H_{16}O$)	178.8	40.0			152.2

[a]T_m is the normal melting point; k_f, the molal freezing point lowering constant; T_b, the normal boiling point; k_b, the molal boiling point elevation constant; M, the molecular weight of the substance.

Solution

Since the molecular weight of glucose is 180, and the density of water is 1.00 g cm^3, 3.00 g glucose per 100 cm^3 corresponds to $30.0/180 = 0.167$ mole per 1000 g water, or to a molality of 0.167 moles kg^{-1}. Then

$$\Delta T = 0.52(0.167) = 0.087°C$$

Hence the boiling point is $100.087°C$.

Freezing Point Lowering

The lowering of the freezing point can be explained like boiling point elevation. In Figure 18-13a, pure liquid B is in equilibrium with pure crystalline solid B, so the molar free energy of B must be identical in the two phases. However, if the solvent, B, is diluted by adding some solute molecules, A, then not every molecule in the liquid phase that strikes the surface of the crystalline solid can stick (Figure 18-13b). Many collisions will be nonproductive. However, the rate of escape of B molecules from the pure solid is unaffected. The free energy or escaping tendency of B will be lower in the liquid than in the crystal, and equilibrium will shift in favor of melting. Some of the solid will begin to melt or dissolve. (This is why $CaCl_2$ or NaCl melts ice when sprinkled on slippery roads in winter.)

To restore equilibrium between crystalline solid and liquid solution, we must lower the temperature until the escaping tendency of the fewer B molecules in the solution again equals that of the greater number of B molecules in the original pure liquid. At the new freezing point, there are fewer molecules of B in the solution to strike the solid, but they are moving more slowly and thus are more likely to be caught.

The derivation of the expression for freezing point lowering is similar to that for boiling point elevation: The combined effect of dilution and temperature lowering on the escaping tendency of solvent molecules must

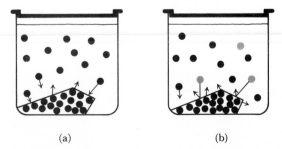

(a)   (b)

Figure 18-13

Freezing point depression. (a) Pure liquid and crystalline solid are in equilibrium. (b) When solute (colored dots) is added to the solvent, equilibrium shifts in favor of melting of the solid. A new equilibrium is restored only by lowering the temperature. Hence addition of solute depresses the freezing point of the solvent.

be zero at the new equilibrium point. The result, likewise, resembles that for boiling point elevation:

$$\Delta T_{\mathrm{f}} = T_{\mathrm{m}} - T_{\mathrm{m}}^{\bullet} = -k_{\mathrm{f}} m_{\mathrm{A}} \tag{18-18}$$

$T_{\mathrm{m}}^{\bullet}$ and T_{m} are the melting or freezing points of pure B and of the solution, m_{A} is the molality of the solute, and k_{f} is the molal freezing point depression constant, given by

$$k_{\mathrm{f}} = \frac{R T_{\mathrm{m}}^{2} M_{\mathrm{B}}}{1000 \, \Delta H_{\mathrm{fus}}} \tag{18-19}$$

ΔH_{fus} is the heat of fusion of the solvent, T_{m} is its normal melting or freezing point, and M_{B} is its molecular weight. Molal freezing point depression constants for some common liquids are given in Table 18-4.

Example 5

What is the freezing point of the glucose solution of Example 4?

Solution   The freezing point is changed by

$$\Delta T_{\mathrm{f}} = -1.86(0.167) = -0.311°$$

The new freezing point is $-0.311°C$.

Example 6

A solution consists of 1.00 g of acetamide (CH_3CONH_2) in 100 cm³ of water. What is its freezing point depression?

Solution   The molecular weight of acetamide is 59.1, so 1.00 g in 100 cm³ of water is equivalent to 10 g or $10/59.1 = 0.169$ moles per kg of water. Hence $m_{\mathrm{A}} = 0.169$, and $\Delta T_{\mathrm{f}} = -1.86(0.169) = -0.315°C$.

Molecular Weight Determinations

Colligative properties can be used to determine molecular weights, since the size of the colligative effect depends on how many particles and hence how many moles of solute are present.

Example 7

A saturated solution of glutamic acid in water has 1.50 g of glutamic acid per 100 g of water. The freezing point of this solution is $-0.189°C$. What is the molecular weight of glutamic acid?

Solution   From the freezing point depression,

$$-0.189 = -1.86\, m_A \quad \text{and} \quad m_A = 0.102 \text{ mole kg}^{-1}$$

The solution was prepared from 1.50 g solute in 100 g water, or 15.0 g kg^{-1} water. Hence 15.0 g glutamic acid is 0.102 mole, and the molecular weight is

$$M_A = \frac{15.0 \text{ g}}{0.102 \text{ mole}} = 147 \text{ g mole}^{-1}$$

Example 8

A compound with a molecular weight of 329.3 is dissolved in water, with 300 mg (300 × 10^{-3} g) of solute in 10 cm^3 of water. What is the freezing point of the solution?

Solution   The expected freezing point is −0.169°C.

Example 9

In the experiment of Example 8, the actual, observed freezing point was −0.676°C, not −0.169°C. What does this fact tell you about what happens when the compound dissolves? [The compound is potassium ferricyanide, K$_3$Fe(CN)$_6$.]

Solution   The freezing point lowering is four times as large as expected, indicating that 1 mole of compound yields 4 moles of particles in solution. Potassium ferricyanide dissolves to produce four ions:

$$\text{K}_3\text{Fe(CN)}_6 \rightarrow 3\text{K}^+ + \text{Fe(CN)}_6^{3-}$$

There are limits to the usefulness of freezing point or boiling point changes in determining molecular weights. For compounds of large molecular weight, it is difficult to get enough material into solution to produce an accurately measurable freezing point change.

Example 10

Two hundred milligrams of the protein cytochrome *c* are dissolved in 10 cm^3 of water. The molecular weight of cytochrome *c* is 12,400. What is the freezing point depression of the solution?

Solution   The solution concentration is equivalent to 20 g protein per kg of water, but because of the high molecular weight, this is only 0.0016 molal. Hence: $T_f = -1.86(0.0016) = 0.003°C$, which is too small for accurate molecular weight determinations.

Neither freezing point lowering nor boiling point elevation is useful to determine molecular weights of macromolecules, but the fourth and last colligative property, osmotic pressure, can be used.

Osmotic Pressure

Many membranes have pores large enough to let some molecules pass through but too small to allow others to pass. These membranes are called **semipermeable membranes**. Some will allow water but not salt ions to pass. Others, with larger pores, will allow water, salts, and small molecules through, but will block proteins or macromolecules with molecular weights of several thousand. Millipore filters can now be manufactured with pore sizes so uniform that they can be used to separate proteins from one another on the basis of size.

Suppose that the beaker shown in Figure 18-14a contains pure water, whereas the thistle tube with the membrane across the bottom contains an aqueous solution of substance A. Furthermore, suppose that water molecules can pass freely through the membrane but molecules of A cannot. The rate of flow of water molecules into the thistle tube from the beaker solution is unimpaired by the presence of A, but the rate of flow from the thistle tube back to the beaker is decreased by the presence of A. The molar free energy or escaping tendency of the water in the thistle tube will be lessened by the presence of solute particles, just as it was for the other colligative properties. More water will flow into the thistle tube than out, so the solution will rise in the tube, as shown in Figure 18-14b.

As the pressure inside the thistle tube increases (this pressure can be measured by the height of the column of solution), the free energy or escaping tendency of water molecules in the tube increases. A point is eventually reached where the increase in free energy of water in the thistle tube because of the increased pressure exactly balances the decrease in free energy of water because of dilution by solute A. There are fewer solvent molecules per cubic centimeter within the tube, but each solvent molecule has a greater tendency to escape through the membrane because of the higher pressure. The net rate of outflow is the same as it would have been without the presence of solute, and equilibrium is reestablished with the water in the open beaker. This extra pressure required to restore equilibrium is called the **osmotic pressure**, and is given the symbol of the Greek pi, π.

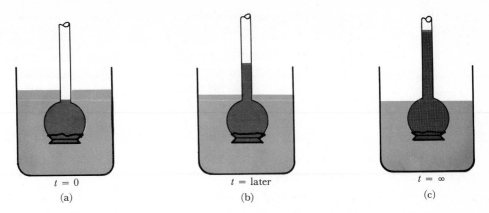

Figure 18-14

An osmotic pressure experiment. When a solute is dissolved in a solvent, and the solution is separated from a reservoir of pure solvent by a membrane that allows solvent but not solute particles through it, the solvent flows into the solution chamber until an extra pressure known as the osmotic pressure builds up to oppose it. (a) Solution (gray) inside thistle tube and pure solvent (colored) outside. (b) Gradual rise of liquid in the tube as solvent flows in. (c) Equilibrium, with the solution in the tube stationary. The osmotic pressure can be calculated from the height of the solution in the tube and the density of the solution.

If the change caused by dilution and the change caused by pressure increase are balanced so that no net tendency to escape across the membrane results, then the following expression obtains:

$$\pi V = n_{A}RT \qquad \text{or} \qquad \pi = c_{A}RT \qquad (18\text{-}20)$$

Here c_{A} is the molarity of the solute, or the number of moles of solute per liter of solution (not per kilogram of solvent, as in molality, m). Notice the resemblance of the osmotic pressure expression to the ideal gas law. An ideal gas is one in which the molecules have *no* attractions for one another; in an ideal solution for which Raoult's law is obeyed and this simple osmotic pressure law holds, all molecules have the *same* attractions for one another.

In the cases of freezing point lowering and boiling point elevation, the disturbance of equilibrium produced by adding solute molecules is compensated by an adjustment of temperature. With osmotic pressure, the restoration of equilibrium is achieved by increasing the pressure instead of changing the temperature. Because the solution is a condensed phase, and relatively incompressible, the pressure required to do this is large and easily measured. Hence osmotic pressure is much more sensitive to the amount of solute present, and more useful with substances of high molecular weights.

Example 11

Calculate the osmotic pressure produced by a solution of 200 mg of cytochrome c in enough water to make a final volume of 10 cm³, if the membrane will pass water molecules but not cytochrome c.

Solution

$$c_A = \frac{0.200 \text{ g}}{0.010 \text{ liter} \times 12{,}400 \text{ g mole}^{-1}} = 0.0016 \text{ mole liter}^{-1}$$

$\pi = 0.0016 \text{ mole liter}^{-1} \times 0.0821 \text{ liter atm K}^{-1} \text{ mole}^{-1} \times 298 \text{ K}$

$\pi = 0.039 \text{ atm}$

A pressure difference of 0.039 atm is easily measured, much more so than a temperature drop of 0.003 K. Osmotic pressure is the preferred colligative property for measuring large molecular weights.

Summary

When a crystalline solid is disordered by temperature at moderate pressure, it normally melts to a liquid, which with increasing temperature becomes a gas. In a crystalline solid the molecules or ions are packed against one another in a regular geometrical array. In a liquid the molecules are still packed against one another, but the regular structure is destroyed because the molecules have enough thermal energy to tumble past one another. In a gas the molecules are no longer in contact; they move in independent trajectories through space, colliding and rebounding.

In a sense, the distinction between liquid and gas is more conventional than real. We are accustomed to seeing two separate phases when a liquid becomes a gas, with a liquid surface or meniscus between them. But at high temperatures and pressures, the density of a gas approaches that of its liquid, and the distinction between phases vanishes. The point at which this occurs is marked by a **critical pressure**, P_c, and a **critical temperature**, T_c. At pressures greater than P_c, the solid melts to a fluid, and there is no second phase transition with increasing temperature.

Melting, evaporation, and sublimation are each characterized by a heat or enthalpy change, and a disorder or entropy change. Heat is always required, but the entropy always increases, in the transitions from solid to liquid to gas. The requirement for heat works against the spontaneity of these processes, and the creation of disorder favors them. At the transition temperatures (T_m, T_b, T_s) these two effects balance, and the free energy of the phase transition is zero.

The **boiling point** of a liquid is the temperature at which the liquid phase is in equilibrium with its vapor at a pressure equal to the surrounding atmospheric pressure. The **normal or standard boiling point** is for 1 standard atmosphere of pressure. At lower temperatures, the equilibrium vapor pressure is less than 1 atm, and is related to the standard free energy of vaporization at that temperature by

$$\Delta G^0 = -RT \ln p_v$$

Equilibria between phases of a pure substance can be represented by a pressure versus temperature phase diagram. Regions where only a single phase is thermodynamically stable appear as areas on the diagram, regions of two-phase equilibrium are curves, and the set of conditions at which all three phases can coexist is a point termed the **triple point**. The **phase rule** predicts that for a pure substance the number of independently adjustable variables such as temperature and pressure is given by $f = 3 - p$, in which p is the number of phases present. Le Chatelier's principle predicts that the slope of each of the two-phase equilibrium curves will be positive, except for the ice–water boundary, whose slope is negative because ice contracts when it melts. We can use the phase diagram to predict both the melting behavior familiar from water and the sublimation behavior of solid carbon dioxide, if we know whether the atmospheric pressure is greater or less than the triple point pressure.

If in a solution the interactions between unlike molecules are the same as those between like molecules of each of the components, then the solution is termed **ideal**. For such solutions, the vapor pressure of a component is proportional to its concentration expressed as a mole fraction: $p_j = X_j p_j^*$. This is **Raoult's law**. If the interactions between unlike molecules are stronger than those between like molecules, then the vapor pressure of each component is less than predicted by Raoult's law, and this is termed **negative deviation** from ideality. If the interactions between unlike molecules are weaker, then the vapor pressure is greater, and the deviation is termed **positive**.

The effects in dilute solution of a nonvolatile solute on the properties of the solvent are called the **colligative properties**. There are four such properties: **vapor pressure lowering**, **boiling point elevation**, **freezing point lowering**, and **osmotic pressure**. In each case the magnitude of the effect is directly proportional to the number of solute molecules or ions present, and is independent of the nature of the particles. The colligative properties are useful in demonstrating ionization in solution, and in determining molecular weights.

Self-Study Questions

1. When a crystalline solid is melted and the temperature thereafter is steadily increased, how does the behavior of the substance differ, depend-

ing on whether the ambient pressure is above or below the critical pressure? What is the critical pressure for H_2O? For CO_2?

2. Can melting ever take place with a decrease in molar enthalpy? A decrease in molar entropy? A decrease in molar volume? Give an example of a real substance for each yes answer.

3. Can vaporization ever take place with a decrease in molar enthalpy? A decrease in molar entropy? A decrease in molar volume? Give an example of a real substance for each yes answer.

4. Are the properties of a substance at its critical point more like those of the liquid, or of the gas, at lower temperature and pressure?

5. If you know the molar entropy of sublimation and of melting at a given temperature, how can you obtain the molar entropy of vaporization? What principle or law is involved in your answer?

6. How can the molar enthalpy of vaporization be obtained from tables such as Appendix 3?

7. What is Trouton's rule, and what is its molecular explanation?

8. How can the boiling point of a liquid be approximated, using only data in Appendix 3? Why is the answer only approximate? What information would be required for an exact solution?

9. What is the equilibrium vapor pressure of a liquid? How is the boiling point related to vapor pressure? What does this mean in terms of the molecular behavior of the boiling liquid?

10. How is the standard free energy of vaporization at a given temperature related to the equilibrium vapor pressure? For what temperature is the standard free energy of vaporization zero?

11. How do the standard enthalpy and entropy of vaporization oppose one another? At what temperature do their effects exactly cancel?

12. How does the boiling point of a liquid change with altitude? What is the advantage of a pressure cooker in preparing food?

13. How is it possible to take a substance from the gas phase to a liquid by condensation, with two phases at intermediate points, and then back to a gas by another route that involves only a single phase at every point and no phase transitions? Illustrate using Figure 18-5.

14. In Figure 18-6 what do the areas and the boundary lines that separate them signify? What is the significance of the point at which the three boundary lines meet?

15. How does Le Chatelier's principle predict that the slope of all three boundary lines in Figure 18-6 will be positive? How does it predict a negative slope for one of the lines in Figure 18-7?

16. At what pressures will solid carbon dioxide melt to a liquid before evaporating to a gas? At what pressures will it sublime directly to a gas?

17. At what pressures will solid H_2O melt to a liquid before evaporating to a gas? At what pressures will it sublime directly to a gas?

18. What is the phase rule? What is its form for a one-component system? Show how the phase rule holds for Figure 18-6.

19. What is an ideal solution? What are the molecular explanations for positive and negative deviation from ideality, and what effect do these deviations have on vapor pressure?

20. What is Raoult's law? For what kind of solution is Raoult's law followed?

21. What are the colligative properties? For what kinds of solutions are the simple expressions as derived in the text valid?

22. Why should the boiling point of a solution increase when a small amount of nonvolatile solute is added? Why should the freezing point decrease?

23. Explain in terms of colligative properties why the antifreeze that one uses in a car radiator also helps to prevent boiling over in hot weather.

24. When a pan of vegetables cooking on the stove is about to boil over, why does throwing in a small amount of table salt save the day?

25. How does an ice skater take advantage of the contrast between Figures 18-6 and 18-7 to lubricate his blades when he skates?

26. Swimming for a long time in salt water makes the skin of one's fingertips wrinkled. What colligative property does this illustrate?

27. How can the colligative properties be used to detect ionization in aqueous solution? How might they be used to detect molecular aggregation in solution?

28. How can the colligative properties be used to measure molecular weights? Which property is most useful with macromolecules, and why are the others less suitable?

Problems

Thermodynamics of phase changes

1. From data in Appendix 3, calculate the changes in enthalpy, entropy, and free energy upon vaporization of BBr_3 at 298 K. What is the approximate boiling point of BBr_3? Check your answers against the true boiling point in a chemical handbook. What is the equilibrium vapor pressure of BBr_3 at 298 K?

2. From data in Appendix 3, calculate the changes in enthalpy, entropy, and free energy of vaporization of formic acid (HCOOH) at 298 K. What is the vapor pressure of formic acid at 298 K? Calculate an approximate boiling point, and compare your answer with the true value from a chemical handbook. How big is the discrepancy?

Free energy and vapor pressure

3. From data in Appendix 3, ascertain which of the following liquids has the largest equilibrium vapor pressure at 298 K: water, benzene (C_6H_6), or formic acid (HCOOH). Explain in terms of molecular interactions.

4. From data in Appendix 3, calculate the equilibrium vapor pressure of boron trichloride at 298 K. What does your answer tell you about the stable form of BCl_3 at that temperature? Calculate the boiling point of BCl_3 from Appendix 3 data, and compare your result with the true value from a chemical handbook.

Trouton's rule

5. Trouton's rule, as originally defined, was an empirical observation that the molar heat of vaporization divided by the boiling temperature was approximately 92 J K^{-1} mole^{-1}. The boiling point of *n*-octane, C_8H_{18}, is 125.7°C. From these data, estimate the molar heat of vaporization of *n*-octane. The standard enthalpy of formation of liquid *n*-octane is −250.0 kJ mole^{-1}, and that of the gas is −208.4 kJ mole^{-1}. Calculate the true heat of vaporization and compare it with your approximate value. What is the percent error?

6. The heat of vaporization of a gram of an unknown hydrocarbon is 367 J g^{-1}, and its boiling point is 69°C. From these data, use Trouton's rule to estimate the molecular weight of the hydrocarbon.

Vapor pressure and boiling point

7. From data in Appendix 3, calculate the vapor pressure of ethyl alcohol, C_2H_5OH, at 298 K, and its approximate boiling point. Assuming that enthalpy and entropy of vaporization are temperature-independent, calculate the boiling point of ethyl alcohol at 12,000 ft altitude, where the pressure is 0.60 atm.

8. Assuming that enthalpy and entropy of vaporization are temperature-independent, calculate the boiling point of water in a pressure cooker at 1 atm over atmospheric pressure, or 2 atm total pressure. Of what value is this in food preparation?

Phase diagrams

9. Sketch the phase diagram for solid–liquid–gas equilibrium for N_2. Show the slope of each curve correctly, locate the triple and critical points, and indicate normal earth atmospheric pressure. Mark where T_m and T_b occur. (Find their values from data in this chapter or from a chemical handbook.) In what pressure range will solid N_2 sublime rather than melt?

10. Sketch the phase diagram for solid–liquid–gas equilibrium in mercury. Show the slope of each curve correctly, locate the critical point, and indicate where the triple point is at least approximately, relative to 298 K and 1 atm pressure. The density of liquid Hg is 13.6 g cm^{-3}. Calculate the molar volume of liquid and gaseous mercury at 298 K, assuming ideal gas behavior, and compare these volumes with the molar volume of the single phase at the triple point, given in Table 18-2. Are the triple point conditions more like the gas or liquid phase at 298 K?

Raoult's law

11. If 10 g each of benzene and toluene are mixed in a beaker at 298 K, what is the vapor pressure of the resultant solution? What contribution to vapor pressure is made by each component?

12. At 363 K the vapor pressures of benzene and toluene are 1.34 atm and 0.534 atm. What is the mole fraction of toluene in a solution that boils at 363 K under atmospheric pressure?

Colligative properties

13. Which of the following substances would you expect to give a 0.1-molal aqueous solution with the lowest freezing point: HNO_3, NaCl, glucose, $CuSO_4$, $BaCl_2$? Why?

14. Which will have a greater effect on the colligative properties of an aqueous solution, 20 g of NaCl or 10 g of $MgCl_2$? Assume complete solubility in each case.

Vapor pressure lowering

15. If 35.5 g of solid chlorine, Cl_2, are dissolved in 32 g of liquid methane, CH_4, at the boiling point of methane, by how much will the vapor pressure of methane be lowered?

16. At 292 K the vapor pressure of carbon tetrachloride, CCl_4, is 0.1128 atm. A solution of 2.182 g of an unknown nonvolatile substance in 100 g of CCl_4 at this temperature has a vapor pressure of 0.1107 atm. Calculate the molecular weight of the unknown. What is its chemical formula, if it analyzes as 94.34% carbon and 5.66% hydrogen?

Boiling point elevation

17. A solution is prepared by mixing 20 g of a nonvolatile solute having a molecular weight of 100 with 500 g of solvent having a molecular weight of 75. The boiling point of the solvent increases from 84.00°C to 85.00°C. Calculate the boiling point elevation constant for this solvent.

18. If 10.0 g of glucose, $C_6H_{12}O_6$, are dissolved in 400 g of ethyl alcohol, the boiling point of the alcohol is increased by 0.1428°. If 2.00 g of an unknown organic substance are dissolved in 100 g of ethyl alcohol, the boiling point increases by 0.1250°. Calculate the molecular weight of the unknown substance.

Freezing point depression

19. Calculate the freezing point of a solution made by dissolving 1.00 g of NaCl in 10.0 g of water. Repeat your calculation using 1.00 g of $CaCl_2$ instead. Which of these salts is more effective as an antifreeze on a weight basis?

20. How many grams of methanol, CH_3OH, must be added to 10.0 kg of water to lower the freezing point of the solution to 263 K? What is the normal boiling point of this solution?

Molecular weight determination

21. A solution is prepared by dissolving 0.400 g of an unknown hydrocarbon in 25.0 g of acetic acid. The freezing point of the solution falls from 16.60°C for pure acetic acid to 16.15°C. The molal freezing point depression constant for acetic acid is 3.60 deg mole^{-1}. What is the molecular weight of the hydrocarbon? Chemical analysis shows it to be 93.76% carbon and 6.25% hydrogen by weight. What is its molecular formula?

22. Benzoic acid is 68.9% carbon, 26.2% oxygen, and 4.96% hydrogen. One gram of the acid in 20.0 g of water freezes at 272.38 K, whereas 1.00 g in 20.0 g of benzene freezes at 277.56 K. What is the apparent molecular formula for benzoic acid in each solvent? Can you explain your results?

23. Thyroxin, a hormone that controls the rate of metabolism in the body, can be isolated from the thyroid gland. If 0.455 g of thyroxin is dissolved in 10.0 g of benzene, the freezing point of the solution is 5.144°C; pure benzene freezes at 5.444°C. What is the molecular weight of thyroxin?

24. A solution of 0.0702 g of an organic compound in 0.804 g of camphor freezes 15.3° lower than pure camphor. What is the molecular weight of the compound? Chemical analysis shows it to be 63.2%

carbon, 8.8% hydrogen, and the rest oxygen by weight. What is the chemical formula?

Osmotic pressure

25. A 4.0-g sample of polyvinyl chloride, with a mean molecular weight of 1.5×10^5, is dissolved in 1.0 liter of dioxane, $C_4H_8O_2$, at a temperature of 298 K. What is the osmotic pressure that develops relative to pure dioxane?

26. When 145 mg of a protein are dissolved in 10.0 cm³ of water at 25°C, the osmotic pressure is 0.00989 atm. What is the molecular weight of the protein?

Suggested Reading

W. J. Moore, *Physical Chemistry*, Prentice Hall, Englewood Cliffs, N.J., 1972, 4th ed. See Chapter 6.

R. M. Rosenberg, *Principles of Physical Chemistry*, Oxford University Press, N.Y., 1977. See Chapter 15.

F. W. Sears, *Thermodynamics*, Addison-Wesley, Reading, Mass., 1955, 2nd ed.

19

Oxidation–Reduction Equilibria and Electrochemistry

Charge!

James Brudenell (Lord Cardigan)
Balaclava, 25 October 1854

The current market prices for ancient coins are not dictated entirely by the value of the gold, silver, or copper in them, or by the relative scarcity of the individual issues. A gold stater of Croesus or a silver teradrachm of Alexander, 2500 years after they were struck, can be miniature art masterpieces in a physical condition that belies their age. In contrast, a typical bronze or copper coin of the same period will be corroded, pitted, and ugly, especially if it has been near moisture or damp earth for a long time. Why does copper corrode, whereas silver and gold do not? The answer lies in the different affinities of the three metals for their electrons.

Under certain circumstances, a person with both gold inlays and silver amalgam fillings in his teeth is in trouble. Every time he bites and brings the two metals into contact, he receives a shock, which makes eating unpleasant. Why the shock? Again, it is a matter of the different affinities that substances have for electrons, and a flow of electrons from where they are not particularly wanted to where they are wanted.

In this chapter we shall discuss ways in which chemical reactions can be divided into two physically separated parts: one that gives up electrons easily, and one that accepts them easily. If we can catch the electrons as they flow "downhill" (to use a gravitational analogy), we may be able to use this flow to do outside work. This is the principle of the **galvanic cell**. Moreover, if we can find ways to push the electrons "uphill" from regions where they are wanted to regions where they are not wanted, then we can either store the energy that this requires, for later use, or make chemical reactions take place that are normally not spontaneous. This is the principle of **electrolytic cells**.

As in previous chapters, we are interested in the concepts of spontaneity and equilibrium. Only this time, we have added a new dimension to spontaneity, that of electron flow.

19-1 HARNESSING SPONTANEOUS REACTIONS

If we open a tank of hydrogen gas at 10 atm pressure and allow the gas to escape, it will do so spontaneously. No useful work is obtained in this experiment. However, if we connect the outlet tube to a plastic bag or a piston we can lift weights, or if we place a pinwheel or windmill in the path of the outrushing gas we can convert the stored energy of the gas into mechanical work (Figure 19-1). The overall reaction is

$$H_2(g, 10 \text{ atm}) \rightarrow H_2(g, 1 \text{ atm})$$

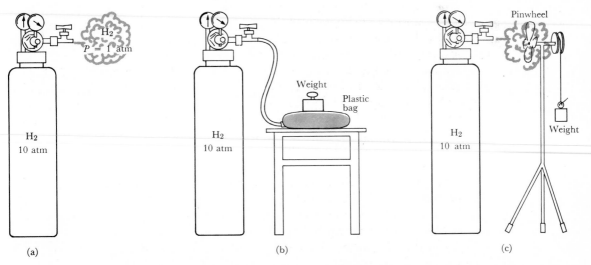

Figure 19-1   Nonproductive and productive use of the free energy stored in hydrogen gas at 10 atm pressure. (a) Expansion to 1 atm with dissipation of energy. (b) Use of part of the available free energy to lift a weight by inflating a plastic bag. (c) Use of part of the free energy to lift a weight by turning a pinwheel or windmill.

and from Chapter 16, we know the maximum free energy obtainable per mole of H_2 to be

$$\Delta G = \Delta G^0 + RT \ln (p_2/p_1)$$
$$\Delta G = 0 + RT \ln (1/10) = -RT \ln (10) = -5.706 \text{ kJ mole}^{-1}$$

(The standard free energy change for the reaction $H_2(g) \rightarrow H_2(g)$ obviously is zero.) This free energy decrease of $5.706 \text{ kJ mole}^{-1}$ represents the *maximum* work obtainable when the process is carried out reversibly, an impractical upper limit that would take forever to reach. For any real, spontaneous, finite-time process, somewhat less than $5.706 \text{ kJ mole}^{-1}$ of useful work will be available.

There is another way to harness the free energy of the escaping gas: through the so-called pressure cell (Figure 19-2). Although it may seem to be a somewhat artificial device, it will lead us directly to other kinds of chemical cells. The pressure cell shown in Figure 19-2 carries out the same reaction in two steps. In the left tank, hydrogen gas at 10 atm is "taken apart" into protons and electrons; in the right tank, hydrogen at 1 atm is "reassembled" from electrons and protons. However, the overall reaction still is

$$H_2(g, \text{ 10 atm}) \rightarrow H_2(g, \text{ 1 atm})$$

and the overall free energy yield likewise must be unchanged.

Since an excess of electrons is produced at the left electrode, and reaction at the right electrode cannot proceed without electrons, a flow of electrons from left to right occurs if the two terminals are connected with a wire. (This is analogous to opening the valve on a tank of compressed gas: high-pressure gas is converted to low-pressure gas without doing any useful work.) This tendency for electrons to flow, or this electron "pressure," is measured by a voltage difference between the two terminals. If we want to prevent electrons from flowing, we must supply an opposing voltage equal to the voltage developed by the pressure cell. Alternatively, we can use this electron "pressure" to carry out some useful task (the pressure-cell analogue of lifting weights with a windmill). In terms of the driving force for doing work, the pressure of compressed hydrogen gas has been converted into a voltage difference between an electron-rich terminal and an electron-poor terminal. For this particular cell the voltage is quite small: 0.0296 volt (V) or 29.6 millivolts (mV). Thus it is not a very useful type of cell.

Example 1

A pressure cell is set up as shown in Figure 19-2, with 10 atm pressure of H_2 on the left and 2 atm pressure on the right. Calculate the free energy change for this reaction. If cell voltage is proportional to the free energy difference, what is the voltage of this pressure cell? What would the voltage be if the H_2 pressures were 5 atm on the left and 1 atm on the right?

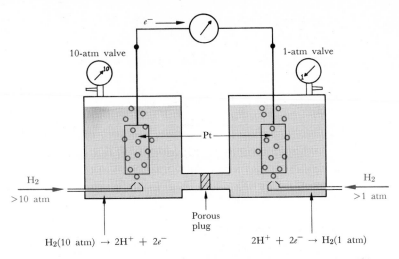

$$H_2(10 \text{ atm}) \rightarrow 2H^+ + 2e^- \qquad 2H^+ + 2e^- \rightarrow H_2(1 \text{ atm})$$

Figure 19-2   A hydrogen pressure cell for converting the free energy of expansion of H_2 at 10 atm into useful work. Two platinum electrodes are immersed in pure water (hydrogen ion concentration of 10^{-7} mole liter^{-1}) in two tanks that are connected by a channel with a porous plug, which permits ion flow but enables a pressure difference to be maintained. Hydrogen gas is bubbled over each electrode, and bleed-off valves and regulators maintain the H_2 pressure at 10 atm in the left tank and 1 atm in the right. In operation, the following reactions occur spontaneously:

Left tank: $H_2(g, 10 \text{ atm}) \rightarrow 2H^+ + 2e^-$
Right tank: $2H^+ + 2e^- \rightarrow H_2(g, 1 \text{ atm})$
Overall: $\overline{H_2(g, 10 \text{ atm}) \rightarrow H_2(g, 1 \text{ atm})}$

Electrons produced in the left-tank reaction flow through the external circuit to the right electrode, where they are used to react with hydrogen ions. (The dial with the arrow in the external electrical circuit symbolizes any current-measuring or work-producing device.) Hydroxide ions diffuse slowly through the porous plug from right to left to maintain electrical neutrality, and combine with the protons produced in the left tank.

Solution

$$\Delta G = \Delta G^0 + RT \ln \left(\frac{p_2}{p_1} \right)$$

$$= 0 + RT \ln \left(\frac{2}{10} \right)$$

$$= -3.99 \text{ kJ mole}^{-1}$$

The cell voltage is

$$\mathscr{E} = \frac{3.99 \text{ kJ}}{5.706 \text{ kJ}} \cdot 29.6 \text{ mV} = 20.7 \text{ mV}$$

Since the pressure ratio in the 5 atm/1 atm cell is the same as in the 10 atm/2 atm cell, the free energy change and voltage also will be the same.

The pressure cell separates the overall reaction into two parts: oxidation and reduction. The H_2 molecule is oxidized to H^+ in the left tank, with a loss of electrons, and H^+ is reduced to H_2 in the right tank, with a gain of electrons. The two terminals or electrodes are named according to whether electrons flow into them or out of them as observed from outside the system. Electrons flow away from the right electrode into the solution during reduction,* therefore this electrode is called the **cathode**. (*Cata-* means "away from," as in *catapult*.) Conversely, during oxidation electrons flow from the solution into the left electrode, which thus is called the **anode**. (*Ana-* means "back." For those of us for whom Greek is not a second tongue, it is easier to remember that *anode* and *oxidation* both begin with a vowel, and *cathode* and *reduction* begin with a consonant.)

Concentration Cells

The pressure cell is a kind of concentration cell; it can generate an external flow of electrons because the concentrations of H_2 gas in the two electrode compartments are different. We can make a similar concentration cell using Cu and $CuSO_4$. If two solutions of different concentrations of copper sulfate are placed in contact, they will mix spontaneously (Figure 19-3a). We can harness this spontaneous reaction by setting up a cell such as that shown in Figure 19-3b. In the left compartment with a dilute solution, the copper electrode will be eroded slowly as copper is oxidized to form more Cu^{2+} ions. Hence, the left electrode is the anode, and it accumulates an excess of electrons. In the right compartment, with a high Cu^{2+} concentration, some of these copper ions will be reduced and Cu will deposit on the copper cathode. If the two electrodes are connected, electrons will flow from right to left through the wire, and sulfate ions will diffuse from left to right through the porous plog to maintain electrical neutrality. The dilute left compartment becomes more concentrated in $CuSO_4$, and the concentrated right compartment becomes more dilute, just as in a free-mixing process. When the two compartments reach equal concentration, electron flow stops.

The overall reaction

$$Cu^{2+}(0.10M) \rightarrow Cu^{2+}(0.01M)$$

---

*In the sense that either a negative ion is formed from a neutral substance, or a positive ion combines with the electrons and is neutralized.

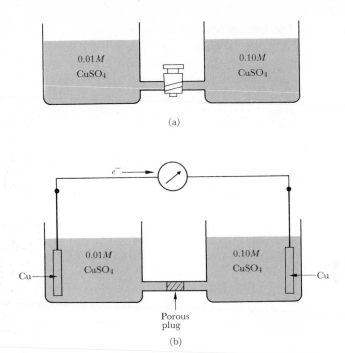

(a)

(b)

Figure 19-3   Copper sulfate concentration cell. (a) When the connecting stopcock is opened, the concentrated and dilute solutions mix without yielding any useful work. (b) Concentration cell for converting the free energy of mixing into useful work. Two copper electrodes are immersed in the 0.01M and 0.10M CuSO$_4$ solutions, which are connected by a porous plug that permits slow ion flow without bulk mixing of the solutions. The following reactions occur spontaneously:

Left compartment: $Cu(s) \rightarrow Cu^{2+}(0.01M) + 2e^-$
Right compartment: $Cu^{2+}(0.10M) + 2e^- \rightarrow Cu(s)$
Overall reaction: $\overline{Cu^{2+}(0.10M) \rightarrow Cu^{2+}(0.01M)}$

Electrons produced in the left compartment flow through the external circuit to the right, where they react with copper ions. The solution in the left compartment gradually becomes more concentrated, and that in the right compartment, more dilute. When the concentrations become equal, no further electron flow occurs.

has a free energy change that can be calculated from the expression

$$\Delta G = \Delta G^0 + RT \ln (c_2/c_1)$$

in which c_2 and c_1 are final and initial concentrations in moles liter^{-1}. For this example,

$$\Delta G = 0 + RT \ln (0.01/0.10) = -5.706 \text{ kJ mole}^{-1}$$

The free energy change is the same as for the 10 atm H_2 pressure cell because the ratio of concentrations is the same. As the reaction proceeds and the concentrations become more nearly equal, the free energy per mole of reaction decreases. (A given mass of water dropped over a 2-ft-high dam can do less work than the same mass of water dropped over a 20-ft dam. See Figure 17-1.) The initial voltage or potential difference between electrodes is 29.6 mV, as in the H_2 cell, and gradually falls to zero as the Cu^{2+} concentrations in the two compartments become equal and the cell runs down.

19-2   ELECTROCHEMICAL CELLS

The two reactions in an electrochemical cell need not be the reverse of one another to produce a useful cell. All that is required is two substances with different tendencies to gain or lose electrons. This difference in affinity for electrons then can be harnessed to accomplish useful work.

Zinc and Copper: The Daniell (Gravity) Cell

If a piece of slightly impure zinc is placed in a solution of copper sulfate, it will slowly be pitted and eroded away. At the same time, copper will be deposited on the zinc surface as a spongy brown coating, and the characteristic blue color of the copper sulfate solution will gradually fade. The zinc spontaneously replaces the copper ions in solution by the reaction

$$Zn(s) + Cu^{2+} \rightarrow Zn^{2+} + Cu(s)$$

because copper ions in solution have a greater affinity for electrons than do zinc ions.

Useful work can be obtained if we separate these two substances in a simple cell, as shown in Figure 19-4a. Zinc is oxidized spontaneously at the anode (left), and copper ions are reduced to the metal, which deposits on the cathode. Electrons flow through the external circuit from anode to cathode with a potential difference of 1.10 V if each of the solutions is $1M$. Anions diffuse left through the porous barrier to maintain electrical neutrality.

With a little reflection it should be apparent that neither the $ZnSO_4$ nor the copper rod is essential. Copper metal will deposit at the cathode on any other good conductor, such as a platinum wire, and the zinc sulfate solution in the anode compartment can be replaced by any other conducting salt that does not react with the zinc anode, such as sodium chloride. The porous barrier has a relatively high resistance to ion diffusion, and hence sets up a relatively high electrical resistance, which cuts down the current that can be drawn from the cell. A better method is to use a **salt bridge**, which is a glass U-tube containing an electrolyte such as KNO_3 mixed with agar or gelatin to hold it in place (Figure 19-4b).

The best setup for a cell that will not be moved is to let gravity separate the solutions, with no internal barrier at all (Figure 19-4c). In this cell a

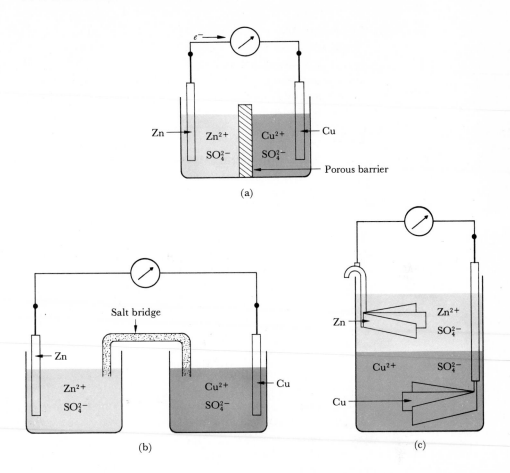

(a)

(b)

(c)

Figure 19-4

Three versions of a simple zinc—copper cell. In each case the oxidation reaction at the anode at the left is

$$Zn(s) \rightarrow Zn^{2+} + 2e^-$$

and the reduction at the cathode at the right is

$$Cu^{2+} + 2e^- \rightarrow Cu(s)$$

The zinc anode is eaten away as copper deposits on the cathode. If $ZnSO_4$ and $CuSO_4$ are present in $1M$ concentrations, this cell develops a potential or emf of $+1.10$ V.
(a) Two solutions separated by a porous barrier. (b) Solutions separated by a salt bridge.
(c) Solutions separated by gravity in a Daniell cell, taking advantage of different solution densities.

dilute solution of zinc sulfate is layered carefully over a concentrated, more dense copper sulfate solution. In the absence of motion or vibration the cell works quite well. Its internal resistance of almost zero makes it possible for large currents to be drawn from it. This Daniell cell was once used widely as a stationary power source in telegraph offices and for home appliances such as doorbells.

The Hydrogen Electrode

Other combinations of metals can be used in a cell similar to that shown in Figure 19-4b. If the metals are nickel and copper, nickel is oxidized at the anode, Cu^{2+} ions are reduced at the cathode, and the cell has a voltage or **electromotive force (emf)** of 0.57 V. If zinc and nickel are used, zinc is oxidized and Ni^{2+} ions are reduced, with an emf of 0.53 V (providing that the ions are at $1M$ concentration). Notice that the cell emf's are additive in the same way that the reactions are:

$$\begin{array}{ll} Ni + Cu^{2+} \rightarrow Ni^{2+} + Cu & \mathscr{E}^0 = +0.57 \text{ V} \\ Zn + Ni^{2+} \rightarrow Zn^{2+} + Ni & \mathscr{E}^0 = +0.53 \text{ V} \\ \hline Zn + Cu^{2+} \rightarrow Zn^{2+} + Cu & \mathscr{E}^0 = +1.10 \text{ V} \end{array}$$

The sum of these two cell reactions is the Zn–Cu reaction of the Daniell cell, and the sum of the two cell potentials gives the potential of the Daniell cell. (We shall draw more conclusions from these observations later.) The symbol $\mathscr{E}^0$, with the zero superscript, indicates a *standard* potential with all reacting ions at $1M$ concentrations* and a temperature of 298 K. The positive sign of the emf means that the cell equation as written is spontaneous from left to right.

An electrode reaction need not involve a metal; metals are merely particularly easy substances to shape and machine. Figure 19-5 shows a cell in which the cathode reaction is the liberation of hydrogen gas:

$$2H^+ + 2e^- \rightarrow H_2(g)$$

For a cell with the overall reaction

$$Zn(s) + 2H^+ \rightarrow Zn^{2+} + H_2(g)$$

the standard emf is $+0.76$ V. If the Zn anode is replaced by Cu, and $ZnSO_4$ by $CuSO_4$, electrons flow in the other direction, from right to left, because copper ions have more of an affinity for electrons than hydrogen ions have. Copper ions are reduced spontaneously at the left electrode, which therefore becomes the cathode:

$$\text{Cathode:} \quad Cu^{2+} + 2e^- \rightarrow Cu(s)$$

---

*Because solutions do not behave ideally when concentrated, the activity (or "effective concentration") really should be used instead of molarity. Standard potentials are actually defined at unit activity, rather than unit molarity. But because our goal is an understanding of basic principles rather than of laboratory technique, we shall ignore the difference between activity and molarity in this chapter.

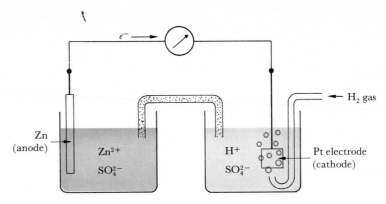

Figure 19-5

Cell with a hydrogen electrode. The two cell reactions are

Anode: $Zn(s) \rightarrow Zn^{2+} + 2e^-$
Cathode: $2H^+ + 2e^- \rightarrow H_2(g)$
Overall: $\overline{Zn(s) + 2H^+ \rightarrow Zn^{2+} + H_2(g)}$

The same overall reaction could be obtained less productively by immersing a strip of zinc in sulfuric acid. The metal would be eaten away (as at the anode in this cell) and bubbles of hydrogen gas would be given off (as at the cathode).

Hydrogen gas is oxidized at the anode on the right to form hydrogen ions:

Anode: $H_2(g) \rightarrow 2H^+ + 2e^-$

The overall spontaneous reaction is

$$Cu^{2+} + H_2(g) \rightarrow Cu(s) + 2H^+$$

and the standard potential of this cell is $\mathscr{E}^0 = +0.34$ V. Once again, notice the additivity of cell potentials. These two hydrogen-electrode reactions can be added to produce the Daniell cell reaction, and the sum of their emf's is the emf of the Daniell cell:

$$
\begin{array}{ll}
Zn(s) + 2H^+ \rightarrow Zn^{2+} + H_2(g) & \mathscr{E}^0 = +0.76 \text{ V} \\
Cu^{2+} + H_2(g) \rightarrow Cu(s) + 2H^+ & \mathscr{E}^0 = +0.34 \text{ V} \\
\hline
Zn(s) + Cu^{2+} \rightarrow Zn^{2+} + Cu(s) & \mathscr{E}^0 = +1.10 \text{ V}
\end{array}
$$

The Dry Cell

Any cell that involves liquids is difficult or impossible to use in a situation involving motion. (Try to imagine a flashlight powered by a Daniell cell!) The dry cell, shown in Figure 19-6, is particularly convenient because its components are either solids or moist pastes, which are sealed tightly from the environment. The anode is the zinc casing of the dry cell itself. Around the carbon-rod cathode is a paste composed of MnO_2, NH_4Cl, and H_2O.

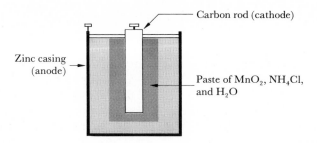

Figure 19-6   Dry cell. The individual electrode reactions are

Anode (zinc casing): $Zn(s) \rightarrow Zn^{2+} + 2e^-$

Cathode (carbon rod): $2MnO_2(s) + 8NH_4^+ + 2e^- \rightarrow 2Mn^{3+} + 4H_2O + 8NH_3$

Overall: $Zn(s) + 2MnO_2(s) + 8NH_4^+ \rightarrow$
$$Zn^{2+} + 2Mn^{3+} + 8NH_3 + 4H_2O$$

All chemical components are either solid or in the form of a paste, and the cell as a whole is sealed (hence the cell's name). Therefore, it is a highly useful cell for flashlights, radios, and other portable units.

At the anode zinc is oxidized to Zn^{2+} ions, and at the cathode MnO_2 is reduced to a mixture of several compounds of Mn in its $+3$ oxidation state. If the cell is used rapidly, ammonia, which is produced by the cathode reaction, can decrease the cell current by forming an insulating layer of gas around the carbon rod. With slower usage, zinc ions from the anode diffuse toward the cathode and combine with the ammonia to form complex ions such as $Zn(NH_3)_4^{2+}$. This is why apparently "dead" flashlight batteries sometimes can recover if allowed to rest.

Reversible Cells: The Lead Storage Battery

Most of the cells mentioned so far are reversible; that is, if a voltage greater than the cell voltage is applied from the outside, the cell reactions can be reversed, and electrical energy can be stored in the cell for withdrawal later. Thus in Figure 19-3b, if electrons are driven from right to left by an outside voltage, the $CuSO_4$ in the left compartment is made even more dilute, and the $CuSO_4$ in the right compartment is made more concentrated. In Figure 19-4c, an external voltage can cause additional Zn to deposit on the left terminal of the Daniell cell, and more copper from the right terminal to go into solution. Thus the free energy of the cell is brought to an even higher state, and at some later time this extra free energy can be released as useful work. Unfortunately, the commercial dry cell cannot be recharged in this way. The Zn^{2+} produced at the anode diffuses away, and the ammonia gas from the cathode reaction complexes with it. There is no convenient mechanism for breaking this complex ion and sending each component back to its original position.

What is needed for rechargeability is a cell in which the electrode products stay in place at the electrodes, ready to be reconverted as the cell is charged. An example of such a cell is the lead storage battery, shown in Figure 19-7. The anode is a spongy lead screen, and when the lead is oxidized to lead sulfate, it remains in place in the screen. Similarly, when lead oxide in the cathode is reduced, also to lead sulfate, the reduction products stay in place. The most noticeable change in the cell as it discharges is a dilution of the initially strong sulfuric acid solution. Therefore, the state of charge of the battery can be measured by a floating hygrometer. This instrument records the density of the battery fluid, and hence the strength of the sulfuric acid solution. The emf of this cell is 2 V, but several cells can be connected in series to produce 6-V and 12-V storage batteries.

When the cell has run down, it can be recharged by applying an external voltage in excess of the normal emf of 2 V per cell. The reactions shown in the caption of Figure 19-7 are reversed, and the lead sulfate is converted to lead and lead oxide. If the lead sulfate fell to the bottom of the tank as the cell discharged, this reverse reaction would be impossible. But it does not; it remains in place on the grid, ready for reconversion. Therefore, the lead storage battery is a convenient device for storing electrical energy in the form of chemical free energy.

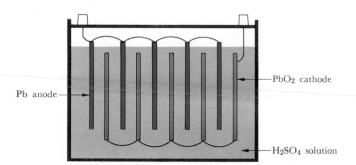

Figure 19-7   Lead storage battery. The electrode reactions are

Anode: $Pb(s) + SO_4^{2-} \rightarrow PbSO_4(s) + 2e^-$
Cathode: $PbO_2(s) + 4H^+ + SO_4^{2-} + 2e^- \rightarrow PbSO_4(s) + 2H_2O$

This battery is reversible (chargeable) since the product of the cell reaction, lead sulfate at both electrodes, adheres to the plates rather than diffusing or falling away. One cell of a lead storage battery, shown here, delivers approximately 2 V, and 6- or 12-V batteries have three or six cells connected in series.

Electrolytic Cells

The cell shown in Figure 19-8 can be used to produce high-purity copper if an external battery or other current source is used to drive the cell in a direction that it would not take spontaneously. An ingot of impure "blister" copper, which is prepared from the reduction of copper oxide (CuO) by coke (C), is suspended in a copper sulfate solution along with a "core" wire of very pure copper. Current is passed through the cell in the direction that makes the ingot the anode, and the pure wire the cathode. The ingot is eaten away, and copper ions plate out as very pure copper metal on the cathode, with the impurities settling to the bottom of the tank below the anode.

Other kinds of electrolytic cells can be used to plate gold or silver over base metals in jewelry, or to make accurate copies of engraving plates for printing. Our U.S. paper money is printed in plates of 12 notes. The engraver makes only one master engraving on steel, which is hardened and then copied by the electroplating process shown in Figure 19-9. The electrodeposited copy, called an *alto* since the engraved lines are now in relief, is then electrocopied to produce a *basso* whose intaglio lines are an exact copy of the grooves on the engraved master plate. These basso copies then are assembled into a 12-note plate that can be used either directly for printing or for making one-piece printing plates by repeating the electrocopying process.

Cells such as these, in which an external current source is used, are called *electrolytic* cells; those such as we discussed previously, which use internal chemical reactions to produce an electric current, are called *galvanic* cells. In both types of cells, the terminal at which oxidation occurs is the anode, and the cathode is the site of reduction.

Figure 19-8   Electrolytic purification of crude copper. Impure copper is oxidized at the anode, and pure copper deposits at the cathode. Impurities accumulate below the anode as "anode slime." The rare metals recovered from the anode slime, such as gold, silver, and platinum, are often valuable enough to pay for the cost of the purification process.

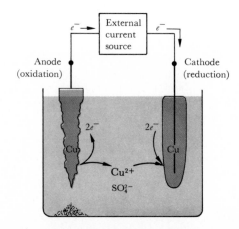

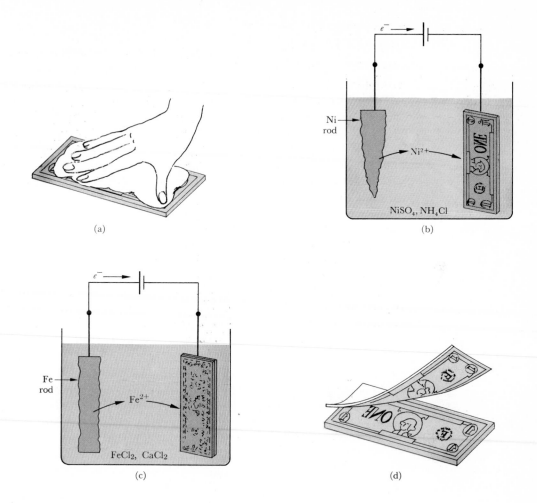

(a)

(b)

(c)

(d)

Figure 19-9   Reproducing engraved plates for U.S. currency by electrodeposition. (a) The engraved steel master plate is rubbed thoroughly with wet graphite powder and washed clean. The minute graphite coating aids electrodeposition and will make it possible to separate the master plate from the plated copy at the end. (b) The master is plated with a 0.025-mm layer of nickel during a 10-hour period. (The vertical tall and short lines at the top are the conventional sign for a battery, with the short line being the electron source—the battery anode. (c) The nickel layer is backed with a 0.1-mm layer of iron for strength, during a 14-hour electrodeposition. (d) The electrodeposited reverse image is peeled away and soldered to a steel plate. The process is repeated, starting with the reverse plate just produced, to yield a copy of the master that can be recopied or used for printing banknotes.

19-3   CELL emf AND FREE ENERGY

When a charge, q, moves spontaneously through a potential drop of $\mathscr{E}$ volts, the external electrical work that can be done on its surroundings is $w_{ext} = q\mathscr{E}$. Therefore, from equation 16-13, the free energy change of the system containing the moving electrical charge is

$$\Delta G = -w_{ext} = -q\mathscr{E} \tag{19-1}$$

Since the electron has a negative charge, it will move spontaneously from a region of low potential to one of high potential. For one electron moving through a potential increase of $\mathscr{E}$, the free energy change is

$$\Delta G = -e\mathscr{E}$$

For a mole of electrons,

$$\Delta G = -Ne\mathscr{E} = -\mathscr{F}\mathscr{E}$$

where N is Avogadro's number and $\mathscr{F} = Ne$ is the charge on 1 mole of electrons, 96,485 coulombs or 1 faraday. In a reaction involving n electrons per molecule of reaction, or n faradays per mole, the free energy change is

$$\Delta G = -n\mathscr{F}\mathscr{E} \tag{19-2}$$

If a galvanic cell is based on a reaction that yields ΔG kJ mole^{-1} of free energy, and if n moles of electrons are transferred through the external circuit per mole of reaction, then the potential difference between terminals, $\mathscr{E} = \mathscr{E}_{cathode} - \mathscr{E}_{anode}$ is given by

$$\mathscr{E}_{cathode} - \mathscr{E}_{anode} = \mathscr{E} = -\Delta G/n\mathscr{F}$$

Since 1 volt-coulomb equals 1 J, equation 19-2 can be written

$$\Delta G(kJ) = -96.5\, n\mathscr{E}\,(volts)$$

The pressure cell discussed in Section 19-1, for which the two-electron reaction had a standard free energy of -5.706 kJ mole^{-1}, therefore has a standard cell potential of

$$\mathscr{E}^0 = -\frac{-5.706 \text{ kJ mole}^{-1}}{2 \times 96.5 \text{ kJ mole}^{-1}\text{ V}^{-1}} = +0.0296 \text{ V}$$

The Daniell cell reaction,

$$Zn(s) + Cu^{2+} \rightarrow Zn^{2+} + Cu(s)$$

has a standard cell potential of $+1.10$ V. Therefore, the standard free energy of this reaction must be

$$\Delta G^0 = -2(96.5)(+1.10) = -212.3 \text{ kJ mole}^{-1}$$

In this expression $n = 2$ because two electrons are transferred per ion. You can calculate the standard free energies of other cells that we have discussed.

Example 2

Calculate the standard free energies of the Ni–Cu, Zn–Cu, and Zn–Ni cells discussed in Section 19-2. Demonstrate that the free energies are additive in the same way that the reactions and the potentials are.

Solution

Ni–Cu cell: $\mathscr{E}^0 = +0.57$ V; $\Delta G = -110$ kJ mole^{-1}

Zn–Ni cell: $\mathscr{E}^0 = +0.53$ V; $\Delta G = -102$ kJ mole^{-1}

Zn–Cu cell: $\mathscr{E}^0 = +1.10$ V; $\Delta G = -212$ kJ mole^{-1}

As you can see, for a cell to deliver even $1\,V$ of potential difference between terminals requires a cell reaction that has a large free energy change. Since the free energies must be additive (by the first law of thermodynamics), it should not be surprising that emf's are additive also, as was explained in Section 19-2.

This additivity of emf's is represented in Figure 19-10. Recall from the discussion of free energy in Chapter 16 that we do not need to tabulate the free energy change for every possible reaction. Having once tabulated the free energy change for a certain kind of reaction, namely, the formation of

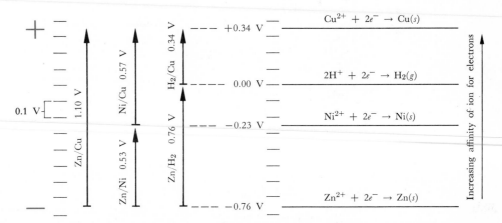

Figure 19-10   Since the potentials for the various cells involving Zn, Cu, Ni, and H$_2$ that we discussed previously are additive, they can be represented as additive distances along a vertical axis measured in volts. The choice of a zero potential along this axis is arbitrary, but once made, all voltages can be specified relative to this zero point. In this figure, the hydrogen electrode has been selected as the reference zero voltage. This is the standard convention in tables, although Ni, Zn, or any other electrode, could have been chosen.

Table 19-1

Standard Reduction Potentials in Acid Solution[a] at 298 K

Half-reaction (couple)	$\mathcal{E}^{\circ}$ (volts)
$F_2 + 2e^- \rightarrow 2F^-$	2.87
$Ag^{2+} + e^- \rightarrow Ag^+$	1.99
$H_2O_2 + 2H^+ + 2e^- \rightarrow 2H_2O$	1.78
$PbO_2 + 4H^+ + SO_4^{2-} + 2e^- \rightarrow PbSO_4 + 2H_2O$	1.69
$MnO_4^- + 4H^+ + 3e^- \rightarrow MnO_2 + 2H_2O$	1.68
$MnO_4^- + 8H^+ + 5e^- \rightarrow Mn^{2+} + 4H_2O$	1.49
$PbO_2 + 4H^+ + 2e^- \rightarrow Pb^{2+} + 2H_2O$	1.46
$Cl_2 + 2e^- \rightarrow 2Cl^-$	1.36
$Cr_2O_7^{2-} + 14H^+ + 6e^- \rightarrow 2Cr^{3+} + 7H_2O$	1.33
$O_2 + 4H^+ + 4e^- \rightarrow 2H_2O$	1.23
$MnO_2 + 4H^+ + 2e^- \rightarrow Mn^{2+} + 2H_2O$	1.21
$Br_2(l) + 2e^- \rightarrow 2Br^-$	1.09
$AuCl_4^- + 3e^- \rightarrow Au + 4Cl^-$	0.99
$NO_3^- + 4H^+ + 3e^- \rightarrow NO + 2H_2O$	0.96
$2Hg^{2+} + 2e^- \rightarrow Hg_2^{2+}$	0.91
$Ag^+ + e^- \rightarrow Ag$	0.80
$Hg_2^{2+} + 2e^- \rightarrow 2Hg$	0.80
$Fe^{3+} + e^- \rightarrow Fe^{2+}$	0.77
$O_2 + 2H^+ + 2e^- \rightarrow H_2O_2$	0.68
$MnO_4^- + e^- \rightarrow MnO_4^{2-}$	0.56
$I_2 + 2e^- \rightarrow 2I^-$	0.54
$Cu^+ + e^- \rightarrow Cu$	0.52

[a]Values from the *Handbook of Chemistry and Physics,* 1978–9, 59th
 ed., Chemical Rubber Co., Cleveland, Ohio.

each compound from elements in their standard states, we can then calculate the free energy change for any reaction involving these compounds, because of the additivity of free energies. Similarly, we do not need to tabulate the potential of every conceivable cell, or of every conceivable combination of anode and cathode reactions. Instead, we need only tabulate voltages of cells in which all electrode reactions are paired with one standard electrode. This amounts to choosing an arbitrary zero in Figure 19-10. We can divide a cell reaction into two half-reactions, one at the anode and the other at the cathode, and we can assign to each half-reaction the voltage that would be observed in a cell if that half-reaction were paired with

$$H_2(g) \rightarrow 2H^+ + 2e^-$$

This is the basis for the scale of reduction potentials in Tables 19-1 and 19-2.

Table 19-1 continued

Standard Reduction Potentials in Acid Solution at 298 K

Half-reaction (couple)	$\mathscr{E}^0$ (volts)
$Cu^{2+} + 2e^- \rightarrow Cu$	0.34
$Hg_2Cl_2 + 2e^- \rightarrow 2Hg + 2Cl^-$	0.27
$AgCl + e^- \rightarrow Ag + Cl^-$	0.22
$SO_4^{2-} + 4H^+ + 2e^- \rightarrow H_2SO_3 + H_2O$	0.20
$Cu^{2+} + e^- \rightarrow Cu^+$	0.16
$2H^+ + 2e^- \rightarrow H_2$	0.00
$Pb^{2+} + 2e^- \rightarrow Pb$	-0.13
$Sn^{2+} + 2e^- \rightarrow Sn$	-0.14
$Ni^{2+} + 2e^- \rightarrow Ni$	-0.23
$PbSO_4 + 2e^- \rightarrow Pb + SO_4^{2-}$	-0.35
$Cd^{2+} + 2e^- \rightarrow Cd$	-0.40
$Cr^{3+} + e^- \rightarrow Cr^{2+}$	-0.41
$Fe^{2+} + 2e^- \rightarrow Fe$	-0.41
$Zn^{2+} + 2e^- \rightarrow Zn$	-0.76
$Mn^{2+} + 2e^- \rightarrow Mn$	-1.03
$Al^{3+} + 3e^- \rightarrow Al$	-1.71
$H_2 + 2e^- \rightarrow 2H^-$	-2.23
$Mg^{2+} + 2e^- \rightarrow Mg$	-2.37
$La^{3+} + 3e^- \rightarrow La$	-2.37
$Na^+ + e^- \rightarrow Na$	-2.71
$Ca^{2+} + 2e^- \rightarrow Ca$	-2.76
$Ba^{2+} + 2e^- \rightarrow Ba$	-2.90
$K^+ + e^- \rightarrow K$	-2.92
$Li^+ + e^- \rightarrow Li$	-3.05

19-4 HALF-REACTIONS AND REDUCTION POTENTIALS

The reduction potentials in Table 19-1 are the potentials that would be observed if the reduction equation as written were paired with the oxidation of hydrogen:

$$H_2(g) \rightarrow 2H^+ + 2e^-$$

A positive sign indicates that the cell reaction will go spontaneously in the direction indicated. A negative sign means that the reverse reaction will occur; that is, the particular substance will be oxidized and protons will be reduced to hydrogen gas. The more positive the reduction potential, the greater the tendency for the substance to accept electrons and become reduced. A large negative reduction potential indicates a strong favoring of

Table 19-2

Standard Reduction Potentials in Basic Solution at 298 K

Half-reaction (couple)[a]	$\mathscr{E}^0$ (volts)
$HO_2^- + H_2O + 2e^- \rightarrow 3OH^-$	0.87
$MnO_4^- + 2H_2O + 3e^- \rightarrow MnO_2 + 4OH^-$	0.59
$O_2 + 4e^- + 2H_2O \rightarrow 4OH^-$	0.40
$Co(NH_3)_6^{3+} + e^- \rightarrow Co(NH_3)_6^{2+}$	0.10
$HgO + H_2O + 2e^- \rightarrow Hg + 2OH^-$	0.10
$MnO_2 + H_2O + 2e^- \rightarrow Mn(OH)_2 + 2OH^-$	−0.05
$O_2 + H_2O + 2e^- \rightarrow HO_2^- + OH^-$	−0.08
$Cu(NH_3)_2^+ + e^- \rightarrow Cu + 2NH_3$	−0.12
$Ag(CN)_2^- + e^- \rightarrow Ag + 2CN^-$	−0.31
$Hg(CN)_4^{2-} + 2e^- \rightarrow Hg + 4CN^-$	−0.37
$S + 2e^- \rightarrow S^{2-}$	−0.51
$Pb(OH)_3^- + 2e^- \rightarrow Pb + 3OH^-$	−0.54
$Fe(OH)_3 + e^- \rightarrow Fe(OH)_2 + OH^-$	−0.56
$Cd(OH)_2 + 2e^- \rightarrow Cd + 2OH^-$	−0.76
$SO_4^{2-} + H_2O + 2e^- \rightarrow SO_3^{2-} + 2OH^-$	−0.92
$Zn(NH_3)_4^{2+} + 2e^- \rightarrow Zn + 4NH_3$	−1.03
$Zn(OH)_4^{2-} + 2e^- \rightarrow Zn + 4OH^-$	−1.22
$Mn(OH)_2 + 2e^- \rightarrow Mn + 2OH^-$	−1.47
$Mg(OH)_2 + 2e^- \rightarrow Mg + 2OH^-$	−2.67
$Ca(OH)_2 + 2e^- \rightarrow Ca + 2OH^-$	−3.02

[a]Couples involving ions not affected by changing pH (such as $Na^+|Na$) have the same potential in acid or base.

the oxidized state. (These are *standard* values, which means a $1M$ concentration for all reacting ions and 1 atm partial pressure for gases at 298 K.)

Example 3

If chlorine gas at 1 atm is bubbled over one platinum electrode in a solution of hydrochloric acid, and hydrogen gas at 1 atm is bubbled over a similar electrode, what will be the overall reaction and what will be the emf of the resulting cell?

Solution   The standard reduction potential for the reaction

$$Cl_2(g) + 2e^- \rightarrow 2Cl^-$$

is $\mathscr{E}^0 = +1.36$ V. This means that when combined with the hydrogen electrode, the chlorine electrode is the cathode (where reduction occurs) and the hydrogen electrode is the anode. The overall reaction

$$Cl_2(g) + H_2(g) \rightarrow 2Cl^- + 2H^+$$

is spontaneous from left to right and has a standard free energy change of

$$\Delta G^0 = -n\mathscr{F}\mathscr{E}^0 = -2(96.5)(+1.36) = -262 \text{ kJ mole}^{-1} \text{ Cl}_2 \text{ or } \text{H}_2$$

Verify this free energy value from the data in Appendix 3.

Example 4

What is the spontaneous reaction when a cadmium electrode and a hydrogen electrode are paired in acid solution? What is the cell potential?

Solution

The standard reduction potential for the half-reaction

$$\text{Cd}^{2+} + 2e^- \rightarrow \text{Cd}(s)$$

is -0.40 V. Thus, the overall reaction

$$\text{Cd}^{2+} + \text{H}_2(g) \rightarrow \text{Cd}(s) + 2\text{H}^+$$

is spontaneous *in reverse,* from right to left. Cadmium will be oxidized spontaneously to ions at the anode, and hydrogen ions will be reduced to hydrogen gas at the cathode. The cell potential is $+0.40$ V.

Example 5

Will a piece of cadmium, dropped into a $1M$ acid solution, produce bubbles of hydrogen gas? What about a piece of silver?

Solution

From the previous example, the reaction

$$\text{Cd}(s) + 2\text{H}^+ \rightarrow \text{Cd}^{2+} + \text{H}_2(g)$$

is spontaneous under standard conditions, so bubbles of hydrogen gas will escape as the cadmium metal is etched away by the acid solution. In contrast, silver has a positive reduction potential and a stronger tendency to remain reduced than hydrogen has. Thus silver, when dropped into a weak acid solution, will remain unaffected. The "nobility" of the noble metals gold, silver, and platinum is primarily a consequence of their large, positive reduction potentials.

Cell Potentials from Reduction Potentials of Half-Reactions

To find the potential of a cell in which a given reaction is occurring, first break the reaction up into its two half-reactions. Choose one of these half-reactions to be a reduction reaction at the cathode, and the other half-reaction to be an oxidation reaction at the anode. Write the second reaction in reverse, as an oxidation. Then find the standard reduction potentials for these two half-reactions and reverse the sign of $\mathscr{E}^0$ for the reaction that has been taken as the oxidation. Add the two half-reactions as a check to be sure that you obtain the original overall reaction, and add the two half-reaction potentials at the same time. After finishing, if you have a positive overall potential, the reaction as written is spontaneous. If the overall potential is negative, you made the wrong assumption about the anode and the reverse reaction is spontaneous.

Example 6

Find the emf of a cell made up of a Zn electrode in $ZnSO_4$ and a Cu electrode in $CuSO_4$. Which is the anode and which is the cathode?

Solution

This is just the Daniell cell, and you know the answer already: the Zn electrode is the anode. But assume that you did not know this, and guessed (wrongly) that the Cu electrode was the anode. The two half-reactions are

$$Zn^{2+} + 2e^- \rightarrow Zn(s) \qquad \mathscr{E}^0 = -0.76 \text{ V}$$
$$Cu^{2+} + 2e^- \rightarrow Cu(s) \qquad \mathscr{E}^0 = +0.34 \text{ V}$$

It is obvious that by reversing the Zn reaction and changing the sign of its emf we obtain a positive overall emf. Nevertheless, let's assume that the copper electrode is the anode. The two reactions to be added then are

$$\begin{array}{ll} Zn^{2+} + 2e^- \rightarrow Zn(s) & \mathscr{E}^0 = -0.76 \text{ V} \\ Cu(s) \rightarrow Cu^{2+} + 2e^- & \mathscr{E}^0 = -0.34 \text{ V} \\ \hline Zn^{2+} + Cu(s) \rightarrow Zn(s) + Cu^{2+} & \mathscr{E}^0 = -1.10 \text{ V} \end{array}$$

The negative potential tells us that the reverse reaction is spontaneous, and that the cell thus set up will have an emf of $+1.10$ V. This method is foolproof against errors in the original assumption about anode and cathode.

Example 7

What is the spontaneous direction in a cell made up of a ferrous–ferric ion electrode and an iodine–iodide ion electrode? What is the cell emf?

Solution

The half-reactions are

$$Fe^{3+} + e^- \rightarrow Fe^{2+} \qquad \mathscr{E}^0 = +0.77 \text{ V}$$
$$I_2 + 2e^- \rightarrow 2I^- \qquad \mathscr{E}^0 = +0.54 \text{ V}$$

It is obvious that the way to obtain a positive overall voltage is to subtract the second half-reaction from the first, thereby reversing the sign of the 0.54-V term. The spontaneous overall reaction is

$$2Fe^{3+} + 2I^- \rightarrow 2Fe^{2+} + I_2 \qquad \mathscr{E}^0 = +0.23 \text{ V}$$

In the preceding example, we had to multiply the first equation by 2 before subtracting the second equation from it because one electron was involved, whereas the iodine equation involved two. If we were subtracting free energies we would multiply the standard free energy change for the iron half-reaction by 2 before subtracting. Should we multiply the potential, $+0.77$ V, by 2 also? No, because the number of electrons has already been taken into account by n in the expressions

$$\Delta G^0 = -n\mathscr{F}\mathscr{E}^0 \qquad \text{and} \qquad \mathscr{E}^0 = -\Delta G^0/n\mathscr{F}$$

Half-cell potentials are already on a "per electron" basis and can be combined directly. To use our hydrostatic analogue again, these potentials are electron "pressures" rather than energy quantities. The pressure of water behind a dam of specified height does not depend on whether we take the water from the bottom in 1- or 2-gallon batches, but the work or energy obtained per batch does.

The factor of 2, for two electrons, will be taken into account as soon as we calculate free energy changes. For the iron half-reaction,

$$2Fe^{3+} + 2e^- \rightarrow 2Fe^{2+}$$
$$\Delta G^0 = -2\mathscr{F}\mathscr{E}^0 = -2(96.5)(+0.77) = -149 \text{ kJ}$$

For the iodine half-reaction,

$$I_2(s) + 2e^- \rightarrow 2I^-$$
$$\Delta G^0 = -2\mathscr{F}\mathscr{E}^0 = -2(96.5)(+0.54) = -104 \text{ kJ}$$

Subtracting the second half-reaction from the first, and the second free energy change from the first, we get

$$\Delta G^0 = -149 \text{ kJ} + 104 \text{ kJ} = -45 \text{ kJ}$$

As a check, verify that the value of this free energy change leads to the overall cell potential, using $\Delta G^0 = -n\mathscr{F}\mathscr{E}^0$.

Any half-reaction with a higher positive reduction potential will dominate over a half-reaction with a lower reduction potential, and send the latter in the reverse direction if the two half-reactions are coupled in a cell. Thus the two half-reactions for the lead storage battery are

$$PbO_2 + 4H^+ + SO_4^{2-} + 2e^- \rightarrow PbSO_4 + 2H_2O \qquad \mathscr{E}^0 = +1.69 \text{ V}$$
$$PbSO_4 + 2e^- \rightarrow Pb + SO_4^{2-} \qquad \mathscr{E}^0 = -0.36 \text{ V}$$

Since the first reaction has the higher reduction potential, it will take place at the cathode, and the reverse of the second reaction will occur at the anode. The overall cell potential will be

$$\mathscr{E}^0 = (+1.69 \text{ V}) - (-0.35 \text{ V}) = +2.04 \text{ V}$$

Ions of any metal in the reduction potential table will become reduced in the presence of a metal lower in the table. Thus silver from a silver nitrate solution will precipitate in the presence of zinc, iron, cadmium, or even copper or mercury. In the presence of Ag^+ will iron be the ferrous ion, Fe^{2+}, or the ferric ion, Fe^{3+}?

The half-reactions from Table 19-1 are

$$(1) \quad Ag^+ + e^- \rightarrow Ag \quad \mathscr{E}^0 = +0.80 \text{ V};$$
$$\Delta G^0 = -1(96.5)(+0.80) = -77 \text{ kJ mole}^{-1}$$

$$(2) \quad Fe^{3+} + e^- \rightarrow Fe^{2+} \quad \mathscr{E}^0 = +0.77 \text{ V};$$
$$\Delta G^0 = -1(96.5)(+0.77) = -74 \text{ kJ mole}^{-1}$$

$$(3) \quad Fe^{2+} + 2e^- \rightarrow Fe(s) \quad \mathscr{E}^0 = -0.41 \text{ V};$$
$$\Delta G^0 = -2(96.5)(-0.41) = +79 \text{ kJ mole}^{-1}$$

But why is no reaction listed for the reduction of Fe^{3+} to metallic iron? The reaction we want,

$$Fe^{3+} + 3e^- \rightarrow Fe(s)$$

can be obtained by adding half-reactions 2 and 3. Can we find the potential by adding the potentials of these two reactions?

$$\mathscr{E}^0 = +0.77 - 0.41 = +0.36 \text{ V}$$

The answer is no; we cannot. The emf of the reduction of Fe^{3+} to $Fe(s)$ is not $+0.36$ V. It is legitimate to subtract one half-cell *potential* from another when the corresponding half-cell *reactions* are subtracted to form a correct overall cell reaction with proper balancing of electron gain and loss. It is not legitimate to add the potentials of a one-electron half-reaction and a two-electron half-reaction to obtain the potential of the resulting three-electron half-reaction.

It is always safe to work with free energies, which is why we bothered to calculate free energies for the three half-reactions. In calculating free energies from cell potentials, explicit counts of electrons involved are made via the factor n in the expression $\Delta G^0 = -n\mathscr{F}\mathscr{E}^0$. Since the desired half-cell reaction is the sum of half-reactions 2 and 3, the overall free energy is the sum of the two free energies:

$$Fe^{3+} + 3e^- \rightarrow Fe(s) \quad \Delta G^0 = -74 + 79 = +5 \text{ kJ mole}^{-1}$$

The standard emf can now be found by dividing by $-3\mathscr{F}$, since this is a three-electron reaction:

$$\mathscr{E}^0 = \frac{+5}{-3(96.5)} = -0.02 \text{ V}$$

But this is not a realistic half-reaction. If both Fe^{3+} and metallic iron were present together, they would combine spontaneously to produce Fe^{2+}, as the free energies indicate:

$$Fe(s) \rightarrow Fe^{2+} + 2e^- \quad \Delta G^0 = -79 \text{ kJ}$$
$$\underline{2Fe^{3+} + 2e^- \rightarrow 2Fe^{2+} \qquad \Delta G^0 = 2(-74) \text{ kJ}}$$
$$Fe(s) + 2Fe^{3+} \rightarrow 3Fe^{2+} \quad \Delta G^0 = -227 \text{ kJ}$$

This is a highly spontaneous reaction with a strongly negative standard free energy change. Since a Fe–Fe^{3+} half-cell is physically unrealistic, its potential is not tabulated.

We have now answered the original question. When silver plates out from an Ag$^+$ solution in the presence of metallic iron, the iron goes into solution as Fe^{2+}. If some Fe^{3+} were present, it immediately would combine with metallic iron by the preceding reactions to make more Fe^{2+}. Thus, the two electrochemical reactions involved are

$$Ag^+ + e^- \rightarrow Ag \qquad \mathscr{E}^0 = +0.80 \text{ V}; \Delta G^0 = -77$$
$$Fe^{2+} + 2e^- \rightarrow Fe(s) \qquad \mathscr{E}^0 = -0.41 \text{ V}; \Delta G^0 = +79$$

Handling of electrode potentials can often be confusing. A safe guideline: *When in doubt, work with free energies.*

Shorthand Notation for Electrochemical Cells

In the usual abbreviated notation for a cell, the reactants and products are listed from left to right in the form

Anode | Anode solution || Cathode solution | Cathode

A single vertical line indicates a change of phase: solid, liquid, gas, or solution. A double vertical line indicates a porous barrier or salt bridge between two solutions. Thus, some of the cells we discussed would be written as follows:

H$_2$ pressure cell: Pt | H$_2$(g, 10 atm) | H$_2$O || H$_2$O | H$_2$(g, 1 atm) | Pt

Cu concentration cell: Cu | Cu^{2+}(0.01M) || Cu^{2+}(0.10M) | Cu

Daniell cell: Zn | Zn^{2+}(xM) || Cu^{2+}(yM) | Cu
(in which x and y
are the molarities of
the ionic solutions)

Lead storage battery: Pb | H$_2$SO$_4$(aq) | PbO$_2$

19-5   THE EFFECT OF CONCENTRATION ON CELL VOLTAGE: THE NERNST EQUATION

The reduction potentials that we have been working with so far all have been standard potentials, that is, with the concentrations of all solutes at 1 mole liter^{-1} and all gases at 1 atm partial pressure,* and at 298 K. Does the emf of a cell change with concentration? It does, for the same reasons that the free energy of the cell reaction changes. We saw special examples of this change at the beginning of this chapter in connection with concentration cells, and now we will derive a more general relationship.

---

*To be absolutely correct, with all substances at unit activity.

The relationship between free energy and concentration was given by equation 16-26:

$$\Delta G = \Delta G^0 + RT \ln Q$$

in which the reaction quotient, Q, is the ratio of products to reactants, each raised to its proper stoichiometric power. We can convert this into a cell-voltage equation, using $\Delta G = -n\mathcal{F}\mathcal{E}$, to yield

$$\mathcal{E} = \mathcal{E}^0 - \frac{RT}{n\mathcal{F}} \ln Q$$

This is called the **Nernst equation** after Walther Nernst, who first proposed it in 1881. In it $\mathcal{E}^0$ is the standard cell potential, Q is the ratio of concentrations under the conditions of a given experiment, and $\mathcal{E}$ is the cell voltage measured under these same conditions. For the copper concentration cell discussed at the beginning of this chapter,

$$\Delta G = 0 + RT \ln (c_2/c_1) = 5.706 \log_{10} (c_2/c_1) \qquad \text{at 298 K}$$

$$\mathcal{E} = 0 - \frac{RT}{2\mathcal{F}} \ln (c_2/c_1) = -\frac{5.706}{2\mathcal{F}} \log_{10} (c_2/c_1)$$

The quantity $RT/\mathcal{F}$ at 298 K is encountered so often that it should be evaluated first:

$$RT/\mathcal{F} = \frac{8.314 \text{ J K}^{-1} \times 298 \text{ K}}{96,485 \text{ coulombs}} = 0.0257 \text{ J coulomb}^{-1}$$

$$RT/\mathcal{F} = 0.0257 \text{ V}$$

$$2.303 \, RT/\mathcal{F} = 0.0592 \text{ V}$$

The general expression then is

$$\mathcal{E} = \mathcal{E}^0 - \frac{0.0592}{n} \log_{10} Q$$

and the emf of the copper concentration cell is

$$\mathcal{E} = -0.0296 \log_{10} (c_2/c_1)$$

As we calculated previously, for a concentration ratio of 1:10 the emf is $+0.0296$ V or 29.6 mV. This value is true for *any* two-electron concentration cell with a 1:10 concentration ratio, no matter what the chemical reaction actually is. Thus, the same voltages were calculated previously for the hydrogen cell and for the copper concentration cell. For a ratio of 1:5 the emf of the cell is

$$\mathcal{E} = +29.6 \log_{10} (5) = +20.7 \text{ mV}$$

Example 8

What are the emf and free energy change for the cell

$$\text{Cd}|\text{Cd}^{2+}(0.0500M)\,||\,\text{Cl}^-(0.100M)|\text{Cl}_2(1 \text{ atm})|\text{Pt}$$

Solution

The half-reactions are

$$Cd^{2+} + 2e^- \rightarrow Cd(s) \qquad \mathscr{E}^0 = -0.40 \text{ V}$$
$$Cl_2(g) + 2e^- \rightarrow 2Cl^- \qquad \mathscr{E}^0 = +1.36 \text{ V}$$

A positive overall voltage will be obtained by taking the $Cd|Cd^{2+}$ electrode to be the anode, where oxidation occurs, and subtracting it from the Cl_2 reaction to yield

$$Cl_2(g) + Cd(s) \rightarrow 2Cl^- + Cd^{2+} \qquad \begin{aligned} \mathscr{E}^0 &= +1.36 - (-0.40) \\ &= +1.76 \text{ V} \end{aligned}$$

The cell voltage at other than standard conditions is found from the expression

$$\mathscr{E} = +1.76 - \frac{0.0592}{2} \log_{10} \frac{[Cl^-]^2[Cd^{2+}]}{1 \cdot 1}$$

Chlorine gas at 1 atm and solid Cd both are in their standard states, thereby giving unit activities in the denominator. Hence,

$$\begin{aligned} \mathscr{E} &= +1.76 - 0.0296 \log_{10} [(0.100)^2(0.0500)] \\ &= +1.76 - 0.0296 \log_{10} (0.000500) \\ &= +1.76 + 0.0296(3.301) = +1.86 \text{ V} \end{aligned}$$

$$\Delta G = -2\mathscr{F}\mathscr{E} = -193(1.86) = -359 \text{ kJ}$$

Single-Electrode Potentials

It is often more convenient to deal with the concentration dependence of each half-reaction separately, and then to combine the results. The Nernst equation can be broken up by imagining that each half-reaction, oxidation and reduction, is coupled with the standard hydrogen reaction,

$$2H^+(1.0M) + 2e^- \rightarrow H_2(g, 1 \text{ atm}) \qquad \mathscr{E}^0 = 0.000 \text{ V}$$

in which both H^+ and H_2 have unit activities. Thus for the $Zn^{2+}|Zn(s)$ half-reaction,

$$\mathscr{E} = -0.76 - 0.0296 \log_{10} \frac{1}{[Zn^{2+}]}$$

since the activity of solid zinc in the numerator is 1. The emf of the hydrogen electrode, $H^+|H_2(g)$, depends upon pH. If the pressure of the hydrogen gas is maintained at 1 atm then this dependence is

$$\mathscr{E} = 0.000 - 0.0592 \text{ pH}$$

(Prove to yourself that this is true.) For the half-reaction $Fe^{3+}|Fe^{2+}$,

$$\mathscr{E} = +0.77 - 0.0592 \log_{10} \frac{[Fe^{2+}]}{[Fe^{3+}]}$$

And for the reaction

$$MnO_4^- + 8H^+ + 5e^- \rightarrow Mn^{2+} + 4H_2O$$

$$\mathscr{E} = +1.49 - 0.0118 \log_{10} \frac{[Mn^{2+}]}{[MnO_4^-][H^+]^8}$$

$$= +1.49 - 0.0118 \log_{10} \frac{[Mn^{2+}]}{[MnO_4^-]} - 0.0947\ pH$$

(Notice the division of 0.0592 by 5 to yield 0.0118, and the remultiplication by 8 at the right.) Therefore, any factors that affect ion concentrations will affect electrode potentials.

These individual half-reaction equations, once written properly with respect to concentrations of ions, then can be combined to form the Nernst equation for the cell as a whole.

Example 9

Write a balanced equation and calculate K_{eq} for the reaction involving the $Fe^{3+}|Fe^{2+}$ and H^+, $MnO_4^-|Mn^{2+}$ couples.

Solution

(a) $5e^- + 8H^+ + MnO_4^- \rightarrow Mn^{2+} + 4H_2O$

$$\mathscr{E} = 1.49 - \frac{0.0592}{5} \log_{10} \frac{[Mn^{2+}]}{[MnO_4^-][H^+]^8}$$

(b) (reversed) $Fe^{2+} \rightarrow Fe^{3+} + e^-$

$$\mathscr{E} = -0.77 - \frac{0.0592}{1} \log_{10} \frac{[Fe^{3+}]}{[Fe^{2+}]}$$

To provide enough electrons for a balanced equation with the manganese reaction, we must multiply the iron reaction by 5:

(c) $5Fe^{2+} \rightarrow 5Fe^{3+} + 5e^-$

$$\mathscr{E} = -0.77 - \frac{0.0592}{5} \log_{10} \frac{[Fe^{3+}]^5}{[Fe^{2+}]^5}$$

Notice how the coefficients in the Fe equations have been treated in the denominator of the 0.0592 term and in the exponents within the logarithm. Recall that

$$\frac{1}{n} \log a^n = \log a$$

Now add (a) and (c):

(d) $8H^+ + MnO_4^- + 5Fe^{2+} \rightarrow Mn^{2+} + 4H_2O + 5Fe^{3+}$

$$\mathcal{E} = 0.72 - \frac{0.0592}{5} \log_{10} \frac{[Mn^{2+}][Fe^{3+}]^5}{[H^+]^8[MnO_4^-][Fe^{2+}]^5}$$

At equilibrium, $\mathcal{E} = 0$, and

$$\mathcal{E}^0_{cell} = 0.72 \text{ V} = \frac{0.0592}{5} \log_{10} K_{eq}$$

$$K_{eq} = 6 \times 10^{60} = \frac{[Mn^{2+}][Fe^{3+}]^5}{[H^+]^8[MnO_4^-][Fe^{2+}]^5}$$

Notice that in considering the half-reaction

$$Fe^{3+} + e^- \rightarrow Fe^{2+}$$

the magnitude of the potential is independent of the number of times we use it in the balanced equation:

$$\mathcal{E} = \mathcal{E}^0 - 0.0592 \log_{10} \frac{[Fe^{2+}]}{[Fe^{3+}]} = \mathcal{E}^0 - \frac{0.0592}{5} \log_{10} \frac{[Fe^{2+}]^5}{[Fe^{3+}]^5}$$

Range of K_{eq} for Oxidation–Reduction Reactions

If we examine the reduction potentials in Tables 19-1 and 19-2, we find that they range from about $+3$ to -3 V. A difference of 6 V in half-reaction potentials corresponds to an equilibrium constant of 10^{100n}, in which n is the number of electrons transferred in a redox process:

$$\mathcal{E}^0 = \frac{0.0592}{n} \log_{10} K_{eq} = 6.0$$

$$\log_{10} K_{eq} \simeq 100n$$
$$K_{eq} \simeq 10^{100n}$$

This is an immensely large equilibrium constant in comparison with the maximum K_{eq} value of about 10^{14} encountered in proton-transfer reactions in aqueous solution. This large range of equilibrium-constant values within a redox potential range of 6 V means that *the chance of two half-cell potentials being close enough to establish equilibrium with significant quantities of both reactants and products present is small.* An equilibrium constant as large as 10^{20} would require only a redox potential difference of 0.59 V for a two-electron reaction. Redox reactions tend to be all-or-nothing processes, with either reactants or products present in significant amounts, but not both. For such reactions, cell emf measurements provide a convenient, practical way of determining equilibrium constants.

19-6   SOLUBILITY EQUILIBRIA AND POTENTIALS

One of the difficulties in measuring solubility products directly is that for slightly soluble salts, such as AgCl, the concentrations at equilibrium are too low to measure accurately. However, the half-reaction

$$\text{Ag}^+ + e^- \rightarrow \text{Ag}$$

for which

$$\mathscr{E} = 0.80 - 0.0592 \log_{10} \frac{1}{[\text{Ag}^+]} = 0.80 + 0.0592 \log_{10} [\text{Ag}^+]$$

can be used to measure solubility. For example, if the silver ion concentration were as low as $10^{-30} M$ the silver electrode potential still would be -1.0 V, which is clearly within the measurable range.

Example 10

Calculate the solubility product constant, K_{sp}, for AgCl at 298 K from appropriate couples in Table 19-1.

Solution

Begin with the two half-reactions:

$$\text{AgCl} + e^- \rightarrow \text{Ag} + \text{Cl}^- \qquad \mathscr{E}^0 = 0.22 \text{ V (reduction)}$$
$$\text{Ag} \rightarrow \text{Ag}^+ + e^- \qquad \mathscr{E}^0 = -0.80 \text{ V (oxidation)}$$

Add the reactions to obtain the equation for the solution of AgCl:

$$\text{AgCl} \rightarrow \text{Ag}^+ + \text{Cl}^- \qquad \mathscr{E}^0 = -0.58 \text{ V}$$

$\mathscr{E}^0$ is negative since the reaction goes spontaneously to the left for $1M$ Ag^+ and Cl^-. For the net reaction,

$$\Delta G^0 = -RT \ln K_{sp} = -n\mathscr{F}\mathscr{E}^0$$

($\mathscr{E}$ in the Nernst equation is zero at equilibrium.)

$$\ln K_{sp} = -\frac{1\mathscr{F}}{RT}(0.58)$$

$$\log_{10} K_{sp} = -\frac{0.58}{0.0592} = -9.8$$

$$K_{sp} = 1.6 \times 10^{-10}$$

Example 11

Calculate K_{sp} for Hg_2Cl_2 at 298 K from data in Table 19-1.

Solution

The solubility equilibrium is

$$Hg_2Cl_2(s) \rightarrow Hg_2^{2+} + 2Cl^-$$
$$K_{sp} = [Hg_2^{2+}][Cl^-]^2$$

The half-reactions for this equilibrium are

$$Hg_2Cl_2 + 2e^- \rightarrow 2Hg + 2Cl^- \qquad \mathscr{E}^0 = +0.27 \text{ V}$$
$$Hg_2^{2+} + 2e^- \rightarrow 2Hg \qquad \mathscr{E}^0 = +0.80 \text{ V}$$

Adding the first half-reaction to the reverse of the second, we obtain

$$Hg_2Cl_2 \rightarrow Hg_2^{2+} + 2Cl^- \qquad \mathscr{E}^0 = 0.27 - 0.80 = -0.53 \text{ V}$$

The solubility product expression is

$$0 = -0.53 - \frac{0.0592}{2} \log_{10} K_{sp}$$

$$\log_{10} K_{sp} = -17.9$$

$$K_{sp} = 1.3 \times 10^{-18}$$

An interesting consequence of the effect of solubility on electrode potentials is revealed when we consider the reaction between silver ion and iodide ion. Table 19-1 "predicts" that silver ion should oxidize iodide ion according to the reaction

$$2Ag^+ + 2I^- \rightarrow 2Ag + I_2$$

If a cell is made of the two half-reactions working separately, silver plates on the cathode and iodine forms at the anode, as predicted. However, if the two ions are mixed directly, the only reaction observed is the formation of insoluble silver iodide:

$$Ag^+ + I^- \rightarrow AgI(s)$$

From the Nernst equation, we can show that since K_{sp} of AgI is about 10^{-16}, the potential of the $Ag^+|Ag$ couple is decreased by 16×0.0592 or nearly 1.0 V in a $1.0M$ I^- solution, whereas the $I_2|I^-$ couple has its potential increased by nearly the same amount in a $1.0M$ Ag^+ solution. Since the $I_2|I^-$ couple is only about 0.25 V below $Ag^+|Ag$, these two changes are more than enough to reverse the position of the couples in the table.

Complex-Ion Formation and Reduction Potentials

Consider the equilibrium between silver ion and cyanide ion:

$$Ag^+ + 2CN^- \rightarrow Ag(CN)_2^-$$

The equilibrium expression is

$$K_{eq} = \frac{[Ag(CN)_2^-]}{[Ag^+][CN^-]^2} \quad \text{or} \quad [Ag^+] = \frac{[Ag(CN)_2^-]}{K_{eq}[CN^-]^2}$$

For a silver electrode immersed in a solution containing $0.01M$ silver ion with an excess of cyanide ion, virtually all of the silver ion is complexed. The potential is

$$\mathscr{E} = 0.80 + 0.0592 \log_{10} \frac{[Ag(CN)_2^-]}{K_{eq}[CN^-]^2}$$

$$= 0.80 - 0.1184 - 0.0592 \log_{10} K_{eq} - 0.1184 \log_{10} [CN^-]$$

Such relationships are used in the study of equilibria involving complex ions.

19-7 REDOX CHEMISTRY GONE ASTRAY: CORROSION

Corrosion of metals is a redox process. For example, iron can be oxidized either by molecular oxygen or by acid, if sufficient moisture is present for the chemical reactions to proceed at an appreciable rate.

Oxidation:	$Fe(s) \rightarrow Fe^{2+} + 2e^-$	$\mathscr{E}^0 = +0.41$ V
Reduction:	$\begin{cases} \frac{1}{2}O_2(g) + H_2O(l) + 2e^- \rightarrow 2OH^- \\ 2H^+ + 2e^- \rightarrow H_2(g) \end{cases}$	$\begin{aligned} \mathscr{E}^0 &= +0.40 \text{ V} \\ \mathscr{E}^0 &= 0.00 \text{ V} \end{aligned}$

When iron rusts, metallic iron is oxidized to the $+2$ state, and it is deposited as flakes of FeO or other iron oxides. Aluminum corrodes even more vigorously:

$$Al(s) \rightarrow Al^{3+} + 3e^- \qquad \mathscr{E}^0 = +1.71 \text{ V}$$

On the reduction-potential scale, aluminum is more susceptible to oxidation than iron. Yet we think of aluminum as relatively inert to corrosion, whereas the rusting of iron and steel is a serious and expensive problem. Why?

We find out why only when we look at the crystal structures of aluminum, iron, and their oxides. The unit-cell or packing distances in aluminum and its oxide are very similar to one another; thus the aluminum oxide formed at the surface of the metal can adhere tightly to the uncorroded aluminum beneath it. The oxidized surface provides a protective layer that prevents oxygen from getting to the metal beneath. "Anodized" aluminum kitchenware has had a particularly tough oxide layer applied to it by placing the aluminum object in a situation where corrosion is especially favored—by making it the anode in an electrochemical reaction.

In contrast, the packing dimensions of metallic iron and FeO are not particularly close; thus there is no tendency for an iron oxide layer to adhere to metallic iron. The curse of rust is not that it forms, but that it constantly flakes off and exposes fresh iron surface for attack (Figure 19-11). One way to prevent rusting is to keep moisture and oxygen away from the surface of an iron object by giving it an artificial coating such as paint. A good paint adheres better than FeO does, but it still is not permanent.

Another, and more effective, method is to make electrochemistry work for you. Just as aluminum can be made to form an oxide film by making it the anode in an electrochemical cell, so iron can be *prevented* from oxidizing by making it the cathode. One way to do this is to coat it or plate it with a more reactive metal, yet one that itself forms a protective oxide coating. Aluminum would be a possible candidate. If iron and aluminum are in contact, the iron will behave as the cathode and the aluminum as the anode, as their reduction potentials indicate:

$$Fe^{2+} + 2e^- \rightarrow Fe(s) \qquad \mathscr{E}^0 = -0.41 \text{ V}$$
$$Al^{3+} + 3e^- \rightarrow Al(s) \qquad \mathscr{E}^0 = -1.71 \text{ V}$$

Aluminum will prevent the iron from becoming oxidized, and its own oxide will protect aluminum from continual destructive corrosion.

But if you are going to aluminum-plate iron, you might as well make the objects out of aluminum to begin with, thereby having the advantage of light weight. Unfortunately, aluminum is expensive. An earlier, and cheaper, alternative has been to "galvanize" iron, that is, to give it a thin coat of zinc. You can see from Table 19-1 that the principle is the same, although the reduction potentials of zinc and iron are closer. A galvanized steel bucket is corrosion-free, not merely because zinc shields iron as paint would, but because zinc electrochemically prevents iron from being re-

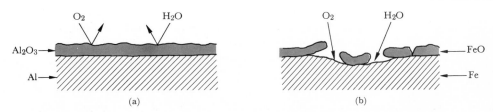

Figure 19-11   (a) A layer of aluminum oxide, once formed, adheres to the surface of the aluminum metal and protects it against further corrosion. (b) Unfortunately, iron oxide does not adhere as well to an iron surface. It continually flakes away as rust and exposes a clean surface to further attack by oxygen and moisture. Iron and steel objects must have their surfaces protected by some artificial method.

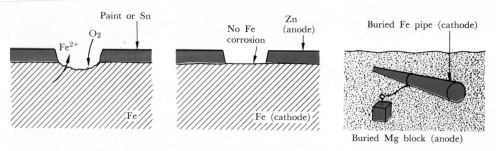

Figure 19-12   One protection for an iron object is an airtight coating of paint or of another metal such as tin. This works as long as the coating is intact, but when a pit or scratch develops, corrosion begins. A zinc coating provides additional electrochemical protection because iron has a higher reduction potential, and tends to remain reduced while zinc is oxidized. Similarly, magnesium can protect iron pipe by corroding in place of it. In these applications zinc and magnesium are called *sacrificial metals*.

duced. Scratch a galvanized pail and the scratched pail will not corrode; in principle the iron object does not even have to be completely covered.

Tin is another story. A "tin can" is tin-plated iron. You can see from Table 19-1 that Sn lies above Fe in reduction potential; thus the Sn^{2+} ion has a greater tendency to become reduced to the metal than does Fe^{2+}. The tin coating actually *encourages* oxidation, and hence corrosion, of the iron. A tin can is protected from corrosion only as long as the entire tin surface is intact. Scratch a tin can and it most certainly will rust. The tin plating functions only as a particularly tough and adhesive "superpaint." This is perhaps fortunate for our environment. Tin cans eventually self-destruct, whereas aluminum cans do not.

The same electrochemical trickery can be used to keep iron pipes from corroding in moist ground, or steel ship hulls from corroding in salt water. Magnesium rods driven into the ground periodically alongside an iron pipeline and connected electrically to it will make the pipe cathodic and prevent it from corroding. The magnesium itself will corrode, but it is easier and cheaper to replace magnesium rods than to dig up and replace the pipeline (Figure 19-12). Blocks of magnesium attached to a ship's hull in seawater perform the same function. The magnesium is called a "sacrificial" metal; its corrosion is acceptable as an alternative to corrosion of the iron object.

Summary

Some atoms and ions attract electrons more strongly than others. When we allow electrons to flow from the less attracting ions or atoms to more attracting ones, a more stable situation develops and energy is released. If no special precautions are taken, this energy is dissipated as heat, or as an increase in disorder (entropy). But if the electron-releasing and electron-accepting half-reactions can be separated physically, then the flow of electrons from one place to another can be harnessed to do electrical work. This is the principle of all electrochemical cells.

The electron-losing substance is **oxidized**, and the electrode where this occurs is the **anode**. The electron-accepting substance is **reduced** at the **cathode**. The "pressure" that these electrons exert, measured between anode and cathode, is the **cell voltage** or **electromotive force (emf)**. A positive cell voltage means that the cell reaction will proceed spontaneously, with electron flow from the anode to the cathode. A negative cell voltage means that the reverse reaction is spontaneous. The cell voltage is related to the free energy of the cell reaction by the expression

$$\Delta G = -n\mathscr{F}\mathscr{E}$$

An overall cell reaction can be separated into two half-reactions that represent the processes at the anode and cathode. Each of these half-reactions can be assigned its own half-cell potential, defined as the potential that would be observed if the half-reaction were paired with the hydrogen half-cell,

$$2H^+ + 2e^- \rightarrow H_2(g)$$

(This amounts to defining the voltage of the hydrogen half-reaction as $\mathscr{E}^0 = 0.000 \ldots$ V.) By convention, half-cell reactions are written as **reductions**, and the corresponding **reduction potential** measures the relative tendency for that reduction to take place. (Some older books use oxidations and oxidation potentials. The oxidation potential is the negative of the reduction potential.) A high positive reduction potential indicates a strong tendency toward reduction, and a low negative reduction potential indicates a strong tendency toward the oxidized state. When one half-reaction is subtracted from another to build a complete cell reaction, the corresponding reduction potentials are subtracted. Although one half-cell reaction may have to be multiplied by a stoichiometric constant to make the total number of electrons cancel in the overall reaction, the corresponding reduction potential is not multiplied by the constant. Reduction potentials are effective "electron pressures," and are already on a one-electron basis. Stoichiometry is taken care of by the quantity n in the expression

$$\Delta G = -n\mathscr{F}\mathscr{E}$$

If the concentrations of all ionic species are $1M$, and all gases are at 1 atm partial pressure, this is the **standard state** for the emf designated $\mathscr{E}^0$ (analogous to the standard free energy, ΔG^0, with a superscript zero). If the reactants and products are not all in their standard concentrations, then the cell voltage is given by the **Nernst equation**,

$$\mathscr{E} = \mathscr{E}^0 - \frac{RT}{n\mathscr{F}} \ln Q$$

which is the electrochemical analogue of the free energy versus concentration equation discussed in Chapters 16 and 17. Just as an overall cell reaction can be divided into two half-reactions, Nernst equations can be written separately for each half-cell process.

One remarkable feature of redox reactions, in contrast to most other chemical reactions, is that they occur over such a wide range of equilibrium-constant values. For a two-electron reaction, a cell voltage of 6 V corresponds to an equilibrium constant of $K_{eq} = 10^{200}$! This means that only rarely will two half-reactions have half-cell potentials so similar that the equilibrium constant for the overall reaction will be of moderate size. Most redox reactions either go to completion (effectively) or do not go at all. However, electrochemical methods can be used to study equilibrium, solubility product, and complex-ion formation in circumstances where one or another component at equilibrium is present in quantities far too small to be detected by standard analytical methods.

Using what we know about electrochemistry, we can design and build cells and batteries that deliver electrical power in small amounts in convenient places, and can use electrical power to bring about desirable chemical reactions. Electroplating and the refining of aluminum are examples. We also can use electrochemical principles to halt corrosion of susceptible metals that have low reduction potentials. What we cannot do yet is produce a cheap, lightweight storage battery with a high-energy density, or an electrochemical fuel cell that will operate with commonly available fuels.

Self-Study Questions

1. Which of the two electrodes, anode or cathode, is associated in all electrochemical cells with oxidation? Which with reduction?
2. What is the driving force that creates a potential in a concentration cell?
3. What would happen if you replaced the semipermeable barrier in a concentration cell by an impermeable one? If you removed the barrier altogether?
4. How is cell potential related to concentrations in a concentration cell?
5. How is a Daniell cell constructed, and what is the source of its electrical potential? Why is it unsuitable for powering flashlights or electric cars?
6. Why is a dry cell more suitable than the Daniell cell for flashlights and electric cars? What chemical reactions take place in a dry cell? Which reaction produces electrons, and how are they used up? Which reaction takes place at the anode, and which at the cathode?
7. How is cell potential related to the free energy of the cell reaction? What is the standard potential?
8. Are the free energies of half-cell reactions always additive? Are the half-cell potentials always additive? Outline the conditions under which

the free energies and potentials will be additive, and the conditions under which they will not.

9. Does a high positive reduction potential for a half-cell reaction indicate a strong tendency for the redox couple to reduce other substances?

10. Given two half-cell reactions with different half-cell potentials, how do we obtain the overall cell voltage? If the resulting cell voltage is positive, what does this indicate about the cell reaction?

11. The two manganese half-reactions with standard reduction potentials of $+1.68$ V and $+1.21$ V, in Table 19-1, can be added to produce a third reaction, listed in that table with a potential of $+1.49$ V. Explain why the reduction potential of this third reaction is not the sum of the other two; that is, not $+1.68$ V $+ 1.21$ V $= +2.89$ V. Account for the observed value of $+1.49$ V in terms of the tabulated potentials of the first two reactions.

12. What is the Nernst equation, and how does it relate cell potentials to concentrations?

13. Why is the first term of the Nernst equation zero for a concentration cell?

14. Why do concentration cells with the same ratio of concentrations always have the same cell potential, no matter what chemical substances are involved in the reaction?

15. How can oxidation–reduction measurements yield the solubility of a substance when the dissolved substance is not present in large enough amounts to be detected by ordinary analytical methods?

16. How is corrosion of iron impeded by a layer of paint? A layer of zinc metal? A layer of tin metal? Why is coating unnecessary with aluminum? What is meant by a "sacrificial metal" in preventing corrosion?

Problems

Data necessary for these problems can be found in Tables 19-1 and 19-2, and in the *Handbook of Chemistry and Physics* or similar references.

Concentration cells

1. A standard $Cl_2|Cl^-$ half-cell has been coupled to a $Cl_2(1.00 \text{ atm})|Cl^-(0.010M)$ half-cell. What is the cell voltage? Determine ΔG_{298} for the reaction.

2. The following cell

$$Ag|Ag^+(0.10M)\|Ag^+(1.0M)|Ag$$

is a concentration cell and is capable of electrical work. (a) What is the cell potential? (b) Which side of the cell is the cathode, and which is the anode? (c) What is ΔG in joules for the spontaneous cell reaction?

3. Two copper electrodes are placed in two copper sulfate solutions of equal concentration and connected to form a concentration cell. What is the cell voltage? One of the solutions is diluted until the concentration of copper ions is one-fifth its original value. What is the cell voltage after dilution?

4. Consider the following perpetual motion device built around a concentration cell: (a) Two copper electrodes are placed in copper sulfate solutions of equal concentration and connected to form a concentration cell. Initially there is no voltage in the cell. Assume that each electrode contains more copper than is present in either solution. (b) Solution A is diluted until its Cu^{2+} concentration is cut in half, at which point the cell has a potential, $\mathscr{E}$. The cell is run and useful work is done on the surroundings until the concentrations in the two solutions are equalized, at which time the cell voltage has fallen to zero again. (c) Solution B is diluted until its Cu^{2+} concentration is halved, at which time the cell has the same potential, $\mathscr{E}$, as before, but in the opposite direction. Again the cell is run, and work is done until the concentrations in solutions A and B are the same. (d) Steps (b) and (c) are repeated, diluting first one solution and then the other by halving its Cu^{2+} concentration after equilibrium has been attained in the previous step. Since neither concentration ever falls to zero by the halving process, we can maintain this process as long as we like and take an infinite amount of work out of the cell. The operation of the cell actually helps us, for it *raises* the concentration of the solution that we had just diluted. What is wrong with this analysis?

Work and free energy

5. Determine the amount of useful work done when a mole of zinc powder is allowed to react with a $1.00M$ solution of $Cu(NO_3)_2$ in a constant-temperature calorimeter. If the reaction were carried out reversibly, how much useful work could be accomplished? ΔH^0_{298} for the reaction is -215 kJ. Calculate the heat liberated when the reaction is carried out reversibly.

Spontaneity

6. Assuming unit activities for all substances, determine which of the following reactions will be spontaneous:
   a) $Zn + Mg^{2+} \rightarrow Zn^{2+} + Mg$
   b) $Fe + Cl_2 \rightarrow Fe^{2+} + 2Cl^-$
   c) $4Ag + O_2 + 4H^+ \rightarrow 4Ag^+ + 2H_2O$
   d) $2AgCl \rightarrow 2Ag + Cl_2$

Half-cell potentials

7. What are the potentials for the following cells or half-cells?
   a) $\mathscr{E}^0$ for the cell
      $$Zn(s)|Zn^{2+}||Cu^{2+}|Cu(s)$$
   b) $\mathscr{E}$ for the half-cell
      $$Zn(s)|Zn^{2+}(0.0010M)$$
   c) $\mathscr{E}$ for the half-cell
      $$Cu^{2+}(10^{-36}M)|Cu(s)$$

8. What are the standard potentials, $\mathscr{E}^0$, for the following half-cells?
   a) $S^{2-}|CuS(s)|Cu(s)$
   b) $NH_3(aq), Zn(NH_3)_4^{2+}|Zn(s)$

Cell potentials and electron flow

9. Consider the cell
   $$Ag(s)|Ag^+(1.0M)||Cu^{2+}(1.0M)|Cu(s)$$
   a) Write the chemical reaction that takes place in this cell. In which direction will the reaction proceed spontaneously?
   b) What is $\mathscr{E}^0$ for the cell?
   c) Do electrons flow from Ag to Cu in the external circuit, or the other way?

10. Consider the following cell:

$$Ni\,|\,Ni^{2+}(0.010M)\,||\,Sn^{2+}(1.0M)\,|\,Sn$$

(a) Predict the direction in which spontaneous reactions will occur. (b) Which metal, Ni or Sn, will be the cathode and which the anode? (c) What is $\mathscr{E}^0$ for the cell? (d) What will $\mathscr{E}$ be for the cell with the specified concentrations at 25°C?

11. Consider the cell

$$Sn\,|\,SnCl_2(0.10M)\,||\,AgCl(s)\,|\,Ag$$

(a) Will electrons flow spontaneously from Sn to Ag, or in the reverse direction? (b) What is the standard potential, $\mathscr{E}^0$, for the cell? (c) What will the cell potential, $\mathscr{E}$, be at 25°C?

12. Use the line notation of the previous problems to represent a cell that uses the following half-reactions:

$$PbO_2 + 4H^+ + 2e^- \rightarrow Pb^{2+} + 2H_2O$$
$$PbSO_4 + 2e^- \rightarrow Pb + SO_4^{2-}$$

a) Which is the reaction at the cathode of the cell? Which way do electrons flow in an external circuit?
b) What is $\mathscr{E}^0$ for this cell?

Half-reaction zero conventions

13. The following two reactions have the $\mathscr{E}^0$ values given:

$$2Ag + Pt^{2+} \rightarrow 2Ag^+ + Pt$$
$$\mathscr{E}^0 = +0.40\ V$$

$$2Ag + F_2 \rightarrow 2Ag^+ + 2F^-$$
$$\mathscr{E}^0 = +2.07\ V$$

If the potential for the reaction $Pt \rightarrow Pt^{2+} + 2e^-$ is assigned a value of zero, calculate the potentials for the following half-reactions:

a) $Ag \rightarrow Ag^+ + e^-$
b) $F^- \rightarrow \frac{1}{2}F_2 + e^-$

Competing reactions

14. Find the missing standard reduction potentials for these following half-reactions from Table 19-1:

Half-reaction	$\mathscr{E}^0$ (V)
$MnO_4^- + 8H^+ + 5e^- \rightarrow$ $Mn^{2+} + 4H_2O$	+1.49
$Au^{3+} + 3e^- \rightarrow Au(s)$	+1.42
$Cl_2 + 2e^- \rightarrow 2Cl^-$	?
$AuCl_4^- + 3e^- \rightarrow$ $Au(s) + 4Cl^-$	?
$4H^+ + NO_3^- + 3e^- \rightarrow$ $NO + 2H_2O$	?

If we assume that all reactants and products are at unit activity, (a) which substance in the half-reactions given is the best oxidizing agent? Which is the best reducing agent? (b) Will permanganate oxidize metallic gold? (c) Will metallic gold reduce nitric acid? (d) Will nitric acid oxidize metallic gold in the presence of Cl^- ion? (e) Will metallic gold reduce pure Cl_2 gas in the presence of water? (f) Will chlorine oxidize metallic gold if Cl^- ion is present? (g) Will permanganate oxidize chloride ion?

Cell reactions and K_{eq}

15. From the data in Tables 19-1 and 19-2: (a) Will Fe reduce Fe^{3+} to Fe^{2+}? (Assume unit activities.) (b) Calculate the equilibrium constant, K_{eq}, for the reaction at 25°C,

$$Fe + 2Fe^{3+} \rightarrow 3Fe^{2+}$$

16. Find the standard reduction potentials for the following half-reactions:

$$SO_4^{2-} + 4H^+ + 2e^- \rightarrow H_2SO_3 + H_2O$$
$$Ag^+ + e^- \rightarrow Ag$$

(a) Write the balanced overall reaction for a successful cell made from these two couples. (b) Write the line notation for the cell. (c) What is $\mathscr{E}^0$ for the cell? (d) What is the equilibrium constant for the cell reaction at 25°C? (e) Calculate the ratio of activities of products and reactants, Q, that will produce a cell voltage of 0.51 V.

Cell potentials

17. Find the standard reduction potentials for the following half-reactions:

$$MnO_4^- + 8H^+ + 5e^- \rightarrow$$
$$Mn^{2+} + 4H_2O$$
$$Al^{3+} + 3e^- \rightarrow Al$$
$$Cl_2 + 2e^- \rightarrow 2Cl^-$$
$$Mg^{2+} + 2e^- \rightarrow Mg$$

(a) Which is the strongest reducing agent? Which is the strongest oxidizing agent? (b) Write the overall reaction for a successful cell made from the Mg and Cl_2 couples. (c) Write the line notation for this cell. (d) Which is the anode of the cell, and which is the cathode? (e) What is $\mathscr{E}^0$ for the cell? (f) What is the equilibrium-constant expression for this cell reaction? Calculate the numerical value of the equilibrium constant at 25°C.

18. Find the standard reduction potentials for the following half-reactions:

$$Hg_2^{2+} + 2e^- \rightarrow 2Hg$$
$$Cu^{2+} + 2e^- \rightarrow Cu$$

(a) Write the overall reaction for a successful cell made from these two half-reactions. (b) Write the line notation for the cell. Which material, Hg or Cu, is the anode? (c) What is $\mathscr{E}^0$ for the cell? (d) What is the equilibrium constant for the cell? (e) What is the voltage of the cell when $[Hg_2^{2+}] = 0.10M$ and $[Cu^{2+}] = 0.010M$?

Le Chatelier's principle

19. For an electrochemical cell in which the spontaneous reaction is

$$3Cu^{2+} + 2Al \rightarrow 2Al^{3+} + 3Cu$$

what will be the qualitative effect on the cell potential if we add ethylenediamine, a ligand that coordinates strongly with Cu^{2+} but not with Al^{3+}?

20. A copper–zinc battery is set up under standard conditions with all species at unit activity. Initially, the voltage developed by this cell is 1.10 V. As the battery is used, the concentration of the cupric ion gradually decreases, and that of the zinc ion increases. According to Le Chatelier's principle, should the voltage of the cell increase or decrease? What is the ratio, Q, of the concentrations of zinc and copper ions when the cell voltage is 1.00 V?

21. A galvanic cell consists of a rod of copper immersed in a 2.0M solution of $CuSO_4$ and a rod of iron immersed in a 0.10M solution of $FeSO_4$. Using the Nernst equation and the reduction potentials for

$$Fe^{2+} + 2e^- \rightarrow Fe$$
$$Cu^{2+} + 2e^- \rightarrow Cu$$

calculate the voltage for the cell as described.

22. What voltage will be generated by a cell that consists of a rod of iron immersed in a 1.00M solution of $FeSO_4$ and a rod of manganese immersed in a $0.10M$ solution of $MnSO_4$?

Concentration and voltage

23. Consider the cell

$$Zn|Zn^{2+}(0.0010M)||Cu^{2+}(0.0010M)|Cu$$

for which $\mathscr{E}^0 = +1.10$ V. Does the cell voltage, $\mathscr{E}$, increase, decrease, or remain unchanged when each of the following changes is made? (a) Excess $1.0M$ ammonia is added to the cathode compartment. (b) Excess $1.0M$ ammonia is added to the anode compartment. (c) Excess $1.0M$ ammonia is added to both compartments at the same time. (d) H_2S gas is bubbled into the Zn^{2+} solution. (e) H_2S gas is bubbled into the Zn^{2+} solution at the same time that excess $1.0M$ ammonia is added to the other solution.

24. Consider the galvanic cell

$$Zn|Zn^{2+}||Cu^{2+}|Cu$$

Calculate the ratio of $[Zn^{2+}]$ to $[Cu^{2+}]$ when the voltage of the cell has dropped to 1.05 V, 1.00 V, and 0.90 V. Notice that a small drop in voltage parallels a large change in concentration. Therefore, a battery that registers 1.00 V is quite "run down."

Competitive reactions

25. Show that hydrogen peroxide is thermo-dynamically unstable and should disproportionate to water and oxygen.

Voltage and pH

26. A silver electrode is immersed in a $1.00M$ solution of $AgNO_3$. This half-cell is connected to a hydrogen half-cell in which the hydrogen pressure is 1.00 atm and the H^+ concentration is unknown. The voltage of the cell is 0.78 V. Calculate the pH of the solution.

Solubility product

27. Using half-reactions, show that Ag^+ and I^- spontaneously form $AgI(s)$ when mixed directly at unit initial activity. Show that K_{sp} for AgI is 10^{-16}.

28. A standard hydrogen half-cell is coupled to a standard silver half-cell. Sodium bromide is added to the silver half-cell, causing precipitation of AgBr, until a concentration of $1.00M$ Br^- is reached. The voltage of the cell at this point is 0.072 V. Calculate the K_{sp} for silver bromide.

Corrosion

29. Chrome-plated automobile trim contains an iron core coated by a thick layer of nickel that is coated by a layer of chromium. Arrange the metals in order of ease of oxidation. What is the purpose of the chromium layer? Of the nickel layer?

Suggested Reading

A. J. Bard, *Chemical Equilibrium,* Harper and Row, New York, 1966.

L. Hepler, "Electrochemistry," in *Chemical Principles,* Chapter 16, Blaisdell, New York, 1964.

W. M. Latimer, *Oxidation Potentials,* 2nd ed., Prentice-Hall, Englewood Cliffs, N.J., 1952.

C. A. VanderWerf, *Oxidation–Reduction,* Van Nostrand Reinhold, New York, 1961.

20

Coordination Chemistry

Key Concepts

20-1 Transition-metal complexes. Ligands. Geometrical isomers. Octahedral geometry, square planar geometry, and tetrahedral geometry. Paramagnetism and diamagnetism. Lability and inertness. Oxidation number and structure. Influence of d-electron structure. Charge transfer.

20-2 Nomenclature for coordination compounds. Monodentate ligands and bidentate ligands. Chelating agents and chelates. Structural isomers and geometrical isomers. Optical isomers or enantiomers. Asymmetric or chiral centers.

20-3 Bonding in coordination complexes. Electrostatic theory. Valence bond theory. d^2sp^3 and sp^3d^2 hybrid orbitals. Outer-orbital complexes and inner-orbital complexes. Crystal field theory. Crystal-field splitting energy. Low-spin complexes and high-spin complexes. Strong-field ligands and weak-field ligands. Molecular orbital theory. Ligand-to-metal (π) interaction. Metal-to-ligand (π) interaction or π back bonding.

20-4 Tetrahedral and square planar coordination. Ligand-field energy levels.

20-5 Complex ion equilibria. Formation constants. The chelate effect.

*The first attempt at generalization seldom
succeeds; speculation anticipates experience,
for the results of observation accumulate
but slowly.*

J. J. Berzelius (1830)

Many compounds of platinum, cobalt, and other transition metals have strange empirical formulas and are often brightly colored. These are called **coordination compounds**. Their major distinguishing feature is the presence of two, four, five, six, and sometimes more chemical groups positioned geometrically around the metal ion. These groups can be neutral molecules, cations, or anions. Each coordinating group can be a separate entity, or all the groups can be connected in one long, flexible molecule that wraps itself around the metal. Coordinating groups significantly change the chemical behavior of a metal. The colors of the compounds provide clues about their electronic energy levels.

For instance, every plant depends on the green magnesium coordination complex known as chlorophyll. The combination of magnesium and its coordinating groups in chlorophyll has electronic properties that the free metal or ion does not have, and can absorb visible light and use the energy for chemical synthesis. Every oxygen-breathing organism requires **cytochromes**, coordination compounds of iron that are essential to the breakdown and combustion of foods and the storage of the energy released by the breakdown. Most larger organisms need hemoglobin, another iron complex in which the coordinating groups enable the iron to bind oxygen molecules without being oxidized. Large areas of biochemistry are really applied transition-metal chemistry. In this chapter we shall look at the structures and properties of some coordination compounds.

20-1   PROPERTIES OF TRANSITION-METAL COMPLEXES

The transition metals are often encountered in highly colored compounds with complex formulas. Although $PtCl_4$ exists as a simple compound, there are other compounds in which $PtCl_4$ is combined with two to six NH_3 molecules or with KCl (Table 20-1). Why should such apparently independent, neutral compounds associate with other molecules, and why should they do so in varying proportions? Measurements of electrical conductivity of solutions, and the precipitation of Cl^- ions by Ag^+, indicate how many ions are present in aqueous solution. This and other evidence lead us to propose the ionic structures listed at the right of the table. These substances that contain ammonia are coordination compounds, in which the NH_3 molecules are arranged around a central Pt^{4+} ion. The Pt(IV) complexes are octahedrally coordinated. In contrast, complexes of Pt(II) are in a square planar coordination, with a coordination number of 4. Complexes of metals with a coordination number of 4 may also be tetrahedral.

Color

Color is a distinctive property of coordination compounds of transition metals. The octahedral complexes of cobalt exist in a wide spectrum of colors, which depend on the groups coordinated to it (Table 20-2). Such coordinated groups are called **ligands**. In solution, color arises from the association of solvent molecules with the metal as ligands, and not from the metal cation itself. In concentrated sulfuric acid (a potent dehydrating

Table 20-1

Platinum Complexes, Numbers of Ions Produced, and Complex Structures

Complex	Molar conductivity (ohm^{-1})	Number of Cl$^-$ ions precipitated by Ag$^+$	Total number of ions	Ions produced
$PtCl_4 \cdot 6NH_3$	523	4	5	$Pt(NH_3)_6^{4+}$; $4Cl^-$
$PtCl_4 \cdot 5NH_3$	404	3	4	$Pt(NH_3)_5Cl^{3+}$; $3Cl^-$
$PtCl_4 \cdot 4NH_3$	229	2	3	$Pt(NH_3)_4Cl_2^{2+}$; $2Cl^-$
$PtCl_4 \cdot 3NH_3$	97	1	2	$Pt(NH_3)_3Cl_3^+$; Cl^-
$PtCl_4 \cdot 2NH_3$	0	0	0	$Pt(NH_3)_2Cl_4^0$
$PtCl_4 \cdot NH_3 \cdot KCl$	109	0	2	K^+; $Pt(NH_3)Cl_5^-$
$PtCl_4 \cdot 2KCl$	256	0	3	$2K^+$; $PtCl_6^{2-}$

Table 20-2

Octahedral Complexes of Co(III), Their Colors, and Estimates of Electronic Transition Energy

Complex[a]	Color	Spectral color absorbed	Approximate wavelength (nm)	Approximate transition energy (wave number, cm^{-1})
$Co(NH_3)_6^{3+}$	Yellow	Indigo	430	23,200
$Co(NH_3)_5NCS^{2+}$	Orange	Blue	470	21,200
$Co(NH_3)_5H_2O^{3+}$	Red	Blue-green	500	20,000
cis-$Co(en)_2(H_2O)_2^{3+}$	Red	Blue-green	500	20,000
$Co(NH_3)_5OH^{2+}$	Pink	Blue-green	500	20,000
$Co(NH_3)_5CO_3^+$	Pink	Blue-green	500	20,000
$Co(NH_3)_5Cl^{2+}$	Purple	Green	530	18,900
$Co(EDTA)^-$	Violet	Yellow	560	17,800
cis-$Co(NH_3)_4Cl_2^+$	Violet	Yellow	560	17,800
$trans$-$Co(en)_2Br(NCS)^+$	Blue	Orange	610	16,400
$trans$-$Co(NH_3)_4Cl_2^+$	Green	Red	680	14,700
$trans$-$Co(en)_2Br_2^+$	Green	Red	>680	<14,700

[a](en) is an abbreviation for ethylenediamine, $NH_2CH_2CH_2NH_2$.

agent) Cu^{2+} is colorless; in water it is aquamarine, and in liquid ammonia it is a deep ultramarine. Complexes of metals in high oxidation states have brilliant colors if they absorb energy in the visible spectrum: CrO_4^{2-} is bright yellow and MnO_4^- is an intense purple.

Whenever a certain energy of visible electromagnetic radiation, E, is absorbed by a compound during the excitation of an electron to a higher quantum state, the wavelength (λ) of light absorbed can be calculated from the expression

$$E = h\nu = hc\bar{\nu} = hc/\lambda$$

where h is Planck's constant, ν is frequency, $\bar{\nu}$ is wave number, and c is the speed of light. If the energy is given in wave numbers, as frequently is done, then the wavelength is simply the reciprocal: $\lambda = 1/\bar{\nu}$. The color that we see in the compound is the *complementary* color to the color absorbed; it is the color that remains in the spectrum after the particular spectral color has been removed. These colors are listed in Table 20-3. If the energy absorbed is so small that it corresponds to wavelengths in the infrared, or so large that it occurs in the ultraviolet (which is usually the case in compounds of representative elements), then the compound will be colorless or white. With transition-metal compounds, interesting things happen during ab-

Table 20-3

Colors of Compounds, Spectral Colors, Wavelengths, and Energies

Color of compound	Spectral color absorbed	Approximate wavelength (nm)	Energy difference between electronic levels (wave number, cm^{-1})
Colorless	Ultraviolet	< 400	> 25,000
Lemon yellow	Violet	410	24,400
Yellow	Indigo	430	23,200
Orange	Blue	480	20,800
Red	Blue-green	500	20,000
Purple	Green	530	18,900
Violet	Lemon yellow	560	17,900
Indigo	Yellow	580	17,300
Blue	Orange	610	16,400
Blue-green	Red	680	14,700
Green	Purple-red	720	13,900
Colorless	Infrared	> 720	< 13,900

sorption in the visible spectrum region of energy. Often you can learn something about trends in chemical behavior by "eyeball spectroscopy."

The colors of transition-metal complexes explain the trick of writing with invisible ink made from $CoCl_2$. If you write with a pale-pink solution of $CoCl_2$, the writing is virtually undetectable on paper. If the paper is heated gently over a candle flame, the message appears in a bright blue. Upon cooling, the writing slowly fades. The pink color is that of the octahedral hydrated cobalt ion: $Co(H_2O)_6^{2+}$. Heating drives away the water and leaves a blue chloride complex with tetrahedral geometry. This compound is **hygroscopic**; that is, it absorbs water from the atmosphere and fades to the pale-pink hydrate again.

Isomers and Geometry

The compound whose empirical formula is $CoCl_3 \cdot 4NH_3$ can be either green or violet. This fact provided transition-metal chemists with convincing evidence that the coordination in this compound is octahedral. Both the green and the violet $CoCl_3 \cdot 4NH_3$ dissociate to produce only one Cl^- ion per molecule, so the cation must be $Co(NH_3)_4Cl_2^+$, with a coordination number of 6. How are the six ligands arranged? We can suggest three possibilities: a flat six-membered ring, a trigonal prism, or an octa-

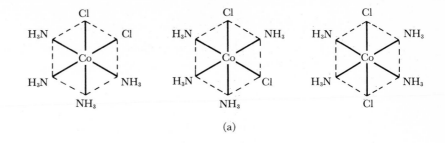

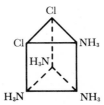 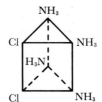 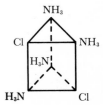

(a)

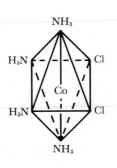 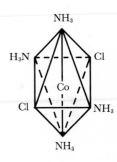

Figure 20-1

The six coordinating groups around Co in $Co(NH_3)_4Cl_2^+$ can be arranged in three possible symmetrical ways: a flat hexagon (a), a triangular prism (b), or an octahedron (c). Three different arrangements of coordinating groups, or geometrical isomers, can be produced in the hexagon and three in the triangular prism, but only two in the octahedron. The octahedral structure is adopted by $Co(NH_3)_4Cl_2^+$ and almost all other six-coordinate complexes.

(b)

(c)

hedron. In each of these three structures there is more than one way of placing the two Cl^- ions among the six coordination positions. Such structures, which differ only in the arrangement of the same ligands around the central metal, are called **geometrical isomers**. As Figure 20-1 shows, the existence of only two geometrical isomers of $Co(NH_3)_4Cl_2^+$ is convincing evidence for the octahedral structure. Octahedral coordination, with a coordination number of 6, is by far the most common structure for such transition-metal compounds.

Fourfold coordination also is found. Is this coordination tetrahedral or square? Again, data on the number of variant forms of a compound with the same empirical formula provide the answer. The compound with the

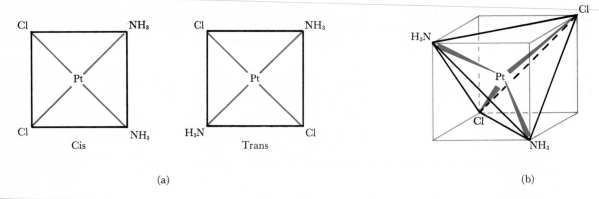

(a)   (b)

Figure 20-2   If the neutral molecule $Pt(NH_3)_2Cl_2$ has square planar coordination, two isomers are possible (a). But if the coordination is tetrahedral, only one can exist (b). Two isomers have been found, so the tetrahedral structure is eliminated. Why is this proof more convincing than that of Figure 20-1?

formula $PtCl_2 \cdot 2NH_3$, or $Pt(NH_3)_2Cl_2$, occurs in two forms that presumably are geometrical isomers. Both isomers are a creamy white, but they differ in solubility and chemical properties. As illustrated in Figure 20-2, there cannot be isomers for the tetrahedral structure, whereas the square planar structure has two. Therefore, the compound $Pt(NH_3)_2Cl_2$ must be square planar. (As an example of a comparable tetrahedral structure, CH_2Cl_2 has only one form and not two.)

Square planar geometry is characteristic of Pd(II), Pt(II), and Au(III), all of whose cations have eight d electrons, or a d^8 structure (Table 20-4). Tetrahedral coordination is encountered most often in transition-metal compounds in which the coordinating group is O^{2-}, as in CrO_4^{2-} and MnO_4^-. Now coordination structures can be examined directly by x-ray crystallography, and the conclusions about geometrical isomers from other experiments have been confirmed.

Magnetic Properties

Some transition-metal complexes are diamagnetic, which indicates no unpaired electrons. Many others are paramagnetic and have one or more unpaired electrons. For example, $Co(NH_3)_6^{3+}$ is diamagnetic, whereas CoF_6^{3-} is paramagnetic and has four unpaired electrons per ion. The ionic charge is not the governing factor, since $Fe(H_2O)_6^{2+}$, which has four unpaired electrons, is paramagnetic, yet $Fe(CN)_6^{4-}$ is diamagnetic. The magnetic properties of several other octahedral complexes are illustrated in Figure 20-3. One of our goals will be to explain this magnetic behavior in terms of electronic arrangement.

Table 20-4

Valence Electronic Configurations of Transition Metals[a]

Configuration	d^1s^2	d^2s^2	d^3s^2	d^4s^2	d^5s^2	d^6s^2	d^7s^2	d^8s^2	d^9s^2	$d^{10}s^2$
Elements	Sc	Ti	V	Cr	Mn	Fe	Co	Ni	Cu	Zn
	Y	Zr	Nb	Mo	Tc	Ru	Rh	Pd	Ag	Cd
	La	Hf	Ta	W	Re	Os	Ir	Pt	Au	Hg
Valence electrons										
Neutral atom	3	4	5	6	7	8	9	10	11	12
M^{2+} ion (*d* electrons)	1	2	3	4	5	6	7	8	9	10
M^{3+} ion (*d* electrons)	0	1	2	3	4	5	6	7	8	9

[a]All configurations are given as d^ns^2, since what interests us here is the number of electrons in the ion and not the electronic configuration of the neutral atom.

Lability and Inertness

A coordination complex that rapidly exchanges its ligands for others is **labile**; a complex that releases its ligands slowly is **inert**. Inertness is not the same as stability in the thermodynamic sense. A complex can be unstable, which means that it is not the most favored state according to the principles of thermodynamics discussed in Chapter 16. Given enough time, the complex will change to some other state. Yet if the transition to the most favored state is extremely slow, the unstable complex is inert. As an example of an inert yet unstable system, H_2 and O_2 can be kept as a mixture for years without an appreciable spontaneous formation of water. However, if a small amount of platinum black (finely divided Pt) is supplied as a catalyst, or if a flame is brought near, the reaction to make the more stable H_2O is sudden, complete, and violent. A mixture of H_2 and O_2 by itself is unstable, yet inert.

Returning to coordination compounds, we note that $Cu(NH_3)_4SO_4$ can be dissolved in water and the $Cu(NH_3)_4^{2+}$ can be allowed to react with dilute acid to produce NH_4^+ and $Cu(H_2O)_6^{2+}$ as fast as the solutions are mixed. In contrast, $Co(NH_3)_6Cl_3$ can be heated in concentrated sulfuric acid to drive off HCl gas and make $[Co(NH_3)_6^{3+}]_2(SO_4^{2-})_3$ without breaking the bonds between Co and NH_3. The copper complex is labile; the cobalt complex is inert. Tripositive ions with three or six *d* electrons form especially inert complexes; these complexes are also remarkably stable in the thermodynamic sense.

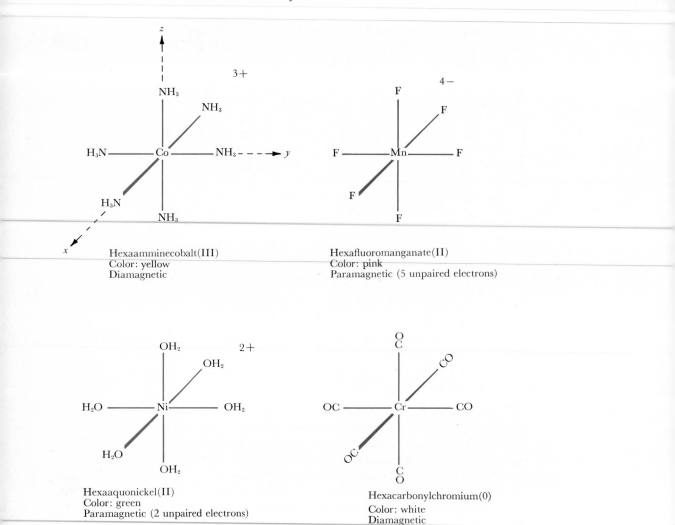

Hexaamminecobalt(III)
Color: yellow
Diamagnetic

Hexafluoromanganate(II)
Color: pink
Paramagnetic (5 unpaired electrons)

Hexaaquonickel(II)
Color: green
Paramagnetic (2 unpaired electrons)

Hexacarbonylchromium(0)
Color: white
Diamagnetic

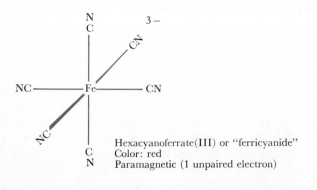

Hexacyanoferrate(III) or "ferricyanide"
Color: red
Paramagnetic (1 unpaired electron)

Figure 20-3

Several octahedral complexes, with systematic names, colors, and magnetic properties.

Oxidation Number and Structure

Coordination number 6 appears to be optimal for ions with oxidation numbers $+2$ and $+3$; these include many transition-metal compounds. An oxidation number of $+1$ is too low to attract six electron-donor groups to build a complex ion. Most complexes of $+1$ ions have smaller coordination numbers, such as 2 for Ag^+ and Cu^+ in $Ag(NH_3)_2^+$ and $CuCl_2^-$. Stable complexes of rather high coordination number do occur with $+1$ ions and neutral atoms. But in most of these instances, such as $Mn(CN)_6^{5-}$ and $Mo(CO)_6$, these ligands have special π-bonding features that transcend simple electron donation.

Complexes of central ions having oxidation numbers greater than $+3$ are rare. They usually exist only with O^{2-} and F^-. We expect stronger bonding as the oxidation number of the central ion increases. However, if the oxidation number becomes too high, the central ion attracts ligand electrons so strongly that they are pulled completely away from the ligand. Then the complex is not stable, and the metal is reduced to a lower oxidation state. For this reason, Fe^{3+} forms no complex with I^-; instead, it oxidizes I^- to I_2. Since O and F are so electronegative, and since O^{2-} and F^- when bound to a central metal are so difficult to oxidize, they can exist in complexes in which the central ion has an oxidation state higher than the usual $+2$ or $+3$.

Influence of the Number of d Electrons

Much of coordination chemistry can be understood in terms of the number of d electrons on the central metal ion. As we have already mentioned, the $+2$ and $+3$ oxidation states are most common. Table 20-4 gives the number of net d electrons for neutral atoms and for $+2$ and $+3$ cations; we shall use this table frequently. In addition to this preference for $+2$ and $+3$ states, ions with the d-shell configurations d^0, d^5, and d^{10} are particularly favored.

■ *Noble-gas shell, d^0.* The noble-gas configuration with no d electrons is especially stable. The ion Sc^{3+} has this configuration, as does Ti(IV) in TiF_6^{2-}. It is increasingly difficult for ions to attain the d^0 structure from left to right across the periodic table. The reason is that the resulting charge on the central metal ion increases. Stabilization is then possible only by coordination to oxide ions. Therefore, we find VO_4^{3-} instead of V^{5+}, CrO_4^{2-} instead of Cr^{6+}, and MnO_4^- instead of Mn^{7+}.

This series of oxide complexes provides a good example of the application of eyeball spectroscopy. Photons of the appropriate energy can excite electrons from the ligand oxygen atoms to the empty d orbitals of the metal. This process is called **charge transfer** and is a common origin for color in transition-metal complexes. The higher the oxidation state of the metal, the easier it is for electrons to transfer, and the lower is the energy of the photons required to bring about the transfer. The required energy in VO_4^{3-} occurs with photons in the ultraviolet region. Therefore, the VO_4^{3-} ion is colorless. In CrO_4^{2-}, the absorption of photons is in the violet region, at approximately 24,000 cm^{-1}; thus the chromate ion in solution appears yellow from the frequencies of light that are *not* absorbed (Table 20-3). (In accordance with standard spectroscopic practice, we express energy in wave numbers, cm^{-1}. See Section 8-2.) The Mn^{7+} ion has the highest oxidation state of all, and absorbs green light (around 19,000 cm^{-1}) for the charge-transfer excitation. Therefore, MnO_4^- appears purple. The colors in these charge-transfer complexes are usually quite intense, which indicates strong absorption. Increasing the size of the central ion makes charge transfer more difficult and moves the absorption into the ultraviolet; thus MoO_4^{2-}, WO_4^{2-}, and ReO_4^- are colorless.

The greater attraction of a large positive charge on the central ion for the negative charge on the ligands is reflected in the decreasing tendency of ligands in the coordination ion to bind to other cations. In the series VO_4^{3-}, CrO_4^{2-}, and MnO_4^-, the vanadate ion is a fairly strong base and will bind H^+ or other cations. The chromate ion is a reasonably strong base also. But the permanganate ion is a weak base; the compound $HMnO_4$ is completely ionized in water. The acid $HMnO_4$ is one of the strongest known (Table 11-2). Reactions of the type

$$2VO_4^{3-} + 2H^+ \rightarrow {}^{2-}O_3V-O-VO_3^{2-} + H_2O$$

occur easily with the vanadate ion, which forms polyvanadates with many $-O-$ bridges, and with the chromate ion, which forms dichromate, $Cr_2O_7^{2-}$, in acid. In contrast, Mn_2O_7 can be made only in concentrated sulfuric acid, which acts as a powerful dehydrating agent. Once formed, it is so unstable that it is a dangerous explosive.

■ *Filled and half-filled shells.* The filled d^{10} structure in Zn^{2+} and Ag^+, and the half-filled d^5 structure in Mn^{2+} and Fe^{3+}, make these ions particularly stable, even though the complexes that Mn^{2+}, Fe^{3+}, and Zn^{2+} form are relatively weak and contribute little to stabilizing the metal. This behavior is another example of the stability of filled and half-filled shells that we have seen so often.

■ *Ions with d^3, d^6, or d^8 structures.* The prominence of the oxidation number $+3$ for $Cr(d^3)$ and for $Co(d^6)$, plus the remarkable inertness of their complexes in chemical reactions—recall $Co(NH_3)_6Cl_3$ in hot sulfuric acid—cannot be explained on the basis of the ideas presented so far. Nor can we account for the special tendency of d^8 ions to adopt square planar rather than octahedral or tetrahedral coordination. To explain these structures and the existence of complexes of metals with oxidation number zero, we must examine how d orbitals participate in bonding with the ligands.

■ *Instability of d^4.* The ion Cr^{2+} (d^4) is a powerful reducing agent that is oxidized to a d^3 arrangement. Also a d^4 ion, Mn^{3+} is an equally powerful oxidant and is reduced to a d^5 ion. And Co^{5+}, likewise with a d^4 structure, does not form any stable compounds at all. Any theory of bonding in coordination complexes will have to interpret this extreme instability of the d^4 configuration.

20-2 NOMENCLATURE FOR COORDINATION COMPOUNDS

Many complex transition-metal salts have common names that were given to them before their chemical identities were known. Some of the names are slightly informative: *potassium ferricyanide* for $K_3Fe(CN)_6$ and *potassium ferrocyanide* for $K_4Fe(CN)_6$. *Luteocobaltic chloride* for $Co(NH_3)_6Cl_3$ and *praseocobaltic chloride* for *trans*-$[Co(NH_3)_4Cl_2]Cl$ are informative only if you know the Latin and Greek for yellow (*luteus*) and green (*praseos*). Luteoiridium chloride, $Ir(NH_3)_6Cl_3$, is not even yellow, and was given that name only because it has the analogous chemical formula to the cobalt salt. And *Reinecke's salt, Erdmann's salt,* and *Zeise's salt* are completely useless names.

Systematic nomenclature is gradually replacing these older names. The following rules are used:

1. In naming the entire complex, the name of the cation is given first and the name of the anion is given second (just as for sodium chloride), whether the cation or the anion is the complex species.

2. In the complex ion, the name of the ligand or ligands precedes that of the central metal atom. Special ligand names are *aquo* for water, *ammine* for NH_3, and *carbonyl* for CO.

3. Ligand names generally end in -*o* if the ligand is negative (*chloro* for Cl^-, *cyano* for CN^-) and -*ium* in the rare cases in which the ligand is positive (*hydrazinium* for $NH_2NH_3^+$). The names are unmodified if the ligand is neutral (*methylamine* for CH_3NH_2, *ethylenediamine* for $NH_2CH_2CH_2NH_2$).

4. A Greek prefix (*mono-, di-, tri-, tetra-, penta-, hexa-,* and so on) indicates the number of each ligand (*mono-* is often omitted for a single ligand of a given type). If the name of the ligand itself contains the terms *mono-, di-,* and so forth (*ethylenediamine*, abbreviated *en; diethylenetriamine*, abbreviated *dien*), then the ligand name is enclosed in parentheses and its number is given with the alternative prefixes *bis-* and *tris-* instead of *di-* and *tri-*. Hence, for example, $Pt(en)_3Br_4$ is tris(ethylenediamine)platinum(IV) bromide.

5. A Roman numeral or a zero in parentheses indicates the oxidation number of the central metal atom.

6. If the complex ion is negative, the name of the metal ends in *-ate.*

7. If more than one ligand is present in a species, the order of ligands in the name is negative, neutral, and positive.

Some examples of systematic nomenclature are

$Pt(NH_3)_6Cl_4$	Hexaammineplatinum(IV) chloride
$[Pt(NH_3)_5Cl]Cl_3$	Chloropentaammineplatinum(IV) chloride
$[Pt(NH_3)_3Cl_3]Cl$	Trichlorotriammineplatinum(IV) chloride
$Pt(NH_3)_2Cl_4$	Tetrachlorodiammineplatinum(IV)
$KPt(NH_3)Cl_5$	Potassium pentachloromonoammine-platinate(IV)
K_2PtCl_4	Potassium tetrachloroplatinate(II)
K_2CuCl_4	Potassium tetrachlorocuprate(II)
$Fe(CO)_5$	Pentacarbonyliron(0)
$[Ni(H_2O)_6](ClO_4)_2$	Hexaaquonickel(II) perchlorate
$K_4Fe(CN)_6$	Potassium hexacyanoferrate(II)
$K_3Fe(CN)_6$	Potassium hexacyanoferrate(III)
$[Pt(en)_2Cl_2]Br_2$	Dichlorobis(ethylenediamine)platinum(IV) bromide
$[Pt(NH_3)_4](PtCl_4)$	Tetraammineplatinum(II) tetrachloro-platinate(II)

Some common ligands are listed in Table 20-5. All these ligands are *monodentate;* that is, each ligand binds to the central ion at only one point.

Table 20-5

Common Monodentate Ligands[a]

Ligand	Name
F^-, Cl^-, Br^-, I^-	Fluoro, chloro, bromo, iodo
$:NO_2^-$ and $:ONO^-$	Nitro and nitrito
$:CN^-$	Cyano
$:SCN^-$ and $:NCS^-$	Thiocyanato and isothiocyanato
$:OH^-$	Hydroxo
$CH_3COO:^-$	Acetato
H_2O	Aquo
NH_3	Ammine
CO	Carbonyl
NO^+	Nitrosyl
py	Pyridine, C_5H_5N

[a]The electron pairs are shown to remind you which atom bonds to the central metal. They ordinarily would not be shown.

Table 20-6

Common Chelating Agents or Multidentate Ligands

Symbol	Ligand name	Formula	Bonds
en	Ethylenediamine	$\overset{..}{N}H_2-CH_2-CH_2-\overset{..}{N}H_2$	2
pn	Propylenediamine	$\overset{..}{N}H_2-CH_2-CH-\overset{..}{N}H_2$ $\qquad\qquad\quad\ \ \vert$ $\qquad\qquad\quad CH_3$	2
dien	Diethylenetriamine	$\overset{..}{N}H_2-CH_2CH_2-\overset{..}{N}H-CH_2CH_2-\overset{..}{N}H_2$	3
trien	Triethylenetetraamine	$\overset{..}{N}H_2-CH_2CH_2-\overset{..}{N}H-CH_2CH_2-\overset{..}{N}H-CH_2CH_2-\overset{..}{N}H_2$	4
EDTA	Ethylenediaminetetraacetate	$^-\!:OOC-CH_2\qquad\qquad CH_2-COO:^-$ $\qquad\quad\vert\qquad\qquad\qquad\quad\vert$ $\qquad\ \ :N-CH_2CH_2-N:$ $\qquad\quad\vert\qquad\qquad\qquad\quad\vert$ $^-\!:OOC-CH_2\qquad\qquad CH_2-COO:^-$	6
ox	Oxalate	$^-\!:OOC-COO:^-$	2

Other ligands are bi-, tri-, or even hexadentate (Table 20-6). Three molecules of ethylenediamine, $NH_2-CH_2-CH_2-NH_2$, can coordinate octahedrally to Pt to produce the cation illustrated in Figure 20-4. The ethylenediaminetetraacetate ion listed in Table 20-6 can wrap itself around a metal ion and coordinate with all six octahedral positions at once (Figure 20-5). Ethylenediaminetetraacetate, or EDTA, is so efficient a scavenger for Ca, Mg, Mo, Fe, Cu, and Zn that it will remove the essential metal atom from an enzyme and completely block its enzymatic activity. EDTA is also a useful scavenger in removing traces of metals from distilled and purified

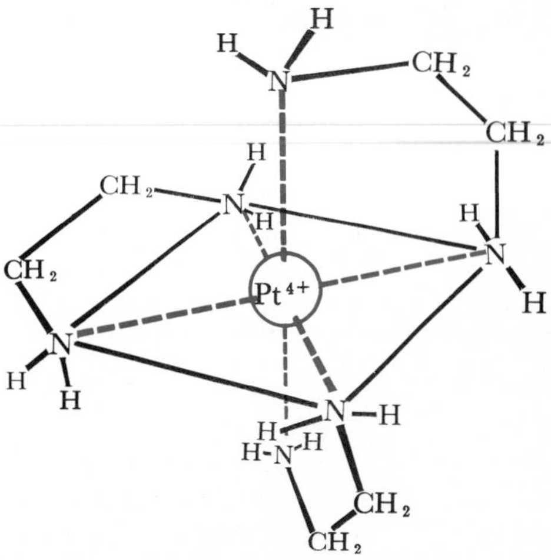

Figure 20-4

The structure of the tris(ethylenediamine)platinum(IV) ion. Each molecule of ethylenediamine, $NH_2-CH_2-CH_2-NH_2$, coordinates to the platinum ion at two points. Such bidentate and multidentate ligands are called *chelating agents*, and the compounds are called *chelates*, from the Greek *chela*, "claw."

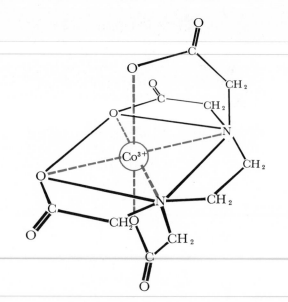

Figure 20-5

One molecule of EDTA, or ethylene-diaminetetraacetate, can completely enclose a metal ion in octahedral coordination. EDTA's attraction for metals is so strong that it will remove metals from enzymes and will inhibit their catalytic activity completely.

water. A molecule or ion that coordinates more than once with a metal ion is called a **chelating agent**, and the total complex is called a **chelate**.

Isomerism

Three types of isomers are found in coordination complexes: structural, geometrical, and optical isomers. **Structural isomers** have the same overall chemical formula but different ways of connecting component parts. Ethyl alcohol (CH_3CH_2OH) and dimethyl ether ($H_3C-O-CH_3$) are structural isomers. The material with the formula $Cr(H_2O)_6Cl_3$ exists in three structural isomers:

$[Cr(H_2O_6]Cl_3$	Hexaaquochromium(III) chloride
$[Cr(H_2O)_5Cl]Cl_2 \cdot H_2O$	Chloropentaaquochromium(III) chloride monohydrate
$[Cr(H_2O)_4Cl_2]Cl \cdot 2H_2O$	Dichlorotetraaquochromium(III) chloride dihydrate

The first of these is violet, the second is light green, and the third is dark green. Their structures can be demonstrated by precipitation of Cl^- with Ag^+ and by elimination of zero, one, or two waters of hydration by drying over H_2SO_4.

As we saw earlier, **geometrical isomers** differ in the arrangement of groups around the same center. The prefix *cis-* indicates that two identical groups are adjacent; *trans-* means that they are across from one another, or at least not adjacent. In Figure 20-6a, the two chlorines in the cis isomer are adjacent to one another along one edge of the octahedron, whereas in the trans isomer they are across a diagonal through the octahedron.

Optical isomers have the same groups connected in the same relative arrangement, but in the reverse sense as your right hand is to your left. Optical isomers arising from the arrangement of groups about a central atom always occur in pairs, one of which is the mirror image of the other. These pairs are called **enantiomers**. An example of two enantiomers is the two $Co(en)_3^{3+}$ complexes shown in Figure 20-6b. A central atom around which such isomers can be formed is called an **asymmetric center** or **chiral center**. Another example of a pair of enantiomers is L- and D-alanine, shown in Figure 21-12. Many optical isomers can be formed when several asymmetric carbon atoms are connected in a chain.

20-3   THEORIES OF BONDING IN COORDINATION COMPLEXES

The maximum number of σ bonds that can be constructed with s and p valence orbitals is 4. Thus, 4 is the highest coordination number commonly encountered in the representative elements in Period 2. These elements do not have filled d orbitals or access to empty d orbitals in the next higher shell. For example, in CH_4 the central carbon atom is "saturated" with four

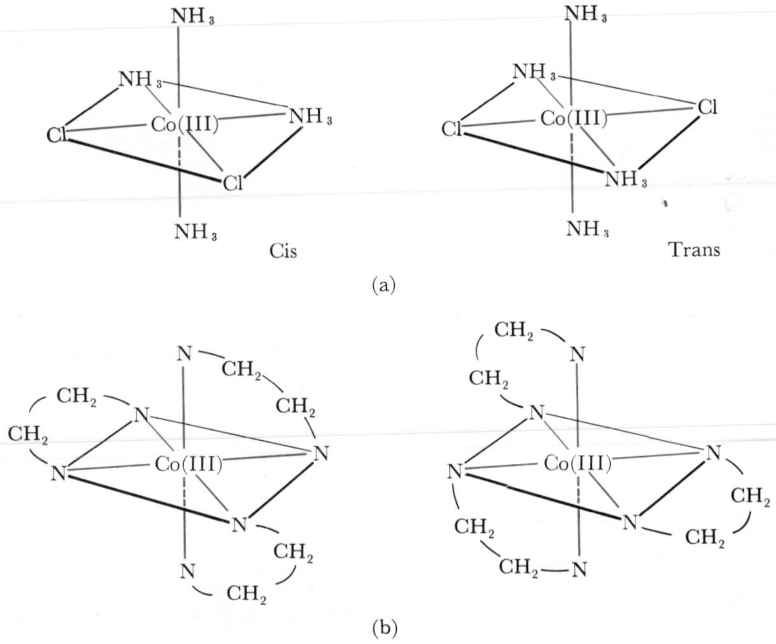

Cis   Trans

(a)

(b)

Figure 20-6   Geometrical and optical isomers of octahedral complexes. (a) *cis*- and *trans*-dichlorotetraamminecobalt(III) ions are geometrical isomers. (b) The two optical isomers of the tris(ethylenediamine)cobalt(III) ion. Can you prove to your own satisfaction that, for each ion, only two such isomers exist?

σ bonds. However, with a first-row transition metal as the central atom, there are five *d* valence orbitals in addition to the four *s* and *p* orbitals.

If the central metal made full use of its *d*, *s*, and *p* valence orbitals in σ bonding, a total of *nine* ligands could be attached. However, because of the bulkiness of most ligands it is extremely difficult to achieve a coordination number of 9. With rhenium (Re), a large third-row transition-metal atom, and H, a small ligand, the coordination number of 9 is found in the complex ReH_9^{2-}. The structure of this interesting complex is illustrated in Figure 20-7.

The bonds in most coordination complexes, however, use fewer than the nine atomic orbitals from the metal. We shall turn now to the theories that have been developed to explain this bonding and the properties of the complexes formed. There have been four stages in the development of transition-metal bonding theory. These are the simple **electrostatic theory**, the **valence bond** or **localized-molecular-orbital theory**, the **crystal field theory**, and the **ligand field** or **delocalized-molecular-orbital** theory. Each of these theories is an improvement on its predecessor. Considered together, they provide a good case study of how bonding ideas develop, and how the same physical facts can be explained by different and seemingly contradictory assumptions.

We shall devote most of the discussion to octahedral coordination because it is both the most common and the easiest to understand. Keep in mind the following questions, which we shall try to answer when developing the theories:

1. How can we explain the difference in absorption of energy (manifested in the color) by a complex as the nature of the ligands is changed (recall Table 20-2)?

2. How can we explain that a complex such as $Co(NH_3)_6^{3+}$ is diamagnetic, yet others such as CoF_6^{3-} are paramagnetic and have one or several unpaired electrons?

Figure 20-7   The structure of the ReH_9^{2-} ion. There are six H atoms at the corners of a trigonal prism, and three more H atoms around the Re atom on a plane halfway between the triangular end faces of the prism.

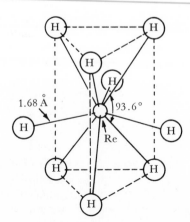

3. The stability of d^0, d^5, and d^{10} electron arrangements can be explained. But why are d^3 and d^6 so stable (recall Cr^{3+} and Co^{3+})?

4. Why do certain ions with the d^8 configuration, such as Pt(II) and Pd(II), prefer square planar geometry to tetrahedral or octahedral?

Electrostatic Theory

The simple electrostatic theory assumes only that the ligands, with negative charges, approach the positively charged central ion. Ligands and central ion attract one another, but ligands repel one another. The electrostatic repulsion between ligands leads to a prediction that a coordination number of 2 will be linear, and three ligands will lie at the corners of an equilateral triangle with the central atom at the center of the triangle. Four ligands will be tetrahedral, and six will be octahedral. This electrostatic theory cannot explain the existence of square planar complexes. Also, it cannot explain why complexes form with neutral molecules (CO, H_2O, NH_3) or with positive ions ($NH_2NH_3^+$). Finally, the theory does not discuss magnetic properties of complexes or their electronic energy levels as revealed by their colors and spectra.

Valence Bond (or Localized-Molecular-Orbital) Theory

One of the first definite advances toward understanding why octahedral geometry occurs was made when Pauling showed, in 1931, that a set of six s, p, and d orbitals could be hybridized in a manner similar to the sp^3 and sp^2 hybridization to produce six equivalent orbitals directed to the vertices of an octahedron. The orbitals required are the s, the three p, and the $d_{x^2-y^2}$ and d_{z^2} orbitals lying either just below or just above these s and p orbitals. The two d orbitals are chosen because they have lobes of maximum density pointing in the six axial directions of an octahedron, as do the three p orbitals. The resulting six octahedrally oriented orbitals are called d^2sp^3 or sp^3d^2 hybrid orbitals, depending on whether the principal quantum number of the d orbitals is one less than or is the same as that of the s and p orbitals.

Each of the hybrid orbitals can be combined with an orbital from a ligand to make a bonding and an antibonding orbital, each with σ symmetry around the metal–ligand bond axis. The lone pair of electrons from each ligand orbital goes into the bonding molecular orbital, and six covalent bonds are produced (Figure 20-8). Similarly, four equivalent hybrid orbitals directed to the corners of a square in the xy plane can be produced from the $d_{x^2-y^2}$, s, p_x, and p_y metal orbitals.

The valence bond theory has not been successful in making quantitative predictions about energies, but at least it gives a rationalization for the magnetic properties of octahedral complexes. Pauling proposed that two types of complexes could be prepared: **outer-orbital** sp^3d^2 complexes in which the d orbitals lie above the s and p orbitals, and **inner-orbital** d^2sp^3

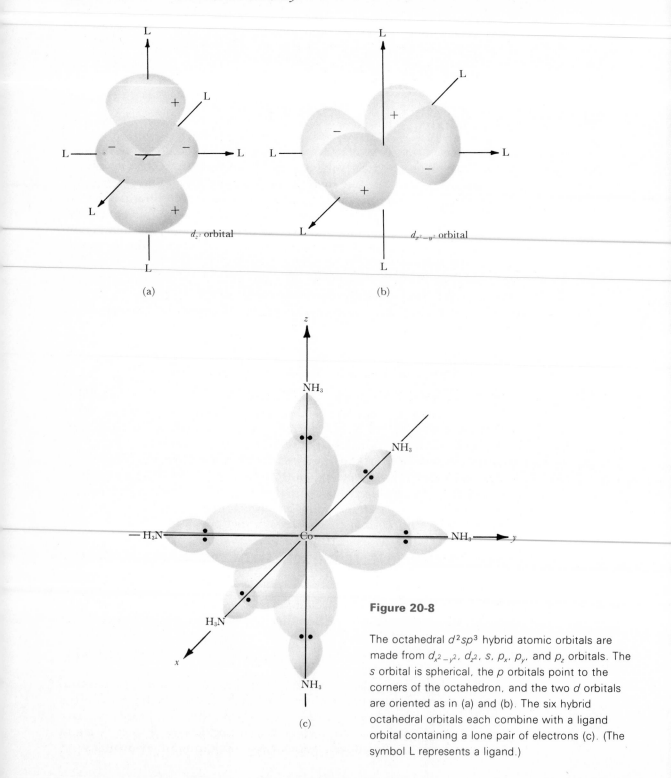

d_{z^2} orbital

(a)

$d_{x^2-y^2}$ orbital

(b)

(c)

Figure 20-8

The octahedral d^2sp^3 hybrid atomic orbitals are made from $d_{x^2-y^2}$, d_{z^2}, s, p_x, p_y, and p_z orbitals. The s orbital is spherical, the p orbitals point to the corners of the octahedron, and the two d orbitals are oriented as in (a) and (b). The six hybrid octahedral orbitals each combine with a ligand orbital containing a lone pair of electrons (c). (The symbol L represents a ligand.)

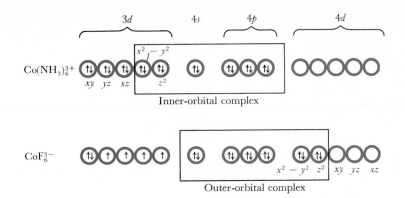

Figure 20-9

The valence bond theory postulates that in inner-orbital complexes of cobalt such as $Co(NH_3)_6^{3+}$, six electrons from the metal are spin-paired in d_{xy}, d_{yz}, and d_{xz} orbitals; the octahedral hybrid orbitals are produced from s, three p, and the two d from the level beneath. In outer-orbital cobalt complexes, all five of the underlevel d orbitals are used for electrons from the metal, now not completely paired. The octahedral hybrids use two d orbitals from the same quantum level as s and p. In either case, lone-pair electrons on the ligands fill the bonding orbitals formed between ligand orbitals and the six metal orbitals of the octahedral hybrid.

complexes in which they lie below (Figure 20-9). In inner-orbital complexes, the number of d orbitals left to hold the d electrons that remain on the metal ion is restricted. Only the d_{xy}, d_{yz}, and d_{xz} are available; the other two are used in octahedral hybridization.

We can use cobalt as an example of the valence bond explanation of magnetic properties. The neutral cobalt atom has nine electrons beyond the Ar noble-gas shell, and can be represented as

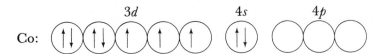

The Co^{3+} ion has six electrons, which by Hund's rule will be distributed among all five $3d$ orbitals:

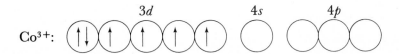

Now let us assume that six ligands, each with an electron pair, are to form six covalent bonds with hybridized metal orbitals that are octahedrally

oriented. If an *outer complex* is formed with $4s$, $4p$, and $4d$ metal orbitals, the electrons in the $3d$ orbitals are undisturbed (Figure 20-9):

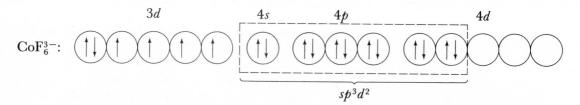

CoF$_6^{3-}$:

sp^3d^2

Four electrons will remain unpaired and, by this theory, CoF$_6^{3-}$ should be paramagnetic, as it is observed to be.

In contrast, if an *inner complex* is formed with $3d$ orbitals in the octahedral hybridization, then only three $3d$ orbitals will be left for the six valence electrons originally present in the Co^{3+} ion:

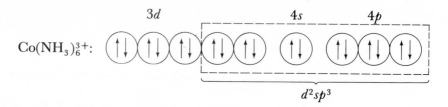

Co(NH$_3$)$_6^{3+}$:

d^2sp^3

Hence we predict Co(NH$_3$)$_6^{3+}$ to be diamagnetic, and it is.

In Figure 20-3, the Mn^{2+} ion in hexafluoromanganate(II) has a d^5 structure:

Mn^{2+}:

If hexafluoromanganate(II) were an inner-orbital complex, its five electrons would be compressed into three d orbitals and one electron would be unpaired:

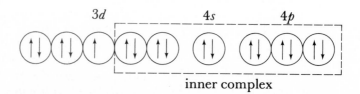

inner complex

Conversely, if it were an outer-orbital complex, all five electrons would be unpaired in the five d orbitals. Both possible complexes would be paramagnetic, but they would differ in the magnitude of the magnetic moment.

Experimental data indicate that the complex has five unpaired spins, so it must be an outer-orbital complex. The Fe^{3+} ion also has a d^5 configuration; however, because magnetic data show that hexacyanoferrate(III) has one unpaired electron, it is described in the valence bond theory as an inner-orbital complex. Ligands such as CN^- and CO tend to form inner-orbital complexes, and ligands such as F^-, Cl^-, Br^-, and I^- usually form outer-orbital complexes.

The valence bond theory produces the correct two alternatives for the number of unpaired electrons, but it offers little help in making the choice between them. It does predict that inner-orbital complexes will be relatively inert. The experimental observation that outer-orbital complexes *are* usually more labile than inner-orbital complexes gives us confidence that the valence bond theory is at least a step in the right direction. It was a landmark at the time that it was proposed; however, it has been supplanted by crystal field theory and a more complete molecular orbital theory.

Crystal Field Theory

From a localized-molecular-orbital theory, the pendulum now swings the other way to a purely electrostatic theory that regards the bonding between metal and ligand as ionic. The simple electrostatic theory predicts that octahedral coordination will arise for the same reason that six unit charges, constrained to move on the surface of a sphere, will adopt an octahedral arrangement as the one of lowest energy. This is simply the electron-pair repulsion idea of Section 11-3.

Crystal field theory is more realistic. With this theory we consider what happens to the five metal d orbitals when six negative charges are brought near the metal in an octahedral array along the three principal axes of the d orbitals. The negative charges represent the lone pairs on the ligands. They are considered to remain with the ligands rather than being involved in any type of covalent bonding with the metal. Therefore, crystal field theory assumes purely *ionic* bonding.

The $d_{x^2-y^2}$ and d_{z^2} orbitals are most affected by the negative charges, which represent the ligands. The orbitals point directly at these charges (Figure 20-10). Any electrons in these d orbitals will respond to the electrostatic repulsion from the ligand lone pairs. Electrons in these two d orbitals will have higher energies than those in the other three. In contrast, the d_{xy}, d_{yz}, and d_{xz} orbitals have their lobes of maximum density directed *between* the ligands (see Figure 8-24). Electrons in these orbitals are more stable. The net result of this electrostatic interaction with the ligands is that the five d orbitals are split into two energy levels separated by a **crystal-field splitting energy,** Δ_o, as shown in Figure 20-11. The lower level is called the t_{2g} level, and the upper, the e_g. The names come from group theory, and their origin need not concern us here.

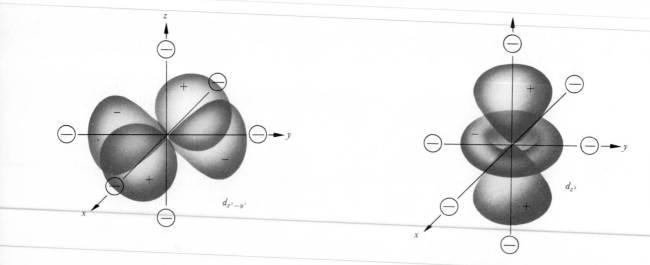

Figure 20-10   By crystal field theory, the six ligands of an octahedral complex may be represented as six negative charges, which point directly at the electron-density lobes of the metal $d_{x^2-y^2}$ and d_{z^2} orbitals. Any electrons in these two d orbitals will be repelled by the negative charges. More energy is required to force electrons into these two d orbitals on the metal than into the d_{xy}, d_{yz}, and d_{xz} metal orbitals, all of which point between the ligands.

The crystal-field splitting energy, Δ_0, is obtained by measuring the energy absorbed when one electron is promoted from the t_{2g} level to the e_g level (Figure 20-12). This splitting energy is crucial in accounting for magnetic properties. If Δ_0 is small, as in CoF_6^{3-}, the six d electrons of Co^{3+} are spread out among all five d orbitals (Figure 20-13). There is a saving of energy if as few electrons as possible are paired. Conversely, if the splitting constant is large enough to overcome the energy of pairing two electrons in the same orbital, the more stable arrangement will be for the three low-lying orbitals of the t_{2g} level to contain one pair of electrons each and for the two upper orbitals to be vacant. This is the situation in $Co(NH_3)_6^{3+}$. Because of the different numbers of unpaired electrons in the two structures, $Co(NH_3)_6^{3+}$ is called a **low-spin complex** and CoF_6^{3-} is called a **high-spin complex**.

Notice how the same facts are explained by two quite different theories, the valence bond and crystal field theories. Both theories state that low-spin octahedral complexes arise when only three d orbitals of low energy are available for electrons originally from the central metal ion. High-spin octahedral complexes occur when there are *five* low-lying d orbitals. However, valence bond theory accounts for the presence of three or five such orbitals in terms of the set of six orbitals used in octahedral hybridization.

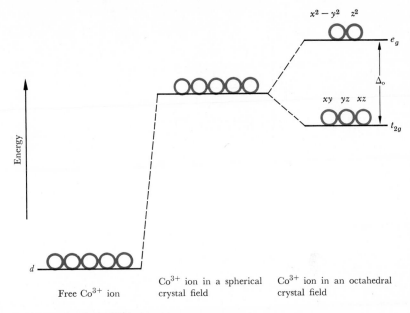

Figure 20-11 Energy-level diagram for the five d orbitals of a metal ion in an octahedral crystal field. On the left is the energy of electrons in the d orbitals of a free ion. In the center is the energy of electrons in the d orbitals if the ion were surrounded by a spherical cloud of negative charges. On the right is the splitting in energies of the d orbitals produced if the negative charges are arranged octahedrally around the metal. The three d orbitals that point between the ligands have lower energies than the two orbitals that point directly at the ligands.

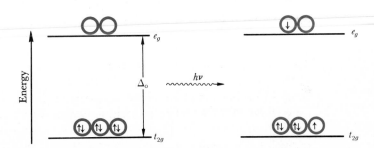

Figure 20-12 When $Co(NH_3)_6^{3+}$ absorbs a photon of violet light and transmits those frequencies that give it its yellow color, the electronic configuration goes from the one at the left to the one at the right.

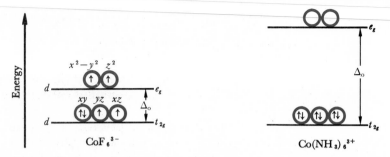

Figure 20-13   Crystal field theory explanations of high-spin and low-spin complexes. The crystal-field splitting produced by the F^- ion is small, and the energy required to place two electrons in the upper level is less than the energy required to pair them with others. Therefore, the high-spin CoF_6^{3-} complex spreads its electrons among all five orbitals, and has four unpaired spins. The NH_3 group produces such a large crystal-field splitting that it is easier to pair electrons in the bottom three orbitals. The low-spin $Co(NH_3)_6^{3+}$ complex has no unpaired electrons.

In contrast, crystal field theory invokes a small or a large energy gap between a low-lying set of three d orbitals and a less stable set of two. In the valence bond theory, the operating factor is hybridization of orbitals from the metal, and the bonds to the ligands are entirely covalent. In the crystal field theory, the operating factor is electrostatic repulsion between ligand electron pairs and electrons on the metal ion, and the bonds to the ligands are entirely ionic. The effects are the same, but the explanations are radically different. Which theory is true?

Some chemists dislike the word *true* and prefer circumlocutions such as "successful in accounting for the facts." But unless the chemist is also a mystic who believes in some sort of inner reality beyond that which can be apprehended by the senses, the two sets of terminology are equivalent. No theory can ever be proven to be true in the absolute sense. All we can say is that one theory is "truer" than another because it can account for more observed properties of its subject than another theory. By this criterion, crystal field theory is better than valence bond theory. The common ligands can be ranked according to the magnitude of crystal-field splitting, Δ_o, that they produce, and this order can be justified to a certain extent.

The stronger the electrostatic field created by the ligand, the greater the splitting should be. Small ions with their lone pairs concentrated in one place, as in F^-, should produce a greater effect than larger groups with electrons diffused over a larger volume, as in Cl^-. Beyond this size-related argument, we can list the ligands in order but cannot explain the order:

$$CO, CN^- > en > NH_3 > -NCS^- > H_2O > OH^-, F^- > Cl^- > Br^- > I^-$$

| strong-field | | intermediate-field | | weak-field |
| ligands | | ligands | | ligands |

We write the isothiocyanate ion as $-NCS^-$ to emphasize that the metal–ligand bond is through the N atom in these cobalt(III) complexes discussed in this chapter.

Without spectroscopes or prisms, we can quickly check the order of ligands in this list merely by looking at the colors of complexes with these ligands. The absorption of visible light during the excitation of metal d electrons from t_{2g} to e_g orbitals is the other important source of color in transition-metal complexes in addition to charge-transfer absorption. For metals in +2 and +3 oxidation states, the charge-transfer absorption is usually in the ultraviolet, and the colors we see are from crystal-field splitting. These colors are not as intense as the charge-transfer absorption colors of CrO_4^{2-} and MnO_4^-. In Table 20-2 is a list of cobalt complexes, their colors, the colors absorbed in the electronic transition of lowest energy, and the approximate wavelengths and energies involved. Replacing even one NH_3 in the complex by $-NCS^-$, H_2O, OH^-, or Cl^- decreases the energy difference between levels, or the transition energy, in the order given. Substituting Br^- for $-NCS^-$ in the ethylenediamine complex lowers the transition energy by approximately 10% and changes the ion from blue to green. Replacing $-NCS^-$ by a halide, Cl^-, in the presence of five NH_3 also lowers the transition energy by 10% and changes the salt from orange to purple.

Why is this list of relative strengths in energy-level splittings as it is? We cannot say from crystal field theory. But if the orbitals on the ligands are taken into account, both those that contain the electron pairs to be shared with the metal and those that contain lone pairs not directly associated with the metal, we can explain more of the order of splitting energies. This extended molecular orbital theory contains both crystal field and valence bond theories as extreme cases, and it is commonly called the **ligand field theory**.

Ligand Field (or Delocalized-Molecular-Orbital) Theory

With ligand field theory we take into account the orbitals on the ligands, and consider the ligands as something more than mere spherical charges. In the delocalized-molecular-orbital treatment, six ligand orbitals, assumed as a first approximation to have σ symmetry around the metal–ligand bond lines, are combined with six of the nine metal s, p, and d orbitals: $d_{x^2-y^2}, d_{z^2}$, s, p_x, p_y, and p_z. These are the same orbitals that Pauling used to synthesize his six hybrid orbitals. Now we shall combine all of them with the six ligand atomic orbitals to produce six delocalized bonding orbitals and six antibonding orbitals (Figure 20-14). The d_{xy}, d_{yz}, and d_{xz} orbitals, having the wrong symmetry for combining with σ-like ligand orbitals, are *nonbonding*.

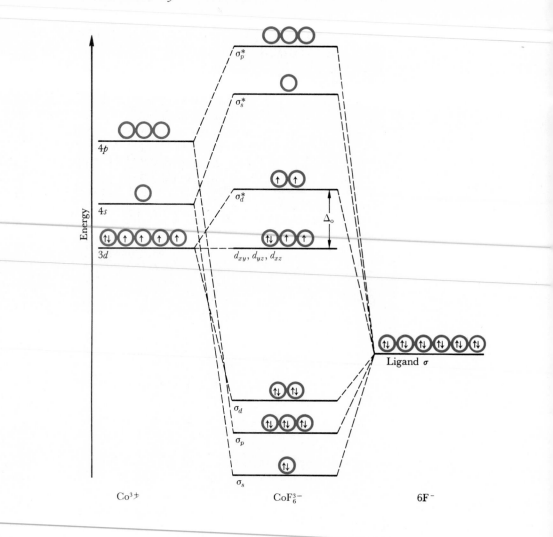

Figure 20-14

In the delocalized-molecular-orbital treatment of octahedral coordination, the same six metal orbitals that were used in the valence bond theory ($d_{x^2-y^2}$, d_{z^2}, s, p_x, p_y, and p_z) now combine with the six lone-pair-containing ligand orbitals to produce six bonding molecular orbitals (σ_s, σ_p, and σ_d) and six antibonding orbitals (σ_d^*, σ_s^*, and σ_p^*). The d_{xy}, d_{yz}, and d_{xz} metal orbitals are nonbonding. The low-lying six bonding orbitals fill with the electron pairs from the ligand to make six electron-pair bonds between metal and ligand. The d electrons of the metal ion are in the nonbonding and lowest antibonding levels, which are separated by the energy Δ_o. These two levels correspond to those in Figure 20-11, but the explanation of their origin is different.

Electron pairs in these orbitals have no effect on holding ligands and metal together, and are described as *metal lone pairs.*

The resulting energy-level diagram appears in Figure 20-14. The six bonding orbitals at the bottom are filled with electron pairs. We can think of them as being the six pairs donated by the ligands, and we can forget about them. The upper four antibonding orbitals are similarly irrelevant; they will be empty except in extreme cases of electronic excitation, which we shall ignore. The nonbonding level and the lowest antibonding level correspond to the two levels, t_{2g} and e_g, produced by crystal-field splitting (Figure 20-13). We shall continue to call them by these names, even in the molecular orbital treatment. But note the difference in the explanation of how this splitting occurs. In crystal field theory, it is the consequence of electrostatic repulsion; in ligand field theory, it is a consequence of the preparation of molecular orbitals. As we saw in Chapter 12 for HF and KF, the same molecular orbital theory can accommodate everything from purely ionic to purely covalent bonding. The choice between these two theories is accordingly a pseudochoice, a consequence of being committed to two extreme models. In CoF_6^{3-} there is a certain ionic character to the bonding, because, as you can see in Figure 20-14, the ligand orbitals are lower than those of the metal and closer in energy to the bonding molecular orbitals. Therefore, the bonding orbitals will have more of the character of the ligand orbitals, and there will be a displacement of negative charge toward the ligands. Thus, the bonds will be partially ionic.

With the molecular orbital theory, we can do a much better job of predicting which ligands will cause large energy differences between the t_{2g} and e_g levels in octahedral coordination, and which will produce small splittings. For this prediction we must look at the interactions of d_{xy}, d_{yz}, and d_{xz} orbitals in the t_{2g} level with atomic orbitals on ligands that have π symmetry around the metal–ligand bond.

The crystal field theory assumes that there are no such ligand orbitals and that each ligand is a featureless sphere of charge. Ligand field theory considers the ligand orbitals that form bonds to the metal ion, and also the two unhybridized p orbitals at right angles to the metal–ligand bond. These unhybridized p orbitals strongly influence the ligand-field splitting energy, Δ_0.

Figure 20-15 depicts four of these chloride p orbitals overlapping one of the three d orbitals in the t_{2g} energy level. If there are electrons in this d orbital, they are repelled by the lone-pair electrons in these p orbitals, and the energy of the t_{2g} level is raised. Any ligand with filled orbitals having such π symmetry around the ligand–metal axis decreases the ligand-field splitting energy, Δ_0. If we retain the crystal field theory terminology, such ligands (OH^-, Cl^-, Br^-, I^-) are called **weak-field ligands.** Fluoride ion is not as efficient at this process because it holds its electrons so tightly. Such an interaction is a ligand-to-metal(π) or $L \rightarrow M(\pi)$ interaction.

Figure 20-15

The lone-pair electrons in the π orbitals of Cl^- repel electrons in the d_{xy}, d_{yz}, and d_{xz} orbitals of the metal (M), thereby making the levels less stable. The t_{2g} level in Figure 20-11 rises and the splitting energy, Δ_o, decreases.

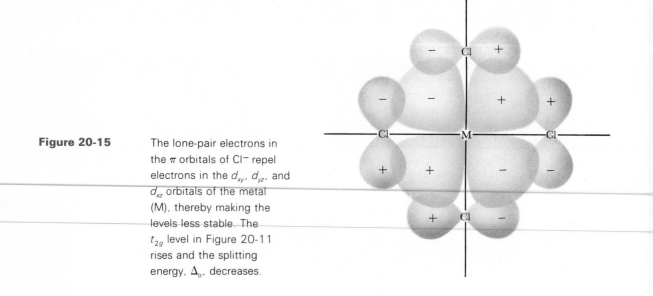

Polyatomic groups that have an unfilled antibonding orbital with π symmetry behave differently. The cyanide ion (Figure 20-16) has a triple bond made from one bonding σ^b orbital and two bonding π^b molecular orbitals. One of these π^b bonding orbitals is shown in Figure 20-16a. This orbital destabilizes or raises the t_{2g} level by a $L \rightarrow M(\pi)$ process just as in Cl^-. But most of the electron density of the π^b orbital lies between the C and the N, *not* in the direction of the metal atom. It is the antibonding π^* orbital (Figure 20-16b) that interacts more with the metal t_{2g} level. Here the effect is the reverse of that in Cl^-. Electrons in the metal t_{2g} orbitals can become partially delocalized and flow into the π^* orbital on the ligand. This

Figure 20-16

The effect of π bonding in cyano complexes. (a) In the CN^- ion, the bonding π^b molecular orbital contains an electron pair, and the antibonding π^* orbital (b) is empty. (c) The metal orbitals of the t_{2g} type are more stable in the presence of simple σ symmetrical ligands because the t_{2g} orbitals do not concentrate their electrons in the directions of the ligands. But if the ligand has filled π orbitals, then these orbitals interact with the metal t_{2g} orbitals and make them less stable. The splitting constant decreases. (d) If the metal has filled t_{2g} orbitals that interact with the empty antibonding π ligand orbitals, then the metal electrons are delocalized, the energy of the orbitals falls, and the splitting energy increases. This last effect predominates in most CN^- complexes, and we say that CN^- produces a large ligand-field splitting.

          ▶

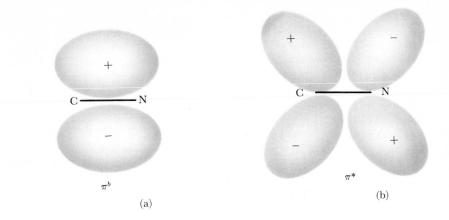

π^b

(a)

π^*

(b)

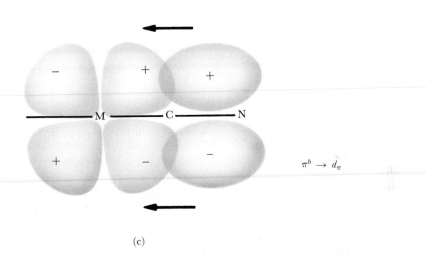

$\pi^b \rightarrow d_\pi$

(c)

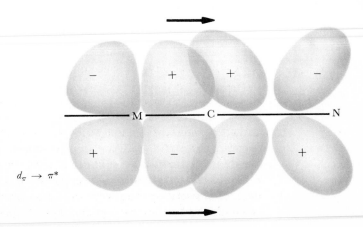

$d_\pi \rightarrow \pi^*$

(d)

delocalization stabilizes the t_{2g} orbital and lowers its energy. Therefore, the splitting energy, Δ_o, increases. This process is a metal-to-ligand (π) or $M \rightarrow L(\pi)$ interaction, and often it is called π **back bonding**. Ligands that increase the splitting of the levels in this way (CO, CN^-, NO_2^-) are called **strong-field ligands** in crystal field terminology. Single atoms with many lone pairs of electrons, such as the halide ions, are weak-field ligands because they donate electrons. Bonded groups of atoms such as CO are more likely to be strong-field ligands because their bonding orbitals of π symmetry are concentrated between pairs of atoms and away from the metal, while the empty antibonding molecular orbitals extend closer to the metal.

The nature of the metal itself also has a large influence on the size of the ligand-field splitting. Metal atoms or ions utilizing $4d$ and $5d$ valence orbitals give rise to much larger splittings than in corresponding complexes involving $3d$-orbital metals. For example, the Δ_o values for $Co(NH_3)_6^{3+}$, $Rh(NH_3)_6^{3+}$, and $Ir(NH_3)_6^{3+}$ are $22,900\ cm^{-1}$, $34,100\ cm^{-1}$, and $40,000\ cm^{-1}$, respectively. Presumably the $4d$ and $5d$ valence orbitals of the ion are more suitable for σ bonding with the ligands than are the $3d$ orbitals, but the reason for this is not well understood. An important consequence of the much larger Δ_o values of $4d$ and $5d$ central metal ions is that *all* second- and third-row metal complexes have low-spin ground states, even complexes such as $RhBr_6^{3-}$, which contain ligands at the weak-field end of the spectrochemical series.

We have discovered that both the magnetic properties and the colors of transition-metal complexes depend on the nature of the ligand and metal by their effects on the ligand-field splitting energy, Δ_o. Thus, two of the questions listed at the beginning of this section have been answered. We can also explain the unusual stability of d^3 and d^6 configurations in complexes with strong-field ligands. The d^3 and d^6 arrangements are half-filled and completely filled t_{2g} levels. When the level splitting is large, these arrangements have the same significance in terms of stability that d^5 and d^{10} configurations do when all five d levels have the same energy. The stability of d^5 and d^{10} arrangements is most noticeable in weak-field complexes, when the ligand-field splitting is small.

20-4   TETRAHEDRAL AND SQUARE PLANAR COORDINATION

Energy levels estimated from ligand field theory for ligands of a given strength in different geometrical arrangements around the metal are compared in Figure 20-17. The relative order of energies in tetrahedral coordination is the reverse of octahedral, and it is not difficult to understand why. Ligands in a tetrahedral complex approach the metal from four of the eight corners of a cube (Figure 20-2b). It is precisely the $d_{x^2-y^2}$ and d_{z^2} orbitals that do *not* point to the corners of the cube around the metal atom. As you

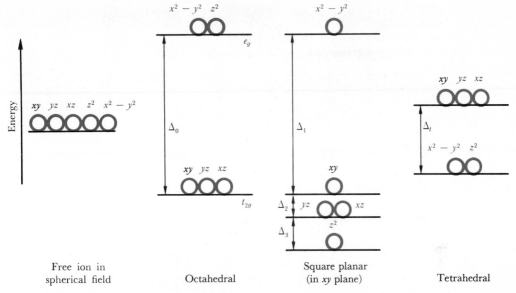

Figure 20-17   Energy levels for the five d orbitals in the free ion in a spherical field of electrical charge and in the three common coordination geometries, all calculated for the same strength ligand. The relative order of levels is explained in the text. The ligand-field splitting energies are represented by Δ_0, Δ_1, Δ_2, Δ_3, and Δ_t.

can verify from Figure 8-24, the density lobes of the d_{xy}, d_{yz}, and d_{xz} orbitals point to the midpoints of the 12 edges of a cube, whereas the other two point to the midpoints of the six faces. The set of three d orbitals, being closer to the tetrahedral ligands, will be less stable, even though the splitting is not as pronounced as for octahedral geometry.

Square planar splitting is almost as straightforward. Since we usually work with d_{z^2} and $d_{x^2-y^2}$ orbitals, let us take the xy plane as the plane of the complex, and assume that the ligands are at equal distance in the $\pm x$ and $\pm y$ directions. The $d_{x^2-y^2}$ orbital then points directly at the four ligands and is least stable. The d_{z^2} orbital points perpendicularly out of the plane of the ligands and is most stable (Figure 20-17). The other three orbitals have intermediate stability; d_{yz} and d_{xz} are more stable than d_{xy} because they are out of the plane of the ligands.

The octahedral arrangement is intrinsically more stable than the square planar because six bonds are formed instead of four. A typical covalent single bond and a typical ionic bond both have a bond energy of 200 to 400 kJ mole^{-1}. This corresponds to 17,000 to 33,000 cm^{-1} in the units in which splitting energies are given in Table 20-7. An octahedral complex,

Table 20-7

Ligand-Field Splitting Energies for Representative Metal Complexes

Octahedral complexes	Δ_o (cm^{-1})	Octahedral complexes	Δ_o (cm^{-1})
$Ti(H_2O)_6^{3+}$	20,300	CoF_6^{3-}	13,000
TiF_6^{3-}	17,000	$Co(H_2O)_6^{3+}$	18,200
$V(H_2O)_6^{3+}$	17,850	$Co(NH_3)_6^{3+}$	22,900
$V(H_2O)_6^{2+}$	12,400	$Co(CN)_6^{3-}$	34,500
$Cr(H_2O)_6^{3+}$	17,400	$Co(H_2O)_6^{2+}$	9,300
$Cr(NH_3)_6^{3+}$	21,600	$Ni(H_2O)_6^{2+}$	8,500
$Cr(CN)_6^{3-}$	26,600	$Ni(NH_3)_6^{2+}$	10,800
$Cr(CO)_6$	32,200	$RhCl_6^{3-}$	22,800
$Fe(CN)_6^{3-}$	35,000	$Rh(NH_3)_6^{3+}$	34,100
$Fe(CN)_6^{4-}$	33,800	$RhBr_6^{3-}$	19,000
$Fe(H_2O)_6^{3+}$	13,700	$IrCl_6^{3-}$	27,600
$Fe(H_2O)_6^{2+}$	10,400	$Ir(NH_3)_6^{3+}$	40,000

Tetrahedral complexes	Δ_t (cm^{-1})
VCl_4	9010
$CoCl_4^{2-}$	3300
$CoBr_4^{2-}$	2900
CoI_4^{2-}	2700
$Co(NCS)_4^{2-}$	4700

Square planar complexes	Δ_1 (cm^{-1})	Δ_2 (cm^{-1})	Δ_3 (cm^{-1})	Total Δ (cm^{-1})
$PdCl_4^{2-}$	23,600	3900	7400	34,900
$PtCl_4^{2-}$	29,700	4700	6800	41,200

with two more bonds than either square planar or tetrahedral, has an intrinsic energy advantage of 35,000 to 65,000 cm^{-1}. Although it appears from Figure 20-17 that square planar coordination is preferable for d^1 through d^6, the extra bond energy causes octahedral coordination to predominate. However, the seventh and eighth electrons are forced into the high-energy e_g orbitals in octahedral coordination, whereas the much more stable d_{xy} is available in square planar. This extra stability is decisive for d^8 configurations in which the ligand-field splitting is large: They are found in square planar coordination. The ligand-field splitting is larger at higher atomic numbers. Hence, Pt(II) and Pd(II) regularly have square planar coordination, whereas Ni(II) is usually octahedral. The ninth and tenth electrons tip the balance back in favor of octahedral because of the extra stability gained from the two additional bonds.

Tetrahedral coordination is seldom preferred and is relatively rare. In addition to the smaller number of bonds in comparison with octahedral, the tetrahedral coordination also suffers from the double disadvantage of a less stable lower level and the necessity of commencing the upper level at the third electron rather than the fourth (in the high-spin complexes).

A selection of measured ligand-field splitting energies for all three coordinations is given in Table 20-7. See whether the octahedral data are compatible with the order of ligand splitting strengths given previously in this section. Also, note how close our guess for the splitting of $Co(NH_3)_6^{3+}$, based purely on color (Table 20-2), really was.

20-5   EQUILIBRIA INVOLVING COMPLEX IONS

When we write Co^{2+} to represent an ion in aqueous solution, we understand implicitly that the bare ion is not present, but that water molecules of hydration are coordinated to the metal. Therefore, the chemistry of complex ions in solution is the chemistry of the substitution of one ligand molecule or ion for another in the coordination shell around a metal. Nevertheless, it is customary, for simplicity, to write the formation of the ammine complex of Co^{2+}, for example, as if it were the addition of NH_3 to dipositive cobalt ions:

$$Co^{2+} + 6NH_3 \rightleftarrows Co(NH_3)_6^{2+} \tag{20-1}$$

We can write an equilibrium constant for this reaction:

$$K_f = \frac{[Co(NH_3)_6^{2+}]}{[Co^{2+}][NH_3]^6} \tag{20-2}$$

Since the equilibrium concerns the formation of a complex, K_f is known as a formation constant. For the formation of hexaamminecobalt(II), $K_f = 1 \times 10^5$.

There is no difference in principle between the mathematics of formation-constant problems and that of dissociation of acids or bases. The parallel would be somewhat more apparent if equation 20-1 were written as a dissociation of $Co(NH_3)_6^{2+}$ rather than as an association, and if a dissociation constant that is the inverse of K_f were used. Formation constants, however, are customary.

As soon as NH_3 is added to a solution of Co^{2+}, some of it combines with Co^{2+} and produces some complex ions. At equilibrium after the addition of NH_3, the concentrations of the complex ion, NH_3, and free Co^{2+} (actually hydrated) can be calculated from equation 20-2.

Example 1

Enough NH_3 is added to a $0.100M$ solution of Ag^+ to make the initial concentration of NH_3 1 mole liter^{-1}. After equilibrium is restored, what will be the concentrations of Ag^+ and of $Ag(NH_3)_2^+$?

Solution

The formation constant for $Ag(NH_3)_2^+$ is given in Table 20-8 as $K_f = 1 \times 10^8$. Therefore, the equilibrium-constant expression is

$$K_f = \frac{[Ag(NH_3)_2^+]}{[Ag^+][NH_3]^2} = 1 \times 10^8$$

Because the formation constant is so large, we can assume that the formation reaction is effectively complete and that the concentration of $Ag(NH_3)_2^+$ is equal to the initial concentration of Ag^+. Since this quantity is appreciable, the concentration of NH_3 remaining at equilibrium is the original concentration less the amount reacted with Ag^+:

$$[Ag(NH_3)_2^+] = 0.100 \text{ mole liter}^{-1}$$
$$[NH_3] = 1.000 - 0.200 = 0.800 \text{ mole liter}^{-1}$$

(Two moles of NH_3 react for every mole of $Ag(NH_3)_2^+$ produced.) Therefore, the concentration of silver ion left at equilibrium is

$$[Ag^+] = \frac{0.100}{(0.800)^2} \times 1 \times 10^{-8} = 2 \times 10^{-9} \text{ mole liter}^{-1}$$

The assumption that the formation reaction is effectively complete is justified by the small Ag^+ concentration.

Example 2

What will be the final concentration of Ni^{2+} hydrated ion if 50 ml of $2.00M$ NH_3 solution are added to 50 ml of $0.200M$ Ni^{2+} solution?

Solution

The formation-constant expression is

$$K_f = \frac{[Ni(NH_3)_6^{2+}]}{[Ni^{2+}][NH_3]^6} = 6 \times 10^8$$

We assume that the concentration of $Ni(NH_3)_6^{2+}$ at equilibrium is approximately 0.100 mole liter^{-1}. The concentration of unbound NH_3 is then $1.00 - 6(0.100) = 0.40$ mole liter^{-1}. Thus we have

$$\frac{[0.100]}{x[0.40]^6} = 6 \times 10^8$$

$$x = 4 \times 10^{-8} \text{ mole liter}^{-1}$$

Table 20-8

Overall Formation Constants for Some Complexes in Aqueous Solution[a] at 298 K

Complex, ML_n	K_f, $[ML_n]/[M][L]^n$	Complex, ML_n	K_f, $[ML_n]/[M][L]^n$
L = NH_3		**L = $H_2NCH_2CH_2NH_2$ (en)**	
$Ag(NH_3)_2^+$	1×10^8	$Mn(en)_3^{2+}$	5×10^5
$Cu(NH_3)_4^{2+}$	1×10^{12}	$Fe(en)_3^{2+}$	4×10^9
$Zn(NH_3)_4^{2+}$	5×10^8	$Co(en)_3^{2+}$	8×10^{13}
$Cd(NH_3)_4^{2+}$	1×10^7	$Ni(en)_3^{2+}$	4×10^{18}
$Ni(NH_3)_6^{2+}$	6×10^8	$Cu(en)_2^{2+}$	1.6×10^{20}
$Co(NH_3)_6^{2+}$	1×10^5	$Zn(en)_3^{2+}$	1.2×10^{13}
L = F^- [b]		**L = Cl^-**	
AlF_6^{3-}	7×10^{19}	$MgCl^+$	4.0
SnF_3^-	8×10^9	$CuCl^+$	1.0
SnF_6^{2-}	10^{25}	$CuCl_4^{2-}$	10^{-5}
ZnF^+	5.0	$AgCl_2^-$	1×10^2
FeF^{2+}	3×10^5	$HgCl_4^{2-}$	1.6×10^{16}
MgF^+	65	$TlCl_4^-$	7.5×10^{18}
HgF^+	10	$BiCl_6^{3-}$	4×10^6
CuF^+	10	$SnCl_4^{2-}$	1.1×10^2
		$PbCl_4^{2-}$	4×10^2
		$FeCl^{2+}$	3.0
		$FeCl_4^-$	6×10^{-2}
L = OH^- [c]		**L = CN^-**	
$Cr(OH)^{2+}$	1×10^{10}	$Fe(CN)_6^{3-}$	10^{31}
$Fe(OH)^{2+}$	1×10^{11}	$Fe(CN)_6^{4-}$	10^{24}
$Co(OH)^{2+}$	1×10^{12}	$Ni(CN)_4^{2-}$	10^{30}
$Al(OH)^{2+}$	2×10^{28}	$Zn(CN)_4^{2-}$	5×10^{16}
$In(OH)_4^-$	1.5×10^{35}	$Cd(CN)_4^{2-}$	6×10^{18}
$Mn(OH)^+$	3×10^4	$Hg(CN)_4^{2-}$	4×10^{41}
$Fe(OH)^+$	1×10^7	$Ag(CN)_2^-$	10^{21}
$Co(OH)^+$	2.5×10^4		
$Ni(OH)^+$	1×10^5		
$Cu(OH)^+$	1×10^7		
$Zn(OH)^+$	1×10^5		
$Ag(OH)$	1×10^3		
$Zn(OH)_4^{2-}$	5×10^{14}		
$Pb(OH)_3^-$	8×10^{13}		

[a] In a strict sense, values should be accompanied by a more detailed description of solvent media and method of measurement. These values are approximate and are useful only for comparisons of similar species. For $n = 1$ assume that three or five water molecules are also in the complex.

[b] Many stable complexes such as SiF_6^{2-} and AsF_6^- form, but they hydrolyze in water to give oxyanions or oxides.

[c] Most polypositive metal ions tend to form poly-nuclear complexes with

$$-O- \quad \text{or} \quad \overset{\displaystyle H}{\underset{\displaystyle}{\overset{|}{O}}}$$

bridges in the presence of OH^-, as in $Fe-O-Fe^{4+}$, $Bi_6(OH)_{12}^{6+}$, $Cr_2(OH)_2^{4+}$, and so on, not to mention extremely insoluble hydroxide precipitates.

Example 3

What will be the final concentration of Ni^{2+} hydrated ion if 50 ml 2.00M ethylenediamine (en) solution are added to 50 ml of 0.200M Ni^{2+} solution?

Solution

The formation-constant expression is

$$K_f = \frac{[Ni(en)_3^{2+}]}{[Ni^{2+}][en]^3} = 4 \times 10^{18}$$

Under the same assumptions made in Example 2, we have

$$\frac{[0.100]}{x[0.70]^3} = 4 \times 10^{18}$$

$$x = 7 \times 10^{-20} \text{ mole liter}^{-1}$$

These exercises illustrate the considerably greater attraction that a chelating agent has for a metal ion as compared with a related ligand. Formation constants for ethylenediamine complexes in Table 20-8 are 8 to 10 orders of magnitude, or about a billion times, as large as formation constants for NH_3 complexes of the same metal ion. The bonding of ammonia and amine chelating agents to the metal is similar; in both cases the lone-pair electrons on an ammonia or amine nitrogen atom interact with the metal. The difference in formation constants between NH_3 and ethylenediamine reflects the increased stability when the bonding atoms of ligands are combined in a chelate molecule. This increased stability is sometimes called the **chelate effect**. However, the cyanide ion, CN^- (which bonds through the carbon), has an intrinsically stronger attraction for metals than does an amine nitrogen atom. As Table 20-8 shows, the formation constants for cyanide complexes are 3 to 13 orders of magnitude greater even than those of the corresponding ethylenediamine complexes!

Because formation constants are usually so large, we can ordinarily assume in complex-ion equilibrium problems that the concentration of the complex is the same as the total concentration of metal ion, as we have in the previous examples. However, for complexes of F^- this approximation is incorrect.

Example 4

What are the final concentrations of F^-, Hg^{2+}, and HgF^+ if 50 ml of a solution 2.00M in F^- are added to 50 ml of a solution 0.200M in Hg^{2+}?

Solution

It is best to begin with a table:

$$F^- \quad + \quad Hg^{2+} \quad \rightleftarrows \quad HgF^+$$

	F^-	Hg^{2+}	HgF^+
Initial conditions:	1.00	0.100	0 mole liter^{-1}
At equilibrium:	$1.00 - x$	$0.100 - x$	x mole liter^{-1}

$$K_f = \frac{x}{(1.00 - x)(0.100 - x)} = 10$$

Solve the equation for x by using the quadratic formula; $x = 0.090$ mole liter^{-1} = $[HgF^+]$; $[F^-] = 0.910$ mole liter^{-1}; and $[Hg^{2+}] = 0.010$ mole liter^{-1}.

Summary

This chapter has been a brief introduction to a rich area of chemistry, that of transition-metal complexes. Much of the richness (and the confusion) in their chemistry results from the presence of closely spaced energy levels involving d orbitals of the metal. The key to understanding transition-metal chemistry is the explanation of how the ligands perturb these metal energy levels. **Valence bond theory** and **crystal field theory** offer partial explanations, but currently the most successful theory is **ligand field theory**.

The story of these theories is an illustration of the dictum, "You can always prove a theory wrong, but you can never prove it right." The success of valence bond theory in accounting for the coordination geometry and magnetic properties is no guarantee that the theory is right, or even that this way of looking at the problem is correct. For example, does the splitting of t_{2g} and e_g levels come about because of the formation of molecular orbitals (ligand field theory), electrostatic repulsion (crystal field theory), or the choice of six orbitals for hybridization (valence bond theory)? Or are all three theories incomplete, and will we some day regard ligand field theory with the same skeptical tolerance with which we now view the old valence bond theory?

For the present, ligand field theory works in many ways and accounts for much of the behavior of transition-metal complexes. Using it, we can explain the absorption of light and the observed magnetic properties of ions. It accounts successfully for the effect of the ligand on the splitting of energy levels. It explains why the d^3 and low-spin d^6 electronic configurations are especially favored in octahedral complexes, and why d^8 leads to square planar geometry.

.Self-Study Questions

1. How can you account for the series of compounds with the formulas $CrCl_3$, $CrCl_3 \cdot 3NH_3$, $CrCl_3 \cdot 4NH_3$, $CrCl_3 \cdot 5NH_3$, and $CrCl_3 \cdot 6NH_3$? Why would you not expect to find the missing members of the series $CrCl_3 \cdot 2NH_3$ and $CrCl_3 \cdot NH_3$?

2. If you found the compound $CrCl_3 \cdot NaCl \cdot xNH_3$, what would you expect x to be?

3. How many different isomers of this compound, $CrCl_3 \cdot NaCl \cdot xNH_3$, would you expect to find?

4. What assumption about the geometry of bonding around the Cr molecule did you make in answering Question 3?

5. How does the number of isomers of a compound distinguish between the possible geometrical arrangements around the central metal ion? Illustrate with tetrahedral and square planar geometry.

6. What is the difference between paramagnetic and diamagnetic compounds? How are these distinguished from one another by experiment?

7. What is the difference between stability and inertness? Can a chemical system be stable yet not be inert? Can it be inert yet not be stable?

8. Why are complexes with electronic configurations of d^5 or d^{10} on the central metal atom stable? Why are complexes with d^3 and d^6 arrangements stable? Which configurations would you predict to be more important for stability in complexes with ligands of large splitting energies? Of small splitting energies?

9. How would you name the following compounds in a systematic way:

   $$Ir(NH_3)_3Cl_3 \qquad Rh(en)_2Cl_2Ir(en)Cl_4$$
   $$Co(NH_3)_6Cl_3 \qquad Rh(en)Cl_4Ir(en)_2Cl_2$$
   $$Rh(en)_3IrCl_6 \qquad RhCl_6Ir(en)_3$$

10. Sketch each of the four Rh–Ir complexes of Question 9.

11. Sketch each of the following complex ions or molecules:

    cis-dichlorotetraamminechromium(III) ion
    trans-dichlorotetraamminechromium(III) ion

    Indicate the charge on each complex.

12. What is the difference between structural, geometrical, and optical isomers? Find examples in Questions 9 and 11 of structural and geometrical isomers.

13. Why do complexes in which the central metal ion has the d^8 electronic configuration exist with square planar geometry?

14. What will be the number of unpaired electrons in $FeCl_6^{3-}$? In $Fe(CN)_6^{3-}$?

15. All octahedral complexes of vanadium(III) have the same number of unpaired electrons, no matter what the nature of the ligand. Why is this so?

16. What is the difference in the way that valence bond theory and crystal field theory explain the magnetic properties of complex ions?

17. How does ligand field theory account for the observed order of ligands in terms of the sizes of their splitting energies?

18. Why, in the crystal field theory, are the five d orbitals on the metal atom divided into two energy levels in the way they are? Where do the corresponding energy levels come from in the molecular orbital theory of complex ion structure?

19. Why are the same groupings of the five d orbitals made in tetrahedral coordination as in octahedral, but with the relative energies of these two groups reversed?

20. What is a chelate? If porphyrin is a tetradentate chelating group, and ethylenediamine is a bidentate chelating group, how would triethylene-tetraamine, diethylenetriamine, and EDTA be described?

21. What is a heme group? How does it function in hemoglobin and in cytochrome c?

Problems

Stoichiometry

1. A student was given 1.00 g of ammonium dichromate for the preparation of a coordination compound. The sample was ignited, thereby producing chromium(III) oxide, water, and nitrogen gas. The chromium(III) oxide was allowed to react at 600°C with carbon tetrachloride to yield chromium(III) chloride and phosgene ($COCl_2$). Upon treatment with excess liquid ammonia, the chromium(III) chloride reacted to produce hexaamminechromium(III) chloride. Calculate the maximum amount of hexaamminechromium(III) chloride that the student could prepare from the 1.00-g sample of ammonium dichromate.

2. When silver nitrate is added to a solution of a substance with the empirical formula $CoCl_3 \cdot 5NH_3$, how many moles of AgCl will be precipitated per mole of cobalt present? Why?

3. Co(III) occurs in octahedral complexes with the general empirical formula $CoCl_m \cdot nNH_3$. What values of n and m are possible? What are the values of n and m for the complex that precipitates 1 mole of AgCl for every mole of Co present?

4. How many ions per mole will you expect to find in solution when a compound with the empirical formula $PtCl_4 \cdot 3NH_3$ is dissolved in water? What about $PtCl_2 \cdot 3NH_3$? Draw diagrams of each of the complex cations.

5. Each of the following is dissolved in water to make a 0.001M solution. Rank the compounds in order of decreasing conductivity of their solutions: K_2PtCl_6, $Co(NH_3)_6Cl_3$, $Cr(NH_3)_4Cl_3$, $Pt(NH_3)_6Cl_4$. Rewrite each compound by using brackets to distinguish the complex ion present in aqueous solution.

Formulas and nomenclature

6. Give the systematic names of $[Co(NH_3)_4Cl_2]Br$, $K_3Cr(CN)_6$, and Na_2CoCl_4.

7. Write the formulas for each of the following compounds by using brackets to distinguish the complex ion from the other ions: (a) hexaaquonickel(II) perchlorate; (b) trichlorotriammineplatinum(IV) bromide; (c) dichlorotetraammineplatinum(IV) sulfate; (d) potassium monochloropentacyanoferrate (III).

8. Write the formula for each of the following by using brackets to distinguish the complex ion: (a) hydroxopentaaquoaluminum(III) chloride; (b) sodium tricarbonatocobaltate(III); (c) sodium hexacyanoferrate(II), (d) ammonium hexanitrocobaltate(III).

Isomers

9. How many isomers are there of the compound $[Cr(NH_3)_4Cl_2]Cl$? Sketch them.

10. Sketch all the geometrical and optical isomers of $PtCl_2I_2(NH_3)_2$.

11. How many geometrical and optical isomers are there of the complex ion $Co(en)_2Cl_2^+$? Of these, how many pairs of isomers are there differing only by a mirror reflection? How many isomers have a plane of symmetry and hence do not exist in pairs of optical isomers?

12. Repeat Problem 11 with propylenediamine substituted for ethylenediamine. Ignore optical isomers from the propylene carbon.

13. How many different *structural* isomers are there of a substance with the empirical formula $FeBrCl \cdot 3NH_3 \cdot 2H_2O$? For each different structural isomer, how many different *geometrical* isomers exist? How many of these can be grouped into right-handed and left-handed pairs of *optical isomers?*

Electronic structure

14. The Co^{2+} ion in aqueous solution is octahedrally coordinated and paramagnetic, with three unpaired electrons. Which one or ones of the following statements follow from this observation: (a) $Co(H_2O)_4^{2+}$ is square planar; (b) $Co(H_2O)_4^{2+}$ is tetrahedral; (c) $Co(H_2O)_6^{2+}$ has a Δ_o that is larger than the electron-pairing energy; (d) the d levels are split in energy and filled as follows: $(t_{2g})^5(e_g)^2$; (e) the d levels are split in energy and filled as follows: $(t_{2g})^6(e_g)^1$.

15. The coordination compound potassium hexafluorochromate(III) is paramagnetic. What is the formula for this compound? What is the configuration of the Cr d electrons?

16. How many unpaired electrons are there in Cr^{3+}, Cr^{2+}, Mn^{2+}, Fe^{2+}, Co^{3+}, Co^{2+} in (a) a strong octahedral ligand field and (b) a very weak octahedral field?

17. A low-spin tetrahedral complex has never been reported, although numerous high-spin complexes of this geometry have been prepared. What conclusion may be drawn regarding the magnitude of Δ_t from this fact?

18. Certain platinum complexes have been found to be active antitumor agents.

Among these are cis-$Pt(NH_3)_2Cl_4$, cis-$Pt(NH_3)_2Cl_2$, and cis-$Pt(en)Cl_2$ (none of the trans isomers is effective). Use valence bond theory to account for the diamagnetism of these complexes. Are these inner or outer complexes? What kinds of hybrid orbitals are used in bonding?

19. What is the d-orbital electronic configuration of $Cr(NH_3)_6^{3+}$? How many unpaired electrons are present? If six Br^- groups were substituted for the six NH_3 groups to give $CrBr_6^{3-}$, would you expect Δ_o to increase or decrease?

20. Diagram the electronic arrangements in $Fe(H_2O)_6^{2+}$ and $Fe(CN)_6^{4-}$ for both the valence bond and crystal field models. Briefly compare these models.

21. For each of the following, sketch the d-orbital energy levels and the distribution of d electrons among them:
a) $Ni(CN)_4^{2-}$ (square planar)
b) $Ti(H_2O)_6^{2+}$ (octahedral)
c) $NiCl_4^{2-}$ (tetrahedral)
d) CoF_6^{3-} (high-spin complex)
e) $Co(NH_3)_6^{3+}$ (low-spin complex)

22. Co(III) can occur in the complex ion $Co(NH_3)_6^{3+}$. (a) What is the geometry of this ion? In the valence bond theory, what Co orbitals are used in making bonds to the ligands? (b) What is the systematic name for the chloride salt of this ion? (c) Using crystal field theory, draw two possible d-electron configurations for this ion. Assign to them the labels *high spin, low spin, paramagnetic, diamagnetic*. Which two labels are correct for the ammine complex? (d) $Co(NH_3)_6^{3+}$ can be reduced to $Co(NH_3)_6^{2+}$ by adding an electron. Draw the preferred d-electron configuration for this reduced ion. Why is it preferred?

23. Pt(II) can occur in the complex ion $PtCl_4^{2-}$. (a) What is the geometry of this ion? In the valence bond theory, what Pt orbitals are used in making bonds to the Cl^- ions? (b) What is the systematic name for the sodium salt of this ion? (c) Using crystal field theory, draw the d-electron configuration for this ion. Is the ion paramagnetic or diamagnetic? (d) Pt(II) can be oxidized to Pt(IV). Draw the d-electron configuration for the chloride complex ion of Pt(IV). Explain the difference between this configuration and that of Pt(II). Is the Pt(IV) chloride complex ion paramagnetic or diamagnetic?

Formation constants

24. A solution is prepared that is $0.025M$ in tetraamminecopper(II), $Cu(NH_3)_4^{2+}$. What will be the concentration of Cu^{2+} hydrated copper ion if the ammonia concentration is 0.10, 0.50, 1.00, and $3.00M$ respectively? What ammonia concentration is needed to keep the Cu^{2+} concentration less than $10^{-15}M$?

25. From the data in Table 20-8, calculate the pH of a $0.10M$ solution of Cr^{3+} ion. *Hint:* Consider the reactions

$$Cr^{3+} + H_2O \rightleftarrows Cr(OH)^{2+} + H^+$$
$$K = ?$$
$$Cr^{3+} + OH^- \rightleftarrows Cr(OH)^{2+}$$
$$K_f = 1 \times 10^{10}$$
$$H^+ + OH^- \rightleftarrows H_2O \qquad K_w = ?$$

26. From the data in Table 20-8, calculate the pH of $0.10M$ solutions of Mn^{2+}, Fe^{2+}, and Ag^+. See Problem 25 if you need help. From the results of these two problems, can you correlate the "acidity" of positive ions with their charge?

27. The ion $Co(NH_3)_6^{3+}$ is very stable, with $K_f = 2.3 \times 10^{34}$. If the hydrolysis constant for the ammonium ion, K_b, is 5×10^{-10}, show that the equilibrium in the reaction

$$Co(NH_3)_6^{3+} + 6H^+ \rightleftharpoons Co^{3+} + 6NH_4^+$$

lies far to the right. Then why does $Co(NH_3)_6^{3+}$ remain intact in hot concentrated sulfuric acid?

28. What is the concentration of chromate ion, CrO_4^{2-}, when solid $BaCrO_4$ is placed in contact with water? What is the chromate ion concentration when solid $BaCrO_4$ is placed in contact with a solution of $0.2M$ Ba^{2+}? $BaCrO_4$ can be dissolved in a solution of pyridine (py), producing the complex $Ba(py)_2^{2+}$, with a formation constant of 4×10^{12}. If $0.10M$ $BaCrO_4$ is dissolved in a solution with a constant pyridine concentration of 1.0 mole liter^{-1}, what is the concentration of Ba^{2+} ion?

29. What is the solubility of $Cu(OH)_2$ in pure water? In buffer at pH 6? Copper(II) forms a complex with NH_3, $Cu(NH_3)_4^{2+}$, with $K_f = 1.0 \times 10^{12}$. What concentration of ammonia must be maintained in a solution to dissolve 0.10 mole of $Cu(OH)_2$ per liter of solution?

30. Calculate the silver ion concentration in a saturated solution of AgCl in water. Silver ions react with an excess of Cl$^-$ as follows:

$$Ag^+ + 2Cl^- \rightleftharpoons AgCl_2^- \quad K_f = 1 \times 10^2$$

Calculate the concentration of $AgCl_2^-$ and show that you were justified in ignoring the complex ion formation in calculating the silver ion concentration at the beginning of the problem.

31. The formation constant for the pyridine complex of silver

$$Ag^+ + 2py \rightleftharpoons Ag(py)_2^+$$

is $K_f = 1 \times 10^{10}$. If a solution is initially $0.10M$ in $AgNO_3$ and $1.0M$ in pyridine, what are the equilibrium concentrations of silver ion, pyridine, and the complex ion?

32. In $0.10M$ NaCl, the concentration of silver ions cannot exceed 10^{-9} mole liter^{-1} because AgCl is so slightly soluble. What concentration of pyridine must be added to dissolve 0.10 mole of AgCl per liter of solution?

Postscript: Coordination Complexes and Living Systems

Since we first realized that we lived on a planet circling one sun among many, rather than being fixed at the center of creation, we have wondered whether we were a one-time miracle (or accident) or part of a general pattern of living things. The astronomer Johannes Kepler (1571–1630) wrote a science fiction novel, *Somnium,* in which he described life on the moon as seen with a new invention, the telescope. He imagined intelligent humanoids and fast-growing plant life that sprouted, matured, and died in the course of one lunar day.

Today we know that any humanoids on the moon or Mars will be immigrants. However, it is possible that we will find the remains of simple life forms or the possible precursors of life forms on Mars, and that these will suggest something about how life evolved on earth. For years, scientists have extracted and analyzed organic matter from meteorites. They have debated whether this organic matter is truly meteoric or only terrestrial contamination, and whether it is of biological origin.

One of the compounds whose presence in meteorite samples is most suggestive of extraterrestrial life is porphin (Figure 20-18), and its derivatives, the *porphyrins*. The porphyrins are flat molecules that can act as tetradentate chelating groups* for metals such as Mg, Fe, Zn, Ni, Co, Cu, and Ag in a square planar complex as in Figure 20-19. The iron complex with the side chains shown in Figure 20-20 is called *heme*. The magnesium complex of porphyrin, with the organic side chain shown in Figure 20-21, is chlorophyll.

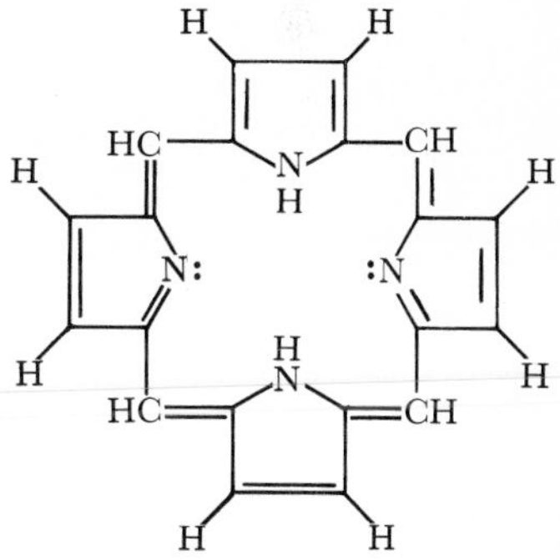

Figure 20-18

The porphin molecule. Porphin molecules with side groups substituted at the eight outermost hydrogen atoms around the ring are called porphyrins. A vertex where several bond lines meet, without a letter symbol, by convention is assumed to be a carbon atom. The four carbon atoms explicitly shown here by the symbol C could have been left out.

---

*The name *chelate* comes from the Greek for "claw"; *tetradentate* literally means "four-toothed." Chelates with twofold, threefold, or fourfold coordination to the metal ion are called *bidentate, tridentate,* or *tetradentate.* It may seem illogical to speak of claws with teeth, but lovers of lobster or crab will appreciate the usage.

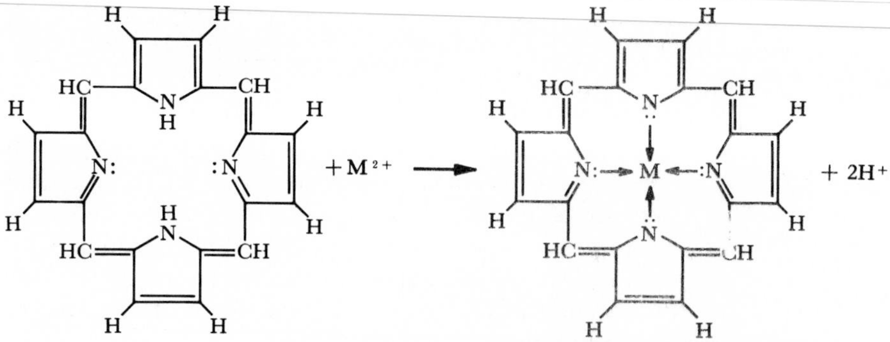

Figure 20-19   A porphyrin molecule can act as a tetradentate chelating group around an ion of a metal such as Mg, Fe, Zn, or Cu.

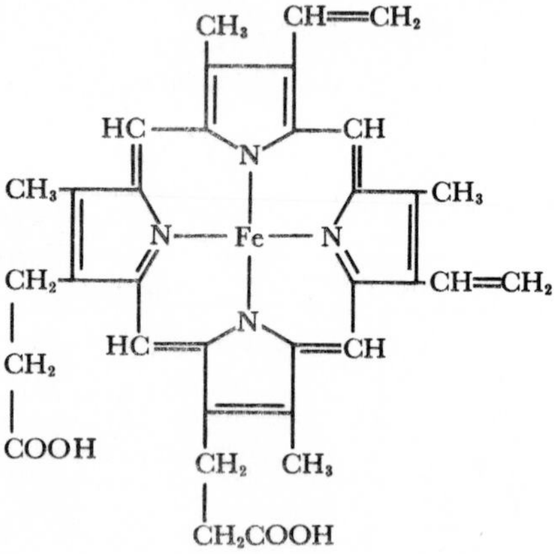

Figure 20-20   The iron—porphyrin complex with the side chains shown here is called a *heme group.*

These two compounds, chlorophyll and heme, are the key components in the elaborate mechanism by which solar energy is trapped and converted for use by living organisms. We have seen that a peculiar feature of transition-metal complexes is their closely spaced d levels, which permit them to absorb light in the visible part of the spectrum and to appear colored. The porphyrin ring around the Mg^{2+} ion in chlorophyll serves the same function. Chlorophyll in plants can absorb photons of visible light and go to an excited electronic state (Figure 20-22). This energy of excitation can initiate a chain of chemical synthesis that ultimately produces sugars from carbon dioxide and water:

$$6CO_2 + 6H_2O \xrightarrow{h\nu} C_6H_{12}O_6 + 6O_2$$

glucose

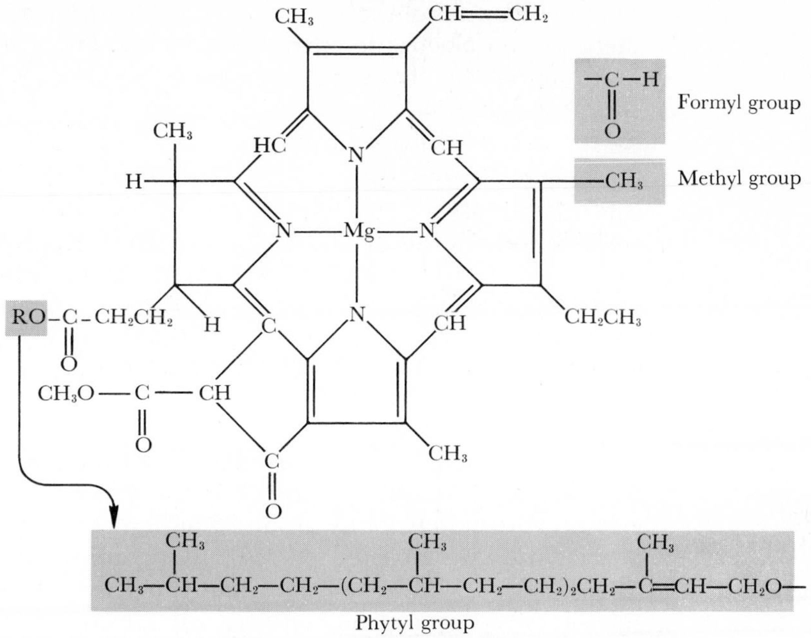

CH₃
CH═CH₂
—C—H ‖ O
Formyl group
CH₃
HC
N
CH
H
—CH₃
Methyl group
N—Mg—N
RO–C–CH₂CH₂
H
C
N
CH
CH₂CH₃
O
CH₃O—C—CH
O
C
O
CH₃
CH₃ CH₃ CH₃
CH₃—CH—CH₂—CH₂—(CH₂—CH—CH₂—CH₂)₂CH₂—C═CH—CH₂O—
Phytyl group

Figure 20-21 The magnesium–porphyrin derivative shown here is called chlorophyll *a*, and is the essential molecule in photosynthesis. Chlorophyll *b* has a formyl group in place of the methyl group.

100
50
Relative intensity of light absorption
Efficiency of
light in
photosynthesis
Spectrum of
chlorophyll a
400
500
600
700
Wavelength (nm)

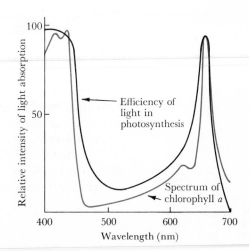

Figure 20-22 Chlorophyll *a* absorbs visible light except in the region around 500 nm (green light), and thus appears green.

Most compounds of the representative elements cannot absorb visible light; there are no electronic energy levels close enough together. Neither can Mg^{2+} alone. But the coordination complex of Mg^{2+} plus its square planar chelating agent has such levels, and chlorophyll is able to trap light and to use its energy in chemical synthesis.

Scientists now believe that life evolved on earth in the presence of a *reducing* atmosphere, an atmosphere with ammonia, methane, water, and carbon dioxide but *no* free oxygen. Free oxygen would degrade organic compounds faster than they could be synthesized by natural processes (electrical discharge, ultraviolet radiation, heat, or natural radioactivity). In the absence of free oxygen, such organic compounds would accumulate in the oceans for eons until finally a packaged, localized bit of chemicals developed that we would call "living."

Living organisms, once developed, would exist by degrading these naturally occurring organic compounds for their energy. The amount of life on the planet would be limited severely if this were the only source of energy. Fortunately for us, around 3 billion years ago, the right combination of metal and porphyrin occurred and an entirely new source of energy was tapped—the sun. The first step that lifted life on earth above the humble role of a scavenger of high-energy organic compounds was an application of coordination chemistry.

Unfortunately, **photosynthesis** (as the chlorophyll photon-trapping process is called) liberates a dangerous by-product, oxygen. Oxygen was not only useless to these early organisms, it competed with them by oxidizing the naturally occurring organic compounds before they could be oxidized within the metabolism of the organisms. Oxygen was a far more efficient scavenger of high-energy compounds than living matter was. Even worse, the ozone (O_3) screen that slowly developed in the upper atmosphere cut off the supply of ultraviolet radiation from the sun and made the natural synthesis of more organic compounds even slower. From all contemporary points of view, the appearance of free oxygen in the atmosphere was a disaster.

As so often happens, life bypassed the obstacle, absorbed it, and turned a disaster into an advantage. The waste products of the original simple organisms had been compounds such as lactic acid or ethanol. These are not nearly so energetic as sugars, but they can release large amounts of energy if oxidized completely to CO_2 and H_2O. Living organisms evolved that were able to "fix" the poisonous O_2 as H_2O and CO_2, and to gain, in the bargain, the energy of combustion of what were once its waste products. Aerobic metabolism had evolved.

Again, the significant development was an advance in coordination chemistry. The central components in the new machinery for aerobic metabolism, by which the combustion of organic molecules was brought to completion, are the **cytochromes**. These are molecules in which an iron atom is complexed with a porphyrin to make a heme (Figure 20-20), and the heme is surrounded with protein. The iron atom changes from iron(II) to iron(III) and back again as electrons are transferred from one component in the chain to another. The entire aerobic machinery is a carefully interlocked set of oxidation–reduction or redox reactions, in which the overall result is the reverse of the photosynthetic process:

$$6O_2 + C_6H_{12}O_6 \rightarrow 6CO_2 + 6H_2O$$

glucose

The energy liberated is stored in the organism for use as needed. The entire, elaborate, chlorophyll–cytochrome system can be regarded as a mechanism for converting the energy of solar photons into stored chemical energy in the muscles of living creatures.

Iron atoms usually exhibit octahedral coordination. What happens to the two coordination positions above and below the plane of the porphyrin ring? In cytochrome c, the heme group sits in a crevice in the surface of the protein molecule (Figure 20-23). From each wall of this crevice, one new ligand extends toward the heme: on one side a nitrogen lone electron pair from a *histidine* side chain on the protein, and on the other side a sulfur lone pair from a *methionine* side chain (Figure 20-24). Therefore, the octahedral coordination positions on the iron are directed to five nitrogen atoms and one sulfur atom.

How does the cytochrome c molecule operate? This is not yet known. The structure of the version with iron(III) was only determined in 1969 by x-ray diffraction, and that of the reduced iron in 1971. The ligands in the complex around the iron, and the protein wrapped around the whole structure, both modify the redox chemistry of the iron atom and ensure that oxidation and reduction are coupled to the earlier and later links in the terminal oxidation chain.

There is one more step in the story of metal–porphyrin complexes. Parkinson might add a subclause to his well-known law: Organisms expand to accommodate the food supplies available. With the guarantee of new energy sources, multicelled organisms evolved. At this point arose the problem, not of obtaining foods or oxygen, but of transporting oxygen to the proper place in the organism. Simple gaseous diffusion through body fluids will work for small organisms but not for large, multicelled creatures. Again, a natural limit was placed on evolution.

For the third time, the way out of the impasse was found with coordination chemistry. Molecules of iron, porphyrin, and protein evolved, in which the iron could *bind* a molecule of oxygen without being oxidized by it. The oxidation of Fe(II) was, in a sense, "aborted" after the first binding step. Oxygen was merely carried along, to be released under the proper conditions of acidity and oxygen scarcity. Two compounds evolved, *hemoglobin*, which carries O_2 in the blood, and *myoglobin*, which receives and stores O_2 in the muscles until it is needed in the cytochrome process.

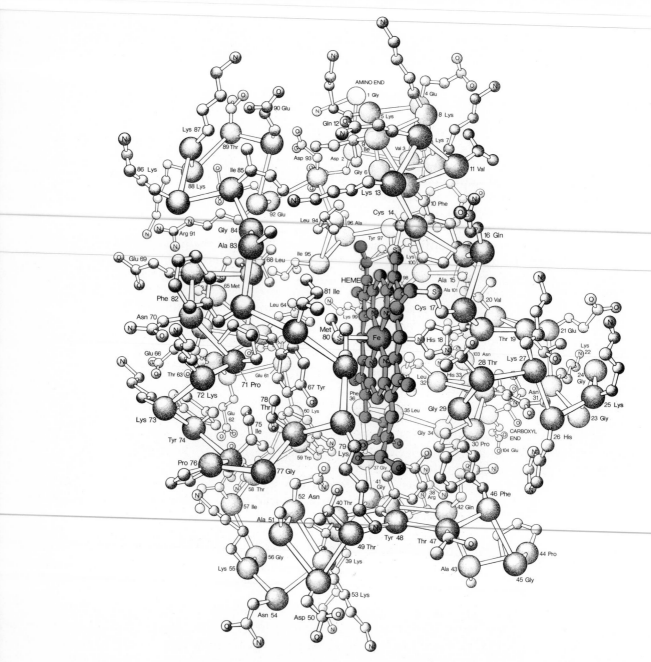

Figure 20-23   Cytochrome *c* is a globular protein with 104 amino acids in one protein chain and an iron-containing heme group. In this schematic drawing, each amino acid is represented by a numbered sphere, and only key amino acid side chains are shown. The heme group is seen nearly edgewise in a vertical crevice in the molecule. Copyright © 1972 R. E. Dickerson and I. Geis: *Scientific American,* April 1972, page 62.

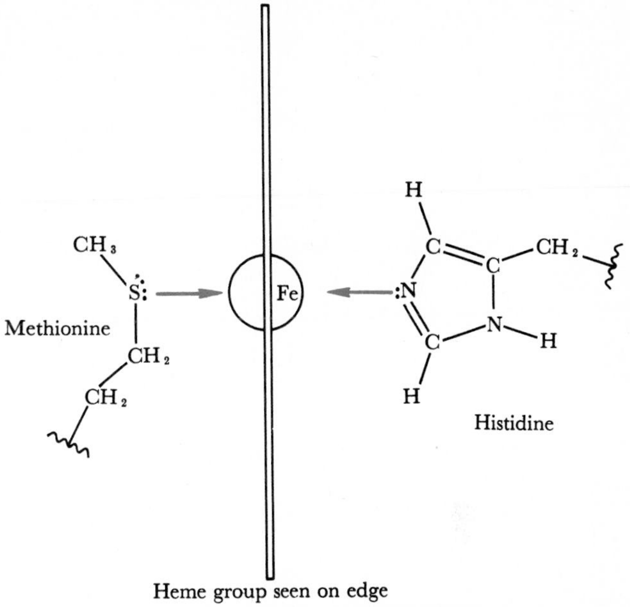

Figure 20-24

The iron atom in cytochrome *c* is octahedrally coordinated through five bonds to nitrogen atoms and one to a sulfur atom. One nitrogen atom and the sulfur atom come from side groups on the protein chain. The other four nitrogen atoms are from the porphyrin ring of the heme.

Methionine

Histidine

Heme group seen on edge

Figure 20-25

The myoglobin molecule is a storage unit for an oxygen molecule in muscle tissue. The heme group is represented by a flat disk, and the iron atom by a ball at the center. The circled W marks the binding site for O_2. The path of the polypeptide chain is shown by double dashed lines.

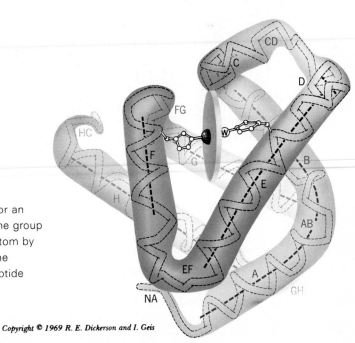

The myoglobin molecule is depicted in Figure 20-25. As in cytochrome *c*, four of the six octahedral iron positions are taken by heme nitrogen atoms. The fifth position has the nitrogen atom of a histidine. However, the sixth position has *no ligand*. This is the place where the oxygen molecule binds, marked by the circled W. In myoglobin, the iron is in the Fe(II) state. If the iron is oxidized, the molecule is inactivated and a water molecule occupies the oxygen position.

Hemoglobin is a package of four myoglobinlike molecules (Figure 20-26). In the past decade, these two structures have been determined by x-ray crystallography. It has become apparent that the four subunits of hemoglobin shift by 7 Å relative to one another when oxygen binds. Hemoglobin and myoglobin now become a model system for transition-metal chemists to study. Why does binding at the sixth ligand site of the iron

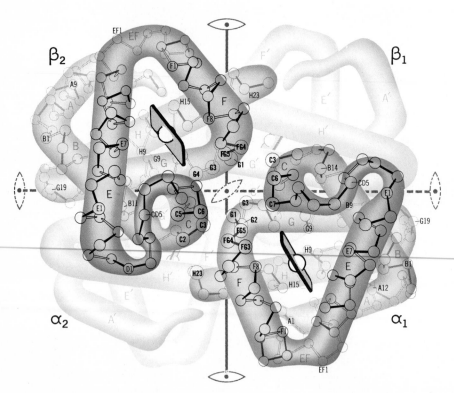

Figure 20-26   The hemoglobin molecule is the carrier of oxygen in the bloodstream. It is built from four subunits, each of which is constructed like a myoglobin molecule. This figure and that of myoglobin are reprinted from R. E. Dickerson and I. Geis, *The Structure and Action of Proteins,* W. A. Benjamin, Menlo Park, Calif., 1969.

complex cause the protein subunits to rearrange? Why does the oxygen molecule fall away from hemoglobin in an acid environment (such as in oxygen-poor muscle tissue)? How is the coordination chemistry of hemoglobin and myoglobin so carefully meshed that myoglobin binds oxygen just as hemoglobin releases it at the tissues?

Heme, or iron porphyrin, is also at the active sites of enzymes such as peroxidase and catalase. Many other transition metals are essential components in enzyme catalysis; we shall discuss some of them in Chapter 21. With the evolution of myoglobin and hemoglobin, the size limitation was removed from living organisms. Thereafter, all the multicelled animals that we ordinarily see around us evolved. In the sense that transition metals and double-bonded organic ring systems such as porphyrin are uniquely suited for absorbing visible light, and their combinations have a particularly rich redox chemistry, life is indeed applied coordination chemistry.

Suggested Reading

F. Basolo and R. Johnson, *Coordination Chemistry,* W. A. Benjamin, Menlo Park, Calif., 1964.

F. A. Cotton and G. Wilkinson, *Basic Inorganic Chemistry,* Wiley, New York, 1977.

R. E. Dickerson, "The Structure and History of an Ancient Protein [cytochrome *c*]," *Scientific American,* April 1972.

R. E. Dickerson, "Chemical Evolution and the Origin of Life," *Scientific American,* September 1978.

H. B. Gray, *Chemical Bonds,* W. A. Benjamin, Menlo Park, Calif., 1973.

H. B. Gray, *Electrons and Chemical Bonding,* W. A. Benjamin, Menlo Park, Calif., 1965.

E. M. Larsen, *Transitional Elements,* W. A. Benjamin, Menlo Park, Calif., 1965.

L. E. Orgel, *An Introduction to Transition-Metal Chemistry,* Methuen, London, 1966, 2nd ed.

E. G. Rochow, *Organometallic Chemistry,* Van Nostrand Reinhold, New York, 1964.

21

The Special Role of Carbon

Organic chemistry just now is enough to drive one mad. It gives me the impression of a primeval tropical forest, full of the most remarkable things, a monstrous and boundless thicket, with no way of escape, into which one may well dread to enter.

Friedrich Wöhler (1835)

The term *organic,* applied to chemistry in Friedrich Wöhler's time, signified "living." Organic chemistry then dealt with the various compounds that were associated with living organisms. Most chemists drew a sharp line between organic and inorganic compounds; they ascribed some special type of "life force" to the former. But in 1828 Wöhler demolished the idea of a vital force in the most direct way possible, by showing that an undoubtedly biological molecule, urea, could be obtained in the laboratory merely by heating an undoubtedly inorganic, nonbiological salt, ammonium cyanate:

$$NH_4OCN \rightarrow NH_2-\overset{\overset{\displaystyle O}{\|}}{C}-NH_2$$

ammonium   urea
cyanate

The synthesis of other organic substances followed rapidly. The term *organic chemistry* gradually came to mean the chemistry of the compounds of carbon, so when chemists in the twentieth century wanted to talk specifically about processes of living organisms, they had to invent another word, *biochemistry.* It is true that the chemistry of life is a subclass of the chemistry of carbon compounds, and it is worth reflecting why this should be so.

Wöhler's complaint about how complicated organic chemistry was can be echoed today, but at an entirely different level of knowledge. We have answered many of the questions that baffled Wöhler, but now we are asking questions that he could never have raised. Charles Darwin once remarked, "It is the merest rubbish, asking about the origin of life; one might as well ask about the origin of matter." It is a measure of how far we have come that scientists today are busy asking both questions, and devising experiments to answer them. One of the goals of the Viking unmanned Martian landers was to search for life on that planet. The results, although not absolutely definitive, were so negative that few people expect future probes to reveal life on Mars. It is interesting that a tacit assumption was made in all the Viking experiments that life would be *carbon-based*. After the fact, no one suggests that this assumption was to be blamed for the negative results, or that Mars is teeming with silicon-based or nitrogen-based creatures. Carbon seems to have a special role in life, and the first part of this chapter will examine why this is so. The rest of the chapter will be an overview of two very large areas of chemistry—organic chemistry and biochemistry. Some of the chapters of this book deal with fundamental and essential techniques, and should be studied intensively. Others, such as this one, are designed to give you a general impression of an area of chemistry. As you read this chapter, try to understand and appreciate, rather than memorize.

21-1   THE SPECIAL TALENTS OF CARBON

The chemistry of carbon puts that of all other elements to shame. The American Chemical Society has maintained a register of chemical compounds mentioned in the literature since 1965. By the middle of 1978 there were $4\frac{1}{4}$ million different chemical substances in the register. Of these, 4 million were compounds based on a carbon backbone, and the remaining quarter million were about evenly divided between alloys and inorganic compounds. Hence, excluding metallic alloys, we know of 32 times as many organic compounds as inorganic! So many carbon compounds exist because carbon can link with itself as no other element can, to make straight chains and branched chains. Some of these compounds are shown in Figure 21-1. Chains made by the repetition of a subunit are called **polymers**, and the repeated unit is called a **monomer**.

Hydrocarbons are compounds containing only C and H atoms. The simplest hydrocarbons are linear polymers of the subunit $-CH_2-$, with the ends of the polymer terminated by hydrogen atoms. Other hydrocarbons have branched chains or rings of connected atoms. Butane is a tetramer (four subunits), and is a gas used for heating and cooking. Five- to 12-carbon polymers are gasolines; heptane (Figure 21-1) is one example. Kerosene is a mixture of molecules with 12 to 16 carbon atoms, and lubricating oils and paraffin wax are mixtures of chains with 17 and more carbons. Polyethylene

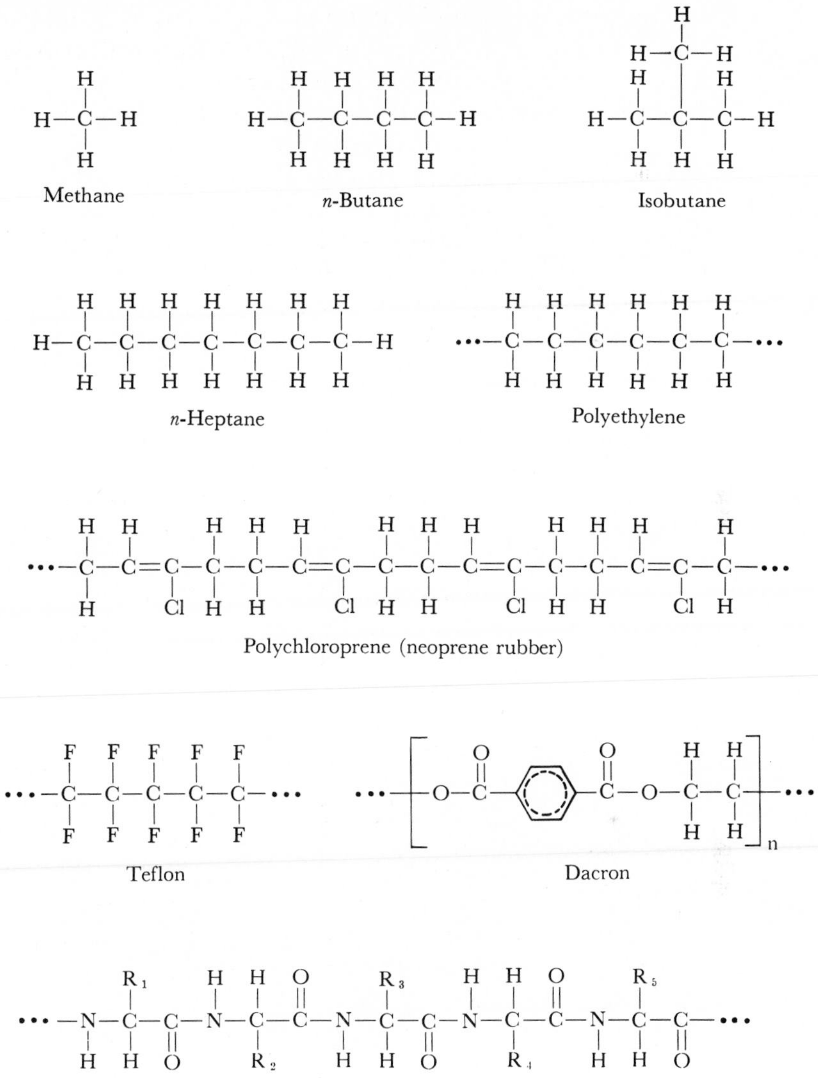

Polypeptides or proteins (silk, wool, hair, collagen, hemoglobin, enzymes, antibodies). R_1, R_2, R_3, and so forth are chemically different side chains.

Figure 21-1 Natural and synthetic chains of carbon atoms, with nitrogen and oxygen. The first two rows are hydrocarbons of increasing chain length from methane, through the commercial heating gases (butane) and gasolines (heptane), to polyethylene plastic. The double bond at every fourth carbon connection in polychloroprene is typical of natural and synthetic rubbers. Dacron shows two kinds of multiple bonding: C=O double bonds of the familiar π^b type, and delocalized benzenelike bonding. Polypeptide chains are cross-linked one to another as in Figure 21-2.

plastic has 5000 to 50,000 —CH_2— monomer units per chain. There are many other organic chains, with more atoms than just C and H. Neoprene rubber, Teflon, and Dacron (Figure 21-1) are synthetic polymers, and the polypeptide chain shown at the bottom of Figure 21-1 is the polymer from which all proteins are built.

Because carbon can make as many as four bonds, branched and cross-linked chains can be built. Isobutane (Figure 21-1) is a branched-chain isomer of C_4H_{10}. Figure 21-2 shows silk and its synthetic analogue, nylon. Both are constructed from parallel, covalently bonded chains that are cross-linked into a sheet by hydrogen bonds. Bakelite and Melmac are hard, inflexible plastics because their monomers are covalently linked in three dimensions.

The other distinguishing feature of carbon is its ability to make double bonds with itself and with other elements, and to do so in the middle of these chains. Neoprene rubber (Figure 21-1) has such double bonds between carbon atoms. Dacron has double bonds between C and O, and it also has the delocalized multiple bonding that we saw in Chapter 13 for benzene. Figure 21-3 depicts some other examples of double bonds in carbon compounds. Since the double bond can often be converted to a single bond by adding an atom at each end of the bond, such double-bond compounds are called **unsaturated**:

$$CH_2{=}CH_2 + H_2 \rightarrow CH_3{-}CH_3$$
ethene   ethane

$$CH_2{=}CH_2 + HCl \rightarrow CH_3{-}CH_2{-}Cl$$
ethene   ethyl chloride

(*Ethene* is the systematic name for C_2H_4; its common name is *ethylene*.) Compounds with rings of atoms having delocalized, benzenelike multiple bonds

Figure 21-2   Three varieties of natural and synthetic polymers. (a) Silk, made from polypeptide chains. ▶ The chains are cross-linked into sheets by hydrogen bonds. (b) Nylon 66 is closely patterned after silk. It was invented, in 1935, by W. H. Carothers at E. I. du Pont de Nemours & Co., Inc. It has hydrogen bonding similar to silk, but at longer intervals down the chains. In both fibers, the fiber axis is horizontal in the figure and parallel to the covalently bonded chains. (c) Bakelite is one of the earliest synthetic plastics, having been invented, in 1909, by L. H. Baekeland, an American chemist who also contributed to the chemistry of photography. Bakelite is one member of a class of phenol—formaldehyde resins that are strong and hard because of their three-dimensional network of covalent bonds.

(a)

(b)

(c)

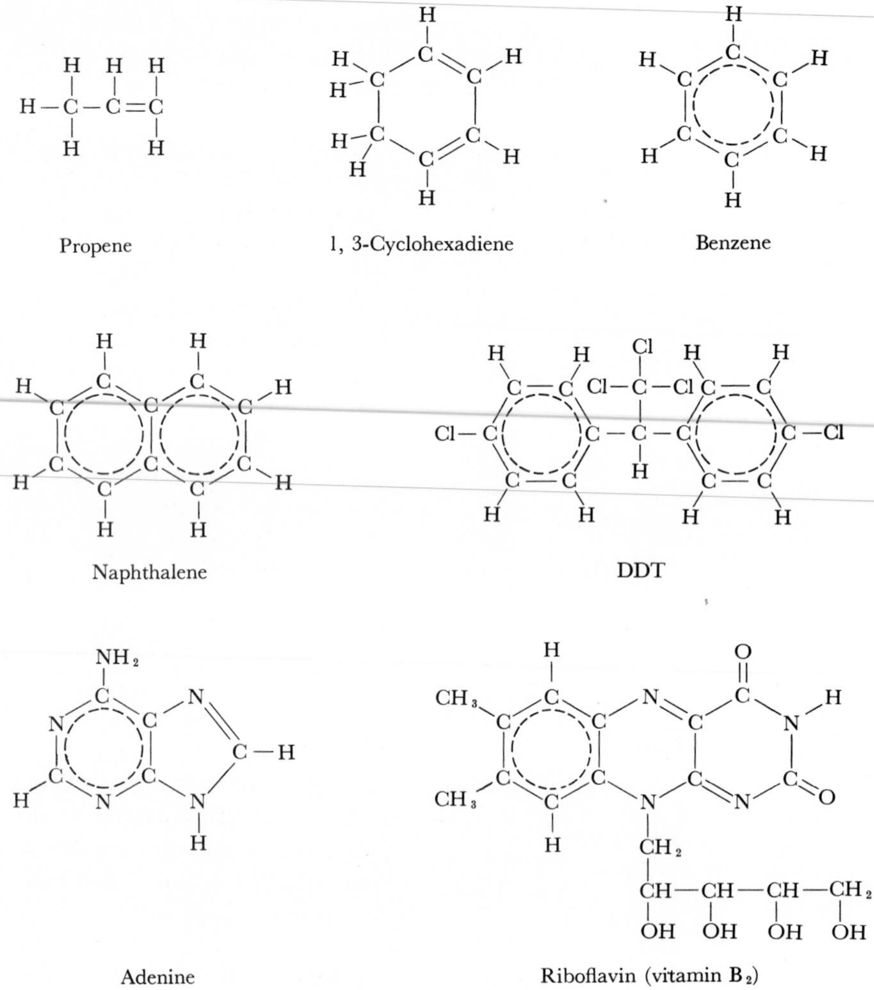

Propene

1, 3-Cyclohexadiene

Benzene

Naphthalene

DDT

Adenine

Riboflavin (vitamin B$_2$)

Figure 21-3   Examples of double bonds and delocalized bonds in organic compounds. Adenine, an essential component of the genetic polymer DNA (deoxyribonucleic acid) and of the energy-storing molecule ATP (adenosine triphosphate), is a pentamer of HCN. It has been prepared from HCN under conditions simulating those of earth in the early stages of the evolution of life. Dashed circles represent delocalized bonds of the type encountered for benzene in Chapter 13.

are called **aromatic** compounds. Dacron (Figure 21-1) and naphthalene, DDT, adenine, and riboflavin (Figure 21-3) all have aromatic components. Adenine and riboflavin also show that carbon can make double bonds to nitrogen, and that nitrogen can participate in a delocalized, aromatic ring.

Much of organic chemistry involves the special properties of aromatic ring systems. Aromatic molecules and transition-metal complexes are the two main classes of compounds in which the energy required to excite an electron falls in the visible part of the spectrum. Hence, these compounds are involved in dyes of all descriptions, and in mechanisms for trapping and transferring photon energy.

The four distinguishing features of organic compounds can be summarized as follows:

1. Long-chain polymers with C—C bonds
2. Branched and cross-linked chains
3. Double and triple bonds
4. Delocalized aromatic bonds

How many of these characteristics are exhibited by the immediate neighbors of carbon—B, N, and Si? What can carbon do that these elements cannot do, and why is this so? What particular combination of electrons and orbitals makes carbon so versatile?

21-2   THE CHEMISTRY OF THE NEIGHBORS OF CARBON

Boron, carbon, and nitrogen all are second-period elements of similar size. They differ in the number of valence electrons that they possess: three electrons in B, four in C, and five in N. Silicon, a third-period element, is like carbon in having four valence electrons, but they are one major energy level farther from the nucleus and have a principal quantum number of 3 instead of 2. Below the valence electrons, Si has 10 inner-orbital electrons: 2 with principal quantum number 1, and 8 with quantum number 2. In contrast, B, C, and N have only 2 electrons below their valence orbitals. All the differences in chemical properties among B, C, N, and Si that will concern us in this chapter come from these two factors: the number of valence electrons and the number of electrons in completed inner orbitals.

Boron has three valence electrons and four valence orbitals per atom. It commonly uses three orbitals in sp^2 hybridization in compounds such as BF_3. Carbon has four valence electrons and four orbitals. Except when involved in multiple bonds, it uses sp^3 hybrid orbitals. Nitrogen has five electrons and four orbitals. It typically makes three bonds to other atoms in tetrahedral configurations, and the fourth sp^3 atomic orbital is occupied by the lone electron pair (Section 13-3). Both carbon and nitrogen can make double and triple bonds involving the π overlap bond discussed in Section 13-4. The bond length for both elements decreases by 13% in a double bond and 22% in a triple bond. The atoms are held more tightly because of the electrons in π^b molecular orbitals derived from overlapping $2p$ atomic orbitals. Conversely, the overlap of these orbitals is too small for significant bonding unless the atoms are closer together. This is the reason why Si and other elements in the third period of the table and beyond cannot form multiple bonds. Silicon has 10 inner-orbital electrons instead of 2 as in C and N. The repulsion between these inner-orbital electrons does not permit two Si atoms to come close enough for p-orbital overlap and the formation of a double bond. Although chemists are actively trying to synthesize compounds with Si$=$Si and Si$=$C double bonds, none have yet been prepared. With one or two exceptions, double and triple bonds are confined to elements in the second period of the periodic table, with no more than two inner-orbital electrons per atom. Exceptions such as S$=$O, P$=$O, and Si$=$O use overlap between p and d orbitals, as we shall see later in this section with Si.

Boron

We can understand why boron is not a good candidate for carbonlike chemistry by looking at the series of boron hydrides. The hydride BH_3 exists only as a short-lived decomposition product of higher hydrides. Other hydrides are known: B_2H_6, B_4H_{10}, B_5H_9, B_5H_{11}, B_6H_{10}, B_6H_{12}, B_8H_{12}, B_9H_{15}, and $B_{10}H_{14}$. The simplest boron hydride, B_2H_6, has eight atoms and only 12 valence electrons. If it were to have an ethanelike structure,

$$
\begin{array}{ccc}
\text{H} & \text{H} \\
| & | \\
\text{H}-\text{B}-\text{B}-\text{H} \\
| & | \\
\text{H} & \text{H}
\end{array}
$$

it would need 14 electrons for the seven covalent bonds. But diborane has only 12 valence electrons: It is an *electron-deficient* compound. Its true structure is

Each B atom has two normal two-center covalent B—H bonds, which use a total of eight electrons. The remaining four electrons are used in two *three-center* B—H—B bonds, in which each of the three atoms contributes an orbital to the bonding molecular orbital. This concept of the three-center bond is enough to explain the structures of all boron hydrides. It also explains why boron cannot do the things that carbon can do.

   For most compounds in which the number of valence electrons is at least as great as the number of valence orbitals, the two-atom chemical bond model is sufficient, and we need consider only two atoms at a time. However, as we learned in the discussion of benzene (Section 13-5), localized molecular orbitals are only an approximation to reality. Sometimes we must construct delocalized molecular orbitals from atomic orbitals contributed by several, or occasionally all, of the atoms in a molecule. In benzene the C—H and the C—C σ bonds can be dealt with individually, but the six p orbitals must be considered together.

To explain the behavior of boron, the smallest unit of bonding that we can consider is sometimes *three atoms*. Three atomic orbitals, one from each atom, can combine to make three molecular orbitals: one bonding, one antibonding, and one nonbonding—the last with virtually the same energy as the original atomic orbitals (Figure 21-4). We have seen nonbonding

Figure 21-4

Three-center orbitals in boron compounds. (a) Three boron atoms each can donate one orbital (two sp^3 and one p) to make a bonding, a nonbonding, and an antibonding orbital. One electron pair in the bonding orbital holds all three atoms together. This arrangement is called an *open three-center bond*. (b) The arrangement of atomic orbitals in a bonding orbital for a B—H—B bridge bond. (c) The arrangement of atomic orbitals in a *closed three-center bond*. Such three-center bonds are found in electron-deficient compounds involving B and Al.

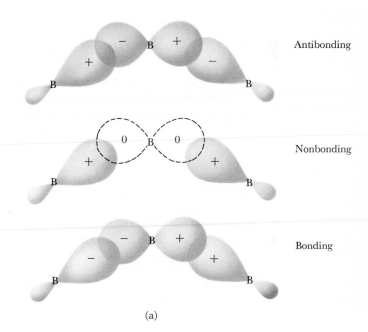

(a)

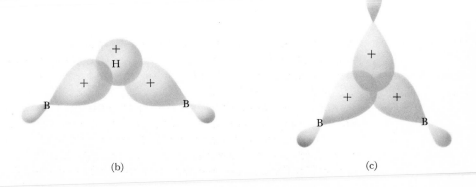

(b)   (c)

orbitals before. In HF (Figures 12-11 and 12-12), the $2p_x$ and $2p_y$ orbitals of F are nonbonding, as are the d_{xy}, d_{yz}, and d_{xz} orbitals of the metal in an octahedral coordination complex.

Two electrons in one of these *bonding* three-center molecular orbitals can hold three atoms together. This economy in bonding helps to compensate for the electron deficiency in boron. However, it also forces a cramped geometry on its compounds that makes boron unsuitable as a rival for carbon. Vast molecular networks can be constructed from the straight- and branched-chain carbon hydrides (hydrocarbons), in which the atoms are connected two at a time. In contrast, the boron hydrides, in which the atoms are connected three at a time, build structures whose boron frameworks are fragments of an icosahedron (Figure 21-5a). The hydride B_4H_{10} is a small fragment of the icosahedron (Figure 21-5b). It has six normal two-center bonds between B and H, one two-center B—B bond, and four three-center B—H—B bonds. Each of these bonds requires one electron pair. In this way, 14 atoms are held together by using 26 atomic orbitals but only 22 electrons. The hydride B_9H_{15} is three-fourths of a complete icosahedron (Figure 21-5c). In this compound, 24 atoms are held together with 51 atomic orbitals and only 42 bonding electrons. The complete B_{12} icosahedron is found in crystalline boron. The manner in which such three-center bonds are used in the larger boron hydrides is shown for B_5H_{11} in Figure 21-6.

In conclusion, boron is an unsuitable candidate for organic chemistry because of its electron deficiency, which leads to three-center bonding and a tendency for boron structures to close in upon themselves. Even worse, the geometrical arrangement produced makes it impossible for p orbitals to lie parallel on adjacent atoms and to form π bonds. In terms of approaching the desirable properties of carbon, boron comes close, but not close enough.

Nitrogen

Nitrogen, like carbon, can make double and triple bonds to itself and to other first- and second-period atoms. But nitrogen suffers from a defect opposite to that of boron; it has too many electrons. Repulsions between lone electron pairs on neighboring nitrogen atoms make the N—N single-bond energy only 161 kJ mole^{-1}, in comparison with 348 kJ mole^{-1} for a C—C bond. In the C—N bond, in which one of these repelling lone electron pairs is absent, the bond energy increases to 292 kJ mole^{-1}.

Some compounds with chains of linked nitrogen atoms exist:

$$H_2N-NH_2 \qquad H-N=\overset{\oplus}{N}=\overset{\ominus}{N} \qquad R_2N-\overset{\overset{\displaystyle R}{|}}{N}-NR_2$$

  hydrazine   hydrazoic acid   triazanes

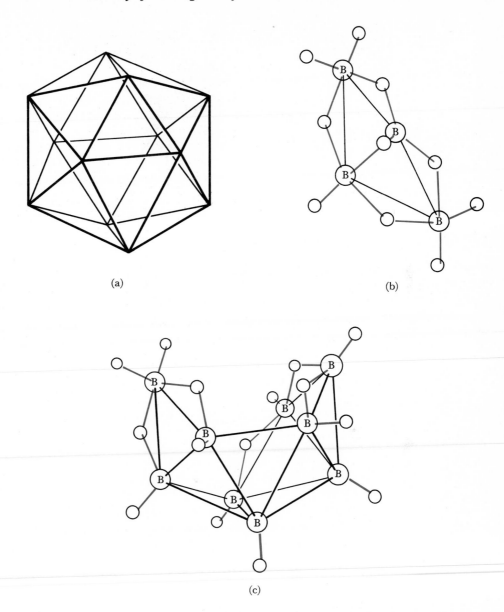

(a)

(b)

(c)

Figure 21-5

(a) The icosahedron is the boron framework for almost all boron hydrides. An icosahedron has 12 vertices and 20 equilateral triangular faces. (b) Tetraborane-10, B_4H_{10}, has its four boron atoms outlining two faces of the icosahedron. Bonds are marked in color Six of the hydrogen atoms make normal two-center covalent bonds to boron; the others participate in four B—H—B bridges. The two central boron atoms are joined by a conventional two-center bond. (c) Enneaborane-15, B_9H_{15}, has a framework that is derived from the icosahedron by removing any three adjacent boron vertices that do not form an equilateral triangle. Ten hydrogen atoms make two-center covalent bonds to the boron atoms; the other five hydrogen atoms participate in B—H—B bridges.

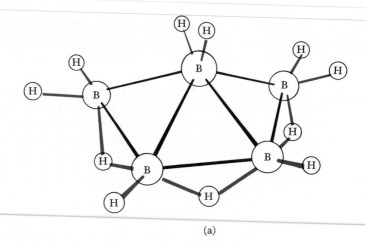

(a)

Figure 21-6

Structure and bonding orbitals in pentaborane-11, B_5H_{11}. Each of the boron atoms has sp^3 hybridization except the central one, which has one unhybridized p orbital and three sp^2 orbitals. The closed three-center bond uses two sp^3 orbitals and one sp^2 orbital. The open three-center bond involving the central B atom uses two sp^3 and one p, as in Figure 21-4a. The entire molecule uses 31 atomic orbitals but only 26 electrons.

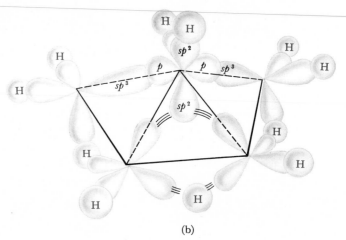

(b)

$$\underset{\text{tetrazanes}}{R_2N-\overset{\overset{\displaystyle R}{|}}{N}-\overset{\overset{\displaystyle R}{|}}{N}-NR_2} \qquad \underset{\text{triazenes}}{RN=N-NR_2} \qquad \underset{\text{tetrazenes}}{R_2N-N=N-NR_2}$$

$$\underset{\text{bisdiazoamines}}{RN=N-\overset{\overset{\displaystyle R}{|}}{N}-N=NR} \qquad \underset{\text{bisdiazohydrazines}}{RN=N-\overset{\overset{\displaystyle R}{|}}{N}-\overset{\overset{\displaystyle R}{|}}{N}-N=NR}$$

$$\underset{\text{octazotrienes}}{RN=N-\overset{\overset{\displaystyle R}{|}}{N}-N=N-\overset{\overset{\displaystyle R}{|}}{N}-N=NR}$$

Hydrazine is used as a rocket fuel. Hydrazoic acid is extremely explosive and toxic. It is sometimes used in detonators for explosives. The higher *hydronitrogens,* as these compounds are called by analogy with the hydrocarbons, can seldom be prepared in the simplest forms, with hydrogen atoms for the R's shown in the preceding structures. Those sufficiently stable even to exist have phenyl groups (benzene rings) for R, or methyl or ethyl groups (CH_3 — or CH_3CH_2 —). They are extremely unstable, and most are explosively so. They decompose rapidly under all conditions. Or as one scientist has said, "They stand on the edge of existence."

An important factor in the instability of nitrogen chains is the unusual stability of the triple bond in the $N \equiv N$ molecule. The N_2 triple bond, whose bond energy is 946 kJ mole^{-1}, is *six times* as strong as the $N-N$ single bond, whereas the $C \equiv C$ triple bond in acetylene is only 2.3 times the strength of the $C-C$ single bond. A long nitrogen chain is far less stable than the system remaining after the chain breaks into a series of N_2 molecules.

Nitrogen participates in chains and rings with carbon and, like carbon, forms double bonds. Diazomethane,

$$H_2C = \overset{\oplus}{N} = \overset{\ominus}{N}$$

is one of the most versatile and useful reagents in organic chemistry, despite the fact that it is highly toxic, dangerously explosive, and cannot be stored without decomposition. A molecule with two or more adjacent nitrogen atoms is rarely stable.

Silicon

The critical difference between Si and C is the greater number of innerorbital electrons in Si, and the consequent inability of two silicon atoms to come close enough together for double and triple bonds. Silicon forms *silanes* analogous to the alkane hydrocarbons to be discussed in Section 21-3. Silanes have the general formula Si_nH_{2n+2}. The longest of these chains that has been prepared is only hexasilane (Figure 21-7). These silanes, like the hydronitrogens, are dangerously reactive. The smallest silanes are stable in a vacuum, but all are spontaneously inflammable in air, and all react explosively with halogens. They are powerful reducing agents.

The silanes are so unstable and susceptible to oxidation because the $Si-O$ bond is much more stable than the $Si-Si$ bond: 369 kJ mole^{-1} versus 177 kJ mole^{-1}. In contrast, with carbon the $C-O$ and $C-C$ bond energies are almost the same: 351 kJ mole^{-1} and 348 kJ mole^{-1} (Table 21-1). Hydrocarbons are oxidized much less easily than are the silanes. Although the reaction

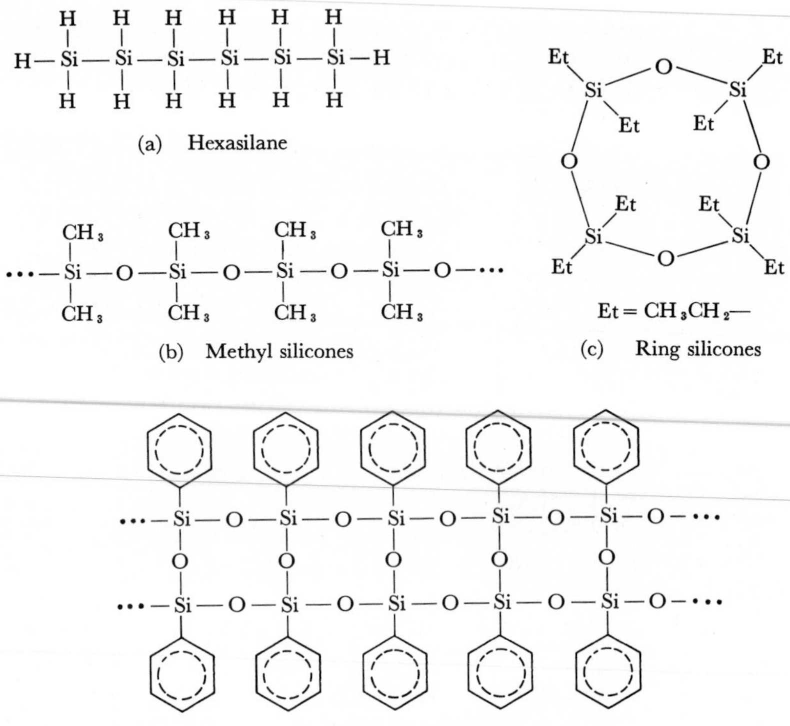

(a)   Hexasilane

(b)   Methyl silicones

(c)   Ring silicones

Et $= CH_3CH_2-$

(d)   "Ladder" silicones

Figure 21-7   Silicon can exist in two types of polymers: the reactive silanes, in which Si atoms are bonded directly, and the inert siloxanes or silicones, in which each connection is through a bridging oxygen atom. The silicones are chemically inert, heat resistant, electrically nonconducting oils and rubbers used as lubricants, insulators, and protective coatings. Three of the four Si bonds are to bridging oxygen atoms in the ladder silicones, which are rubbery or plastic materials. When all four Si bonds are involved in oxygen bridges, the silicate minerals result.

$$H-Si-Si-Si-Si-H + 6\tfrac{1}{2}O_2 \rightarrow 4SiO_2 + 5H_2O$$

is explosively spontaneous, the analogous reaction with butane

$$H-C-C-C-C-H + 6\tfrac{1}{2}O_2 \rightarrow 4CO_2 + 5H_2O$$

Table 21-1

Relative Bonding Abilities of Carbon and Its Neighbors

Element, R	B	C	N	Si
Valence electrons	3	4	5	4
Usual coordination	Threefold (sp^2) (or fourfold with 3-center bonds)	Fourfold (sp^3)	Fourfold (sp^3) (including lone pair)	Fourfold (sp^3)
Single bond energies (kJ mole^{-1})				
R—R		348	161	177
R—C		348	292	290
R—N		292	161	—
R—O		351	~230	369
R—H		413	391	295
Electron-to-orbital ratio and bonding behavior	Electron-deficient; 3-center bonds (in multiple-B compounds)	Electron match; 2-center bonds	Electron surfeit; 2-center bonds. Lone-pair repulsions	Electron match; 2-center bonds
Linkage of like atoms	Icosahedral shells	Extensive chains	Chains, limited extent	Chains, limited extent
Double and triple bonds, π bonding	π bonding impossible with 3-center bonds in icosahedral framework	Good π overlap in double and triple bonds. Entire bonding capacity *cannot* be satisfied with one other like atom. Builds networks	Good π overlap in double and triple bonds. Entire bonding capacity can be satisfied with one other like atom. Builds N_2 molecules	Double and triple bonds impossible. Builds networks, principally using the stable Si—O bond rather than the less stable Si—Si bond

must be ignited by heat, and continues under ordinary conditions only because the heat released by the reaction keeps the reactants at a high temperature.

Part of this difference in oxidation of C and Si compounds arises because the Si—Si bond is weaker than the C—C bond. This is to be expected from the greater size of Si. The bonding electrons are farther from each nucleus, and the bond is not as strong. The same effect gives Si a lower ionization energy than C and makes it less electronegative (Table 9-1). But an even more important factor in the difference in the behavior of C and Si

Figure 21-8   The strength of the Si—O bond is due to an unusual partial double-bond character. One of the filled $2p$ lone-pair orbitals of oxygen shares its electrons with an empty $3d$ orbital of Si that has similar energy. For this reason, the Si—O bond energy is 369 kJ mole^{-1}; whereas the comparable silicon bond with C, which lacks the lone electron pairs, is only 290 kJ mole^{-1}.

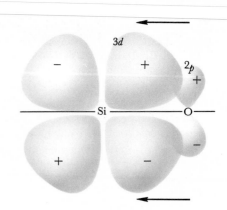

is the anomalously high strength of the Si—O bond. In carbon, the empty $3d$ orbitals have a much higher energy than the filled lone-pair $2p$ orbitals of oxygen. There is no interaction between them. However, in silicon the added nuclear charge lowers the energy of the empty $3d$ atomic orbitals closer to the energy of the oxygen $2p$ orbitals. Oxygen can then share part of its lone-pair electrons with Si (Figure 21-8) in a back bonding similar to the $L \rightarrow M(\pi)$ and $M \rightarrow L(\pi)$ sharing in coordination complexes discussed in Section 20-3. Since the d_{xy}-type orbitals of Si extend farther toward O than an Si p orbital of a π bond, Si and O need not come as close as if they formed a $p\pi-p\pi$ double bond. The result of this sharing of oxygen lone electron pairs is that, although the Si—Si bond is 171 kJ mole^{-1} weaker than the C—C bond, the Si—O bond is 18 kJ mole^{-1} *stronger* than the C—O bond.

These results suggest that compounds in which Si atoms are linked by bridging oxygen atoms might be stable. This is so, and these compounds are the silicones. As shown in Figure 21-7, silicones can exist as straight chains, as rings, or as "ladder" compounds with two parallel linked chains. The silicones are extremely inert compounds. The silanes are much *more* reactive than the hydrocarbons; the silicones are much *less* reactive.

Comparison of Boron, Nitrogen, and Silicon

Each of the neighboring elements of carbon is unable to do the things that make carbon so important: to build long, stable chains with branching, cross-linking, and double bonds, and rings with delocalized electrons. The relative behavior of these elements is summarized in Table 21-1. Boron is forced into an unfavorable geometry by its deficiency of electrons and cannot overlap p orbitals to make double bonds. Although N can occasionally replace C in carbon rings and chains, and can form double bonds as easily as carbon can, long chains of nitrogen atoms are unstable. Silicon is hampered by the weakness of its Si—Si bond in comparison with the Si—O bond and by its inability to make double bonds.

Carbon, then, is the fortunate combination of a small atom that has as many valence electrons as valence orbitals, and a bond to itself that is as strong as a bond to oxygen. Science fiction writers have long speculated on totally alien extraterrestrial life based on nonaqueous chemistry and an element other than carbon. Silicon has been the favorite element, and Mars has been the favorite homeland for rock-metabolizing, silicone-putty-fleshed monsters. But the more we learn about what carbon compounds do in terrestrial living creatures, the less easy it is to imagine silicon compounds performing even remotely similar roles. Carbon *is* special, and its properties can be duplicated by no other element.

21-3 SATURATED HYDROCARBONS OR ALKANES

Compounds that contain only carbon and hydrogen are, as we have seen, called **hydrocarbons**, and those in which all carbon atoms form four single bonds to other atoms are called **saturated hydrocarbons**, **paraffins**, or **alkanes**. The word *paraffin* originates from the Greek word for "little reactivity," and the chemical properties of these paraffins are in marked contrast to those of the silanes and hydronitrogens.

Examples of several alkanes are given in Table 21-2. The first four have common names; those with 5 through 19 carbon atoms are commonly described by a Greek prefix indicating the total number of carbon atoms, and the standard suffix *-ane*. If there are more than 19 carbon atoms the chemical formula is usually employed as the name. Each carbon atom has four tetrahedrally oriented bonds either to another carbon atom or to a hydrogen.

The general chemical formula for the noncyclic alkanes is C_nH_{2n+2}. Alkanes exhibit a regular increase in melting point and boiling point with increasing molecular weight. Methane, ethane, propane, and butane are gases; pentane through $C_{20}H_{42}$ are liquids; and $C_{21}H_{44}$ and heavier compounds are waxy solids.

If there are four or more carbon atoms, there can be more than one way of connecting the carbons. Consequently, structural isomers can exist, in which the same numbers of each kind of atom are present, but the atoms are connected in different ways. The five isomers of hexane have the following carbon skeletons and systematic names (CH_3— is called the *methyl group*):

$$C-C-C-C-C-C \qquad C-\overset{\displaystyle C}{\underset{\displaystyle |}{C}}-C-C-C \qquad C-C-\overset{\displaystyle C}{\underset{\displaystyle |}{C}}-C-C$$

hexane   2-methylpentane   3-methylpentane

Table 21-2
Some Common Saturated Hydrocarbons

Formula	Common name	Systematic name
CH_4	Methane	Methane
CH_3-CH_3	Ethane	Ethane
$CH_3-CH_2-CH_3$	Propane	Propane
$CH_3-CH_2-CH_2-CH_3$	*n*-Butane	Butane
$CH_3-CH-CH_3$ with CH_3 below	Isobutane	Methylpropane
$CH_3-CH_2-CH_2-CH_2-CH_3$	*n*-Pentane	Pentane
$CH_3-CH_2-CH-CH_3$ with CH_3 below	Isopentane	Methylbutane
$CH_3-\underset{CH_3}{\overset{CH_3}{C}}-CH_3$	Neopentane	Dimethylpropane
$CH_3-\underset{CH_3}{\overset{CH_3}{C}}-CH_2-\overset{CH_3}{CH}-CH_3$	Isooctane	2,2,4-Trimethylpentane
(cyclohexane ring structure)	Cyclohexane	Cyclohexane

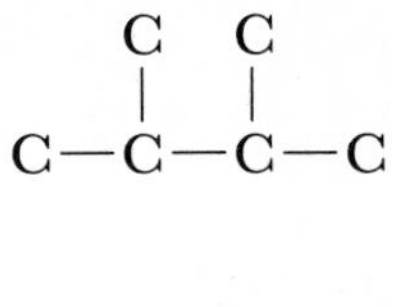

$$C-C-C-C \quad C-C-C-C$$

2,3-dimethylbutane 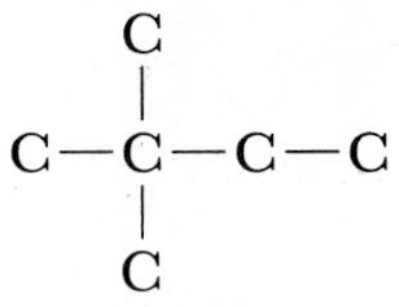   2,2-dimethylbutane

The old labels of *normal-* (or *n-*) for straight chain, *iso-* for a branched chain, and *neo-* for a third isomer rapidly become confusing as the number of carbon atoms increases, and the systematic nomenclature of the right column in Table 21-2 must be used. With the systematic nomenclature, the compound is given the name corresponding to the *longest carbon chain* that can be traced through the molecule. The molecule is stretched along this longest chain, and carbon atoms are counted beginning with the end that has the nearest branch point. The side chains are then identified, and located by giving the number of the carbon to which they are attached on the main chain. Hydrocarbon side chains are named by analogy with the hydrocarbons: CH_3-, methyl; CH_3CH_2-, ethyl; $CH_3CH_2CH_2-$, propyl; and

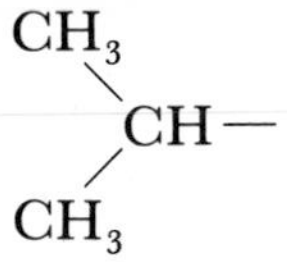

$$\begin{matrix} CH_3 \\ \diagdown \\ CH- \\ \diagup \\ CH_3 \end{matrix} \quad \text{isopropyl}$$

Thus, neopentane in systematic nomenclature is dimethylpropane, and not trimethylethane or even tetramethylmethane, because the longest continuous carbon chain has three carbon atoms as in propane.

Example 1

Draw the carbon skeletons and give the systematic names for the nine isomers of C_7H_{16}.

Solution

You will find one heptane (7-carbon straight chain), two substituted hexanes (6-carbon chain with side branches), five pentanes, and one butane. The last compound is 2,2,3-trimethylbutane.

Example 2

Give the systematic name of the following compound:

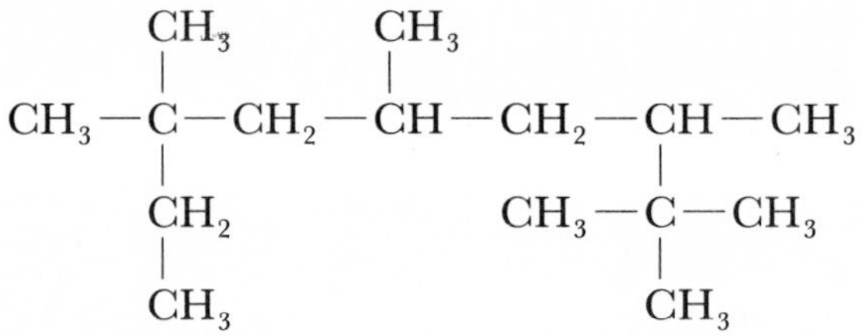

$$
\begin{array}{c}
\quad\;\; \overset{\displaystyle CH_3}{\underset{\displaystyle |}{}} \qquad\qquad \overset{\displaystyle CH_3}{\underset{\displaystyle |}{}} \\
CH_3 - \underset{\displaystyle |}{\overset{\displaystyle |}{C}} - CH_2 - CH - CH_2 - CH - CH_3 \\
\quad\;\; \underset{\displaystyle |}{CH_2} \qquad\qquad\quad CH_3 - \underset{\displaystyle |}{\overset{\displaystyle |}{C}} - CH_3 \\
\quad\;\; CH_3 \qquad\qquad\qquad\quad CH_3
\end{array}
$$

The structure has been written to suggest the name 2,4-dimethyl-2-ethyl-6-isopropylheptane. But a chain longer than seven C atoms can be found, and the proper name should be 2,2,3,5,7,7-hexamethylnonane.

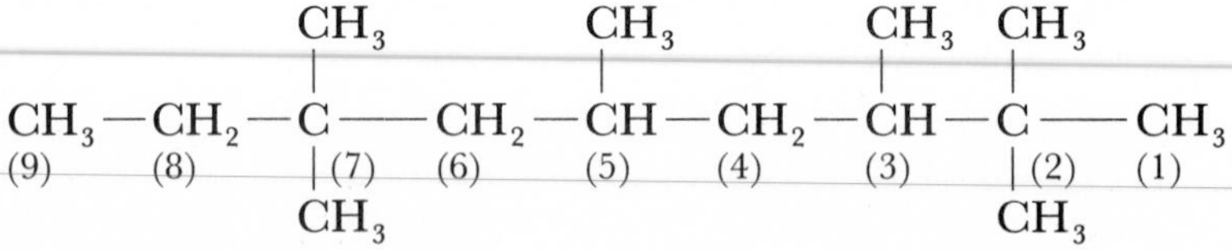

$$
\begin{array}{c}
\qquad\qquad CH_3 \qquad\qquad CH_3 \qquad\; CH_3\; CH_3 \\
\qquad\qquad\;\; |\qquad\qquad\quad |\qquad\qquad |\quad\; | \\
CH_3 - CH_2 - C - CH_2 - CH - CH_2 - CH - C - CH_3 \\
\;(9)\qquad (8)\quad |\,(7)\quad (6)\quad (5)\quad (4)\quad (3)\;\; |\,(2)\;\; (1) \\
\qquad\qquad CH_3 \qquad\qquad\qquad\qquad\;\; CH_3
\end{array}
$$

It is customary to begin the numbering at the end of the chain that is nearest the first branch point.

Hydrocarbons can form rings as well as chains. The smallest is the three-carbon ring of cyclopropane:

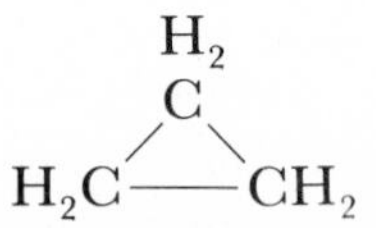

$$
\begin{array}{c}
H_2 \\
C \\
H_2C \underline{\qquad} CH_2
\end{array}
$$

This ring is highly strained, and the strain energy was calculated in Table 15-2. The optimum bond angle is 109° (the tetrahedral angle), but the angles in this three-membered ring are 60°. Cyclobutane and cyclopentane are less strained, and six-membered rings with the cyclohexane structure are extremely common. Cyclohexane can have two different structures, called the **boat** and the **chair** forms (Figure 21-9). The boat form is less stable because of the close approach of two hydrogen atoms across the top of the ring. Sugars and other substances whose molecules have a cyclohexanelike ring almost always occur in the chair form.

Reactions of Alkanes

As an example of the chemical unreactivity of the alkanes, the compound *n*-hexane is not attacked by boiling HNO_3, concentrated H_2SO_4, the strong oxidizing agent $KMnO_4$, or molten NaOH. The inertness of the alkanes makes them useful as lubricating oils, plastic films, and solid plastics for tubing and containers. Polyethylene is a familiar example. Virtually the only chemical reactions of the alkanes are combustion, dehydrogenation, and halogenation.

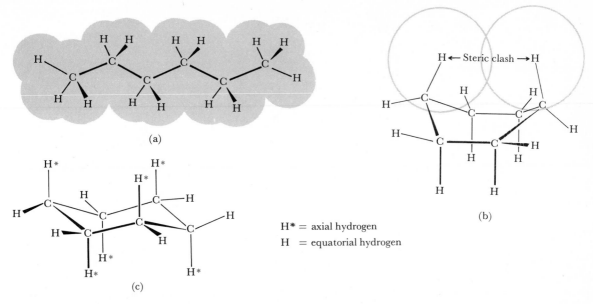

H* = axial hydrogen

H = equatorial hydrogen

Figure 21-9 *n*-Hexane, C_6H_{14}, and cyclohexane, C_6H_{12}. (a) The straight-chain hydrocarbon *n*-hexane. (b) The boat configuration of cyclohexane. The two top hydrogen atoms at the prow and stern of the boat are too close; this configuration is less stable than (c), the chair form. This same type of ring occurs in hexose sugars, which also adopt the chair form. These drawings emphasize the tetrahedral bond geometry around each carbon atom. (Adapted from R. E. Dickerson and I. Geis, *Chemistry, Matter, and the Universe*, W. A. Benjamin, Menlo Park, Calif., 1976.)

Combustion makes the alkanes useful as fuels:

$$CH_3-CH_2-CH_2-CH_3 + 6\tfrac{1}{2}O_2 \rightarrow 4CO_2 + 5H_2O(l)$$

butane gas

$$\Delta H^0_{298} = -2878 \text{ kJ}$$

Propane and butane gas, gasolines, and kerosenes all are alkanes whose value lies in their combustibility.

Dehydrogenation is the removal of atoms of hydrogen and the creation of double or triple bonds. This process usually occurs at high temperatures and in the presence of a catalyst such as Cr_2O_3:

$$H_3C-CH_3 \xrightarrow[\text{Cr}_2\text{O}_3 \text{ catalyst}]{500°C} H_2C=CH_2 + H_2$$

ethane   ethene

These dehydrogenated products are called *alkenes* or *olefins*. We shall discuss them further in the next section.

Halogenation is the reaction of a hydrocarbon with F_2, Cl_2, or Br_2 (I_2 is too inert under ordinary conditions) and the replacement of one or more H atoms by halogen atoms:

$$\begin{array}{ccc} & H & & H \\ & | & & | \\ H-C-H + Cl-Cl & \rightarrow & H-C-Cl + H-Cl \\ & | & & | \\ & H & & H \end{array}$$

methane   methyl chloride
                             or monochloromethane

These halogenated hydrocarbons are the gateway to a great many other chemical reactions.

21-4   UNSATURATED HYDROCARBONS

Dehydrogenation turns saturated hydrocarbons or **alkanes** into unsaturated hydrocarbons or **alkenes** and **alkynes**:

$$CH_3-CH_2-CH_3 \xrightarrow{\text{heat}} CH_2=CH-CH_3 + H_2$$

propane   propene

In the cracking process for petroleum, heat and catalysts break long-chain hydrocarbons into saturated hydrocarbons in the gasoline size range, and unsaturated alkenes such as propene, ethene, and butadiene. Double bonds can also be produced by removing HCl from alkyl halides with KOH in alcohol, or by removing H_2O from alcohols with concentrated H_2SO_4:

$$CH_3-CH_2-Cl + KOH \xrightarrow{\text{alcohol}} CH_2=CH_2 + KCl + H_2O$$

ethyl chloride   ethene

$$CH_3-CH_2-OH \xrightarrow{\text{acid}} CH_2=CH_2 + H_2O$$

Triple bonds also can be formed, as in ethyne or acetylene, $HC\equiv CH$, but these are not as important or as widespread as double bonds. By analogy with the alkanes, compounds with double bonds are called *alkenes,* and those with triple bonds are *alkynes* in the systematic International Union of Pure and Applied Chemistry (IUPAC) nomenclature. The common names of ethane, ethene, and ethyne are ethane, ethylene, and acetylene.

Because of the double bond, rotation around the central bond is restricted, and *geometrical isomers* are the result. Thus, $CH_3CH=CHCH_3$, 2-butene, can exist as two isomers:

$$\begin{array}{cc} H_3C \quad\quad CH_3 & H_3C \quad\quad H \\ \diagdown \quad\quad \diagup & \diagdown \quad\quad \diagup \\ C=C & C=C \\ \diagup \quad\quad \diagdown & \diagup \quad\quad \diagdown \\ H \quad\quad H & H \quad\quad CH_3 \end{array}$$

cis-2-butene   *trans*-2-butene

The double bond forces the two central C atoms, and the C and H attached directly to them, to lie in a plane. As with the isomers of coordination complexes, the prefix *cis-* indicates adjacent positioning of similar groups, and *trans-* means "across" or, at least, not adjacent. The *trans*-2-butene molecule is slightly more stable than the cis form because its bulky methyl groups are farther apart. We shall find that **steric hindrance**, or the bumping of bulky groups, plays a significant role in determining the structures of organic and biological molecules.

In the longer paraffins, dehydrogenation leads to a mixture of several products with the double bond in different places. The straight-chain isomer of butane, *n*-butane, can be dehydrogenated to two structural isomers of butene with one double bond, and two isomers of butadiene with two double bonds:

$$CH_2{=}CH{-}CH_2{-}CH_3 \qquad \text{1-butene}$$
$$CH_3{-}CH{=}CH{-}CH_3 \qquad \text{2-butene}$$
$$CH_2{=}CH{-}CH{=}CH_2 \qquad \text{1,3-butadiene}$$
$$CH_2{=}C{=}CH{-}CH_3 \qquad \text{1,2-butadiene}$$

The numbers 1, 2, and 3 locate the positions of the double bonds.

Addition reactions can occur at the double bonds with H_2, HCl, or Cl_2. For example,

$$CH_2{=}CH{-}CH_2{-}CH_3 + Cl_2 \rightarrow \overset{\displaystyle Cl}{\overset{\displaystyle |}{CH_2}}{-}\overset{\displaystyle Cl}{\overset{\displaystyle |}{CH}}{-}CH_2{-}CH_3$$

1-butene   1,2-dichlorobutane

The corresponding 1,3-butadiene addition reaction is peculiar; the addition takes place at the extreme ends of the two double bonds, in what appears to be a simultaneous (concerted) process. One double bond disappears in the reaction, and the other moves to the center of the molecule:

$$CH_2{=}CH{-}CH{=}CH_2 + Cl_2 \rightarrow \overset{\displaystyle Cl}{\overset{\displaystyle |}{CH_2}}{-}CH{=}CH{-}\overset{\displaystyle Cl}{\overset{\displaystyle |}{CH_2}}$$

1,3-butadiene   1,4-dichloro-2-butene

This unusual behavior occurs because the double bonds in the 1,3-butadiene molecule are delocalized. Such an alternating arrangement of double and single bonds $(-C{=}C{-}C{=}C{-})$ is called a **conjugated system**. When such conjugated double bonds occur in flat, closed rings with all atoms in a plane, we call the compounds **aromatic**. (See Section 21-6.)

21-5 DERIVATIVES OF HYDROCARBONS: FUNCTIONAL GROUPS

In a **chlorination** reaction, one or more hydrogen atoms can be replaced by Cl, and many isomers are possible. Some examples with their systematic names, are

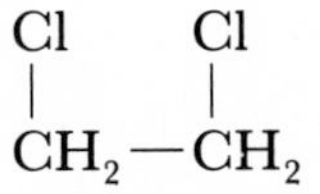

$$\begin{matrix} Cl & Cl \\ | & | \\ CH_2 & CH_2 \end{matrix}$$

1,2-dichloroethane

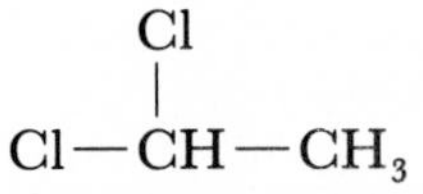

$$\begin{matrix} Cl \\ | \\ Cl-CH-CH_3 \end{matrix}$$

1,1-dichloroethane

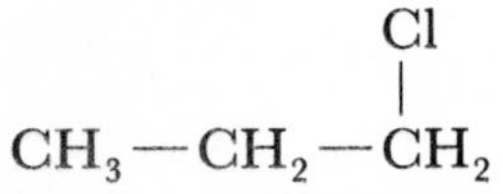

$$\begin{matrix} Cl \\ | \\ CH_3-CH_2-CH_2 \end{matrix}$$

1-chloropropane

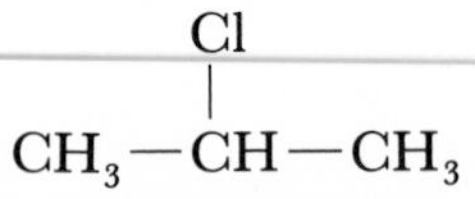

$$\begin{matrix} Cl \\ | \\ CH_3-CH-CH_3 \end{matrix}$$

2-chloropropane

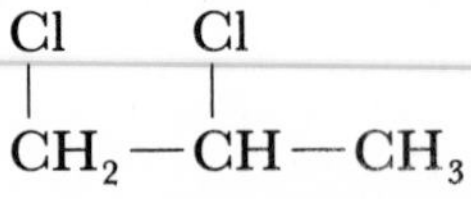

$$\begin{matrix} Cl & Cl \\ | & | \\ CH_2-CH-CH_3 \end{matrix}$$

1,2-dichloropropane

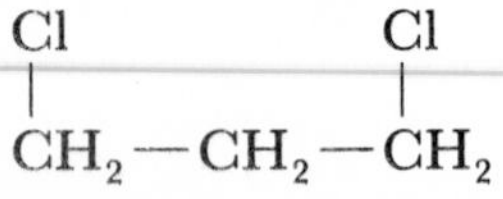

$$\begin{matrix} Cl & Cl \\ | & | \\ CH_2-CH_2-CH_2 \end{matrix}$$

1,3-dichloropropane

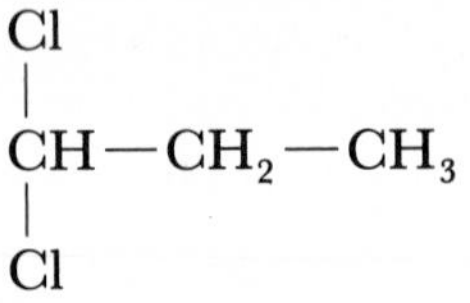

$$\begin{matrix} Cl \\ | \\ CH-CH_2-CH_3 \\ | \\ Cl \end{matrix}$$

1,1-dichloropropane

$$\begin{matrix} Cl \\ | \\ CH_3-C-CH_3 \\ | \\ Cl \end{matrix}$$

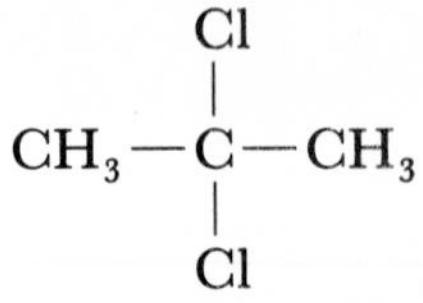

2,2-dichloropropane

Example 3

How many different isomers are there of trichloropropane, and what are they?

Solution   Five. 1,2,3; 1,2,2; 1,1,3; 1,1,2; 1,1,1.

These chlorinated hydrocarbons are the starting materials for the preparation of many classes of compounds that cannot be prepared directly from the hydrocarbons. Their chemical reactivity lies in the C—Cl bond, and the rest of the molecules act as a unit in many reactions. Therefore, it is convenient to think of the hydrocarbon part of the molecules as a **radical** attached to a **functional group**. Ethyl chloride, CH_3CH_2—Cl, behaves chemically like the combination of an ethyl radical, CH_3CH_2— or C_2H_5—, and a chloride group, —Cl. Many **replacement reactions** can occur, given the proper temperatures and catalysts:

$$C_2H_5-Cl + H_2O \rightarrow C_2H_5-OH + HCl$$

ethyl chloride   ethyl alcohol

$$C_2H_5-Cl + H_2S \rightarrow C_2H_5-SH + HCl$$

ethyl mercaptan

$$C_2H_5-Cl + NH_3 \rightarrow C_2H_5-NH_2 + HCl$$

ethylamine

$$C_2H_5-Cl + AgCN \rightarrow C_2H_5-CN + AgCl$$

ethyl cyanide

In subsequent reactions of these products, the ethyl group usually remains intact, while chemical activity takes place at the bond between the ethyl radical and the functional group.

Several common functional groups are listed in Table 21-3 and are shown in three-dimensional skeletal models in Figures 21-10 and 21-11. The **alcohols** are good solvents for organic materials, and the lower-molecular-weight alcohols are soluble in water. Methanol, or "wood alcohol," is a toxic alcohol that produces blindness and death when ingested. It attacks the nervous system by dissolving fatty material at nerve endings. The less toxic ethanol, or "grain alcohol," is the end product of energy extraction in anaerobic (non-oxygen-using) organisms such as yeasts:

$$C_6H_{12}O_6 \xrightarrow[\text{enzymes}]{\text{yeast}} 2C_2H_5OH + 2CO_2$$

Methanol and ethanol are employed in vast quantities both as solvents and as raw materials for chemical syntheses. Methanol is synthesized commer-

Figure 21-10

Models of hydrocarbon derivatives, showing typical functional groups. (a) Methyl alcohol, with the —OH group; (b) acetaldehyde (named as a derivative of acetic acid, CH_3COOH), with the —CHO aldehyde group; (c) dimethyl ether, with the —O— ether bridge; and (d) dimethyl ketone, or acetone, with the ketone linkage,

Table 21-3

Hydrocarbon Derivatives and Functional Groups

Derivative	Functional group	General formula	Examples
Halides	$-Cl, -Br$	$R-Cl$	CH_3-CH_2-Cl ethyl chloride (chloroethane) $Cl-CH_2-CH_2-Cl$ 1,2-dichloroethane
Alcohols	$-OH$	$R-OH$	CH_3-OH methanol CH_3-CH_2-OH ethanol
Ethers	$-O-$	R_1-O-R_2	CH_3-O-CH_3 dimethyl ether $CH_3-O-CH_2-CH_3$ methyl ethyl ether
Ketones	$\overset{O}{\overset{\|}{-C-}}$	$R_1-\overset{O}{\overset{\|}{C}}-R_2$	$CH_3-\overset{O}{\overset{\|}{C}}-CH_3$ dimethyl ketone or acetone
Aldehydes	$-\overset{O}{\overset{\|}{C}}-H$	$R-\overset{O}{\overset{\|}{C}}-H$	$H-\overset{O}{\overset{\|}{C}}-H$ formaldehyde $CH_3-\overset{O}{\overset{\|}{C}}-H$ acetaldehyde
Acids	$-\overset{O}{\overset{\|}{C}}-OH$	$R-\overset{O}{\overset{\|}{C}}-OH$	$H-\overset{O}{\overset{\|}{C}}-OH$ formic acid $CH_3-\overset{O}{\overset{\|}{C}}-OH$ acetic acid
Esters	$-\overset{O}{\overset{\|}{C}}-O-$	$R_1-\overset{O}{\overset{\|}{C}}-O-R_2$	$CH_3-\overset{O}{\overset{\|}{C}}-O-CH_2-CH_3$ ethyl acetate
Amines	$-NH_2$	$R-NH_2$	CH_3-NH_2 methyl amine $(CH_3)_2-NH$ dimethyl amine
Amino acids	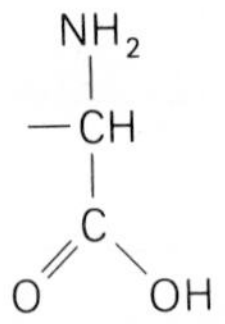 $\begin{array}{c}NH_2\\\|\\-CH\\\|\\C\\O\diagup\diagdown OH\end{array}$	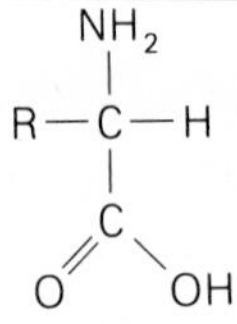 $\begin{array}{c}NH_2\\\|\\R-C-H\\\|\\C\\O\diagup\diagdown OH\end{array}$	H_2N-CH_2-COOH glycine 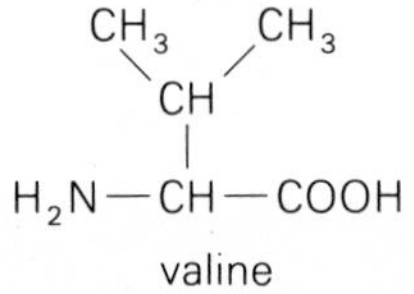 $H_2N-\overset{CH_3}{\overset{\|}{CH}}-COOH$ alanine

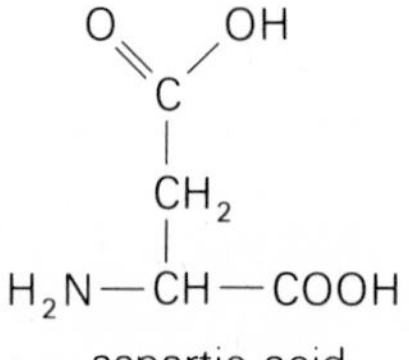

For amino acids examples (continued):

valine: $H_2N-\overset{\overset{\displaystyle CH_3\quad CH_3}{\diagdown\diagup}}{\overset{\displaystyle CH}{\underset{}{CH}}}-COOH$

aspartic acid: $\begin{array}{c}O\diagdown\ \diagup OH\\C\\\|\\CH_2\\\|\\H_2N-CH-COOH\end{array}$

Figure 21-11

Organic acids and bases, and their derivatives. (a) Acetic acid, shown with its carboxyl group ionized. (b) Methyl acetate, with the characteristic

$$\begin{matrix} & O \\ & \| \\ -C&-O- \end{matrix}$$

ester linkage. (c) Methylamine, with the amine $-NH_2$ group. The model shows the amine in its ionic form, $-NH_3^+$.

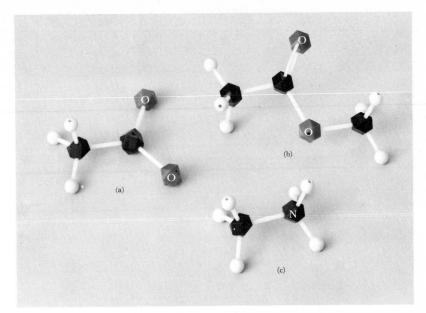

cially from carbon dioxide and hydrogen:

$$CO_2 + 3H_2 \rightarrow CH_3OH + H_2O$$

and ethanol is produced from ethene:

$$CH_2{=}CH_2 + H_2O \rightarrow CH_3CH_2OH$$

(For the names of some alcohols and other hydrocarbon derivatives, see Table 21-4.)

Ethers are relatively volatile compounds obtained when alcohols condense in the presence of concentrated sulfuric acid to eliminate water:

$$CH_3CH_2{-}O{-}H + H{-}O{-}CH_2CH_3 \xrightarrow{\ H_2SO_4\ }$$

ethyl alcohol \qquad\qquad ethyl alcohol

$$CH_3CH_2{-}O{-}CH_2CH_3 + H_2O$$

diethyl ether

Diethyl ether is the familiar ether used as an anesthetic. Ethers are valuable as solvents for waxes, fats, and other water-insoluble organic substances.

Aldehydes and **ketones** are the first step in the oxidation of alcohols:

$$CH_3CH_2OH + \tfrac{1}{2}O_2 \rightarrow CH_3{-}\overset{\displaystyle O}{\overset{\displaystyle \|}{C}}{-}H + H_2O$$

ethanol \qquad\qquad acetaldehyde

Table 21-4

Names of Hydrocarbon Derivatives[a]

R	Alcohols R—OH	Aldehydes R—CHO	Acids R—COOH	Esters CH_3—COO—R	Esters R—COO—CH_3
H—	Water	Formaldehyde (methanal)	Formic	Acetic acid	Methyl formate
CH_3—	Methyl (methanol)	Acetaldehyde (ethanal)	Acetic (ethanoic)	Methyl acetate	Methyl acetate (methyl ethanoate)
C_2H_5—	Ethyl (ethanol)	Propionaldehyde (propanal)	Propionic (propanoic)	Ethyl acetate	Methyl propionate (methyl propanoate)
C_3H_7—	Propyl (propanol)	Butyraldehyde (butanal)	Butyric (butanoic)	Propyl acetate	Methyl butyrate (methyl butanoate)
C_4H_9—	Butyl (butanol)	Valeraldehyde (pentanal)	Valeric (pentanoic)	Butyl acetate	Methyl valerate (methyl pentanoate)
C_5H_{11}—	Pentyl (pentanol)	Caproaldehyde (hexanal)	Caproic (hexanoic)	Pentyl acetate	Methyl caproate (methyl hexanoate)
C_6H_{13}—	Hexyl (hexanol)	Heptaldehyde (heptanal)	Heptanoic (heptanoic)	Hexyl acetate	Methyl heptylate (methyl heptanoate)
C_7H_{15}—	Heptyl (heptanol)	Octaldehyde (octanal)	Caprylic (octanoic)	Heptyl acetate	Methyl caprylate (methyl octanoate)
C_8H_{17}—	Octyl (octanol)	Pelargonic aldehyde (nonanal)	Pelargonic (nonanoic)	Octyl acetate	Methyl pelargonate (methyl nonanoate)
$C_{11}H_{23}$—	Undecyl (undecanol)	Lauric aldehyde (dodecanal)	Lauric	Undecyl acetate	Methyl laurate
$C_{15}H_{31}$—	Pentadecyl (pentadecanol)	Palmitic aldehyde (hexadecanal)	Palmitic	Pentadecyl acetate	Methyl palmitate
$C_{17}H_{35}$—	Heptadecyl (heptadecanol)	Stearic aldehyde (octadecanal)	Stearic	Heptadecyl acetate	Methyl stearate
$CH_3(CH_2)_7CH$=$CH(CH_2)_7$— (cis isomer)			Oleic		Methyl oleate

[a]The problem of names in organic chemistry is formidable. There are two parallel systems: the common names, and systematic names agreed upon by the International Union of Pure and Applied Chemistry (IUPAC). Common names are generally shorter and more convenient, but are only labels. From the systematic name you usually can determine most of the molecule's structure. Systematic names are given in parentheses in this table.

Common names are based on two series: those of the alkanes and those of the acids. The alkane series begins with arbitrary names but quickly shifts to the Greek prefixes indicating the number of carbon atoms: methyl, ethyl, propyl, butyl, pentyl, hexyl, and so forth. Unfortunately, the acid series retains its nonnumerical names, which usually reflect the source of the material.

Note that the numerical prefixes for acids are one place out of step with the alcohols because the carbon atom of the carboxyl group is included in the counting. Thus, C_5H_{11}COOH is *hexa*noic acid and not pentanoic.

Aldehydes and the carbon-linked part of esters use the acid nomenclature. Alcohols, ethers, ketones, amines, and the oxygen-linked part of esters use the alkane nomenclature.

You should know the names through C_4, and should understand the principles of systematic nomenclature beyond this point.

This reaction occurs at moderately high temperatures in the presence of a catalyst such as finely divided silver, or a mixture of powdered iron and molybdenum oxide. The second step in oxidation leads to a **carboxylic acid**, an acid with the carboxyl group,

$$
\begin{array}{c}
O \\
\parallel \\
-C-OH
\end{array}
$$

For example,

$$
\begin{array}{ccc}
O & & O \\
\parallel & & \parallel \\
CH_3-C-H + \tfrac{1}{2}O_2 & \rightarrow & CH_3-C-OH \\
\text{acetaldehyde} & & \text{acetic acid}
\end{array}
$$

Aldehydes and ketones are used as solvents and as raw materials for chemical syntheses. Formaldehyde,

$$
\begin{array}{c}
O \\
\parallel \\
H-C-H
\end{array}
$$

is the starting point for phenyl–formaldehyde resins such as Bakelite. Acetone,

$$
\begin{array}{c}
O \\
\parallel \\
CH_3-C-CH_3
\end{array}
$$

is one of the most common laboratory solvents.

The carboxylic acids are relatively weak acids; they dissociate to a limited extent in aqueous solution. When the carboxyl group does dissociate, the negative charge is spread over both oxygen atoms. The three p orbitals on the two oxygen atoms and the carbon atom connecting them are combined into one delocalized molecular orbital:

$$
CH_3-C\overset{\displaystyle O}{\underset{\displaystyle O-H}{\big<}} \rightarrow CH_3-C\overset{\displaystyle O}{\underset{\displaystyle O^-}{\big<}} \quad \text{or} \quad CH_3-C\overset{\displaystyle O}{\underset{\displaystyle O}{\big<}} \; -
$$

Both carbon–oxygen bonds in the ionized carboxyl group have the *same length*. The negative charge is spread over all three atoms. (The middle structure in the preceding equation can be considered as one of the two resonance structures contributing to the true carboxyl ion structure. What would the other resonance structure look like?) With metal hydroxides and carbonates, the carboxylic acids react as any other acid would to make salts:

$$
\underset{\text{propionic acid}}{C_2H_5COOH} + NaOH \rightarrow \underset{\text{sodium propionate}}{C_2H_5COONa} + H_2O
$$

Sodium propionate is dissociated in aqueous solution into sodium ions and propionate ions, and is obtained as a salt only on drying.

Formic acid, HCOOH, is the main irritant in insect stings. Acetic acid, CH_3COOH, is the acid in vinegar. The acids from butyric (C_4) to heptanoic (C_7) have acrid odors that are encountered in rancid butter and strong cheese.

Esters are made by allowing acids and alcohols to react:

$$CH_3\overset{\overset{\displaystyle O}{\|}}{C}-OH + C_4H_9OH \rightarrow CH_3\overset{\overset{\displaystyle O}{\|}}{C}-OC_4H_9 + H_2O$$

acetic acid   *n*-butanol   butyl acetate

Although they are named as though they were salts, esters are not ionized. Many are volatile liquids with pleasing, fruity odors. Butyl acetate gives bananas their odor and therefore is called banana oil. Ethyl butyrate, $C_3H_7COOC_2H_5$, has the odor of pineapples, and octyl acetate, $CH_3COOC_8H_{17}$, the odor of oranges. Oils such as linseed, cottonseed, and olive oil, and fats such as butter, lard, and tallow, are esters of the trihydroxyl alcohol glycerol,

$$\begin{array}{ccc} CH_2 & -CH- & CH_2 \\ | & | & | \\ OH & OH & OH \end{array}$$

with large molecular weight acids such as palmitic, $C_{15}H_{31}COOH$; stearic, $C_{17}H_{35}COOH$; and oleic, $C_{17}H_{33}COOH$.

Soluble soaps are the alkali metal salts of these fatty acids, obtained by treating animal fats with alkali metal hydroxides, especially NaOH:

$$(C_{17}H_{35}COO)_3C_3H_5 + 3NaOH \rightarrow 3C_{17}H_{35}COONa + C_3H_5(OH)_3$$

glyceryl stearate   sodium stearate   glycerol
(from animal fat)   (a soap)

In aqueous solution, a soap molecule has a hydrocarbon end and a charged end. Soaps "lift" dirt into solution by surrounding a small amount of grease with many molecules; all their hydrocarbon tails point in toward the grease and their carboxyl groups point out. The soap molecules thus "package" the grease in droplets or **micelles** that can be taken up into the solution and washed away.

The most common organic bases are called **amines** and can be thought of as derivatives of ammonia:

$$CH_3-NH_2 \qquad CH_3CH_2-NH_2 \qquad CH_3-NH-CH_3$$

methylamine   ethylamine   dimethylamine

$$\begin{array}{c} CH_3 \\ | \\ CH_3-N-CH_3 \end{array} \qquad CH_3-NH-C_2H_5$$

trimethylamine   methylethylamine

$$H_2N-CH_2-CH_2-NH_2 \qquad HO-NH_2$$

ethylenediamine   hydroxylamine

They are called **primary**, **secondary**, or **tertiary amines**, depending on how many of the hydrogen atoms of NH_3 are replaced by organic radicals. These organic bases are about as strong as ammonia, and add a proton to produce the ionic form:

$$CH_3CH_2-NH_2 + H^+ \rightarrow CH_3CH_2-NH_3^+$$

Methylamine is shown in Figure 21-11c in its ionic form.

The amines as a group have fishy odors and are generally toxic. Triethylamine in moderate concentrations has a choking odor of rotting fish. In toxic concentrations, the olfactory receptors are saturated, and only the ammonia smell is sensed.

The **amino acids** are an important combination of carboxylic acid and amine in one molecule. They have the general formula

amine   acid   ionized form

in which $-R$ indicates the side group that gives each amino acid its identity and chemical properties (Figure 21-12). The carbon from which the side group branches off is called the α carbon. As Figure 21-12 shows, the α carbon is an asymmetric carbon, and there are two optical isomers or enantiomorphs of an amino acid.

In solution, both the amine end and the carboxyl end of an amino acid are ionized, and the charged molecule is known as a **zwitterion**. All proteins are built from polymers of amino acids in which water is removed and a **peptide bond** is formed:

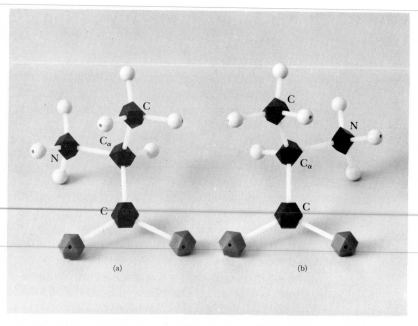

Figure 21-12   The two optical isomers of an amino acid, showing the side chain branching in opposite directions from the central carbon atom, which is called the α carbon. (a) L-Alanine, and (b) D-alanine. Both the carboxyl group and the amine group are shown in their ionized form

$$\begin{array}{c} CH_3 \\ | \\ {}^+H_3N-CH-COO^- \end{array}$$

The proteins of all living organisms are built from only L-amino acids. Their mirror images, D-amino acids, are found in small amounts in bacterial cell walls and in antibiotics produced by some microorganisms. One of the problems in explaining the evolution of life is accounting for this asymmetry of the components of living organisms. Any carbon atom that has four different atoms bonded to it will have two different possible configurations which are mirror images of one another. Such a carbon atom is called an asymmetric carbon.

Once the peptide bond is made, electrons in the C=O double bond become delocalized and the peptide C—N bond acquires a partial double bond character. The peptide unit (Figure 21-13) is thus forced to remain planar. This unit is the cornerstone of all protein structure and is one of the most important examples of delocalization of π bonding in chemical systems.

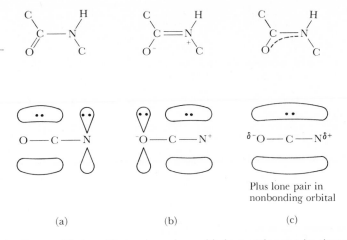

Plus lone pair in
nonbonding orbital

(a)   (b)   (c)

Figure 21-13 A structure for the peptide bond between amino acids in proteins can be drawn so that a double bond connects C and O, and the peptide bond itself is a single C—N bond (a). Another structure can be drawn with a single bond from C to O and a double peptide bond (b). This structure places charges on O and N, and is therefore less favorable. The true situation can be represented by combining p atomic orbitals from O, C, and N to make bonding, nonbonding, and antibonding delocalized molecular orbitals. The delocalized bonding orbital extends over all three atoms (c), and is therefore more stable. The nonbonding orbital is not shown. The extra stability of the delocalized electrons more than compensates for the slight charge separation at O and N. The partial double bond character of the the C—N peptide bond prevents rotation around the bond and keeps the peptide unit planar.

21-6   AROMATIC COMPOUNDS

Aromatic compounds are ring molecules with delocalized electrons. The simplest of these is benzene, C_6H_6. The delocalized electrons give aromatic compounds the special properties that differentiate them from the **aliphatic** compounds we have been examining so far. The benzene ring is commonly written as one of the Kekulé structures,

although a better representation of delocalization is

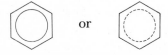

The delocalization can extend over more than one adjacent ring as in naphthalene:

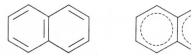

and anthracene:

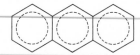

Coronene has seven adjoining rings:

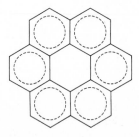

The ultimate limit of this process is graphite, with its sheets of hexagonal rings and delocalization over the entire sheet. Because of these delocalized electrons, graphite is a good conductor of electricity, whereas diamond is not. In a sense, graphite is a two-dimensional metal whose electron mobility is restricted to the individual stacked sheets.

Benzene is surprisingly unreactive in comparison to alkenes such as butene. In its lack of reactivity it is more like the saturated alkanes. It does not undergo addition reactions at a double bond; if it did so, the extent of the delocalization of electrons would be reduced. Because of this delocalization, benzene is 166 kJ mole^{-1} more stable than we would expect a compound with three single and three double bonds to be (Figure 15-9). In general, the larger the region in a molecule over which delocalization occurs, the more stable the molecule is.

Rather than addition, the typical reaction of aromatic rings is **substitution:**

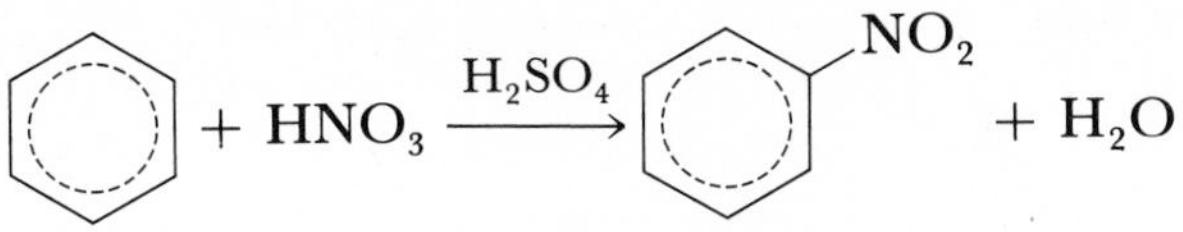

$$\text{benzene} + HNO_3 \xrightarrow{H_2SO_4} \text{nitrobenzene-}NO_2 + H_2O$$

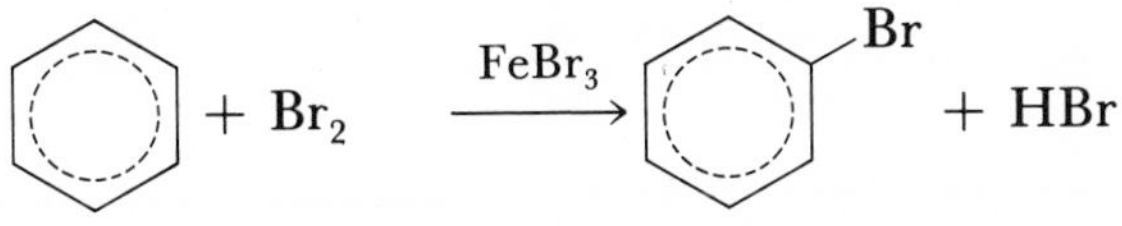

nitrobenzene

$$\text{benzene} + Br_2 \xrightarrow{FeBr_3} \text{bromobenzene-}Br + HBr$$

bromobenzene

$$\text{benzene} + RCl \xrightarrow{AlCl_3} \text{alkylbenzene-}R + HCl$$

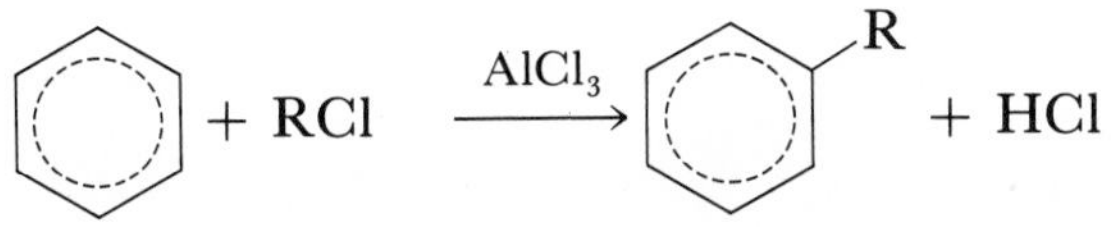

alkyl benzene

(In writing structures of this sort, it is common practice to omit the H's bonded to the ring carbons; each apex of a hexagonal ring represents a C to which one H is attached.) In the first of these reactions, sulfuric acid aids the reaction by converting HNO_3 to NO_2^+, the species that attacks the benzene ring. Sulfuric acid also acts as a dehydrating agent to remove the water formed as a product. The compounds $FeBr_3$ and $AlCl_3$ are **catalysts**. To see why they are necessary we must look at the mechanism of the reaction. Aromatic rings are particularly susceptible to attack by **electrophilic groups**, or Lewis acids, which have a strong affinity for electron pairs. In the bromination reaction, Br_2 is not electrophilic; in the absence of the $FeBr_3$ catalyst, no reaction takes place within a reasonable time. However, $FeBr_3$ itself has an attraction for another Br^- ion with its electron pair and will tear a Br_2 molecule into Br^- and Br^+ ions:

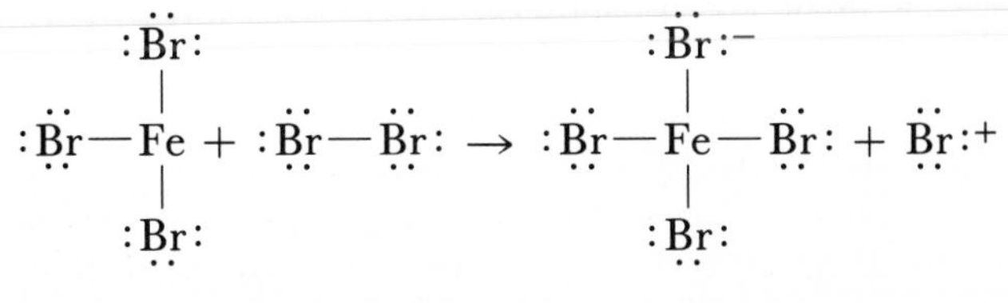

$$:\!\ddot{B}r\!-\!Fe + :\!\ddot{B}r\!-\!\ddot{B}r: \ \rightarrow \ :\!\ddot{B}r\!-\!Fe\!-\!\ddot{B}r: + \ \ddot{B}r:^+$$

The electrophilic Br^+ then attacks the aromatic ring and attracts an electron pair to make a C—Br bond. This intermediate compound is unstable and dissociates either by ejecting the Br^+ to make the starting material again, or by ejecting a hydrogen ion to make the end product, bromobenzene:

reactant   unstable intermediate   product

The liberated H^+ reacts with the $FeBr_4^-$ to produce HBr and the original $FeBr_3$. The NO_2^+ in the nitrobenzene reaction also is an electrophilic group, and that reaction proceeds by a similiar mechanism.

Two important aspects of chemical reactions are illustrated by this mechanism: the lack of completeness of most reactions, and the use of catalysts. Not every molecule of unstable intermediate that decomposes produces bromobenzene; many molecules break down to yield the reactant again. The result of most syntheses is a mixture in which the desired end product is one component (hopefully a major one) of a range of possible products. One of the challenges of chemical synthesis is to devise procedures and synthetic pathways that maximize the yield of the desired product. Often the long way around is better than an obvious one-step synthesis because the more involved synthesis produces essentially a single product.

As we have seen, a catalyst is a substance that accelerates a chemical reaction by providing an easier pathway, without being used up itself in the reaction. This does not mean that it is uninvolved. The $FeBr_3$ plays an important part in the stepwise mechanism previously outlined. But at the end of the reaction the $FeBr_3$ is regenerated in its original form. This is the general and defining behavior of a catalyst. A mixture of H_2 and O_2 can remain for years at room temperature without appreciable reaction, but the introduction of a small amount of platinum black causes an instant explosion. Platinum black has the same effect on butane gas or alcohol vapor and O_2. (Cigarette lighters with platinum black instead of a wheel and flint once were manufactured, but they soon ceased to operate because of the poisoning of the catalytic surface by impurities in the butane gas. Tetraethyl lead similarly poisons the catalytic converters that control auto emission, which is why such cars must be run on lead-free gasoline.) Platinum black acts as a catalyst by aiding the dissociation of diatomic gas molecules adsorbed on its surface. These dissociated atoms (e.g., H or O) are much more reactive than the original molecules in the gas phase. A catalyst does not affect the *overall* energy of a reaction or enter irreversibly into the reaction. It only provides an easier mechanism or pathway that makes the reaction go more rapidly.

Many catalysts, but not all, are surface-acting agents like platinum black. The substances catalyzed, called the **substrates**, bind to the surface of the catalyst. If chemical groups on the surface of the catalyst weaken a bond in a substrate, the cleavage of the substrate becomes easier. This is what happens with platinum catalysis. The finely divided black powder of Pt is a more efficient catalyst than a block of Pt only because it has much more surface exposed.

The compound $AlCl_3$ plays a catalytic role in the alkylation reaction to produce benzene derivatives with alkyl side chains. An important class of biological catalysts are the protein molecules called **enzymes**. These molecules have regions on their surface, called **active sites**, where catalysis occurs. Transition metals are frequently bound to the enzymes at their active sites, and are essential participants in catalysis. We shall look at an example of enzymatic catalysis in Section 21-10.

Several of the derivatives of benzene are shown in Figure 21-14. Phenol is weakly acidic, unlike the alcohols of which it appears to be an aromatic analogue. This ability of phenol and its derivatives to lose the hydroxyl proton arises because electrons of the oxygen become partially involved in the delocalization. The bond from ring to oxygen attains a partial double bond character, and hydrogen, robbed of some of its bonding electron pair, dissociates easily. However, the acidity of the phenols is generally less than that of the carboxylic acids.

For the same reason, aniline is a weaker base than ammonia or the aliphatic amines. The nitrogen lone electron pair that would have attracted a proton is partially involved with the aromatic ring and is less able to attract the proton and to ionize the molecule.

21-7 AROMATIC COMPOUNDS AND THE ABSORPTION OF LIGHT

Aromatic ring compounds with delocalized electrons, like transition-metal complexes with d orbitals, frequently have energy levels close enough together to absorb visible light. Hence, these two classes of compounds are often brightly colored. When a photon of energy is absorbed, one electron in a π^b bonding orbital (Figure 13-26) is promoted to the lowest π^* anti-bonding molecular orbital. Therefore, this absorption is called a $\pi \rightarrow \pi^*$ transition. In benzene and in naphthalene the levels are too far apart for the absorption to be in the visible spectrum, and these compounds are colorless. But if two nitro groups are added to make 1,3-dinitronaphthalene, the electronic-level spacing falls below 25,000 cm^{-1} and the compound appears pale yellow. (Table 20-3 with its spectral and complementary colors will be helpful throughout this section.) This phenomenon happens because the delocalized electron system has been enlarged to include the two nitro groups, and the energy levels (and the spacings between them)

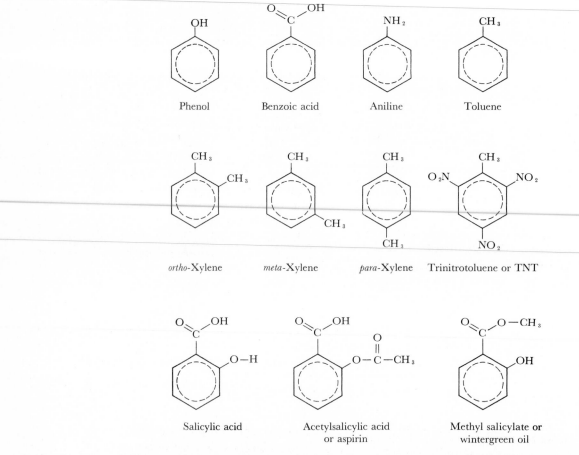

Figure 21-14   Some representative derivatives of benzene. Salicylic acid can form an ester in two ways: by using either its acid group in methyl salicylate or its hydroxyl group in aspirin. Unlike alcohols, phenols are acids, although they are usually far weaker acids than the carboxylic acids. Aromatic amines such as aniline are weaker bases than aliphatic amines. *Ortho-, meta-,* and *para-* (frequently abbreviated *o-, m-,* and *p-*) denote relative positions of groups attached to the benzene ring, as for the three isomers of xylene.

have fallen accordingly. The effect is continued in Martius Yellow, a common dye for wool and silk. An added hydroxyl group enlarges the conjugated system even more, and the energy of a $\pi \to \pi^*$ transition decreases. The color of the compound changes to yellow-orange. The light actually absorbed by the three compounds is ultraviolet for naphthalene, violet for 1,3-dinitronaphthalene, and blue for Martius Yellow.

naphthalene   1,3-dinitronaphthalene   Martius Yellow, or
2,4-dinitro-1-naphthol

Many substances such as phenolphthalein, methyl orange, and litmus have different colors in acidic or basic solutions; hence, they are useful **acid–base indicators** (Figure 5-7). *p*-Nitrophenol is a poor indicator because its color change is not vivid, but it is a simple molecule for showing what happens when an indicator changes color. Since phenols are weak acids, the reaction in solution is

acidic solutions   basic solutions

p-Nitrophenol in basic solution is a deep yellow. Its maximum absorption occurs at a wavelength of 400 nm. In the basic form, the oxygen atom and the nitro group can combine with the aromatic ring to make one large delocalized system:

Such an extended delocalized structure is called a *quinone structure,* after the yellow benzoquinone:

In the acidic form of *p*-nitrophenol, the negative charge on the oxygen atom is lacking. It is not as easy to involve the oxygen lone electron-pairs in delocalization; therefore, the energy level of the first excited electronic state is not lowered as much. Absorption occurs with a maximum just inside the ultraviolet, at 320 nm, and the compound appears a pale greenish-yellow. Phenolphthalein, which is colorless in acid and pink in base, is a more complicated molecule that works by exactly the same principle.

A particularly good way of expanding a delocalized system is illustrated by the azo dyes, which have two aromatic rings bridged by the $-N{=}N-$ group. Methyl orange, another acid–base indicator (Figure 5-6), is an azo dye:

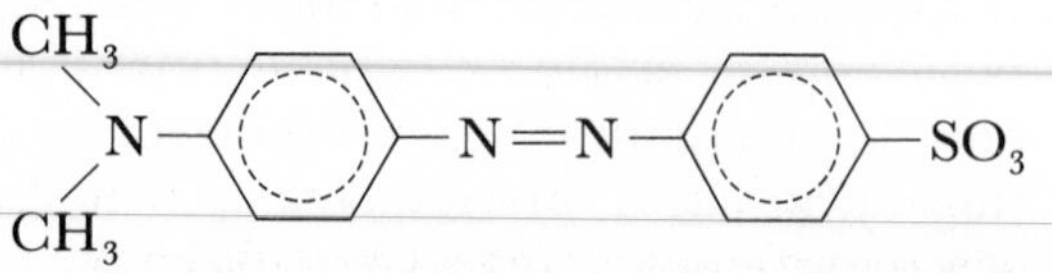

It is red in acid and yellow in base. (In which conditions are its electronic energy levels more widely spaced? Can you figure out why?)

An important example of delocalization and energy absorption is chlorophyll, which was discussed in the Postscript to Chapter 20. The aromatic ring surrounding the Mg^{2+} ion is an extended delocalized system derived from porphyrin (Figure 20-19). The electronic energy levels are such that one absorption occurs in the violet, at 430 nm, and a second in the red, at 690 nm, (Figure 20-22). When light is absorbed by chlorophyll molecules, the energy excites an electron to a higher energy level, thereby enabling it to reduce the Fe^{3+} ions in *ferredoxin,* which is a protein of molecular weight 13,000 that has two iron atoms coordinated to sulfur. The reoxidation of ferredoxin supplies the energy to drive other reactions that eventually lead to the splitting of water, the reduction of carbon dioxide, and ultimately the synthesis of glucose, $C_6H_{12}O_6$.

21-8   CARBOHYDRATES

The sugar glucose, produced in the leaves of green plants, is a **carbohydrate**. The name *carbohydrate* comes from an early misconception about the structures of these compounds. The formula for glucose, $C_6H_{12}O_6$, can be written as $(C \cdot H_2O)_6$. Substances whose formulas could be represented by equal amounts of carbon and water were called carbohydrates.

The glucose molecule is polymerized in chains of thousands of monomer units in plants to make cellulose, and in a slightly different way to make starch. A close relative of glucose, *N*-acetylglucosamine (NAG), is polymerized to form chitin, the material from which the shells of insects are made. NAG and a close variant, *N*-acetylmuramic acid (NAM), are copolymerized in alternating sequence in chains that make up part of the walls of bacterial cells. Glucose is decomposed in a stepwise fashion to produce the energy that a living organism requires. Excess glucose is carried in the bloodstream to the liver and is converted into the animal starch glycogen, which is reconverted to glucose when needed. Glucose, cellulose, starch, and glycogen all are carbohydrates.

Carbohydrates, in the form of starch, are the primary sources of energy from foods. To obtain this energy, we either eat the grains in which the starch is stored, or feed the grains to animals and let them synthesize meat protein before we eat *them*. In either case, the energy that we obtain ultimately originates from starch, the polymerized product of photosynthesis. We encounter cellulose fibers in cotton and linen, and in the artificial products cellulose acetate and rayon. The shelter over our heads probably is cellulose in the form of wood. This book is a processed cellulose called paper. Even our money, having ceased to be made from noble metals, is well on its way to becoming notarized cellulose. In this section we shall look very briefly at what carbohydrates are and how they are used.

The most fundamental unit of a carbohydrate is a **monosaccharide**, or simple sugar. Such sugars can have three, four, five, or six carbon atoms, in which case they are called **trioses**, **tetroses**, **pentoses**, or **hexoses**. We shall look only at hexoses, and especially at the most common one, D-glucose. The structure of D-glucose is depicted in Figure 21-15, parts a–c. Figure 21-15a shows the numbering of the six carbon atoms, and the Fischer convention of writing formulas to indicate the structure around an asymmetric carbon.

An **asymmetric carbon atom** is one that is bonded to four different groups, as are carbon atoms 1 through 5 in glucose. As we saw for the α carbon of an amino acid (Figure 21-12), each such asymmetric carbon atom has two different arrangements of the four groups, which are related by a mirror reflection. With five asymmetric carbon atoms, and two different configurations around each, there are a total of $2^5 = 32$ different isomers of the hexose sugars.

By the Fischer convention, the bonds to the right and left in Figure 21-15a lead from the central atom to atoms that lie above the plane of the page. Bonds extending up or down from the central atom go to atoms below the plane of the diagram. A change in configuration at any asymmetric carbon atom in the hexose is produced by exchanging the —H and —OH groups right for left in the Fischer diagram. This asymmetry is easier to see in the flat hexagon representation of the same molecule in Figure 21-15b. The actual shape of the molecule, with its tetrahedral geometry at the carbon atoms, is depicted more accurately by Figure 21-15c. Glucose has the chair conformation, which we first saw with cyclohexane, rather than the boat form.

In the 32 isomers of hexose that arise from the 32 possible interchanges of arrangement at carbon atoms 1 through 5, the positions —H and —OH at carbon atom 1 are indicated by the prefixes α and β. In α-hexoses the hydroxyl group points down as in part b or c of Figure 21-15; in β-hexoses it points up as in Figure 21-15d. A complete mirror reversal of a hexose at all five asymmetric carbon atoms simultaneously produces an L-hexose from

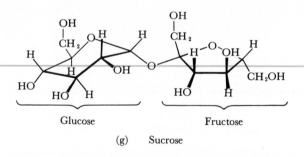

(a) α-D-Glucose

(b) α-D-Glucose

(c) α-D-Glucose

(d) β-D-Glucose

(e) α-D-Galactose

(f) α-D-Mannose

(g)   Sucrose

Figure 21-15   (a) α-D-Glucose in the Fischer representation, (b) in the flat hexagon diagram, and (c) in a form that most closely represents its actual shape. (d) β-D-Glucose, which differs from the α form only at carbon 1. (e) α-D-Galactose, which differs from glucose at carbon 4. β-D-Galactose is produced by exchanging —H and —OH at carbon 1. (f) α-D-Mannose, which differs from α-D-glucose only at carbon 2. (g) Sucrose, a dimer of α-D-glucose and β-D-fructose.

a D-hexose. Therefore, for each type of hexose there are four variants: α-D, α-L, β-D, and β-L. There must be $32/4 = 8$ different named types of hexose. However, only three of these occur naturally: glucose, galactose, and mannose. These three sugars differ at carbon atoms 2 and 4, and are compared in Figures 21-15d, e, and f. Galactose occurs in the milk sugar lactose, and mannose is a plant product (named for the Biblical *manna*). However, the most common hexose by far is glucose.

Of the hexose sugars with a five-membered ring, the most common is *fructose*. Fructose occurs naturally in honey and fruit (hence its name), and is combined with glucose as the common table sugar *sucrose* Figure 21-15g.

Polysaccharides

Cellulose is the structural fiber in trees and plants. It is found in wood, cotton, and linen, and, in a modified form, in paper. Cellulose is a polymer of β-D-glucose, with a typical chain length of about 3000 monomer units. The connection from one β-glucose to another, shown in Figure 21-16a, is called a **β-glucoside link**.

The hydroxyl groups of glucose can form esters; the treatment of cellulose with acetic anhydride, acetic acid, and a small amount of sulfuric acid produces the derivative cellulose acetate. The chains are broken to a length of 200–300 monomers, and an average of two acetate groups attach to each monomer. Cellulose acetate is the material for photographic film backing; it is also dissolved in acetone and extruded through fine holes in a metal cup to form threads of rayon.

Cellulose is not a source of food. With the exception of termites, and ruminants such as cows, both of which carry cellulose-digesting microorganisms in their stomachs, animals are incapable of breaking the β-glucosidic bond. The cleavage is an enzymatically catalyzed process, and we lack the enzymes. In 1967, a process was developed for degrading cellulose to produce an artificial flour that, although usable in baking like starch flour, had no nutritive value. It was touted briefly as a dieting aid, but rapidly sank into obscurity. (*Life* magazine referred to it as "non-food," and suggested that its inventor be paid in non-money.) However, it has been suggested quite seriously that if man could in some way learn to live in a symbiotic relationship with cellulose-digesting microorganisms in his intestines as ruminants do, his food problems would be resolved for many centuries.*

---

*This is one of those superficially attractive suggestions with possibly disastrous social consequences. It is an illustration of the unpleasant fact that a blind application of science and technology to isolated problems often creates more problems than it solves. What would be the most likely consequences if cellulose suddenly became an apparently unlimited source of cheap food? We can list a few of the results.

1. A rapid diminution of interest in the crisis of overpopulation, and an upsurge in the total population of the planet. (footnote continued on page 830)

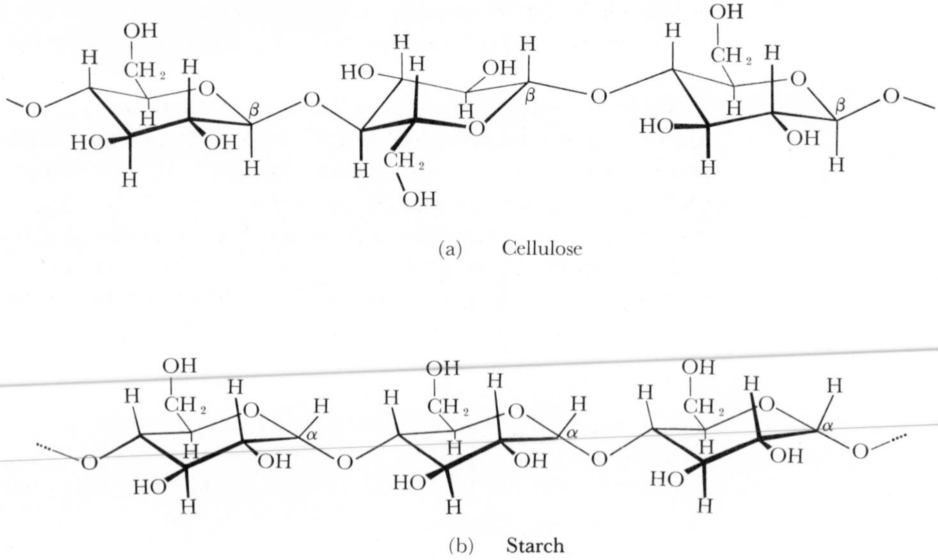

(a)   Cellulose

(b)   Starch

Figure 21-16   (a) Cellulose, a polymer of β-D-glucose. (b) Starch, a polymer of α-D-glucose.

Starch also is a polymer of glucose, but with the α linkage of Figure 21-16b. Starch is the standard storage medium for glucose to be used as a food supply in plants, and is our chief source of trapped solar energy. It is stored in plant stalks, leaves, roots, seeds, and grains. All organisms possess the enzymes necessary to digest starch. The first step in fermentation, whether it takes place in the stomach or in the brewer's vat, is the breakdown of starch to glucose. A piece of bread held in your mouth eventually will taste sweet because the enzymes in the saliva can digest bread starch to sugar.

Polymers of hexose derivatives are the structural materials in insect shells (chitin) and bacterial cell walls. In insect chitin, a hexose derivative called N-acetylglucosamine is polymerized without cross-linking. One layer

---

2. Great changes in standards and modes of living in the face of the population increases made possible by "unlimited" cheap food.

3. Deforestation of large areas of the world where people are starving. This deforestation would lead to flooding, erosion of topsoil, crop failures of conventional foods, and probable starvation again.

The introduction of edible cellulose, *by itself,* would be as shortsighted and disastrous a step as finding an efficient way of diffusing the smog of Los Angeles over the farms of Iowa. It would not avoid the eventual crisis; it would merely delay its arrival. Our planet is so carefully structured that almost any major technological or scientific change introduced without thinking is likely to lead to trouble. Who could have convinced Henry Ford that his mass transportation machines would eventually become a curse on the landscape?

in the walls of bacteria is a polymer of hexose derivatives, cross-linked with short chains of four amino acids for strength. We and all other higher organisms have evolved an enzyme, lysozyme, to protect us by lysing or dissolving this polysaccharide wall structure in invading bacteria. Lysozyme is found in most external secretions such as sweat and tears. One of the few places where D-amino acids exist in nature is in the walls of certain bacteria. One view is that they have been placed there by the bacteria simply to keep them out of the way, but it also has been speculated that they might have evolved in the structure of the wall as a defensive measure against attack by enzymes (not lysozymes, which attack the β-glucoside link) that operate most effectively against the common L-amino acids.

21-9 PROTEINS AND ENZYMES

A protein is a folded polymer of amino acids. Such a polymer is shown at the bottom of Figure 21-1, and a model of a single amino acid appears in Figure 21-12. Enzymes are one class of proteins, and perhaps the most glamorous class. They are approximately globular molecules with molecular weights from 10,000 to several million and diameters of 20 Å and more. They serve as the catalysts that control biological reactions. Other globular protein molecules such as myoglobin and hemoglobin are carriers and storage units for molecular oxygen (Figures 20-25, 20-26). The cytochromes are oxidation–reduction proteins that serve as intermediate links in the extraction of energy from foods (Figure 20-23). The gamma globulins are antibody molecules with a molecular weight of 160,000. They attach themselves to viruses, bacteria, or other foreign bodies, and precipitate them from body fluids to protect their host. All of these proteins are **globular proteins**.

The other large class of proteins is the **fibrous proteins**. These are mainly structural materials. **Keratin**, found in skin, hair, wool, nails, and beaks, is a fibrous protein. **Collagen** in tendons, the underlayers of skin, and the cornea of the eye is another type of fibrous protein, as are silks and many kinds of insect fibers. Proteins in combination with carbohydrates, and with lipids (long-chain fats and fatty acids), are the structural materials of all living organisms.

The chief distinction between a protein chain and a chain of polyethylene or Dacron is that not all side chains in a protein are alike. In fibrous proteins, it is the repetition of the sequence of side groups that gives a particular fibrous protein—silk or hair or collagen—its special mechanical properties. Globular proteins are even more intricate. These molecules typically have 100 to 500 amino acids polymerized in one long chain, and the complete sequence of side groups is the same in every molecule of the same globular protein. The side groups can be hydrocarbonlike, acidic, basic, or neutral but polar. Both the folding of the protein chain to make a compact globular molecule and the chemical behavior of the molecule once

Table 21-5

Side Groups of the Twenty Common Amino Acids

Symbol	Name	Side group

All of the amino acids given except Pro can be represented by the formula

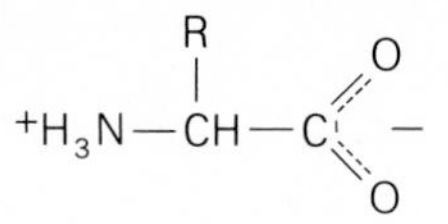

$$^+H_3N-CH-C\overset{\displaystyle O}{\underset{\displaystyle O}{\diagdown}}-$$

with R above the CH

in which the side groups, —R, are listed below. The entire molecule is shown for Pro.

Acidic side groups

Asp	Aspartic acid	$-CH_2COOH$
Glu	Glutamic acid	$-CH_2CH_2COOH$

Basic side groups

Lys	Lysine	$-CH_2CH_2CH_2CH_2NH_2$
Arg	Arginine	$-CH_2CH_2CH_2NH-C(NH)NH_2$
His	Histidine	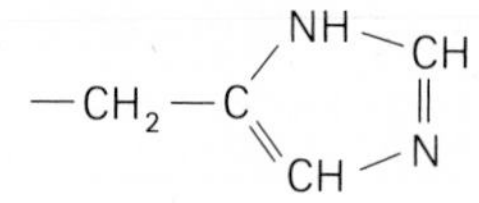

$$-CH_2-C\underset{CH}{\overset{NH}{\diagup}}\underset{N}{\overset{CH}{\diagdown}}$$

Uncharged but polar side groups

Asn	Asparagine	$-CH_2-CO-NH_2$
Gln	Glutamine	$-CH_2CH_2-CO-NH_2$
Ser	Serine	$-CH_2OH$
Thr	Threonine	$-CH(OH)-CH_3$
Gly	Glycine	$-H$

Sulfur-containing side groups

Cys	Cysteine	$-CH_2SH$
Met	Methionine	$-CH_2CH_2-S-CH_3$

Aliphatic side groups

Ala	Alanine	$-CH_3$

it is folded are determined by the kind and sequence of amino acid side groups.

Only 20 different kinds of amino acid side groups are ordinarily found in living organisms. These side groups are shown in Table 21-5. Some of them are hydrocarbonlike, such as Val, Leu, Ile, and Phe. The **hydro-**

Table 21-5 (continued)

Side Groups of the Twenty Common Amino Acids

Symbol	Name	Side group
Val	Valine	$-CH-CH_3$ $\quad\ \ \mid$ $\quad\ \ CH_3$
Leu	Leucine	$-CH_2-CH-CH_3$ $\qquad\ \ \mid$ $\qquad\ \ CH_3$
Ile	Isoleucine	$-CH-CH_2-CH_3$ $\quad\ \ \mid$ $\quad\ \ CH_3$
Pro	Proline (entire amino acid shown)	
Aromatic side groups		
Phe	Phenylalanine	
Tyr	Tyrosine	
Trp	Tryptophan	

phobic groups are more stable if they can be removed from an aqueous environment. Hence a protein chain in aqueous solution tends to fold into a molecule with these groups *inside*. Other side groups are charged: The acids Asp and Glu are ionized and negatively charged, and the bases Lys and Arg are positively charged at pH 7. Others, such as Asn, Gln, and Ser, although uncharged, are compatible with an aqueous environment. One of the most important factors in determining how a protein chain will fold into a globular molecule is the stability that results if hydrophobic groups are buried on the inside of the molecule and charged groups are outside. Although either optical isomer shown in Figure 21-12 might seem equally probable, all of the amino acids in proteins are L-amino acids (Figure 21-12a).

A protein chain is particularly stable when coiled into a right-handed α helix (Figure 21-17). In this structure, the side groups point away from the axis of the helix, and the $C{=}O$ groups of one turn of the helix are bonded to the $H{-}N$ groups of the next higher turn by a hydrogen bond. Hydrogen bonds form between especially electronegative atoms, such as F or O, and hydrogen atoms with a slight local excess of partial positive charge. Such bonds are mainly electrostatic and depend on the two atoms' being able to approach closely. Because they are small, O and F can make such bonds; the larger Cl ordinarily cannot. Hydrogen bonds are quite common in proteins because of the presence of the carbonyl oxygen and amine hydrogen on the polypeptide chain. As Figure 21-13 shows, the partial double bond character of the $C{-}N$ peptide bond not only keeps the linkage planar, it also makes the oxygen atom slightly negative and the nitrogen atom with its hydrogen atom slightly positive. These are favorable conditions for hydrogen bonding.

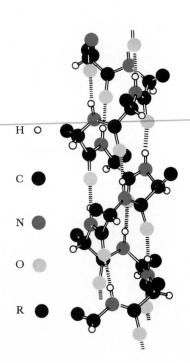

Figure 21-17   The α helix, a type of protein chain folding found in both fibrous and globular proteins. The α helix was proposed first by L. Pauling and R. B. Corey from model-building experiments based on the bond lengths and bond angles determined in x-ray analyses of individual amino acids and polymers of two and three amino acids. The structure since has been discovered in hair and wool, in skin keratin, and in globular proteins such as myoglobin and hemoglobin.

H   O

C   ●

N   ●

O   ●

R   ●

In hair, wool, and other keratins, α helices are twisted into threads, strands, and cables to make the fibers that we can see and manipulate. In silk, the chains are stretched full length rather than in an α helix, and are cross-connected by hydrogen bonds into the sheets illustrated in Figure 21-2a. In the globular proteins, the strands can be neither fully extended nor fully α-helical; there must be some bending back and forth to keep the molecule compact. In myoglobin (Figure 20-25), the 153 amino acids in the protein chain are coiled into eight lengths of α helix (lettered A through H), which then are folded back and forth to make a compact molecule. Helices E and F form a pocket into which the heme group fits, and the oxygen molecule binds to the iron of this heme. Hemoglobin is constructed along similar lines and has four such myoglobinlike units (Figure 20-26). The small protein cytochrome c (Figure 20-23) has less room for α helices. Its 103 amino acids are spun around its heme group like a cocoon, leaving only one edge exposed. In larger enzymes such as trypsin (223 amino acids) and carboxypeptidase (307 amino acids), there are regions in the center of the molecule where the protein chain zigzags back and forth in several parallel and antiparallel strands held together by hydrogen bonds very much as in silk.

The purpose of an enzyme is to provide a surface to which its *substrates* (the molecules acted upon) can bind, and to facilitate the formation or rupture of bonds in these molecules. The site on the surface of the enzyme where these activities take place is called the **active site**. The enzyme has two functions: **recognition** and **catalysis**. If it indiscriminately bound every molecule that came along, it would spend only a small proportion of its time catalyzing the reaction that it was supposed to catalyze. Conversely, even if it bound the right molecules, it would be useless if it could not assist in making or breaking the proper bonds. Enzymes recognize their true substrates by having at their active site properly positioned amino acid side chains that can interact with the substrate molecule by way of charge interaction, hydrogen bonding, or the attraction of hydrophobic groups. This selection of molecules that an enzyme will or will not bind is called its **specificity**.

Once bound, the substrate is subject to attack by groups on the enzyme. Many enzymes involved in bond-breaking reactions use metals such as Zn, Mg, Mn, or Fe. Sometimes one part of the substrate will coordinate to the metal; in other cases, the metal draws electrons from the substrate and weakens a bond. Both are illustrated in the catalytic action of trypsin discussed in the next section.

Only recently have we known the molecular structures of proteins. The first x-ray analysis of a protein, that of myoglobin, was completed in 1959. That of the first enzyme, lysozyme, was accomplished in 1964. Research on larger enzymes, electron carriers, and antibodies is progressing rapidly. We now know the detailed molecular framework of more than 90 proteins. Biochemistry, in these areas, merges imperceptibly with its sister field of molecular biology.

21-10   THE MECHANISM OF ACTION OF AN ENZYME

One of the most studied families of enzymes is that of the serine proteases. These are all designed to cut polypeptide chains of proteins by a mechanism involving a serine amino acid side chain ($-CH_2-OH$) at the active site. Three of these proteases, trypsin, elastase, and chymotrypsin, are synthesized in the pancreas and secreted into the intestines, where they digest food proteins into amino acids that can be absorbed through the intestinal walls. Because they are easy to isolate and relatively stable, they were studied intensively by chemical means before the days of protein crystal structure analysis. Today, the combination of biochemistry and x-ray diffraction has led to an especially clear picture of how these enzymes operate, illustrating the two aspects of enzyme action: **catalytic mechanism** and **substrate specificity**.

 The reaction that trypsin, chymotrypsin, and elastase catalyze is the hydrolysis or cleavage of a peptide bond in a protein:

$$\cdots -\underset{\underset{R'}{|}}{CH}-\underset{\overset{\|}{O}}{C}-\underset{\underset{H}{|}}{N}-\underset{\overset{|}{R''}}{CH}-\cdots + H_2O \rightarrow \cdots -\underset{\underset{R'}{|}}{CH}-\underset{\overset{\|}{O}}{C}-OH + H_2N-\underset{\overset{|}{R''}}{CH}-\cdots \qquad (21\text{-}1)$$

(R′, R″ are amino acid side chains.) The cleavage is carried out in two steps. In the first, the right half of the chain as shown in equation 21-1 is cut away, and the left half is attached to the serine side chain on the enzymes:

$$\cdots -\underset{\underset{R'}{|}}{CH}-\underset{\overset{\|}{O}}{C}-\underset{\underset{H}{|}}{N}-\underset{\overset{|}{R''}}{CH}-\cdots + HO-CH_2-\boxed{enzyme} \rightarrow$$

substrate   enzyme

$$\cdots -\underset{\underset{R'}{|}}{CH}-\underset{\overset{\|}{O}}{C}-O-CH_2-\boxed{enzyme} + H_2N-\underset{\overset{|}{R''}}{CH}-\cdots \qquad (21\text{-}2)$$

acyl enzyme   product 1

This intermediate is called the **acyl enzyme**. If the reaction stopped here, then trypsin or chymotrypsin would be not a catalyst but a reactant. The essence of catalysis is that the catalyst provides an easier, and hence faster, pathway for reaction, but comes through the process unscathed at the end. The enzyme is restored in a second step by the introduction of a molecule of water:

$$\cdots-\underset{\underset{R'}{|}}{CH}-\overset{\overset{O}{\|}}{C}-O-CH_2-\boxed{enzyme} + H-O-H \rightarrow$$

acyl enzyme   water

$$\cdots-\underset{\underset{R'}{|}}{CH}-\overset{\overset{O}{\|}}{C}-OH + HO-CH_2-\boxed{enzyme} \qquad (21\text{-}3)$$

product 2   enzyme

The left half of the original substrate molecule falls away, the serine side chain on the enzyme is restored, and the enzyme is ready to repeat the process. Notice that equation 21-3 is the reverse of equation 21-2, but with a water molecule substituting for the right half of the substrate chain.

Nothing happens in enzymatic catalysis that could not take place without the catalyst. A water molecule *could* come up to the protein chain in equation 21-1 and split it apart, donating its —OH to the left half and its H— to the right half of the separated chain. The activation-energy barrier to this direct reaction would be formidable, however, so the reaction itself would be very slow. The two-step reaction assisted by the serine side chain on the enzyme does not require as high an energy for any intermediate form, hence the reaction takes place more quickly. Since the effect of the enzyme is a lowering of the activation-energy barrier, the reverse reaction is catalyzed to the same extent. The same ultimate equilibrium conditions are reached with or without the catalyst; the enzyme only makes the approach to equilibrium more rapid.

The part that the enzyme plays in making the reaction easier is better seen in the eight-step diagram of Figure 21-18. The first four steps, acylation of the enzyme, correspond to equation 21-2, and the last four, deacylation, to equation 21-3. Besides the key serine side chain at the active site, the serine proteases also have histidine and aspartic acid side chains that are directly involved in the catalytic mechanism. Before the substrate chain binds, the serine is hydrogen bonded to the histidine, which in turn is hydrogen bonded at the other side of its five-membered ring to the aspartic acid (step 1). In step 2, the protein chain to be cut binds to the active site of the enzyme, which is especially shaped to receive it. In step 3, the serine oxygen

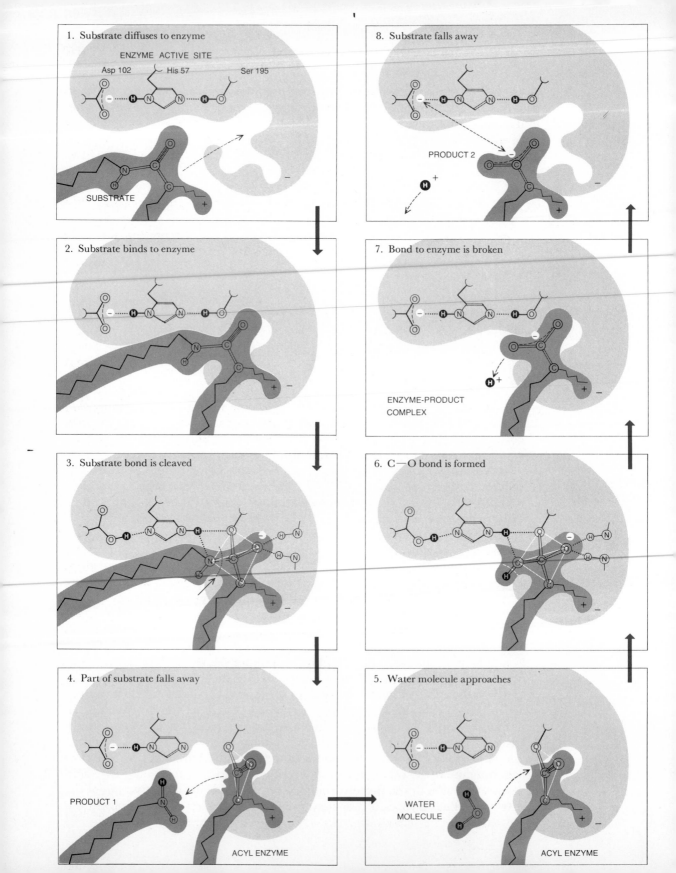

1. Substrate diffuses to enzyme

ENZYME ACTIVE SITE

Asp 102   His 57   Ser 195

SUBSTRATE

2. Substrate binds to enzyme

3. Substrate bond is cleaved

4. Part of substrate falls away

PRODUCT 1

ACYL ENZYME

5. Water molecule approaches

WATER
MOLECULE

ACYL ENZYME

6. C—O bond is formed

7. Bond to enzyme is broken

ENZYME-PRODUCT
COMPLEX

8. Substrate falls away

PRODUCT 2

has passed its hydrogen to the histidine nitrogen, and formed a bond with the carboxyl carbon of the substrate chain. This carbon now has a tetrahedral arrangement of bonds to four other atoms. The carbonyl double bond to oxygen at the right has become a single bond, and the extra electron pair has been pushed onto the carbonyl oxygen, giving it a negative charge. The enzyme helps to stabilize this negative charge by means of two amide hydrogen atoms from the enzyme protein chain. (Remember that the NH in a protein chain has a partial positive charge.) Step 3 is called the **tetrahedral intermediate**.

At the same time that the bond between serine oxygen and carbonyl carbon is being formed, the bond between the carbonyl carbon and the amide nitrogen is being weakened, and this weakening is assisted by the nearby presence of the formerly serine hydrogen atom, now on the histidine nitrogen. As the peptide N—C bond breaks, this hydrogen attaches to the N to make a completed —NH$_2$ group on the end of the departing chain, labeled product 1 in step 4. Half of the substrate chain has now fallen away, and the other half is attached to the serine side chain of the enzyme. Bonding around the carbonyl carbon atom once again is trigonal and planar, with a C=O double bond.

Steps 5 through 8 (equation 21-3) are just the reverse of steps 1 through 4, with the water molecule replacing product 1. The water molecule donates its —OH to a new tetrahedral intermediate (step 6) and its H— to the histidine, which then passes it on to the serine (step 7) as the acyl bond is broken. The second half of the substrate chain, or product 2, is then free to fall away, and the enzyme is ready for another cycle (step 8).

So far nothing has been said about the specificity of the enzymes. If trypsin, chymotrypsin, and elastase use identical catalytic mechanisms, then how do they differ from one another? The answer is that they are selective about the kind of side chain next to which they cleave a peptide bond. The critical side chain is marked R′ in equations 21-1 through 21-3, just before the carbonyl group of the bond to be cut. Each of the three enzymes has a pocket on its surface into which this side chain is inserted when the substrate binds. The specificity pocket in trypsin is long and deep, with a negative charge from an ionized aspartic acid at the bottom (Figure 21-19a). Hence trypsin favors cleavage next to the positively charged basic side chains of lysine or arginine. In chymotrypsin the specificity pocket is wider (Figure 21-19b) and is lined entirely with hydrophobic side chains, so chymotrypsin favors cleavage next to a bulky aromatic side chain such as phenylalanine or tryptophan. In elastase, the specificity pocket is quite shallow (Figure 21-19c). Elastase is less selective than the other two enzymes, but tends to favor cleavage next to small side chains such as alanine or valine.

◀ **Figure 21-18**   Catalytic mechanism of the enzymes trypsin, chymotrypsin, and elastase. See text for explanation. (Drawing by Irving Geis, from R. M. Stroud, "A Family of Protein-Cutting Proteins," *Scientific American*, July 1974. Copyright Stroud, Dickerson, and Geis.)

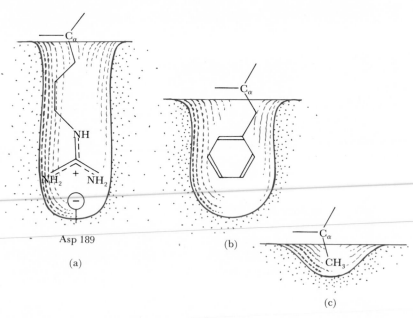

Figure 21-19   Specificity pockets of trypsin (a), chymotrypsin (b), and elastase (c). The size of each pocket, and the nature of the side chains lining it, determine what kind of amino acid chain it will hold best. This in turn determines at which position along a substrate chain cleavage will occur. (From Dickerson and Geis, *Chemistry, Matter, and the Universe,* W. A. Benjamin, Menlo Park, Calif., 1976.)

Hence both the catalytic mechanism and the substrate specificity can be explained by the folding of the polypeptide chain of the enzyme and the positioning of side chains along it. The actual folding of the protein chain in trypsin can be seen in Figure 21-20. The enzyme is built from one continuous polypeptide chain of 223 amino acids. (The conventional chain numbering has gaps and insertions to harmonize with chymotrypsin and elastase.) The molecule is roughly spherical and 45 Å in diameter, with a bowl-shaped depression or dimple on one side for the active site. The essential aspartic acid, histidine, and serine in the active site are shown by black stippling in Figure 21-20. The protein chain to be cut is shown in color with black outlining, and the arrow indicates the bond to be cleaved. The dashed lines at both ends of the substrate indicate that the chain stretches on for a considerable length in both directions. The specificity pocket for side chain R' is indicated in light-colored stippling, and since the molecule illustrated is trypsin, an arginine side chain is shown inserted into the pocket, attracted by the negative charge on aspartic acid 189 at the bottom of the pocket.

When enzyme chemists of only a generation ago spoke of the "mechanism" of enzymatic catalysis, they could only do so figuratively, even though they had a literal meaning in mind. Now we can diagram a mechanism that is every bit as literal and mechanical as the mechanism of a combination

Figure 21-20   Main-chain skeleton of the trypsin molecule. The α-carbon atoms are shown by shaded spheres, with certain of them given residue numbers for identification. For simplicity, the connecting —CO—NH— amide groups are represented only by straight lines. A portion of the polypeptide chain substrate appears in dark color with black outline. The specificity pocket is shaded, with an asparagine side chain from the substrate inserted. The catalytically important Asp, His, and Ser are poised for cleavage of the peptide bond, as marked by an arrow. (Drawing by Irving Geis, from R. M. Stroud, "A Family of Protein-Cutting Proteins," *Scientific American,* July 1974. Copyright Stroud, Dickerson, and Geis.)

lock. Although we may not yet know the mechanism of a particular enzyme of interest, there is no longer anything mysterious about the basis of enzymatic catalysis, or the principles involved. Enzymes are superb examples of molecular engineering.

21-11   ENERGY AND METABOLISM IN LIVING SYSTEMS

To mountaineers who take their hobby seriously, a particularly challenging operation is a "dynamic traverse." This is a traverse across a difficult piece of terrain where, at each instant, the climber is in an unstable situation, and where he is prevented from a disastrous fall only by his momentum. In a sense, every living organism is engaged continually in a dynamic traverse. One of the most important generalizations in science, the second law of thermodynamics, states that in any process taking place in a closed system (i.e., the object studied plus its entire environment with which it exchanges matter or energy), the disorder of the system as a whole *increases* (Chapter 16). A living organism is an intricate chemical machine, evolved to a high level of complexity. It is faced constantly with the dilemma that every chemical reaction that takes place inside it increases the disorder and reduces its complexity. A constant supply of energy is needed from outside sources, not only to provide the power to do physical work, but also to keep down the level of disorder, or *entropy,* within. If this supply of energy fails, then death and the breakdown of the chemical machine is only a matter of time: a day for the fast-living shrew, a few weeks for man. Just as momentum saves the climber from falling, so a constant influx of energy keeps the living machine from collapsing. The degradation of high-energy fuels and the extraction of their energy is called **metabolism**. In this section we shall trace briefly the outlines of the metabolism that is common to all oxygen-using living organisms. Since glucose is so important a metabolite, we shall use it as an illustration.

The Combustion of Glucose

If 180.16 g of glucose, or 1 mole, are burned, 2816 kJ of heat are liberated. Such an uncontrolled combustion is wasteful; only a small fraction, if any, of the energy stored in glucose is put to good use. It is more efficient to feed glucose to horses and use them to pull a load than it is to burn glucose and operate a locomotive with it. This is because in the horse's metabolism glucose is broken down in a series of small steps. The energy released at each step is stored in the chemical bonds of a special molecule, adenosine triphosphate (ATP), and is available for use in other chemical reactions that make muscles do work. Combustion in the horse is controlled and efficient; combustion in the locomotive is less controlled and wasteful.

The standard free energy change in the combustion of glucose,

$$C_6H_{12}O_6 + 6O_2 \rightarrow 6H_2O + 6CO_2$$

is $\Delta G^0 = -2870$ kJ mole^{-1}. This represents the chemical drive available from glucose combustion. Of this, 2816 kJ come from the heat of combustion ($\Delta H^0 = -2816$ kJ mole^{-1}) and 54 kJ come from the increased disorder of the products ($\Delta S^0 = +181$ J K^{-1} mole^{-1}, and $-T \Delta S^0 = -54$ kJ mole^{-1}). In the horse, 40% of the liberated free energy is saved by using it to synthesize 38 molecules of ATP for every molecule of glucose. It is this orderly breakdown process that we shall examine.

The Three-Step Process in Metabolic Oxidation

There are three parts to the combustion process in oxygen-using organisms. In the first part, **anaerobic fermentation** or **glycolysis**, all foods, no matter what their chemical nature, are degraded to pyruvic acid,

$$CH_3-\overset{\overset{\displaystyle O}{\|}}{C}-\overset{\overset{\displaystyle O}{\|}}{C}-OH$$

Little energy is obtained from this process. Its main purpose is to reduce everything to a common set of chemicals and prepare for the real energy-producing steps. In the second part, called the **citric acid cycle**, pyruvate is oxidized to CO_2, and the hydrogen atoms from pyruvate are transferred to the carrier molecules NAD$^+$ (nicotinamide adenine dinucleotide) and FAD (flavin adenine dinucleotide). Again, only a small amount of free energy is stored as ATP in this cycle. Its principal purpose is to break up the 1142 kJ mole^{-1} of free energy in pyruvate into four smaller and more easily handled packages, of around 220 kJ mole^{-1} each, in the form of 4 moles of reduced carrier molecules. The third part of the process, the **respiratory chain**, accepts these reduced carrier molecules. It reoxidizes them, uses the hydrogen atoms obtained during the oxidation to reduce O_2 to water, and uses the free energy obtained to synthesize ATP.

We can see two objectives to this three-part machinery: to reduce the thousands of different possible foods to a common set of chemical reactions as rapidly as possible, and to break the inconveniently large packages of free energy into several smaller ones that can be handled by the machinery for synthesizing ATP. Let us now look more closely at each of these three parts.

Step 1: Glycolysis

The first part of this combustion process does not require oxygen. It is common to all living organisms, and is known as anaerobic fermentation or glycolysis ("glucose breakdown"). If O_2 is present, the end product is pyruvic acid, as we have stated. But in other organisms that do not use oxygen, or in some oxygen-using microorganisms deprived of oxygen, other compounds are produced. Yeast cells produce ethanol under anaerobic conditions, certain types of bacteria produce acetone, and human muscle cells make lactic acid:

$$C_6H_{12}O_6 \rightarrow 2CH_3CH(OH)COOH \qquad \Delta G^0 = -198 \text{ kJ}$$

glucose   lactic acid

It is this accumulation of lactic acid in our muscles that produces muscle cramps during sudden exertion when the oxygen supply in the muscles is exhausted. As more oxygen is brought to the muscles, the lactic acid is reconverted to pyruvate, the normal product:

$$2CH_3CH(OH)COOH + O_2 \rightarrow 2CH_3COCOOH + 2H_2O$$

lactic acid   pyruvic acid

$$\Delta G^0 = -388 \text{ kJ}$$

This first part of metabolism occurs in 11 chemical steps, in which glucose is degraded to fructose and then to two three-carbon glyceraldehyde derivatives. Only in the last step or two does the process branch into separate pathways to produce pyruvic acid, lactic acid, ethanol, or acetone. Each step of the breakdown is controlled by its own catalyst, an enzyme with a molecular weight ranging from 30,000 to 500,000.

Glycolysis probably is a chemical "fossil," dating from the time before oxygen existed in the atmosphere, when one-celled organisms lived by degrading naturally occurring organic molecules. When organisms increased in size and complexity and in their energy requirements, and when oxygen appeared in our atmosphere, the more complex and much more energetic biochemical process known as the citric acid cycle evolved. Before we explore this process, we must look at the universal method of storing chemical energy in every kind of living organism.

Energy Storage and Carrier Molecules

The structure of the key molecule in the energy storage process, adenosine triphosphate (ATP), is illustrated in Figure 21-21. It is built from adenine (Figure 21-3), ribose (a five-carbon sugar), and three linked phosphate groups. The terminal phosphate group in ATP can be hydrolyzed, or split off with the addition of OH^- and H^+ from water, to yield phosphate and adenosine diphosphate (ADP). ADP can be decomposed even further to produce another phosphate group and adenosine monophosphate (AMP). Finally, the last phosphate group can be removed to make adenosine. The first two cleavages liberate 30.5 kJ mole^{-1} of free energy each, whereas the third cleavage liberates only 8 kJ mole^{-1}. It is this substance, and more particularly the first phosphate bond (farthest left in the figure), that is the

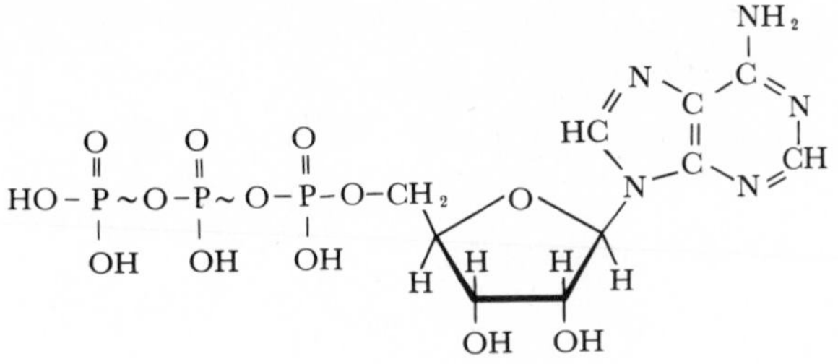

Figure 21-21

The structure of adenosine triphosphate (ATP). The bonds marked by wavy lines in the phosphate groups liberate an unusually large amount of energy when they are cleaved by hydrolysis, and are the means by which chemical energy is stored in the ATP molecule.

principal means of energy storage in any living cell. Every time a molecule of glucose is degraded biochemically to two molecules of pyruvate, eight molecules of ATP are formed from 8ADP:

$$\text{Glucose} + 8\text{ADP} + 8 \text{ phosphate} \rightarrow 2\text{CH}_3\text{COCOOH} + 8\text{ATP}$$

This results in the storage of $8 \times 30.5 = 244$ kJ mole^{-1} of free energy. The enzymes that control all of the steps of the breakdown ensure that the energy released at a step of this process is used to synthesize an ATP molecule rather than wasted as heat.

There are two other carriers that we should examine before going on to the citric acid cycle. One is nicotinamide adenine dinucleotide, whose structure is shown in Figure 21-22. It resembles ATP in having an adenine group, ribose, and phosphate. However, the essential part is a nicotine ring that can be reduced and oxidized. This molecule is a redox carrier. When a metabolite is oxidized at one step in the citric acid cycle, the oxidized form of nicotinamide adenine dinucleotide, NAD$^+$, can accept two H atoms and be reduced to NADH and H$^+$. The other important carrier is FAD or flavin adenine dinucleotide, which is reduced to FADH$_2$. Both of these carriers feed into the last production line of the energy storage factory, the terminal oxidation chain or respiratory chain. This is a four-step pathway, involving the cytochrome enzymes, in which the reduced electron carriers, NADH and FADH$_2$, are reoxidized. In this process, oxygen is reduced to water, and the energy released is stored as ATP. Each time a reduced carrier molecule is reoxidized, the energy released by this oxidation is conserved by the synthesis of several molecules of ATP.

Many of the essential vitamins are the semifinished components for energy carriers such as these. Small amounts of these vitamins must be supplied from outside sources because we have lost the ability to synthesize them ourselves. For example, the chemical group that is oxidized and reduced in FAD is a flavin. Some organisms can synthesize flavins; perhaps

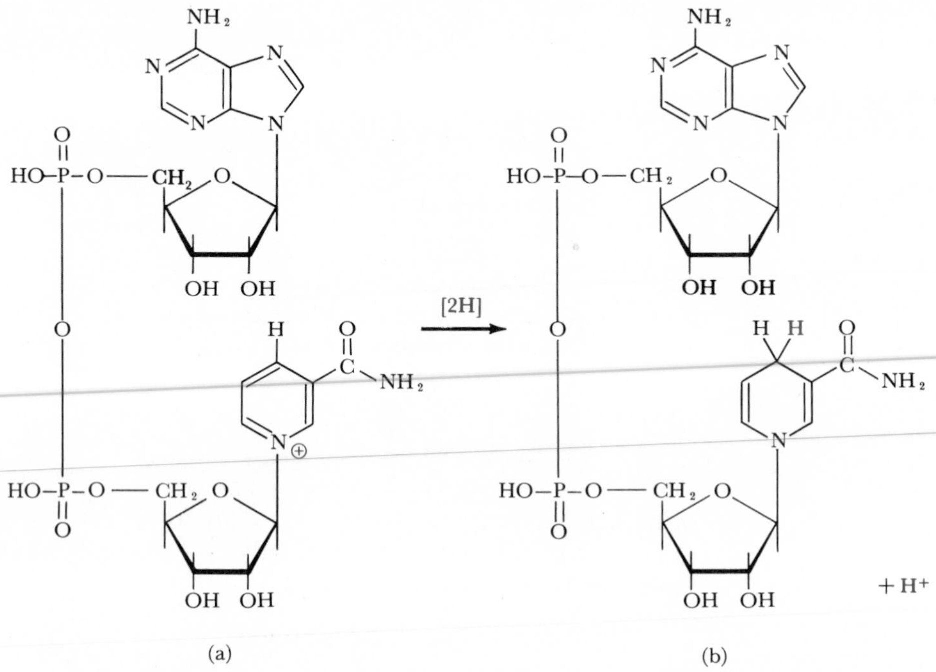

Figure 21-22   The structure of nicotinamide adenine dinucleotide (a) oxidized, NAD+, and (b) reduced, NADH. Note the similarity between the top half of the molecule as drawn here and ATP. Reduced NADH is a carrier molecule that can pass on its stored energy in chemical syntheses in a similar way as reduced ferredoxin does in photosynthesis.

we or our ancestor species also could synthesize them at one time, but we cannot do so now. To keep the energy-transfer system operating and to replace the FAD molecules as they are gradually worn out, we need small amounts of *riboflavin*, or vitamin B_2, supplied from external sources (Figure 21-3). Similarly, the nicotinamide ring of NAD+ is made from nicotinic acid, or *niacin*, which we must obtain in small quantities in our diet.

Step 2: Citric Acid Cycle

Now we are ready to look at the second part of the three-part energy-retrieval process of metabolism. The main role of this part, the citric acid cycle, is to convert the more than 1050 kJ mole^{-1} of free energy contained in pyruvic acid molecules into four packages of 220 kJ mole^{-1}, which are in the form of reduced NADH and FADH$_2$. The "primer" step before the cycle begins is the combination of pyruvic acid with a molecule called reduced coenzyme A (CoA—SH) to form acetyl coenzyme A. Acetyl coenzyme A is the raw material for the citric acid cycle, which is diagrammed in Figure 21-23:

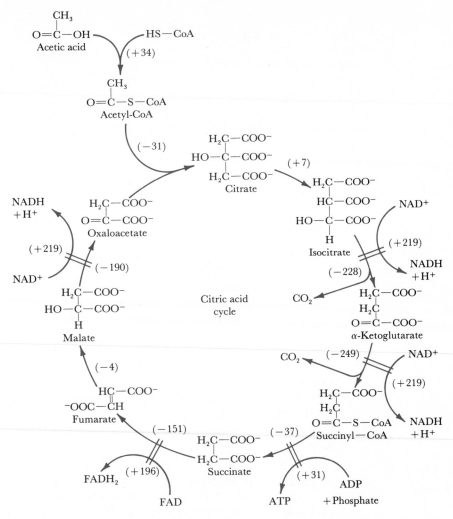

Figure 21-23   The citric acid cycle, also called the Krebs cycle or the tricarboxylic acid cycle. The numbers in parentheses are the standard free energies in kilojoules for the reactions shown. A double bar indicates that the oxidation in the cycle and the reduction of a carrier molecule are made to occur together by an enzyme. The reduced carrier molecules feed into the terminal oxidation chain, where they are oxidized again, and O_2 is reduced to H_2O.

$$CH_3COCOOH + NAD + CoA—SH \rightarrow$$

pyruvic acid

$$CH_3CO—S—CoA + CO_2 + NADH_2$$

acetyl coenzyme A

In the cycle, the two-carbon acetate group is combined first with four-carbon oxaloacetate to make the six-carbon citrate ion. Then citrate is degraded in seven steps to release two of its carbon atoms as CO_2 and to restore the oxaloacetate again. Each of the steps in the citric acid cycle is either an oxidation (isocitrate to α-ketoglutarate, malate to oxaloacetate) or a rearrangement in preparation for the next oxidation (citrate to isocitrate). In four of the oxidation steps, the liberated energy is saved by using it to reduce a carrier molecule: NAD^+ or FAD.

Step 3: Terminal Oxidation or Respiratory Chain

Each NADH molecule, no matter what its origin, funnels into the third part of the metabolic process, the terminal oxidation chain or respiratory chain, and produces three molecules of ATP. Each $FADH_2$ arrives midway through the same chain and produces only two ATP. The respiratory chain is a series of flavin-containing proteins and cytochromes (Figure 20-23), along which the hydrogen atoms and electrons from NADH and $FADH_2$ are passed, until they ultimately reduce O_2 to H_2O. The components of the respiratory chain are shown in Figure 21-24. When NADH is reoxidized, the two hydrogens are used to reduce a flavoprotein, and the free energy is saved by synthesizing a molecule of ATP from ADP and phosphate. The flavoprotein is reoxidized by reducing a small organic quinone molecule, known as ubiquinone or coenzyme Q. At this point, the electrons and protons of the reducing hydrogen atoms go their separate ways. The electrons are used to reduce the iron atom in a cytochrome b from Fe^{3+} to Fe^{2+}, and the protons go into solution. Cytochrome b reduces a cytochrome c_1, and the electrons go from c_1 to c to a to a_3, all of these being iron-containing cytochrome molecules related to the molecule shown in Figure 20-23. Two more molecules of ATP are synthesized along the way. Finally, the electrons from cytochrome a_3 and the protons from solution recombine to reduce an atom of oxygen to H_2O.

When succinate is oxidized to fumarate in the citric acid cycle (Figure 21-23), the two H atoms are given to a molecule of FAD, which is actually bound to the enzyme succinate dehydrogenase. From there the hydrogen atoms are passed to the same pool of coenzyme Q molecules used by NADH, and the process continues, but with only two ATP produced.

With every NADH molecule reduced during the citric acid cycle being "worth" three ATP, and every $FADH_2$ worth two, you can keep score on energy storage during one turn of the cycle. During the degradation of 1 mole of acetyl coenzyme A, 883 kJ of free energy are released, and 12 ATP-equivalents are saved: 12×30.5 kJ $= 366$ kJ. In the total degradation of 1 mole of glucose—glycolysis to 2 moles of pyruvate, conversion to 2 moles of acetate, and two turns of the citric acid cycle—the balance sheet looks like this (for simplicity, the conversion of pyruvate to acetyl CoA has been included in glycolysis):

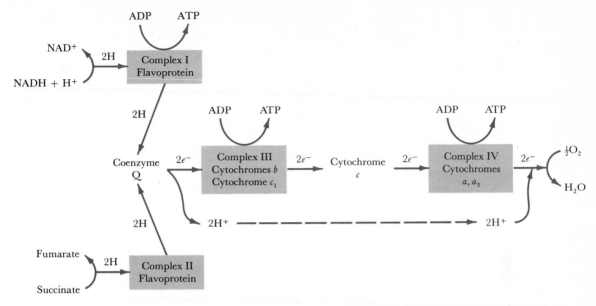

Figure 21-24   The terminal respiratory chain. The components are organized on the inner surface of the inner mitochondrial membrane into four macromolecular complexes, containing cytochromes, flavoproteins, and other non-heme iron proteins. Coenzyme Q, or ubiquinone, and cytochrome c act as conduits from one complex to the next. Reduction is by transfer of hydrogen atoms until coenzyme Q is reached, after which it occurs by transfer of electrons, with protons going into solution. Electrons and protons are reunited at the end when oxygen is reduced to water. Free energy is stored as ATP at three of the four complexes.

Degradation

$$C_6H_{12}O_6 + 2O_2 \rightarrow 2CH_3COOH + 2H_2O + 2CO_2 \qquad \Delta G^0 = -1105 \text{ kJ}$$
$$\underline{2CH_3COOH + 4O_2 \rightarrow 4CO_2 + 4H_2O \qquad\qquad\qquad \Delta G^0 = -1765 \text{ kJ}}$$
$$C_6H_{12}O_6 + 6O_2 \rightarrow 6CO_2 + 6H_2O \qquad\qquad\qquad \Delta G^0 = -2870 \text{ kJ}$$

Synthesis

$$14 \text{ ADP} + 14 \text{ phosphate} \rightarrow 14 \text{ ATP} \qquad \Delta G^0 = 14 \times 30.5 \text{ kJ} = +427 \text{ kJ}$$
$$\underline{24 \text{ ADP} + 24 \text{ phosphate} \rightarrow 24 \text{ ATP} \qquad \Delta G^0 = 24 \times 30.5 \text{ kJ} = +732 \text{ kJ}}$$
$$38 \text{ ADP} + 38 \text{ phosphate} \rightarrow 38 \text{ ATP} \qquad \Delta G^0 = \qquad\qquad\qquad +1159 \text{ kJ}$$

The overall efficiency of energy conversion from glucose as food to ATP stored in the muscles is $1159/2870 = 0.40$.

Anaerobic fermentation or glycolysis, the citric acid cycle, and the respiratory chain are part of the common heritage of all life on earth above the level of bacteria. Some aerobic, or oxygen-using, bacteria also have this same machinery to oxidize glucose or similar metabolite completely to carbon dioxide and water. Other anaerobic, or non-oxygen-using, bacteria make do only with fermentation: ingesting glucose or other energy-rich molecules, breaking them down into smaller molecules such as propionic acid, acetic acid, or ethanol, and using the relatively small quantity of free energy that is released. The anaerobic way of life is wasteful; when yeasts are given a plentiful supply of air they use the entire machinery described in this section to produce 38 moles of ATP from each mole of glucose, but when deprived of oxygen and forced to ferment glucose anaerobically to ethanol, they obtain only 2 moles of ATP per mole of glucose.

If O_2 respiration yields 19 times as much energy per gram of food as anaerobic fermentation, then why should *any* life form exist in an anaerobic lifestyle? The bacteria that are responsible for botulism in improperly canned foods, or gangrene in wounds, are two kinds of *Clostridia,* which not only cannot use oxygen, but are poisoned by it. Why are they not oxygen-breathers like the rest of us?

The answer probably lies in the way that life evolved on earth. The primitive earth is believed to have had a reducing atmosphere, with gases such as H_2, CH_4, NH_3, H_2O, and H_2S, but little or no free O_2. Under these reducing conditions, organic molecules that were formed by nonbiological means would not be degraded by oxidation as they are today, but would accumulate over the millennia. The first life forms were probably scavengers, evolving from this chemical "soup" in the oceans, and obtaining energy by degrading naturally occurring high-free-energy compounds. The *Clostridia* and their relatives today most likely are living fossils: descendants of these early anaerobic fermenters that retreated into the odd anaerobic pockets of the world when the atmosphere as a whole accumulated large quantities of free O_2 and acquired an oxidizing character.

What caused the atmosphere to change so drastically? The changeover appears to be a by-product of the invention of a new kind of energy trapping, photosynthesis, which gave its possessors an enormous advantage over the purely fermentative energy-scavengers. The organisms that developed this new talent could use the energy of sunlight to make their own high-free-energy molecules, rather than depending on what they could find in their surroundings. They became the ancestors of all green plants. Living organisms today are divided metabolically into two categories: those that can make their own food from sunlight, and those that cannot. Since the creatures in the second category subsist by eating those in the first, energy storage by means of photosynthesis is the driving source that makes all life on earth possible.

Winding the Mainspring of Life: Photosynthesis

The overall reaction of green-plant photosynthesis is the reverse of that of combustion of glucose:

$$6CO_2 + 6H_2O \rightarrow C_6H_{12}O_6 + 6O_2 \qquad \Delta G^0 = +2870 \text{ kJ}$$

Water is split apart as the source of hydrogen atoms to reduce CO_2 to glucose, and the unwanted oxygen gas is released into the atmosphere. The energy for this highly nonspontaneous process comes from sunlight. The earliest kinds of bacterial photosynthesis used substances other than water as sources of reducing hydrogens—H_2S, organic matter, or hydrogen gas itself—but the easy availability of water made this form enormously successful, and it is used today by all algae and green plants. The simplest organisms that have O_2-liberating photosynthesis are the blue-green algae, which should more properly be called by their modern name of *cyanobacteria,* since they are indeed bacteria that have learned the trick of making their own food with CO_2, H_2O, and sunlight.

Green-plant photosynthesis can be divided into two separate processes: the **photo** reactions and the **synthesis** reactions, or as they are more commonly known, the **light** reactions and the **dark** reactions. In the dark reactions, CO_2 is reduced to glucose using hydrogen atoms from NADPH (NADP$^+$ is NAD$^+$ with a phosphate on one ribose $-$OH), with ATP as driving energy:

$$6CO_2 + 12NADPH + 12H^+ \rightarrow$$
$$C_6H_{12}O_6 + 6H_2O + 12NADP^+ \qquad \Delta G^0 = +226 \text{ kJ}$$
$$18ATP + 18H_2O \rightarrow 18ADP + 18P_i \qquad \Delta G^0 = -548 \text{ kJ}$$

(P_i is a shorthand notation for inorganic phosphate.) The first reaction by itself would be nonspontaneous by 226 kJ per mole of glucose, but the second reaction provides energy to spare, giving the process a net spontaneous drive of 322 kJ. The dark reactions are indifferent as to the source of the NADPH and ATP to keep them running. Although the source in green plants today is sunlight via the light reactions, it may be that the dark reactions are older, and originally were driven by NADPH and ATP from a different source. The machinery for the dark reactions is known as the **Calvin–Benson cycle**, and is similar in logic to the citric acid cycle. Carbon dioxide is combined first with a carrier molecule, ribulose diphosphate. After a series of steps, some of which look like glycolysis run in reverse, part of the organic material is ejected as glucose, $C_6H_{12}O_6$, while the remainder is used to make ribulose diphosphate for the next turn of the cycle.

The driving energy for the Calvin–Benson cycle comes from the light reactions. The light-absorbing elements are molecules of chlorophyll (Figure 20-21), a conjugated ring of carbon atoms with delocalized electrons surrounding a magnesium atom. One type of chlorophyll molecule sits at the photocenter or trap where chemical reaction actually occurs, but other chlorophylls and related conjugated molecules surround the photocenter and act as "antennas," absorbing photons of light and passing electronic excitation on to the photocenter molecules.

The light energy absorbed by chlorophyll is used to drive an inherently unlikely reaction: reduction of $NADP^+$ using *water* as the reducing agent:

$$NADP^+ + H_2O \xrightarrow[\text{energy}]{\text{light}} NADPH + H^+ + \tfrac{1}{2}O_2$$

Oxygen gas is released into the atmosphere as a by-product. Energy also is stored by making ATP, and this ATP and NADPH provide the fuel for the dark reactions.

A more detailed comparison of glycolysis, respiration, and photosynthesis, and the various enzymes that control them, suggests that the energy-managing machinery of life evolved on our planet in a series of steps:

1. **Anaerobic fermentation, especially glycolysis.** The first living organisms were scavengers, obtaining energy by breaking down high-free-energy molecules that were not biologically formed into smaller molecules, without oxidizing them. During this distant era, the atmosphere of our planet was reducing in character, with no free oxygen.

2. **Dark reactions for glucose synthesis, leading eventually to the Calvin–Benson cycle.** Any organism that could use external energy sources to make its own high-free-energy molecules for later use would be at an enormous advantage over its less talented brothers. The first external energy source need not have been solar radiation.

3. **Light reactions for production of ATP and NADPH to drive the dark reactions.** Sunlight was the most widespread and abundant energy source on the primitive earth, as it is today. The by-product of the light reactions as carried out by green plants is free O_2, and this, poured into the atmosphere for more than a billion years, gradually changed our atmosphere from reducing to what we know today, with 20% free oxygen.

4. **Oxygen respiration.** With free O_2 available, it was perhaps inevitable that some line of organisms would learn how to use it to extract far more energy from a given amount of food. The advantages of combustion with O_2 were so great that the vast majority of life forms—plant and animal—use O_2 respiration today.

As far as we can learn, the planet itself was formed approximately 4.6 billion years ago, and simple fermenting one-celled life forms were already present 3.5 billion years ago. They may have been photosynthetic as early as 3.1 billion years ago, but geological evidence about the oxidation state of sedimentary iron deposits indicates that the atmosphere only became oxidizing around 1.8 to 1.4 billion years ago. Multicelled life forms, which perhaps were dependent on the energy surplus that only O_2 respiration could provide, made their appearance roughly between 1000 and 700 million years ago, and the pattern for the later evolution of higher organisms was set. The most revolutionary single step, after the evolution of life itself, was the tapping of an extraplanetary energy source, the sun. This step turned life from a thinly spread scavenger of naturally occurring high-free-energy molecules into a force capable of transforming the surface of the planet and ultimately leaving it altogether. Whether, from a planetary standpoint, life proves to be a beneficial or a pathological phenomenon remains to be seen.

Summary

We have come a long way in this chapter from speculating about the relative chemistries of B, C, N, and Si. Carbon undoubtedly has a special role, conferred upon it by its adequate supply of electrons for orbitals, absence of repelling lone electron pairs, and ability to make double and triple bonds. The simple **alkanes**, or single-bonded compounds of hydrogen and carbon, illustrate the diversity of compounds that carbon can build because of its ability to make long, stable chains. The **alkyl halides** are the bridges from the relatively unreactive alkanes to the wealth of hydrocarbon derivatives: **alcohols, ethers, aldehydes, ketones, esters, acids, amines, amino acids**, and others that have not been discussed in this chapter. The ability of carbon to make double and triple bonds has been seen in the **alkenes** and **alkynes**, and is especially important in **conjugated** and **aromatic** molecules.

The aromatic compounds are special because a certain number of their electrons are delocalized, or not confined to the region between any two bonded atoms. In general, the greater the delocalization, the more stable the molecule will be. Benzene is 166 kJ mole^{-1} more stable than a calculation of the bond energies of a Kekulé structure would lead one to expect. Moreover, the first unfilled orbitals in many delocalized-electron molecules are lower and closer to the filled orbitals than in other molecules. The energy required to excite an electron often falls in the visible part of the spectrum, making these compounds colored. The ability to absorb visible light is crucial in molecules of chlorophyll, because the energy of the light absorbed is used to synthesize glucose. Without photosynthesis, life on this planet would have remained a rare and irrelevant scavenger, breaking down once again those organic molecules that had been synthesized naturally by non-biological means.

The molecule that is synthesized as an energy-storage agent in photosynthesis is **glucose**, a **carbohydrate**. Besides being storehouses for chemical energy, carbohydrates are important structural materials in plants: wood, cotton, woody stalk tissue in softer plants. Glucose is polymerized into **cellulose**, which is the basis of the structural carbohydrates, and cannot be redigested, and **starch**, which is stored in seeds, grains, and roots, and is destined to be degraded later for its glucose.

When needed as an energy source, the glucose is not burned in one step. Instead, it is degraded in a series of more than 25 separate steps. During many of these steps the energy released is saved by synthesizing molecules of ATP. **Anaerobic fermentation** or **glycolysis** provides the initial breakdown of glucose to pyruvate, and the **citric acid cycle** completes the oxidation of carbon to CO_2. The hydrogens are transferred to carrier molecules, NAD^+ and FAD. These are reoxidized in the **respiratory chain**, where energy is stored as more ATP, and the hydrogen atoms are used to reduce O_2 to H_2O.

Photosynthesis is the effective reverse of the foregoing processes. The **dark reactions** use NADPH and ATP to reduce CO_2 to glucose, and the **light reactions** employ the energy of absorbed photons to generate the necessary NADPH and ATP.

Proteins are polymers of amino acids. The **fibrous** proteins have structural roles in hair, skin, nails, muscle, and tendons. In these structures the protein chains are coiled into cables or cross-connected by hydrogen bonds into sheets. The **globular** proteins include enzymes, carriers, and antibodies. In these the chains may be in helices or sheets, but these substructures are folded back on themselves to form compact, discrete molecules.

The most important class of globular proteins is made up of the biological catalysts or **enzymes**. These are characterized by a **catalytic mechanism**, by which they hasten the approach of a given reaction toward thermodynamic equilibrium, and by a **substrate specificity**, by which they choose between potential substrate molecules, acting on some and refusing to accept others. The region on the surface of the enzyme where catalysis occurs is called the **active site**. The mechanism of catalysis may involve charged groups, electron or hydrogen atom donors and acceptors, and metal atoms at the active site. Enzymatic specificity arises from the shape of the surface of the enzyme, and interactions with the substrate such as hydrogen bonding, charge interaction, and hydrophobic attractions. The enzyme and its substrate are chemically tailored for one another.

Trypsin, chymotrypsin, and elastase are three related enzymes that share the same catalytic mechanism but have different substrate specificities. They act on a protein chain substrate by forming an acyl intermediate between a serine on the enzyme and half the substrate chain, and then cleaving away this half of the chain with a water molecule. The specificity arises from a pocket adjacent to the active site, into which a substrate side chain fits.

A living organism as found on earth (and this is the only kind we know of now) is a complex collection of carbon-based molecules, tailored by evolution for survival, and using solar energy directly or indirectly as a means of driving otherwise unfavorable reactions and keeping the entropy within the organism low. As long as this can be maintained, the organism survives. When the machinery ultimately breaks down, the individual organism goes over into that state of low energy and high entropy that traditionally is known as death.

Self-Study Questions

1. What are hydrocarbons, and what size-range of hydrocarbon molecules is found in oils, paraffin waxes, gasolines, heating gas, and kerosene?

2. Illustrate branched chains in (a) hydrocarbons and (b) a synthetic fiber. How does cross-linking affect the physical properties of the fiber or plastic?

3. Why are hydrocarbons with double bonds between carbons referred to as "unsaturated" hydrocarbons?

4. Why is the chemical bonding in boron compounds so different from that for carbon?

5. In what way does the chemical bonding in B_2H_6 differ from that in C_2H_6? How is this evidenced in the geometry of the two molecules?

6. What is a three-center bond in boron compounds? How many electrons are used, and how many atoms are held together?

7. Does the three-center bond violate the principle, enunciated in Chapters 12 and 13, that a combination of a given number of atomic orbitals leads to the *same number* of molecular orbitals? In a three-center bond, how many molecular orbitals remain unfilled by electrons?

8. The carbon hydrides, or hydrocarbons, are straight-chain, cyclic, or branched-chain molecules with carbon atom backbones. How does the geometry of the boron hydrides differ from this pattern?

9. Why is it difficult for nitrogen to form long $N-N-N-N-N$ chains analogous to those of carbon? How is this a consequence of the electronic structures of the atoms?

10. To what extent do nitrogen hydrides exist? What is their single most striking property?

11. How does the bond energy of N_2 contribute to the instability of long-chain nitrogen compounds?

12. Why is silicon different in chemical behavior from carbon? Why is SiO_2 a high-melting solid, whereas CO_2 is a gas?

13. What is the backbone chain structure in silicones? In silanes? How do these two classes compare in reactivity with hydrocarbons?

14. Why can't silicon form double bonds with other silicon atoms? Then how can it form a double bond with oxygen?

15. In view of the difficulty of forming long chains of nitrogen atoms, why can nitrogen fit so easily into carbon chains?

16. What is the distinction between alkanes, alkenes, and alkynes?

17. How are branched-chain alkanes given systematic names? How is the framework for naming the alkane chosen? How are the branching side chains identified and located?

18. What are structural isomers? Show with diagrams that there are only nine structural isomers of heptane.

19. Write the structures of the following alkane radicals: methyl, ethyl, isopropyl.

20. Why are cyclopropane and cyclobutane termed *strained* molecules? Why is cyclohexane not strained?

21. Why is the chair conformation of cyclohexane more stable than the boat form?

22. How are alkene molecules constrained, in a way that alkanes are not?

23. What do the numbers in 1-butene and 2-butene signify? Why is there no 3-butene?

24. How many geometrical isomers are there of 1-butene? Of 2-butene? Draw them and give their proper systematic names.

25. What is a diene?

26. What types of chemical reactions do alkanes most commonly undergo? What are the principal reactions of alkenes?

27. What are functional groups in organic molecules? In what ways do they determine the chemical behavior of the molecules?

28. How can alcohols be prepared from halogenated hydrocarbons? How else are alcohols obtained?

29. What is an ether, and how is it made from alcohols?

30. How are aldehydes made from alcohols? How are ketones made?

31. What compounds are produced when aldehydes are oxidized? Can ketones be oxidized under similar conditions?

32. How does delocalization of electrons influence the structure of a carboxylate ion? Where is the negative charge after a carboxylic acid has ionized?

33. What is a fatty acid?

34. What are the salts of fatty acids and alkali metal cations called? What makes such salts useful to us, and how do they work?

35. How do soaps operate as cleaning agents?

36. How are fats related to fatty acids? What is the alcohol that is esterified in fats?

37. Why are organic amines bases? How can they be bases if they have no hydroxyl groups to dissociate? What inorganic base do they most resemble?

38. What are primary, secondary, and tertiary amines? Illustrate with methyl groups. Why can there be quaternary ammonium ions, but not quaternary amines?

39. What is an amino acid? In what sense is it both an acid and a base? What is the zwitterion form of an amino acid, and under what conditions is it present?

40. What is meant by an asymmetric carbon atom in an amino acid?

41. How are amino acids combined to form a protein molecule? What molecule is released during this combining process?

42. What are the two stereoisomers of an amino acid called? Which one is found almost exclusively in living organisms?

43. What is a conjugated molecule? An aromatic molecule? In what sense are they examples of electron delocalization?

44. What effect does delocalization of electrons have on the spacings of electronic energy levels? Why does it follow that many of these molecules are colored?

45. How do certain conjugated molecules function as "antennas" for light, and what is done with the absorbed light energy?

46. Why do aromatic molecules not undergo ethenelike addition reactions across double bonds? What reactions occur instead?

47. In what sense can graphite be considered a relative of benzene? What substance bears the same logical relationship to ethane?

48. Why were sugars given the name "carbohydrates," and why is this a misleading name?

49. What is the molecular formula of glucose? What is its molecular weight?

50. What is the difference between α- and β-D-glucose? Why does an equilibrium mixture exist in aqueous solution?

51. What is the difference between the intersubunit bonds in cellulose and starch? Why does no confusion arise in a plant regarding making and breaking of these two different kinds of bonds? What enzymes are involved?

52. What use is made by a plant of the two polymers, starch and cellulose? What substances play the corresponding roles in the human body?

53. How can cows and termites digest cellulose, when they cannot make the necessary enzyme, cellulase?

54. What is the difference between globular and fibrous proteins? What are the main functions of each?

55. How many different amino acid side chains are encountered in most living organisms? Give an example of one that is (a) positively charged; (b) negatively charged; (c) hydrophobic and aliphatic; (d) hydrophobic and aromatic; (e) uncharged but polar.

56. What is an α helix? Is it found in fibrous, or globular, proteins? What makes it a particularly stable structure for a protein chain to adopt?

57. How do hydrogen bonds influence the structure of a globular protein? How do hydrophobic side chains influence its structure?

58. Which atom in an amino acid is responsible for the existence of optical isomers? How many optical isomers are there? How many ordinarily are found in proteins of living organisms?

59. Is there any inherent reason why life on earth should be based on L-amino acids exclusively, rather than D-amino acids? Would there be any conceivable advantage to the use of either L- or D-amino acids, as opposed to a mixture of the two?

60. What is the active site of an enzyme? Is it involved with enzymatic catalysis, or with substrate specificity?

61. How do enzymes recognize their proper substrates and reject other molecules?

62. What is the acyl enzyme in the trypsin cleavage of a protein chain?

63. What is the tetrahedral intermediate in protein digestion by trypsin or chymotrypsin? How many tetrahedral intermediates are formed in one complete cycle of catalysis?

64. Where does the hydrogen atom come from that is added to the amino end of the severed substrate protein chain during protein digestion by trypsin? Where does the hydrogen come from to restore the enzyme to its initial state at the end of one cycle of catalysis?

65. How does the enzyme help to stabilize the tetrahedral intermediate during cleavage by trypsin?

66. How do trypsin, chymotrypsin, and elastase discriminate between their respective substrates?

67. What parts do histidine and aspartic acid play at the active site during trypsin cleavage?

68. What are NAD^+, $NADP^+$, and FAD, and what biological roles do they play? In what sense are they carriers of free energy, and how much does each carry? How does their energy compare with the amount of free energy carried by a mole of ATP?

69. In what sense are NAD^+, $NADP^+$, and FAD carriers of reducing power? Can ATP also function as such a carrier?

70. What is a vitamin? What vitamins are necessary for the synthesis of NAD^+? Of FAD? Why is no vitamin necessary for the synthesis of ATP?

71. Is a molecule that is a vitamin for one organism necessarily a vitamin for every other organism? Why, or why not?

72. How much free energy from combustion of a mole of glucose comes from enthalpy, and how much from entropy?

73. Why is the combustion of glucose in a cat more efficient than combustion of the same glucose in a Bunsen burner?

74. What is glycolysis? In what sense is it anaerobic? What is the end product of glycolysis in human muscles deprived of oxygen? What is the end product if ample oxygen is present?

75. In what sense is glycolysis probably a chemical fossil?

76. How does anaerobic fermentation compare with aerobic respiration, in terms of free energy produced per gram of glucose?

77. What is the citric acid cycle? What metabolite goes into the cycle, and what carbon compounds come out? What happens to the hydrogen atoms?

78. Where do the hydrogen atoms from the citric acid cycle eventually go? How is energy stored in this process? How much energy is stored per 2 moles of hydrogen atoms?

79. Why do we suspect that the citric acid cycle and respiratory chain are younger than the glycolytic pathway?

80. How is the respiratory chain linked to the citric acid cycle? What molecules are oxidized by the respiratory chain? What molecules are

reduced by the respiratory chain at its other end? How is energy stored along the chain?

81. Why does the reoxidation of $FADH_2$ produce less ATP than reoxidation of NADH?

82. How many moles of ATP are synthesized during the breakdown of a mole of glucose? How much free energy is stored in this way? How much free energy is released by oxidizing the glucose, and what percentage is saved via ATP?

83. Which component of energy metabolism of higher organisms is common to all life forms? When the component operates without the other components, how many moles of ATP result from 1 mole of glucose? How many times more productive is the entire system?

84. What are the two metabolic components of photosynthesis? Which is believed to be older, and what basis is there for such a hypothesis?

85. What gross effect has green-plant photosynthesis had on our planetary atmosphere?

Problems

Molecular geometry

1. Which of the following isomers are identical?

a)
```
     H   H   H   H
     |   |   |   |
 H — C — C — C — C — H
     |   |   |   |
     H   H   H   Cl
```

b)
```
     H   H   Cl  H
     |   |   |   |
 H — C — C — C — C — H
     |   |   |   |
     H   H   H   H
```

c)
```
     H   H   H
     |   |   |
 H — C — C — C — H
     |       |
     H       Cl
         |
     H — C — H
         |
         H
```

d)
```
     H   H   H   Cl
     |   |   |   |
 H — C — C — C — C — H
     |   |   |   |
     H   H   H   H
```

e)
```
      H   H   H   H
      |   |   |   |
Cl — C — C — C — C — H
      |   |   |   |
      H   H   H   H
```

f)
```
     H   H   H   H
     |   |   |   |
 H — C — C — C — C — H
     |   |   |   |
     H   Cl  H   H
```

g)
```
     H   H   H   H   H
     |   |   |   |   |
 H — C — C — C — C — C — Cl
     |   |   |   |   |
     H   H   H   H   H
```

2. Consider the following six molecules:

a)
```
    H  H  H  H  H  H  H
    |  |  |  |  |  |  |
H — C— C— C— C— C— C— C— H
    |  |  |  |  |  |  |
    H  H  H  H  H  H  H
```

b)
```
    H  H  H  H
    |  |  |  |
H — C— C— C— C— H
    |  |  |  |
    H  |  H  H
       |
       H  H
       |  |
H — C— C— C— H
    |  |  |
    H  H  H
```

c)
```
    H  H  H
    |  |  |
H — C— C— C— H
    |  |  |
    H  H  |
          |
       H— C— H
          |
          H  H
          |  |
       H— C— C— C— H
          |  |  |
          H  H  H
```

d)
```
          H              H
          |              |
      H — C ———————— C — H
          |              |
    H  H  |
    |  |  |
H — C— C— C— H   H — C — H
    |  |  |          |
    H  H  H      H — C — H
                     |
                     H
```

e)
```
    H  H              H  H  H
    |  |              |  |  |
H — C— C— H    H — C— C— C— H
    |  |              |  |  |
    H  |              H  H  H
       |
    H— C ———————— C — H
       |              |
       H              H
```

f)
```
              H
              |
          H — C — H
              |
    H         H  H
    |         |  |
H — C— C— C— C— H
    |  |  |  |
    H  |  H  H
       |
       H
       |
    H— C— C— H
       |  |
       H  H
```

Of molecules b–f, which are identical with *n*-heptane (molecule a)? What are the systematic names for all the molecules that are not identical with a?

Isomers and nomenclature

3. Draw all the possible structures for compounds having the following molecular formulas. Give their systematic names.
   a) C_3H_8
   b) C_3H_4
   c) C_4H_8
   d) C_3H_5Cl

4. Draw all the possible structural isomers of C_5H_{10}. Give the systematic name for each isomer.

Isomers

5. Which of the following compounds are isomers of one another?
   a) $CH_3CH_2CH_2OH$
   b) $CH_3CHClCH_3$
   c) $CH_3CH_2CH_3$
   d) 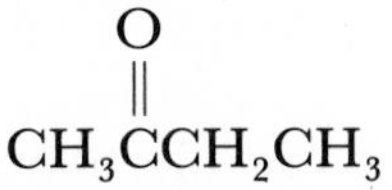 O
        $\parallel$
        $CH_3CCH_2CH_3$
   e) $CH_3CH_2CH_2Cl$

Geometry and bonding

6. Sketch the structure of each of the following molecules and indicate the hybridization around each carbon atom.
   a) C_2Cl_4
   b) CBr_4
   c) C_2Cl_2
   d) CH_2Cl_2
   e) C_2F_6

Systematic nomenclature

7. What is the systematic name of each of the following compounds?
   a) $CH_3CH_2CH=CH_2$
   b) $CH_3CH_2C\equiv CH$
   c) $CH_2=CF_2$
   d) $(CH_3)_2C=CHCH_3$

e) $CH_3CH=CCl_2$
f) $CH_2=CHCH_2CH_2CH=CH_2$
g) 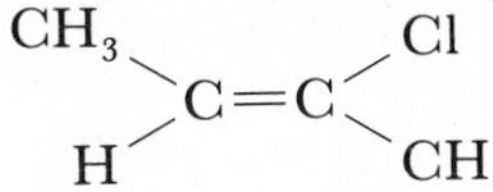 CH_3 \qquad Cl
        $\diagdown$ \qquad $\diagup$
          $C=C$
        $\diagup$ \qquad $\diagdown$
      H \qquad\qquad CH_3
h) $CH_2=CH-CBr=CH-CH_3$
i) $(CH_3)_2C=CHCH_2CH(CH_3)_2$
j) $CH_3CHClC\equiv CCH_3$
k)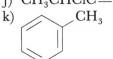

Nomenclature

8. Give the systematic names of the following substances:
   a) $CH_3CH_2CH_2COOCH_3$
   b) $CH_3CHOHCH_2CH_2CH_2CH_2CH_3$
   c) $CH_3CH_2CH_2OCH_2CH_3$
   d) $CH_3COCH(CH_3)_2$

Structural formulas

9. Draw the structural formula for each of the following compounds:
   a) *trans*-2-hexene
   b) *cis*-2,3-dichloro-2-butene
   c) 1-methylcyclopentene
   d) *trans*-1,2-dibromocyclohexane
   e) 4-ethyl-1-octene
   f) 3-hexyne
   g) *cis*-diiodoethylene
   h) 2-methyl-2-butene
   i) 2-bromo-1,3-butadiene

Hydrocarbon structure

10. Write the structural formulas of
    a) five different, simple alkanes
    b) five different, simple alkenes (only *one* double bond)

c) five different, simple alkynes (only *one* triple bond)

d) five different, simple cycloalkanes (only *one* ring)

Show that they conform to the general formulas C_nH_{2n+2}, C_nH_{2n}, C_nH_{2n-2}, and C_nH_{2n}, respectively.

Isomers of hydrocarbons

11. A hydrocarbon has a molecular weight of approximately 60 and contains 17.2% hydrogen. What is the molecular formula of the hydrocarbon? Write all of the structural isomers that have this formula.

12. A hydrocarbon has a molecular weight of 56.0 and contains 85.7% carbon. What is its molecular formula? What structural and geometric isomers could it have?

Alcohol structures

13. Write all the structural formulas of the isomeric alcohols with molecular composition $C_6H_{13}OH$. Give each its proper systematic name.

Reactions of alkenes

14. Outline a scheme for synthesizing a butanol from 1-butene. What intermediate compound might be formed? Would the product be 1-butanol or 2-butanol, and what is the principle that tells you which?

Isomers and structure

15. How many structural isomers are there of $C_6H_{14}O$ that are ethers? What are their systematic names? What other kinds of compounds in addition to ethers can be structural isomers of $C_6H_{14}O$?

16. Name the following substances. What kind of chemical compound are they? Are c and d isomers?
a) $C_6H_5OC_2H_5$
b) $(CH_3)_2CHOCH(CH_3)_2$
c) $CH_3OCH_2C_2CH_3$
d) $CH_3CH_2CH_2OCH_3$

17. Name the following substances. What kind of chemical compound are they? Are c and d isomers?
a) $C_6H_5COOC_2H_5$
b) $C_2H_5COOC_6H_5$
c) $CH_3COOCH_2CH_2CH_3$
d) $CH_3CH_2CH_2COOCH_3$

Oxidation of alcohols

18. What compounds are obtained by subjecting 1-propanol to moderate (i.e., nondestructive) oxidizing conditions? The intermediate and the final product in this reaction have quite different vapor pressures at room temperature. Explain this in terms of molecular structure. How could this help you to design experimental conditions to maximize the yield of either intermediate or final product?

Molecular structures

19. Draw the structural formula for each of the following:
a) sodium propionate
b) *m*-bromobenzoic acid
c) ethyl benzoate
d) isobutyryl chloride
e) methyl formate
f) diethylamine
g) tri-*n*-propyl amine
h) benzyl amine
i) *m*-bromoaniline
j) tetraethylammonium hydroxide
k) alanine

Suggested Reading

N. L. Allinger and J. Allinger, *Structures of Organic Molecules,* Prentice-Hall, Englewood Cliffs, N.J., 1965.

"The Biosphere," *Scientific American,* September 1970. An issue devoted to a common theme: the chemistry of our planet and of life.

R. Breslow, "The Nature of Aromatic Molecules," *Scientific American,* August 1972.

R. E. Dickerson and I. Geis, *Chemistry, Matter, and the Universe,* W. A. Benjamin, Menlo Park, Calif., 1976. Chapters 18–26 provide an elementary introduction to organic and biochemistry that is the next logical step in difficulty beyond this chapter.

R. E. Dickerson and I. Geis, *The Structure and Action of Proteins,* W. A. Benjamin, Menlo Park, Calif., 1969. Elementary introduction, abundantly illustrated.

E. Frieden, "The Chemical Elements of Life," *Scientific American,* July 1972. Discussion of the essential elements for life, and the roles that they play.

W. Herz, *The Shape of Carbon Compounds,* W. A. Benjamin, Menlo Park, Calif., 1963. Elementary introduction to organic chemistry, with emphasis on reaction mechanisms.

R. C. Johnson, *Introductory Descriptive Chemistry,* W. A. Benjamin, Menlo Park, Calif., 1966. Useful information on boron and nitrogen chemistry.

L. Stryer, *Biochemistry,* W. H. Freeman, San Francisco, 1975. Excellent introductory textbook.

22

Rates and Mechanisms of Chemical Reactions

Chemical phenomena must be treated as if they were problems in mechanics.

Lothar Meyer (1868)

The goal of every chemist, no matter what types of chemical compounds he or she works with, is to understand how and why chemicals react and change. Yet this is the most difficult task of all. It is not enough to know the structures of all the reactants and products, although such knowledge is a vital starting point. We must also know how the molecules approach one another and with what energies and with what orientations they interact. The concepts of energy and entropy are important in understanding chemical reactions. In this chapter we shall look at some of the problems that face us because we cannot examine individual molecular events. We shall see how to explain complicated experimental expressions for the rate of reaction in terms of a mechanism of reaction or a sequence of simple reactions leading to the overall chemical change. We shall examine two theories for predicting the rates of such simple reactions and compare their success or lack of it. We shall look at the two factors that often make reactions slow—energy and entropy—and see how catalysts can overcome these factors and accelerate chemical changes. Although we cannot present a complete theory of chemical reaction (no one can do this yet), we can outline the foundations on which this theory will someday be constructed.

22-1   WHAT HAPPENS WHEN MOLECULES REACT?

Let's suppose that we can watch what happens when two molecules react. We can take as an example the reaction of a molecule of thioacetamide, $CH_3-CS-NH_2$, with water to yield acetamide, $CH_3-CO-NH_2$, and H_2S (Figure 22-1).

In the original thioacetamide molecule, the central carbon atom is bound to C and to N by σ bonds, and by a σ, π double bond to S (Figure 22-1a). Since S and N are more electronegative than C, the electron pairs of their bonds to C are slightly displaced toward S and N. These two atoms bear a small negative charge, and the central C has a small positive charge. All four atoms lie in a plane.

The most favorable direction of approach of a water molecule is perpendicularly from either side of the plane of the four heavy atoms. The most favorable orientation of the incoming water molecule is as shown in Figure 22-1a. Here a lone electron pair from the water is attracted to the positive charge on the central C. As the water molecule approaches this C atom, the lone-pair electrons are drawn to it and begin to form a partial bond. This partial bond formation has two effects: It weakens the bond between C and S by repelling the electrons even more toward S, and it simultaneously weakens the $O-H$ bonds in water by pulling electrons from these bonds toward O as the O lone-pair electrons are attracted toward C. This intermediate state appears in Figure 22-1b. The central carbon atom now has two single bonds to C and N, and two partial bonds to S and O.

This intermediate state is not stable. If the water molecule falls away again and the situation reverts to that of Figure 22-1a (and there is no reason why this could not happen), then we see no net reaction. The water molecule rebounds from a collision with thioacetamide and goes its separate way. It also could happen that the *sulfur* atom falls away, as in Figure 22-1c. In this eventuality, the two protons released by O as it makes a double bond with C are attracted by the sulfur atom with four electron pairs, and a molecule of H_2S results. The reaction

$$CH_3-CS-NH_2 + H_2O \rightarrow CH_3-CO-NH_2 + H_2S$$

is complete. The reverse reaction also can occur; a molecule of H_2S can collide with one of acetamide and produce water and thioacetamide. We would be less likely to see such an event if we could watch reactions at the molecular level, simply because there are very few H_2S molecules in comparison to the number of water molecules.

What factors might affect the reaction of thioacetamide? One factor is certainly the geometry of approach of the water molecule. If the water molecule approached *in the plane* of the thioacetamide molecule, it would find its entry blocked by sulfur lone electron pairs and hydrogen atoms (to a greater extent than is apparent in the skeletal drawings of Figure

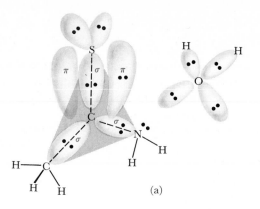

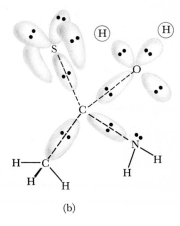

(a)

Figure 22-1

A mechanism for the reaction of thioacetamide, $CH_3—CS—NH_2$, with water to make acetamide, $CH_3—CO—NH_2$, and H_2S. (a) The thioacetamide molecule has all four heavy atoms in one plane, with S, C, and N in a triangular arrangement around a central C. The central C has sp^2 hybridization and makes a σ single bond to C and N and a σ, π double bond to S. Molecular orbitals and lone-pair orbitals are in color. The orbitals of the double bond are distorted toward S to represent its greater electronegativity. Orbitals that play no part in the reaction (C—H, N—H, N lone-pair) are not drawn. (b) Intermediate or transition state, with partial bonds from C to both S and O. The former O—H bonding electron pairs are becoming lone pairs. (c) Products of the reaction: acetamide and H_2S. The tetrahedral geometry of the transition state has reverted to trigonal planar geometry as the S atom leaves. Bonding around the S atom is shown with unhybridized s and p orbitals, in contrast to the sp^3 hybridization in water (a). This is probably close to the truth, since the bond angle in H_2S is 92° in contrast to the 105° of H_2O.

(b)

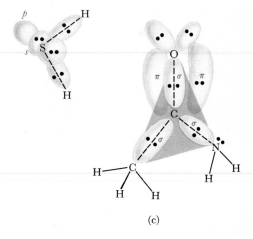

(c)

22-1). Moreover, if the water molecule approached with a *hydrogen atom,* instead of a lone electron pair, pointed at the central C, it would not be as attracted to the thioacetamide molecule and would be more likely to re-bound without reacting. If we could watch every collision, we might see that only 1 collision in 10, or in 100, had both molecules properly oriented for reaction.

A second factor is the energy of the two molecules. In the simplest theories this is expressed only as the relative *speed* of the two molecules upon collision. If the relative speed of the two molecules is small upon impact, the intermediate state will be more likely to revert to the starting mol-ecules. A slowly moving water molecule can bounce harmlessly off the thioacetamide. In contrast, a water molecule that slams forcefully against the thioacetamide has more of a chance of driving away the sulfur atom, thereby producing acetamide and H_2S. We might find that we could plot a curve of the probability of reaction as a function of the velocity of approach of the two molecules along a line connecting their centers.

Unfortunately, the kinds of observations we have been describing for a reaction are an unattainable dream. We must try to find out what is hap-pening during a reaction in a more indirect way. Frequently the most that we can say about a proposed mechanism of reaction is that it is not incom-patible with the data. There is always the lingering possibility that some other mechanism of reaction might explain the same data just as well. A classic example of this ambiguity is the reaction of H_2 and I_2. In 1893, Max Bodenstein, in Germany, studied the reaction

$$H_2 + I_2 \rightarrow 2HI$$

This was the first comprehensive kinetic study of a reaction occurring in the gas phase. From that time until 1967, virtually every kinetics text and treatise used this reaction as the ideal example of a two-body collision mechanism. One gas molecule of H_2 collides with a molecule of I_2, they reshuffle atoms, and two molecules of HI are the result. But in 1967, J. H. Sullivan showed that this reaction does not take place by a two-body colli-sion at all, but by a complicated chain reaction. We shall see later why the data measured before 1967 could be explained with equal ease with the two-body or a three-body model.

Not only are we unable to watch individual molecules, we cannot choose the orientation of the molecules upon collision. The best we can do is to estimate the probability of the molecules' being suitably oriented and then modify our calculations of rates of reaction by a suitable factor. Such a correction sometimes is used and is called a **steric factor**.

In a gas reaction or a reaction in solution, we cannot even choose the velocity of approach of the reacting molecules. The molecules in a sample of gas will have a distribution of velocities. We can shift the distribution of velocities by varying the temperature of the gas. As illustrated in Figure 3-11, in nitrogen gas the fraction of all the molecules having a velocity greater than some value such as 1000 m sec^{-1} increases as the temperature increases. At 273 K, only 0.44% of the N_2 molecules have velocities of 1000 m sec^{-1} or greater; at 1273 K, 35% have this velocity or greater; at 2273 K, this fraction increases to 55%. However, nothing that we can do to the system will give us *one specific* velocity.

We can remove the velocity distribution for certain reactions by using the method of crossed molecular beams (Figure 22-2). Instead of reactions occurring between molecules dispersed in a solution or a gas, beams of molecules or ions are passed through one another in an evacuated chamber

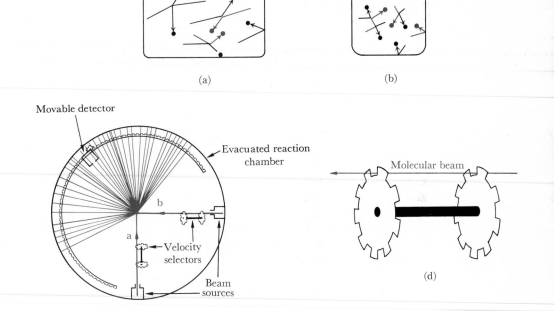

Figure 22-2   The two main classes of experiments in chemical kinetics. In bulk reactions in the gas (a) or liquid phase (b), the orientation of reacting molecules is uncontrolled, and a distribution of molecular velocities exists. In a crossed-molecular-beam experiment (c), the orientations are still uncontrolled, but only molecules or ions with certain velocities are used. A typical velocity selecting device (d) is a pair of wheels with suitably spaced sectors cut out so only molecules that take a specified time to travel the distance from one wheel to the next can pass through the open sectors on both wheels. The beam sources typically are ovens that emit a stream of gas molecules, and electrostatic fields that accelerate ions.

with negligible amounts of other molecules present. The molecules in the crossed beams react with one another and are scattered from the beams. The products of the reaction, and the unreacted initial molecules, can be observed as a function of angle of scattering by using a movable detector mounted inside the chamber. This arrangement has the great advantage that the velocity selectors can limit the beam to molecules with velocities in a chosen small range. A knowledge of the products of the reaction as a function of angle of deflection or scatter provides much more information about the process of reaction. The orientation problem remains in a molecular-beam experiment, but one can imagine experiments in which this factor is controlled as well. Intense magnetic or electric fields placed just before the beams intersect might give the majority of the molecules in the beam one preferred orientation in space if the molecules had magnetic moments or dipole moments.

Some of the reactions that have been studied with crossed molecular beams are

$$K + HBr \rightarrow KBr + H$$
$$K + CH_3I \rightarrow CH_3 + KI$$
$$K + C_2H_5I \rightarrow C_2H_5 + KI$$

The reactants, beams of K atoms, HBr, CH_3I, and C_2H_5I molecules, are emitted from heated ovens within the evacuated chamber. The detector is a heated wire filament called a *surface ionization detector,* which is sensitive to alkali metals or compounds of alkali metals.

The disadvantage of molecular-beam experiments is that not all chemical reactions are suitable for study with molecular beams in evacuated chambers. Molecular-beam methods remain a special tool for making complete studies of certain special reactions. The majority of chemical reactions must be studied by bulk methods, in gas mixtures, solutions, and (less frequently) solids.

22-2 MEASUREMENT OF REACTION RATES

The rate of reaction is usually followed in bulk methods by watching the disappearance of a reactant or the appearance of a product in a given time. If the chemical reaction is

$$A + 2B \rightarrow 3C$$

then the rate of appearance of product C in a time interval Δt is

$$\frac{\Delta[C]}{\Delta t} \tag{22-1}$$

in which the concentration of C, [C], is usually expressed in moles liter^{-1}. This is the average rate of appearance of C during the time interval Δt.

The limit of this average rate as the time interval becomes smaller is called the *rate of appearance* of C at time *t*. It is the slope of the curve of [C] versus *t* at time *t*. This instantaneous slope or rate is written $d[C]/dt$. Since one molecule of A disappears for every three of C that are produced, and two molecules of B disappear during the same process, the rates of disappearance and appearance of chemical species are related by the expression

$$-\frac{d[A]}{dt} = -\frac{1}{2}\frac{d[B]}{dt} = +\frac{1}{3}\frac{d[C]}{dt}$$

The rate of a chemical reaction will depend on the concentrations of the reactants, although not always in the way that might be expected from the overall chemical equation. For the reaction of hydrogen gas with gaseous iodine to produce HI,

$$H_2 + I_2 \rightarrow 2HI$$

the relationship between rates is

$$\frac{d[HI]}{dt} = -2\frac{d[H_2]}{dt} = -2\frac{d[I_2]}{dt}$$

and as you might intuitively expect, the rate equation is

$$\frac{d[HI]}{dt} = k[H_2][I_2] \tag{22-2}$$

The rate of reaction is proportional to the concentrations of H_2 and I_2, and dependent on the first power of each concentration. This does not mean that the reaction proceeds by a collision of one H_2 molecule with one I_2; since 1967 we have had evidence that it does not. We must distinguish clearly between the order of a reaction and the molecularity of the reaction. The constant of proportionality in equation 22-2 is called the **rate constant**.

The **order** of a reaction is the sum of all the exponents of the concentration terms in the rate equation. The HI reaction is first order in each of the reactant concentrations and is second order overall. Order is a purely experimental parameter and describes what is observed about the rate equation rather than implying anything about the mechanism of reaction.

The **molecularity** of a simple one-step reaction is the number of individual molecules that interact in the reaction. Molecularity requires a knowledge of the reaction mechanism. A reaction such as that of hydrogen and iodine actually may take place as a series of half a dozen individual reactions for which we could specify the molecularity of each. The concept of the molecularity of an overall reaction that occurs in a series of steps has no meaning. Most simple one-step reactions are unimolecular (spontaneous decay) or bimolecular (collision). True trimolecular reactions are rare, as three-body collisions are unlikely. Tetramolecular and higher reactions are virtually unheard of. Reactions that from their stoichiometry appear to be trimolecular or higher, are after careful study usually seen to be the sum of a series of simple unimolecular and bimolecular steps. One of the challenges of chemical kinetics is to determine the true set of reactions in such a case.

The reaction of hydrogen gas with bromine is in complete contrast to that with iodine. The overall reaction is similar:

$$H_2 + Br_2 \rightarrow 2HBr$$

but the experimental dependence of the rate of production of HBr upon concentrations of reactants and products is utterly different from equation 22-2:

$$\frac{d[HBr]}{dt} = \frac{k[H_2][Br_2]^{1/2}}{1 + k'([HBr]/[Br_2])} \tag{22-3}$$

This expression has two experimental rate constants, k and k'. We cannot talk about the molecularity of the reaction, because the overall process is the result of an elaborate chain of reactions that we shall come back to later. Even the order is a puzzle. At the start of a reaction of H_2 with Br_2, when little HBr is present, the second term in the denominator can be neglected. Then the reaction is effectively $1\frac{1}{2}$ order: first order in H_2 and one-half order in Br_2. As the product, HBr, accumulates, it slows down the rate of production of more HBr. Therefore, HBr is called an **inhibitor** of the reaction.

The formation of HCl is even more complicated. The production of HCl is accelerated by light of intensity I and is inhibited by the presence of oxygen gas, even at low oxygen concentrations. For years the difficulty of purifying the H_2 and Cl_2 gases and eliminating all traces of O_2 led to erroneous conclusions about the kinetics of this reaction. The best experimental rate equation for the appearance of HCl is

$$\frac{d[HCl]}{dt} = \frac{k_1 I[H_2][Cl_2]}{k_2[Cl_2] + [O_2]([H_2] + k_3[Cl_2])} \tag{22-4}$$

Notice that, in the limit of the complete absence of oxygen gas, the rate is proportional to the concentration of H_2 gas and not dependent on the concentration of Cl_2 gas at all! (The second term in the denominator of equation 22-4 is zero, and the remaining Cl_2 concentrations in the numerator and denominator cancel.) The reaction is complicated by side reactions that take place on the surfaces of the reacting vessels. The results obtained sometimes depend on the size and shape of the reaction container. All of this is a far cry from the simplicity of the HI system. There are side reactions in the HI system, too, but they are not important below 800 K.

Following the Course of a Reaction

How do we measure concentrations of reactants and products during a reaction to find rate equations such as we have been examining? If the total number of moles of gas changes during a gas reaction, the course of the reaction can be measured from the change in pressure at constant volume or the change in volume at constant pressure. These are examples of *physical* measurements that can be performed on the system while it is reacting. They have the advantage of not disturbing the reacting system, and usually they are rapid. With automatic recording devices, we can monitor a physical quantity continuously during the reaction.

Other physical measurements often used in kinetic studies include optical methods such as the rotation of light by a solution (useful if reactants and products have different abilities to rotate polarized light), changes in refractive index of a solution, color, and absorption spectra. Common electrical methods include the electrical conductivity of a solution (especially useful when ions are being produced or consumed), electrical potential in a cell, and mass spectrometry. Thermal conductivity, viscosity of a polymerizing solution, heats of reaction, and freezing points also have been employed. The disadvantage of all such methods is that they are indirect. The property observed must be calibrated in terms of concentrations of reactants and products. The calibration is subject to systematic errors, especially if side reactions are occurring.

Chemical methods are more straightforward and yield concentrations directly. With such methods, a small sample is extracted from the reacting mixture, and the reaction is halted by dilution or cooling long enough to measure concentrations. The serious disadvantages are that we are removing a part of the reacting system and thereby gradually changing it. Moreover, if the reaction cannot be stopped in the sample removed for analysis, then the analysis is that much less accurate. In the gas-phase reactions between H_2 and Cl_2, Br_2 or I_2, there is no change in the number of moles of gas before and after reaction, so pressure- or volume-change methods cannot be used. To study these reactions, samples are taken, and the gas mixtures are analyzed chemically for their compositions.

A First-Order Rate Equation and the Decay of ^{14}C

In a first-order process, the rate of disappearance of reactant is proportional to the amount of reactant present. Each reactant molecule has the same probability of breakdown in a given time interval, and the total rate of breakdown simply depends on how many molecules are present. The expression is

$$\frac{dn}{dt} = -kn$$

with n being the total number of molecules present. This rate expression can be integrated to yield the concentration as a function of time:

$$n = n_0 e^{-kt}$$

This rate equation is used in the example of dating with ^{14}C in Section 23-5, where the expressions in terms of the concentration of ^{14}C are

$$\frac{d[^{14}C]}{dt} = -k[^{14}C] \tag{22-5}$$

$$[^{14}C] = [^{14}C]_0 e^{-kt} \tag{22-6}$$

The integrated equation 22-6 is plotted in Figure 23-3. Taking the logarithm of both sides of equation 22-6 yields

$$\ln[^{14}C] = \ln[^{14}C]_0 - kt \tag{22-7}$$

This is the equation of Figure 23-8. The plot is a straight line, with a negative slope equal to the rate constant, k. The slope of the plot of equation 22-6 at any time t, as shown in Figure 23-3, is proportional to the concentration of ^{14}C remaining at that time. This, in words, is what equation 22-5 means.

Decomposition of N_2O_5

In Section 16-5, we encountered the decomposition of solid N_2O_5 as an example of a reaction that is spontaneous yet strongly endothermic. Now we shall look at the decomposition of N_2O_5 dissolved in carbon tetrachloride as an example of a first-order chemical reaction. Solid N_2O_5 and one product, NO_2, are soluble in CCl_4; the other product, O_2, is not. The reaction

$$N_2O_5 \rightarrow 2NO_2 + \tfrac{1}{2}O_2(g)$$

can be followed by measuring the total volume of oxygen gas that bubbles out of the solution.

The data for this reaction are given in Table 22-1, after O_2 volume measurements have been converted to concentrations of N_2O_5 left in the solution. These data are plotted in Figure 22-3 as an example of the way in which concentration data are treated. In this figure are the concentration of N_2O_5 at any time, the rate of change in this concentration, and this rate of change divided by the concentration. This last quantity is equal to the rate constant. That the rate of change divided by the concentration is constant (within the limits of experimental error in the data in Table 22-1) demonstrates that the reaction is indeed first order.

Stoichiometry and Rate Expressions

The reaction

$$2NO(g) + O_2(g) \rightarrow 2NO_2(g)$$

has an observed rate equation of the form

Table 22-1

Decomposition of N_2O_5 in CCl_4 Solution at $45°C^{a,b}$

Time, t (sec)	$[N_2O_5]$ (mole liter^{-1})	$N_2O_5 \rightarrow 2NO_2 + \frac{1}{2}O_2(g)$		$-\dfrac{\Delta[N_2O_5]}{\Delta t}$ (mole liter^{-1} sec^{-1})	$k = -\dfrac{1}{[N_2O_5]}\dfrac{\Delta[N_2O_5]}{\Delta t}$
		$\Delta[N_2O_5]$	Δt		
0	2.33				
		-0.25	184	1.36×10^{-3}	6.2×10^{-4}
184	2.08				
		-0.17	135	1.26×10^{-3}	6.3×10^{-4}
319	1.91				
		-0.24	207	1.16×10^{-3}	6.5×10^{-4}
526	1.67				
		-0.32	341	0.94×10^{-3}	6.2×10^{-4}
867	1.35				
		-0.24	331	0.72×10^{-3}	5.9×10^{-4}
1198	1.11				
		-0.39	679	0.57×10^{-3}	6.2×10^{-4}
1877	0.72				

[a]From H. Eyring and F. Daniels, *J. Am. Chem. Soc.* **52**, 1472 (1930).

[b]In the last column, $[N_2O_5]$ = average of concentrations at beginning and end of this time interval. For the first entry, $[N_2O_5] = (2.33 + 2.08)/2$.

$$\frac{d[NO_2]}{dt} = k[NO]^2[O_2]$$

The reaction is second order in NO and first order in O_2, and is third order overall. The rate equation happens to agree with the stoichiometry of the chemical reaction; this agreement suggests (but does not prove) that this may be a simple one-step reaction involving three molecules.

In contrast, ethanol and decaborane react in solution according to the equation

$$30C_2H_5OH + B_{10}H_{14} \rightarrow 10B(OC_2H_5)_3 + 22H_2$$

One might naively expect this to have a thirty-first order rate expression. In fact, the reaction is second order, first order in each of the two reactants. For the rate of disappearance of ethanol,

$$-\frac{d[C_2H_5OH]}{dt} = k[C_2H_5OH][B_{10}H_{14}]$$

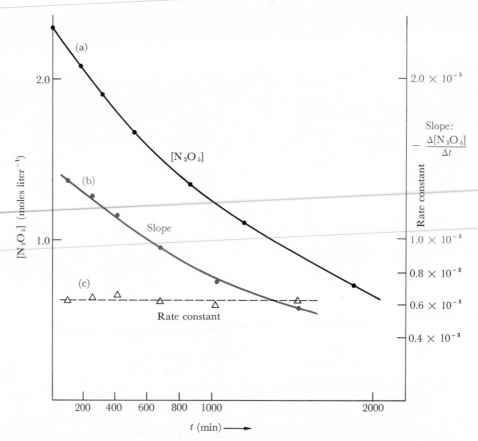

Figure 22-3　The kinetics of the reaction $N_2O_4 \rightleftarrows 2NO_2 + \frac{1}{2}O_2$ from the data in Table 22-1. (a) Plot of concentration of N_2O_5 as a function of time. (b) Plot of the negative of the slope of curve in (a) as a function of time, or of $-\Delta[N_2O_5]/\Delta t$ versus Δt. (c) Plot of the slope at any time divided by the concentration at that time, or of $-(\Delta([N_2O_5]/\Delta t)/[N_2O_5]$ $= k$, the first-order rate constant. The constancy of this number proves that the reaction follows a first-order rate law.

The Goals of Chemical Kinetics

Some chemical processes are simple one-step reactions involving one, two, or occasionally three molecules. Many more processes are the combination of several such simple reactions. One of the goals of chemical kinetics is to find out what the true molecular mechanism of a complex process is. Why do HI, HBr, and HCl have such different experimental rate equations for a reaction that looks superficially the same in all three cases? To a kineticist the question, "What is the mechanism of the reaction?" means this: "What is the sequence of simple reactions that produces the observed kinetics and stoichiometry of the overall reaction?" To this question organic and inorganic chemists have added, "What is the geometry of the reaction for each simple step in the overall process?" The goal of this inquiry is to predict why the simple reactions proceed as they do and to predict the rates at which they occur. The theories that have been developed to calculate the rate constants for simple unimolecular and bimolecular reactions are our next topic.

22-3 CALCULATING RATE CONSTANTS FROM MOLECULAR INFORMATION

Let us assume a simple bimolecular reaction,

$$A + B \rightarrow C + D \tag{22-8}$$

with a rate expression,

$$-\frac{d[A]}{dt} = k[A][B]$$

How far can we go in calculating k from the molecular properties of A, B, C, and D? One of the earliest observations was that k varies with temperature; the rate constant is larger, and the rate of reaction is faster, at higher temperatures.

Arrhenius' Activation Energy

If we plot the logarithm of the rate constant against the reciprocal of temperature, we usually obtain a straight line. Although Arrhenius was not the first person to do this, he developed the idea and gave it an explanation. Therefore, such a plot is called an Arrhenius plot. What does it mean in terms of reaction mechanisms?

Van't Hoff and others had been working, in the late 1800s, on the variation of free energy change of reaction and of the equilibrium constant with temperature. They discovered that the equilibrium constant, K_{eq}, varies with absolute temperature, T, and with the heat of reaction in the following way:

$$\frac{d \ln K_{eq}}{dT} = \frac{\Delta H^0}{RT^2} \tag{22-9}$$

This expression can be derived from the Gibbs–Helmholtz equation mentioned in connection with Figure 17-3, and ultimately can be derived rigorously from thermodynamics. During the same period, G. M. Guldberg and P. Waage found that they could derive the equilibrium constant from kinetic arguments. If the forward reaction in equation 22-8 has the rate

$$\text{Rate}_1 = k_1[\text{A}][\text{B}]$$

and the reverse reaction has the rate

$$\text{Rate}_2 = k_2[\text{C}][\text{D}]$$

then (they assumed) *equilibrium* is the state in which forward and reverse rates are equal, so no net change in the reacting system is occurring with time:

$$[\text{A}][\text{B}]k_1 = k_2[\text{C}][\text{D}]$$

$$K_{\text{eq}} = \frac{k_1}{k_2} = \frac{[\text{C}][\text{D}]}{[\text{A}][\text{B}]}$$

The equilibrium constant, in this argument, is the ratio of the rate constants for the forward and reverse reactions.

This is an erroneous derivation. It is valid only when the reaction is a simple one-step process in which the stoichiometry of the reaction is reflected in the coefficients of the concentration terms in the rate equation. Nevertheless, it is valid for the kind of reactions we are considering here: simple bimolecular reactions. If the equilibrium constant is the ratio of forward and reverse rate constants, equation 22-9 suggests that the enthalpy of reaction might be the difference between two energies, E_1 and E_2:

$$K_{\text{eq}} = \frac{k_1}{k_2} \qquad \Delta H^0 = E_1 - E_2$$

$$\frac{d \ln (k_1/k_2)}{dT} = \frac{E_1 - E_2}{RT^2}$$

$$\frac{d \ln k_1}{dT} = \frac{E_1}{RT^2}$$

$$\frac{d \ln k_2}{dT} = \frac{E_2}{RT^2}$$

Or, in general,

$$\frac{d \ln k}{dT} = \frac{E_a}{RT^2} \tag{22-10}$$

The quantity E_a is called the **Arrhenius energy of activation**. Equation 22-10 can be rearranged to yield

$$\frac{d \ln k}{d(1/T)} = -\frac{E_a}{R} \tag{22-11}$$

If the Arrhenius energy of activation is not a function of temperature, equation 22-11 predicts that a plot of $\ln k$ against the reciprocal of the absolute temperature will generate a straight line. This is true for many reactions, and the activation energy is one of the standard experimental parameters by which a chemical reaction is described. If E_a is not a function of temperature, equation 22-11 can be integrated to yield

$$k = Ze^{-(E_a/RT)} \qquad\qquad (22\text{-}12)$$

in which Z would be the rate constant if there were no activation energy required.

The activation energy is a barrier that the colliding molecules must surmount if they are to react rather than recoil from one another. We already have used this idea in the thioacetamide reaction of Section 22-1. We postulated that if the thioacetamide and water molecules do not collide head-on with sufficient energy the redistribution of bonds in Figure 22-1b and c will never occur. Water will recoil from the thioacetamide molecule and no reaction will take place. Now we have experimental evidence, in the form of the temperature dependence of k, that some such threshold energy *is* involved in chemical reactions. Arrhenius' explanation of activation energies assumes that every pair of molecules with energy less than E_a will not react, and every pair with energy greater than E_a will react. The theory is certainly too simple, but it is a beginning.

Nothing is changed in the thermodynamics of the overall reaction, as Figure 22-4 shows. The activation barriers to forward and reverse reactions, E_1 and E_2, are such that their difference, $\Delta H^0 = E_1 - E_2$, is the thermodynamic heat of reaction. The higher the barrier, E_1, the slower the forward reaction will be. However, since E_2 must rise by the same amount as E_1 if their difference is fixed, the reverse reaction is slowed by the same amount.

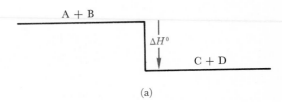

(a)

Figure 22-4   (a) The enthalpy or heat of a reaction is the change in enthalpy when reactants become products. (b) The activation energy for the reaction of A and B, E_1, is the energy necessary before A and B will react instead of rebound. The reverse reaction, in which C and D restore A and B, also has an activation energy, E_2. The difference between these two activation energies is the enthalpy of the reaction.

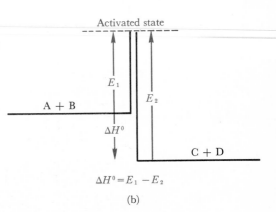

The point of equilibrium is affected not by the individual numerical values of the activation energies for forward and reverse reactions, but only by the difference between them, which is ΔH^0.

Collision Theory of Bimolecular Gas Reactions

The next logical step is to construct a collision theory for gas reactions. A reaction between two molecules occurs, in this theory, when the molecules collide with energy in excess of E_a. A theory could hardly be simpler. There are two questions to be answered before the rate constant can be calculated:

1. How often do two molecules collide per cubic centimeter of gas mixture?

2. In what fraction of these collisions does the combined energy of the two molecules exceed E_a?

The collision frequency can be calculated from the simple kinetic theory of gases with the methods that were introduced in Chapter 3. The frequency depends on the concentrations of the two reacting gases, and also on their molecular weights, the distance between the molecular centers on collision, and on the square root of the temperature. Since the molecules move more rapidly at higher temperatures, they collide more often. The fraction of pairs of molecules having energy equal to or greater than E_a upon collision (if we assume the type of Boltzmann distribution of molecular velocities that we saw in Figure 3-11) is

$$e^{-(E_a/RT)}$$

According to simple collision theory, the rate of reaction then is

Rate = collision frequency $\times$ probability that $E \geq E_a$

$$-\frac{d[A]}{dt} = (Z[A][B]) \times e^{-(E_a/RT)}$$

$$-\frac{d[A]}{dt} = Ze^{-(E_a/RT)}[A][B]$$

The rate of a reaction is greater at higher temperatures because collisions are more frequent and because the probability that a colliding pair will have an energy greater than E_a is also higher. The constant, Z, can be calculated from the molecular weights and the diameters of the reacting molecules by approximating them with spheres. The bimolecular rate constant, k, then is

$$k = Ze^{-(E_a/RT)} \tag{22-13}$$

This theory is tested in the data in Table 22-2. The Arrhenius activation energy is tabulated for six bimolecular gas reactions, along with the observed preexponential factor, Z, and its theoretical value as calculated

Table 22-2

Calculation of Rate Constants for Bimolecular Reactions

| | | A + B $\rightarrow$ products $\text{Rate} = Ze^{-(E_a/RT)}[A][B]$ | | | |
| | | | $\log_{10} Z$ | | |
Reaction	E_a (kJ)[a]	Observed	Collision theory	Absolute rate theory	ΔH° (kJ)[a]
$NO + O_3 \rightarrow NO_2 + O_2$	10.5	11.9	13.7	11.6	-200
$NO + Cl_2 \rightarrow NOCl + Cl$	84.9	12.6	14.0	12.1	$+83.7$
$NO_2 + CO \rightarrow NO + CO_2$	132	13.1	13.6	12.8	-226
$2NO_2 \rightarrow 2NO + O_2$	111	12.3	13.6	12.7	$+113$
$2NOCl \rightarrow 2NO + Cl_2$	103	13.0	13.8	11.6	$+75.7$
$2ClO \rightarrow Cl_2 + O_2$	0.0	10.8	13.4	10.0	-138

[a]Per mole of reactants.

from the collision theory and the absolute rate theory that we will discuss in the next sections. Keep in mind that these are the *logarithms* of Z that are tabulated, so a disagreement between theory and experiment of 1.0 means an error by a factor of 10 in the rate constant. The agreement is generally encouraging for so simple a theory that has no assumptions other than those of the kinetic theory of gases. There are discrepancies; for example, the ClO reaction rate is incorrect by a factor of 400. When discrepancies occur, the absolute rate theory usually does a better job of predicting Z than the collision theory does.

In the right column of Table 22-2 are the standard enthalpies or heats of reaction. The relative enthalpy of reactants and products, and the activation barrier between them, are plotted for these reactions in Figure 22-5. Some reactions, such as $NO_2 + CO$, must surmount a considerable activation barrier. For other reactions, the barrier is nonexistent, as with 2ClO. For others such as $2NO_2$, the barrier to reaction is only the heat of reaction itself, and the reverse reaction has a zero activation energy. The most general case is diagrammed at the bottom of Figure 22-5.

Activated Complexes

Before we consider the absolute rate theory, we must look more closely at the state of the reactant molecules as they cross the activation barrier. In the reaction

$$2ClO \rightarrow Cl_2 + O_2$$

the Cl and O atoms are bonded at the start, and the two ClO molecules are too far apart to exert any influence on one another. At the end of the

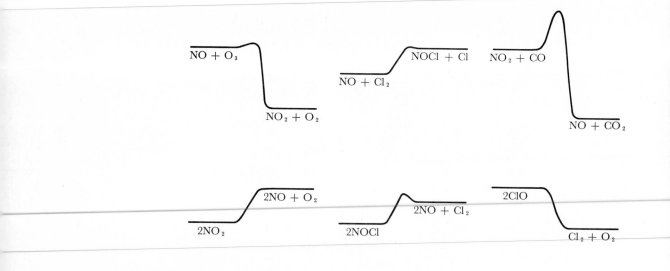

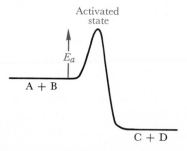

Figure 22-5   Activation-energy barriers for the six reactions tabulated in Table 22-2. Some of these reactions have appreciable barriers; others such as the 2ClO reaction have none at all. The general activation-energy diagram is shown at the bottom.

reaction, the Cl atoms are 1.99 Å apart in a Cl_2 molecule, the O atoms are 1.21 Å apart in an O_2 molecule, and these two molecules are far apart. What is the intermediate, activated state?

The activated complex is diagrammed in Figure 22-6b. All four atoms are an unspecified distance apart, somewhat farther away than if they were bonding in the stable sense of the term. We can be sure that the activated complex is *not* one in which all four atoms are so far apart that they exert no influence on one another; some sort of a loose complex must exist. The basis for this assertion is a knowledge of the bond energies of the three molecules, and that the activation energy for the 2ClO reaction is zero. The bond energies, or the energies required to separate completely the atoms in a diatomic molecule, are

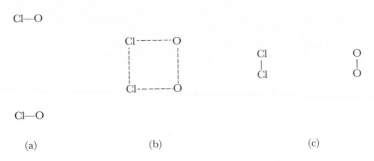

Figure 22-6   (a) The initial state in the reaction $2ClO \rightarrow Cl_2 + O_2$, with two ClO molecules infinitely far apart. (b) A possible transition state. (c) The final state with Cl_2 and O_2 molecules infinitely far apart.

Molecule	Bond energies
ClO	270 kJ mole^{-1}
O$_2$	494 kJ mole^{-1}
Cl$_2$	239 kJ mole^{-1}

If during the reaction the two ClO molecules were first pulled apart, and then the isolated atoms were combined into Cl_2 and O_2, the activation energy for this reaction would be twice the bond energy of ClO, or 540 kJ per 2 moles of ClO. Instead, the activation energy is zero. The activated complex must be a combination of the four atoms such that whatever instability created as Cl and O separate is immediately compensated by the stabilizing influence of associations between Cl and Cl and between O and O.

We can think of the activated complex as an unstable "molecule," with many of the properties of a molecule, except that it decomposes spontaneously either to reactants or to products. The thioacetamide and water molecules in Figure 22-1b are in an activated complex, and the energy of this complex is greater than either that of thioacetamide and water or that of acetamide and hydrogen sulfide.

Potential Energy Surfaces

The 2ClO reaction suggests that in principle we should be able to calculate the total potential energy of a collection of atoms as a function of their positions in space. This calculation would produce a potential energy surface with hills and plateaus of high energy, and valleys of low energy. Any region of a minimum of potential energy in this plot will represent a stable molecule. Even with four atoms as in the 2ClO reaction we would need, unfortunately, six variables to describe the arrangement of atoms: the bond lengths from each atom to the other three, for example. Our potential energy plot would have to be in seven-dimensional space. This is difficult to visualize and impossible to construct.

We need an example with only two variables so the map can be plotted in three-dimensional space. One of the first maps to be calculated, by Henry Eyring in 1935, is the potential energy surface for the reaction

$$H + H_2 \rightleftarrows H_2 + H$$

in which all three atoms are constrained to lie on a straight line. The only variables are the distances from the central hydrogen atom to the other two, r_1 and r_2:

H—H—H

$\quad r_1 \quad r_2$

The potential energy of the three-atom system as a function of r_1 and r_2 is shown in Figure 22-7a. The actual arrangements of the three atoms at the six numbered points marked in color are drawn in Figure 22-8. Sections through this potential energy surface at fixed values of r_1 are shown in Figure 22-7b.

If either r_1 or r_2 is large, the three hydrogen atoms exist as an H_2 molecule and an isolated H atom. The potential energy section at constant r_1, for r_1 greater than 3 Å in Figure 22-7b, is the same as that for an isolated H_2 molecule in Figure 12-2. As an atom approaches an H_2 molecule from the right (points 1 and 2 of Figures 22-7 and 22-8), the first noticeable effect is an increase in the potential energy of the system of three atoms. The

Figure 22-7

The potential energy of three hydrogen atoms in a straight line, plotted in kilojoules as a ▶ function of the separation of the two outside hydrogen atoms from the central one, r_1 and r_2. Contours of equal potential energy in (a) are numbered in kilojoules. The shape of the potential surface is that of two deep valleys parallel to the r_1 and r_2 axes, with sheer walls rising to these axes and with less steep walls rising to a plateau at the upper right corner of (a). The two valleys are connected by a path over a pass or saddle, with the crest of the pass at $r_1 = 0.8$ Å $= r_2$. The calculations for the three-atom system were carried out by semiempirical methods, in 1935, by Henry Eyring and his co-workers. Recently, more exact quantum mechanical calculations have indicated that the depression at the summit of the pass may not be real. Several sections through this potential surface are shown in (b) for different values of r_1. At sufficiently large values of r_1 (over 3 Å), the potential energy curve is that of a diatomic H_2 molecule virtually unaffected by the third H atom. Compare the potential energy curve for $r_1 > 3$ Å with Figure 12-2. The axes in this plot have been skewed so a marble rolling down a model of this surface would represent accurately the vibrations of the three-atom system. Points 1 through 6, marked in color, correspond to the atomic arrangements shown in Figure 22-8).

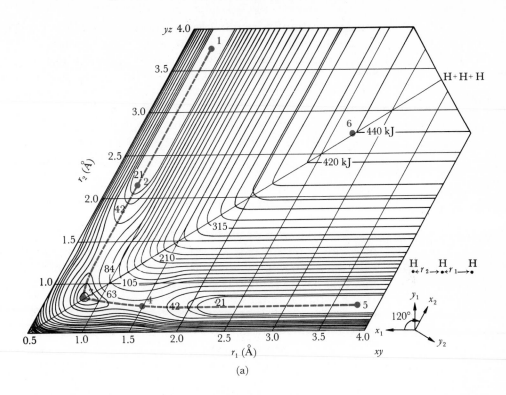

(a)

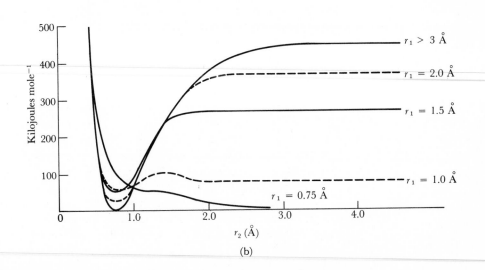

(b)

incoming atom is repelled by the molecule, and a more stable situation results if the atom rebounds and moves away again. If the atom has enough kinetic energy to keep approaching the H_2 molecule, it will begin to weaken the H—H bond in the H_2 molecule. At point 3, both outer atoms are slightly farther from the central one than a normal H—H bond length, but the potential energy of the system of atoms is 25 kJ mole^{-1} higher than that of isolated H_2 and H. Point 3 is the activated complex for the reaction.

The activated complex can decompose either to products or to reactants. There is no reason why the three atoms in the state of point 3 cannot return to point 1 as well as proceed to point 5. What is certain is that the activated complex is unstable and must decompose. Points 1 through 5 in Figure 22-7 are connected by a colored dashed line called a **reaction pathway**. If we plot potential energy along this pathway, an activation energy barrier curve such as those in Figure 22-5 results. Notice that the reaction pathway at all times is a path along a valley between steep walls. It may take 25 kJ of energy to build the activated complex of point 3, but it takes over 400 kJ to separate the atoms as at point 6.

Similar potential energy surfaces have been calculated for other systems of atoms; the one for $H_2 + Br \rightleftarrows H + HBr$ is shown in Figure 22-9. Now the shape of the surface is altered because H_2 is more stable than HBr. As the Br atom approaches H_2, it pushes the H atoms apart. The activated complex (point 2) has the two H atoms twice as far apart as in

Figure 22-8

Relative positions of the three hydrogen atoms for the six points indicated in color in Figure 22-7a. Points 1 through 5 represent stages in the reaction of a hydrogen molecule with a hydrogen atom. Point 6 represents the three atoms 2.5 Å apart along a straight line. This state is about 390 kJ less stable than any stage along the reaction pathway. The potential energies in kilojoules are given at the right.

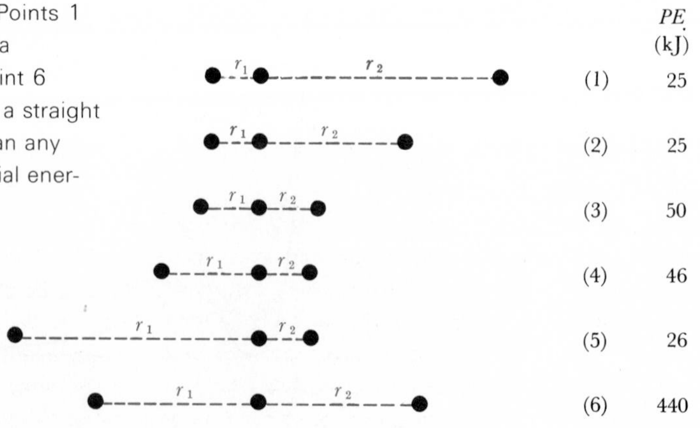

		PE (kJ)
	(1)	25
	(2)	25
	(3)	50
	(4)	46
	(5)	26
	(6)	440

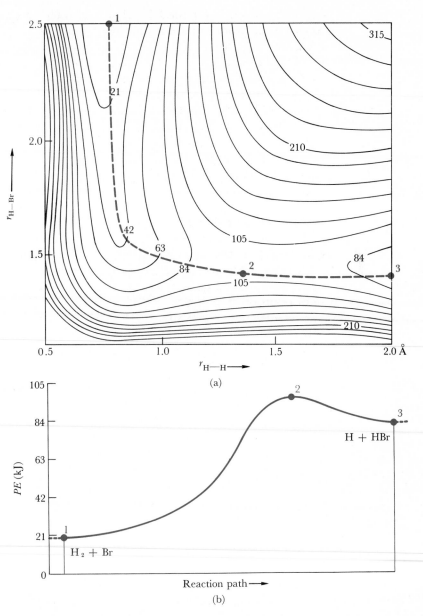

(a)

(b)

Figure 22-9 (a) Potential energy surface for the linear array of H — — — — —H— — — — —Br. Contours of equal potential energy are marked in kilojoules per mole of three-atom sets. The reaction pathway for the reaction $H_2 + Br \rightleftharpoons H + HBr$ is marked by a dashed colored line. (b) A potential energy profile of the reaction pathway. Points 1, 2, and 3 correspond to the pathways on the potential energy surface. Point 2 is the activated complex for the reaction. You can verify from the data in Appendix 3 on standard heats of formation of atoms and molecules that the difference in potential energy at points 1 and 3 is approximately correct.

the H_2 molecule, but has H and Br at nearly their final bond distance in HBr. The activated complex is almost the same as an HBr molecule, and indeed it is only about 15 kJ mole^{-1} less stable than HBr. Calculated potential energy surfaces such as Figure 22-9 are the basis for the common drawings of reaction barriers such as Figure 22-5. A profile of potential energy such as Figure 22-9b is still useful even with reactions of molecules so complicated that we cannot calculate or even plot their complete multi-dimensional potential energy surface.

Absolute Rate Theory

In the absolute rate theory, reaction takes place when an activated complex breaks down into products. Therefore, the rate of reaction is the product of three factors:

1. The concentration of activated complexes per cubic centimeter
2. The rate of breakdown of individual complexes or their rate of passage over the activation energy barrier
3. The probability that a breakdown will form products and not reactants again

Since the activated complex represents an unstable state of transition between reactants and products, it is often called a **transition state**. The absolute rate theory is also called the **transition-state theory**. We shall use these terms interchangeably.

The transition-state theory assumes an equilibrium between reactants and the activated complex, usually represented by a superscript double dagger:

$$A + B \rightleftarrows AB^{\ddagger}$$
$$K^{\ddagger} = \frac{[AB^{\ddagger}]}{[A][B]}$$

Hence, the concentration of the activated complex is given by

$$[AB^{\ddagger}] = K^{\ddagger}[A][B]$$

The rate of decomposition is more complicated to calculate, but it turns out to be a universal constant for all bimolecular reactions at a given temperature:

$$\text{Rate of decomposition} = \frac{kT}{h}$$

in which k = Boltzmann's constant, and h = Planck's constant. The probability that a breakdown will be to products and not to reactants is the transmission coefficient, κ. It can be estimated only as having a value between 0.5 and 1.0 in most reactions. Therefore, the overall rate of reaction is

$$-\frac{d[A]}{dt} = \kappa\frac{kT}{h}K^{\ddagger}[A][B]$$

and the rate constant, k_2, is

$$k_2 = \kappa\frac{kT}{h}K^{\ddagger} \tag{22-14}$$

(The symbol k_2 is used here instead of k to represent a bimolecular rate constant to avoid confusion with Boltzmann's constant, k.)

It is possible to calculate the equilibrium constant, $K^{\ddagger}$, between reactants and the activated complex from molecular properties by using statistical mechanics. We shall not even attempt this calculation here; instead we shall look at the thermodynamic interpretation of this rate-constant expression. The equilibrium constant is related to the standard free energy of formation of the activated complex from reactants, and this in turn is related to the standard enthalpy and entropy of the formation of the activated complex:

$$-RT\ln K^{\ddagger} = \Delta G^{0\ddagger} = \Delta H^{0\ddagger} - T\,\Delta S^{0\ddagger} \tag{22-15}$$

Thus, the bimolecular rate constant, k_2, can be written

$$k_2 = \kappa\frac{kT}{h}e^{-(\Delta G^{0\ddagger}/RT)} = \kappa\frac{kT}{h}e^{+(\Delta S^{0\ddagger}/R)}e^{-(\Delta H^{0\ddagger}/RT)} \tag{22-16}$$

The enthalpy of activation, $\Delta H^{0\ddagger}$, is nearly the same quantity as the activation energy, E_a. The difference is irrelevant is this discussion. Equation 22-16 indicates that the rate of reaction is slower if the activation energy is large. This result was already obtained in the collision theory; if the activation energy is large, only a small fraction of the molecules will have enough energy to surmount the barrier and to react rather than rebound upon collision. Equation 22-16 also suggests that the reaction rate is faster if the *entropy* of activation is large. If the activated complex is much more disordered than the reactants, the reaction is enhanced because the equilibrium constant for formation of the complex is large, and more of the complex is present. In contrast, if the reactants are severely constrained when they combine to make the activated complex, then the reaction is inhibited. We might guess that the entropy of activation in the thioacetamide-plus-water reaction is negative since the two molecules combine to form one unit in the complex. Both molecules are limited severely in their initial orientations if they are to build the activated complex in Figure 22-1b successfully. The entropy of activation in bimolecular reactions is almost always large and negative because the two reactants lose entropy when they combine in the complex. Often the most useful application of absolute rate theory is not to calculate the rate constant directly, but to use the observed rate constant and equation 22-16 to calculate the entropy of activation. The entropy of activation provides information about the structure of the activated complex. For example, if the calculated entropy of activation is positive, then any mechanism that leads through a tightly organized activated complex must be rejected.

As an example, in the next section we will consider two reactions of the type

$$R_3C—Br + OH^- \rightarrow R_3C—OH + Br^-$$

which proceed by different mechanisms, depending on the nature of the R groups. In one mechanism, the Br^- is driven away as OH^- approaches, in the same fashion as the thioacetamide reaction. The activated complex is then a combination of $R_3C—Br$ and OH^-:

$$R_3C—Br + OH^- \rightarrow \begin{bmatrix} & R \quad R & \\ Br\cdots C\cdots OH \\ & | & \\ & R & \end{bmatrix}^- \rightarrow Br^- + R_3C—OH \qquad (22\text{-}17)$$

This is called an associative or S_N2 reaction, meaning that it is a *substitution* of one group for another, that the groups are *nucleophilic* (donating electrons and attracting nuclei: Lewis bases, in fact), and that *two* molecules are involved in the reaction. The other mechanism is for the $R_3C—Br$ molecule to dissociate spontaneously into Br^- and what is known as a *carbonium ion*, R_3C^+, and for the OH^- ions to react rapidly in a separate step with any free carbonium ions. The activated complex or transition state in this mechanism will be the reactant $R_3C—Br$ just before dissociation:

$$R_3C—Br \rightarrow [R_3C—Br]^{\ddagger} \rightarrow R_3C^+ + Br^-$$
$$R_3C^+ + OH^- \rightarrow R_3C—OH \qquad (22\text{-}18)$$

This is called a dissociative or S_N1 mechanism since it is a nucleophilic substitution in which the slowest step involves dissociation of a single molecule. One should be able to distinguish between these two mechanisms by their entropies of activation, calculated from equation 22-16, and the measured rate constants. The S_N2 mechanism will have a large negative entropy of activation since the activated complex is formed by combining two molecules. In contrast, the S_N1 mechanism will have virtually a zero entropy of activation because the activated complex differs only slightly from the reactant molecule.

Comparison of Theories

Both the collision theory and the absolute rate theory build upon the idea of an energy of activation that acts as a barrier to reaction. Both, to this extent, are based on the older Arrhenius explanation of the variation of rate constants with temperature. The collision theory focuses on the collision of two reactant molecules, whereas the absolute rate theory deals more with the complex formed after collision and assumes an equilibrium between this complex and the reactants. The collision theory uses the activation energy concept by stating that all molecular pairs that do not have this energy on collision will rebound instead of react. The absolute rate theory postulates instead that a high enthalpy of activation to the complex means that the equilibrium constant and hence the concentration of complexes will be small. If you think of the complex as that which is formed when two molecules have the energy demanded by the collision theory, then the two theories are seen for what they really are: different viewpoints of the same phenomenon.

As was mentioned earlier, the equilibrium constant for activated complex formation, $K^{\ddagger}$, can be calculated from the properties of the reactant molecules and the presumed properties of the complex. This means that the rate constant, k (or k_2), can be calculated from first principles just as it can be in the collision theory. The calculated values from these two theories are compared with observed values in Table 22-2. As you can see, the absolute rate theory is usually a little better in predicting rate constants than is the collision theory. The collision theory is not wrong; it is just too simple a picture of what happens when chemical reactions occur.

22-4 COMPLEX REACTIONS

Most chemical reactions are not simple unimolecular or bimolecular reactions, but combinations of these. This is why such complicated rate equations as equations 22-3 or 22-4 arise. Even the hydrogen–iodine reaction, which has been used for over half a century as the classic example of a simple bimolecular reaction (equation 22-2), is complex.

The Hydrogen–Iodine Reaction

For the reaction

$$H_2 + I_2 \rightarrow 2HI \tag{22-19}$$

the observed rate equation is

$$-\frac{d[H_2]}{dt} = k[H_2][I_2] \tag{22-20}$$

Above 800 K, side reactions with different mechanisms occur, yet these reactions can be neglected at moderate temperatures. Both N. N. Semenov and Henry Eyring have suggested that the true mechanism might be not that of equation 22-19, but a two-step mechanism involving the reversible dissociation of I_2 to $2I$, followed by the trimolecular reaction of I and H_2:

$$I_2 \rightleftarrows 2I$$
$$H_2 + 2I \rightarrow 2HI \tag{22-21}$$

The rate expression for the reaction of one H_2 molecule with two I atoms is

$$-\frac{d[H_2]}{dt} = k'[H_2][I]^2 \tag{22-22}$$

If the dissociation of I_2 is reversible and at equilibrium, we can write an equilibrium-constant expression:

$$K = \frac{[I]^2}{[I_2]} \qquad [I]^2 = K[I_2] \tag{22-23}$$

However, substituting for the concentration of I atoms in equation 22-22 by using the equilibrium of equation 22-23,

$$-\frac{d[H_2]}{dt} = k'K[H_2][I_2] \tag{22-24}$$

produces the *same* rate expression as if the mechanism were really one of bimolecular collision. We thus have two different mechanisms with the same rate expression. How can we choose between them?

The two mechanisms have the same rate equation *so long as* the dissociation of I_2 is at thermal equilibrium, and the amount of I atoms present is given by the thermal equilibrium constant of equation 22-23. At higher temperatures, more I_2 dissociates, thereby producing the same effect that would have resulted from the greater bimolecular rate constant in the bimolecular mechanism. J. H. Sullivan decided to test the two theories by making the concentration of iodine atoms different from what it normally is in thermal dissociation of I_2. He did this by dissociating I_2 with 578-nm light from a mercury vapor lamp. This light should have little effect if the reaction is bimolecular, aside from a slight decrease in I_2 concentration. Conversely, if the trimolecular reaction with I atoms is correct, the rate of reaction should increase with the intensity of irradiating light since more I atoms are being produced.

Sullivan calculated the concentration of I atoms present at several intensities of irradiating light and found that the rate of appearance of HI is proportional to the *square* of the I atom concentration. Therefore, the mechanism of equation 22-21 is the correct one. The classical $H_2 + I_2$ reaction is a trimolecular reaction masquerading as a simpler bimolecular reaction because of the thermal equilibrium that normally exists between I_2 and 2I. (At least, until someone even more ingenious designs an experiment that proves that it is a more complicated reaction masquerading as trimolecular. As Sullivan points out [*J. Chem. Phys.* **46**, 73 (1967)], the trimolecular reaction

$$H_2 + 2I \rightarrow 2HI$$

can be replaced by two bimolecular steps:

$$H_2 + I \rightleftarrows H_2I$$
$$H_2I + I \rightarrow 2HI$$

If the first of these is fast and reversible, so that reactants and products are in equilibrium, then the rate expression is the same as for the trimolecular process, and the two mechanisms cannot be distinguished by reaction rates.

This example makes a point that must be kept in mind at all times. We can never prove that a proposed mechanism is right; we can only prove that it has not yet been shown to be wrong. There is always the chance that a more subtle experiment, such as Sullivan's upsetting of thermal equilibrium with light, may uncover the weaknesses in an accepted mechanism. When two theories are presented, the temptation and usually the wiser choice is to opt for the simpler one, at least until the data compel you to do otherwise. But you should always be prepared to change your mind when new data demand it.

Rates and Mechanisms of Substitution Reactions

The reaction of *tert*-butyl bromide with OH^-,

$$(CH_3)_3CBr + OH^- \rightleftarrows (CH_3)_3COH + Br^- \qquad (22\text{-}25)$$

has the experimental rate expression

$$-\frac{d[(CH_3)_3CBr]}{dt} = k[(CH_3)_3CBr] \qquad (22\text{-}26)$$

The rate does not appear to depend on the OH^- concentration at all. In contrast, the similar reaction with a less highly substituted carbon atom in ethyl bromide,

$$CH_3CH_2Br + OH^- \rightleftarrows CH_3CH_2OH + Br^- \qquad (22\text{-}27)$$

has the rate expression that we might expect from the chemical equation:

$$-\frac{d(CH_3CH_2Br)}{dt} = k[CH_3CH_2Br][OH^-] \qquad (22\text{-}28)$$

Why should these two similar reactions proceed by different mechanisms and have different rate equations? And how is it that the rate in equation 22-26 can be independent of concentration of one of the reactants?

Reaction 22-25 takes place by the S_N1 mechanism of equation 22-18. The *tert*-butyl bromide first dissociates in a slow reaction, and the carbonium ion that is formed reacts immediately with OH^-. Whenever a process takes place by a series of rapid steps with one quite slow step, the overall rate of reaction will be controlled by the slow step. The rate in this S_N1 reaction depends entirely on how fast the molecules of $(CH_3)_3CBr$ decompose. The capacity of reacting with carbonium ions by OH^- presumably far exceeds the amount of carbonium ions supplied by the dissociation. The total amount of OH^- present is unimportant.

Why do the reactions go with different mechanisms? The S_N2 scheme is possible for ethyl bromide because there is room for three substituents of the C atom (CH_3 and two H), plus OH^- and Br^-. The activated complex

$$\left[\begin{array}{c} H \qquad H^- \\ \diagdown \quad \diagup \\ Br \cdots C \cdots OH \\ | \\ CH_3 \end{array} \right]^{\ddagger}$$

is sterically possible. In contrast, in *tert*-butyl bromide the groups attached to the carbon atom (three CH_3) are large enough that OH^- and Br^- cannot bind at the same time. The activated complex of the S_N2 reaction is impossible. No reaction can occur until a molecule of *tert*-butyl bromide spontaneously dissociates. The dissociated carbonium ion is then subject to attack, either by Br^- to form the reactants again, or by OH^- to form the product. If Br^- is present only as a result of previous reaction of *tert*-butyl bromide, its concentration is probably much smaller than that of OH^-, and most of the carbonium ions will be converted to *tert*-butyl alcohol, $(CH_3)_3COH$.

In general, a rate expression that disagrees with the stoichiometry of the overall reaction is an indication that the reaction is proceeding by a series of steps. The problem, then, often is to find a set of steps including a slow step that accounts for the observed rate law.

The difference in reaction mechanisms encountered in *tert*-butyl bromide and ethyl bromide also is found in octahedral and square planar complexes of transition metals. Square planar complexes of Pt(II) and other metals can react with new ligands by associative (S_N2) mechanisms because the metal atom is accessible from either side of the plane. The S_N2 mechanism of the reaction

$$Pt(NH_3)_3Cl^+ + Br^- \rightarrow Pt(NH_3)_3Br^+ + Cl^-$$

can be written

The activated complex is five-coordinated platinum, which breaks down rapidly to products. The rate of the overall reaction depends on the rate of formation of the activated complex. This rate is influenced strongly by the nature of the entering group (Br^- in this example). Ligands capable of forming strong bonds with the central atom are the best entering groups; that is, they displace the leaving group (Cl^- in this example) most rapidly. The ions CN^- and I^- are good entering groups for Pt(II) complexes, whereas NH_3 and H_2O are relatively poor.

It is much more difficult for the six-coordinated octahedral complexes to react by an S_N2 mechanism because six ligands around a central metal, such as Co(III), leave little or no room for the attachment of an entering group in a transition state. Studies of substitution reactions of octahedral Co(III) complexes have established that the important or rate-determining step involves the dissociation of the bond between the Co(III) and the leaving group. The entering group is not involved in this initial dissociation step. For example, in aqueous solution, H_2O displaces Cl^- in the complex $Co(NH_3)_5Cl^{2+}$, thereby producing $Co(NH_3)_5H_2O^{3+}$. The mechanism most consistent with rate studies of this and similar reactions is the dissociative or S_N1 mechanism, which can be written

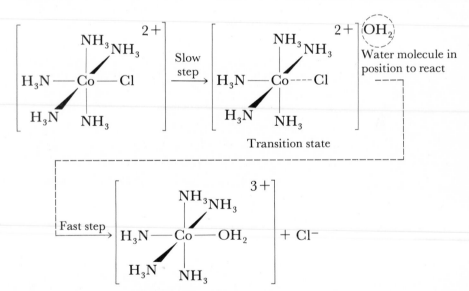

For such a mechanism the entering group plays no significant role in the creation of the transition state and hence in the rate of the overall reaction. A characteristic of most octahedral substitution reactions is the *lack* of influence of entering groups on the rate of reaction.

Chain Reactions

The reaction $H_2 + Br_2 \rightleftarrows 2HBr$ has the strange rate equation that we have already seen,

$$\frac{d[HBr]}{dt} = \frac{k[H_2][Br_2]^{1/2}}{1 + k'([HBr]/[Br_2])} \qquad (22\text{-}3)$$

For 13 years after this rate law was discovered, no one could account for it. Then, three groups did so almost simultaneously, those of Henry Eyring, K. F. Herzfeld, and Michael Polanyi. They proposed that the reaction proceeds by a chain mechanism involving two *chain-propagating steps:*

(1) $H_2 + Br \xrightarrow{k_1} H + HBr$

(2) $H + Br_2 \xrightarrow{k_2} HBr + Br$

When a molecule breaks apart into uncharged fragments having unpaired electrons, the fragments are called **radicals**. The unpaired electrons (e.g., in H and Br) make the fragments chemically reactive. The atomic product of each of these steps is one reactant for the other step, and they both produce HBr. Thus, HBr results, not from a bimolecular collision, but from an endless chain of reactions 1 and 2. The first of these two steps is the reaction of Figure 22-9. But where do these atoms of Br and H come from? The Br atoms are postulated to come initially from a **chain-initiating step**:

(3) $Br_2 \xrightarrow{k_3} Br + Br$

Why is the dissociation of H_2 not included also? The real reason is that the explanation of equation 22-3 does not require it, and that if we add it, we obtain the wrong rate expression. We can justify this omission in another way: The dissociation energy of H_2 is 432 kJ mole^{-1}, whereas that of Br_2 is only 190 kJ mole^{-1}.

A large concentration of HBr inhibits the reaction, as we can see from the HBr term in the denominator of equation 22-3. Moreover, a large concentration of Br_2 counteracts this inhibition. So HBr and Br_2 are evidently competing for the same chemical substance. What might that substance be?

The most likely candidate is hydrogen atoms; the inhibiting reaction would be

(4) $H + HBr \xrightarrow{k_4} H_2 + Br$

This is a *chain-inhibiting reaction* and is counteracted if an excess of Br_2 makes reaction 2 go rapidly, as the rate equation 22-3 predicts. Finally, the chain is *terminated* by the recombination of Br:

(5) $Br + Br \xrightarrow{k_5} Br_2$

How do we obtain equation 22-3 from these five reactions? If we can do so, this will be a strong argument for the correctness of the chain mechanism, although *not* an absolute proof, as we have seen with HI.

The rate of appearance of HBr is given by

$$\frac{d[\text{HBr}]}{dt} = +k_1[\text{H}_2][\text{Br}] + k_2[\text{H}][\text{Br}_2] - k_4[\text{H}][\text{HBr}] \qquad (22\text{-}29)$$

since HBr appears as a result of reactions 1 and 2, and disappears in reaction 4. The rates of production of H and Br atoms are given by

$$\frac{d[\text{H}]}{dt} = k_1[\text{H}_2][\text{Br}] - k_2[\text{H}][\text{Br}_2] - k_4[\text{H}][\text{HBr}] \qquad (22\text{-}30)$$

$$\frac{d[\text{Br}]}{dt} = -k_1[\text{H}_2][\text{Br}] + k_2[\text{H}][\text{Br}_2] + k_4[\text{H}][\text{HBr}]$$
$$+ 2k_3[\text{Br}_2] - 2k_5[\text{Br}]^2 \qquad (22\text{-}31)$$

The coefficients of 2 in front of k_3 and k_5 arise because each unit of reaction 3 produces two Br atoms, and each unit of reaction 5 removes two Br atoms.

At this point an essential simplification must be made. The actual amount of H and Br atoms present at any time must be small because they are consumed almost at the same rate that they are produced. Soon after the reaction begins, the concentrations of H and Br will reach a *steady state* and will remain constant so long as the reaction continues with a plentiful supply of reactants. Then each of the rate equations in 22-30 and 22-31 can be set equal to zero:

$$0 = k_1[\text{H}_2][\text{Br}] - k_2[\text{H}][\text{Br}_2] - k_4[\text{H}][\text{HBr}] \qquad (22\text{-}32)$$
$$0 = -k_1[\text{H}_2][\text{Br}] + k_2[\text{H}][\text{Br}_2] + k_4[\text{H}][\text{HBr}] + 2k_3[\text{Br}_2] - 2k_5[\text{Br}]^2 \qquad (22\text{-}33)$$

Adding these two equations yields

$$2k_5[\text{Br}]^2 = 2k_3[\text{Br}_2]$$
$$[\text{Br}] = \left(\frac{k_3}{k_5}\right)^{1/2}[\text{Br}_2]^{1/2} \qquad (22\text{-}34)$$

This calculation gives us a steady-state concentration for Br atoms in terms of the concentration of Br_2 molecules.

The HBr rate equation can be rewritten as

$$\frac{d[\text{HBr}]}{dt} = k_1[\text{H}_2][\text{Br}] + \{k_2[\text{Br}_2] - k_4[\text{HBr}]\}[\text{H}] \qquad (22\text{-}35)$$

We can eliminate the H concentration by expressing it in terms of Br concentration from equation 22-32:

$$[\text{H}] = \left(\frac{k_1[\text{H}_2]}{k_2[\text{Br}_2] + k_4[\text{HBr}]}\right)[\text{Br}] \qquad (22\text{-}36)$$

Substituting equation 22-36 into 22-35, placing everything over a common denominator, and canceling terms yields

$$\frac{d[\text{HBr}]}{dt} = \frac{2k_1 k_2 [\text{H}_2][\text{Br}_2][\text{Br}]}{k_2[\text{Br}_2] + k_4[\text{HBr}]} \tag{22-37}$$

Dividing top and bottom by $[\text{Br}_2]$ and then eliminating $[\text{Br}]$ with equation 22-34 yields

$$\frac{d[\text{HBr}]}{dt} = \frac{2k_1(k_3/k_5)^{1/2}[\text{H}_2][\text{Br}_2]^{1/2}}{1 + (k_4/k_2)\dfrac{[\text{HBr}]}{[\text{Br}_2]}} \tag{22-38}$$

This is exactly the experimental rate law, in which the experimental rate constants are related to those for the individual reactions in the chain by

$$k = 2k_1 \left(\frac{k_3}{k_5}\right)^{1/2}$$

$$k' = \frac{k_4}{k_2}$$

Now that we know what these two experimental constants mean in terms of the individual reactions, we can give a much fuller interpretation to the rate law, equation 22-38. Suppose that we could vary the individual rate constants, k_1 to k_5, at will. What effects would these changes have on the overall rate? The overall rate of production of HBr is accelerated if rate constants k_1, k_2, and k_3 are large, or if reactions 1, 2, and 3 are fast. The first two of these reactions produce HBr; the third prepares the way by making more Br atoms. The production of HBr is slowed if k_4 and k_5 are large, or if the chain-inhibiting and chain-terminating reactions are fast. So long as k_3 and k_5 change together, there is no change in the overall rate of reaction. Reactions 3 and 5 are the opposing initiating and terminating steps. Similarly, so long as k_2 and k_4 change together, the rate is unaffected. This, too, is sensible; for reactions 2 and 4 are similar in that they both consume an H and produce a Br, but differ in producing HBr in reaction 2 and removing it in reaction 4. Inhibition by HBr occurs because reaction 4 is enhanced, and inhibition is lessened by Br_2 because reaction 2 is enhanced.

22-5 CATALYSIS

A mixture of hydrogen and oxygen gas can be kept for years without appreciable reaction to produce water. But if a small amount of platinum black is introduced, the mixture explodes. The Pt is a catalyst for the reaction. As we learned earlier, a catalyst is a substance that accelerates the attainment of thermodynamic equilibrium without itself being consumed in the process. It does this by providing an alternative mechanism or pathway for the reaction, with a lower activation energy. If the activation energy for the forward reaction (E_1 in Figure 22-4) is lowered, that of the reverse reaction (E_2) must be lowered by the same amount if the heat of reaction is to be unchanged. A catalyst accelerates both forward and reverse reactions. It does not change the conditions of equilibrium in a reaction; it changes only the speed of getting there. The Pt catalyst dissociates H_2 gas into hydrogen atoms on the metal surface. These H atoms then react far more rapidly with O_2 molecules that meet them at the metal surface than H_2 molecules do with O_2 in the gas phase.

This is an example of **heterogeneous catalysis**, involving a gas or liquid phase and a surface. Even more common is **homogeneous catalysis**, in which both the reactants and the catalyst are in solution.

Homogeneous Acid Catalysis

The ionization of methanol in aqueous solution,

$$CH_3OH + H_2O \rightarrow CH_3O^- + H_3O^+ \tag{22-39}$$

proceeds rapidly (but to a small extent), and in fact too rapidly to measure by bulk sampling methods. In contrast, the analogous reaction with dimethyl ether,

$$CH_3-O-CH_3 + H_2O \rightarrow CH_3O^- + CH_3OH_2^+ \tag{22-40}$$

does not occur to a measurable extent. This is because the hydroxyl hydrogen atom on the alcohol is so small and so exposed that it can be attacked by a nucleophilic reagent such as water (Figure 22-10), whereas the analogous $-CH_3$ group in the ether blocks the approach of H_2O. **Steric hindrance**—the bumping together and mutual repulsion of nonbonded atoms—is the greatest single factor in determining activation energies of reacting molecules. The H^+ ion, being a lone proton, is free from such steric hindrance. It is so small that such barriers to reaction can be avoided, and proton-transfer reactions are usually quite fast. If an alternative mechanism to a given reaction involves proton-transfer steps, it is likely that the reaction will be catalyzed by acids, which provide a source of protons. This is why acid catalysis is so important to chemistry.

The reaction

$$CH_3COOCH_3 + H_2O \rightleftharpoons CH_3COOH + CH_3OH$$

has the rate equation

$$-\frac{d[CH_3COOCH_3]}{dt} = k'[H_2O][CH_3COOCH_3]$$

$$= k[CH_3COOCH_3]$$

Although this rate equation is really second order, it is effectively first order since the water concentration in aqueous solution is effectively constant

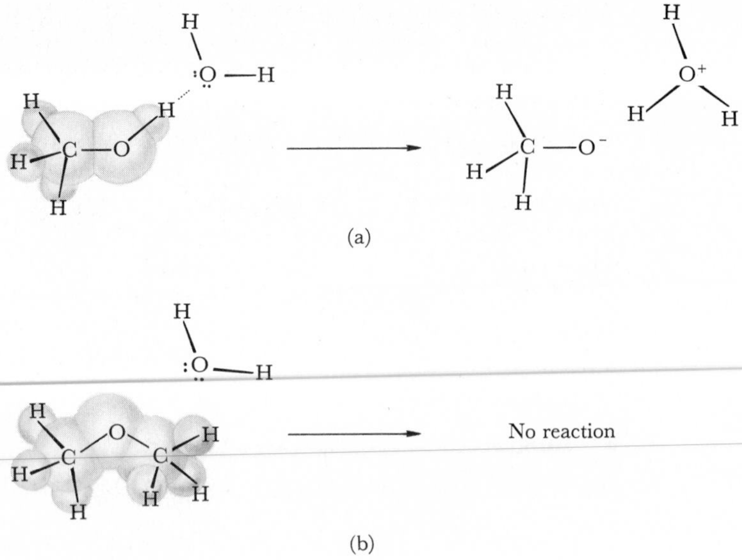

(a)

(b)

Figure 22-10   When methanol is dissolved in water, the proton on the methanol hydroxyl group is small enough that a water molecule can approach it and subject it to a nucleophilic attack. The result is that the O—H bond in methanol is broken and the proton is bound to the solvent. (b) In dimethyl ether, the proton is replaced by a CH_3 group. This group is so large and bulky that the water molecule cannot approach close enough to attack it as with methanol. The O—C bond in dimethyl ether does not break, and no CH_3O^- is formed. The colored outlines mark the approximate relative sizes of the atoms as measured by van der Waals contact distances between nonbonded atoms.

during the reaction. This reaction is catalyzed by acids, and was studied by Ostwald before the turn of the century. Ostwald determined that the rate constant was increased by a factor of 300 by HCl, and that other acids gave the increases shown in Table 22-3. You can see a correlation between the rate constant and the dissociation constant for the acid. We know now that the effectiveness of the acid as a catalyst (as reflected in the rate constant) arises from the concentration of H^+ ion produced by the acid (as reflected in the dissociation constant). All the completely dissociated strong acids produce approximately the same rate constant for the reaction.

Summary

This chapter has been a short introduction to a very large field. It is impossible in this small space to give you more than a brief outline of the problems

Table 22-3

Increase in Rate Constant by Acid Catalysis for the Reaction

$$CH_3COOCH_3 + H_2O \rightleftarrows CH_3COOH + CH_3OH^a$$

Acid	$-\dfrac{d[CH_3COOCH_3]}{dt} = k[CH_3COOCH_3]$	
	k/k_{HOAc}	K_a
Acetic, CH_3COOH (formed in reaction)	1.0	1.76×10^{-5}
Formic, HCOOH	3.8	1.77×10^{-4}
Dichloroacetic, $CHCl_2COOH$	66.7	3.32×10^{-2}
Trichloroacetic, CCl_3COOH	198.0	0.20
Sulfuric, H_2SO_4	214.0	—
Nitric, HNO_3	267.0	—
Hydrobromic, HBr	284.0	—
Hydrochloric, HCl	290.0	—

[a]Column 2 gives the ratio of the rate constant, k, for a given acid to that for acetic acid at the same concentration.

involved in finding out how molecules react and some of the methods for solving them. Chemical kinetics is presently less well developed than structural chemistry because the problems are basically more difficult. Kinetics needs every shred of information about structure that can be found, but this information is only the starting point for proposing mechanisms of reaction and designing experiments to test them.

The essential question to be answered is, "How does one molecule react with another molecule?" However, because we cannot study individual molecules, we are forced to work with large numbers of molecules, with a distribution of energies, and with unknown relative orientations of the molecules. The energy distribution can be avoided for certain special reactions by **molecular-beam methods**, but the orientation problem remains.

The **rate law** or **rate expression** for the appearance of products does not necessarily agree with what would have been expected from the number of moles of each reactant in the overall equation. When it does agree, this is suggestive evidence that the reaction proceeds in a one-step process as written (although there are pitfalls, as in the HI reaction). When the rate law and overall equation do not agree, as with HBr, this suggests that the overall reaction really proceeds in a series of simpler steps. If one of these steps is much slower than the others, the kinetics of the overall reaction is controlled by this **rate-determining step**

The **order** and the **molecularity** of a reaction are two quite different quantities that reflect the difference between overall stoichiometry and mechanism. The order of a reaction is simply the sum of the exponents of all the concentration terms in an expression that is a product of such terms. The molecularity of a simple reaction is the number of molecules or ions that collide in that step. An overall, multistep reaction has no molecularity, although its order may be well defined. But the rate expression for the HBr reaction is so complicated that even the concept of reaction order has no meaning, except at low HBr concentrations.

Two theories of simple bimolecular reactions exist: the **collision theory** and the **absolute rate theory**. Both are based on Arrhenius' interpretation of the dependence of rate constant on temperature in terms of an energy of activation. The collision theory focuses on the collision that precedes reaction; the absolute rate theory concentrates on the assembly of atoms just after collision but before separation into products. Both give a reasonable explanation of observed rate constants; the absolute rate theory is somewhat better.

Both the **enthalpy of activation** and the **entropy of activation** are important in determining the size of the barrier to reaction. Reaction is favored if the enthalpy barrier is low and the entropy of activation is large and positive (or at least not negative). If the activated complex is much more ordered than the reactants, the entropy of activation is large and negative, and reaction is slowed.

A **catalyst** accelerates a reaction by providing an alternative reaction pathway or mechanism with a lower free energy of activation. It can do this by providing energy for a dissociation or by assisting in the ordering of reactants in the complex. In the first pathway, the enthalpy of activation is lowered (as with H_2 on a Pt surface). The second pathway increases the probability that the reactants will be ordered more properly than would be the case by chance in solution. In either case, reaction is more rapid because $\Delta G^{0\ddagger}$ is lower and $K^{\ddagger}$ is larger.

Thermodynamics indicates nothing about the time involved in attaining equilibrium, as we have emphasized several times before. Thermodynamics is concerned only with comparing the initial and final states of a reacting system for quantities such as T, P, V, E, H, S, and G, which are state functions. The change in these quantities is the same whether the reaction takes place in a nanosecond (10^{-9} sec) or an eon (10^9 years), and whether the reaction takes place in one step or in a thousand, so long as the starting conditions and the final conditions are the same. In contrast, kinetics deals with how fast reactions occur. A rock rolling down a hillside will come to a halt and remain stationary forever if it meets a barrier that is even a small fraction of the height of the hill. If the rock is given random disturbances by passersby, the probability that it will be knocked over the obstacle and will continue down the hill within a specified time depends on the height of the barrier (among other factors). The task of chemical kineticists is to investigate these barriers to chemical reaction, to see what effect they have in slow-ing reactions, and to find ways to avoid them, either by surmounting them by proper chemical conditions or by circumventing them by catalysis.

Self-Study Questions

1. Why does the probability of reaction of thioacetamide and water in Figure 22-1 depend on the relative orientations of the two molecules as they approach each other?

2. Why does the probability of the reaction mentioned in Question 1 depend on the relative velocity of approach of the two molecules?

3. Why does the approach of the water molecules to thioacetamide weaken the C—S bond? Why does it weaken the O—H bonds?

4. Why are the molecules of H_2O (Figure 22-1a) and H_2S (Figure 22-1c) not drawn the same way?

5. Which of the two factors important in determining the rate of reaction, energy and entropy, is controlled in molecular-beam experiments? How is this done?

6. Why are molecular-beam experiments unsuitable for studying the acid–base reactions discussed in Chapter 5? What kinds of reactions can be studied by molecular-beam methods?

7. How is the rate constant for a reaction defined? In the HI rate equation (equation 22-2), what is the meaning of the d's in numerator and denominator on the left side? What do the chemical symbols in brackets mean?

8. What is the difference between order and molecularity in a chemical reaction? What is the overall order of reactions of equations 22-25 and 22-27? What is their molecularity? Under what conditions are order and molecularity different? For what kinds of reactions is the concept of molecularity meaningless? For what kinds of reactions is the concept of order meaningless?

9. What physical methods can be used to follow the course of a reaction? What are the relative advantages of physical and chemical techniques for following reaction kinetics?

10. How can you tell when a chemical reaction is first order?

11. Can you suggest an explanation of why the reaction between ethanol and decaborane in Section 22-2 is not thirty-first order? Can you think of a possible mechanism that would account for the observed rate equation?

12. What physical evidence leads to the concept of an energy of activation for a reaction? What is the interpretation of this energy of activation in terms of the reaction mechanism?

13. What is questionable about the derivation of the equilibrium constant in terms of forward and reverse reactions occurring at the same rates? When is this derivation valid?

14. At constant temperature, how does the enthalpy of a reaction change as the activation energy of the forward reaction changes? How can the activation energy be altered?

15. How is the activation energy used in the collision theory of reaction? In this theory, what factors affect the rate of reaction? In what two ways does temperature influence the rate of reaction in the collision theory?

16. Why, from the evidence in Table 22-2 and Figure 22-5, can we say that the activated complex for the reaction

$$2ClO \rightarrow Cl_2 + O_2$$

is not a complex with four atoms at a great distance from each other?

17. If three hydrogen atoms were spaced at intervals of 2.0 Å along a straight line, what would be the potential energy (expressed as kJ mole^{-1} of such triplets of atoms) as given by Figure 22-7? How much less stable is this state than the highest point in the reaction path from $H + H_2$ to $H_2 + H$?

18. What is an activated complex? What is the activated complex in Figure 22-1? In the absolute rate theory, what assumption is made about the amount of activated complex present?

19. Does a large positive enthalpy of activation favor rapid reaction? Does a large positive entropy of activation favor rapid reaction?

20. Why is it useful to calculate the entropy of activation from measured rate constants and the absolute rate theory? What information do such data give us about reaction mechanisms?

21. What do the symbols mean in S_N1 and S_N2 mechanisms? What is the difference between these mechanisms? What is the activated complex in each mechanism? In which mechanism is the nature of the entering group more important? What factors determine whether a reaction will proceed by S_N1 or S_N2 mechanism?

22. Does the reaction of thioacetamide with water use S_N1 or S_N2 mechanism?

23. Which of the two mechanisms of the preceding two questions will have a larger entropy of activation? What effect will this have on the rate of reaction?

24. Which of the two mechanisms in Questions 21 and 22 will have a larger enthalpy of activation, other factors being the same? What effect will this have on the rate of reaction?

25. In view of the answers to the preceding two questions, why can't we make dogmatic statements about relative rates of reactions that proceed by S_N1 and S_N2 mechanisms?

26. What is the evidence for the assertion that the reaction of H_2 with I_2 to produce HI is not a simple bimolecular reaction, as had been believed for such a long time?

27. Which mechanism, S_N1 or S_N2, is more likely to be encountered with octahedral complex ions? With square planar ions? Why?

28. Why is the rate equation for the reaction of H_2 with Br_2 so much more complicated than that for H_2 and I_2?

29. What is the distinguishing feature of a chain reaction? What is a chain-initiating step? What is a chain-inhibiting step? What chain-inhibiting step is used in nuclear power reactors? (See Section 23-4.)

30. What is the steady-state assumption in solving rate expressions?

31. Why is acid catalysis so common and so effective?

32. Why is the hydrolysis reaction for methyl acetate in aqueous solution a first-order reaction, as commonly measured?

33. In terms of the theories of reaction rates of this chapter, how does a catalyst work? What is the activated complex in a catalytic reaction? How can the catalyst affect the enthalpy of activation? The entropy of activation?

Problems

Rate equations

1. For the hypothetical reaction

$$2A + 3B \rightarrow 3C + 2D$$

the following rate data were obtained in three experiments at the same temperature:

Initial [A] (mole liter^{-1})	Initial [B] (mole liter^{-1})
0.10	0.10
0.20	0.10
0.20	0.20

Initial rate (moles of A consumed liter^{-1} sec^{-1})
0.10
0.40
0.40

(a) Determine the experimental rate equation for the reaction. (b) Calculate the specific rate constant, k. (c) What is the rate of this reaction when $[A] = 0.30M$ and $[B] = 0.30M$?

2. For the hypothetical reaction

$$2A + B \rightarrow 2C$$

the following data were collected in three experiments at 25°C:

Initial [A] (mole liter^{-1})	Initial [B] (mole liter^{-1})
0.10	0.20
0.30	0.40
0.30	0.80

Initial rate (moles of A consumed liter^{-1} sec^{-1})
300
3600
14400

(a) What is the experimental rate equation for this reaction? (b) Calculate the specific rate constant for this reaction.

3. In the reaction

$$2NO + Cl_2 \rightarrow 2NOCl$$

the reactants and products are gases at the temperature of the reaction. The following rate data were measured for three experiments:

Initial p_{NO} (atm)	Initial p_{Cl_2} (atm)
0.50	0.50
1.0	1.0
0.50	1.0

Initial rate (atm sec^{-1})
5.1×10^{-3}
4.0×10^{-2}
1.0×10^{-2}

(a) From these data, write the rate equation for this gas reaction. What order is the reaction in NO, Cl_2, and overall? (b) Calculate the specific rate constant for this reaction.

4. The reaction $2NO + O_2 \rightarrow 2NO_2$ is first order in oxygen pressure and second order in the pressure of nitric oxide. Write the rate expression.

5. The indicated rate expressions have been obtained for each of the reactions listed below. If these rate expressions hold, even at equilibrium, what are the rate expressions for the reverse reactions at equilibrium?
a) $C_2H_2 + H_2 \rightarrow C_2H_4$
  Rate $= k[H_2]/[C_2H_2]$
b) $C_2H_4 + H_2 \rightarrow C_2H_6$
  Rate $= k[H_2]$
c) $2H_2 + O_2 \rightarrow 2H_2O$
  Rate $= k[H_2][O_2]^{4/3}$
d) $N_2O_5 \rightarrow 2NO_2 + \frac{1}{2}O_2$
  Rate $= k[N_2O_5]$

6. Given the following data for the reaction

trans   cis

Time (min):	0	10	20	30
Moles trans:	1.00	0.90	0.81	0.73

What is the order of the reaction? How long will it take for half of the trans compound to decompose?

7. The reaction

$$A + B + C \rightarrow D + F$$

was found to be zero order with respect to A. A solution of reactants A, B, and C was prepared with the following initial concentrations: $0.2M$ of A, $0.4M$ of B, and $0.6M$ of C. The concentration of A in this solution dropped to essentially zero in 5 min. A second solution was prepared with the following initial concentrations: $0.03M$ of A, $0.4M$ of B, and $0.6M$ of C. How long will it take for A to disappear?

8. For the reaction

$$2NO + H_2 \rightarrow N_2O + H_2O$$

the following experimental rate data are collected in three successive experiments at the same temperature:

Initial [NO] (molar)	Initial [H_2] (molar)
0.60	0.37
1.20	0.37
1.20	0.74

Initial rate (mole
liter^{-1} min^{-1})

0.18
0.72
1.44

Using these experimental data, write the rate expression for the reaction.

9. The reaction

$$2HCrO_4^- + 3HSO_3^- + 5H^+ \rightarrow$$
$$2Cr^{3+} + 3SO_4^{2-} + 5H_2O$$

follows the rate equation

$$\text{Rate} = k[HCrO_4^-][HSO_3^-]^2[H^+]$$

Why isn't the rate proportional to the numbers of ions of each kind that are shown by the equation?

10. The reaction

$$I^- + OCl^- \rightarrow Cl^- + OI^-$$

has the experimental rate equation

$$\text{Rate of disappearance of } OCl^- =$$
$$k[I^-][OCl^-]$$

How would you describe the order of this reaction?

11. The rate constant in Problem 10 depends on hydroxide ion concentration. For hydroxide ion concentrations of $1.00M$, $0.50M$, and $0.25M$, k is 60, 120, and 240 liters mole^{-1} sec^{-1}, respectively. What is the order of the reaction with respect to hydroxide ion concentration?

12. The reaction $SO_2Cl_2 \rightarrow SO_2 + Cl_2$ is a first-order reaction with the rate constant $k = 2.2 \times 10^{-5}$ sec^{-1} at 320°C. What fraction of SO_2Cl_2 is decomposed on heating at 320°C for 90 min?

13. For the first-order reaction A → C, $k = 5$ min^{-1}. When the first-order reaction D → B occurs, only 10% of D decomposes in the same length of time that it takes for 50% of A to decompose in the first reaction. Calculate k for the second reaction.

14. For the decomposition of ammonia, the following data were measured. The top line indicates time in seconds, and the bottom line indicates the corresponding concentration of ammonia, in moles liter^{-1}.

0	1	2	$t_{1/2}$
2.000	1.993	1.987	1.000

For this first-order reaction, write an expression for the rate of reaction and calculate the rate constant, k.

15. For the decomposition of N_2O_5 in CCl_4, a plot of $\log[N_2O_5]$ against time is a straight line. The rate constant for the reaction is 6.2×10^{-4} sec^{-1} at 45°C. If one begins with 1 mole of N_2O_5 in a 1-liter flask, how long will it take for 20% of the N_2O_5 to decompose? How long for 50%?

Activation energy

16. It often is said that, near room temperature, a reaction rate doubles if the temperature is increased by 10°C. Calculate the activation energy of a reaction whose rate exactly doubles between 27°C and 37°C.

17. What is the activation energy for a reaction for which an increase in temperature from 20°C to 30°C exactly triples the rate constant?

18. The following data give the temperature dependence of the rate constant for the reaction $N_2O_5 \rightarrow 2NO_2 + \frac{1}{2}O_2$. Plot the data and calculate the activation energy of the reaction.

$T(\text{K})$	$k(\text{sec}^{-1})$
273	7.87×10^{-7}
298	3.46×10^{-5}
308	1.35×10^{-4}
318	4.98×10^{-4}
328	1.50×10^{-3}
338	4.87×10^{-3}

19. For the decomposition of CH_3I at 285 K, the energy of activation is 180 kJ mole^{-1}. Assuming that the energy of activation is constant, calculate the percent increase in the fraction of molecules with energy greater than E_a when the temperature is increased to 300 K.

20. The rate constant for the decomposition of N_2O_5 in carbon tetrachloride is 6.2×10^{-4} sec^{-1} at 45°C. Calculate the rate constant at 100°C if the activation energy is 103 kJ mole^{-1}.

21. Why does it take longer to boil an egg on the top of Mt. San Jacinto (11,000 ft) than in Pasadena (750 ft)? (Smog is not the answer.)

Surface reactions

22. For the reaction $Ni + \frac{1}{2}O_2 \rightarrow NiO$, $\Delta H^0 = -248$ kJ at 298 K, and the value does not change drastically with temperature. Suppose that the reaction takes place on the surface of Ni so rapidly that all the heat is used to heat the remaining nickel. If at 25°C one oxygen atom reacts with each 10 Å^2 of surface, what will be the final temperature when a cube of nickel 1.00 cm on an edge reacts with oxygen? The density of nickel is about 9 g cm^{-3}. If the 1-cm cube is ground to form 10^{15} equal cubes and reaction occurs with these, what will be the final temperature? Ignore the change in heat capacity when nickel oxide is formed. Assume that the law of Dulong and Petit (Chapter 6) is valid.

23. If a cube of NaCl 1.00 cm on an edge is dissolved in an enormous quantity of water stirred in a tank, it takes 6 hours before solution is complete. If the cube is ground to a fine powder containing 10^{15} equal spheres, what will be the time required for solution if this time is inversely proportional to the initial area of contact between the NaCl and the water?

Rates and mechanisms

24. The reaction

$$H_2 + Cl_2 \rightarrow 2HCl$$

occurs explosively in the presence of light. Assume that this explosion takes place by a chain reaction and is initiated by the formation of (a) hydrogen atoms or (b) chlorine atoms. Using the bond energies of H_2 and Cl_2, calculate the wavelength of light needed in each case.

25. Calculate the standard free energy change of the reaction

$$2C_6H_6(g) \rightarrow 3CH_4(g) + 9C(gr)$$

at 298 K, and calculate the equilibrium constant, K_{eq}. Other typical reactions that you may have encountered are the precipitation of AgCl from solution,

$$Ag^+ + Cl^- \rightarrow AgCl(s)$$
$$K_{eq} \simeq 10^{10}$$

and the formation of the diammine complex of silver in aqueous solution,

$$Ag^+ + 2NH_3 \rightarrow Ag(NH_3)_2^+$$
$$K_{eq} \simeq 10^8$$

All three reactions are highly spontaneous. Yet in the laboratory the preceding two reactions proceed essentially instantaneously, whereas the decomposition of benzene to methane and carbon apparently does not proceed at all.

Account for this fantastic difference in reaction rates.

26. The decomposition of gaseous N_2O_5 occurs according to the reaction

$$N_2O_5 \rightarrow 2NO_2 + \tfrac{1}{2}O_2$$

The experimental rate equation is

$$\frac{-d[N_2O_5]}{dt} = k[N_2O_5]$$

and a proposed reaction mechanism is

1) Equilibrium: $N_2O_5 \underset{}{\overset{K}{\rightleftarrows}} NO_2 + NO_3$
2) Slow reaction:

$$NO_2 + NO_3 \xrightarrow{k_2} NO_2 + O_2 + NO$$

3) Fast reaction: $NO + NO_3 \xrightarrow{k_3} 2NO_2$

(a) Show that this mechanism is consistent with the observed rate equation. (b) If $k = 5 \times 10^{-4}$ sec^{-1}, how long does it take for the concentration of N_2O_5 to fall to one-tenth its original value?

27. Consider the reaction

$$CH_4 + Cl_2 \xrightarrow{\text{light}} CH_3Cl + HCl$$

The mechanism is a chain reaction involving Cl atoms and CH_3 radicals. Which of the following steps does not terminate this chain reaction?
a) $CH_3 + Cl \rightarrow CH_3Cl$
b) $CH_3 + HCl \rightarrow CH_4 + Cl$
c) $CH_3 + CH_3 \rightarrow C_2H_6$
d) $Cl + Cl \rightarrow Cl_2$

28. Assume that the reaction

$$5Br^- + BrO_3^- + 6H^+ \rightarrow 3Br_2 + 3H_2O$$

proceeds by the mechanism

1) Fast reaction:

$$BrO_3^- + 2H^+ \xrightarrow{k_1} H_2BrO_3^+$$

2) Fast reaction:

$$H_2BrO_3^+ \xrightarrow{k_{-1}} BrO_3^- + 2H^+$$

3) Slow reaction:

$$Br^- + H_2BrO_3^+ \xrightarrow{k_2} Br\!-\!BrO_2 + H_2O$$

4) Fast reaction:

$$Br\!-\!BrO_2 + 4H^+ + 4Br^- \xrightarrow{k_3} 3Br_2 + 2H_2O$$

Deduce the rate equation that agrees with this mechanism; express the rate constant for the overall reaction in terms of rate constants for the individual steps. The rate equation depends on the concentrations of H^+, Br^-, and BrO_3^-.

29. Consider the reaction

$$5H^+ + [Co(NH_3)_5Cl]^{2+} + [Cr(H_2O)_6]^{2+} \xrightarrow{H_2O} [Co(H_2O)_6]^{2+} + [Cr(H_2O)_5Cl]^{2+} + 5NH_4^+$$

When this reaction is carried out in the presence of radioisotopically labeled chloride ions, it is found that the radioactive ions do not appear in the product. Keeping in mind that the rate of exchange of ligands bound to Cr^{2+} and Co^{2+} is quite rapid, whereas for Cr^{3+} and Co^{3+} it is quite slow, postulate a mechanism for the reaction [see *J. Am. Chem. Soc.* **75**, 4118 (1953)].

Suggested Reading

J. O. Edwards, *Inorganic Reaction Mechanisms: An Introduction,* W. A. Benjamin, Menlo Park, Calif., 1965.

H. Eyring and E. M. Eyring, *Modern Chemical Kinetics,* Reinhold, New York, 1963.

A. A. Frost and R. G. Pearson, *Kinetics and Mechanisms,* Wiley, New York, 1961, 2nd ed.

E. L. King, *How Chemical Reactions Occur,* W. A. Benjamin, Menlo Park, Calif., 1963.

23

Nuclear Chemistry

Let us take the Road;
   Hark! I hear the Sound of Coaches!
   The Hour of Attack approaches,
To Arms, brave Boys, and load.

See the Arms we hold!
   Let the Chymists toil like Asses,
   Our Fire, their Fire surpasses,
And turns all our Lead to Gold!

**Chorus of Highwaymen, John Gay,
The Beggars' Opera (1728)**

Transmutation of the elements—specifically of base metals into gold—was to medieval scientists what the search for extraterrestrial life is to our generation: the half-mystical but theoretically attainable goal that captures the imagination of the nonscientist as well as the professional. The alchemist's fire was the symbol of arcane knowledge, with both benefits and hazards, rather like the nuclear reactor or the genetic engineering laboratory today. We have solved the problem of transmutation of the elements, but the other great goal of medieval chemistry (or alchemy), indefinite prolongation of life and youth, is still the subject of intense research activity.

It was the physicist and not the chemist who ultimately succeeded in bringing about transmutation of one element into another; the alchemist's crucible gave way to the nuclear power reactor. At first, attention was focused on the immense quantities of energy released during nuclear reactions. The fact that uranium was being transmuted into barium and other light elements was of secondary interest. But it did not take chemists long to realize that radioactive isotopes of common elements were uniquely valuable. A radioactive atom acts as a label or tag; it can be added at one point in a reaction and followed thereafter, to unravel the complex skein of chemical steps. Carbon isotope tracer studies, for example, are responsible for our present understanding of the reactions that take place during photosynthesis, and it is hard to imagine that the same information could have been obtained without them. Isotopes, radioactive and stable, have enabled us to find answers to chemical problems that are solvable in no other way. Radioisotopes have also given us a means of accurately dating past events, both historical and geological. And they have revealed comparative ages for the earth and the moon that have disproved some old theories about the origin of the moon.

23-1 THE NUCLEUS

As was mentioned in Chapter 1, the nucleus of an atom is composed of two main fundamental particles, protons and neutrons, both of which are called **nucleons**. It has a positive charge equal to the number of protons it contains, and this number, Z, is the **atomic number**. A neutral atom has the same number of electrons around the nucleus as protons within the nucleus. Since these outer electrons give each atom its chemical properties, all neutral atoms with the same number of electrons and protons are classified as the same element. Hence, the atomic number identifies an element. The total number of protons and neutrons in a nucleus is its **mass number, A**.

Atoms with the same number of protons but different numbers of neutrons are called **isotopes**. In the representation of an isotope, the atomic number, Z, is written as a left subscript to the symbol of the element, and the mass number, A, is written as a left superscript. Thus, the mercury isotope with 80 protons and 116 neutrons is written $^{196}_{80}\text{Hg}$ (80 + 116 = 196). The mass of the nucleus, in atomic mass units (amu), is nearly equal to its mass number, A. One amu is defined as exactly one-twelfth the mass of one atom of carbon in the $^{12}_{6}\text{C}$ isotope. One amu is 1.66057×10^{-24} g. Other mass conversions are given in Appendix 2.

Elements found in nature are usually mixtures of several isotopes. For example, hydrogen has three isotopes, $^{1}_{1}\text{H}$, $^{2}_{1}\text{H}$, and $^{3}_{1}\text{H}$, of which the first two occur in nature in the proportions indicated in Table 23-1. The nuclei of these three isotopes contain a proton, a proton and a neutron, and a proton and two neutrons, respectively. Mercury has isotopes that contain from 189 to 206 neutrons and protons, or from 109 to 126 neutrons. The seven isotopes listed in Table 23-1 occur naturally in the relative amounts listed. The observed atomic weight, 200.59, is the weighted average of the atomic weights of the individual naturally occurring isotopes.

Size and Shape

Rutherford made the first measurements of the size of a nucleus during his α-particle scattering experiments. Better measurements can be made with neutron scattering because neutrons are not deflected by electrostatic repul-

Table 23-1

Natural Isotopes of Hydrogen and Mercury[a]

Isotope	Mass (amu)	Percent natural abundance	Name
${}^{1}_{1}H$	1.0078	99.985	Hydrogen
${}^{2}_{1}H$ or D	2.0140	0.015	Deuterium
${}^{3}_{1}H$ or T	3.0161	—	Tritium
Natural mixture	1.0080		
${}^{196}_{80}Hg$	195.9658	0.15	Mercury-196
${}^{198}_{80}Hg$	197.9668	10.02	Mercury-198
${}^{199}_{80}Hg$	198.9683	16.84	Mercury-199
${}^{200}_{80}Hg$	199.9683	23.13	Mercury-200
${}^{201}_{80}Hg$	200.9703	13.22	Mercury-201
${}^{202}_{80}Hg$	201.9706	29.80	Mercury-202
${}^{204}_{80}Hg$	203.9735	6.85	Mercury-204
Natural mixture	200.59		

[a]Values from the *Handbook of Chemistry and Physics,* Chemical Rubber Co., Cleveland, Ohio, 1978–9, 59th ed.

sion. Numerous neutron-scattering experiments have shown that the radius of a nucleus is proportional to the cube root of the number of nucleons within it, or that the volume is proportional to the number of nucleons:

$$r = 1.33 \times 10^{-13} A^{1/3} \text{ cm}$$

The isotope ${}^{1}_{1}H$ has a radius of 1.33×10^{-13} cm, and ${}^{238}_{92}U$ has a radius of 8.25×10^{-13} cm. The atomic radii, including electron clouds, are 20,000 times the size of the nuclei.

If a collection of charges is not perfectly spherical, then it can have what is called an **electric quadrupole moment** even though it does not have a dipole moment. These quadrupole moments can be measured, although they will not concern us here. Such measurements have revealed that many nuclei are spherical, and most of those that are not spherical are elongated like a football in which the ratio of longest to shortest diameter never exceeds 1.2.

Binding Energy

The mass number, A, and the mass of a nucleus in amu are not the same, partly because the mass of a proton or a neutron is not exactly 1 amu. The mass of a proton (Appendix 2) is 1.007276 amu, and the mass of a neutron is 1.008665 amu. There is another reason as well: An atom of a stable isotope weighs *less* than the sum of masses of electrons, protons, and neutrons from which it is built.

Example 1

What is the total mass of the particles of which an atom of $^{200}_{80}\text{Hg}$ is composed?

Solution

The mass of a neutron is 1.008665 amu, and that of a proton *and* the electron in the electron cloud that neutralizes the proton's charge is 1.007825 amu. (Masses of neutral atoms are usually tabulated, rather than masses of nuclei.) The given isotope of mercury has 80 protons and $200 - 80 = 120$ neutrons. The total mass of the component particles then is

$$80 \times 1.007825 \text{ amu} = \underline{80.62600 \text{ amu of protons and electrons}}$$
$$120 \times 1.008665 \text{ amu} = \underline{121.03980 \text{ amu of neutrons}}$$
$$\text{Total} = 201.66580 \text{ amu of component parts of } ^{200}_{80}\text{Hg}$$

Although the components of $^{200}_{80}\text{Hg}$ weigh 201.6658 amu, the observed atomic mass of $^{200}_{80}\text{Hg}$ is only 199.9683 amu. What has happened to the other 1.6975 amu of matter? The missing mass has been converted to energy that is given off, energy that is needed before the nucleus breaks apart again. This **binding energy** of the nucleus can be calculated from Einstein's famous relationship between mass and energy:

$$E = mc^2$$
$$\text{(joules)} \quad \text{(g)(m}^2 \text{ sec}^{-2})$$

Here c is the speed of light, 2.998×10^8 m sec^{-1}. The energy into which 1.0000 amu of matter can be converted is

$$E = 1.0000 \text{ amu} \times \frac{1.66057 \times 10^{-27} \text{ kg}}{1 \text{ amu}} \times (2.998 \times 10^8)^2 \text{ m}^2 \text{ sec}^{-2}$$

$$= 1.4925 \times 10^{-10} \text{ J} \times \frac{1 \text{ eV}}{1.60219 \times 10^{-19} \text{ J}} = 9.315 \times 10^8 \text{ eV}$$

(The conversions from atomic mass units to kilograms, and from joules to electron volts, are given in Appendix 2.) It is useful to remember that 1 amu of mass is equivalent to 931.5 MeV (million electron volts) of energy. Although electron volts are not strict SI units, they are common in nuclear physics, and are useful since a joule is an impractically large unit of energy for breakdown of a single atom. Scientists discuss nuclear energies in electron volts per atom, or joules per mole of atoms.

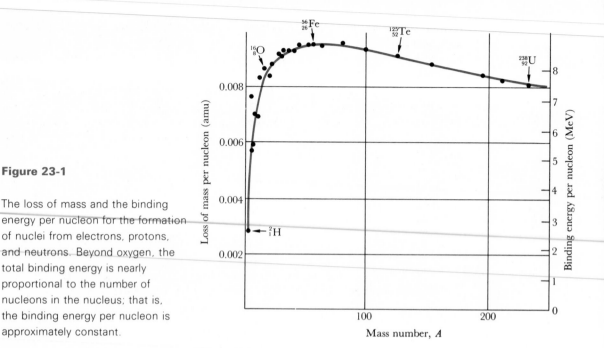

Figure 23-1

The loss of mass and the binding energy per nucleon for the formation of nuclei from electrons, protons, and neutrons. Beyond oxygen, the total binding energy is nearly proportional to the number of nucleons in the nucleus; that is, the binding energy per nucleon is approximately constant.

$$1 \text{ eV atom}^{-1} = 96{,}485 \text{ J mole}^{-1}$$

(This latter quantity is Faraday's constant. Can you explain why?)

In making $^{200}_{80}\text{Hg}$ from electrons, protons, and neutrons, the loss in mass *per nucleon* is 1.6975 amu/200 nucleons = 0.00849 amu nucleon^{-1}. This mass corresponds to a binding energy of 0.00849 amu nucleon^{-1} × 931 MeV amu^{-1} = 7.90 MeV nucleon^{-1}. The binding energy per nucleon is plotted against mass number, A, for the entire range of elements in Figure 23-1. In the first few elements, the binding energy per nucleon is low and is approximately proportional to the number of nucleons. Beyond oxygen, the binding energy per nucleon is almost constant at 8 MeV per nucleon. This fact means that forces between nucleons are quite short range, and that each nucleon essentially interacts only with its immediate neighbors. If nuclei were held together by longer-range forces so every nucleon in the nucleus interacted with every other one, the *total* binding energy would increase with the square of the number of nucleons and not with the first power, as it does. The initial increase of binding energy per nucleon with number of nucleons is reasonable when we realize that in such small nuclei each nucleon does not have its full complement of neighbors. Therefore, each new nucleon increases the binding of a preexisting nucleon by increasing its packing coordination number. Protons and neutrons are not simply closely packed in the nucleus like marbles in a box. (We shall soon see evidence for structure within the nucleus.) But with regard to binding energy, they behave as if they were.

Why have we waited until now to introduce the idea that energy changes in a reaction are accompanied by mass changes? If the total binding energy of a mercury nucleus is equivalent to 1.6975 amu of matter, why have we not made comparable calculations for the bond energies of molecules? The answer lies in the enormously greater energies involved in nuclear reactions. As an example, let us calculate the amount of mass that is equivalent to the entire bond energy of a Cl_2 molecule.

Example 2

The enthalpy of the following reaction is 242.8 kJ mole^{-1} of Cl_2:

$$Cl_2(g) \rightarrow 2Cl(g)$$

What is the mass equivalent of this energy, in amu molecule^{-1}?

Solution

$$\Delta H = 242.8 \text{ kJ mole}^{-1} \times \frac{1 \text{ eV molecule}^{-1}}{96.485 \text{ kJ mole}^{-1}} = 2.516 \text{ eV molecule}^{-1}$$

$$\Delta m = 2.516 \text{ eV} \times \frac{1 \text{ amu}}{9.315 \times 10^8 \text{ eV}} = 2.70 \times 10^{-9} \text{ amu molecule}^{-1}$$

Or, more directly from Einstein's $E = mc^2$, we have:

$$\Delta m = \frac{\Delta H}{c^2} = \frac{2.428 \times 10^5 \text{ J mole}^{-1}}{(3.00 \times 10^8 \text{ m sec}^{-1})^2} = 2.70 \times 10^{-12} \text{ kg mole}^{-1}$$

$$\Delta m = 2.70 \times 10^{-9} \text{ g mole}^{-1} \text{ or amu molecule}^{-1}$$

since $1J \equiv 1 \text{ kg m}^2 \text{ sec}^{-2}$. This calculation illustrates the simplicity of SI units, since the answer follows from the definition of the joule rather than from remembered conversion factors between energy units.

The total energy involved in the dissociation of a molecule of Cl_2 amounts to a mass of only five millionths of an electron. Chemical reactions usually involve energies of a few electron volts, whereas nuclear energies are in the million electron volt range. One MeV per molecule equals 96.5 *million* kJ mole^{-1}, which is quite outside the range of most chemical reactions. Therefore, in chemical reactions we can work with the separate principles of conservation of mass and of energy. Interconversions are not detectable. In contrast, for nuclear reactions the interconversion of mass and energy is normal; here we must use the more general principle of the conservation of mass–energy. In any nuclear reaction, the total of *mass and energy* of the reacting species and their surroundings does not change during the course of the reaction.

23-2   NUCLEAR DECAY

Many nuclei do not decay. These are called the **stable isotopes**. Others break down spontaneously to new elements; they are the **radioactive isotopes**. The radiation that Becquerel observed in 1896 was α particles ($_2^4$He nuclei) from the decay of uranium:

$$_{92}^{238}\text{U} \rightarrow \, _{90}^{234}\text{Th} + \, _2^4\text{He}$$

One single isotope of polonium, found in uranium ore by the Curies in 1898, decays in three different ways:

$$_{84}^{207}\text{Po} \rightarrow \, _{82}^{203}\text{Pb} + \, _2^4\text{He} \qquad (\alpha \text{ emission})$$
$$_{84}^{207}\text{Po} \rightarrow \, _{83}^{207}\text{Bi} + \, _{+1}^{0}e \qquad (\beta^+ \text{ emission})$$
$$_{84}^{207}\text{Po} \rightarrow \, _{83}^{207}\text{Bi} \qquad\qquad (\text{electron capture})$$

In the first reaction, an α particle is emitted and polonium changes to lead. In the second and third reactions, one proton in the nucleus changes to a neutron. This is accomplished in the second reaction by emitting a **positron** (β^+), a particle with the mass of an electron but with a unit positive charge:

$$_{+1}^{1}p \rightarrow \, _0^1n + \, _{+1}^{0}e$$

In the third reaction, the change occurs by capturing one of the electrons from the orbitals around the nucleus:

$$_{+1}^{1}p + \, _{-1}^{0}e \rightarrow \, _0^1n$$

(The two nuclear reactions of a proton just written do not occur in the simple manner implied here. The overall result is as shown, but the actual mechanism is more complicated. Such a simplification is analogous to that involved in the way we write most chemical reactions.) In both of these last two reactions, the atomic number decreases by one, but the atomic mass remains the same. Radium-228 decays by another mechanism, with the emission of an electron:

$$_{88}^{228}\text{Ra} \rightarrow \, _{89}^{228}\text{Ac} + \, _{-1}^{0}e \qquad (\beta^- \text{ emission})$$

These examples illustrate the four types of decay: β^- emission, electron capture, β^+ emission, and α-particle emission.

β^- or Electron Emission

In spontaneous β^- decay, one of the neutrons in a nucleus becomes a proton and an electron that is emitted from the nucleus. Energy accompanies the stream of emitted electrons from a sample undergoing decay, and the calculation of this energy is relatively straightforward. Consider the decay of carbon-14 to nitrogen-14:

$$^{14}_{6}C \rightarrow \ ^{14}_{7}N + \ _{-1}^{0}e$$

During this reaction, a carbon nucleus and six electrons are converted into a nitrogen ion having six electrons (one too few), and one electron as a β^- particle. Thus, the masses of reactants and products are simply the masses of neutral atoms of C and N:

Reactants: mass of $^{14}_{6}C$ atom: 14.003242 amu

Products: mass of $^{14}_{7}N$ atom: 14.003074 amu

Loss of mass in reaction: 0.000168 amu

The energy equivalent to this loss of mass is

$$E = 0.000168 \text{ amu} \times 931.5 \text{ MeV amu}^{-1} = 0.156 \text{ MeV}$$

Therefore, the emitted electron has an energy of 0.156 MeV. In β^- decay, the atomic number always increases by one unit.

Orbital Electron Capture, EC

In orbital electron capture, one electron from the cloud surrounding the nucleus is captured by the nucleus and combines with a proton to form a neutron. An example is the decay of beryllium-7 to lithium-7:

$$^{7}_{4}Be \xrightarrow{\text{EC}} \ ^{7}_{3}Li$$

In this reaction, a nucleus and four orbital electrons are changed to a nucleus that is heavier by one electron, and three orbital electrons. Again, the masses of reactants and products are just the masses of Be and Li atoms

Reactants: mass of $^{7}_{4}Be$ atom: 7.0169 amu

Products: mass of $^{7}_{3}Li$ atom: 7.0160 amu

Loss of mass in reaction: 0.0009 amu or 0.84 MeV

In electron capture, the atomic number always decreases by one.

β^+ or Positron Emission

Carbon-11 decays by the emission of positrons:

$$^{11}_{6}C \rightarrow \ ^{11}_{5}B + \ _{+1}^{0}e$$

The energy calculation contains a trap. Carbon and its six electrons are converted to boron with six electrons (one too many) *plus* the mass of the positron. In the mass balance, a neutral carbon atom is converted to a neutral boron atom, an electron, and a positron of the same mass as an electron:

Products:	one $^{11}_{5}$B atom:	11.009305 amu
	two electrons:	0.001098 amu
		11.010403 amu
Reactants:	one $^{11}_{6}$C atom:	11.011443 amu
Loss of mass in reaction:		0.001040 amu or 0.968 MeV

In positron emission, as in electron capture, the atomic number decreases by one.

α-particle Emission

For the heavier elements, above $A = 200$, a common mode of decay is the emission of a helium nucleus or an α particle:

$$^{232}_{90}\text{Th} \rightarrow {}^{228}_{88}\text{Ra} + {}^{4}_{2}\text{He}$$

In α-particle emission, the atomic number decreases by 2, and the mass by 4.

γ Emission During α Decay

High-energy electromagnetic radiation can also be emitted during a nuclear decay. In the uranium-238 reaction,

$$^{238}_{92}\text{U} \rightarrow {}^{234}_{90}\text{Th} + {}^{4}_{2}\text{He}$$

α particles of two different energies are liberated, 4.195 MeV and 4.147 MeV. Electromagnetic radiation accompanies these with an energy equal to the difference in α-particle energies, 0.048 MeV. Radiation of such energy occurs at the extreme right of the electromagnetic spectrum, as shown in Figure 8-5; this is called γ radiation.

Example 3

What is the wavelength of 0.048-MeV γ radiation?

Solution

$$E = 0.048 \times 10^6 \text{ eV} \times \frac{1.602 \times 10^{-19} \text{ J}}{1 \text{ eV}} = 0.077 \times 10^{-13} \text{ J}$$

$$= h\nu = hc/\lambda \qquad (\lambda = hc/E)$$

$$\lambda = \frac{6.626 \times 10^{-34} \text{ J sec} \times 3.00 \times 10^8 \text{ m sec}^{-1}}{0.077 \times 10^{-13} \text{ J}}$$

$$= 0.26 \times 10^{-10} \text{ m} = 0.26 \text{ Å}$$

Radiation in this wavelength region is called either x rays or γ rays, depending on the source of the radiation. That from nuclear decay traditionally is called γ radiation, and that obtained by bombarding a metal anode with a

Figure 23-2

A nuclear energy-level diagram showing the ground state and an excited state for $^{234}_{90}$Th. This diagram explains the occurrence of α particles with 4.195 and 4.147 MeV energy, and γ radiation with 0.048 MeV energy, in the decay of $^{238}_{92}$U. More complicated patterns of γ radiation during decay reveal even more about nuclear energy levels.

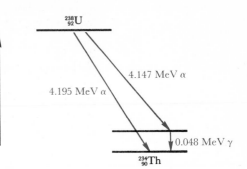

beam of electrons is called x radiation. Both are electromagnetic radiation.

The 4.13-MeV α particles are emitted when ^{238}U is converted to ^{234}Th in an *excited* nuclear state. When ^{234}Th drops to its *ground* state, 0.05-MeV γ radiation is liberated. The two energy levels for ^{234}Th are shown in Figure 23-2. Thus, a form of "nuclear spectroscopy" is possible, analogous to atomic spectroscopy with the hydrogen atom. In the future, such studies may reveal the substructure of the nucleus.

Stability and Half-Life

The number of nuclei that decay in a given time in a sample of material is always proportional to the amount of material left to decay. Nuclear decay is an independent, nucleus-by-nucleus process. The probability that a given nucleus will decay in a given time is constant and independent of the surroundings of the nucleus. Therefore, the total number of nuclei that decay in a given time is proportional to the total number of nuclei present. If the number of nuclei is n, and if the change in this number is Δn during a time Δt, then the decay is represented by

$$\text{Nuclei decaying per unit of time} = -\frac{\Delta n}{\Delta t} = kn \qquad (23\text{-}1)$$

This is known as a *first-order* rate equation since the rate depends on the first power of a concentration term, n. Using calculus, we can convert this rate equation into an expression relating the number of nuclei, n, that remain at time t to the number originally present, n_0, at time $t = 0$. This integrated expression is

$$n = n_0 e^{-kt} \qquad (23\text{-}2)$$

This expression is plotted in Figure 23-3 for the decay of $^{14}_{6}$C, starting with $n_0 = 1$ g of carbon-14. Although this is a plot of equation 23-2, you can verify that it is compatible with equation 23-1 by noting that the slope of the curve, which is approximately equal to $\Delta n/\Delta t$, is proportional to the amount of carbon left at every point on the curve.

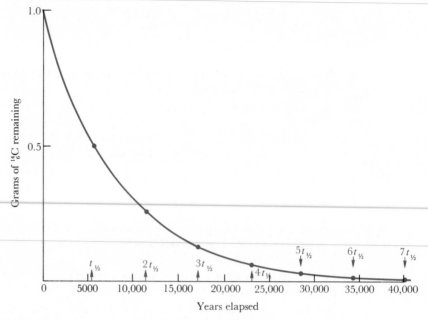

Figure 23-3

The nuclear decay curve for carbon-14: $^{14}_{6}C \rightarrow ^{14}_{7}N + ^{0}_{-1}e$. Every 5570 years, the amount of carbon remaining is halved. This half-life of 5570 years is given the symbol $t_{1/2}$. The regular decay of carbon-14 is the basis for dating carbon-containing objects that ceased to be alive within the last 20,000 years.

An important property of first-order decay is that the time required for any amount of material to decay to half that amount is constant and independent of the amount of material present. If this time is defined as the **half-life**, $t_{1/2}$, then

$$\frac{n_0}{2} = n_0 e^{-kt_{1/2}}$$

$$2 = e^{+kt_{1/2}}$$

$$t_{1/2} = \frac{\ln 2}{k} = \frac{0.693}{k}$$

If either the half-life, $t_{1/2}$, or the first-order rate constant, k, is known, the other can be calculated. If the half-life of a radioactive substance is 10 days, the same time (10 days) is required for 1 g of material to decay to a half gram, for 16 g to decay to 8 g, for 10 tons to decay to 5 tons, or for 2 mg to decay to 1 mg. This may seem paradoxical, until you recall that 10 tons of radioactive material contain many more atoms than 2 mg, so it is natural that many more should decay in the 10-day period if *each atom* has precisely the same probability of decaying within 10 days' time.

The half-lives for the reactions that we have examined vary enormously. The half-life for the reaction

$$^{238}_{92}U \rightarrow ^{234}_{90}Th + ^4_2He$$

is 4.51 billion years. The half-life for polonium decay, according to the reaction

$$^{207}_{84}Po \rightarrow ^{203}_{82}Pb + ^4_2He$$

is 5.7 hours. The decay of carbon-14 in Figure 23-3 has a half-life of 5570 years, and the decay of astatine to bismuth,

$$^{216}_{85}At \rightarrow ^{212}_{83}Bi + ^4_2He$$

has a half-life of only 3×10^{-4} sec.

23-3 STABILITY SERIES

In Figure 23-4 the stable isotopes for the elements are plotted in color; the radioactive isotopes are plotted in black. Notice that in stable isotopes beyond H and He the number of protons is never greater than the number of neutrons, and that the most stable isotopes have an excess of neutrons. The neutrons "dilute" the positive charges of protons and help to stabilize the nucleus against charge repulsion.

Note also the zigzag appearance of the band of stable isotopes, and the predominance of isotopes with even numbers of protons, or neutrons, or both. This is suggestive of some type of pairing and of substructure within the nucleus. This preference for even numbers of each type of nucleon is even more apparent in Table 23-2.

Isotopes to the right of and below the region of greatest stability in Figure 23-4 can reach this region by losing electrons. β^- emission is the rule when such nuclei decay. Isotopes to the left and above the stable region can decay to stable isotopes by electron capture or positron emission. Above the region of approximately $Z = 80$, α-particle emission predominates. β^- emission moves the isotope diagonally up to the left by one square; either electron capture or positron emission proceeds in the reverse direction, down and to the right by one square. α-particle emission takes the nucleus down and to the left by *two* squares, almost along the line of greatest stability. This mode of decomposition is used by atoms beyond the end of the stable region in Figure 23-4.

Figure 23-4 shows only the *existence* of stable (nonradioactive) isotopes, and not their degree of nuclear stability or their abundance. Nuclei are particularly stable and abundant when they have either Z or n (the number of neutrons) equal to 2, 8, 20, 28, 50, 82, or 126. These have been called **magic numbers**. Although they convey information about the shell structure of the nucleus, we do not yet have a theory that will explain them. They are comparable to the set of magic numbers of 2, 10, 18, 36, 54, and

Table 23-2

Frequency of Occurrence of Stable Nuclei with Even and Odd Numbers of Protons and Neutrons

	Neutrons	
	Even	Odd
Protons		
Even	166	53
Odd	57	8

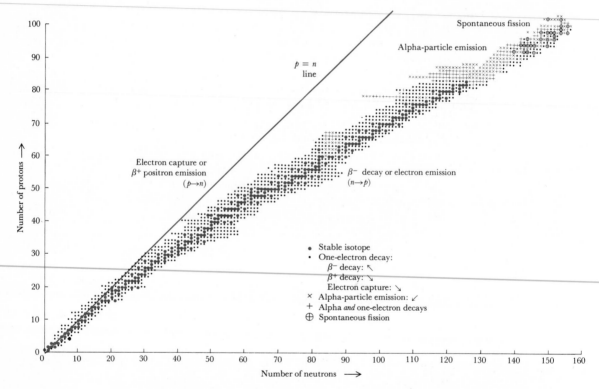

Figure 23-4   A plot of stable isotopes (colored dots) and radioactive isotopes (black dots) as a function of their number of protons, p or Z, and number of neutrons, n. The band of stable isotopes is bordered on both sides by radioactive isotopes. On this plot, radioisotopes lying above the band of stable isotopes decay to stable isotopes by electron capture (EC) or positron emission (β^+). Radioisotopes below the stable band decay by electron (β^-) emission. The stable band ends at $^{209}_{83}$Bi, but there are naturally occurring isotopes with quite long half-lives beyond this point. For example, $^{232}_{90}$Th has a half-life of 13.9 billion years, and $^{238}_{92}$U has a half-life of 4.51 billion years. (From Dickerson and Geis, *Chemistry, Matter, and the Universe.*)

86 for the atomic numbers of the particularly stable noble gases. There must be an explanation in terms of nuclear shell structure, and these nuclear quantum shells must exist independently for protons and neutrons. A magic number of either protons or neutrons bestows stability upon the nucleus; atoms such as $^{208}_{82}Pb$ with magic numbers of each are exceptionally stable. Such nuclei are also nearly spherical, as determined from their quadrupole moments. The shell theories of the nucleus that have been proposed have led to some useful predictions, but with nuclear structure we are presently where Bohr and Sommerfeld were with electronic structures of atoms. Someone in the next generation of scientists is needed to give us the equivalent of the Schrödinger equation and wave mechanics for nuclei.

Natural Radioactive Series

The heavy, unstable elements at the upper right end of the stability curve in Figure 23-4 decay in a set of four pathways. Two of these, starting with $^{238}_{92}U$ and $^{235}_{92}U$, and ending with stable isotopes of lead, are illustrated in Figure 23-5. This enlarged fragment of Figure 23-4 shows more clearly the changes produced on the plot by α emission, β^- emission, and electron capture or β^+ emission. A third such series begins with $^{232}_{90}Th$ and ends, after 10 steps, at $^{208}_{82}Pb$. A fourth begins at the artificial $^{241}_{94}Pu$ and ends at $^{209}_{83}Bi$.

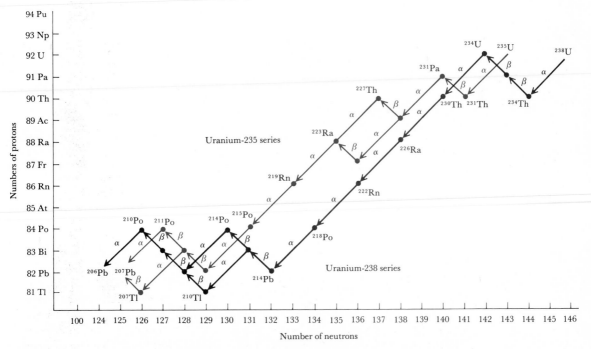

Figure 23-5   Uranium-235 and uranium-238 each decay by a series of α particle and β emissions to a stable isotope of lead. Thorium-232 and plutonium-241 have similar multistep decay chains. (From Dickerson and Geis, *Chemistry, Matter, and the Universe.*)

23-4 NUCLEAR REACTIONS

In 1919, Rutherford carried out the first transmutation of one element into another by bombardment. He used α particles to change nitrogen into oxygen:

$$^{14}_{7}N + ^{4}_{2}He \rightarrow ^{17}_{8}O + ^{1}_{1}H$$

He demonstrated the occurrence of this reaction by detecting the protons emitted. In this reaction, the α particle fuses with the nitrogen nucleus to form an unstable and excited intermediate, $^{18}_{9}F$, which then decomposes to oxygen and the proton. In nuclear reactions such as Rutherford's it is difficult to force a charged particle close enough to the nucleus to react. One of the chief goals of the development of particle accelerators such as the linear accelerator and the cyclotron has been to produce beams of positively charged nuclei having enough energy to make them react with target nuclei.

Neutron beams need not be as energetic, because the neutrons themselves are not electrostatically repelled by the target nuclei. For example, neutron beams from atomic reactors are used to prepare tritium, $^{3}_{1}H$, for medical or for chemical tracer work:

$$^{10}_{5}B + ^{1}_{0}n \rightarrow ^{3}_{1}H + 2\,^{4}_{2}He$$
$$^{6}_{3}Li + ^{1}_{0}n \rightarrow ^{3}_{1}H + ^{4}_{2}He$$

Radioactive cobalt-60, which is used in cancer therapy, is prepared from the stable isotope by neutron bombardment:

$$^{59}_{27}Co + ^{1}_{0}n \rightarrow ^{60}_{27}Co$$

Artificial Elements

One of the most interesting applications of high-energy accelerators has been the preparation of new transuranium elements. Elements 93 through 105 have been prepared by the following bombardment reactions:

$$^{238}_{92}U + ^{2}_{1}H \rightarrow ^{238}_{93}Np + 2\,^{1}_{0}n \qquad \text{(Neptunium)}$$
$$^{238}_{92}U + ^{4}_{2}He \rightarrow ^{239}_{94}Pu + 3\,^{1}_{0}n \qquad \text{(Plutonium)}$$
$$^{239}_{94}Pu + ^{4}_{2}He \rightarrow ^{240}_{95}Am + ^{1}_{1}p + 2\,^{1}_{0}n \qquad \text{(Americium)}$$
$$^{239}_{94}Pu + ^{4}_{2}He \rightarrow ^{242}_{96}Cm + ^{1}_{0}n \qquad \text{(Curium)}$$
$$^{244}_{96}Cm + ^{4}_{2}He \rightarrow ^{245}_{97}Bk + ^{1}_{1}p + 2\,^{1}_{0}n \qquad \text{(Berkelium)}$$
$$^{238}_{92}U + ^{12}_{6}C \rightarrow ^{246}_{98}Cf + 4\,^{1}_{0}n \qquad \text{(Californium)}$$
$$^{238}_{92}U + ^{14}_{7}N \rightarrow ^{247}_{99}Es + 5\,^{1}_{0}n \qquad \text{(Einsteinium)}$$
$$^{238}_{92}U + ^{16}_{8}O \rightarrow ^{249}_{100}Fm + 5\,^{1}_{0}n \qquad \text{(Fermium)}$$
$$^{253}_{99}Es + ^{4}_{2}He \rightarrow ^{256}_{101}Md + ^{1}_{0}n \qquad \text{(Mendelevium)}$$

$$^{246}_{96}\text{Cm} + ^{13}_{6}\text{C} \rightarrow ^{254}_{102}\text{No} + 5\,^{1}_{0}n \qquad \text{(Nobelium)}$$

$$^{252}_{98}\text{Cf} + ^{10}_{5}\text{B} \rightarrow ^{257}_{103}\text{Lr} + 5\,^{1}_{0}n \qquad \text{(Lawrencium)}$$

$$^{249}_{98}\text{Cf} + ^{12}_{6}\text{C} \rightarrow ^{257}_{104}\text{XX} + 4\,^{1}_{0}n \qquad \text{(Unnamed)}$$

$$^{249}_{98}\text{Cf} + ^{15}_{7}\text{N} \rightarrow ^{260}_{105}\text{XX} + 4\,^{1}_{0}n \qquad \text{(Unnamed)}$$

As you can surmise from the names given these new elements, many of them have been produced at the Lawrence Radiation Laboratory at the University of California, Berkeley. (When announcing the naming of elements 97 and 98, the *New Yorker* magazine commented sarcastically that the Lawrence Laboratory had missed a good thing in not using the names "Universitium" and "Ofium" first.) Perhaps now that Mendeleev has been honored, it would be appropriate to use the name *Newlandium* for element 105, since 105 is evenly divisible into octaves (which does *not* mean divisible by eight).

What lies ahead for the synthesis of transuranium elements? Will there be more radioactive and extremely short-lived species such as 97 through 105? It now appears that there is a chance of reaching a new zone of stability that might even include some nonradioactive elements. Calculations with nuclear-shell models have led to the expectation that element $^{298}_{114}\text{XX}$, with 114 protons and 184 neutrons (both magic numbers in the nuclear shell theory) would be on an island of stability in a sea of instability. Figure 23-4 is replotted with a third dimension in Figure 23-6, whose vertical axis is a measure of nuclear stability. If some means can be found to reach the elements in the neighborhood of $^{298}_{114}\text{XX}$, we may have a set of relatively long-lived species. Some attempts have been made at Berkeley, and the following reactions are possibilities:

$$^{248}_{96}\text{Cm} + ^{40}_{18}\text{Ar} \rightarrow ^{284}_{114}\text{XX} + 4\,^{1}_{0}n$$

$$^{244}_{94}\text{Pu} + ^{48}_{20}\text{Ca} \rightarrow ^{288}_{114}\text{XX} + 4\,^{1}_{0}n$$

The first reaction has been unsuccessful, probably because its product nuclei are neutron-deficient and hence unstable; they lie just to the left of the island of stability in Figure 23-6. The second reaction is more promising, but could not be tried until recently for lack of suitable heavy-ion accelerators.

With the realization that it may be possible to jump to a new region of stability, it is of interest to extend the periodic table further. Figure 23-7 extends the table through the end of the presently partially complete seventh period and a new eighth period. In this period, for the first time, there appear *g* orbitals, the 5*g*. There may be some initial uncertainties about the order of filling the 5*g*, 6*f*, and 7*d*. However, recent calculations at Los Alamos have indicated that after the first one or two electrons, the remainder fill the 5*g* orbitals in an orderly manner. These elements might be called the *hypotransition metals,* from *hypo,* meaning "deeply buried" or "below."

Nuclear Chemistry

Figure 23-6 The relative stabilities of isotopes as a function of the number of protons, p, and neutrons, n. The axes in the horizontal plane are the ones of Figure 23-4. The vertical axis indicates the relative stability of the nuclei, with stable nuclei rising above an imaginary ''sea level'' in a long peninsula. Nuclei that have a magic number of either protons or neutrons (28, 50, 82, 126) are less susceptible to breakdown than their neighbors. These correspond to ridges on the ocean floor. The doubly magic $^{208}_{82}Pb$ nucleus is shown as a mountain, and the suspected region of stability around $^{298}114$ as a hill above the waterline. [Redrawn from G. T. Seaborg and J. L. Bloom, ''The Synthetic Elements: IV,'' Copyright © April 1969 by *Scientific American*. Also see S. G. Thompson and C. F. Tsang, ''Superheavy Elements,'' *Science* **178,** 1047 (1972).]

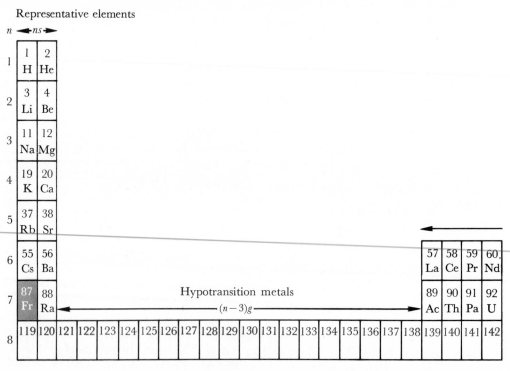

Figure 23-7 The hyperlong form of the periodic table, extended to include the eighth period and the insertion of the 5*g* orbitals to create a series of hypotransition metals. Artificial elements are printed in color. (Francium and astatine do exist in nature as short-lived intermediates in nuclear decay processes, but it is estimated that there is never more than an ounce of either element in the earth's crust at any given moment.) Elements with colored atomic numbers have not yet been prepared. The new artificial transuranium elements from 95 to

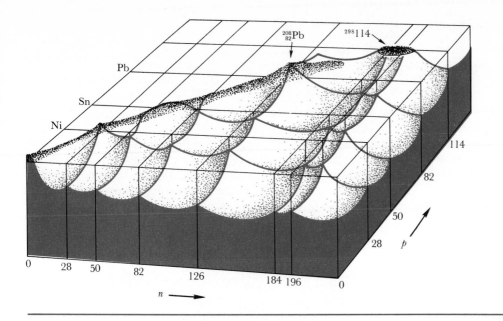

Representative elements

																5 B	6 C	7 N	8 O	9 F	10 Ne	2
																13 Al	14 Si	15 P	16 S	17 Cl	18 Ar	3
							21 Sc	22 Ti	23 V	24 Cr	25 Mn	26 Fe	27 Co	28 Ni	29 Cu	30 Zn	31 Ga	32 Ge	33 As	34 Se	35 Br	36 Kr	4
							39 Y	40 Zr	41 Nb	42 Mo	43 Tc	44 Ru	45 Rh	46 Pd	47 Ag	48 Cd	49 In	50 Sn	51 Sb	52 Te	53 I	54 Xe	5
61 Pm	62 Sm	63 Eu	64 Gd	65 Tb	66 Dy	67 Ho	68 Er	69 Tm	70 Yb	71 Lu	72 Hf	73 Ta	74 W	75 Re	76 Os	77 Ir	78 Pt	79 Au	80 Hg	81 Tl	82 Pb	83 Bi	84 Po	85 At	86 Rn	6
93 Np	94 Pu	95 Am	96 Cm	97 Bk	98 Cf	99 Es	100 Fm	101 Md	102 No	103 Lr	104	105														7
																										8

Transition metals
$(n-1)d$

Inner transition metals
$(n-2)f$

105 have completed the second inner-transition series and are beginning a new fourth transition-metal series. The island of stability that Seaborg and his co-workers at Berkeley hope to find lies below lead in the new seventh-period representative elements. The g orbitals being filled in the first hypotransition metals would be so deeply buried that the elements would be extremely difficult to characterize and separate.

Fission

One of the most famous (or infamous) of nuclear reactions is

$$^{235}_{92}\text{U} + ^{1}_{0}n \longrightarrow ^{139}_{56}\text{Ba} + ^{94}_{36}\text{Kr} + 3\,^{1}_{0}n$$

This is the reaction of the ^{235}U atomic bomb. This reaction was produced first in the late 1930s by Enrico Fermi and his colleagues in Rome, and by Otto Hahn, Lise Meitner, and Fritz Strassman at the Kaiser Wilhelm Institute in Berlin. Both groups were trying to produce transuranium elements by the methods used later by the Berkeley group and others. No one expected **fission** (or fragmentation) into two approximately equal pieces to occur. In 1938, Hahn identified one of the decomposition products as barium, thereby indicating that splitting or fission was taking place. He passed the news privately by letter to Lise Meitner, who had been forced to leave Nazi Germany and was working in Scandinavia. Otto Frisch, her cousin, discovered that tremendous quantities of energy were emitted during the reaction, and he realized that fission might have military applications. At the time (early 1939), Frisch was working in Niels Bohr's laboratory in Copenhagen. On a visit to the United States in January of 1939, Bohr related this information about nuclear fission to physicists in the United States. Fermi had fled the Fascists in Italy and was at Columbia University. He verified Bohr's news, as did others at Berkeley.

The Hungarian physicist Leo Szilard, one of many European scientists who sought political asylum in England and the U.S. in the late 1930s, acted as a spokesman for many concerned physicists. Foreseeing the significance of fission, he persuaded Albert Einstein to write a letter to President Roosevelt in August of 1939. In the now-famous letter, Einstein described the military potential of nuclear fission and mentioned the grave possibility that the Nazis might develop it as a weapon in the war that was brewing in Europe. Roosevelt responded by establishing the Manhattan Project and supporting the enormous research effort that led, in 1945, to the first bomb test at Trinity Flats, New Mexico, and the two bombs dropped on Hiroshima and Nagasaki. The atomic bomb, more than any other development, made scientists aware that they had to be concerned with the political and social impact of their discoveries, and led to the increased involvement of scientists in public affairs that has typified the post-World War II era.

The uranium fission reaction is a potential chain reaction because it produces three neutrons for every one used to initiate a fission of a nucleus. Naturally occurring uranium has only a small amount of ^{235}U dispersed among the more abundant ^{238}U. However, if ^{235}U is purified, and enough of it is brought together, each fission of a nucleus will liberate neutrons that will cause the fission of *more* than one nucleus. Thus, a branched-chain explosion will occur. If the piece of ^{235}U is too small, the losses of neutrons to the surroundings will prevent a chain reaction. The mass at which the rate of loss of neutrons to the surroundings becomes smaller than the rate of production of neutrons is the **critical mass** of uranium.

In a nuclear reactor, the reaction takes place rapidly enough to produce usable heat, but not rapidly enough to become a branched-chain explosion. The control of such a reactor is achieved by means of cadmium rods extended through the atomic pile to absorb neutrons. When the rods are pushed into the pile, so many neutrons are absorbed by the cadmium rods that fission occurs at a very low level. As the rods are withdrawn, fewer neutrons are absorbed by the cadmium and more fission occurs.

Fusion

One of the reactions by which the sun produces energy is the **fusion** of hydrogen nuclei to form helium:

$$^2_1H + ^3_1H \rightarrow \,^4_2He + ^1_0n + 17.6\,\text{MeV}$$

To overcome electrostatic repulsion between hydrogen nuclei, the collision energy must be about 0.02 MeV. This energy can be reached in accelerators on a small scale, but the energy obtained is far less than that needed to operate the accelerators. If we could discover some way to make these reactions work on a large scale and in a controlled manner, large quantities of power would be available. Unfortunately, military applications again are farther advanced than peaceful ones. Temperatures of more than 200 million degrees are required to start the fusion process; these temperatures are attained in hydrogen bombs by using an atom bomb as a kind of match. It is much more difficult to carry out fusion in a controlled manner, and the successful fusion reactor has not yet been designed. At the necessary temperatures for fusion, there are no solids, and the normal concepts of containers for reactants are irrelevant. The most promising approach is to utilize magnetic lines of force to keep the hot ionized atoms constrained and away from the walls of the reactor. However, so far no approach has succeeded.

23-5 APPLICATIONS OF NUCLEAR CHEMISTRY AND ISOTOPES

Isotopes are useful in studying nonisotopic and nonradioactive chemical reactions because they are a tag by which the pieces of a reactant can be identified in the products of a reaction. The most common nonradioactive isotopes that are used as chemical markers are ^2H, ^{15}N, and ^{18}O. Radioactive isotopes have the additional advantage that they can be detected and their concentrations measured by counting their radioactivity, rather than by chemical analysis. Some radioactive tracer nuclei are ^3H, ^{14}C, ^{32}P, and ^{35}S.

Chemical Markers

As an example of the usefulness of chemical tags in determining the mechanism of a reaction, consider the cleavage of ATP to make ADP and phosphate:

$$
\text{Adenosine}-O-\underset{\underset{OH}{|}}{\overset{\overset{O}{\|}}{P}}-O-\underset{\underset{OH}{|}}{\overset{\overset{O}{\|}}{\underset{a}{P}}}-O-\underset{\underset{OH}{|}}{\overset{\overset{O}{\|}}{\underset{b}{P}}}-OH + H-O-H \rightarrow
$$

adenosine triphosphate (ATP)

$$
\text{Adenosine}-O-\underset{\underset{OH}{|}}{\overset{\overset{O}{\|}}{P}}-O-\underset{\underset{OH}{|}}{\overset{\overset{O}{\|}}{P}}-OH + H_3PO_4
$$

adenosine diphosphate (ADP)

Is the bond marked *a* or the bond marked *b* cleaved by hydrolysis? Without isotopes there is no way to find out. But if the reaction is run in water enriched with ^{18}O, the question is answered by seeing where the ^{18}O occurs in the products. If cleavage is at bond *a*, the hydroxide from water will bond to ADP, and ADP will contain the ^{18}O. If cleavage occurs at *b*, the ^{18}O will appear in the phosphate. This experiment has been carried out, and cleavage is at *b*.

Melvin Calvin and his colleagues at the University of California, Berkeley, determined the molecular mechanism of photosynthesis with isotope tracers by using $^{14}CO_2$ as the starting material.

Radiometric Analysis

Radiometric analysis is often faster and more accurate than conventional chemical analysis. In the analysis for small quantities of Zn(II), the Zn is precipitated by an excess of $(NH_4)_2HPO_4$, in which the P is the ^{32}P radioisotope. The insoluble $Zn(NH_4)PO_4$ precipitate is washed, and its radioactivity measured. From the known radioactivity of pure ^{32}P, we can calculate the concentration of the phosphate precipitate and hence of Zn. This method is faster than conventional weighing analysis. No weighings are required, and the product need not be pure, so long as all of the radioactive $(NH_4)_2HPO_4$ reagent is washed away.

Isotope-Dilution Methods

Isotope dilution is used when it is difficult to separate all of the substance to be analyzed from a complex mixture. By this method, a small amount of the component to be analyzed in the mixture is added *to* the mixture.

However, the added material has 100% (or at least a known percentage) radioactive species for some atom in the compound. Radioactivities are calculated as **specific activities**, or radioactive disintegrations per second per gram of substance. The added substance is thoroughly mixed. Then the component to be analyzed is isolated by a method that yields, not quantitative separation, but a small amount of extremely pure compound. The dilution of the specific activity of the added compound by the nonradioactive version of the same compound in the mixture indicates the amount of the compound in the initial mixture. For example, if the specific activity of the purified sample is the *same* as that of the added compound, then there was no such compound in the mixture; we are detecting only what we put into the mixture. If the specific activity is half that of the added compound, the compound was present in the mixture in an amount equal to the amount that we added. If the specific radioactivity has fallen to a tenth of its initial value, there must have been nine times as much compound in the mixture as was added. In general, if

W_m = weight of a compound in a mixture

W_a = weight of a radioactive version of the same compound added to the mixture

A_i = specific activity or radioactive decomposition per second per gram of the added compound

A_f = specific activity of the final purified sample

then the weight of the compound in the mixture can be calculated from the expression

$$\frac{W_m}{W_a} = \frac{A_i}{A_f} - 1$$

Example 4

At the end of a synthesis of a benzoic acid (C_6H_5COOH), 10.0 mg of pure benzoic acid with a specific activity for ^{14}C of 1600 counts per minute per milligram (cpm mg^{-1}) are added to the products of reaction and stirred well. A sample of 40 mg of pure benzoic acid is extracted and shows a specific activity for ^{14}C decay of 190 cpm mg^{-1}. How much benzoic acid was produced in the reaction?

Solution

From the expression given, with W_a = 10.0 mg, A_i = 1600 cpm mg^{-1}, and A_f = 190 cpm mg^{-1}

$$W_m = 10.0 \text{ mg} \left(\frac{1600}{190} - 1\right) = 74.2 \text{ mg benzoic acid}$$

Example 5

A 1.0-ml sample of an aqueous solution containing 2.00×10^6 counts sec^{-1} of tritium is injected into the bloodstream of an animal. After time is allowed for complete circulatory mixing, a 1.00-ml sample of blood is withdrawn and found to have an activity of 1.50×10^4 counts sec^{-1}. Calculate the blood volume of the animal.

Solution

The blood volume is 133 ml. This is both more accurate and less harmful than the old method of draining the beast dry.

Radiocarbon Dating

Reactions in the upper atmosphere involving cosmic neutrons maintain a constant supply of $^{14}_{6}C$ in the atmosphere according to the reaction

$$^{14}_{7}N + ^{1}_{0}n \rightarrow {}^{14}_{6}C + {}^{1}_{1}H$$

So long as an organism is alive, it maintains a constant ratio of carbon-14 to the stable carbon-12. Carbon is being lost continually by a living organism in the form of CO_2 and organic waste products, and new carbon is ingested in foods. In plants, the intake of carbon from the atmosphere is direct, by means of photosynthesis. Animals that eat plants, or eat the protein from other animals that have eaten plants, also maintain a steady intake of carbon from the photosynthetic process. Since the turnover of carbon atoms in the food chain is so rapid compared with the half-life of decay of ^{14}C, the isotope composition of carbon in a living organism is the *same* as that in the atmosphere around it. But as soon as the organism dies, it ceases to equilibrate its carbon isotopes with the atmosphere. Carbon-14 decay by the reaction

$$^{14}_{6}C \rightarrow {}^{14}_{7}N + {}^{0}_{-1}e$$

is no longer compensated by more carbon of the atmospheric ratio. Thus, the proportion of carbon-14 in the dead organism decreases. The ratio of ^{14}C to ^{12}C can be used to reveal the age of a sample of a carbon-containing substance (or more precisely, at what time in the past it ceased to be alive).

Living organisms have mixtures of carbon-14 and carbon-12 that produce 15.3 ± 0.1 disintegrations of carbon-14 atoms per minute per gram of carbon. The number of ^{14}C atoms in a sample is proportional to the disintegration rate: $n \propto d$. From the disintegration equation, equation 23-2, we can write

$$kt = \ln \left(\frac{n_0}{n} \right) = \ln \left(\frac{d_0}{d} \right)$$

in which d is the disintegration rate in counts per minute per gram of carbon. Since the rate constant, k, is related to the half-life, $t_{1/2}$, by $t_{1/2} = 0.693/k$, we can write

$$t = \frac{1}{k} \ln \left(\frac{d_0}{d}\right) = \frac{t_{1/2}}{0.693} \ln \left(\frac{d_0}{d}\right) = t_{1/2} \times \frac{2.303 \log_{10} (d_0/d)}{0.693}$$

$$= t_{1/2} \times 3.33 \log_{10} \left(\frac{d_0}{d}\right) \tag{23-3}$$

For $^{14}_{6}C$, which has a half-life of 5570 years and an initial decay rate of 15.3 cpm g^{-1}, equation 23-3 can be written

$$t = 18{,}600 \log_{10} \left(\frac{15.3}{d}\right) \text{years} \tag{23-4}$$

The results of several analyses are shown in Figure 23-8. Carbon-14 dating has been of the utmost importance in straightening out the chronology of prehistoric cultures of Europe and the Middle East. The last ice age ended around 10,000 years ago, and the resulting climatic change in the Middle East led to the invention of farming, the domestication of animals, and the beginning of settled village life—in short, the Neolithic Revolution. It is fortunate for archaeologists that a common element such as carbon has an isotope with a convenient half-life of 5570 years, suitable for dating events in the past 10,000 years.*

The Age of the Earth and the Moon

The ^{238}U radioactive series shown in Figure 23-5 can be used to date the rocks of the earth by measuring the ratio of ^{238}U to the stable end product, ^{206}Pb (assuming that all of this isotope of lead came from a decomposition of ^{238}U). Since the half-life of ^{238}U, 4.5 billion years, is 20,000 times that of the next longest half-life, we can assume that the time required is for a decay of ^{238}U to ^{234}Th, and that the rest of the chain is relatively instantaneous. Then equation 23-3 applies, and $t_{1/2} = 4.5 \times 10^9$ years.

Example 6

In a sample of uranium ore, there are 0.277 g of ^{206}Pb for every 1.667 g ^{238}U. How old is the ore?

*For some of the results, and some of the difficulties in applying carbon-14 dating to archaeology and prehistory, see C. Renfrew, "Carbon-14 and the Prehistory of Europe," *Scientific American,* October 1971, and R. Protsch and R. Berger, "Earliest Dates for Domestication of Animals," *Science* **179**, 235 (1973).

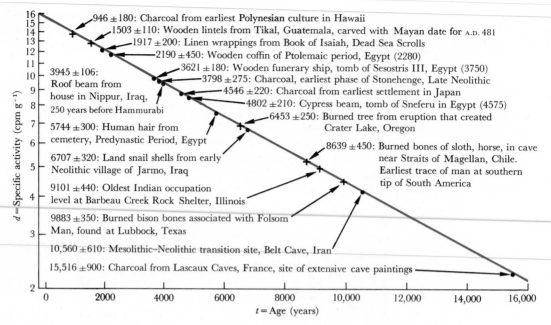

946 ±180: Charcoal from earliest Polynesian culture in Hawaii
1503 ±110: Wooden lintels from Tikal, Guatemala, carved with Mayan date for A.D. 481
1917 ±200: Linen wrappings from Book of Isaiah, Dead Sea Scrolls
2190 ±450: Wooden coffin of Ptolemaic period, Egypt (2280)
3621 ±180: Wooden funerary ship, tomb of Sesostris III, Egypt (3750)
3798 ±275: Charcoal, earliest phase of Stonehenge, Late Neolithic
3945 ±106: Roof beam from house in Nippur, Iraq, 250 years before Hammurabi
4546 ±220: Charcoal from earliest settlement in Japan
4802 ±210: Cypress beam, tomb of Sneferu in Egypt (4575)
5744 ±300: Human hair from cemetery, Predynastic Period, Egypt
6453 ±250: Burned tree from eruption that created Crater Lake, Oregon
6707 ±320: Land snail shells from early Neolithic village of Jarmo, Iraq
8639 ±450: Burned bones of sloth, horse, in cave near Straits of Magellan, Chile. Earliest trace of man at southern tip of South America
9101 ±440: Oldest Indian occupation level at Barbeau Creek Rock Shelter, Illinois
9883 ±350: Burned bison bones associated with Folsom Man, found at Lubbock, Texas
10,560 ±610: Mesolithic–Neolithic transition site, Belt Cave, Iran
15,516 ±900: Charcoal from Lascaux Caves, France, site of extensive cave paintings

Figure 23-8 Plot of carbon-14 decay rate (in counts per minute per gram of carbon from the sample) against the age of the sample in years. This plot is prepared from the equation $t = 18{,}600 \log_{10} (15.3/d)$. It is what is called a *semilog plot*. Although the vertical axis is marked in units of d, it is really measured in equal units of $\log_{10} d$. Such historically datable points as the Ptolemaic period and the period of Sneferu in Egypt permit us to check the whole concept of radiocarbon dating. The agreement is excellent.

Solution If the lead is entirely a decay product of ^{238}U, then to produce 0.277 g lead, the following amount of ^{238}U had to decay:

$$0.277 \text{ g} \times \frac{238}{206} = 0.320 \text{ g uranium}$$

Hence, the initial amount of uranium was 0.320 g + 1.667 g = 1.987 g. Applying equation 23-3, and realizing that the ratio of disintegrations is the same as the ratio of grams, we find that

$$t = 4.5 \times 10^9 \times 3.33 \log_{10} \frac{1.987}{1.667} \text{ years}$$

$$= 1.13 \times 10^9 \text{ years}$$

We can correct for our simplifying assumption that all of the ^{206}Pb came from ^{238}U by looking at the ratio of different lead isotopes in materials containing and lacking ^{238}U. The oldest rocks yet found, granite from the west coast of Greenland, have an age of slightly more than 3.7×10^9 years. We know from dating stony meteorites that the earth was formed by accretion of rocky matter and dust approximately 4.6 billion years ago. (Independent checks of this figure are obtained by rubidium–strontium and potassium–argon dating.) The first billion years of our planet's history have been erased by the continual processes of weathering, erosion, and rebuilding that are typical of the earth.

The moon tells us a different story. Radioisotope dating of rock and soil samples brought back from the moon by the Apollo 11–17 expeditions has transformed our ideas of the moon's origin and history. Before the Apollo program, astronomers argued about whether the lunar craters all had arisen from meteoritic impact, or whether volcanic activity had helped to shape the surface. Two rival theories were proposed for the origin of the moon: that it grew by accretion of dust and rock in the same way and at the same time that the earth did, or that it split off from the earth much later, perhaps leaving a scar that we now see as the Pacific basin.

Apollo 11 and 12 landed in two "maria" or flat lowland regions of the moon in 1969: the Sea of Tranquility and the Ocean of Storms. In both regions the astronauts found a terrain made up of two kinds of material: crystalline rocks of volcanic origin resembling basalt, and breccia, a conglomerate rock produced by shattering and reforming of fragments and dust over the eons. The volcanic basalt samples brought back to earth were dated by three isotope-ratio methods: potassium–argon, rubidium–strontium, and uranium–thorium–lead. All gave the same remarkably great age: between 3.6 billion and 4.2 billion years.* It became apparent that the maria or "seas" on the moon are the result of outfloodings of lava from the interior, which took place in the first billion years of the moon's 4.6-billion-year existence. The cratering in the maria arose from meteorite impact after the lava hardened.

There is even more to be learned from isotope dating. The lava on the floor of the Ocean of Storms is younger by 300 million years than that on the Sea of Tranquility. This is not surprising, since the Ocean of Storms is less heavily cratered by meteorites and therefore appears younger. But the difference in age is not as large as the difference in cratering suggests. To reconcile cratering and age, we must conclude that meteorite collisions were more frequent during the moon's first billion years, and gradually have diminished. This makes very good sense in terms of a picture of the moon initially nucleating out along with the earth and the other planets from a mass of dust and debris around the sun. The rate of meteoritic collision

*The variation is in the rock samples, not the dating methods.

would have been very high during the accretion period, and then would have fallen slowly as most of the solid matter in the vicinity was swept up into the moon or earth.

In early 1971, Apollo 14 landed near Fra Mauro, on a blanket of material that was thrown out during the meteoritic collision that produced the Mare Imbrium. Later that year, Apollo 15 visited the floor of the Mare Imbrium, near Hadley Rille. Isotope dating by Rb–Sr, K–Ar, and ^{40}Ar–^{39}Ar methods again produced a surprise. Basalt samples were found at both sites, but the floor of the Mare Imbrium near Hadley Rille proved to be younger than the debris thrown out by impact: 3.3 billion versus 3.8 billion years. This could only mean that subsequent lava flow occurred years after the mare was formed. Not only did meteorite craters pockmark lava beds, lava flows occurred later in the bottoms of large craters. Both processes transpired continuously to shape the moon.

The "Genesis rock" that Scott and Irwin were so excited about during the Apollo 15 moonwalk was anorthosite, a calcium aluminum silicate of a type that is found in the very oldest terrestrial rocks, crystallized from a molten magma. The astronauts had been trained to look for this type of rock as possible samples of the original lunar crust, and their Genesis rock proved to be 4.2 billion years old. The Apollo 16 and 17 missions to the lunar highlands found mainly this same ancient anorthosite, not covered over by subsequent lava flow as the maria had been. These minerals were compressed into breccias at least 200 meters deep, which are the lunar records of a surface plowed up, crushed, remelted and compacted by 4.2 billion years of meteoritic collisions.

The Antiquity of Life on Earth

Terrestrial geological strata in which fossils of invertebrates first appear in profusion are classified in the Cambrian era; the beginning of this era has been dated by radioisotopic methods to 600 million years ago. Yet almost every present-day phylum of invertebrates is already represented in Cambrian fossils. Where are the fossils of their predecessors during the period of differentiation of phyla?

These pre-Cambrian fossils are more difficult to find; for the animals were soft-bodied, and the rocks into which their sediment hardened have undergone metamorphoses that tended to destroy the fossil record. Fossil plant remains have been unearthed in the Nonesuch shale deposits of northern Michigan, and rubidium–strontium methods date this deposit at about 1.05 billion years. Fossil plants resembling blue-green algae, and others resembling no known species, have been discovered in the Gunflint iron formations of Ontario; these have been dated by potassium–argon methods as about 2 billion years old. The oldest fossil remains of what are believed to be living organisms are fossil bacteria in the Fig Tree deposits in the Transvaal, South Africa. These have been dated by the rubidium–

strontum method as 3.1 billion years old. While the floors of the great lunar maria were being blasted out by meteorites and covered by lava flows, the first traces of life were beginning to appear in the oceans of our planet.

Summary

We are approximately at the same position with nuclei and nuclear structure now that we were with atoms and atomic structure in 1925. We can measure, describe, and classify nuclei, but we do not have a general theory to explain them. Nuclei are made of protons and neutrons that pack together and interact most strongly only with their immediate neighbors in the nucleus. In some ways (as binding energy) they act like packed droplets of uniform particles, and in other ways (preference for even nucleons and **magic numbers**) they act as if they had shell structures similar to electronic shell structures. Energy-level diagrams can be drawn for nuclei from their γ-radiation spectrum in nuclear transformations. There are ground states and excited states for nuclei, as well as for electrons around the nucleus.

Nuclei decay spontaneously by **electron capture**, **electron** or **positron emission**, and **α-particle emission**. This decay is a **first-order rate process**. The **half-life**, or time for the amount of material to decrease to half its initial value, is independent of the amount of material present.

Nuclear reactions can occur when target nuclei are bombarded with other nuclei that are accelerated to sufficient speed to overcome electrostatic repulsion between the positively charged nuclei. Neutrons interact more easily because they have no charge. One important use for nuclear reactions is the preparation of isotopes for chemistry, industry, and medicine. Another is the synthesis of new transuranium elements. Elements to $Z = 105$ have been prepared, and there is hope that those around 114 might be more stable than those prepared recently.

Nuclear fission, having had its beginnings in war, is now being adapted as a practical source of peacetime power. **Fusion** has yet to be made practical as a power source, but remains promising.

Isotopes, radioactive or not, are of great utility in chemistry as a means of tagging reactants and following them through the reaction to products. Radioactive isotopes are especially useful in chemical analyses.

In 1650, Archbishop Ussher made one of the first serious calculations of the age of the earth. On the basis of biblical genealogies, he placed Creation Day at 4004 B.C.; this gave the earth an age of 5654 years. More recent values based on nuclear genealogies set the figure closer to 4.5 billion years. For those who like coincidences, Ussher's original age of the earth is almost exactly one half-life of carbon-14. The first living organisms of which we have any fossil remains are fossil bacteria with an age of approximately 3.4 billion years. Somewhere in the first billion years of this planet's existence, chemical evolution developed to the point at which bacterialike organisms existed. From these organisms, during the following 3.4 billion years, evolved the vast diversity of living organisms that we now find around (and including) us.

Self-Study Questions

1. How are isotopes distinguished when chemical symbols are written?
2. How does the size of a nucleus compare with that of the entire atom?
3. What is the binding energy of a nucleus? How is it calculated, and what does it signify? Where does the energy come from?
4. What is the difference between β^+ emission and electron capture if both result in the decrease of atomic number by one?
5. How can the γ radiation accompanying α emission give us information about the nucleus?
6. What is the half-life of a radioactive element?
7. Why does it take as long for 10 g of a radioactive substance to decay to 5 g as for 1 g to decay to $\frac{1}{2}$ g? If in the first reaction, 10 g decay to 5 g in a year, why won't 7 g decay to 2 g (also a loss of 5 g) in the same time?
8. What are the magic numbers in nuclear physics, and what do they imply about the nucleus? Are there analogous numbers in atomic structure? What do they represent?
9. How does fission differ from decay?
10. What is a nuclear chain reaction?
11. What is the difference between nuclear fission and fusion?
12. Why is it useful to use radioactive isotopes, or at least unusual isotopes, in studying chemical reactions? What advantages do ^{32}P and ^{35}S have that ^{15}N and ^{18}O do not have?
13. How does radiometric analysis work?
14. With isotope-dilution methods, you can determine how much of a chemical substance is present in a sample *without* doing a true quantitative separation and analysis. How can you do this?
15. What experimental measurements are made in the radiocarbon dating process? How are these converted to an age for the sample?
16. A scientist interested in the age of a Greek temple investigated the ^{14}C content of wood from the roof beams and $CaCO_3$ from the marble walls, and found radically different results. Why? Which result would be a better means of dating the age of the temple?
17. What inferences might you draw about the history of the moon from the information that the moon contains a greater amount of elements with high melting and boiling points than does the earth?

Problems

The masses of isotopes needed for many of these problems are tabulated in the *CRC Handbook of Chemistry and Physics* and other standard physics reference books.

Nuclear size

1. Determine which naturally occurring isotope has a nuclear radius that is most nearly three times the radius of the hydrogen nucleus. Which isotope has a radius most nearly four times that of the hydrogen nucleus?

Nuclear reaction balancing

2. Complete the following nuclear equations and supply symbols or values for X or x:

a) $^x_{88}\text{Ra} \rightarrow {}^4_2\text{X} + {}^{222}_x\text{X}$
b) $^{14}_x\text{C} \rightarrow {}^x_x\text{N} + {}^0_{-1}e$
c) $^x_x\text{Ne} \rightarrow {}^{19}_x\text{F} + {}^0_{+1}e$
d) $^{73}_x\text{As} + {}^0_{-1}e \rightarrow {}^x_{32}\text{X}$
e) $^{176}_x\text{Lu} \rightarrow {}^x_x\text{X} + {}^x_{-1}e$
f) $^{235}_x\text{U} + {}^1_0n \rightarrow x^1_0n + {}^{94}_{36}\text{X} + {}^{139}_x\text{X}$
g) $^{59}_x\text{Co} + {}^2_1\text{X} \rightarrow {}^{60}_x\text{Co} + {}^x_x\text{X}$
h) $^{19}_9\text{X} + {}^x_x\text{H} \rightarrow {}^{16}_8\text{X} + {}^4_x\text{X}$
i) $^{16}_8\text{X} + {}^2_1\text{X} \rightarrow {}^{14}_7\text{X} + {}^4_x X$
j) $^{26}_x\text{Mg} + {}^1_0\text{X} \rightarrow {}^4_x\text{He} + {}^x_x\text{X}$

Nuclear equations

3. Write equations for each of the following nuclear processes: (a) positron emission by $^{120}_{51}\text{Sb}$; (b) electron emission by $^{35}_{16}\text{S}$; (c) α-particle emission by $^{226}_{88}\text{Ra}$; (d) electron capture by ^7_4Be.

4. The decay of $^{234}_{92}\text{U}$ ultimately leads to $^{206}_{82}\text{Pb}$. The process proceeds via the sequence of α and β^- decay steps: α, α, α, α, α, β^-, α, β^-, β^-, β^-, α. Write the symbol for each of the isotopes produced in the decay process.

Mass–energy equivalence

5. (a) The heat of combustion of $CH_4(g)$ is -495.0 kJ mole^{-1}. Calculate the mass equivalent of this energy in grams. (b) When ^{14}C decays to $^{14}_7\text{N} + {}^0_{-1}e$, a mass loss of 0.000168 amu occurs. How many moles of $CH_4(g)$ would have to be burned to produce the same amount of energy as would be produced from the decay of 1 mole of ^{14}C?

6. When two ^1_1H nuclei and two neutrons combine to form ^4_2He, the mass of the product helium nuclei is not the same as the sum of the masses of the reactant particles. Calculate the energy, in joules per mole of helium atoms, that is equivalent to the change in mass during the reaction. If the energy per helium atom were released as a single photon, what would be its wavelength? How does this wavelength compare with the approximate radius of the helium nucleus?

7. Calculate the binding energy per nucleon for the following nuclear species: (a) $^{12}_6\text{C}$ ($m = 12.0000$ amu); (b) $^{37}_{17}\text{Cl}$ ($m = 36.96590$ amu); (c) $^{208}_{82}\text{Pb}$ ($m = 207.9766$ amu); (d) $^{32}_{16}\text{S}$ ($m = 31.97207$ amu); (e) $^{16}_8\text{O}$ ($m = 15.99491$ amu).

8. The masses of $^{22}_{11}\text{Na}$ and $^{22}_{10}\text{Ne}$ atoms are 21.994435 amu and 21.991385 amu, respectively. In terms of energy, is it possible for ^{22}Na to decay to ^{22}Ne by positron emission?

Mass change and wavelength

9. The α particles emitted by radium have energies of 4.795 MeV and 4.611 MeV. What is the wavelength of the γ radiation accompanying the decay?

10. When an electron and a positron meet, they are annihilated and two photons of equal energy result. What are the wavelengths of these photons?

Radioactive decay

11. The only stable isotope of fluorine is ^{19}F. What radioactivity would you expect from the isotopes ^{17}F, ^{18}F, and ^{21}F?

12. The half-lives of ^{235}U and ^{238}U are 7.1×10^8 years and 4.5×10^9 years, respectively. Presently, uranium is 99.28% ^{238}U and 0.72% ^{235}U. Calculate the percent natural abundance of each uranium isotope when the earth was formed 4.5 billion years ago.

13. The human body contains about 18% carbon by weight. Of that carbon, $1.56 \times 10^{-10}\%$ is ^{14}C. Calculate the mass of ^{14}C present in a 150-lb body. Calculate the number of disintegrations occurring per minute in a body of this weight. (The half-life of ^{14}C is 5570 years.)

14. Radioactive Na with a half-life of 14.8 hours is injected into an animal in a tracer experiment. How many days will it take for the radioactivity to fall to 0.10 of its original intensity?

Half-lives

15. If 1 g of ^{99}Mo decays by β emission to $\frac{1}{8}$ g in 200 hours, what is the half-life of ^{99}Mo?

16. The number of α particles emitted per second by 1 g of radium is 3.608×10^{10}. Determine the decay constant and the half-life of radium.

Chemical analysis

17. Ten grams of a protein are digested and decomposed into amino acids. To this mixture 100 mg of pure deuterium-substituted alanine, $H_2N-CH(CD_3)-COOH$, are added. After thorough mix-

ing, some of the alanine is separated from the mixture and purified by crystallization. This crystalline alanine contains 1.03% deuterium by weight. How many grams of alanine were originally present in the protein digest?

18. A man weighing 70.8 kg was injected with 5.09 ml of water containing tritium (9×10^9 cpm). After 3 hours, the tritiated water had equilibrated with the body water of the patient. A 1-ml sample of plasma water then showed an activity of 1.8×10^5 cpm. Estimate the weight percent of water in the human body.

Radiocarbon dating

19. A sample of wood from an Egyptian mummy case gives 9.4 cpm g^{-1} of carbon of ^{14}C disintegrations. How old is the mummy case?

20. In the skeleton of a fish caught in the Pacific Ocean in 1960, the ^{14}C disintegration rate is 17.2 cpm g^{-1} of carbon. How can this be? What does this imply for Pacific archaeology?

Radioisotope dating

21. In a sample of uraninite ore from the Black Hills of South Dakota, the weight of Pb is 22.8% the weight of the U present. Estimate a minimum age for the earth from this information.

Fission vs. combustion

22. One atom of ^{235}U evolves about 200 MeV when it undergoes fission. How does this heat of fission compare on a weight basis with the heat evolved in the combustion of 1 g of carbon? How many tons of coke (carbon) will give as much heat on combustion as 1 lb (454 g) of ^{235}U evolves on fission?

Suggested Reading

G. Choppin, *Nuclei and Radioactivity, Elements of Nuclear Chemistry,* W. A. Benjamin, Menlo Park, Calif., 1964. A more extensive introduction to nuclear chemistry than this chapter, but at the same level.

R. Jungk, *Brighter Than a Thousand Suns,* Harcourt Brace Jovanovich, New York, 1958. An exceedingly well-written narrative of the rise of the nuclear age, from quantum mechanics in the 1920s through the development of the atom bomb and the rise of the cold war. Raises some ethical issues that are relevant now.

D. H. Kenyon and G. Steinman, *Biochemical Predestination,* McGraw-Hill, New York, 1969. Ignore the title; the book is good. An introduction to what we know about the chemical evolution of life on the primitive earth. Chapter 2, on radioisotope dating and the antiquity of terrestrial life, is particularly relevant.

W. F. Libby, *Radiocarbon Dating,* University of Chicago Press, Chicago, 1955, 2nd ed. Interesting monograph with a discussion of methods of dating objects with ^{14}C, and a summary of the findings in Western Hemisphere and Near Eastern archaeology.

R. Protsch and R. Berger, "Earliest Dates for Domestication of Animals," *Science* **179**, 235 (1973).

C. Renfrew, "Carbon-14 and the Prehistory of Europe," *Scientific American,* October 1971.

G. T. Seaborg, *Man-Made Transuranium Elements,* Prentice-Hall, Englewood Cliffs, N.J., 1963.

S. C. Thompson and C. F. Tsang, "Superheavy Elements," *Science* **178**, 1047 (1972).

Appendix 1:
The Système Internationale (SI) of Units

In 1960 the International Bureau of Weights and Measures established the International System of Units, commonly known as "SI units," to simplify communication among world scientists. SI units are a streamlining of the metric system, in which certain traditional units that arose for historical reasons have been replaced with more rational units in cases where they can be defined easily in terms of simpler, fundamental quantities. Some of the remaining arbitrariness of even the metric system has thus been removed. With familiarity, the SI units are even easier to use than their metric predecessors.

The International System has seven fundamental units:

Physical quantity	Unit	Abbreviation
Length	meter	m
Mass	kilogram	kg
Time	second	s (or sec)
Electric current	ampere	A
Thermodynamic temperature	kelvin	K
Amount of substance	mole	mol
Luminous intensity	candela	cd

All other units can be derived from these, although it is sometimes convenient to create a special name for one of the derived units. For example, volume can be expressed in cubic meters, or m³. But the *liter,* now defined as a cubic decimeter or one-thousandth of a cubic meter, is too convenient a laboratory volume unit to be abandoned.

It is also convenient to define special units for force and work. By Newton's first law of motion, the force that must be exerted on a mass, m, to give it an acceleration, a, is $F = ma$. If a mass of m kilograms is given an acceleration of a meters per second per second, then the force required is $F = ma$ kg m sec⁻². The SI unit of force is the kilogram meter per second per second. By itself such a unit is clumsy, so it is given the name of a *newton* of force after the famous British physicist.

$$1 \text{ newton (N)} = 1 \text{ kg m sec}^{-2} \quad (\text{since } F = ma)$$

Similarly, if a force of F newtons is exerted over a distance of s meters, the work done is $w = Fs$ newton meters. Again, for convenience, the unit of newton meters is defined as a joule, J, after the British thermodynamicist. Hence

$$1 \text{ joule (J)} = 1 \text{ newton meter (N m)} = 1 \text{ kg m}^2 \text{ sec}^{-2} \quad (\text{since } w = Fs)$$

The traditional unit of heat, work, and energy has been the calorie, which was based empirically on the amount of heat required to increase the temperature of a gram of water by 1° kelvin (or simply "1 kelvin" in SI units). Although according to thermodynamics heat, work, and energy are equivalent, the unit of calories has no immediately obvious connection with mass and acceleration. The choice of units tended to obscure a physical relationship. The joule as a measure of heat has the great advantage of making the connection between heat, work, and energy apparent by its very definition. Although most of the older thermodynamic literature has been expressed in calories, the logical simplicity of the joule will eventually ensure its adoption, just as the gallon and yard have given way to the liter and meter in most of the enlightened countries of the world.

Some of the most common derived units in the SI are the following:

Physical quantity	Unit	Abbreviation	Definition in fundamental units
Volume	liter	l	m³/1000
Force	newton	N	m kg sec⁻²
Energy	joule	J	m² kg sec⁻² = newton meters
Power	watt	W	m² kg sec⁻³ = joules per second
Pressure	pascal	Pa	m⁻¹ kg sec⁻² = newtons per square meter
Electric charge	coulomb	C	A sec = ampere seconds
Electric potential	volt	V	m² kg sec⁻³ A⁻¹ = watts per ampere
Electric resistance	ohm	Ω	m² kg sec⁻³ A⁻² = volts per ampere
Frequency	hertz	Hz	sec⁻¹

Multiples or decimal fractions of the basic SI units are designated by prefixes:

Fraction	Prefix	Symbol		Multiple	Prefix	Symbol
10^{-1}	deci	d		10	deka	da
10^{-2}	centi	c		10^2	hecto	h
10^{-3}	milli	m		10^3	kilo	k
10^{-6}	micro	μ		10^6	mega	M
10^{-9}	nano	n		10^9	giga	G
10^{-12}	pico	p		10^{12}	tera	T
10^{-15}	femto	f				
10^{-18}	atto	a				

Hence a centimeter, cm, is one-hundredth of a meter, and a kilometer (km) is 1000 meters. The liter is retained as a derived unit because it is a convenient size for laboratory measurements, and it would be clumsy to refer to it either as a milli-(cubic meter) or a cubic decimeter.

There are other cases where auxiliary units are employed because the straightforward primary SI units are the wrong size. The charge on an electron, or on any ion that has gained or lost one electron, has a magnitude of 1.602189×10^{-19} coulomb. Yet in ordinary chemical discussions, we do not say that the sodium ion has a charge of this magnitude, or even of 0.16022 attocoulomb (aC), although this would be perfectly correct within the framework of the SI. Instead, we define an *auxiliary unit* of charge, equal in magnitude to the charge on the electron, and we express charges on other ions in terms of electronic charges. This is such an unconscious process that we are not aware that we are doing it, and do not think of the electronic charge as an auxiliary unit. If we were absolute purists about SI units (and few working scientists would be), we would have to reject the electronic charge and express the charge on an aluminum ion as $+0.4807$ attocoulomb, rather than simply as $+3$ (electronic charges).

By the same token, the pascal, or newton per square meter, is the logical SI unit of pressure, but is of an inconvenient magnitude for measuring gas pressures comparable to that of the atmosphere of our planet. The standard sea-level atmospheric pressure on this planet is 101,325 pascals. Although we could carry out gas-law calculations in megapascals (MPa), with the standard atmospheric pressure as 0.101325 MPa, convenience and past custom argue for the retention of the *atmosphere* (atm) as an auxiliary unit of pressure, like the electron charge for measuring charge on ions. We have used the atmosphere for gas calculations in this textbook.

One more concession to convenience has been made in this book. Beyond 10^{-2}, there are SI prefixes for fractional powers of 10 only if the exponent is divisible by 3: 10^{-3}, 10^{-6}, 10^{-9}, 10^{-12}. Yet typical bond lengths between atoms lie in the range of one or two times 10^{-10} m. We could have referred to the carbon–carbon single bond length as 0.154 nanometer (nm), and this is done in many textbooks. Yet it seemed to be more convenient to make the unit match the quantity measured, and retain the angstrom unit (Å), now defined as 10^{-10} m. The carbon–carbon single bond length is expressible as 1.54×10^{-10} m, 0.154 nm, or 1.54 Å, and the fact that as an exponent 10 is not integrally divisible by 3 did not seem a sufficiently good reason to abandon a familiar and convenient unit. Most spectroscopic wavelengths in the ultraviolet and visible range are reported in nanometers, however, and you should be able to go back and forth between nanometers and angstroms (1nm = 10 Å) with as much ease as between meters and centimeters.

More information on the SI units can be found in *National Bureau of Standards Special Publication 330,* "The International System of Units (SI)," U.S. Government Printing Office, Washington, D.C., 1972, and in Martin A. Paul, "International System of Units (SI)," *Chemistry* **45**, 14 (1972).

Appendix 2:
Physical Constants
and Conversion Factors

Physical Constants

Acceleration of gravity at sea level and equator	$g = 9.806$ m sec^{-2}
Atomic mass unit	1 amu $= 1.660566 \times 10^{-24}$ g $= (6.022045 \times 10^{23})^{-1}$ g
Avogadro's number	$N = 6.022045 \times 10^{23}$ mole^{-1} or amu g^{-1}
Bohr radius	$a_0 = 0.529177$ Å $= 5.29177 \times 10^{-11}$ m
Boltzmann's constant	$k = 1.38066 \times 10^{-23}$ J K^{-1} molecule^{-1}
Electron charge	$e = 1.602189 \times 10^{-19}$ C
Faraday's constant	$\mathscr{F} = 96{,}484.6$ C mole^{-1}
Gas constant	$R = 8.31441$ J K^{-1} mole^{-1}
	$R = 0.082057$ liter atm K^{-1} mole^{-1}

Masses of fundamental particles:

Electron	$m_e = 9.109534 \times 10^{-31}$ kg $= 0.0005486$ amu
Proton	$m_p = 1.672649 \times 10^{-27}$ kg $= 1.007276$ amu
Neutron	$m_n = 1.674954 \times 10^{-27}$ kg $= 1.008665$ amu
Proton plus electron	$m_{p+e} = 1.673560 \times 10^{-27}$ kg $= 1.007825$ amu

Planck's constant	$h = 6.626176 \times 10^{-34}$ J sec
	$h/2\pi = \hbar = 1.054589 \times 10^{-34}$ J sec
Rydberg constant	$R_\infty = 109{,}737.32$ cm^{-1}
Speed of light	$c = 2.9979246 \times 10^8$ m sec^{-1}

These values have been taken from E. R. Cohen and B. N. Taylor, *J. Phys. Chem. Ref. Data* **2** (4), 741 (1973), and CODATA Task Group on Fundamental Constants Bulletin No. 11, December 1973.

Conversion Factors

1 electron volt (eV) $= 1.602189 \times 10^{-19}$ J

1 J $= 4.184$ calories

1 liter atm $= 101.325$ J

$2.303\,RT = 5.7061$ kJ at 298 K

1 kJ mole^{-1} $= 83.593$ cm^{-1} (from $E = h\nu = hc\bar{\nu}$, with $\bar{\nu}$ in cm^{-1})

1 amu $= 1.49244 \times 10^{-10}$ J $= 931.502 \times 10^6$ eV $= 931.502$ MeV
     (from $E = mc^2$)

Appendix 3:
Standard Enthalpies and Free Energies of Formation, and
Standard Third-Law Entropies, at 298 K

This table gives the standard enthalpies (ΔH^0) and free energies (ΔG^0) of formation of compounds from elements in their standard states in kilojoules per mole, and the thermodynamic or third-law entropies (S^0) of compounds in joules per kelvin per mole, all at 298 K. The state of the compound is specified by the following symbols: (g) = gas; (l) = liquid; (s) = solid; (aq) = aqueous solution. Occasionally the crystal form of the solid is also specified. Compounds are arranged by the group number of a principal element, with metals taking precedence over nonmetals and O and H being considered least important.

   This table is an abbreviated version of a more complete one in calories, in R. E. Dickerson, *Molecular Thermodynamics*, W. A. Benjamin, Menlo Park, Calif., 1969. Other convenient tabulations are found in the *Chemical Rubber Company Handbook of Chemistry and Physics*, and in *Lange's Handbook of Chemistry*. When using tables in calories, remember that 1 calorie = 4.184 joules.

Substance	ΔH^0_{298} (kJ mole^{-1})	ΔG^0_{298} (kJ mole^{-1})	S^0_{298} (J K^{-1} mole^{-1})
H(g)	217.94	203.24	114.61
H$^+$(aq)	0.0	0.0	0.0
H$_3$O$^+$(aq)	−285.84	−237.19	69.940
H$_2$(g)	0.0	0.0	130.59

	Substance	ΔH^0_{298} (kJ mole^{-1})	ΔG^0_{298} (kJ mole^{-1})	S^0_{298} (J K^{-1} mole^{-1})
IA	Li(g)	155.1	122.1	138.67
	Li(s)	0.0	0.0	28.0
	Li$^+$(aq)	−278.46	−293.8	14
	LiF(s)	−612.1	−584.1	35.9
	LiCl(s)	−408.8	−383.6	55.2
	LiBr(s)	−350.3	−339.9	69.0
	LiI(s)	−271.1	−268	—
	Na(g)	108.7	71.11	153.62
	Na(s)	0.0	0.0	51.0
	Na$^+$(aq)	−239.66	−261.87	60.2
	Na$_2$(g)	142.1	104.0	230.2
	NaO$_2$(s)	−259	−195	—
	Na$_2$O(s)	−416	−377	72.8
	Na$_2$O$_2$(s)	−504.6	−430.1	66.9
	NaF(s)	−569.0	−541.0	58.6
	NaCl(s)	−411.00	−384.03	72.4
	NaBr(s)	−359.95	−347.7	—
	NaI(s)	−288.0	−237	—
	Na$_2$CO$_3$(s)	−1131	−1048	136
	K(g)	90.00	61.17	160.23
	K(s)	0.0	0.0	63.6
	K$^+$(aq)	−251.2	−282.2	103
	KCl(s)	−435.87	−408.32	82.68
	KCl(g)	−219	−235	239.5
	Rb(g)	85.81	55.86	169.99
	Rb(s)	0.0	0.0	69.5
	Rb$^+$(aq)	−246	−282.2	125
	RbF(s)	−551.9	−520.0	114
	RbCl(s)	−430.58	−405	—
	RbBr(s)	−389.2	−378.1	108.3
	RbI(s)	−328	−326	118.0
	Cs(g)	78.78	51.21	175.49
	Cs(s)	0.0	0.0	82.8
	Cs$^+$(aq)	−248	−282.0	133
	CsF(s)	−530.9	−500.0	—
	CsCl(s)	−433.0	−404.1	—
	CsBr(s)	−394	−383	121
	CsI(s)	−337	−333	130
IIA	Be(g)	320.6	282.8	136.17
	Be(s)	0.0	0.0	9.54
	Be^{2+}(aq)	−390	−356	—
	Mg(g)	150	115	148.55
	Mg(s)	0.0	0.0	32.5
	Mg^{2+}(aq)	−461.96	−456.01	−118
	MgCl$_2$(s)	−641.83	−542.32	89.5

	Substance	ΔH^0_{298} (kJ mole^{-1})	ΔG^0_{298} (kJ mole^{-1})	S^0_{298} (J K^{-1} mole^{-1})
	$MgCl_2 \cdot 6H_2O(s)$	-2499.6	-2115.6	366
	$Ca(g)$	192.6	158.9	154.8
	$Ca(s)$	0.0	0.0	41.6
	$Ca^{2+}(aq)$	-542.96	-553.04	-55.2
	$CaCO_3(s, \text{ calcite})$	-1206.9	-1128.8	92.9
	$CaCO_3(s, \text{ aragonite})$	-1207.0	-1127.7	88.7
	$Sr(g)$	164	110	164.53
	$Sr(s)$	0.0	0.0	54.4
	$Sr^{2+}(aq)$	-545.51	-557.3	-39
	$Ba(g)$	175.6	144.8	170.28
	$Ba(s)$	0.0	0.0	67
	$Ba^{2+}(aq)$	-538.36	-560.7	13
	$BaCl_2(s)$	-860.06	-810.9	126
	$BaCl_2 \cdot H_2O(s)$	-1165	-1059	167
	$BaCl_2 \cdot 2H_2O(s)$	-1461.7	-1296	203
IVB	$Ti(g)$	469	423	100.20
	$Ti(s)$	0.0	0.0	30.3
	$TiO_2(s, \text{ rutile III})$	-912.1	-852.7	50.25
	$TiO^{2+}(aq)$	—	-577	—
	$Ti_2O_3(s)$	-1536	-1448	78.78
	$Ti_3O_5(s)$	-2443	-2300	129.4
VIB	$W(g)$	843.5	801.7	173.85
	$W(s)$	0.0	0.0	33
VIII	$Fe(g)$	404.5	358.8	100.3
	$Fe(s)$	0.0	0.0	27.2
	$Fe^{2+}(aq)$	-87.9	-84.93	-113
	$Fe^{3+}(aq)$	-47.7	-10.6	-293
	$Fe_2O_3(s, \text{ hematite})$	-822.1	-741.0	90.0
	$Fe_3O_4(s, \text{ magnetite})$	-1121	-1014	146
IB	$Cu(g)$	341.0	301.4	166.28
	$Cu(s)$	0.0	0.0	33.3
	$Cu^+(aq)$	51.9	50.2	-26
	$Cu^{2+}(aq)$	64.39	64.98	-98.7
	$CuSO_4(s)$	-769.86	-661.9	113
	$Ag(g)$	289.2	250.4	172.892
	$Ag(s)$	0.0	0.0	42.702
	$AgCl(s)$	-127.03	-109.72	96.11
	$AgNO_2(s)$	-44.371	19.85	128.1
	$AgNO_3(s)$	-123.1	-37.2	140.9
IIB	$Zn(s)$	0.0	0.0	41.6
	$ZnO(s)$	-348.0	-318.2	43.9
	$Hg(g)$	60.84	31.8	174.9

Substance	ΔH^0_{298} (kJ mole^{-1})	ΔG^0_{298} (kJ mole^{-1})	S^0_{298} (J K^{-1} mole^{-1})
Hg(l)	0.0	0.0	77.4
HgCl$_2$(s)	-230	-186	144
Hg$_2$Cl$_2$(s)	-264.9	-210.7	196

	Substance	ΔH^0_{298} (kJ mole^{-1})	ΔG^0_{298} (kJ mole^{-1})	S^0_{298} (J K^{-1} mole^{-1})
IIIA	B(g)	406	363	153.34
	B(s)	0.0	0.0	6.53
	B$_2$O$_3$(s)	-1264	-1184	54.02
	B$_2$H$_6$(g)	31	82.8	232.9
	B$_5$H$_9$(g)	62.8	166	275.6
	BF$_3$(g)	-1110	-1093	254.0
	BF$_4^-$(aq)	-1527	-1435	167
	BCl$_3$(g)	-395	-380	289.9
	BCl$_3$(l)	-418.4	-379	209
	BBr$_3$(g)	-187	-213	324.2
	BBr$_3$(l)	-221	-219	229
	Al(g)	314	273	164.44
	Al(s)	0.0	0.0	28.32
	Al^{3+}(aq)	-524.7	-481.2	-313
	Al$_2$O$_3$(s)	-1669.8	-1576.4	50.986
	AlCl$_3$(s)	-653.4	-636.8	167
	TlI(g)	33	-13	274
	TlI(s)	-124	-124	123

	Substance	ΔH^0_{298} (kJ mole^{-1})	ΔG^0_{298} (kJ mole^{-1})	S^0_{298} (J K^{-1} mole^{-1})
IVA	C(g)	718.384	672.975	157.99
	C(s, diamond)	1.896	2.866	2.439
	C(s, graphite)	0.0	0.0	5.694
	CO(g)	-110.523	-137.268	197.91
	CO$_2$(g)	-393.513	-394.383	213.64
	CO$_2$(aq)	-412.9	-386.2	121
	CH$_4$(g)	-74.848	-50.794	186.2
	C$_2$H$_2$(g)	226.75	209	200.82
	C$_2$H$_4$(g)	52.283	68.124	219.5
	C$_2$H$_6$(g)	-84.667	-32.88	229.5
	C$_3$H$_8$(g)	-103.8	-23.5	269.9
	n-C$_4$H$_{10}$(g)	-124.7	-15.7	310.0
	i-C$_4$H$_{10}$(g)	-131.6	-18.0	294.6
	C$_6$H$_6$(g)	82.927	129.65	269.2
	C$_6$H$_6$(l)	49.028	124.50	172.8
	n-C$_8$H$_{18}$(l)	-250.0	6.48	360.8
	HCOOH(g)	-362.6	-335.7	251
	HCOOH(l)	-409.2	-346.0	129.0
	HCOOH(aq)	-410.0	-356.1	164
	HCOO$^-$(aq)	-410.0	-334.7	91.6
	H$_2$CO$_3$(aq)	-698.7	-623.4	191
	HCO$_3^-$(aq)	-691.1	-587.1	95.0
	CO$_3^{2-}$(aq)	-676.3	-528.1	-53.1
	CH$_3$COOH(l)	-487.0	-392	160

Substance	ΔH^0_{298} (kJ mole^{-1})	ΔG^0_{298} (kJ mole^{-1})	S^0_{298} (J K^{-1} mole^{-1})
$CH_3COOH(aq)$	−488.453	−399.6	—
$CH_3COO^-(aq)$	−488.871	−372.5	—
$(COOH)_2(s)$	−826.8	−697.9	120
$(COOH)_2(aq)$	−818.3	−697.9	—
$HC_2O_4^-(aq)$	−818.8	−698.7	154
$C_2O_4^{2-}(aq)$	−824.2	−674.9	51.0
$HCHO(g)$	−116	−110	218.7
$HCHO(aq)$	—	−130	—
$CH_3OH(g)$	−201.3	−161.9	236
$CH_3OH(l)$	−238.64	−166.3	127
$CH_3OH(aq)$	−245.9	−175.2	132.3
$C_2H_5OH(g)$	−235.4	−168.6	282
$C_2H_5OH(l)$	−277.63	−174.8	161
$CH_3OCH_3(g)$	−184.05	−112.92	267.1
$CH_3CHO(g)$	−166.4	−133.7	266
$CH_3CHO(l)$	−195	—	—
$CH_3CHO(aq)$	−208.7	—	—
$CH_3NH_2(g)$	−28	28	241.5
$CH_3SH(g)$	−12.4	0.88	254.8
$C_6H_{12}O_6(s)$	−1260	−919.2	288.9
$Si(g)$	368.4	323.9	167.86
$Si(s)$	0.0	0.0	18.7
$SiO(g)$	−111.8	−137.1	206.1
$SiO_2(s, \text{quartz})$	−859.4	−805.0	41.84
$Ge(g)$	328.2	290.8	167.8
$Ge(s)$	0.0	0.0	42.42
$Sn(g)$	300	268	168.4
$Sn(s, \text{gray})$	2.5	4.6	44.8
$Sn(s, \text{white})$	0.0	0.0	51.5
$Pb(g)$	193.9	161.0	175.27
$Pb(s)$	0.0	0.0	64.89
$Pb^{2+}(aq)$	1.6	−24.3	21
$PbO(s, \text{red})$	−219.2	−189.3	67.8
$PbO(s, \text{yellow})$	−217.9	−188.5	69.5
$PbO_2(s)$	−276.6	−219.0	76.6

VA

Substance	ΔH^0_{298} (kJ mole^{-1})	ΔG^0_{298} (kJ mole^{-1})	S^0_{298} (J K^{-1} mole^{-1})
$N(g)$	472.646	455.512	153.195
$N_2(g)$	0.0	0.0	191.49
$N_3^-(aq)$	252	325	134
$NO(g)$	90.374	86.688	210.62
$NO_2(g)$	33.85	51.840	240.5
$NO_2^-(aq)$	−106	−34.5	125
$NO_3^-(aq)$	−206.57	−110.6	146
$N_2O(g)$	81.55	103.6	220.0
$N_2O_2^{2-}(aq)$	−10.8	138	28
$N_2O_4(g)$	9.661	98.286	304.3

	Substance	ΔH^0_{298} (kJ mole^{-1})	ΔG^0_{298} (kJ mole^{-1})	S^0_{298} (J K^{-1} mole^{-1})
	$N_2O_5(s)$	−41.8	134	113
	$NH_3(g)$	−46.19	−16.64	192.5
	$NH_3(aq)$	−80.83	−26.6	110
	$NH_4^+(aq)$	−132.8	−79.50	112.8
	$NH_4Cl(s)$	−315.4	−203.9	94.6
	$N_2H_4(l)$	50.42	—	—
	$NH_2-CO-NH_2(s)$	−333.2	−197.2	104.6
	$(NH_4)_2SO_4(s)$	−1179.3	−900.35	220.3
	$P(g)$	314.5	279.1	163.1
	$P(s, \text{white})$	0.0	0.0	44.4
	$P(s, \text{red})$	−18	−14	29
	$P_4(g)$	54.89	24.4	279.9
	$PCl_3(g)$	−306.4	−286.3	311.7
	$PCl_5(g)$	−398.9	−324.6	353
	$As(g)$	253.7	212.3	174.1
	$As(s, \text{gray metal})$	0.0	0.0	35
	$As_4(g)$	149	105	289
VIA	$O(g)$	247.52	230.09	160.95
	$O_2(g)$	0.0	0.0	205.03
	$O_3(g)$	142	163.4	238
	$OH(g)$	42.09	37.4	183.63
	$OH^-(aq)$	−229.94	−157.30	−10.5
	$H_2O(g)$	−241.83	−228.59	188.72
	$H_2O(l)$	−285.84	−237.19	69:940
	$H_2O_2(l)$	−187.6	−113.97	92
	$H_2O_2(aq)$	−191.1	−131.67	—
	$S(g)$	222.8	182.3	167.72
	$S(s, \text{rhombic})$	0.0	0.0	31.9
	$S(s, \text{monoclinic})$	0.30	0.096	32.6
	$S^{2-}(aq)$	41.8	83.7	—
	$SO(g)$	79.58	53.47	221.9
	$SO_2(g)$	−296.9	−300.4	248.5
	$SO_3(g)$	−395.2	−370.4	256.2
	$H_2S(g)$	−20.15	−33.02	205.6
	$H_2S(aq)$	−39	−27.4	122
VIIB	$F(g)$	76.6	59.4	158.64
	$F^-(aq)$	−329.1	−276.4	−9.6
	$F_2(g)$	0.0	0.0	203
	$HF(g)$	−269	−271	173.5
	$Cl(g)$	121.39	105.40	165.09
	$Cl^-(aq)$	−167.46	−131.17	55.2
	$Cl_2(g)$	0.0	0.0	222.95
	$ClO^-(aq)$	—	−37	43.1
	$ClO_2(g)$	103	123	249
	$ClO_2^-(aq)$	−69.0	−10.7	101

Substance	ΔH^0_{298} (kJ mole^{-1})	ΔG^0_{298} (kJ mole^{-1})	S^0_{298} (J K^{-1} mole^{-1})
$ClO_3^-(aq)$	−98.32	−2.6	163
$ClO_4^-(aq)$	−131.4	−8	182
$Cl_2O(g)$	76.15	93.72	266.5
$HCl(g)$	−92.312	−95.265	186.68
$HCl(aq)$	−167.46	−131.17	55.2
$HClO(aq)$	−116.4	−79.956	130
$ClF_3(g)$	−155	−114	278.7
$Br(g)$	111.8	82.38	174.913
$Br^-(aq)$	−120.9	−102.82	80.71
$Br_2(g)$	30.7	3.14	245.35
$Br_2(l)$	0.0	0.0	152.3
$HBr(g)$	−36.2	−53.22	198.48
$I(g)$	106.61	70.149	180.68
$I^-(aq)$	−55.94	−51.67	109.4
$I_2(g)$	62.241	19.4	260.58
$I_2(s)$	0.0	0.0	117
$I_2(aq)$	21	16.43	—
$I_3^-(aq)$	−51.9	−51.51	174
$HI(g)$	26	1.3	206.32
$ICl(g)$	18	−5.52	247.4
$ICl_3(s)$	−88.3	−22.6	172
$IBr(g)$	40.8	3.8	259
$He(g)$	0.0	0.0	126.1
$Ne(g)$	0.0	0.0	144.1
$Ar(g)$	0.0	0.0	154.7
$Kr(g)$	0.0	0.0	164.0
$Xe(g)$	0.0	0.0	169.6
$Rn(g)$	0.0	0.0	176.1

0

Appendix 4:
Significant Figures and
Exponential [Scientific] Notation

SIGNIFICANT FIGURES*

Calculations in chemistry (or any other science for that matter) involve numbers that owe their origin to an experimental measurement. For example, our problem may be to calculate the volume of a particular gas, given its weight, pressure, and temperature. The data are measured experimentally, and each measurement contains some error. Obviously, this error will be reflected in calculated value for the volume of the gas. There is a natural tendency to "play it safe" and carry out such a calculation to a greater number of decimal points than is justified by the experimental accuracy. Not only does the answer then misrepresent the correct volume, but much effort probably has been wasted in arriving at most of the digits. The temptation to carry too many digits is stronger with a pocket calculator than it was with a slide rule with its intrinsically limited precision. A scientist indicates how "good" a number is by including in it only those digits that are known with certainty *plus* one more. The known digits plus the doubtful one constitute *significant figures*. For example, recording the volume of a gas as 48.12 ml implies four significant figures, of which 4, 8, and 1 are known with certainty, but 2 is doubtful.

We should consider what factors determine the error in a measurement such as the volume of the gas referred to in the preceding paragraph.

*The authors wish to thank Professor Wilbert Hutton for his courtesy in granting permission to reprint the material in Appendix 4 from *A Study Guide to Chemical Principles, Second Edition.*

The error in the volume measured is a composite of the *accuracy* and the *precision* of the measurement. Accuracy concerns the absolute truth of a measurement, whereas precision involves the detail with which a measurement is made. For instance, suppose that the volume of the gas were measured in a 50-ml gas burette. The stated volume of 48.12 ml indicates that the chemist has made this measurement and that he can reproduce it in the same burette to within 0.01 ml (the last digit is doubtful).* His precision, implied in the number 48.12 ml, can be expressed as 48.12 $\pm$ 0.01 ml, or 48.12 $\pm$ 0.02% ml, because

$$\frac{0.01}{48.12} \times 100 = 0.02\%$$

However, the burette itself may be inaccurate; that is, its scale markings might be in error, temperature fluctuations in the room might alter the capacity of the burette from what it was when the markings were etched on the glass, or the volume-indicating liquid in the burette might not have drained completely when the measurement was made. In any of these events, the volume may be measured with great precision, but it would not be very accurate. Of course, all chemists hope that their instruments are calibrated accurately so that the validity of a measurement depends only on the precision with which the chemist is capable of making the measurement.

Although we may not be justified on some occasions, we will assume that all the instruments used for obtaining data have an accuracy comparable to the measurements. Therefore, so far as we are concerned, all numbers in problems contain significant figures; our objective is to make sure that when we manipulate these numbers we do not distort the information by throwing precision away or apparently adding to it. Some simple rules will enable you to do this.

Addition and Subtraction

The reason for a rule in computations involving addition and subtraction can be understood from the following example. If an empty beaker weighs 64 g and you put a sample of NaCl weighing 0.176 g into the beaker, what is the total weight of the beaker and the salt?

Without thinking, you might follow the natural tendency simply to add the two numbers, 64 + 0.176, and report the final weight as 64.176 g. If you do this, you are wrong. Remember that we should write numbers with significant figures only. Stating that the combined weight of the beaker and NaCl is 64.176 g means that you are certain of the digits 6, 4, 1, and 7, and are doubtful of the last 6. You are, in fact, saying that

---

*A less experienced chemist may be capable of reproducing a reading on this burette to within ± 0.02 ml. In such a case, a notation such as 48.12 $\pm$ 0.02 ml should be used; otherwise, one will assume a precision of ± 1 unit in the doubtful significant figure.

the combined weight is known to ±0.001 g, that is, to within plus or minus one part in 64,176—about one part in 64,000 or ±0.0015%. Clearly this is nonsense. The weight of the empty beaker was stated as 64 g, which implies that its weight is known to be 64 ± 1 g. Not only are you in doubt as to the digit 4—it could be 3 or 5, for instance—you haven't been given the slightest indication as to what the values are of the digits to the right of the 4. Consequently, any digits to the right of the decimal point are uncertain, and you should not report them. To do so would be indicating that you have information that you really do not possess.

The correct answer can be computed easily by indicating the uncertain digits by question marks:

$$\begin{array}{r} 64.??? \\ + \ 0.176 \\ \hline 64.??? \end{array}$$

It is obvious that even though we know the digits 1, 7, and 6 in the second line, when they are added to the corresponding uncertain digits in the first line, the resulting values for the digits to the right of the decimal point also will be uncertain. Therefore, you should report 64 g as the correct weight of the beaker plus the salt.

The conclusions we reached in this example can be expressed as the following rule for addition and subtraction:

Round off all the numbers in the group of numbers to be added or subtracted so each has the same number of digits to the right of the decimal point *as that number in the series having the smallest number of digits to the right of the decimal point.* Then add or subtract the resulting series of rounded off numbers.

For example, consider the addition of the numbers

$$\begin{array}{r} 119.2 \\ 204.12 \\ 1.75 \\ 260.3734 \end{array}$$

The number 119.2 has the smallest number of digits to the right of the decimal point: one. Therefore, round off all numbers in the series so each has one digit to the right of the decimal point, then add:

$$\begin{array}{r} 119.2 \\ 204.1 \\ 1.8 \\ 260.4 \\ \hline 585.5 \end{array}$$

Two points need further discussion before proceeding.

1. The convention for rounding off a number depends on the value of the digit to its right. If the digit to the right is greater than 5, the number is increased by one; thus, 260.3734 rounds off to 260.4. If the digit to the right is less than 5, the number is unchanged in rounding it off; thus, 204.12 rounds off to 204.1. If the number to the right is 5, the convention is to increase the value of the digit being rounded off by one if the digit is odd and to leave it unchanged if the digit is even; thus, 1.75 rounds off to 1.8, whereas 1.85 rounds off to 1.8. This latter convention, although somewhat arbitrary, can be justified by reasoning that the chance of there being an odd number to round off is as great as the chance of there being an even number. Therefore, by applying the convention stated here, on the average we will increase as many numbers in rounding off as we will leave unchanged. Any error introduced in rounding off to a number that is too large will eventually be compensated by an error in rounding off a number to a value that is too small.

2. We can alter the procedure stated in the rule and add or subtract the numbers as they are. Then we can round off the answer so it has the same number of digits to the right of the decimal point as does that number in the series with the smallest number of digits to the right of its decimal point. Occasionally, a slightly different answer will result, depending on which procedure you follow. Don't be concerned; remember that the last digit is doubtful anyway. Do the following exercises for practice, and express the answer to the correct number of significant figures.

Exercises

(1) 4.72 + 203.6 + 121.780 + 55
(2) 3.1416 + 2.73 + 5.921 + 3.83
(3) 297.64 − 31.279
(4) 32.7945 + 121.5 − 326.73
(5) 49378.2 + 25.98 − 33

Answers   (1) 385; (2) 15.62; (3) 266.36; (4) −172.4; (5) 49371.

Zeros further complicate the matter of significant figures because they serve two purposes in a number. A zero may indicate that a given decimal place has been measured to be zero; thus, it is significant. Second, a zero may be used to indicate the location of the decimal point, in which case it is not a significant figure. As examples, consider the following numbers: (a) 0.0123; (b) 2027.3; (c) 0.1072; (d) 0.200. The first number has three significant figures, the 1, the 2, and the 3. The zero between the decimal point and the 1 merely locates the decimal; that is, it indicates only that the number is 123 ten-thousandths and not 123, 123 thousandths, and so forth. Thus, it is not counted as a significant figure. The second number has five significant figures. The zero here does not locate the decimal point; it is a necessary digit in the number. The same applies to the zero in (c), which has four significant figures. The last number represents an interesting case. The fact that the number two-tenths can be expressed just as well either as 0.2 or 0.200 indicates that the two zeros to the right of 2 must be significant; otherwise, they would not be written as part of the number. Consequently, there are three significant figures in (d), and we can presume that the measurement was made with a device that is accurate to ± 0.001.

One method often used to avoid the confusion presented by zeros is to write the number as a power of 10. In this form, the exponent locates the decimal point, and only significant figures are included in the base of the number. (If you cannot remember the meaning of exponents, refer to the next section.) The numbers in the preceding example are written as powers of 10 (often called scientific notation) in the following way:

(a) $0.0123 = 1.23 \times 10^{-2}$
(b) $2027.3 = 2.0273 \times 10^3$
(c) $0.1072 = 1.072 \times 10^{-1}$
(d) $0.200 = 2.00 \times 10^{-1}$

As a test of your understanding, indicate the number of significant figures in the following and express the numbers as powers of 10.

Exercises

(1) 2305.0
(2) 0.00007062
(3) 21.070
(4) 0.02003
(5) 900.0
(6) 1000 apples when you are sure of the exact number of apples.
(7) 0.7020 ± 0.001

Answers   (1) five significant figures, 2.3050×10^3; (2) four significant figures, 7.062×10^{-5}; (3) five significant figures, 2.1070×10; (4) four significant figures, 2.003×10^{-2}; (5) four significant figures, 9.000×10^2; (6) four significant figures, 1.000×10^3; (7) three significant figures, 7.02×10^{-1}. The ± 0.001 indicates the uncertainty is in the third decimal place; the digit 2 is uncertain.

Multiplication and Division

The evaluation of the uncertainty in an answer acquired from a sequence of multiplication and division operations is more complicated than it is when the computations involve only addition and subtraction. A precise evaluation requires one to determine the uncertainty in each of the factors and then sum these together to find the uncertainty in the answer. The answer then is written in significant figures, in which the uncertainty appears in the last digit. This procedure is too time consuming to use in practice, and a more rapid, if less precise, procedure is preferable. Such a procedure is stated in the following rule:

> Express the answer to multiplication and/or division such that the answer has the same number of significant figures as does the factor having the *smallest* number of significant figures.

Note that the emphasis is on the number of *significant figures* in multiplication and division. It is not on the number of decimal places in the measurements, as was the case in addition and subtraction. The rule is based on the logical principle that the reliability of an answer determined from the combination of a sequence of numbers will not be any greater than the least reliable number in the sequence. Therefore, since the last digit in a number containing significant figures is doubtful, the uncertainty in the number can be approximated by the number of significant figures; that is, the more significant figures in a number, the more precisely is that number known. A number with four significant figures is known to at least 1 part in 1000, three significant figures to at least 1 part in 100, and so on. We assume, of course, that the uncertainty in the doubtful digit is plus or minus one unit. This assumption will be considered valid for the data in problems.

Consider the following sequences of multiplications and divisions.

Example

$2.760/5.46 = ?$

Solution

The answer, carried to four decimal places, is 0.5055. To determine where the answer should be rounded off, we observe that there are four significant figures in the factor 2.760 (if the zero was not significant, it would not have been included as part of the number) and three significant figures in the factor 5.46. Consequently, the answer should be rounded off to three significant figures and written correctly as 0.506.

Example

$$\frac{(2.56)(1.9)(3.725)}{(6.02 \times 10^{23})(0.0071)} = ?$$

Solution

The numbers of significant figures in the various factors are three in 2.56, two in 1.9, four in 3.725, three in 6.02×10^{23}, and two in 0.0071. The smallest number of significant figures in any of the factors is two, so the answer should be rounded off to two significant figures and is written correctly as 4.2×10^{-21}.

Occasionally a complication occurs such as the following.

Example

$$\frac{(276)(9.9)}{2497.3} = ?$$

Solution

The answer to four decimals is 1.0941. Our rule dictates that we round off this number to two significant figures, that is, 1.1, since 9.9 is the factor known with the least precision (9.9 ± 0.1, or one part out of 99, or about 1%).

But something is not quite right in the preceding solution. The answer (1.1) indicates a precision of one part out of 11, or within about 10%. This is a precision less than that of the least precise factor. In a sense, we cheat ourselves a bit in expressing the answer this way because the most uncertain number that we have in the data is known with almost 10 times the precision indicated by the answer, 1.1. On this basis, we are justified in adding an additional significant figure and expressing the answer as 1.09. This procedure would indicate that the answer is known to 1.09 ± 0.01 (i.e., within one part out of 109 or about 1%, which is a more honest estimate of what we know than is the 1.1 value).

Let us compute the volume of a sphere from the relationship $V = \frac{4}{3}\pi r^3$. The measured quantity is r, and the number of significant figures in the value for r will determine the correct answer. What about $\frac{4}{3}\pi$? Let us think about these numbers for a moment. Pi (π) has an established value that can be determined to any desired number of significant figures, 3.141592653589793. We need only employ in our calculations a value having more significant figures than are known for r. The 4 and the 3 in the fraction $\frac{4}{3}$ each are exact numbers. Although by convention they are not written as such, they are known to an infinite number of significant figures (4.0000000 . . .). You will use many exact numbers in problems and should recognize that since these numbers are exact, they need not be considered in significant-figure decisions.

Example

Suppose that we wish to calculate the volume of a sphere with a diameter, d, of 4.00 cm.

Solution

Since $d = 2r$

$$r = \frac{d}{2} = \frac{4.00 \text{ cm}}{2} = 2.00 \text{ cm}$$

Since 2 is an exact number, the number of significant figures in the radius is determined by the three digits in the value for the diameter. Therefore,

$$V = \tfrac{4}{3}\pi r^3 = \tfrac{4}{3}\pi(2.00 \text{ cm})^3$$

Because there are three significant figures in r, there will be three significant figures in the answer, provided that we use a value for π expressed to at least three significant figures. The correct answer is 33.5 cm^3.

To test your understanding, state the answers to the following exercises in the correct number of significant figures.

Exercises

1. $\dfrac{(2.75)(0.01267)}{(3.1416)} = 0.0110906$

2. $(4.00 \times 10^2)^3 = 64000000$

3. $\dfrac{(105.2)(3.21)}{(1.007)(3.1 \times 10^3)} = 108.176$

4. $\dfrac{(55.2)(0.90)}{(0.4557)} = 109.01$

5. Three samples of ore are weighed on different balances with the uncertainties indicated

    376.6 ± 0.5 g 273.17 ± 0.02 g 0.1725 ± 0.0001 g

    What is the average percent uncertainty in the measurements?

Answers

(1) 0.0111; (2) 6.40 $\times$ 10^7; (3) 1.1 $\times$ 10^2; (4) 109; (5) The three percent uncertainties are 0.13%, 0.007%, and 0.058%; the average percent uncertainty is 0.06%.

EXPONENTIAL NUMBERS OR "SCIENTIFIC NOTATION"

Not only does writing a number in exponential form enable us to express significant-figure information with minimal confusion, it avoids writing many zeros for either small or large numbers. On numerous occasions you will find it convenient to use this notation.

We use exponential numbers to express quantities as multiple powers of 10. An exponential number consists of two parts: a coefficient (chosen to be between 1 and 10) and a power of 10. For example, Avogadro's number is written 6.02×10^{23}; 6.02 is the coefficient and 23 is the power of 10.

A positive exponent, n, indicates that the coefficient should be multiplied by 10 n times; that is, the decimal should be moved n places to the right of its position in the coefficient. A negative exponent, $-m$, indicates that the coefficient should be divided by 10 m times; that is, the decimal should be moved m places to the left. For example,

$$0.0000000192 = 1.92 \times 10^{-8}$$
$$1 \text{ thousand} = 1 \times 10^{3}$$
$$96500 = 9.65 \times 10^{4}$$

To *add* or *subtract* exponential numbers, we must be certain that the powers of 10 are the same. Otherwise, the operation would be like adding different things: $2x + 2y = ?$, whereas $2x + 2x = 4x$. Put another way, 2 hundreds plus 2 thousands does not equal 4 hundreds or 4 thousands. But 2 hundreds plus 20 hundreds (2 thousands) equals 22 hundreds. Thus, before adding or subtracting quantities, the units (in this case, the relative position of the decimal point) must be the same. This requirement may force you to rewrite the exponential number. The revision is easy if you remember that each time the power of 10 becomes more positive by one unit it is the same as multiplying the number by 10, or moving the decimal in the coefficient one place to the right. Similarly, if the power of 10 is made more negative, it is the same as moving the decimal in the coefficient to the left. For example,

$$6.022 \times 10^{23} + 7.65 \times 10^{21} = ?$$

Rewrite both numbers so they have the same power of 10; for example, 21. To write 6.022×10^{23} as a multiple of 10^{21} (the exponent has been decreased by two powers of 10) requires that the coefficient be *increased* by two powers of 10. Therefore, its decimal point should be moved two places to the right:

$$6.022 \times 10^{23} = 602.2 \times 10^{21}$$

Now the numbers can be added:

$$602.2 \ \times 10^{21}$$
$$+7.65 \times 10^{21}$$
$$609.8 \ \times 10^{21} \text{ or } 6.098 \times 10^{23}$$

Do the following exercises to test your understanding.

Exercises

1. Add 2.46×10^{-9} cm to 2.46×10^{-8} cm.

2. Subtract 1.625×10^{-1} cm from 2.234×10^2 cm.

3. Add 4.0075×10^3 ml to 6.23×10^2 ml.

4. Subtract 1.725×10^{-1} g from 2.1623×10^1 g.

Answers   (1) 2.71×10^{-8} cm; (2) 2.232×10^2 cm; (3) 4.630×10^3 ml; (4) 2.1450×10 g.

In *multiplication,* you need only multiply the coefficients together and then multiply the powers of 10 together (add their exponents) to obtain the coefficient and power of 10 for the answer. For example,

$$6.02 \times 10^{23} \times 1.76 \times 10^{-2} = ?$$

The product of the coefficients to the proper number of significant figures is $6.02 \times 1.76 = 10.6$. The product of the powers of 10 (exponentials) are $10^{23} \times 10^{-2} = 10^{[23+(-2)]} = 10^{21}$. The answer to the multiplication is 10.6×10^{21} or, writing the coefficient in the preferred way as a number between 1 and 10, 1.06×10^{22}.

In *division,* the coefficients are divided separately and then the exponentials are divided. Recall that in division of exponential numbers, the exponent of the divisor (the denominator) is subtracted from the exponent of the dividend (the numerator). For example,

$$\frac{6.022 \times 10^{23}}{5.976 \times 10^{27}} = ?$$

Dividing 6.022 by 5.976 gives 1.008 to the correct number of significant figures. Dividing the exponentials gives $10^{23}/10^{27} = 10^{(23-27)} = 10^{-4}$. Therefore the answer is 1.008×10^{-4}.

The same general procedure is followed when raising an exponential to a power. The coefficients are done first, the exponentials next, and the results of these computations combined for the answer. Thus,

$$(6 \times 10^3)^3 = 216 \times 10^9 = 2 \times 10^{11} \qquad \text{(if only one significant figure is justified)}$$
$$(5.1 \times 10^{-2})^2 = 26 \times 10^{-4} = 2.6 \times 10^{-3}$$

To avoid fractional exponents when extracting a root, we must adjust the power of 10 so it becomes a whole number factor of 2 if a square root is to be extracted, a whole number factor of 3 if a cube root is to be extracted, and so forth. Therefore, to extract the cube root of Avogadro's number $(6.02 \times 10^{23})^{1/3}$ you first must rewrite the number so the power of 10 is a whole-number multiple of three. Since $3 \times 7 = 21$ and $3 \times 8 = 24$, either 10^{21} or 10^{24} is a suitable exponential. Let us rewrite the number as a coefficient times 10^{21} by moving the decimal in the coefficient two places to the right and decreasing the exponent of 10 by two units: $(602 \times 10^{21})^{1/3}$. The cube root of 602 is 8.45; the cube root of 10^{21} is 10^7. The answer is 8.45×10^7.

As a self-test of your understanding, do the following exercises.

Exercises

1. $\dfrac{5.23 \times 10^{27}}{9.76 \times 10^3} =$

2. $\dfrac{3.42 \times 10^{-29}}{6.704 \times 10^5} =$

3. $\dfrac{(2.46 \times 10^3)(1.7 \times 10^{-5})}{3.25 \times 10^4} =$

4. $(5.2 \times 10^{-3})^3 =$

5. $(7.5 \times 10^{-5})^{1/2} =$

Answers

(1) 5.36×10^{23}; (2) 5.10×10^{-35}; (3) 1.3×10^{-6}; (4) 1.4×10^{-7}; (5) 8.7×10^{-3}.

Appendix 5:
A More Exact Treatment
of Acid-Base Equilibria

In Chapter 5, we introduced some simple acid–base equilibria, with equilibrium expressions uncomplicated enough to solve with the quadratic formula or with approximation methods. In many cases, these expressions were only valid for dilute solutions, or situations in which one component was present in small amounts. These methods are ordinarily good enough, except in unusual circumstances. In this appendix we shall derive the exact expressions that must be used when the approximations of Chapter 5 fail. We shall see how to handle equilibrium problems when the contribution of protons or hydroxide ions from the dissociation of water cannot be neglected. We shall carry out an exact derivation of the equilibrium equations for weak acids and their salts, and see that the weak-acid equilibria of Section 5-7 and the hydrolysis equilibria of Section 5-8 really are only special cases of the same general process. The exact equations then will permit us to calculate the titration behavior of a weak acid titrated by a strong base, and to compare these results with the strong-acid titration of Section 5-5.

STRONG AND WEAK ACIDS: THE CONTRIBUTION FROM DISSOCIATION OF WATER

In Section 5-5, we stated that when a strong acid is added to water, the effect is that of adding the same amount of hydrogen ions, since the acid is totally dissociated. But the acid is not the only source of hydrogen ions; water itself dissociates:

$$H_2O \rightleftharpoons H^+ + OH^-$$

with a dissociation constant or ion product of

$$[H^+][OH^-] = K_w = 10^{-14}$$

Is this source of protons important?

For a $0.01M$ solution of nitric acid, the answer is no. The hydrogen ion concentration from the acid is 10^{-2} mole liter^{-1}, and $[H^+]$ from water dissociation even in pure water is only 10^{-7} mole liter^{-1}, one hundred-thousandth as much. Since the added H^+ from the acid will repress water dissociation, the actual contribution of water to the $[H^+]$ will be smaller yet. We can find the concentration of hydroxide ion (which comes only from water dissociation) from the equilibrium expression:

$$[OH^-] = \frac{K_w}{[H^+]_t} = \frac{10^{-14}}{10^{-2}} = 10^{-12}$$

Here $[H^+]_t$ is the total supply of hydrogen ions from all sources.

Protons have no labels. For every hydroxide ion produced by dissociation of water, one hydrogen ion is produced. Thus the concentration of hydrogen ions *coming from the dissociation of water alone* is

$$[H^+]_w = [OH^-] = 10^{-12} \text{ mole liter}^{-1}$$

and this is entirely negligible in comparison with 10^{-2} mole liter^{-1} of hydrogen ions from the nitric acid.

Example 1

What is the pH of $10^{-6}M$ HCl?

Approximate Solution

$$[H^+] = 10^{-6} \text{ mole liter}^{-1} \quad \text{and} \quad pH = 6.0$$

As a check on the dissociation of water:

$$[H^+]_w = [OH^-] = \frac{10^{-14}}{10^{-6}} = 10^{-8} \text{ mole liter}^{-1}$$

The water contribution to the hydrogen ion concentration still is only 1% as large as that from the added acid. Neglecting $[H^+]_w$ leads to only a 1% error in total hydrogen ion concentration.

Exact Solution

An exact solution involves two simultaneous equations, each relating the total hydrogen ion concentration, $[H^+]_t$, to that provided by water dissociation alone, $[H^+]_w$.

Mass balance: $[H^+]_t = 10^{-6} + [H^+]_w$

Water equilibrium: $\quad [H^+]_w = [OH^-] = \dfrac{10^{-14}}{[H^+]_t}$

The mass-balance equation simply states that the total mass of protons is the sum of that provided from each of two sources—acid and water. To solve these two equations, let y represent the total hydrogen ion concentration from all sources, $[H^+]_t$ and eliminate $[H^+]_w$:

$$y = 10^{-6} + \frac{10^{-14}}{y}$$

$$y^2 - 10^{-6}y - 10^{-14} = 0$$
$$y = 1.01 \times 10^{-6} \text{ mole liter}^{-1}$$

The true hydrogen ion concentration is 1% higher, because of the contribution from dissociating water, than our approximate calculation predicted. This correction is trivial for strong acids, but it can become important when we deal with weak acids.

Weak Acids and Water Dissociation

In Section 5-6, we used the simple dissociation-constant expression

$$K_a = 1.76 \times 10^{-5} = \frac{y^2}{0.0010 - y}$$

to calculate the hydrogen ion concentration of $0.0010M$ acetic acid:

$$[H^+] = y = 1.24 \times 10^{-4} \text{ mole liter}^{-1}$$

Nothing was said about any contribution from the other equilibrium present:

$$K_w = [H^+][OH^-]$$

Are we wrong to neglect this component? We can make a quick test to see. Since every dissociation of a water molecule produces one proton and one hydroxide ion, we can follow the concentration of protons from water dissociation by calculating the hydroxide ion concentration. The protons coming from dissociation of acid will repress the water equilibrium, and the hydroxide ion concentration will be only:

$$[OH^-] = \frac{10^{-14}}{1.24 \times 10^{-4}} = 8.1 \times 10^{-11} \text{ mole liter}^{-1}$$

Because this is also the amount of H^+ that comes from the dissociation of water, we are clearly not wrong in omitting this in comparison with H^+ from the acid, 1.24×10^{-4} mole liter^{-1}. For a weaker acid such as HCN the story can be different.

Example 2

Calculate the pH and percent dissociation of HCN in solutions that are $10^{-2}M$, $10^{-3}M$, and $10^{-7}M$.

Approximate Solution

HCN is an extremely weak acid, with a dissociation constant of $K_a = 4.93 \times 10^{-10}$. The answers, using approximate equation 5-34, are:

Concentration (moles liter^{-1})	10^{-2}	10^{-3}	10^{-7}
pH:	5.65	6.15	8.17
Percent dissociation:	0.02%	0.07%	6.8%

The preceding example yields the surprising result that by making an acid solution sufficiently dilute ($10^{-7}M$), we apparently can make its solution basic! This cannot be true. The flaw is that we finally have reduced the hydrogen ion concentration from the acid to the point where it is close to that coming from dissociation of water. The simple equilibrium expression that we have been using is no longer good enough. In the proper treatment for the general acid HA, there are four unknown concentrations, $[H^+]$, $[HA]$, $[A^-]$, and $[OH^-]$, and four equations that relate these unknowns:

Acid dissociation:
$$K_a = \frac{[H^+][A^-]}{[HA]} \tag{A-1}$$

Water dissociation:
$$K_w = [H^+][OH^-] \tag{A-2}$$

Mass balance on the acid anion: $c_0 = [HA] + [A^-]$ (A-3)

Charge balance: $[H^+] = [A^-] + [OH^-]$ (A-4)

The key to solving what might seem to be a complicated set of equations is to understand what the charge-balance equation means physically. Since $[OH^-]$ also equals the amount of hydrogen ion produced from dissociation of water, the charge-balance equation shows that the acid anion concentration, $[A^-]$, is less than the total hydrogen ion concentration, $[H^+]$, by just that amount of H^+ that came from water, rather than from HA:

Charge balance: $[A^-] = [H^+] - [OH^-]$

If we represent the desired hydrogen ion concentration by y as before, and use the water-dissociation equation A-2 to eliminate $[OH^-]$, then

Charge balance: $[A^-] = y - \dfrac{K_w}{y}$ (A-5)

The rest of the derivation is as before, with the undissociated acid concentration being the initial overall concentration less that which has dissociated:

Mass balance: $[HA] = c_0 - [A^-] = c_0 - y + \dfrac{K_w}{y}$ (A-6)

These two concentrations, $[HA]$ and $[A^-]$, now can be replaced in the acid-dissociation equilibrium expression:

$$K_a = \frac{y(y - K_w/y)}{c_0 - y + K_w/y}$$ (A-7)

The expression for K_a that we have just derived looks more complicated than it really is. All we have to do to regain the simpler equation 5-34 is to strike out the two terms in numerator and denominator that represent the contributions of water dissociation, K_w/y. However, solving equation A-7 is tedious because it leads to a cubic equation, so it is important to know when you do not need to use it. The easiest approach is to carry out an approximate solution using equation 5-34, and then verify that K_w/y is significantly smaller than y, or that $10^{-14}/[H^+]$ is negligibly small compared with $[H^+]$. For the calculations of Example 2:

Original c_0 of HCN:	10^{-2}	10^{-3}	10^{-7}
Approximate $[H^+]$:	2.22×10^{-6}	7.02×10^{-7}	6.78×10^{-9}
$10^{-14}/[H^+]$:	4.50×10^{-9}	1.42×10^{-8}	1.48×10^{-6}

The last line is the contribution to $[H^+]$ from water dissociation. It certainly is negligible for $10^{-2}M$ HCN, on the borderline for $10^{-3}M$, and clearly *not* negligible for $10^{-7}M$ HCN, which is why the approximate answer went astray. If you do use equation A-7 for $10^{-7}M$ HCN, the exact answer is

$$[H^+] = 1.0025 \times 10^{-7} \quad \text{and} \quad pH = 6.999$$

or pH 7 within the accuracy of K_a itself. HCN is such a weak acid and is dissociated so little that its own contribution to the pool of hydrogen ions at this concentration can be neglected in comparison with that from water molecules!

It is very seldom that you will need the full equation A-7. The usual practice is to solve a problem with approximate equation 5-34, and then to verify as we have just done that the full treatment is not called for.

WEAK ACIDS AND THEIR SALTS: EXACT TREATMENT

In Section 5-6 we calculated the pH of a solution of a weak acid, and in Section 5-8 we calculated the pH of a solution of the salt of such an acid with a strong base. These and the buffers mentioned in Section 5-7 are only different extremes of the same problem: that of finding the pH of a solution of c_a moles liter^{-1} of a weak acid and c_s moles liter^{-1} of its salt with a strong base, where both c_a and c_s can range from zero to appreciable concentrations. In the weak acid problems, of course, $c_s = 0$. In hydrolysis problems, $c_a = 0$. When dealing with buffers, we assume that c_a and c_s are of similar magnitude. Each of these special cases is relatively easy to handle mathematically because we can make approximations by neglecting small quantities when they are added to or subtracted from larger terms. In the regions between these extremes, the mathematics, but not the chemistry, becomes more difficult. It is worth setting up the general expressions, however, to show how the simpler expressions are derived from them, and how all three special cases—weak acids, hydrolysis, and buffers—are examples of the same phenomenon.

The heart of the problem still is the equilibrium expression for a weak acid:

$$K_a = \frac{[H^+][A^-]}{[HA]} \quad \text{(acid dissociation)} \tag{A-8}$$

The water-dissociation expression is always valid:

$$K_w = [H^+][OH^-] \quad \text{(water dissociation)} \tag{A-9}$$

even though it is sometimes unimportant. One mass-balance statement for the acid anion is that the total supply of material, either as anion or as undissociated acid, must equal the sum of the starting concentrations of weak acid and salt:

$$c_a + c_s = [HA] + [A^-] \quad \text{(mass balance: anion)} \tag{A-10}$$

Another mass-balance statement is that the total positive ion from the salt (let us assume that it is sodium ion) equals the starting salt concentration, since this ion takes no part in the reactions:

$$c_s = [Na^+] \quad \text{(mass balance: positive ion)} \tag{A-11}$$

The next equation is a charge balance, which states that the overall solution is electrically neutral:

$$[H^+] + [Na^+] = [OH^-] + [A^-] \quad \text{(charge balance)} \tag{A-12}$$

For simplicity let us represent the hydrogen ion concentration by y, and the hydroxide ion concentration by z. The goal in solving these equations is to eliminate $[A^-]$ and $[HA]$ from the acid-equilibrium expression. The charge-balance equation can be rearranged to

$$[A^-] = [Na^+] + y - z = c_s + y - z$$

The first mass-balance equation then can be converted to

$$[HA] = c_a + c_s - [A^-] = c_a + c_s - y + z - c_s = c_a - y + z$$

Substitution into the equilibrium expression completes the derivation:

$$K_a = \frac{y(c_s + y - z)}{(c_a - y + z)} \tag{A-13}$$

How does this general expression reduce to equations we have used previously? Under acid conditions, the hydroxide ion concentration will be unimportant, so z can be neglected in comparison with y, the hydrogen ion concentration. The general expression then becomes

$$K_a = \frac{y(c_s + y)}{c_a - y} \tag{A-14}$$

which we have already used in the weak acid–salt and buffer problems of Section 5.7.

If no salt is present, then $c_s = 0$, and the general expression becomes

$$K_a = \frac{y(y - z)}{c_a - y + z} = \frac{y(y - K_w/y)}{c_a - y + K_w/y} \tag{A-15}$$

But this is just the expression that we derived in the previous section of this appendix, when we took water dissociation into account in the dissociation of weak acids.

If no salt is present, and if the conditions are sufficiently acid so that we can neglect water dissociation, then $c_s = 0$, $z = 0$, and the general expression becomes

$$K_a = \frac{y^2}{c_a - y} \tag{A-16}$$

This is the first and simplest form of the dissociation constant for weak acids, which we used in Section 5-6.

Our last special equation, the hydrolysis expression, arises under basic conditions where $z = [OH^-]$ is appreciable and $y = [H^+]$ is negligible by comparison:

$$K_a = \frac{y(c_s - z)}{c_a + z} = \frac{K_w(c_s - z)}{z(c_a + z)}$$

or

$$K_b = \frac{K_w}{K_a} = \frac{z(c_a + z)}{c_s - z} \tag{A-17}$$

(Notice that we can neglect y only when it is compared with a larger quantity, not when it stands alone as a multiplier.) If the acid concentration, c_a, is zero, the preceding equation becomes the hydrolysis equilibrium expression that we used in Section 5-8 (y was used for $[OH^-]$):

$$K_b = \frac{z^2}{c_s - z} \tag{A-18}$$

In short, every acid–base equilibrium expression that we have used in Chapter 5 is contained in the general equation A-13. The secret is not to try to force through a solution from this equation, but rather to understand the physical reality of the particular problem in hand, to see what quantities are negligible in comparison with others, and to make the right simplification for the mathematical solution.

TITRATION OF A WEAK ACID BY A STRONG BASE

In Section 5-5 we plotted the results of titrating a strong acid with a strong base, but deferred the more difficult problem of titration of weak acids, where equilibria are involved. The general expression that we have just derived, equation A-13, makes it possible to calculate this kind of titration.

Let us titrate 50 ml of a $0.10M$ solution of acetic acid with a $0.10M$ solution of a strong base, NaOH, and calculate the pH as a function of the volume of base added. The effect of adding sodium hydroxide is to turn some of the acetic acid into sodium acetate by the neutralization reaction:

$$HAc + NaOH \rightarrow H_2O + NaAc$$

or, more accurately,

$$HAc + OH^- \rightarrow H_2O + Ac^-$$

The calculations for various amounts of added NaOH solution are listed in Table A-1, and the results are plotted in Figure A-1. Point a on the plot was obtained from the simple weak-acid equilibrium expression of Section 5-6:

Table A-1

Titration of 50 ml of 0.10M Acetic Acid by 0.10M Sodium Hydroxide

Point:	a	b	c	d	e
v = ml base added:	0	10	20	25	30
V = total volume:	50	60	70	75	80
[HAc] as added:	0.1000	0.0833	0.0714	0.0667	0.0625
[NaOH] as added:	0.0000	0.0167	0.0286	0.0333	0.0375
$[HAc]_{net} = c_a$:	0.1000	0.0667	0.0428	0.0333	0.0250
$[NaAc]_{net} = c_s$:	0.0000	0.0167	0.0286	0.0333	0.0375
$[H^+]$:	1.33×10^{-3}	7.04×10^{-5}	2.64×10^{-5}	1.76×10^{-5}	1.17×10^{-5}
pH	2.88	4.15	4.58	4.76	4.93

f	g
40	50
90	100
0.0555	0.0500
0.0444	0.0500
0.0111	0.0000
0.0444	0.0500
4.4×10^{-5}	1.89×10^{-9}
4.36	8.68

$$K_a = \frac{y^2}{c_a - y} \tag{5-34}$$

Points b through f came from the weak acid–salt equilibrium expression of Section 5-7:

$$K_a = \frac{y(c_s + y)}{(c_a - y)} \tag{5-42}$$

The equivalence point, g, at equal concentrations of acid and base, came from the hydrolysis expression of Section 5-8:

$$K_b = \frac{K_w}{K_a} = \frac{y^2}{c_s - y} \tag{5-50}$$

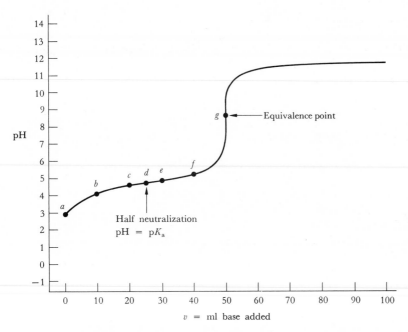

Figure A-1

Titration curve for a typical weak acid and strong base, in this case acetic acid and sodium hydroxide. Data for this plot appear in Table A-1. Compare this titration curve with that in Figure 5-5. The pH in this curve increases after initial addition of base because even after partial neutralization, the acetate ion continues to repress dissociation of the remaining acetic acid. Neutralizing a certain portion of acetic acid with NaOH is the same as removing the acetic acid and replacing it with sodium acetate, which is why this figure and Figure 5-8 are similar. For the completely dissociating strong acid HNO_3 in Figure 5-5 the nitrate anion has no such repressing effect, and the titration curve initially rises slowly. In this figure, the curve beyond the equivalence point is similar to that for a strong acid.

All these equations are simplifications for special conditions of the general equation A-13 derived in Section A-2. Beyond the equivalence point, where we are in effect adding more base to a sodium acetate solution, the titration curve is little different from that of a strong acid and base. Only in the region between points *f* and *g* would we have to struggle with the complete general expression.

The half-neutralization point is equivalent to having identical concentrations of acetic acid and sodium acetate (half of the initial acetic acid has been neutralized by sodium hydroxide). Therefore, the pH at this point is the pK_a of acetic acid. The equivalence point, *g*, does not occur within such a wide range of pH values as in Figure 5-5, and it becomes more important to choose an indicator with a pK_a near 8 or 9. For example, methyl orange could not be used to locate the equivalence point in this acetic acid titration (see Figure 5-7), although phenolphthalein or thymol blue would be ideal.

Appendix 6:
Answers to Odd-Numbered Problems

Problems at the ends of chapters in this book have generally been paired, with first an odd-numbered and then an even-numbered problem to illustrate the same point or technique. Answers for the odd-numbered problems are given here so you can use them for self-testing, whereas the even-numbered problems are not provided with answers so they can be used for class testing and homework assignments. If you are ever given an even-numbered problem as an assignment, and find that you cannot work it, try the preceding problem first and check your answer against that given below.

Chapter 1

1. <u>RA</u>dium vs. <u>Rad</u>o<u>N</u>; <u>RH</u>odium vs. <u>Rh</u><u>E</u>nium; <u>RU</u>thenium vs. <u>Ru</u><u>B</u>idium

3. 13 protons; $27 - 13 = 14$ neutrons; 13 electrons. Expected atomic weight = 27.223 amu. Observed = 26.982. Mass converted to energy when nucleus is formed.

5. 7 e, 7 p, 7 n; $^{14}_{7}N$; $^{15}_{7}N$; No effect; $^{15}_{8}O$; One more electron.

7. Always 30 electrons. (a) 34 n, ^{64}Zn; (b) 36 n, ^{66}Zn; (c) 37 n, ^{67}Zn; (d) 38 n, ^{68}Zn; 65.366 amu.

9. 64.923 amu

11. 1 g boron has more atoms: 5.57×10^{22} atoms to 2.23×10^{22} atoms per g of aluminum. Same number of atoms per mole.

13. 2.44×10^{19} atoms

15. 331.6 amu; 1.54×10^{22} atoms

17. 90.03 amu; 26.7 g carbon

19. 88.8% Cu; 888 g Cu; 42.0 g C; 153.8 g CO_2 or 3.49 moles.
1000 g Cu_2O = 42.0 g C = 1042 g
888 g Cu + 153.8 g CO_2 = 1042 g

21. Li_2S; $Zn_3(PO_4)_2$; $CaSO_4$; $UO_2C_2O_4$; $Zn_3[Fe(CN)_6]_2$; $Fe_2(SO_4)_3$; CrF_3

23. 0.0246 faradays or 2370 coulombs

25. 0.0871 moles Al

27. 40.2 hours

29. 1.92 g lead

31. 1.6×10^6 kg day^{-1}; Roughly 2 dams (1.86)

Chapter 2

1. 26.58% K; 35.35% Cr; 38.07% O; 294.29 g mole^{-1}

3. Ca_3SiO_5

5. P_4O_{10}; 283.88 g mole^{-1} or amu molecule^{-1}

7. 51.9 amu atom^{-1}

9. Zinc reacts with sulfuric acid to form hydrogen and zinc sulfate. 65.37 g Zn and 98.08 g H_2SO_4 produce 2.02 g H_2 and 161.43 g $ZnSO_4$. Same numbers of atoms of each element on each side of equation.

11. a) $Na + H_2O \rightarrow NaOH + \frac{1}{2}H_2$
b) $Ca(OH)_2 + CO_2 \rightarrow CaCO_3 + H_2O$
c) $CO + 3H_2 \rightarrow CH_4 + H_2O$
d) $Al(NO_3)_3 + 3NH_3 + 3H_2O \rightarrow Al(OH)_3 + 3NH_4NO_3$

13. 366 g CO_2

15. (a) Yes, balanced; (b) 0.540 mole CO_2; 23.8 g CO_2

17. $Ca_3P_2 + 6H_2O \rightarrow 3Ca(OH)_2 + 2PH_3$; 0.653 g PH_3

19. 0.244 millimoles XeF_6; 59.8 mg XeF_6; 2.6 mg F_2 remaining unused.

21. (a) No reactants left unused; (b) 1.37 g O_2 left unused

23. 0.054 mole liter^{-1}

25. Molality = 1.133 mole kg^{-1}; Molarity = 1.087 mole liter^{-1}; Normality = 2.174 equiv liter^{-1}; Density = 1.066 g ml^{-1}

27. 300 ml water added

29. $0.700M$ H_2SO_4

31. 0.000372 mole liter^{-1}; 0.000744 equiv liter^{-1}

33. 17.3 ml nitric acid solution

35. $3Mg(OH)_2 + 2H_3PO_4 \rightarrow Mg_3(PO_4)_2 + 6H_2O$
37. 6.0 ml H_2SO_4
39. 35 ml HCl
41. 0.179 equiv liter^{-1} and 0.0895 mole liter^{-1}
43. Pyruvic acid, $CH_3COCOOH$
45. 2.96 mg formic acid
47. $0.237M$ NaOH
49. 104 g equiv^{-1}
51. 135.3 g equiv^{-1}; $CH_3—C_6H_4—COOH$
53. 785 newtons; 2.35 kilojoules; 42.4 mg methane
55. (a) -52 kJ; (b) -26 kJ; (c) $+10$ kJ; (d) $+52$ kJ; (e) $+161$ kJ;
(f) $+597$ kJ; (g) -52 kJ.
57. $+0.30$ kJ mole^{-1}; -297.2 kJ mole^{-1}
59. -321 kJ mole^{-1}
61. -2510 kJ; Heat is given off.
63. -457 kJ

Chapter 3

1. 760 mm high for any cross-sectional area
3. 4.68
5. (a) 15.0 ml; (b) 750 ml; (c) 7600 ml
7. 5.28 liters
9. 495 in^3
11. 3.75×10^{14} molecules
13. $p_T = 1.67$ atm; $p_{CO_2} = 1.23$ atm; $p_{N_2} = 0.44$ atm
15. $p_{Ar} = 0.10$ atm; $p_{He} = 1.01$ atm
17. $p_{CO} = 0.020$ atm
19. 10.94 g liter^{-1} at STP; 13.0 g liter^{-1}
21. 4.00 g mole^{-1}; He
23. 42.2 g mole^{-1}; C_3H_6
25. $FeO \cdot (SO_3)_2 \cdot H_2O \cdot (NH_3)_2 \cdot (H_2O)_6$ or perhaps:
$FeSO_4 \cdot (NH_4)_2SO_4 \cdot 6H_2O$
27. (a) 1.000 atm; (b) 0.593 atm; (c) 1.593 atm; (d) 0.0201 moles water;
(e) 0.626 ml water remaining as liquid
29. 3.72 kJ mole^{-1}; About 1% of bond energy; Thermal breaking
of bonds
31. 3.49 g liter^{-1}; $V_1 = 148$ A^3 molecule^{-1}; $V_g = 37,200$ A^3 molecule^{-1};
Factor of 252
33. 482 m sec^{-1}; 29,800 K; 100 times
35. 348 m sec^{-1}; Faster in He; 308 m sec^{-1}
37. 0.989 liter; 1.003 liter
39. 64 g mole^{-1}; SO_2
41. 22.412 liter mole^{-1}; 0.9990 atm; 0.1% difference

Chapter 4

1. (a) $K_{eq} = \dfrac{[H_2]^2[S_2]}{[H_2S]^2}$ (b) $K_{eq} = \dfrac{[H_2]^2}{[H_2S]^2}$ (c) $K_{eq} = \dfrac{[PCl_5]}{[PCl_3][Cl_2]}$

(d) $K_{eq} = [CO_2]$ (e) $K_{eq} = \dfrac{[NO]^2[O_2]}{[NO_2]^2}$

3. (a) $K_{eq} = \dfrac{[NO_2]^2}{[N_2O_4]}$ (b) $K_{eq} = \dfrac{[N_2][O_2]^2}{[N_2O_4]}$ (c) $K_{eq} = \dfrac{[N_2O_4]}{[NO_2]^2}$

(d) $K_{eq} = \dfrac{[N_2O_4]^{1/2}}{[NO_2]}$ (e) $K_{eq} = [NO_2]^2[O_2]^{1/2}$

5. (a) $atm^{-1/2}$; (b) $K_p = 1.6 \times 10^{-11}$ atm; (c) $K_c = 1.3 \times 10^{-13}$ mole liter^{-1};
(d) $[O_2] = 1.6 \times 10^{-11}$ atm $= 1.3 \times 10^{-13}$ mole liter^{-1}

7. K_{eq} unitless. (a) reverse; (b) reverse; (c) forward

9. $K_{p_1} = 1.86$ atm$^{-1/2}$; $K_{p_2} = 0.289$ atm; $K_{p_2} = 1/K_{p_1}^2$

11. (a) $NO + \frac{1}{2}Cl_2 \rightarrow ClNO$;
(b) $K_c = 13.6$ mole$^{-1/2}$ liter$^{1/2}$;
$K_p = 2.11$ atm$^{-1/2}$

13. (a) $K_c = 0.0488$ mole liter^{-1}; (b) $[N_2O_4] = 1.85$ mole liter^{-1}

15. No units; $K_p = K_c$; $p_{H_2} = 0.98$ atm; $p_{H_2O} = 9.02$ atm

17. moles liter^{-1}; (a) $[CO_2] = 0.0060$ mole liter^{-1}; (b) 0.060 moles;
(c) 60% converted; (d) $p_{CO_2} = 0.50$ atm

19. Endothermic by Le Chatelier's principle, since dissociation favored by higher temperature.

21. (a) products; (b) no effect; (c) products; (d) reactants; (e) products

Chapter 5

1. pH = 12

3. (a) pH = 3; (b) pH = 1.7; (c) pH = 11; (d) pH = 11.3

5. pH = 2.8; pOH = 11.2

7. (a) pH lower; (b) little or no change; (c) pH higher;
(d) no change; (e) pH higher

9. pH = 2.9; $K_a = 1.71 \times 10^{-5}$

11. $K_{diss} = 1.79 \times 10^{-5}$; No; $NH_3 + H_2O \rightarrow NH_4^+ + OH^-$; pH = 10.6

13. pH = 5.1; pOH = 8.9

15. $[NO_2^-] = 1.05 \times 10^{-2}$ mole liter^{-1}; pH = 2.0; 4.2% ionization

17. 3.97×10^{-5} mole liter^{-1}; pH = 9.6

19. pH = 5.0

21. $OBr^- + H_2O \rightarrow HOBr + OH^-$

$$K_b = \frac{[HOBr][OH^-]}{[OBr^-]} = 5.05 \times 10^{-5}$$

$K_a = 1.98 \times 10^{-9}$

23. $K_b = 2.00 \times 10^{-11}$; $K_a = K_w/K_b = 5.00 \times 10^{-4}$
25. pH = 9.31
27. pH = 4.75
29. pH = 2.0; pH = 4.78
31. pH = 9.07; pH = 9.01
33. (a) pH = 10.5; (b) pH = 5.5; (c) yes
35. $K_a = 1.4 \times 10^{-6}$
37. $[H^+] = [HCO_3^-] = 1.2 \times 10^{-4}$ mole liter^{-1}
$[CO_3^{2-}] = 5.6 \times 10^{-11}$ mole liter^{-1}
$[CO_2] = 0.034$ mole liter^{-1}
pH = 3.9
39. $[H^+] = [HSO_3^-] = 2.11 \times 10^{-2}$ mole liter^{-1}
$[H_2SO_4] = 0.0289$ mole liter^{-1}
$[SO_3^{2-}] = 1.02 \times 10^{-7}$ mole liter^{-1}
41. $K_{sp} = 1.77 \times 10^{-18}$; Sol. = 1.77×10^{-15} mole liter^{-1}
43. Solubilities = 2.7×10^{-3} mole liter^{-1} and 3.2×10^{-5} mole liter^{-1}
45. $[Ag^+] = 1.6 \times 10^{-4}$ mole liter^{-1} and 4.4×10^{-6} mole liter^{-1}
47. No
49. pH = 0.7
51. Sol. = 2.0×10^{-7} mole liter^{-1}
53. At pH = 2, Sol. = 8.9×10^{12} mole liter^{-1} (or infinitely soluble);
at pH = 12, Sol. = 8.9×10^{-8} mole liter^{-1}
55. (a) 8.5×10^{-15} mole liter^{-1}; (b) 2.0×10^{-7} mole liter^{-1};
(c) 2.3×10^{-24} mole liter^{-1}; answer c is best.

Chapter 6

1. CO_2: 44.007 g   CO: 28.009 g
H_2: 2.016 g H_2O: 18.014 g
46.023 g   46.023 g

Volume happens to be conserved at 110°C, but would not be at
a temperature where water is liquid. Mass is more fundamental
than volume.
3. 23.0 g; 46.0 g per mole O; 23.0 g mole^{-1}; sodium
5. 10 liters H_2O vapor; 5 liters unused O_2
7. 80.6 g mole^{-1}; CH_3X, CH_2X_2, CHX_3, CX_4; X = bromine
9. 63.6 g mole^{-1}
11. 238.1 g mole^{-1}; combining capacity = 3
13. CH_2O; 60.1 g mole^{-1}; $C_2H_4O_2$
15. C_7H_8

Chapter 7

1. Au (like 111); Hg (like 112); Rn (like 118)
3. (a) N, VA; (b) Al, IIIA; (c) Cl, VIIA; (d) Rb, IA
5. Decrease
7. $HI + H_2O \rightarrow H_3O^+ + I^-$
9. CaH_2, H_2Te, GeH_4, WH_6; CaH_2 will be ionic. The hydrogens will behave as cations in H_2Te, and as anions in CaH_2, GeH_4, and WH_6. Aqueous solution of CaH_2 will be most basic $[CaH_2 + 2H_2O \rightarrow Ca(OH)_2 + 2H_2]$.

Chapter 8

1. $\lambda = 2.5 \times 10^2$ nm; 8.0×10^{-19} J photon^{-1}; 4.8×10^2 kJ mole^{-1}; ultraviolet light
3. 6.626×10^{-28} J photon^{-1}; 3.990×10^{-7} kJ mole^{-1}; $\lambda = 2.998 \times 10^4$ cm. The energy of radio waves is much less than that of a C—C bond. Therefore, such waves would not produce a chemical reaction.
5. $E = h\nu = -\dfrac{k}{n^2}$ and $k = \dfrac{2\pi^2 m_e e^4 Z^2}{h^2}$

Since $h\nu$ is proportional to Z^2, a plot of Z against the square root of ν should give a straight line.
7. $\lambda = 4.5 \times 10^2$ nm; 4.4×10^{-19} J photon^{-1}; $\bar{\nu} = 2.2 \times 10^4$ cm^{-1}
9. 0.544 eV atom^{-1}
11. $\lambda = \dfrac{18\ ch^3}{7\pi^2 m_e e^4} = 4.70 \times 10^{-7}$ m (or 4.70×10^2 nm)
13. 0.66 eV atom^{-1}
15. -0.85 eV atom^{-1}

17.

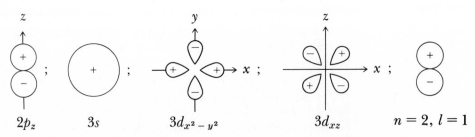

$2p_z$; $3s$; $3d_{x^2-y^2}$; $3d_{xz}$; $n = 2, l = 1$

19. $n = 3$, $l = 2$, $m = 2,1,0,-1,-2$.

21. $n = 7, 8$, etc. $\rightarrow n = 6$ series in Be^{3+}; lowest-energy line is $n = 7 \rightarrow n = 6$; for $n = 7$, 98 combinations; for $n = 6$, 72 combinations.

23. $n = 4$; $l = 3$; $m = 3, 2, 1, 0, -1, -2, -3$; $s = +\frac{1}{2}, -\frac{1}{2}$

25. $\lambda = 1.27 \times 10^{-11}$ m; $\Delta p = 0.53 \times 10^{-23}$ kg m sec^{-1}; $p = 5.2 \times 10^{-23}$ kg m sec^{-1}; the uncertainty in the momentum is about one-tenth the actual momentum.

Chapter 9

1. (a) Ca, $1s^2 2s^2 2p^6 3s^2 3p^6 4s^2$; (b) Mg^{2+}, $1s^2 2s^2 2p^6$

3. C, 2 unpaired electrons; F, 1; Ne, 0

5. $1s^2 2s^2 2p^4$ for both $^{18}_8 O$ and $^{16}_8 O$

7. (a) impossible (violates the Pauli principle); (b) impossible (violates the Pauli principle); (c) ground state (Ti); (d) impossible (violates the Pauli principle); (e) excited state (Fe); (f) excited state (Li); (g) excited state (Ca); (h) ground state (O); (i) impossible ($2d$ orbital does not exist)

9. Li, $1s^2 2s^1$; Lu, $1s^2 2s^2 2p^6 3s^2 3p^6 4s^2 3d^{10} 4p^6 5s^2 4d^{10} 5p^6 6s^2 4f^{14} 5d^1$; La, $1s^2 2s^2 2p^6 3s^2 3p^6 4s^2 3d^{10} 4p^6 5s^2 4d^{10} 5p^6 6s^2 5d^1$; Lr, $1s^2 2s^2 2p^6 3s^2 3p^6 4s^2 3d^{10} 4p^6 5s^2 4d^{10} 5p^6 6s^2 4f^{14} 5d^{10} 6p^6 7s^2 5f^{14} 6d^1$; Li^+, $1s^2$; Lu^{3+}, $1s^2 2s^2 2p^6 3s^2 3p^6 4s^2 3d^{10} 4p^6 5s^2 4d^{10} 5p^6 4f^{14}$; La^{3+}, $1s^2 2s^2 2p^6 3s^2 3p^6 4s^2 3d^{10} 4p^6 5s^2 4d^{10} 5p^6$; Lr^{3+}, $1s^2 2s^2 2p^6 3s^2 3p^6 4s^2 3d^{10} 4p^6 5s^2 4d^{10} 5p^6 6s^2 4f^{14} 5d^{10} 6p^6 5f^{14}$

11. The first two noble gases would have atomic numbers 2 and 6.

13. $1s^2 2s^2 2p^6$; relative sizes $N^{3-} > O^{2-} > F^- > Ne > Na^+$

15. (a) 18; (b) $5g^7$ and $5g^{11}$; (c) at least 9, $5g^9$, and probably 10, in the configuration $5g^9 7d^1$ (in either case a new record); (d) 0.544 eV atom^{-1}; this will be smaller than the *IE* of a $5g$ electron in any one of the hypotransition elements, because in the latter the effective nuclear charge will be much larger than $+1$ (atomic hydrogen).

17. (a) The plot has a slope of about 4.9×10^2 kJ χ^{-1}; (b) χ_{Ne} is approximately 4; Ne—F bonds would be covalent; (c) The electron affinity of Sr is less than that of Rb because the $5s$ orbital is filled for Sr but not for Rb (an electron added to Sr would go into a higher energy level). From Y to Ru and Rh, the general trend is the expected increase in *EA* (the effective nuclear charge increases as the $4d$ orbitals are being filled). Pd is $4d^{10}$, so it is reasonable that its *EA* should drop, but the reason for the additional decrease from Pd to Ag ($4d^{10} 5s^1$) is not clear. The sharp decrease from Ag to Cd ($4d^{10} 5s^2$) is expected, as an electron added to Cd would have to occupy a $5p$ orbital. The *EA* of In ($4d^{10} 5s^2 5p^1$) is larger than that of Cd, as its effective nuclear charge is greater.

Chapter 10

1. $+8(XeO_4)$; $+2(XeF_2)$; $+6(XeO_3)$; $+4(XeF_4)$; $+6(XeF_6)$
3. $+3(Co)$; $+4(Pt)$; $+4(C)$; $+4(S)$
5. $+2(Pt)$
7. $+5(V)$; $+5(P)$; $-3(P)$; $+3(N)$; $-1(O)$; $-1(H)$; $-3(N)$; $+3(N)$; $+5(I)$; $+1(Ag)$
9. (a) MnO_4^-, SO_3^{2-}, SO_3^{2-}, MnO_4^-; (b) $Cr_2O_7^{2-}$, HI, HI, $Cr_2O_7^{2-}$; (c) Cl_2 for all
11. $2MnO_4^- + 6H^+ + 5H_2S \rightarrow 2Mn^{2+} + 8H_2O + 5S$; $+7(Mn)$; $S(-2)$; $Mn(+7)$; reduced
13. (a) $2MnO_2 + 4KOH + O_2 \rightarrow 2K_2MnO_4 + 2H_2O$;
    (b) $CuCl_4^{2-} + Cu \rightarrow 2CuCl_2^-$;
    (c) $NO_3^- + 4Zn + 10H^+ \rightarrow NH_4^+ + 4Zn^{2+} + 3H_2O$;
    (d) $2ClO_2 + 2OH^- \rightarrow ClO_2^- + ClO_3^- + H_2O$;
    (e) $6Fe^{2+} + Cr_2O_7^{2-} + 14H^+ \rightarrow 6Fe^{3+} + 2Cr^{3+} + 7H_2O$;
    (f) $3Cu + 2NO_3^- + 8H^+ \rightarrow 3Cu^{2+} + 2NO + 4H_2O$
15. $0.0759M$; 52.68 g equiv^{-1}, $0.228N$; 158.04 g equiv^{-1}, $0.0759N$; 39.51 g equiv^{-1}, $0.304N$; 31.61 g equiv^{-1}, $0.380N$
17. 52.3 g equiv^{-1}
19. (a) 0.50 liter; (b) 0.50 liter; (c) 0.17 liter; (d) 0.25 liter
21. (a) $0.02N$; (b) $0.02N$
23. 3.13×10^{-4} mole
25. $0.0125M$
27. (a) $+7(Mn)$ in MnO_4^-; $+2(Mn)$ in Mn^{2+}; $+3(C)$ in $(COOH)_2$, $+4(C)$ in CO_2;
    (b) $2MnO_4^- + 5(COOH)_2 + 6H^+ \rightarrow 2Mn^{2+} + 10CO_2 + 8H_2O$;
    (c) $KMnO_4$, 158.04 g mole^{-1}, 31.61 g equiv^{-1}, $(COOH)_2$, 90.04 g mole^{-1}, 45.02 g equiv^{-1}; (d) 0.004 mole; (e) 0.04 equiv; (f) $0.500N$; (g) $0.05M$; (h) 1.34 g

Chapter 11

1. $Na\cdot$; $\overset{\cdot\cdot}{C}\cdot$; $\overset{\cdot\cdot}{Si}\cdot$; $:\overset{\cdot\cdot}{Cl}\cdot$; Ca^{2+}; K^+; $:\overset{\cdot\cdot}{Ar}:$; $:\overset{\cdot\cdot}{Cl}:^-$; $:\overset{\cdot\cdot}{S}:^{2-}$
3. $:\overset{\cdot\cdot}{Cl}-\overset{\cdot\cdot}{Cl}:$; $:N\equiv N:$; $\cdot\overset{\cdot\cdot}{N}=\overset{\cdot\cdot}{O}:$; $H-\overset{\cdot\cdot}{Cl}:$; NO has one unpaired electron and should be paramagnetic.

5. $[:\overset{\cdot\cdot}{Cl}:]^-[Ca]^{2+}[:\overset{\cdot\cdot}{Cl}:]^-$; $\quad H-\underset{\underset{H}{|}}{\overset{\overset{H}{|}}{Si}}-H$; $\quad :\overset{\cdot\cdot}{S}=C=\overset{\cdot\cdot}{S}:$;

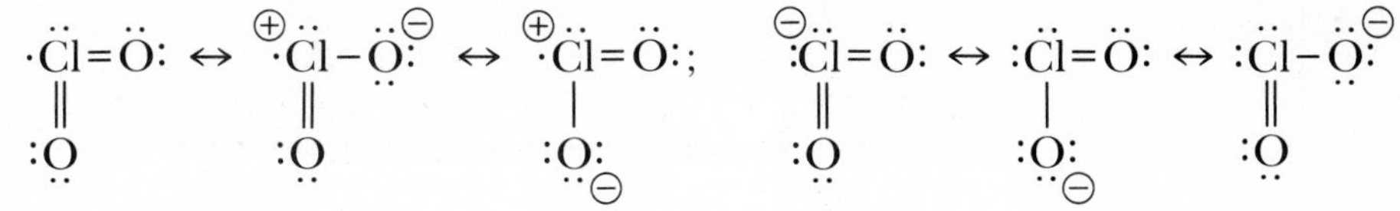

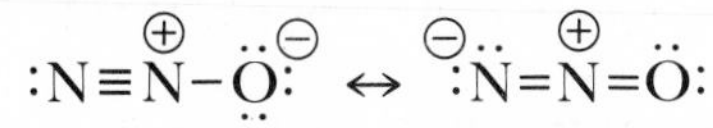

7. 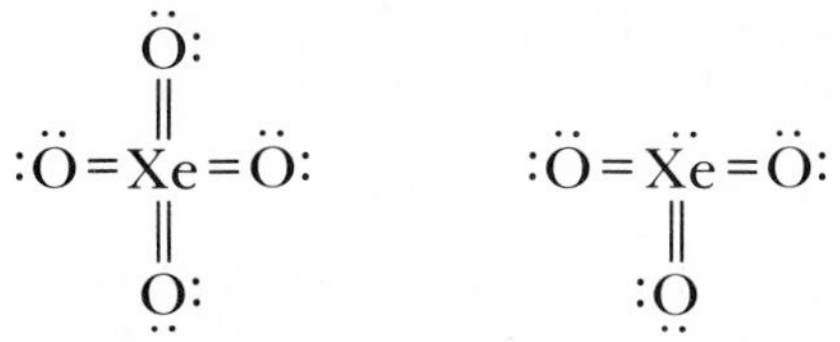

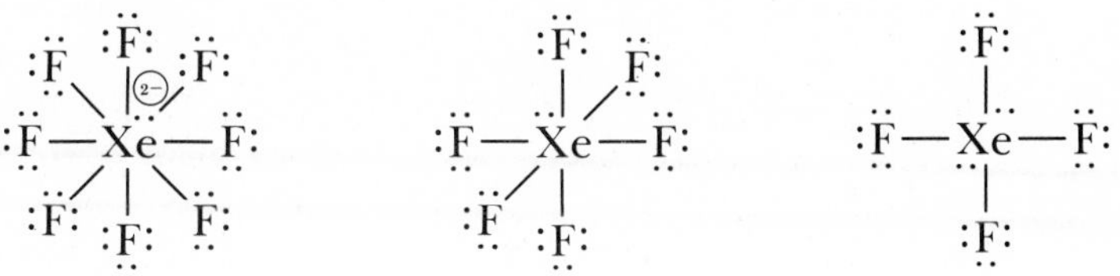

Xe—F bond length in XeF_4 will be shorter than I—F in IF_4^-.

9. 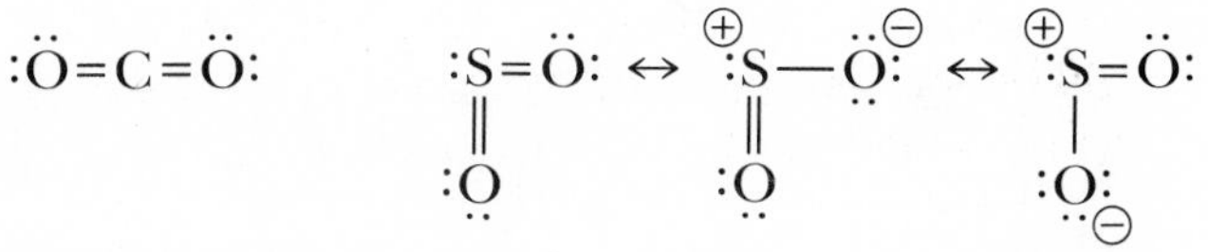

The C—O bonds are primarily covalent; CO_2 should be linear; SO_2 should be angular.

11. 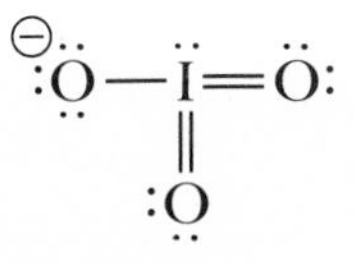 (plus 2 equivalent resonance structures, each I—O has a bond order of $1\frac{2}{3}$);

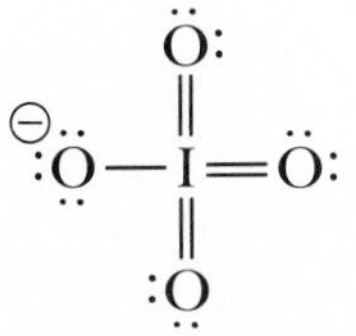 (plus 3 equivalent resonance structures, each I—O has a bond order of $1\frac{3}{4}$);

(plus 5 equivalent resonance structures, each I—O has a bond order of $1\frac{1}{6}$)

The relative I—O bond lengths are predicted to be $IO_4^- < IO_3^- < IO_6^{5-}$.

13. :F—B—F: (plus a small contribution from 3 equivalent

resonance structures of the type $:\overset{\ominus}{F}—B=\overset{\oplus}{F}:$);

BF_3 and NO_3^- are isoelectronic species.

15. (a) :F—Br: (b) :S=S: (c) :Cl—Cl: (d) :P≡P:

(e) :N≡C—$\overset{\ominus}{O}$: ↔ $\overset{\ominus}{:N}$=C=O:

(f) $\overset{\ominus}{:C}$≡$\overset{\oplus}{N}$—$\overset{\ominus}{O}$: ↔ $\overset{2\ominus}{:C}$=$\overset{\oplus}{N}$=O: (the latter resonance

structure does not contribute substantially because the formal charge separation is highly unfavorable [−2 on C])

(g) [:Cl:]⁻[Be]²⁺[:Cl:]⁻ (h) :S=C=S: (i) see 13

(j) see text

(k) $\overset{\ominus}{:O}$—C=O: ↔ $\overset{\ominus}{:O}$—C—$\overset{\ominus}{O}$: ↔ :O=C—$\overset{\ominus}{O}$:

(o) $:\ddot{C}l-\overset{\displaystyle|}{\underset{\displaystyle |}{\ddot{N}}}-\ddot{C}l:$ (p) $:\ddot{F}-\overset{\displaystyle |}{\underset{\displaystyle |}{\ddot{P}}}-\ddot{F}:$ (q) $H-\overset{\displaystyle\ominus}{\underset{\displaystyle |}{\ddot{C}}}-H$ (r) $:\ddot{S}-\ddot{F}:$

$:\ddot{C}l:$ $:\ddot{F}:$ H $:\ddot{F}:$

(s) $:\ddot{O}=\ddot{Xe}=\ddot{O}:$ (t) see 9   (u) $:\ddot{F}-\overset{\displaystyle:\ddot{F}:}{\underset{\displaystyle:\ddot{F}:}{\overset{\displaystyle\diagup}{\underset{\displaystyle\diagdown}{S}}}}-\ddot{F}:$ (v) $[Na]^+[:\ddot{O}:]^{2-}[Na]^+$

$\|$
$:\ddot{O}:$

(w) see 5   (x) $:\ddot{F}-\ddot{N}=\ddot{N}-\ddot{F}:$ (y) $[Cs]^+[:\ddot{F}:]^-$ (z) $[Sr]^{2+}[:\ddot{O}:]^{2-}$

17. $:\ddot{O}=\overset{\displaystyle\oplus}{N}=\ddot{O}:$ (bond order of 2);

$\overset{\displaystyle\oplus}{N}-\ddot{O}:{}^{\ominus} \leftrightarrow \overset{\displaystyle\oplus}{N}=\ddot{O}:$ (bond order of $1\frac{1}{2}$);
$\|$ $|$
$:\ddot{O}$ $:\ddot{O}:{}_{\ominus}$

$\ddot{N}-\ddot{O}:{}^{\ominus} \leftrightarrow \ddot{N}=\ddot{O}:$ (bond order of $1\frac{1}{2}$);
$\|$ $|$
$:\ddot{O}$ $:\ddot{O}:{}_{\ominus}$

relative N—O bond lengths are predicted to be
$NO_2^+ < NO_2 \approx NO_2^-$.

19. $[{}^{\ominus}:C\equiv N:]^-$; $[:\ddot{S}=C=\ddot{N}:{}^{\ominus}]^-$; $[{}^{\ominus}:\ddot{N}=\overset{\displaystyle\oplus}{N}=\ddot{N}:{}^{\ominus}]^-$;

$[:\ddot{S}=C=\ddot{O}:]$

21. (a) Ag^+, Lewis acid; NH_3, Lewis base; (b) $HF(H^+)$, Lewis acid; $C_2H_3O_2^-$, Lewis base.

23. $\begin{array}{c} H \\ | \\ H-C-C \\ | \\ H \end{array} \overset{\ddot{O}\cdot}{\underset{\ddot{O}:{}^{\ominus}}{<}} \leftrightarrow \begin{array}{c} H \\ | \\ H-C-C \\ | \\ H \end{array} \overset{\ddot{O}:{}^{\ominus}}{\underset{\ddot{O}\cdot}{<}}$

25. $H-\ddot{O}-\overset{\displaystyle\ddot{O}:}{\underset{\displaystyle\ddot{O}:}{\overset{\displaystyle\|}{I}}}-\ddot{O}-H$;   the I—O bond is shorter because it has
double bond ($I=O$) character.

27. (a) linear; (b) trigonal planar; (c) T-shaped; (d) trigonal planar; (e) tetrahedral; (f) tetrahedral; (g) angular; (h) octahedral; (i) trigonal pyramidal; (j) angular; (k) square pyramidal; (l) angular; (m) angular

29. Br (trigonal pyramidal); C (tetrahedral); Cl (tetrahedral); S (angular).

$$\begin{matrix} & & O \\ & & \| \end{matrix}$$

31. H_3C—(tetrahedral); —C—(trigonal planar); C—OH will be longer than C=O.

Chapter 12

1. $KK(\sigma_s^b)^1$

3. $KK(\sigma_s^b)^2(\sigma_s^*)^2(\pi_{x,y}^b)^4(\sigma_z^b)^2(\pi_{x,y}^*)^1$; NO has one unpaired electron in π_x^* (or π_y^*) and should be paramagnetic; Lewis structure and MO configuration agree (both predict one unpaired electron); bond energy of NO should be less than that of NO^+ ($2\frac{1}{2}$ bonds in NO *vs.* 3 in NO^+).

5. $KLKL$ $(\sigma^b)^2(\sigma_s^*)^2(\pi_{x,y}^b)^4(\sigma_z^b)^2$; no unpaired electrons in P_2; bond order = 3; bond energy of P_2 should be greater than that of S_2 (bond order of $S_2 = 2$); P_2 should have a shorter bond length than that of S_2.

7. NF^+ should have the larger bond energy (NF^+ bond order = $2\frac{1}{2}$; NF bond order = 2).

9. . . . $(\pi_{x,y}^*)^4(\sigma_z^*)^1$; Br_2^- has one unpaired electron and is paramagnetic.

11. PO has one unpaired electron [. . . $(\pi_{x,y}^*)^1$]; PO bond order = $2\frac{1}{2}$, SO bond order = 2, therefore the bond energy of PO should be greater than that of SO; the bond length of PO should be smaller than that of SO.

13. The $(\sigma^b)^1(\sigma^*)^1$ state is lower, as promotion of two electrons to give $(\sigma^*)^2$ requires much more energy; the lower state has two unpaired electrons and is paramagnetic.

15. π_x (or π_y) nonbonding molecular orbital; HF^+ is . . . $(\sigma^b)^2(\pi_{x,y})^3$; the π_x (or π_y) orbital is localized on F [it is $2p_x$ (or $2p_y$)]; HF^- is . . . $(\sigma^b)^2(\pi_{x,y})^4(\sigma^*)^1$; HF^- bond order = $\frac{1}{2}$; HF^+ bond order = 1; HF^+ should have a larger bond energy than HF^-; HF^- has one unpaired electron and should be paramagnetic.

17. CO (2.1% ionic)

19. HCl (17% ionic); CsCl (75% ionic); TlCl (29% ionic); CsCl should have the largest percent ionic character because the alkali metal atoms have very low electronegativities (and the *s*

valence electron in a heavy Group IA atom such as Cs is far
from the nucleus, thus the *IE* is very low); the percent ionic
character in these molecules increases as the electronegativity
of the atom attached to Cl decreases: $\chi_H = 2.20$; $\chi_{Tl} = 2.04$;
$\chi_{Cs} = 0.79$.

21. CsH has greater ionic character than LiH ($\chi_{Cs} = 0.79$; $\chi_{Li} = 0.98$);
CsH should have the larger dipole moment.

Chapter 13

1. sp^3

3. sp^2; trigonal planar

5. sp^2

7.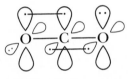

C—O σ bonds are
$sp(C) + p(O)$, and the $2s^2(O)$
lone pairs are not shown; the
same orbital picture can be
drawn for NO_2^+;

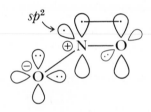

N—O σ bonds are
$sp^2(N) + p(O)$, and the $2s^2(O)$
lone pairs are not shown;
NO_2^- and SO_2 can be represented in the same way as NO_2,
but in these cases the nonbonding sp^2 orbital on N (or S)
is fully occupied.

b) CO_2 (linear, nonpolar); NO_2^+ (linear); NO_2 (angular, polar);
NO_2^- (angular); SO_2 (angular, polar).

c) CO_2 (0); NO_2^+ (0); NO_2 (1); NO_2^- (0); SO_2 (0).

d) NO_2^+ (180°); NO_2^- (120°).

e) Relative N—O bond lengths are $NO_2^+ < NO_2 \approx NO_2^-$.

9.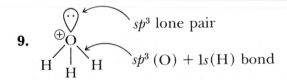

11. See NO_2^- or SO_2, problem 7; the central O in O_3 is sp^2
hybridized; $\frac{1}{2}\pi$ bond (avg) for each O—O.

13.

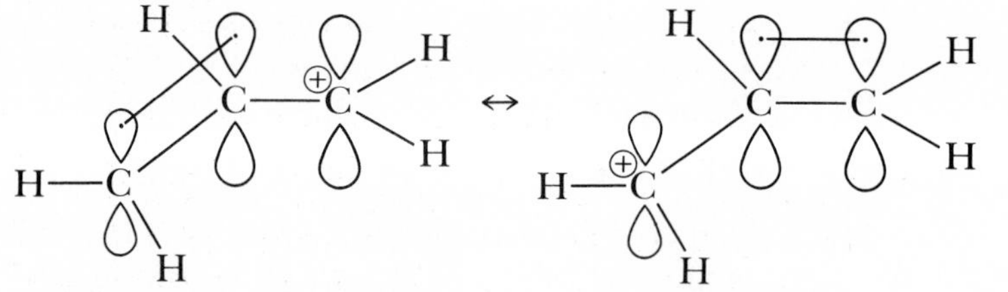

Planar allyl cation; each C is sp^2; the empty p orbital in $C_3H_5^+$ takes 1 electron in the allyl radical, C_3H_5, and 2 electrons in the allyl anion. The π bonding network requires all three species ($C_3H_5^+$, C_3H_5, $C_3H_5^-$) to be planar.

15.

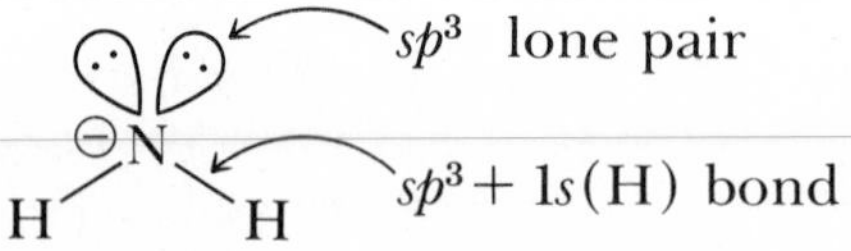

; NH_2^- is angular

17. COS has a higher dipole moment than CS_2; CS_2 is linear and the $\overset{\delta+}{C}$—$\overset{\delta-}{S}$ bond dipoles exactly cancel; the $\overset{\delta+}{C}$—$\overset{\delta-}{O}$ bond dipole is larger than $\overset{\delta+}{C}$—$\overset{\delta-}{S}$ in linear COS, so there is a net dipole moment (O—C—S).

19. The electronegativity difference between H and O is relatively large, so the bond dipoles $\overset{\delta+}{H}$—$\overset{\delta-}{O}$ are large, the electro-negativities of O and F do not differ as much, so the $\overset{\delta-}{F}$—$\overset{\delta+}{O}$ bond dipoles are small.

21. Se is much less electronegative than O; thus $\overset{\delta+}{H}$—$\overset{\delta-}{O}$ bond dipoles are larger than $\overset{\delta+}{H}$—$\overset{\delta-}{Se}$ ones.

23. IE (ethylene) < IE (acetylene).

25. Benzene will absorb light of lower energy than ethylene because the spacing between the highest filled and lowest unoccupied π orbitals is smaller in benzene (see the text discussions of the energy levels in benzene, sect. 13-5, and ethylene, sect. 13-7).

Chapter 14

1. Attractive van der Waals interactions are greater between the heavier halogen atoms.

3. 1.4 kJ mole^{-1}; 4.0 Å (Kr—Kr).

5. Each Na^+ in $NaCl(s)$ is surrounded by 6 Cl^- ions; Cl^-—Cl^- repulsion in the crystal prevents the Na^+Cl^- distance from being as short as in $NaCl(g)$, where no Cl^-—Cl^- repulsions exist.

7. (a) nonmetallic covalent; (b) van der Waals; (c) carbon layers can slip across each other in graphite, as they are bonded weakly; melting requires drastic rearrangement of the strong intralayer carbon bonds, so the melting point is high as in diamond.

(d) $\leftrightarrow$ $\leftrightarrow$ etc.;

(e) The π electrons are mobile and are responsible for electrical conduction.

9. See text discussion (section 14-6).

11. BF_3 should have a higher melting temperature and it does ($-127°C$); NF_3 melts at $-207°C$.

13. For Si: $\nu = 2.63 \times 10^{14}$ Hz; $\lambda = 1.14 \times 10^3$ nm; $\bar{\nu} = 8.77 \times 10^3$ cm^{-1}; infrared; for Ge: $\nu = 1.5 \times 10^{14}$ Hz; $\lambda = 2.0 \times 10^3$ nm; $\bar{\nu} = 5.0 \times 10^3$ cm^{-1}; infrared

Chapter 15

1. Motion of shot converted to atomic vibrations within metal.

3. a) $H_2(g) + \frac{1}{2}O_2(g) \rightarrow H_2O(l)$; Heat is given off.
   b) -285.8 kJ mole^{-1}

5. a) $CH_3CHO(l) + 2\frac{1}{2}O_2(g) \rightarrow 2CO_2(g) + 2H_2O(l)$
   b) -1164 kJ; -582 kJ; -466 kJ
   c) -26.4 kJ g^{-1}
   d) -193 kJ mole^{-1}

7. a) $CH_3OH(l) + 1\frac{1}{2}O_2(g) \rightarrow CO_2(g) + 2H_2O(l)$
   b) -726.6 kJ mole^{-1} methanol

9. a) $C_8H_{18}(l) + 12\frac{1}{2}O_2(g) \rightarrow 8CO_2(g) + 9H_2O(l)$;
      $\Delta H^0_{298} = -5471$ kJ mole^{-1} octane
   b) $C_8H_{18}(l) + 8\frac{1}{2}O_2(g) \rightarrow 8CO(g) + 9H_2O(l)$;
      $\Delta H^0_{298} = -3207$ kJ mole^{-1} octane
   c) $8CO(g) + 4O_2(g) \rightarrow 8CO_2(g)$; $\Delta H^0_{298} = -2264$ kJ

d) -3207 kJ $- 2264$ kJ $= -5471$ kJ. Additivity of enthalpies, or first law of thermodynamics.

11. $\Delta H^0_{298} = +90$ kJ mole^{-1} NO

13. $\Delta H^0_{298} = +50.5$ kJ mole^{-1} N$_2$H$_4$;
$N_2H_4(l) + O_2(g) \rightarrow N_2(g) + 2H_2O(l)$;
$\Delta H^0_{298} = -622$ kJ mole^{-1} N$_2$H$_4$

15. a) $\Delta H^0_{298} = +131.31$ kJ mole^{-1} C
   b) $CO(g) + \frac{1}{2}O_2(g) \rightarrow CO_2(g)$; $H_2(g) + \frac{1}{2}O_2(g) \rightarrow H_2O(l)$;
      $\Delta H^0_{298} = -568.83$ kJ
   c) 1163 kJ

17. a) $\Delta H^0_{298} = -184$ kJ mole^{-1}
   b) $\Delta H_{bond} = -188$ kJ mole^{-1}

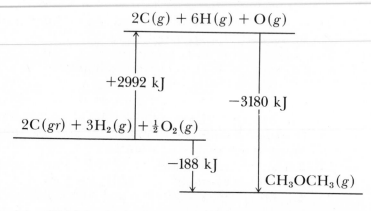

19. (a) -82 kJ mole^{-1}; (b) -24 kJ mole^{-1}; (c) -33 kJ mole^{-1}

21. a) $+93.6$ kJ mole^{-1}
   b) $+51.4$ kJ mole^{-1}. Less heat required since already gaseous.
   c) $+47$ kJ mole^{-1}; 4.4 kJ difference.
   d) C—C and O—H bond in reactants replaced by C—O and C—H bond in products. C—C and C—O comparable strength, but O—H bond more stable than C—H, requiring extra energy from outside to break O—H.

23. $\Delta H^0_{298} = +75$ kJ mole from thermodynamic tables.
   $= +105$ kJ mole^{-1} from bond energy calculations.
   Resonance energy $= 30$ kJ mole^{-1}.
   Resonance structures:

Chapter 16

1. (a) cards spread out; (b) unassembled parts; (c) raw materials
3. (a) positive, increase; (b) positive, increase; (c) negative, decrease; (d) positive, increase. Rank: c < b < a < d.
5. a) negative: 3 moles of gas going to 2 moles of gas.
   b) negative: Gas disappearing
      (solid entropies approximately equal).
   c) positive: Solid converted to liquid.
   d) positive: Solid converted to gas.
   e) negative: 3 moles gas converted to 1 mole gas and 2 of liquid.
7. One mole liquid Br_2 replaced by one mole gaseous HBr.
9. (a) $\Delta S^0 = -25.5$ J K^{-1} mole^{-1} = -25.5 e.u. mole^{-1}; (b) Products more ordered than reactants.
11. $\Delta S^0_{298} = +11.6$ e.u. mole^{-1}
13. $\Delta G^0_{298} = +329.2$ kJ mole^{-1}; $\Delta H^0_{298} = +490.5$ kJ mole^{-1}; $\Delta S^0_{298} = +541.0$ e.u. mole^{-1} or J K^{-1} mole^{-1};

$$490.5 - \frac{541.0 \times 298}{1000} = 329.3$$

15. $\Delta G^0_{298} = -33.28$ kJ mole^{-1}

$$K_{eq} = \frac{[NH_3]^2}{[H_2]^3[N_2]} = 6.8 \times 10^5$$

17. Na(s) < NaCl(s) < $N_2O_4(s)$ < $Br_2(l)$ < $Br_2(g)$
19. (a) $\Delta G^0_{298} = -818$ kJ; Very spontaneous; (b) Rate of the spontaneous reaction is slow.
21. a) No. $K_{eq} = 2.3 \times 10^{-37}$. Reactants very strongly favored.
    b) $\Delta G^0_{298} = -375.2$ kJ mole^{-1}. Thermodynamically much better, with large K_{eq}. (But in fact acetylene would be only one of a great many products formed.)
23. $\Delta H^0_{273} = +6.01$ kJ mole^{-1};
    $\Delta S^0_{273} = +22.0$ e.u. mole^{-1} = $+1.22$ e.u. g^{-1}; $\Delta G^0_{273} = 0$
25. a) $\Delta H^0_{353} = +30.54$ kJ mole^{-1};
       $\Delta S^0_{353} = +86.5$ e.u. mole^{-1}; $\Delta G^0_{353} = 0$
    b) $\Delta E^0_{353} = +27.6$ kJ mole^{-1}
    c) $\Delta H^0_{298} = +33.90$ kJ mole^{-1}; $\Delta G^0_{298} = +5.15$ kJ mole^{-1}. Reaction is favored at higher temperature because it is heat-absorbing.

Chapter 17

1. $K_{eq} = \dfrac{p_{NH_3}^2}{p_{N_2} p_{H_2}^3} = 0.50$

$K_{eq} = \dfrac{p_{N_2}^{1/2} p_{H_2}^{3/2}}{p_{NH_3}} = 1.41$

3. $K_{tot} = \dfrac{K_{eq} K_{sp}^3}{K_{diss}^3} = 5.1 \times 10^{59}$; Yes; $\Delta G^0 = -341$ kJ

5. $K_{eq} = \dfrac{a_{H_2}^4}{a_{H_2O}^4}$

7. Exothermic ($\Delta H_{298}^0 = -113$ kJ); K_{eq} decreases as T rises; K_{eq} unchanged by pressure or catalyst.

9. $\Delta H_{298}^0 = -58.04$ kJ; $\Delta G_{298}^0 = -5.394$ kJ; $K_{eq} = 8.82$. Less $\dot{N}_2O_4$ at higher T; more at higher P; less at larger V. Less conversion to N_2O_4 if Na_2O added, but K_{eq} is unchanged.

11. $K_{eq} = 8.22 \times 10^{-2}$; $\Delta G_{498}^0 = +10.3$ kJ; $\Delta G_{298}^0 = +16.8$ kJ (Remember that I_2 is a gas, not a solid here.); $K_{eq(298)} = 1.15 \times 10^{-3}$. Reaction is more favored at higher T because it is endothermic.

13. $\Delta G_{298}^0 = +80.5$ kJ

15. $K_{eq} = 1.78 \times 10^{-4}$; K_a from table $= 1.77 \times 10^{-4}$; pH $= 2.45$

17. (a) No shift, K_{eq} same; (b) Shift left, K_{eq} same; (c) Shift left, K_{eq} same; (d) Shift right, K_{eq} increases; (e) No change in gases or K_{eq}.

19. $p_{N_2O_4} = 0.034$ atm

Chapter 18

1. $\Delta H_{298}^0 = +34$ kJ mole^{-1}; $\Delta G_{298}^0 = +6$ kJ mole^{-1}; $\Delta S_{298}^0 = +95$ e.u. mole^{-1}; 358 K; 364 K; 0.089 atm

3. C_6H_6 has largest p_v since smallest free energy of vaporization. No hydrogen bonds to hold benzene molecules against one another.

5. 37 kJ mole^{-1}; 41.6 kJ mole^{-1}; 11% error

7. 0.082 atm; 349 K; 337 K

9.

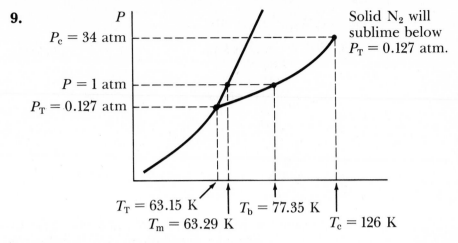

$P_c = 34$ atm

$P = 1$ atm

$P_T = 0.127$ atm

Solid N_2 will sublime below $P_T = 0.127$ atm.

$T_T = 63.15$ K

$T_m = 63.29$ K

$T_b = 77.35$ K

$T_c = 126$ K

11. 0.083 atm; $X_B = 0.807$; $X_T = 0.193$
13. $BaCl_2$, since three ions
15. $\Delta P = 0.20$ atm
17. $k_b = 2.5$ deg mole^{-1}
19. NaCl: $\Delta T = 6.36°$ (better)
   $CaCl_2$: $\Delta T = 5.03°$
21. 128 g mole^{-1}; $C_{10}H_8$
23. 776 g mole^{-1}
25. 6.52×10^{-4} atm

Chapter 19

1. $+0.118$ V; -22.8 kJ mole^{-1} Cl_2
3. 0 V; $+0.021$ V
5. $w_{irr} = 0$; $w_{rev} = 212$ kJ; $q = -3$ kJ
7. (a) $+1.10$ V; (b) $+0.85$ V; (c) -0.72 V
9. (a) $2Ag + Cu^{2+} \rightarrow 2Ag^+ + Cu$; Spontaneous in reverse;
   (b) -0.46 V; (c) Flow from Cu to Ag
11. (a) Flow from Sn to Ag; (b) $+0.36$ V; (c) $+0.39$ V
13. (a) $+0.40$ V; (b) -1.67 V
15. (a) Yes, Fe will reduce Fe^{3+} to Fe^{2+} ($\mathscr{E}^0 = +1.18$ V);
   (b) $K_{eq} = 7.9 \times 10^{39}$
17. a) Mg is strongest reducing agent; MnO_4^- strongest oxidizing agent
   b) $Mg + Cl_2 \rightarrow Mg^{2+} + 2\ Cl^-$
   c) $Mg|Mg^{2+}\|Cl^-|Cl_2$
   d) Mg is anode. Cathode is inert electrode such as Pt, with Cl_2 absorbed on it.
   e) $\mathscr{E}^0 = +3.73$ V

f) $K_{eq} = 1.0 \times 10^{126} = \dfrac{[Mg^{2+}][Cl^-]^2}{[Cl_2]}$

19. $\mathscr{E}$ decreases as Cu^{2+} is removed by complexing

21. $\mathscr{E} = +0.79$ V

23. a) $\mathscr{E}$ decreases as $Cu(NH_3)_4^{2+}$ is formed
b) $\mathscr{E}$ increases as $Zn(NH_3)_4^{2+}$ is formed
c) $\mathscr{E}$ decreases since Cu complex is more stable than Zn complex
d) $\mathscr{E}$ increases as ZnS is formed, removing Zn^{2+}
e) Opposing influences. Difficult to predict, but $\mathscr{E}$ probably increases

25. $2H_2O_2 \rightarrow 2H_2O + O_2$; $\mathscr{E}^0 = +0.55$ V

27. $Ag^+ + I^- \rightarrow AgI$; $\mathscr{E}^0 = +0.95$ V; $K_{sp} = 9 \times 10^{-17}$

29. Increasing ease of oxidation: Ni < Fe < Cr. Cr provides protection in the form of an inert chromium oxide layer. Ni protects iron passively like paint, or tin on a tin can.

Chapter 20

1. 2.07 g

3. $n = 3,4,5,6$; $m = 3$; $n = 4$; $m = 3$

5. $Pt(NH_3)_6Cl_4 > Co(NH_3)_6Cl_3 > K_2PtCl_6 > Cr(NH_3)_4Cl_3$

7. (a) $[Ni(H_2O)_6](ClO_4)_2$; (b) $[Pt(NH_3)_3Cl_3]Br$; (c) $[Pt(NH_3)_4Cl_2]SO_4$; (d) $K_3[Fe(CN)_5Cl]$.

9. Two isomers:

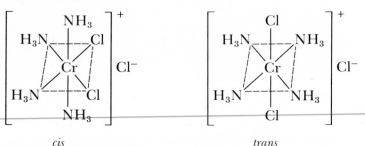

        cis   *trans*

11. Three isomers; one pair (*cis*); one (*trans*)

13. Assuming octahedral coordination about Fe(II), there are two structural isomers: $[Fe(NH_3)_3(H_2O)_2Cl]Br$ and $[Fe(NH_3)_3(H_2O)_2Br]Cl$. Each has three geometrical isomers. If the octahedral directions are identified by coordinates North–East–South–West in a plane and Up–Down perpendicular to it, then these isomers are (a) H_2O at North and East, halide at South, (b) H_2O at North and East, halide Up and (c) H_2O at North and South, halide at East. (NH_3 are understood to be at the three unspecified positions.) All other isomers can be obtained from these three by tumbling the ion. Each of these three contains a mirror plane, so there are no optical isomers.

15. K_3CrF_6; $(t_{2g})^3$

17. Δ_t is smaller than the energy required to pair two electrons.

19. $(t_{2g})^3$; 3; decrease

21. (a) $(d_{z^2})^2(d_{xz},d_{yz})^4(d_{xy})^2$; (b) $(t_{2g})^2$; (c) $(e)^4(t_2)^4$; (d) $(t_{2g})^4(e_g)^2$; (e) $(t_{2g})^6$

23. (a) square planar, $5d_{x^2-y^2}$, $6s$, $6p_x$, $6p_y(dsp^2)$; (b) sodium tetrachloroplatinate(II); (c) $(d_{z^2})^2(d_{xz},d_{yz})^4(d_{xy})^2$, diamagnetic; (d) $(t_{2g})^6$ (two electrons have been removed from the d_{z^2} orbital, and $d_{xz},d_{yz},d_{xy}(t_{2g})$ are degenerate in octahedral $PtCl_6^{2-}$), diamagnetic.

25. $pH = 2.5$

27. For $Co(NH_3)_6^{3+} + 6H^+ \rightleftharpoons Co^{3+} + 6NH_4^+$; $K = 3 \times 10^{21}$; $Co(NH_3)_6^{3+}$ remains intact because it is kinetically inert.

29. $3.4 \times 10^{-7}M$; $1.6 \times 10^{-3}M$; $13M$

31. $2 \times 10^{-11}M$; $0.8M$; $0.10M$

Chapter 21

1. $a = d = e$ and $b = f$

3. a)

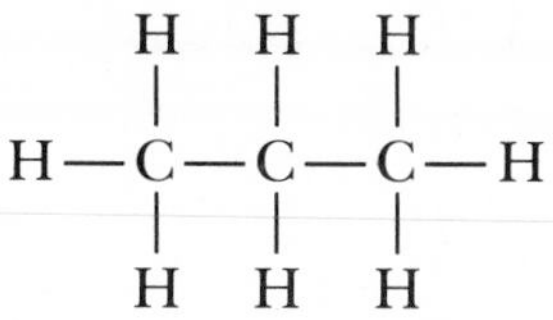

propane

b)

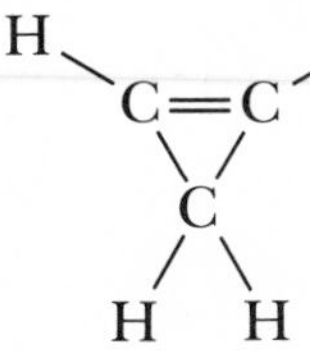

cyclopropene 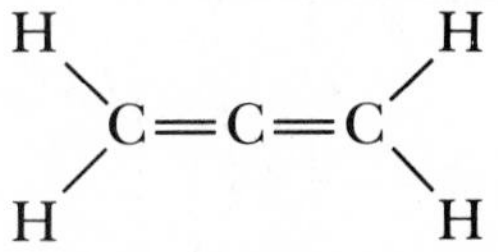   1,2-propadiene   propyne

c)

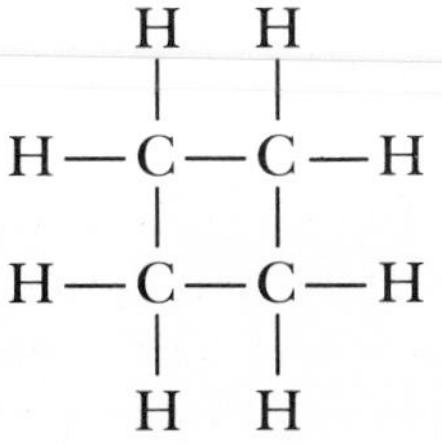

cyclobutane 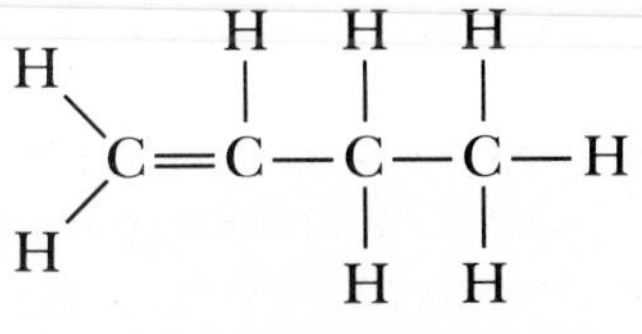   1-butene 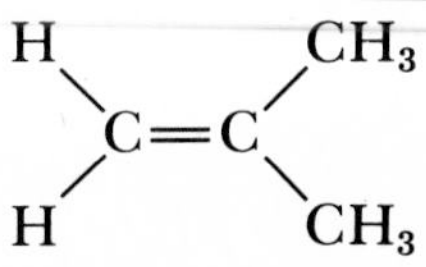   2-methyl propene

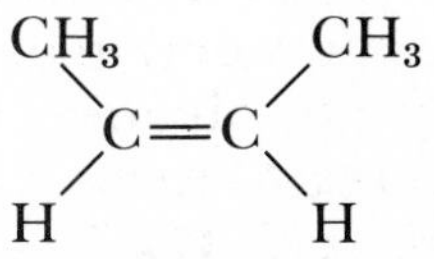

cis-2-butene 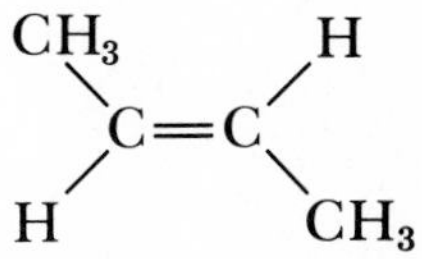   trans-2-butene

d)

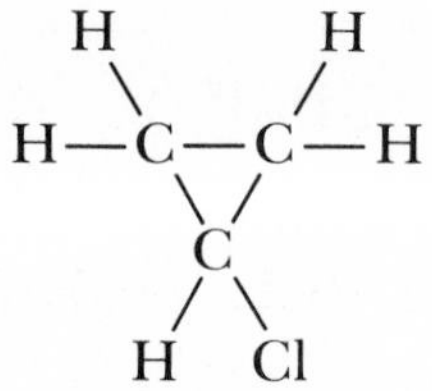

chlorocyclopropane

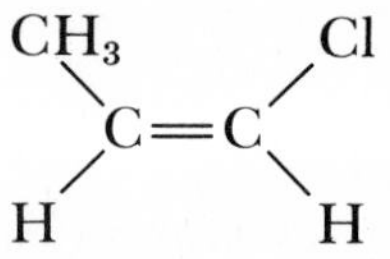

cis-1-chloropropene

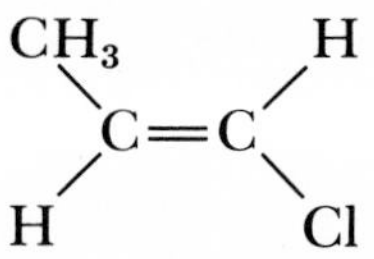

trans-1-chloropropene

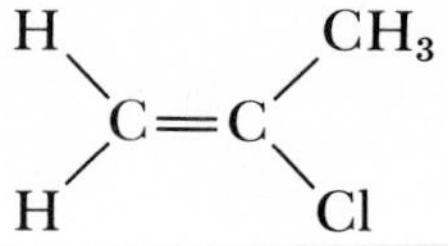

2-chloropropene

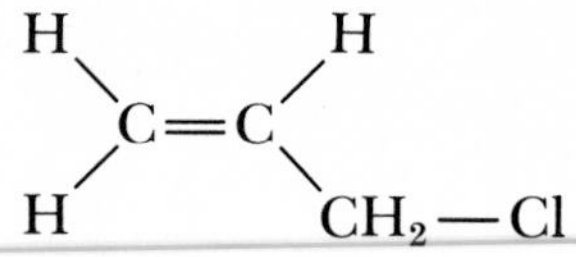

3-chloropropene

5. b and e

7. a) 1-butene
   b) 1-butyne
   c) 1,1-di/uoroethene
   d) 2-methyl-2-butene
   e) 1,1-dichloropropene
   f) 1,5-hexadiene
   g) *trans*-2-chloro-2-butene
   h) 3-bromo-1,3-pentadiene
   i) 2,5-dimethyl-2-hexene
   j) 4-chloro-2-pentyne
   k) methyl benzene or toluene

9. a) 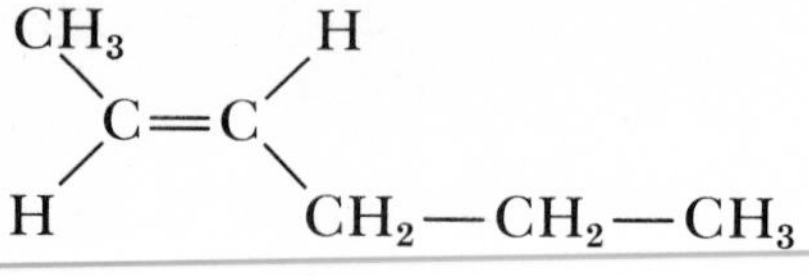   b)

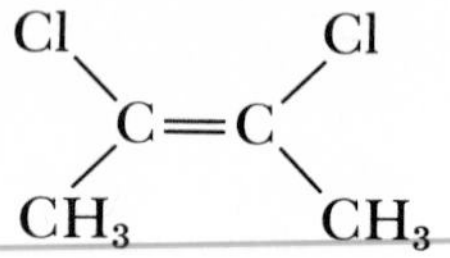

c) 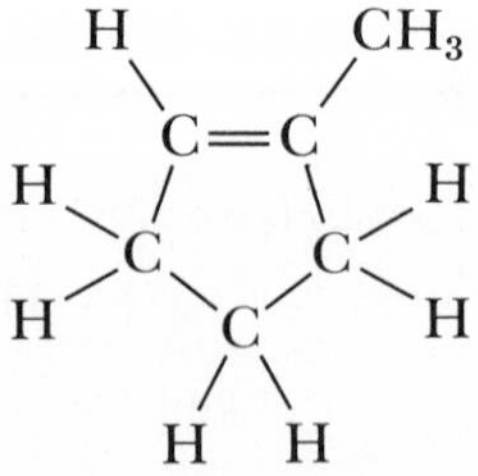   d)

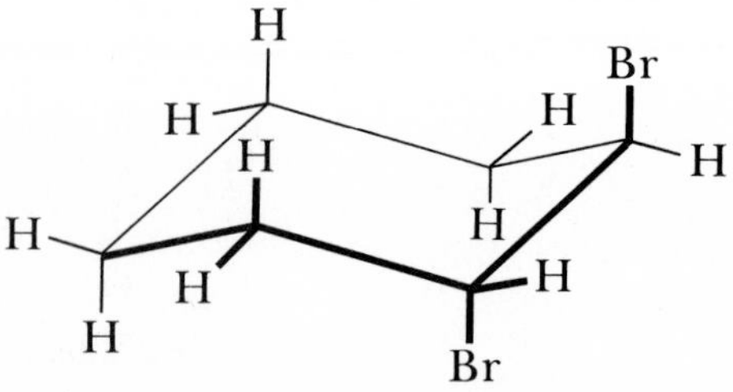

$$CH_3$$
$$|$$
$$CH_2$$
$$|$$

e) 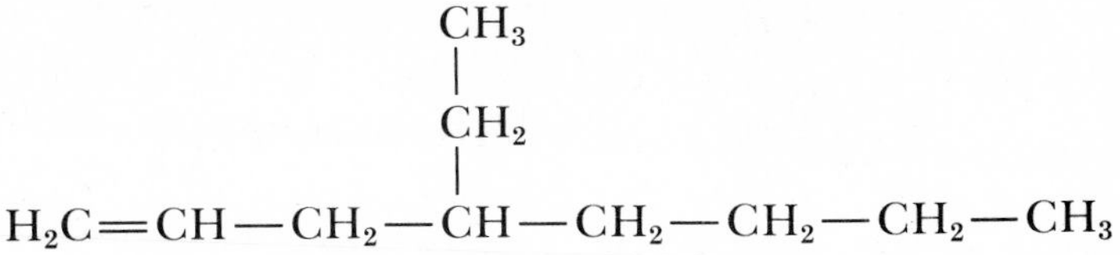 $H_2C{=}CH{-}CH_2{-}CH{-}CH_2{-}CH_2{-}CH_2{-}CH_3$

f) $CH_3{-}CH_2{-}C{\equiv}C{-}CH_2{-}CH_3$

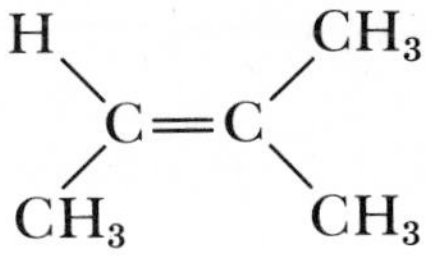

g) 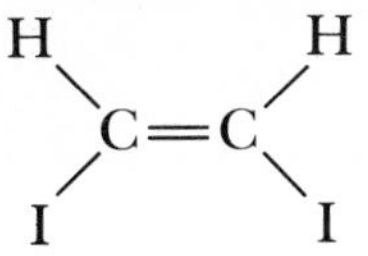
$$\begin{array}{ccc} H & & H \\ & C{=}C & \\ I & & I \end{array}$$

h)
$$\begin{array}{ccc} H & & CH_3 \\ & C{=}C & \\ CH_3 & & CH_3 \end{array}$$

i)
$$Br$$
$$|$$
$$H_2C{=}C{-}CH{=}CH_2$$

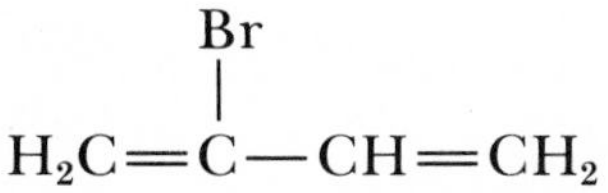

11. C_4H_{10}; Two isomers: $CH_3{-}CH_2{-}CH_2{-}CH_3$ and
$$\begin{array}{c} CH_3 \\ | \\ CH_3{-}CH{-}CH_3 \end{array}$$

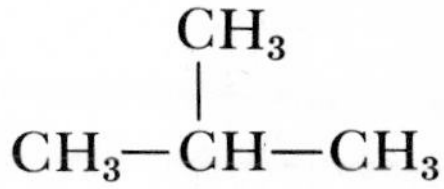

13. $CH_3{-}CH_2{-}CH_2{-}CH_2{-}CH_2{-}CH_2{-}OH$
1-hexanol

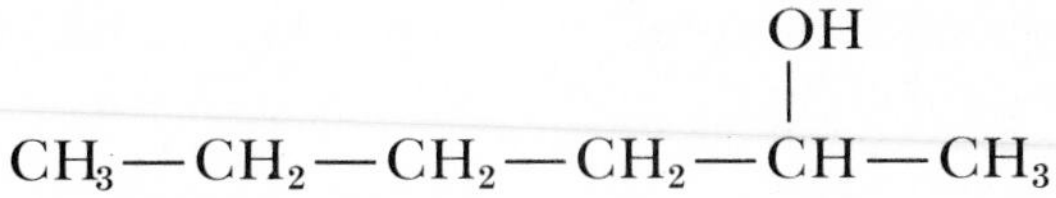

$$OH$$
$$|$$
$$CH_3{-}CH_2{-}CH_2{-}CH_2{-}CH{-}CH_3$$
2-hexanol

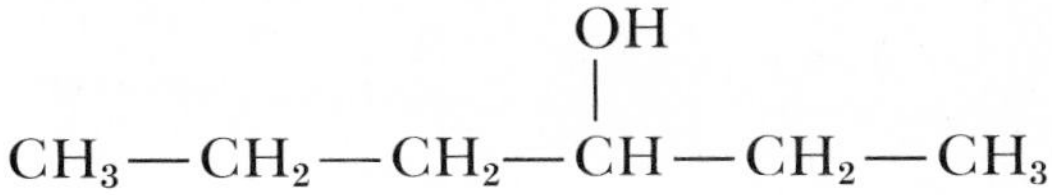

$$OH$$
$$|$$
$$CH_3{-}CH_2{-}CH_2{-}CH{-}CH_2{-}CH_3$$
3-hexanol

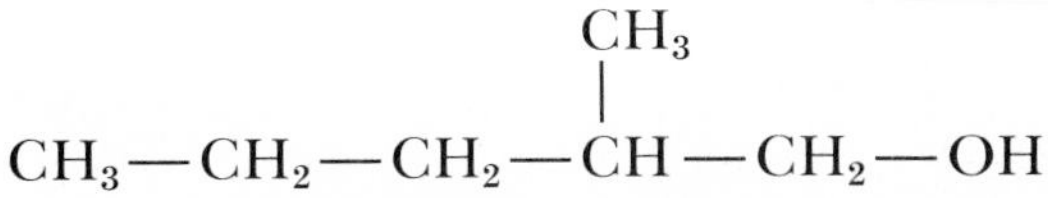

$$CH_3$$
$$|$$
$$CH_3{-}CH_2{-}CH_2{-}CH{-}CH_2{-}OH$$
2-methyl-1-pentanol

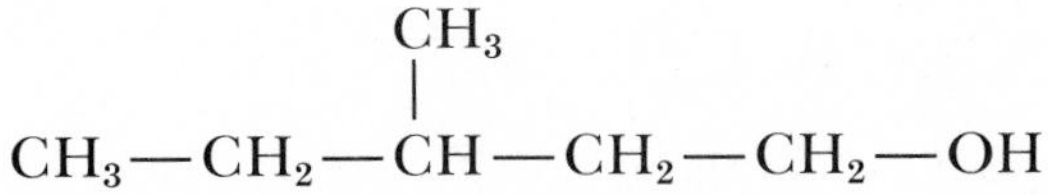

$$CH_3$$
$$|$$
$$CH_3{-}CH_2{-}CH{-}CH_2{-}CH_2{-}OH$$
3-methyl-1-pentanol

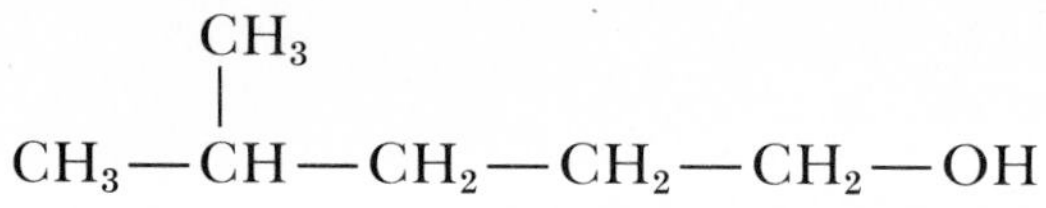

$$CH_3$$
$$|$$
$$CH_3-CH-CH_2-CH_2-CH_2-OH$$

4-methyl-1-pentanol

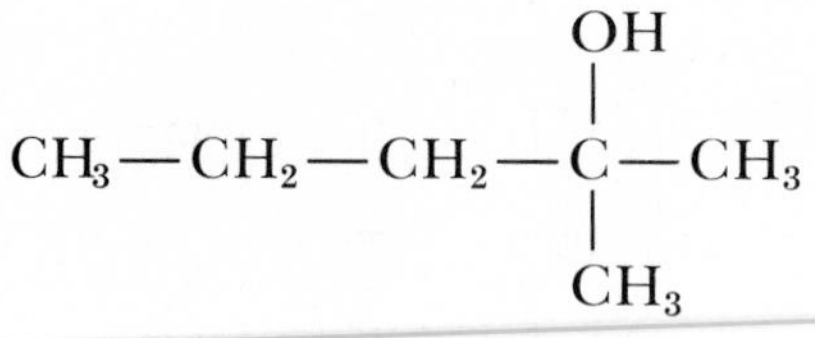

$$OH$$
$$|$$
$$CH_3-CH_2-CH_2-C-CH_3$$
$$|$$
$$CH_3$$

2-methyl-2-pentanol

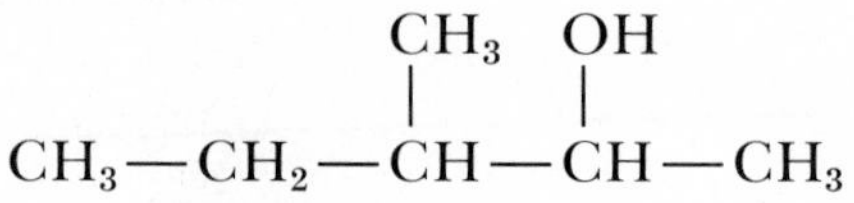

$$CH_3 \quad OH$$
$$| \qquad |$$
$$CH_3-CH_2-CH-CH-CH_3$$

3-methyl-2-pentanol

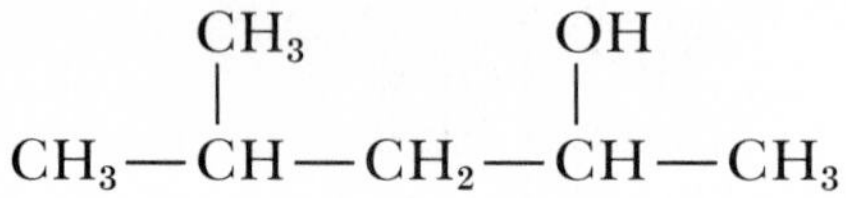

$$CH_3 \qquad OH$$
$$| \qquad\qquad |$$
$$CH_3-CH-CH_2-CH-CH_3$$

4-methyl-2-pentanol

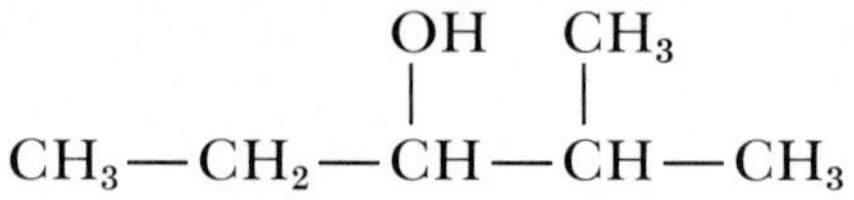

$$OH \quad CH_3$$
$$| \qquad |$$
$$CH_3-CH_2-CH-CH-CH_3$$

2-methyl-3-pentanol

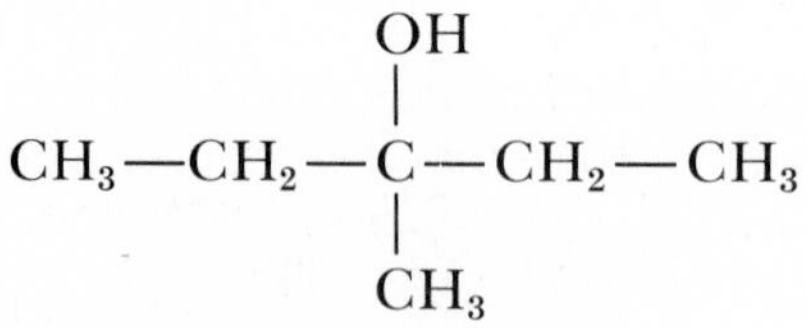

$$OH$$
$$|$$
$$CH_3-CH_2-C-CH_2-CH_3$$
$$|$$
$$CH_3$$

3-methyl-3-pentanol

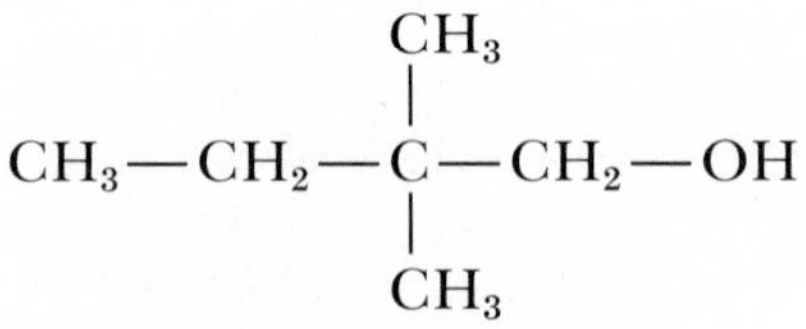

$$CH_3$$
$$|$$
$$CH_3-CH_2-C-CH_2-OH$$
$$|$$
$$CH_3$$

2,2-dimethyl-1-butanol

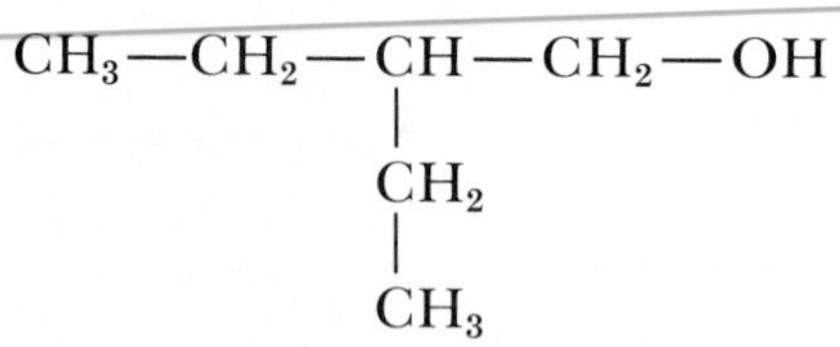

$$CH_3-CH_2-CH-CH_2-OH$$
$$|$$
$$CH_2$$
$$|$$
$$CH_3$$

2-ethyl-1-butanol

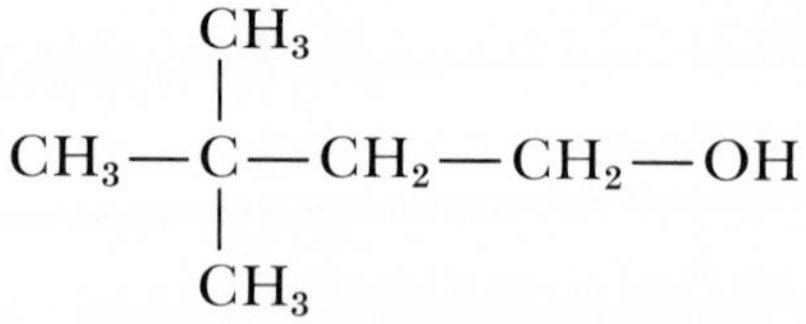

$$CH_3$$
$$|$$
$$CH_3-C-CH_2-CH_2-OH$$
$$|$$
$$CH_3$$

3,3-dimethyl-1-butanol

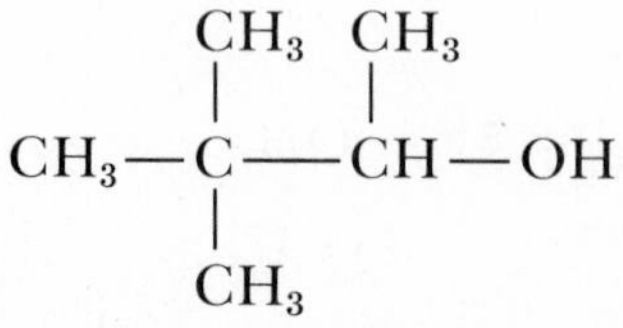

$$CH_3 \quad CH_3$$
$$| \qquad |$$
$$CH_3-C-CH-OH$$
$$|$$
$$CH_3$$

3,3-dimethyl-2-butanol

$$CH_3 \quad CH_3$$
$$| \qquad |$$
$$CH_3-CH_2-C-OH$$
$$|$$
$$CH_3$$

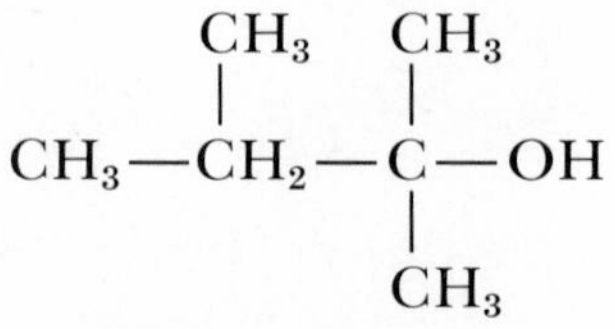

2,3-dimethyl-2-butanol

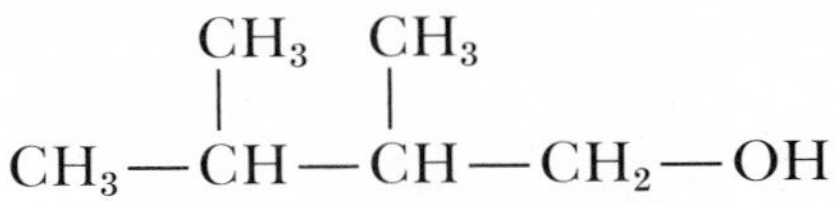

$$CH_3-CH-CH-CH_2-OH$$

with CH_3 and CH_3 substituents

2,3-dimethyl-1-butanol

15. Fifteen isomeric ethers:
   dipropyl ether
   di-1-methylethyl ether
   1-methylethyl propyl ether
   butyl ethyl ether
   1-methylpropyl ethyl ether
   2-methylpropyl ethyl ether
   pentyl methyl ether
   1-methylbutyl methyl ether
   2-methylbutyl methyl ether
   3-methylbutyl methyl ether
   1,1-dimethylpropyl methyl ether
   1,2-dimethylpropyl methyl ether
   2,2-dimethylpropyl methyl ether
   1,1-dimethylethyl ethyl ether
   1-ethylpropyl methyl ether
   Other possible types of isomers: Alcohols (see problem 13).
   Aldehydes, ketones, and carboxylic acids are all impossible
   because of the number of H atoms present.

17. (a) ethyl benzoate; (b) benzyl propionate; (c) *n*-propyl acetate;
   (d) methyl-*n*-butyrate. These compounds are all esters, and c
   and d are isomers.

19. a)   b)

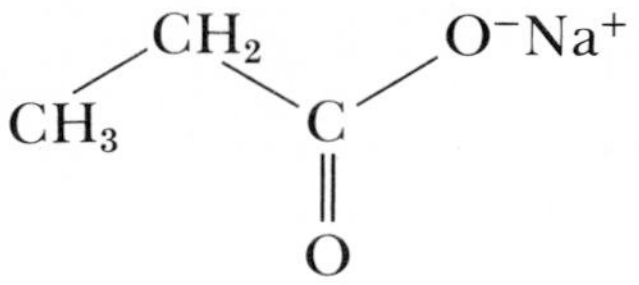

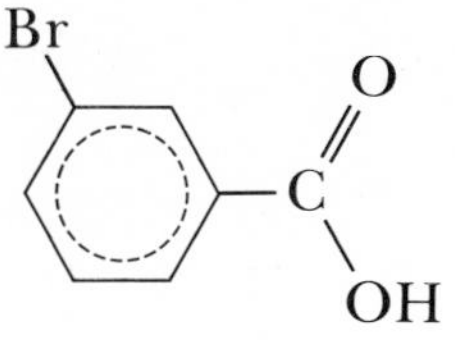

c) 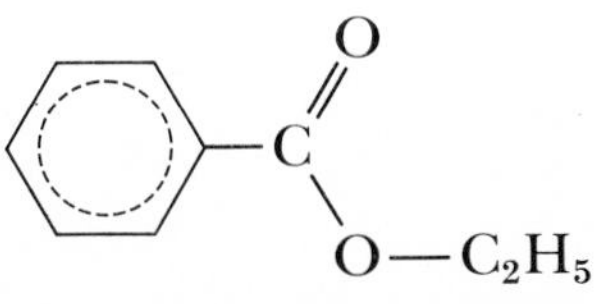   d)

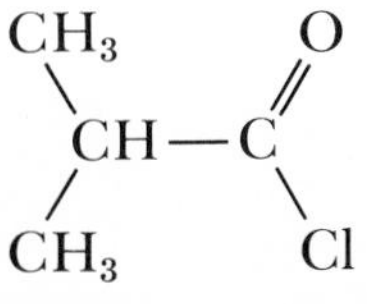

e) 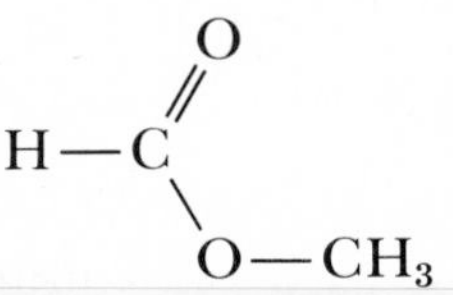   f)

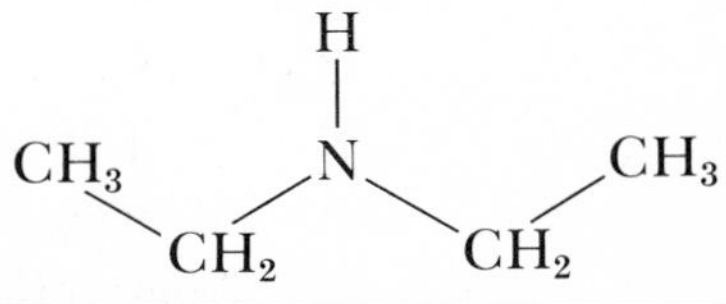

g)

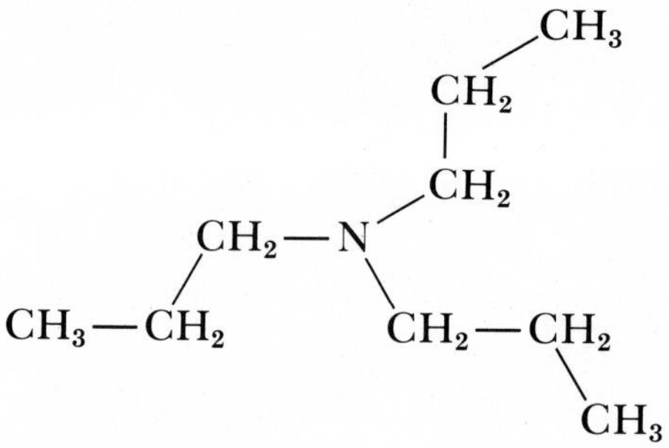

$$CH_3-CH_2-CH_2-N(CH_2-CH_2-CH_3)(CH_2-CH_2-CH_3)$$

The structure g) shows:

CH₃
|
CH₂
|
CH₂
|
CH_2-N
|
CH_3-CH_2 ... CH_2-CH_2
|
CH₃

h)

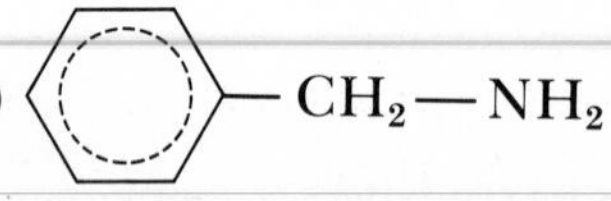

$$-CH_2-NH_2$$

i)

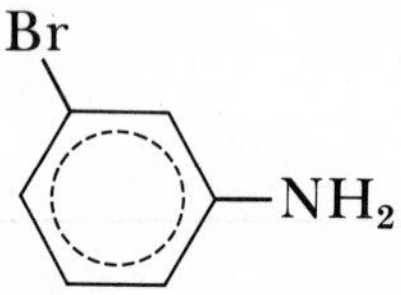

Br (on ring), $-NH_2$

j)

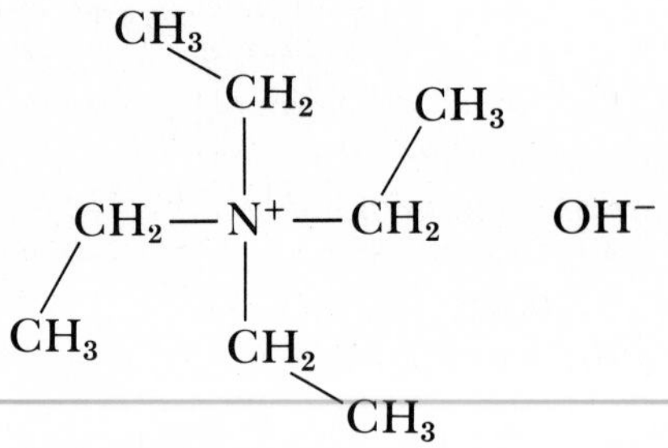

CH₃
\
CH₂   CH₃
| /
$CH_2-N^+-CH_2$ OH^-
/ |
CH₃   CH₂
\
CH₃

k)

CH₃
|
$H_2N-CH-COOH$

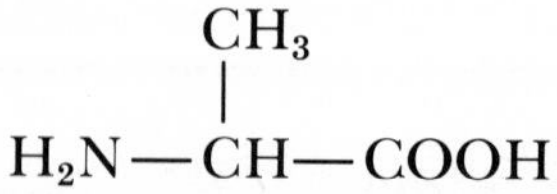

Chapter 22

1. (a) Rate $= -\dfrac{d[A]}{dt} = k[A]^2$; (b) $k = 10$ liters mole^{-1}sec^{-1};

(c) 0.90 mole liter^{-1}sec^{-1}

3. (a) Rate $= kp_{NO}^2 p_{Cl_2}$ (second order in NO, first order in Cl$_2$, third order overall); (b) $k = 4.0 \times 10^{-2}$ atm^{-2}sec^{-1}

5. (a) Rate $= k_2 \dfrac{[C_2H_4]}{[C_2H_2]^2}$; (b) Rate $= k_2 \dfrac{[C_2H_6]}{[C_2H_4]}$;

   (c) Rate $= k_2 \dfrac{[H_2O]^2[O_2]^{1/3}}{[H_2]}$; (d) Rate $= k_2 [NO_2]^2[O_2]^{1/2}$

7. 45 sec

9. A balanced chemical equation for a reaction gives no information about the rate equation.

11. Inverse first order in OH^- concentration, or $[OH^-]^{-1}$

13. $k = 1$ min^{-1}

15. 3.6×10^2 sec; 1.1×10^3 sec

17. 81.1 kJ mole^{-1}

19. 4400%

21. Water boils at a lower temperature at a higher elevation. Consequently, according to the Arrhenius equation, the rate constant for the coagulating reaction will be lower (and the rate will be slower) at the higher elevation.

23. 0.269 sec

25. $\Delta G^0 = -411.7$ kJ; $K_{eq} = 10^{72}$; the decomposition of benzene involves breaking carbon–carbon bonds, and the activation energy for this process is very large.

27. Step b

29. The mechanism involves displacement of a water molecule on $Cr(H_2O)_6^{2+}$ by the Cl^- on $Co(NH_3)_5Cl^{2+}$, giving an activated complex $Co(NH_3)_5Cl^{2+}$---$Cr(H_2O)_5^{2+}$, followed by dissociation of the bridged activated complex to give $Co(NH_3)_5^{2+}$ and $Cr(H_2O)_5Cl^{2+}$ (i.e., a Cl atom is transferred from Cr to Co). In aqueous solution, $Co(NH_3)_5^{2+}$ loses its NH_3 ligands to give $Co(H_2O)_6^{2+}$. The $Cr(H_2O)_5Cl^{2+}$ is kinetically inert (it contains Cr^{3+}) and does not exchange its Cl^- for the radioisotopically labeled Cl^- ions in solution.

Chapter 23

1. ^{27}Al; ^{64}Zn or ^{64}Cu

3. a) $^{120}_{51}Sb \rightarrow {}^{120}_{50}Sn + {}^{0}_{+1}e$

   b) $^{35}_{16}S \rightarrow {}^{35}_{17}Cl + {}^{0}_{-1}e$

   c) $^{226}_{88}Ra \rightarrow {}^{222}_{86}Rn + {}^{4}_{2}He$

   d) $^{7}_{4}Be \xrightarrow{\text{EC}} {}^{7}_{3}Li$

5. (a) 5.51×10^{-12} kg; (b) 30,500 moles CH_4

7. (a) 7.65 MeV nucleon^{-1}; (b) 8.51 MeV nucleon^{-1}; (c) 7.84 MeV nucleon^{-1}; (d) 8.46 MeV nucleon^{-1}; (e) 7.94 MeV nucleon^{-1}

9. 0.0677 Å

11. ^{17}F: positron emission or electron capture
    ^{18}F: positron emission or electron capture
    ^{21}F: electron emission
13. 1.91×10^{-8} g; 194,600 disintegrations min^{-1}
15. 66.7 hours
17. 0.537 g
19. 3940 years old
21. 1.50×10^9 years

Index

Atomic Weights of Elements Referred to $^{12}C = 12$ (exactly)

Name	Symbol	Atomic Number	Atomic Weight	Name	Symbol	Atomic Number	Atomic Weight
Actinium	Ac	89	227.028	Erbium	Er	68	167.26
Aluminum	Al	13	26.98154	Europium	Eu	63	151.96
Americium	Am	95	(243)	Fermium	Fm	100	(257)
Antimony	Sb	51	121.75	Fluorine	F	9	18.9984
Argon	Ar	18	39.948	Francium	Fr	87	(223)
Arsenic	As	33	74.9216	Gadolinium	Gd	64	157.25
Astatine	At	85	~210	Gallium	Ga	31	69.72
Barium	Ba	56	137.33	Germanium	Ge	32	72.59
Berkelium	Bk	97	(247)	Gold	Au	79	196.9665
Beryllium	Be	4	9.01218	Hafnium	Hf	72	178.49
Bismuth	Bi	83	208.9806	Helium	He	2	4.00260
Boron	B	5	10.81	Holmium	Ho	67	164.9303
Bromine	Br	35	79.904	Hydrogen	H	1	1.0079
Cadmium	Cd	48	112.41	Indium	In	49	114.82
Calcium	Ca	20	40.08	Iodine	I	53	126.9045
Californium	Cf	98	(251)	Iridium	Ir	77	192.22
Carbon	C	6	12.011	Iron	Fe	26	55.847
Cerium	Ce	58	140.12	Krypton	Kr	36	83.80
Cesium	Cs	55	132.9054	Lanthanum	La	57	138.9055
Chlorine	Cl	17	35.453	Lawrencium	Lr	103	(260)
Chromium	Cr	24	51.996	Lead	Pb	82	207.2
Cobalt	Co	27	58.9332	Lithium	Li	3	6.941
Copper	Cu	29	63.546	Lutetium	Lu	71	174.967
Curium	Cm	96	(247)	Magnesium	Mg	12	24.305
Dysprosium	Dy	66	162.50	Manganese	Mn	25	54.9380
Einsteinium	Es	99	(254)	Mendelevium	Md	101	(258)